Contents

General Catalogue Excerpt (2014-2015)

College of Natural Sciences and Mathematics Curriculum Components

Students in these programs must complete the Basic Curriculum, the appropriate Major Core Curriculum, and the Major Curriculum.

The Basic Curriculum

I. Written Composition: 6 hours

A. Six semester hours from one of the following sequences:

EH 101 - Written English I (3)
EH 102 - Written English II (3)
OR
EH 103 - Honors English I (3)
EH 104 - Honors English II (3)

II. Humanities and Fine Arts: 15 hours

A. Choose one of the following sequences:

EH 221 - British Literature I (3)
EH 222 - British Literature II (3)

EH 231 - American Literature I (3)
EH 232 - American Literature II (3)

EH 213 - Honors Literature I (3)
EH 214 - Honors Literature II (3)
OR
Choose one of the following:
EH 221 - British Literature I (3)
EH 222 - British Literature II (3)
EH 231 - American Literature I (3)
EH 232 - American Literature II (3)
AND
SP 102 - Introductory Spanish II (3) (or higher)
OR
FR 102 - Introductory French II (3) (or higher)
OR
PL 100 - An Introduction to Philosophy: Humans and Society (3)
OR
PL 204 - Medical Ethics (3)

B. Any two of the following:

AT 100 - Introduction to Art (3)
MU 100 - Introduction to Music (3)
TH 100 - Introduction to Theatre (3)
OR
For Honors students, any one of the above AND
HR 100 - Honors Forum (1)
HR 200 - Honors Special Topics: Interdisciplinary (2)

C. Three semester hours from the following:

SH 100 - Principles of Public Speaking (3)
SH 150 - Professional Speaking (3)

III. Natural Sciences and Mathematics: 11 hours

BY 101 - Principles of Biology (4)
OR
BY 103 - Honors Biology (4)

MH 114 - Precalculus Trigonometry (3)

For majors in Chemistry, Biological or Environmental Sciences:

CH 111 - General Chemistry I (4)

For majors in Mathematics and Mathematics-Computer Information Systems:

PH 211 - Technical Physics I (4)

IV. History, Social, and Behavioral Sciences: 12 hours

A. Six semester hours from one of the following sequences:

HY 101 - History of Western Civilization I (3)
HY 102 - History of Western Civilization II (3)
OR
HY 103 - Honors Western Civilization (3)
HY 104 - Honors Western Civilization II (3)
OR
HY 211 - American History I (3)
HY 212 - American History II (3)

B. Six semester hours from the following:

AN 100 - Introduction to Anthropology (3)
EC 202 - Principles of Macroeconomics (3) *
EC 201 - Principles of Microeconomics (3) *
PS 110 - American Government (3)
PY 100 - General Psychology (3)

SY 100 - Principles of Sociology (3)
OR
SY 110 - Social Problems (3)

Note(s):

*Required for MH and MH-CIS majors.

The Major Core Curriculum

Choose one of the following sequences:

I. Biological and Environmental Sciences and Marine Biology Basic Curriculum: 24 hours

BY 212 - General Botany (4)
BY 222 - General Zoology (4)
CH 112 - General Chemistry II (4)
CH 241 - Organic Chemistry I (4)
CH 242 - Organic Chemistry II (4)
MH 246 - Introduction to Biostatistics (4)

II. Chemistry Major Basic Curriculum: 24 hours

CH 112 - General Chemistry II (4)
MH 121 - Calculus I (4)
MH 122 - Calculus II (4)
MH 246 - Introduction to Biostatistics (4)

PH 201 - College Physics I (4)
PH 202 - College Physics II (4)
OR
PH 211 - Technical Physics I (4)
PH 212 - Technical Physics II (4)

III. Mathematics Major Basic Curriculum: 19 hours

CS 205 - Microcomputer Applications (3) (for non-CIS majors or minors)
OR
CS 210 - Introduction to CIS (3) **
OR
* ED 405 - Technology and Education (3)

MH 121 - Calculus I (4)
MH 122 - Calculus II (4)
MH 223 - Multivariable Calculus (4)
PH 212 - Technical Physics II (4)

Note(s):

*Required for teacher certification students.
**For Mathematics/CIS majors or CIS minors

The Major Curriculum

Choose one of the following sequences:

I. Biology, Biology Comprehensive (both tracks), Marine Biology, and Biology (teacher certification) Majors: 24 hours

- BY 308 - Seminar in Biology (1)
- BY 340 - Microbiology (4)
- BY 380 - Genetics (4)
- BY 400 - Senior Seminar (1)
- BY 450 - Ecology (4)
- BY 472 - Cell Biology (4)
- BY 490 - Evolutionary Theory (3)
- BY 495 - Research in Biology (3)

A. Choose one of the following sequences:

1. Biology. 36 hours

This program must be completed in conjunction with a minor in another field.

Electives from the following list (must include at least 3 hours in botanical sciences and 3 hours in zoological sciences): 12 hours

- BY 307 - Independent Study in Biology (1-8)
- BY 309 - Biological Science Internship (1-4)
- BY 314 - Trees and Shrubs of Alabama (4)
- BY 320 - Invertebrate Zoology (4)
- BY 330 - Hematology (4)
- BY 331 - Immunology (4)
- BY 367 - Independent Study in Marine Biology (1-8)
- BY 392 - History of Life on Earth (4)
- BY 393 - Paleontology (4)
- BY 404 - Research Design and Data Analysis (3)
- BY 410 - Field Botany (4)
- BY 413 - Advanced Plant Biology (4)
- BY 420 - Field Zoology (4)
- BY 428 - Vertebrate Zoology (4)
- BY 429 - Entomology (4)
- BY 431 - Histology (4)
- BY 441 - Environmental Toxicology (4)
- BY 453 - Appalachian Ecology (4)
- BY 456 - Ecological Restoration (3)
- BY 458 - Subtropical Ecology (4)
- BY 461 - Aquatic Biology (4)
- EN 307 - Independent Study in Environmental Sciences (1-8)
- EN 360 - Environmental Chemistry (3)
- EN 480 - Environmental Law (3)

2. Biology Comprehensive (Conservation and Field Biology): 60 hours

MH 121 - Calculus I (4)
PH 201 - College Physics I (4)
PH 202 - College Physics II (4)

Electives from the following list (must include at Least 16 hours of field courses and at least 4 hours of Zoological sciences and 4 hours of botanical sciences): 24 hours

BY 307 - Independent Study in Biology (1-8)
BY 309 - Biological Science Internship (1-4)
* BY 314 - Trees and Shrubs of Alabama (4)
BY 320 - Invertebrate Zoology (4)
* BY 392 - History of Life on Earth (4)
* BY 393 - Paleontology (4)
BY 404 - Research Design and Data Analysis (3)
* BY 410 - Field Botany (4)
BY 413 - Advanced Plant Biology (4)
* BY 420 - Field Zoology (4)
BY 428 - Vertebrate Zoology (4)
* BY 429 - Entomology (4)
BY 441 - Environmental Toxicology (4)
* BY 453 - Appalachian Ecology (4)
BY 456 - Ecological Restoration (3)
* BY 458 - Subtropical Ecology (4)
* BY 461 - Aquatic Biology (4)
BY 495 - Research in Biology (3)
EN 307 - Independent Study in Environmental Sciences (1-8)
EN 360 - Environmental Chemistry (3)
EN 480 - Environmental Law (3)
* GE 370 - Environmental Geology (4)

Note(s):

* Designates a field course.

3. Biology Comprehensive (Medical Track): 60 hours

BY 231 - Human Anatomy and Physiology I (4)
BY 232 - Human Anatomy and Physiology II (4)
BY 471 - Biochemistry I (4)
MH 121 - Calculus I (4)
PH 201 - College Physics I (4)
PH 202 - College Physics II (4)

Electives from the following list: 12 hours

BY 307 - Independent Study in Biology (1-8)
BY 309 - Biological Science Internship (1-4)
BY 330 - Hematology (4)
BY 331 - Immunology (4)
BY 404 - Research Design and Data Analysis (3)
BY 428 - Vertebrate Zoology (4)
BY 429 - Entomology (4)
BY 431 - Histology (4)
BY 441 - Environmental Toxicology (4)
BY 474 - Human Physiology (4)
CH 321 - Analytical Chemistry (4)
CH 472 - Biochemistry II (3)

4. Marine Biology Comprehensive: 60 hours

GE 102 - Physical Geology (4)
GE 370 - Environmental Geology (4)
PH 201 - College Physics I (4)
PH 202 - College Physics II (4)

Electives from the following list (At least 12 hours From courses taught at the Dauphin Island Sea Lab (DISL): 20 hours

BY 367 - Independent Study in Marine Biology (1-8) *
BY 413 - Advanced Plant Biology (4)
BY 414 - Marine Botany (4) *
BY 420 - Field Zoology (4)
BY 424 - Marine Invertebrate Zoology (4) *
BY 425 - Marine Vertebrate Zoology (4) *
BY 426 - Coastal Ornithology (4) *
BY 427 - Marine Behavioral Ecology (4) *
BY 428 - Vertebrate Zoology (4)
BY 451 - Marine Ecology (4) *
BY 452 - Marsh Ecology (4) *
BY 460 - Oceanography (4) *
EN 340 - Coastal Zone Management (2) *

Note(s):

* Indicates courses taught at the Dauphin Island Sea Lab

5. Biology (Teacher Certification): 33 hours

BY 389 - Advanced Laboratory Practicum in Biology (1)
BY 410 - Field Botany (4)
BY 420 - Field Zoology (4)

Note(s):

Students pursuing teacher certification in Biology should follow the pattern listed above. Students should refer to the College of Education section of the Catalogue for a listing of additional requirements for teacher certification.

II. Core Curriculum for Environmental Sciences Comprehensive and General Science (Teacher Certification) Majors: 23 hours

BY 308 - Seminar in Biology (1)
BY 380 - Genetics (4)
BY 490 - Evolutionary Theory (3)
BY 400 - Senior Seminar (1)
BY 450 - Ecology (4)
EN 100 - Introduction to Environmental Sciences (4)
EN 404 - Research Design and Data Analysis (3)
EN 495 - Research in Environmental Sciences (3)

A. Choose one of the following sequences:

1. General Science (Teacher Certification): 45 hours

BY 389 - Advanced Laboratory Practicum in Biology (1)
ES 100 - Introduction to Geology (4)
PH 201 - College Physics I (4)
PH 202 - College Physics II (4)

Electives from the following list in at least two Different areas: 9 hours

BY 307 - Independent Study in Biology (1-8)
BY 314 - Trees and Shrubs of Alabama (4)
BY 393 - Paleontology (4)
BY 410 - Field Botany (4)
BY 420 - Field Zoology (4)
BY 428 - Vertebrate Zoology (4)
BY 429 - Entomology (4)
BY 457 - Natural History of the Black Belt (3)
BY 461 - Aquatic Biology (4)

CH 321 - Analytical Chemistry (4)
CH 422 - Instrumental Analysis (4)
CH 480 - Forensic Chemistry (4)

EN 307 - Independent Study in Environmental Sciences (1-8)
EN 453 - Appalachian Ecology (4)
EN 454 - Conservation Biology (3)
EN 456 - Ecological Restoration (3)
EN 458 - Subtropical Ecology (4)
EN 480 - Environmental Law (3)

GE 370 - Environmental Geology (4)

Note(s):

Students pursuing teacher certification in Science should follow the pattern listed above. Students should refer to the College of Education section of the Catalogue for a listing of additional requirements for teacher certification.

2. Environmental Sciences Comprehensive: 56 Hours

EN 409 - Environmental Sciences Internship (12)
GE 102 - Physical Geology (4)

Electives from the following list: 17 Hours

BY 307 - Independent Study in Biology (1-8)
BY 314 - Trees and Shrubs of Alabama (4)
BY 340 - Microbiology (4)

BY 367 - Independent Study in Marine Biology (1-8)
BY 393 - Paleontology (4)
BY 410 - Field Botany (4)
BY 420 - Field Zoology (4)
BY 428 - Vertebrate Zoology (4)
BY 429 - Entomology (4)
BY 461 - Aquatic Biology (4)
CH 321 - Analytical Chemistry (4)
CH 422 - Instrumental Analysis (4)
CH 480 - Forensic Chemistry (4)
EN 307 - Independent Study in Environmental Sciences (1-8)
EN 360 - Environmental Chemistry (3)
EN 441 - Environmental Toxicology (4)
EN 453 - Appalachian Ecology (4)
EN 454 - Conservation Biology (3)
EN 456 - Ecological Restoration (3)
EN 458 - Subtropical Ecology (4)
EN 480 - Environmental Law (3)
GE 370 - Environmental Geology (4)
GE 392 - History of Life on Earth (4)
TY 338 - Geographic Information Systems (3)
TY 351 - Managing Occupational Safety and Health (3)
TY 352 - Hazardous Waste Operations and Emergency Response (3)

Faculty

Department of Biological & Environmental Sciences

Name	Station	Ext	Building	Room	Email
Beaird, Janis	07	3710	Bibb Graves Hall	207A	jbeaird@uwa.edu
Burnes, Brian	07	3442	Bibb Graves Hall	214D	bburnes@uwa.edu
Keener, Brian	07	3796	Bibb Graves Hall	105	bkeener@uwa.edu
McCall, John	07	3724	Bibb Graves Hall	107A	jmccall@uwa.edu
Merida, Jeffery	07	3771	Bibb Graves Hall	101A	jmerida@uwa.edu
Morse, Kevin	07	3804	Bibb Graves Hall	201C	kmorse@uwa.edu
Morsy, Mustafa	07	5541	Bibb Graves Hall	214I	mmorsy@uwa.edu
Rindsberg, Andrew	07	3416	Bibb Graves Hall	214D	arindsberg@uwa.edu
Rundles, Joan	07	3862	Bibb Graves Hall	212B	jrundles@uwa.edu
Shumaker, Ketia	07	3406	Bibb Graves Hall	201B	kshumaker@uwa.edu
Stanton, Lee	07	3415	Bibb Graves Hall	101B	lstanton@uwa.edu

1

The Science of Biology

Learning Objectives

Biology and the Living World

1.1 The Diversity of Life
1. List the six kingdoms of life.
2. Identify the kingdom to which you belong.

1.2 Properties of Life
1. Name the five basic properties shared by all living things.
2. Explain why complexity, movement, and response to stimulation are not properties that define life.

1.3 Organization of Life
1. List the 13 hierarchal levels of the organization of life.
2. Factor them into three general levels of complexity.
3. Define emergent property, and describe one for each of the three general levels of life's complexity.

1.4 Biological Themes
1. List the five general themes that unify biology as a science.
2. Describe the flow of energy among living organisms.
3. Define symbiosis.
4. Identify the key ability that allows a complex body like yours to maintain homeostasis.

The Scientific Process

1.5 How Scientists Think
1. Differentiate between deductive and inductive reasoning.
2. Identify the form of reasoning used in most scientific studies.

1.6 Science in Action: A Case Study
1. Describe the mechanism producing the ozone "hole" over Antarctica.
2. Explain why this ozone depletion is dangerous to humans.

1.7 Stages of a Scientific Investigation
1. State the six stages of a scientific investigation.

1.8 Theory and Certainty
1. Define hypothesis.
2. Relate hypothesis to theory.
3. Contrast how scientists and the public use the word theory.
4. Appraise the so-called "scientific method."

Author's Corner: Where Are All My Socks Going?

Core Ideas of Biology

1.9 Four Theories Unify Biology as a Science
1. Identify the four major theories that unite biology as a science.
2. Describe the cell theory.
3. Define gene.
4. Identify in what way the chromosomal theory of inheritance extends Mendel's ideas.
5. Explain how Darwin's theory of evolution is related to the gene theory.

Inquiry & Analysis: Does the Presence of One Species Limit the Population Size of Others?

These Antarctic Gentoo penguins share many properties with you and all living things. Their bodies are made up of cells, just as yours is. They have families, with children that resemble their parents, just as your parents did. They grow by eating, as you do, although their diet is limited to fish and krill they catch in the cold Antarctic waters. The sky above them shields them from the sun's harmful UV radiation, just as the sky above you shields you. Not in the summer, however. In the Antarctic summer an "ozone hole" appears, depleting the ozone above these penguins and exposing them to the danger of UV radiation. Scientists are analyzing this situation by a process of observation and experimentation, rejecting ideas that do not match their data. Proceeding in this way they are learning more and more about what is going on. The study of biology is a matter of observing carefully and asking the right questions. When a possible answer—what a scientist calls a *hypothesis*—was proposed, that destruction of Antarctic ozone is the result of leakage of industrial chemicals containing chlorine into the world's atmosphere, scientists carried out experiments and further observations in an attempt to prove this hypothesis wrong. Nothing they have learned so far leads them to reject the hypothesis. It appears human activities far to the north are having a serious impact on the environment of these penguins. This chapter begins your study of biology, the science of life, of penguins and people. Its study helps us to better understand ourselves, our world, and our impact on it.

Biology and the Living World

1.1 The Diversity of Life

In its broadest sense, biology is the study of living things—the science of life. The living world is teeming with a breathtaking variety of creatures—whales, butterflies, mushrooms, and mosquitoes—all of which can be categorized into six groups, or **kingdoms,** of organisms. Representatives from each kingdom can be seen in figure 1.1. All organisms that are placed into a kingdom possess similar characteristics with all other organisms in that same kingdom and are very different from organisms in the other kingdoms.

Biologists study the diversity of life in many different ways. They live with gorillas, collect fossils, and listen to whales. They isolate bacteria, grow mushrooms, and examine the structure of fruit flies. They read the messages encoded in the long molecules of heredity and count how many times a hummingbird's wings beat each second. In the midst of all this diversity, it is easy to lose sight of the key lesson of biology, which is that all living things have much in common.

Key Learning Outcome 1.1 The living world is very diverse, but all things share many key properties.

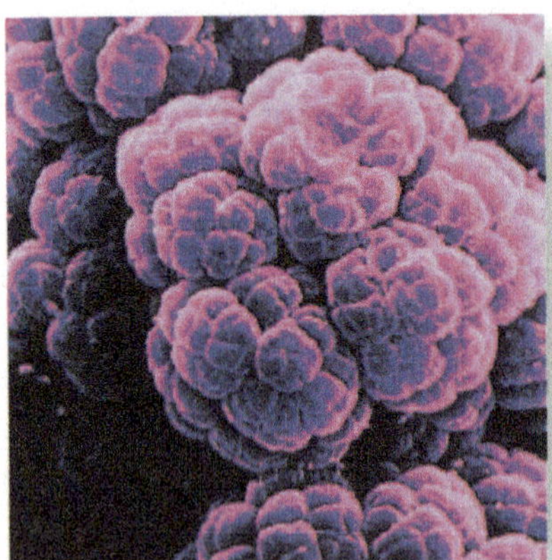

Archaea. This kingdom of prokaryotes (the simplest of cells that do not have nuclei) includes this methanogen, which manufactures methane as a result of its metabolic activity.

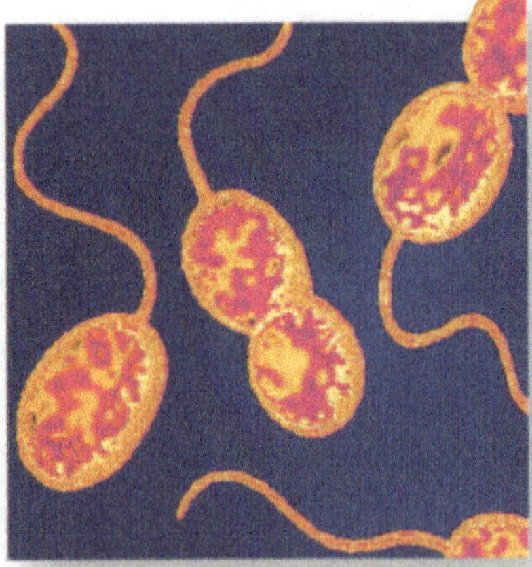

Bacteria. This group is the second of the two prokaryotic kingdoms. Shown here are purple sulfur bacteria, which are able to convert light energy into chemical energy.

Protista. Most of the unicellular eukaryotes (those whose cells contain a nucleus) are grouped into this kingdom, and so are the multicellular algae pictured here.

Fungi. This kingdom contains nonphotosynthetic organisms, mostly multicellular, that digest their food externally, such as these mushrooms.

Plantae. This kingdom contains photosynthetic multicellular organisms that are terrestrial, such as the flowering plant pictured here.

Animalia. Organisms in this kingdom are nonphotosynthetic multicellular organisms that digest their food internally, such as this ram.

Figure 1.1 The six kingdoms of life.
Biologists assign all living things to six major categories called *kingdoms*. Each kingdom is profoundly different from the others.

1.2 Properties of Life

Biology is the study of life—but what does it mean to be alive? What are the properties that define a living organism? This is not as simple a question as it seems because some of the most obvious properties of living organisms are also properties of many nonliving things—for example, *complexity* (a computer is complex), *movement* (clouds move in the sky), and *response to stimulation* (a soap bubble pops if you touch it). To appreciate why these three properties, so common among living things, do not help us to define life, imagine a mushroom standing next to a television: The television seems more complex than the mushroom, the picture on the television screen is moving while the mushroom just stands there, and the television responds to a remote control device while the mushroom continues to just stand there—yet it is the mushroom that is alive.

All living things share five basic properties, passed down over millions of years from the first organisms to evolve on earth: *cellular organization, metabolism, homeostasis, growth and reproduction,* and *heredity.*

1. **Cellular organization.** All living things are composed of one or more cells. A cell is a tiny compartment with a thin covering called a *membrane.* Some cells have simple interiors, while others are complexly organized, but all are able to grow and reproduce. Many organisms possess only a single cell, like the paramecia in figure 1.2; your body contains about 10–100 trillion cells (depending on how big you are)—that's how many centimeters long a string would be wrapped around the world 1,600 times!
2. **Metabolism.** All living things use energy. Moving, growing, thinking—everything you do requires energy. Where does all this energy come from? It is captured from sunlight by plants and algae through photosynthesis. To get the energy that powers our lives, we extract it from plants or from plant-eating animals. That's what the kingfisher is doing in figure 1.3, eating a fish that ate algae. The transfer of energy from one form to another in cells is an example of *metabolism.* All organisms require energy to grow, and all organisms transfer this energy from one place to another within cells using special energy-carrying molecules called ATP molecules.

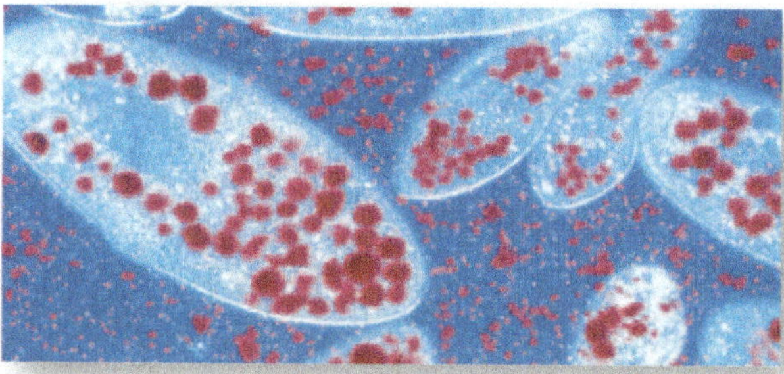

Figure 1.2 Cellular organization.
These paramecia are complex single-celled protists that have just ingested several yeast cells. Like these paramecia, many organisms consist of just a single cell, while others are composed of trillions of cells.

Figure 1.3 Metabolism.
This kingfisher obtains the energy it needs to move, grow, and carry out its body processes by eating fish. It metabolizes this food using chemical processes that occur within cells.

3. **Homeostasis.** All living things maintain stable internal conditions so that their complex processes can be better coordinated. While the environment often varies a lot, organisms act to keep their interior conditions relatively constant; a process called *homeostasis.* Your body acts to maintain an internal temperature of 37°C (98.6°F), however hot or cold the weather might be.
4. **Growth and reproduction.** All living things grow and reproduce. Bacteria increase in size and simply split in two as often as every 15 minutes, while more complex organisms grow by increasing the number of cells and reproduce sexually (some, like the bristlecone pine of California, have reproduced after 4,600 years).
5. **Heredity.** All organisms possess a genetic system that is based on the replication and duplication of a long molecule called *DNA (deoxyribonucleic acid).* The information that determines what an individual organism will be like is contained in a code that is dictated by the order of the subunits making up the DNA molecule, just as the order of letters on this page determines the sense of what you are reading. Each set of instructions within the DNA is called a *gene.* Together, the genes determine what the organism will be like. Because DNA is faithfully copied from one generation to the next, any change in a gene is also preserved and passed on to future generations. The transmission of characteristics from parent to offspring is a process called *heredity.*

Key Learning Outcome 1.2 All living things possess cells that carry out metabolism, maintain stable internal conditions, reproduce themselves, and use DNA to transmit hereditary information to offspring.

1.3 Organization of Life

The organisms of the living world function and interact with each other at many levels, from the very small and simple to the large and complex.

A Hierarchy of Increasing Complexity

A key factor in organizing these interactions is the degree of complexity. We will examine the complexity of life at three levels: cellular, organismal, and populational.

Cellular Level Following down the first section of figure 1.4, you can see that structures get more and more complex—that there is a *hierarchy* of increasing complexity within cells.

1. **Atoms.** The fundamental elements of matter are atoms.
2. **Molecules.** Atoms are joined together into complex clusters called molecules.
3. **Macromolecules.** Large complex molecules are called macromolecules. DNA, which stores the hereditary information in all living organisms, is a macromolecule.
4. **Organelles.** Complex biological molecules are assembled into tiny compartments within cells called organelles, within which cellular activities are organized. The nucleus is an organelle within which the cell's DNA is stored.
5. **Cells.** Organelles and other elements are assembled in the membrane-bounded units we call cells. Cells are the smallest level of organization that can be considered alive.

Organismal Level At the organismal level, in the second section of figure 1.4, cells are organized into four levels of complexity.

6. **Tissues.** The most basic level is that of tissues, which are groups of similar cells that act as a functional unit.

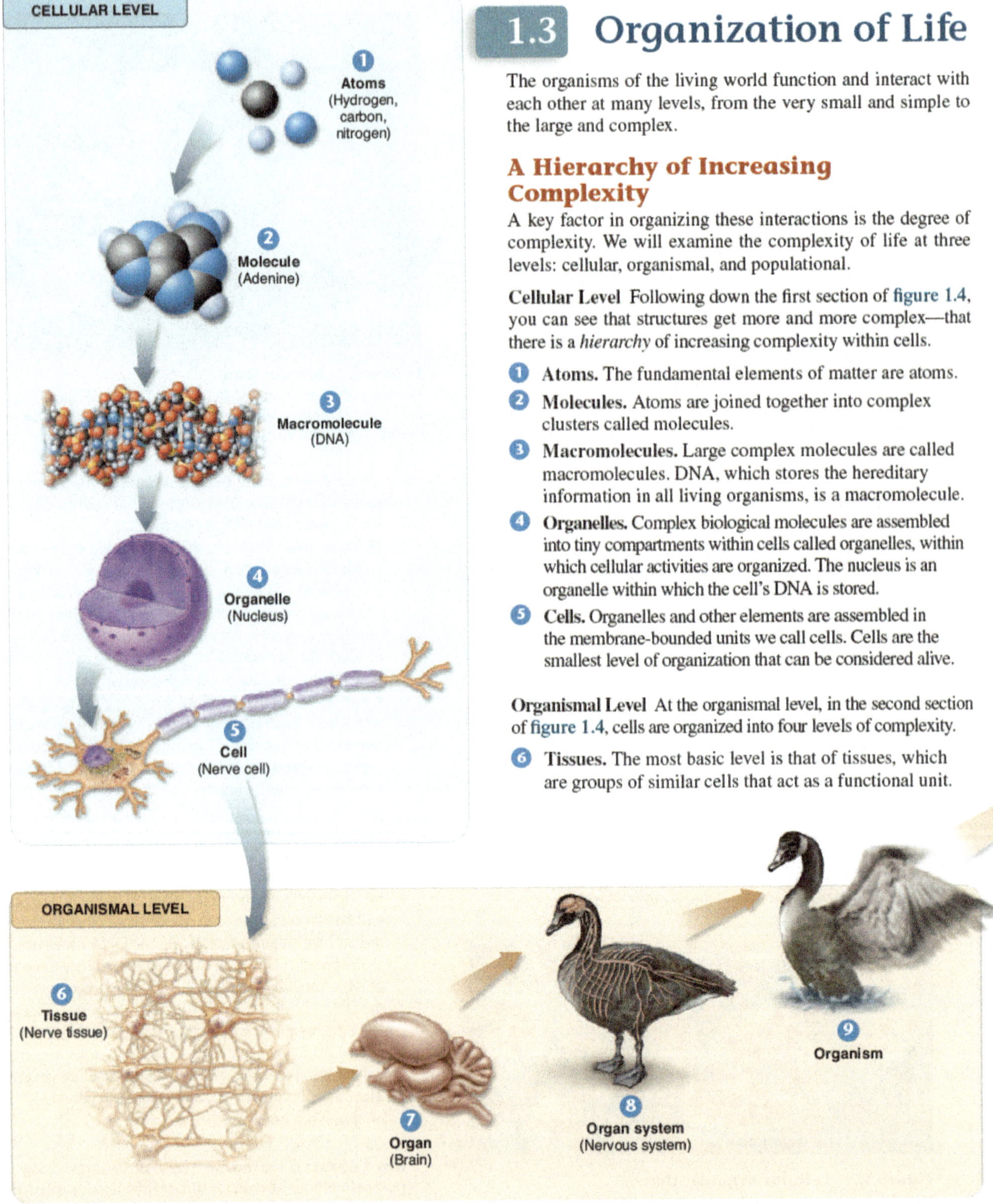

Figure 1.4 Levels of organization.
A traditional and very useful way to sort through the many ways in which the organisms of the living world interact is to organize them in terms of levels of organization, proceeding from the very small and simple to the very large and complex. Here we examine organization within the cellular, organismal, and populational levels.

Nerve tissue is one kind of tissue, composed of cells called neurons that are specialized to carry electrical signals from one place to another in the body.

7. **Organs.** Tissues, in turn, are grouped into organs, which are body structures composed of several different tissues grouped together in a structural and functional unit. Your brain is an organ composed of nerve cells and a variety of connective tissues that form protective coverings and distribute blood.
8. **Organ systems.** At the third level of organization, organs are grouped into organ systems. The nervous system, for example, consists of sensory organs, the brain and spinal cord, neurons that convey signals to and from them, and supporting cells.
9. **Organism.** Separate organ systems function together to form an organism.

Populational Level Organisms are further organized into several hierarchical levels within the living world, as you can see in the third section of figure 1.4.

10. **Population.** The most basic of these is the population, which is a group of organisms of the same species living in the same place. A flock of geese living together on a pond is a population.

11. **Species.** All the populations of a particular kind of organism together form a species, its members similar in appearance and able to interbreed. All Canada geese, whether found in Canada, Minnesota, or Missouri, are basically the same, members of the species *Branta canadensis*. Sandhill cranes are a different species.
12. **Community.** At a higher level of biological organization, a community consists of all the populations of different species living together in one place. Geese, for example, may share their pond with ducks, fish, grasses, and many kinds of insects. All interact in a single pond community.
13. **Ecosystem.** At the highest tier of biological organization, a biological community and the soil and water within which it lives together constitute an ecological system, or ecosystem.

Emergent Properties

At each higher level in the living hierarchy, novel properties emerge, properties that were not present at the simpler level of organization. These **emergent properties** result from the way in which components interact, and often cannot be guessed just by looking at the parts themselves. You have the same array of cell types as a giraffe, for example. Yet, examining a collection of its individual cells gives little clue of what your body is like.

The emergent properties of life are not magical or supernatural. They are the natural consequence of the hierarchy, or structural organization, that is the hallmark of life. Water, which makes up 50%–75% of your body's weight, and ice are both made of H_2O molecules, but one is liquid and the other solid because the H_2O molecules in ice are more organized.

Functional properties emerge from more complex organization. Metabolism is an emergent property of life. The chemical reactions within a cell arise from interactions between molecules that are orchestrated by the orderly environment of the cell's interior. Consciousness is an emergent property of the brain that results from the interactions of many neurons in different parts of the brain.

Key Learning Outcome 1.3
Cells, multicellular organisms, and ecological systems each are organized in a hierarchy of increased complexity. Life's hierarchical organization is responsible for the emergent properties that characterize so many aspects of the living world.

1.4 Biological Themes

Just as every house is organized into thematic areas such as bedroom, kitchen, and bathroom, so the living world is organized by major *themes,* such as how energy flows within the living world from one part to another. As you study biology in this text, five general themes will emerge repeatedly, themes that serve to both unify and explain biology as a science (table 1.1):

1. evolution;
2. the flow of energy;
3. cooperation;
4. structure determines function;
5. homeostasis.

Evolution

Evolution is genetic change in a species over time. Charles Darwin was an English naturalist who, in 1859, proposed the idea that this change is a result of a process called **natural selection.** Simply stated, those organisms whose characteristics make them better able to survive the challenges of their environment live to reproduce, passing their favorable characteristics on to their offspring. Darwin was thoroughly familiar with variation in domesticated animals (in addition to many nondomesticated organisms), and he knew that varieties of pigeons could be selected by breeders to exhibit exaggerated characteristics, a process called **artificial selection.** You can see some of these extreme-looking pigeons pictured in table 1.1 under the heading "evolution." We now know that the characteristics selected are passed on through generations because DNA is transmitted from parent to offspring. Darwin visualized how selection in nature could be similar to that which had produced the different varieties of pigeons. Thus, the many forms of life we see about us on earth today, and the way we ourselves are constructed and function, reflect a long history of natural selection. Evolution will be explored in more detail in chapter 14.

The Flow of Energy

All organisms require energy to carry out the activities of living—to build bodies and do work and think thoughts. All of the energy used by most organisms comes from the sun and is passed in one direction through ecosystems. The simplest way to understand the flow of energy through the living world is to look at who uses it. The first stage of energy's journey is its capture by green plants, algae, and some bacteria by the process of photosynthesis. This process uses energy from the sun to synthesize sugars that photosynthetic organisms like plants store in their bodies. Plants then serve as a source of life-driving energy for animals that eat them. Other animals, like the eagle in table 1.1, may then eat the plant eaters. At each stage, some energy is used for the processes of living, some is transferred, and much is lost, primarily as heat. The flow of energy is a key factor in shaping ecosystems, affecting how many and what kinds of animals live in a community.

Cooperation

The ants cooperating in the upper right photo in table 1.1 protect the plant on which they live from predators and from shading by other plants, while this plant returns the favor by providing the ants with nutrients (the yellow structures at the tips of the leaves). This type of cooperation between different kinds of organisms has played a critical role in the evolution of life on earth. For example, organisms of two different species that live in direct contact, like the ants and the plant on which they live, form a type of relationship called **symbiosis.** Animal cells possess organelles that are the descendants of symbiotic bacteria, and symbiotic fungi helped plants first invade land from the sea. The coevolution of flowering plants and insects—where changes in flowers influenced insect evolution and, in turn, changes in insects influenced flower evolution—has been responsible for much of life's great diversity.

Structure Determines Function

One of the most obvious lessons of biology is that biological structures are very well suited to their functions. You will see this at every level of organization: Within cells, the shape of the proteins called enzymes that cells use to carry out chemical reactions are precisely suited to match the chemicals the enzymes must manipulate. Within the many kinds of organisms in the living world, body structures seem carefully designed to carry out their functions—the long tongue with which the moth in table 1.1 sucks nectar from deep within a flower is one example. The superb fit of structure to function in the living world is no accident. Life has existed on earth for over 2 billion years, a long time for evolution to favor changes that better suit organisms to meet the challenges of living. It should come as no surprise to you that after all this honing and adjustment, biological structures carry out their functions well.

Homeostasis

The high degree of specialization we see among complex organisms is only possible because these organisms act to maintain a relatively stable internal environment, a process introduced earlier called homeostasis. Without this constancy, many of the complex interactions that need to take place within organisms would be impossible, just as a city cannot function without rules to maintain order. Maintaining homeostasis in a body as complex as yours or the hippo's in table 1.1 requires a great deal of signaling back-and-forth between cells.

As already stated, you will encounter these biological themes repeatedly in this text. But just as a budding architect must learn more than the parts of buildings, so your study of biology should teach you more than a list of themes, concepts, and parts of organisms. Biology is a dynamic science that will affect your life in many ways, and that lesson is one of the most important you will learn. It is also a great deal of fun.

Key Learning Outcome 1.4 The five general themes of biology are (1) evolution, (2) the flow of energy, (3) cooperation, (4) structure determines function, and (5) homeostasis.

TABLE 1.1 BIOLOGICAL THEMES

Evolution Charles Darwin's studies of artificial selection in pigeons provided key evidence that selection could produce the sorts of changes predicted by his theory of evolution. The differences that have been obtained by artificial selection of the wild European rock pigeon (*top*) and such domestic races as the red fantail (*middle*) and the fairy swallow (*bottom*), with its fantastic tufts of feathers around its feet, are indeed so great that the birds probably would, if wild, be classified in different major groups.

Cooperation Latin American ants live within the hollow thorns of certain species of acacia trees. The nectar at the bases of the leaves and at the tips of the leaflets provide food. The ants supply the trees with organic nutrients and protection.

The Flow of Energy Energy passes from the sun to plants to plant-eating animals to animal-eating animals, such as this eagle.

Homeostasis Homeostasis often involves water balance to maintain proper blood chemistry. All complex organisms need water—some, like this hippo, luxuriate in it. Others, like the kangaroo rat that lives in arid conditions where water is scarce, obtain water from food and never actually drink.

Structure Determines Function With its long tongue, this moth is able to reach the nectar deep within these flowers.

The Scientific Process

1.5 How Scientists Think

Deductive Reasoning

Science is a process of investigation, using observation, experimentation, and reasoning. Not all investigations are scientific. For example, when you want to know how to get to Chicago from St. Louis, you do not conduct a scientific investigation—instead, you look at a map to determine a route. In other investigations, you make individual decisions by applying a "guide" of accepted general principles. This is called **deductive reasoning.** Deductive reasoning, using general principles to explain specific observations, is the reasoning of mathematics, philosophy, politics, and ethics; deductive reasoning is also the way a computer works. All of us rely on deductive reasoning to make everyday decisions—like whether you need to slow down while driving along a city street, as in figure 1.5. We use general principles as the basis for examining and evaluating these decisions.

Inductive Reasoning

Where do general principles come from? Religious and ethical principles often have a religious foundation; political principles reflect social systems. Some general principles, however, are not derived from religion or politics but from observation of the physical world around us. If you drop an apple, it will fall whether or not you wish it to and despite any laws you may pass forbidding it to do so. Science is devoted to discovering the general principles that govern the operation of the physical world.

How do scientists discover such general principles? Scientists are, above all, observers: They look at the world to understand how it works. It is from observations that scientists determine the principles that govern our physical world.

This way of discovering general principles by careful examination of specific cases is called **inductive reasoning.** Inductive reasoning first became popular about 400 years ago, when Isaac Newton, Francis Bacon, and others began to conduct experiments and from the results infer general principles about how the world operates. The experiments were sometimes quite simple. Newton's consisted simply of releasing an apple from his hand and watching it fall to the ground. This simple observation is the stuff of science. From a host of particular observations, each no more complicated than the falling of an apple, Newton inferred a general principle—that all objects fall toward the center of the earth. This principle was a possible explanation, or *hypothesis,* about how the world works. You also make observations and formulate general principles based on your observations, like forming a general principle about the timing of traffic lights in figure 1.5. Like Newton, scientists work by forming and testing hypotheses, and observations are the materials on which they build them.

Key Learning Outcome 1.5 Science uses inductive reasoning to infer general principles from detailed observation.

DEDUCTIVE REASONING

An Accepted General Principle

When traffic lights along city streets are "timed" to change at the time interval it takes traffic to pass between them, the result will be a smooth flow of traffic.

DEDUCTIVE REASONING

Using a General Principle to Make Everyday Decisions

Traveling at the speed limit, you approach each intersection anticipating that the red light will turn green as you reach the intersection.

INDUCTIVE REASONING

Observations of Specific Events

Driving down the street at the speed limit, you observe that the red traffic light turns green just as you approach the intersection.

Maintaining the same speed, you observe the same event at the next several intersections: the traffic lights turn green just as you approach the intersections. When you speed up, however, the light doesn't change until after you reach the intersection.

INDUCTIVE REASONING

Formation of a General Principle

You conclude that the traffic lights along this street are "timed" to change in the time it takes your car, traveling at the speed limit, to traverse the distance between them.

Figure 1.5 Deductive and inductive reasoning. A driver who assumes that the traffic signals are timed can use deductive reasoning to expect that the traffic lights will change predictably at intersections. In contrast, a driver who is not aware of the general control and programming of traffic signals can use inductive reasoning to determine that the traffic lights are timed as the driver encounters similar timing of signals at several intersections.

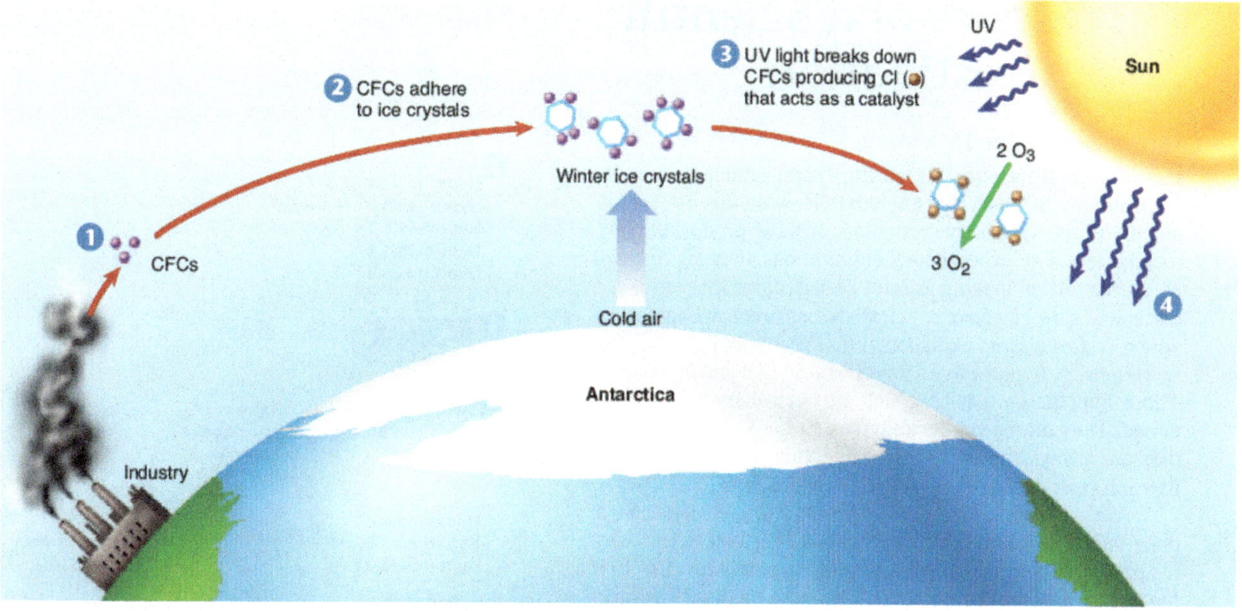

Figure 1.6 How CFCs attack and destroy ozone.

CFCs are stable chemicals that accumulate in the atmosphere as a by-product of industrial society ❶. In the intense cold of the Antarctic, these CFCs adhere to tiny ice crystals in the upper atmosphere ❷. UV light causes the breakdown of CFCs, producing chlorine (Cl). Cl acts as a catalyst, converting O_3 into O_2 ❸. As a result, more harmful UV radiation reaches the earth's surface ❹.

1.6 Science in Action: A Case Study

In 1985 Joseph Farman, a British earth scientist working in Antarctica, made an unexpected discovery. Analyzing the Antarctic sky, he found far less ozone (O_3, a form of oxygen gas) than should be there—a 30% drop from a reading recorded five years earlier in the Antarctic!

At first it was argued that this thinning of the ozone (soon dubbed the "ozone hole") was an as-yet-unexplained weather phenomenon. Evidence soon mounted, however, implicating synthetic chemicals as the culprit. Detailed analysis of chemicals in the Antarctic atmosphere revealed a surprisingly high concentration of chlorine, a chemical known to destroy ozone. The source of the chlorine was a class of chemicals called **chlorofluorocarbons (CFCs).** CFCs (the purple balls ❶ in figure 1.6) have been manufactured in large amounts since they were invented in the 1920s, largely for use as coolants in air conditioners, propellants in aerosols, and foaming agents in making Styrofoam. CFCs were widely regarded as harmless because they are chemically unreactive under normal conditions. But in the atmosphere over Antarctica, CFCs condense onto tiny ice crystals ❷; in the spring, the CFCs break down and produce chlorine, which acts as a catalyst, attacking and destroying ozone, turning it into oxygen gas without the chlorine being used up ❸.

The thinning of the ozone layer in the upper atmosphere 25 to 40 kilometers above the surface of the earth is a serious matter. The ozone layer protects life from the harmful ultraviolet (UV) rays from the sun that bombard the earth continuously. Like invisible sunglasses, the ozone layer filters out these dangerous rays. So when ozone is converted to oxygen gas, the UV rays are able to pass through to the earth ❹. When UV rays damage the DNA in skin cells, it can lead to skin cancer. It is estimated that every 1% drop in the atmospheric ozone concentration leads to a 6% increase in skin cancers.

The world currently produces less than 200,000 tons of CFCs annually, down from 1986 levels of 1.1 million tons. As scientific observations have become widely known, governments have rushed to correct the situation. By 1990, worldwide agreements to phase out production of CFCs by the end of the century had been signed. Production of CFCs declined by 86% in the following 10 years.

Nonetheless, most of the CFCs manufactured since they were invented are still in use in air conditioners and aerosols and have not yet reached the atmosphere. As these CFCs move slowly upward through the atmosphere, the problem can be expected to continue. Ozone depletion is still producing major ozone holes over the Antarctic.

But the worldwide reduction in CFC production is having a major impact. The period of maximum ozone depletion will peak in the next few years, and researchers' models predict that after that the situation should gradually improve, and that the ozone layer will recover by the middle of the 21st century. Clearly, global environmental problems can be solved by concerted action.

Key Learning Outcome 1.6 **Industrially produced CFCs catalytically destroy ozone in the upper atmosphere.**

1.7 Stages of a Scientific Investigation

How Science Is Done

How do scientists establish which general principles are true from among the many that might be? They do this by systematically testing alternative proposals. If these proposals prove inconsistent with experimental observations, they are rejected as untrue. After making careful observations concerning a particular area of science, scientists construct a hypothesis, which is a suggested explanation that accounts for those observations. A hypothesis is a proposition that might be true. Those hypotheses that have not yet been disproved are retained. They are useful because they fit the known facts, but they are always subject to future rejection if—in the light of new information—they are found to be incorrect.

We call the test of a hypothesis an experiment. Suppose that a room appears dark to you. To understand why it appears dark, you propose several hypotheses. The first might be, "The room appears dark because the light switch is turned off." An alternative hypothesis might be, "The room appears dark because the light bulb is burned out." And yet another alternative hypothesis might be, "I am going blind." To evaluate these hypotheses, you would conduct an experiment designed to eliminate one or more of the hypotheses. For example, you might reverse the position of the light switch. If you do so and the light does not come on, you have disproved the first hypothesis. Something other than the setting of the light switch must be the reason for the darkness. Note that a test such as this does not prove that any of the other hypotheses are true; it merely demonstrates that one of them is not. A successful experiment is one in which one or more hypotheses is demonstrated to be inconsistent with the results and is thus rejected.

As you proceed through this text, you will encounter a great deal of information, often accompanied by explanations. These explanations are hypotheses that have withstood the test of experiment. Many will continue to do so; others will be revised as new observations are made. Biology, like all science, is in a constant state of change, with new ideas appearing and replacing old ones.

The Scientific Process

Joseph Farman, who first reported the ozone hole, is a practicing scientist, and what he was doing in Antarctica was science. Science is a particular way of investigating the world, of forming general rules about why things happen by observing particular situations. A scientist like Farman is an observer, someone who looks at the world in order to understand how it works.

Scientific investigations can be said to have six stages as illustrated in figure 1.7: ❶ observing what is going on; ❷ forming a set of hypotheses; ❸ making predictions; ❹ testing them and ❺ carrying out controls, until one or more of the hypotheses have been eliminated; and ❻ forming conclusions based on the remaining hypothesis.

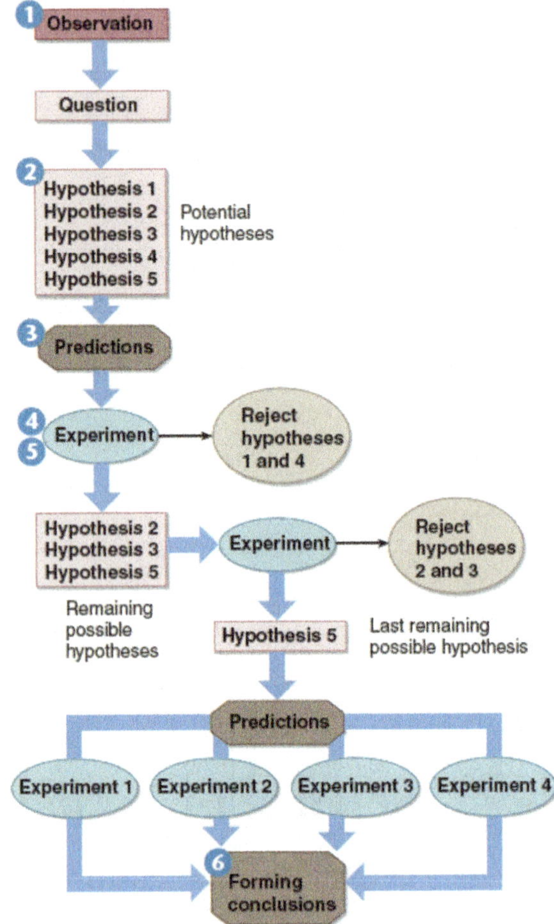

Figure 1.7 The scientific process.
This diagram illustrates the stages of a scientific investigation. First, observations are made that raise a particular question. Then a number of potential explanations (hypotheses) are suggested to answer the question. Next, predictions are made based on the hypotheses, and several rounds of experiments (including control experiments) are carried out in an attempt to eliminate one or more of the hypotheses. Finally, any hypothesis that is not eliminated is retained. Further predictions can be made based on the accepted hypothesis and tested with experiments. If it is validated by numerous experiments and stands the test of time, a hypothesis may eventually become a theory.

1. **Observation.** The key to any successful scientific investigation is careful **observation.** Farman and other scientists had studied the skies over the Antarctic for many years, noting a thousand details about temperature, light, and levels of chemicals. You can see an example in figure 1.8, where the purple colors represent the lowest levels of ozone that the scientists recorded. Had these scientists not kept careful records of what they observed, Farman might not have noticed that ozone levels were dropping. Observations usually generate questions, such as: Why were ozone levels dropping?

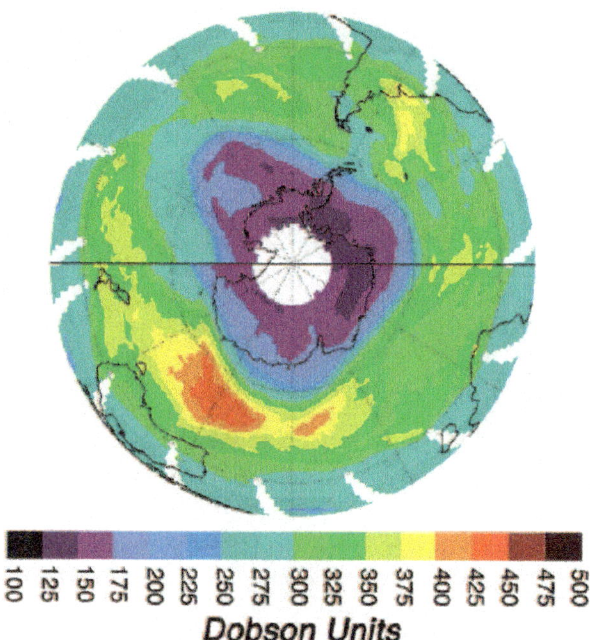

Figure 1.8 The ozone hole.

The swirling colors represent different concentrations of ozone over the South Pole as viewed from a satellite on September 15, 2001. As you can easily see, there is an "ozone hole" (the *purple* areas) over Antarctica covering an area about the size of the United States. (The color *white* indicates areas where no data were available.)

2. **Hypothesis.** When the unexpected drop in ozone was reported and questioned, environmental scientists made a guess to answer their questions—perhaps something was destroying the ozone; maybe the culprit was CFCs. Of course, this was not a guess in the true sense; scientists had some working knowledge of CFCs and what they might be doing in the upper atmosphere. We call such a guess a **hypothesis.** A hypothesis is a guess that might be true. What the scientists guessed was that chlorine from CFCs was reacting chemically with ozone over the Antarctic, converting ozone (O_3) into oxygen gas (O_2) and in the process removing the ozone shield from our earth's atmosphere. Often, scientists will form **alternative hypotheses** if they have more than one guess about what they observe. In this case, there were several other hypotheses advanced to explain the ozone hole. One suggestion explained it as the result of convection: the ozone spun away from the polar regions much as water spins away from the center as a clothes washer moves through its spin cycle. Another hypothesis was that the ozone hole was a transient phenomenon, due perhaps to sunspots, and would soon disappear.
3. **Predictions.** If the CFC hypothesis is correct, then several consequences can reasonably be expected. We call these expected consequences **predictions.** A prediction is what you expect to happen if a hypothesis is true. The CFC hypothesis predicts that if CFCs are responsible for producing the ozone hole, then it should be possible to detect CFCs in the upper Antarctic atmosphere as well as the chlorine released from CFCs that attack the ozone.
4. **Testing.** Scientists set out to test the CFC hypothesis by attempting to verify some of its predictions. We call the test of a hypothesis an **experiment.** To test the hypothesis, atmospheric samples were collected from the stratosphere over 6 miles up by a high-altitude balloon. Analysis of the samples revealed CFCs, as predicted. Were the CFCs interacting with the ozone? The samples contained free chlorine and fluorine, confirming the breakdown of CFC molecules. The results of the experiment thus support the hypothesis.
5. **Controls.** Events in the upper atmosphere can be influenced by many factors. We call each factor that might influence a process a **variable.** To evaluate alternative hypotheses about one variable, all the other variables must be kept constant so that we do not get misled or confused by these other influences. This is done by carrying out two experiments in parallel: In the first experimental test, we alter one variable in a known way to test a particular hypothesis; in the second, called a **control experiment,** we do *not* alter that variable. In all other respects, the two experiments are the same. To further test the CFC hypothesis, scientists carried out control experiments in which the key variable was the amount of CFCs in the atmosphere. Working in laboratories, scientists reconstructed the atmospheric conditions, solar bombardment, and extreme temperatures found in the sky far above the Antarctic. If the ozone levels fell without addition of CFCs to the chamber, then CFCs could not be what was attacking the ozone Carefully monitoring the chamber, however, scientists detected no drop in ozone levels in the absence of CFCs.
6. **Conclusion.** A hypothesis that has been tested and not rejected is tentatively accepted. The hypothesis that CFCs released into the atmosphere are destroying the earth's protective ozone shield is now supported by a great deal of experimental evidence and is widely accepted. While other factors have also been implicated in ozone depletion, destruction by CFCs is clearly the dominant phenomenon. A collection of related hypotheses that have been tested many times and not rejected is called a **theory.** A theory indicates a higher degree of certainty; however, in science, nothing is "certain." The theory of the ozone shield—that ozone in the upper atmosphere shields the earth's surface from harmful UV rays by absorbing them—is supported by a wealth of observation and experimentation and is widely accepted. The explanation for the destruction of this shield is still at the hypothesis stage.

Key Learning Outcome 1.7 Science progresses by systematically eliminating potential hypotheses that are not consistent with observation.

1.8 Theory and Certainty

A theory is a unifying explanation for a broad range of observations. Thus we speak of the theory of gravity, the theory of evolution, and the theory of the atom. Theories are the solid ground of science, that of which we are the most certain. There is no absolute truth in science, however, only varying degrees of uncertainty. The possibility always remains that future evidence will cause a theory to be revised. A scientist's acceptance of a theory is always provisional. For example, in another scientist's experiment, evidence that is inconsistent with a theory may be revealed. As information is shared throughout the scientific community, previous hypotheses and theories may be modified, and scientists may formulate new ideas.

Very active areas of science are often alive with controversy, as scientists grope with new and challenging ideas. This uncertainty is not a sign of poor science but rather of the push and pull that is the heart of the scientific process. The hypothesis that the world's climate is growing warmer due to humanity's excessive production of carbon dioxide (CO_2), for example, has been quite controversial, although the weight of evidence has increasingly supported the hypothesis.

The word theory is thus used very differently by scientists than by the general public. To a scientist, a theory represents that for which he or she is most certain; to the general public, the word theory implies a *lack* of knowledge or a guess. How often have you heard someone say, "It's only a theory!"? As you can imagine, confusion often results. In this text the word theory will always be used in its scientific sense, in reference to a generally accepted scientific principle.

Figure 1.9 Nobel Prize winner.
Sherwood Rowland, along with Mario Molina and Paul Crutzen, won the 1995 Nobel Prize in Chemistry for discovering how CFCs act to catalytically break down atmospheric ozone in the stratosphere, the chemistry responsible for the "ozone hole" over the Antarctic.

The Scientific "Method"

It was once fashionable to claim that scientific progress is the result of applying a series of steps called the **scientific method;** that is, a series of logical "either/or" predictions tested by experiments to reject one alternative. The assumption was that trial-and-error testing would inevitably lead one through the maze of uncertainty that always slows scientific progress. If this were indeed true, a computer would make a good scientist—but science is not done this way! If you ask successful scientists like Farman how they do their work, you will discover that without exception they design their experiments with a pretty fair idea of how they will come out. Environmental scientists understood the chemistry of chlorine and ozone when they formulated the CFC hypothesis, and they could imagine how the chlorine in CFCs would attack ozone molecules. A hypothesis that a successful scientist tests is not just any hypothesis. Rather, it is a "hunch" or educated guess in which the scientist integrates all that he or she knows. The scientist also allows his or her imagination full play, in an attempt to get a sense of what *might* be true. It is because insight and imagination play such a large role in scientific progress that some scientists are so much better at science than others (figure 1.9)—just as Beethoven and Mozart stand out among composers.

The Limitations of Science

Scientific study is limited to organisms and processes that we are able to observe and measure. Supernatural and religious phenomena are beyond the realm of scientific analysis because they cannot be scientifically studied, analyzed, or explained. Supernatural explanations can be used to explain any result, and cannot be disproven by experiment or observation. Scientists in their work are limited to objective interpretations of observable phenomena.

It is also important to recognize that there are practical limits to what science can accomplish. While scientific study has revolutionized our world, it cannot be relied upon to solve all problems. For example, we cannot pollute the environment and squander its resources today, in the blind hope that somehow science will make it all right sometime in the future. Nor can science restore an extinct species. Science identifies solutions to problems when solutions exist, but it cannot invent solutions when they don't.

Key Learning Outcome 1.8 **A scientist does not follow a fixed method to form hypotheses but relies also on judgement and intuition.**

Author's *Corner*

Where Are All My Socks Going?

All my life, for as far back as I can remember, I have been losing socks. Not pairs of socks, mind you, but single socks. I first became aware of this peculiar phenomenon when, as a young man, I went away to college. When Thanksgiving rolled around that first year, I brought an enormous duffle bag of laundry home. My mother, instead of braining me, dumped the lot into the washer and dryer, and so discovered what I had not noticed—that few of my socks matched anymore.

That was over 40 years ago, but it might as well have been yesterday. All my life, I have continued to lose socks. This last Christmas I threw out a sock drawer full of socks that didn't match, and took advantage of sales to buy a dozen pairs of brand-new ones. Last week, when I did a body count, three of the new pairs had lost a sock!

Enough. I set out to solve the mystery of the missing socks. How? The way Sherlock Holmes would have, scientifically. Holmes worked by eliminating those possibilities that he found not to be true. A scientist calls possibilities "hypotheses" and, like Sherlock, rejects those that do not fit the facts. Sherlock tells us that when only one possibility remains unrejected, then—however unlikely—it must be true.

Hypothesis 1: It's the socks. I have four pairs of socks bought as Christmas gifts but forgotten until recently. Deep in my sock drawer, they have remained undisturbed for five months. If socks disappear because of some intrinsic property (say the manufacturer has somehow designed them to disappear to generate new sales), then I could expect at least one of these undisturbed ones to have left the scene by now. However, when I looked, all four pairs were complete. Undisturbed socks don't disappear. Thus I reject the hypothesis that the problem is caused by the socks themselves.

Hypothesis 2: Transformation, a fanciful suggestion by science fiction writer Avram Davidson in his 1958 story "Or All the Seas with Oysters" that I cannot get out of the quirky corner of my mind. I discard the socks I have worn each evening in a laundry basket in my closet. Over many years, I have noticed a tendency for socks I have placed in the closet to disappear. Over that same long period, as my socks are disappearing, there is something in my closet that seems to multiply—COAT HANGERS! Socks are larval coat hangers! To test this outlandish hypothesis, I had only to move the laundry basket out of the closet. Several months later, I was still losing socks, so this hypothesis is rejected.

Hypothesis 3: Static cling. The missing single socks may have been hiding within the sleeves of sweatshirts or jackets, inside trouser legs, or curled up within seldom-worn garments. Rubbing around in the dryer, socks can garner quite a bit of static electricity, easily enough to cause them to cling to other garments. Socks adhering to the outside of a shirt or pant leg are soon dislodged, but ones that find themselves within a sleeve, leg, or fold may simply stay there, not "lost" so much as misplaced. However, after a diligent search, I did not run across any previously lost socks hiding in the sleeves of my winter garments or other seldom-worn items, so I reject this hypothesis.

© Eric Lewis/The New Yorker Collection/www.cartoonbank.com

Hypothesis 4: I lose my socks going to or from the laundry. Perhaps in handling the socks from laundry basket to the washer/dryer and back to my sock drawer, a sock is occasionally lost. To test this hypothesis, I have pawed through the laundry coming into the washer. No single socks. Perhaps the socks are lost after doing the laundry, during folding or transport from laundry to sock drawer. If so, there should be no single socks coming out of the dryer. But there are! The singletons are first detected among the dry laundry, before folding. Thus I eliminate the hypothesis that the problem arises from mishandling the laundry. It seems the problem is in the laundry room.

Hypothesis 5: I lose them during washing. Perhaps the washing machine is somehow "eating" my socks. I looked in the washing machine to see if a sock could get trapped inside, or chewed up by the machine, but I can see no possibility. The clothes slosh around in a closed metal container with water passing in and out through little holes no wider than a pencil. No sock could slip through such a hole. There is a thin gap between the rotating cylinder and the top of the washer through which an errant sock might escape, but my socks are too bulky for this route. So I eliminate the hypothesis that the washing machine is the culprit.

Hypothesis 6: I lose them during drying. Perhaps somewhere in the drying process socks are being lost. I stuck my head in our clothes dryer to see if I could see any socks, and I couldn't. However, as I look, I can see a place a sock could go—behind the drying wheel! A clothes dryer is basically a great big turning cylinder with dry air blowing through the middle. The edges of the turning cylinder don't push hard against the side of the machine. Just maybe, every once in a while, a sock might get pulled through, sucked into the back of the machine.

To test this hypothesis, I should take the back of the dryer off and look inside to see if it is stuffed with my missing socks. My wife, knowing my mechanical abilities, is not in favor of this test. Thus, until our dryer dies and I can take it apart, I shall not be able to reject hypothesis 6. Lacking any other likely hypothesis, I take Sherlock Holmes' advice and tentatively conclude that the dryer is the culprit.

Core Ideas of Biology

1.9 Four Theories Unify Biology as a Science

The Cell Theory: Organization of Life

As was stated at the beginning of this chapter, all organisms are composed of cells, life's basic units. Cells were discovered by Robert Hooke in England in 1665. Hooke was using one of the first microscopes, one that magnified 30 times. Looking through a thin slice of cork, he observed many tiny chambers that reminded him of monks' cells in a monastery. Not long after that, the Dutch scientist Anton van Leeuwenhoek used microscopes capable of magnifying 300 times, and discovered an amazing world of single-celled life in a drop of pond water like you see in figure 1.10. He called the bacterial and protist cells he saw "wee animalcules." However, it took almost two centuries before biologists fully understood their significance. In 1839, the German biologists Matthias Schleiden and Theodor Schwann, summarizing a large number of observations by themselves and others, concluded that all living organisms consist of cells. Their conclusion forms the basis of what has come to be known as the **cell theory.** Later, biologists added the idea that all cells come from other cells. The cell theory, one of the basic ideas in biology, is the foundation for understanding the reproduction and growth of all organisms. The nature of cells and how they function is discussed in detail in chapter 4.

Figure 1.10 Life in a drop of pond water.
All organisms are composed of cells. Some organisms, including these protists, are single-celled, while others, such as plants, animals, and fungi, consist of many cells.

The Gene Theory: Molecular Basis of Inheritance

Even the simplest cell is incredibly complex, more intricate than a computer. The information that specifies what a cell is like—its detailed plan—is encoded in a long cablelike molecule called **DNA (deoxyribonucleic acid).** Researchers James Watson and Francis Crick discovered in 1953 that each DNA molecule is formed from two long chains of building blocks, called nucleotides, wound around each other. You can see in figure 1.11 that the two chains face each other, like two lines of people holding hands. The chains contain information in the same way this sentence does—as a sequence of letters. There are four different nucleotides in DNA (symbolized as A, T, C, and G in the figure), and the sequence in which they occur encodes the information. Specific sequences of several hundred to many thousand nucleotides make up a *gene,* a discrete unit of hereditary information. A gene might encode a particular protein, or a different kind of unique molecule called RNA, or a gene might act to regulate other genes. All organisms on earth encode their genes in strands of DNA. This prevalence of DNA led to the development of the **gene theory.** Illustrated in figure 1.12, the gene theory states that the proteins and RNA molecules encoded by an organism's genes determine what it will be like. The entire set of DNA instructions that specifies a cell is called its **genome.** The sequence of the human genome, 3 billion nucleotides long, was decoded in 2001, a triumph of scientific investigation. How genes function is the subject of chapter 12. In chapter 13 we explore how detailed knowledge of genes is revolutionizing biology and having an impact on the lives of all of us.

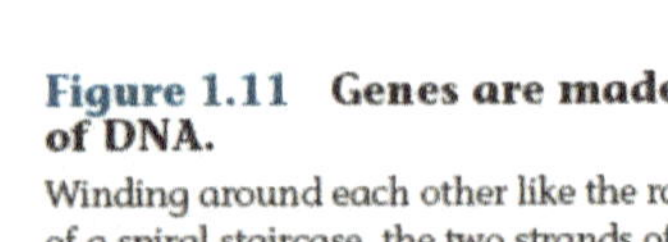

Figure 1.11 Genes are made of DNA.
Winding around each other like the rails of a spiral staircase, the two strands of a DNA molecule make a double helix. Because of its size and shape, the nucleotide represented by the letter A can only pair with the nucleotide represented by the letter T, and likewise G can only pair with C.

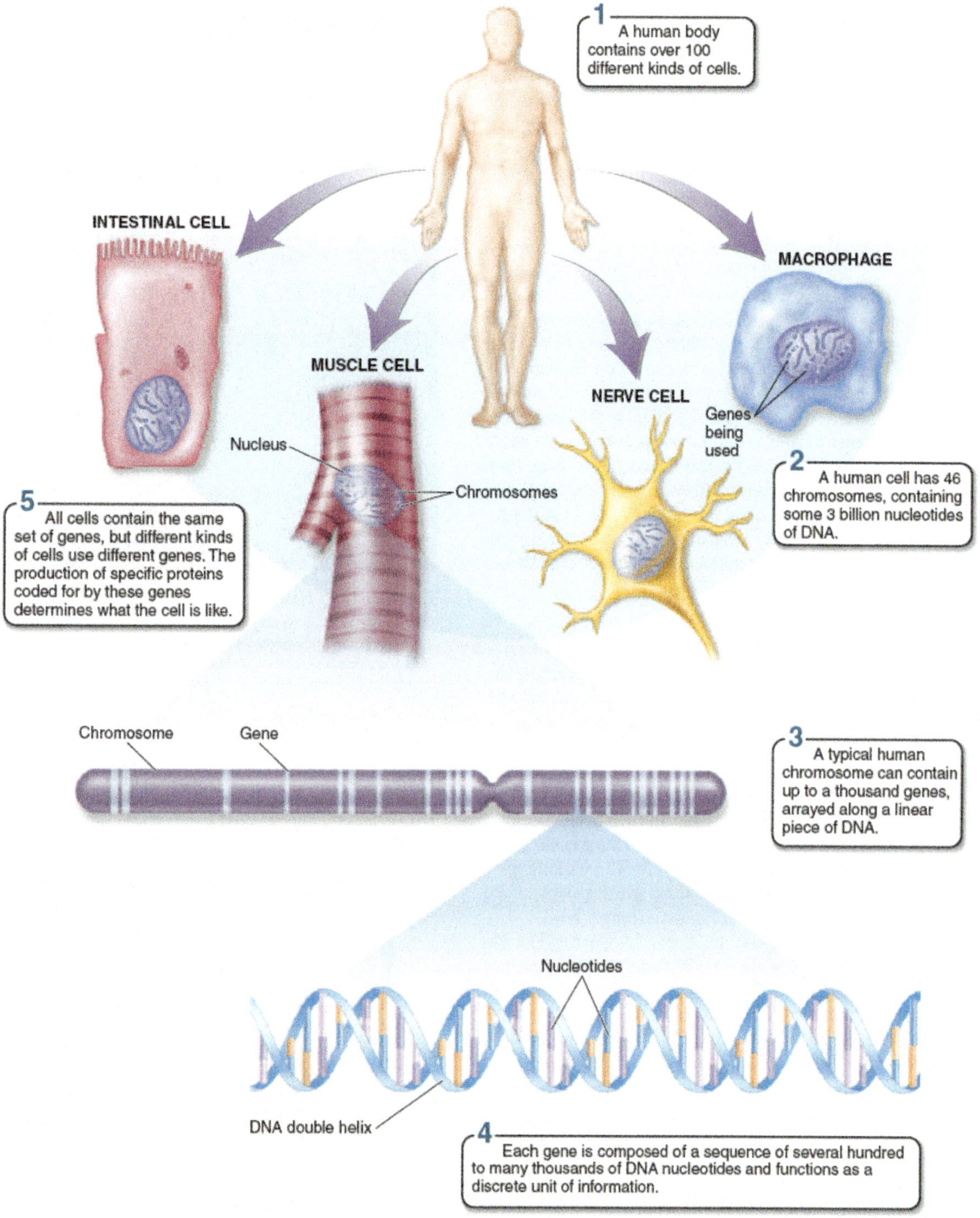

Figure 1.12 The gene theory.

The gene theory states that what an organism is like is determined in large measure by its genes. Here you see how the many kinds of cells in the body of each of us are determined by which genes are used in making each particular kind of cell.

The Theory of Heredity: Unity of Life

The storage of hereditary information in genes composed of DNA is common to all living things. The **theory of heredity,** first advanced by Gregor Mendel in 1865, states that the genes of an organism are inherited as discrete units. A triumph of experimental science developed long before genes and DNA were understood, Mendel's theory of heredity is the subject of chapter 10. Soon after Mendel's theory gave rise to the field of genetics, other biologists proposed what has come to be called the **chromosomal theory of inheritance,** which in its simplest form states that the genes of Mendel's theory are physically located on chromosomes, and that it is because chromosomes are parceled out in a regular manner during reproduction that Mendel's regular patterns of inheritance are seen. In modern terms, the two theories state that genes are a component of a cell's chromosomes (like the 23 pairs of human chromosomes you see in figure 1.13), and that the regular duplication of these chromosomes during sexual reproduction is responsible for the pattern of inheritance we call Mendelian segregation. Sometimes a character is conserved essentially unchanged in a long line of descent, reflecting a fundamental role in the biology of the organism, one not easily changed once adopted. Other characters might be modified due to changes in DNA.

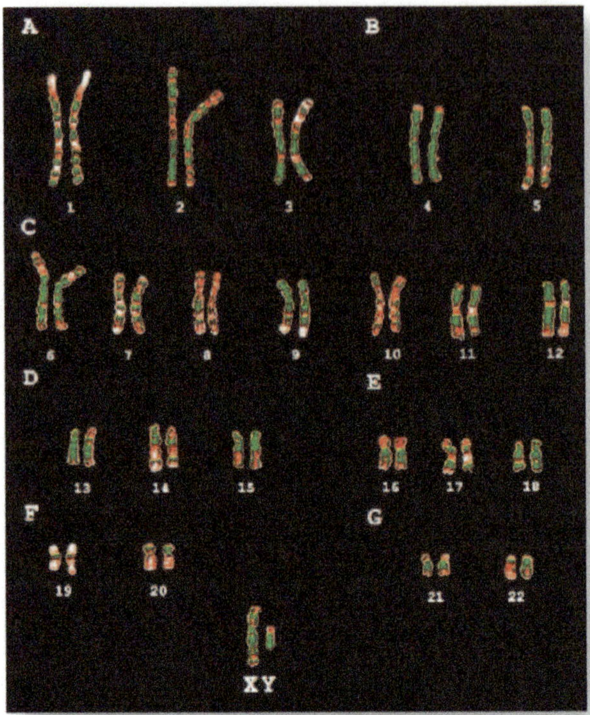

Figure 1.13 Human chromosomes.
The chromosomal theory of inheritance states that genes are located on chromosomes. This human karyotype (an ordering of chromosomes) shows banding patterns on chromosomes that represent clusters of genes.

The Theory of Evolution: Diversity of Life

The unity of life, which we see in the retention of certain key characteristics among many related life-forms, contrasts with the incredible diversity of living things that have evolved to fill the varied environments of earth. These diverse organisms are sorted by biologists into six kingdoms, as you learned in section 1.1. Organisms placed in the same kingdom have in common some general characteristics. In recent years, biologists have added a classification level above kingdoms, based on fundamental differences in cell structure. The six kingdoms are each now assigned into one of three great groups called *domains:* Bacteria, Archaea, and Eukarya (figure 1.14).

The **theory of evolution,** advanced by Charles Darwin in 1859, attributes the diversity of the living world to natural selection. Those organisms best able to respond to the challenges of living will leave more offspring, he argued, and thus their traits become more common in the population. It is because the world offers diverse opportunities that it contains so many different life-forms.

Today scientists can decipher many of the thousands of genes (the genome) of an organism. One of the great triumphs of science in the century and a half since Darwin is the detailed understanding of how Darwin's theory of evolution is related to the gene theory—of how changes in life's diversity can result from changes in individual genes (figure 1.15).

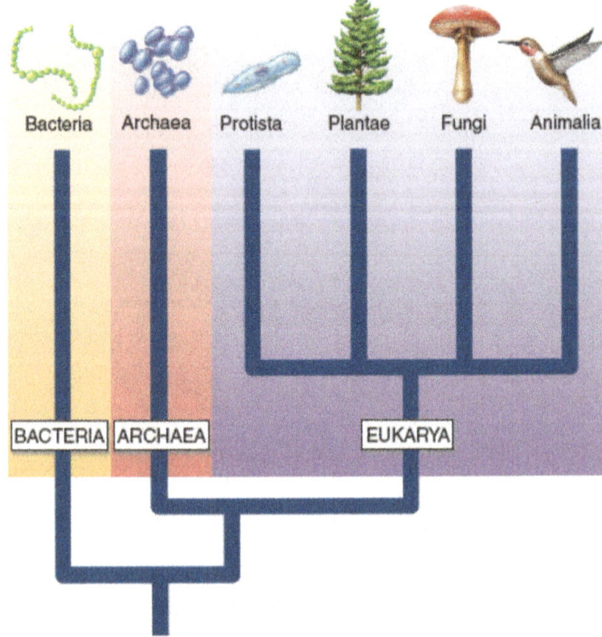

Figure 1.14 The three domains of life.
Biologists categorize all living things into three overarching groups called domains: Bacteria, Archaea, and Eukarya. Domain Bacteria contains the kingdom Bacteria, and domain Archaea contains the kingdom Archaea. Domain Eukarya is composed of four more kingdoms: Protista, Plantae, Fungi, and Animalia.

Key Learning Outcome 1.9 **The theories uniting biology state that cellular organisms store hereditary information in DNA. Sometimes DNA alterations occur, which when preserved result in evolutionary change. Today's biological diversity is the product of a long evolutionary journey.**

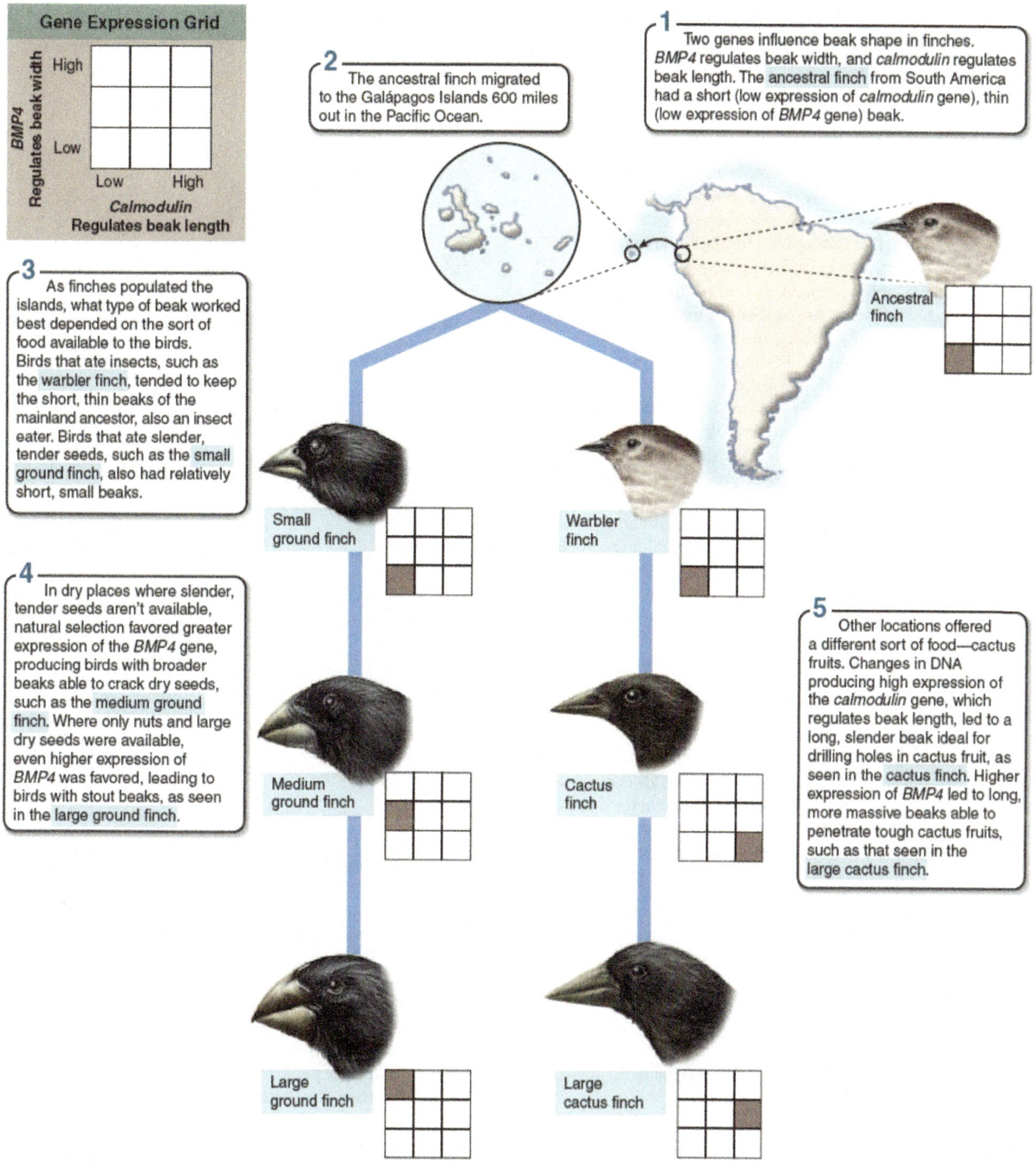

Gene Expression Grid

BMP4 Regulates beak width: High / Low

Calmodulin Regulates beak length: Low / High

1 Two genes influence beak shape in finches. *BMP4* regulates beak width, and *calmodulin* regulates beak length. The ancestral finch from South America had a short (low expression of *calmodulin* gene), thin (low expression of *BMP4* gene) beak.

2 The ancestral finch migrated to the Galápagos Islands 600 miles out in the Pacific Ocean.

3 As finches populated the islands, what type of beak worked best depended on the sort of food available to the birds. Birds that ate insects, such as the warbler finch, tended to keep the short, thin beaks of the mainland ancestor, also an insect eater. Birds that ate slender, tender seeds, such as the small ground finch, also had relatively short, small beaks.

4 In dry places where slender, tender seeds aren't available, natural selection favored greater expression of the *BMP4* gene, producing birds with broader beaks able to crack dry seeds, such as the medium ground finch. Where only nuts and large dry seeds were available, even higher expression of *BMP4* was favored, leading to birds with stout beaks, as seen in the large ground finch.

5 Other locations offered a different sort of food—cactus fruits. Changes in DNA producing high expression of the *calmodulin* gene, which regulates beak length, led to a long, slender beak ideal for drilling holes in cactus fruit, as seen in the cactus finch. Higher expression of *BMP4* led to long, more massive beaks able to penetrate tough cactus fruits, such as that seen in the large cactus finch.

Ancestral finch

Small ground finch

Warbler finch

Medium ground finch

Cactus finch

Large ground finch

Large cactus finch

Figure 1.15 The theory of evolution.

Darwin's theory of evolution proposes that many forms of a gene may exist among members of a population, and that those members with a form better suited to their particular habitat will tend to reproduce more successfully, and so their traits become more common in the population, a process Darwin dubbed "natural selection." Here you see how this process is thought to have worked on two pivotal genes that helped generate the diversity of finches on the Galápagos Islands, visited by Darwin in 1831 on his round-the-world voyage on HMS *Beagle*.

Does the Presence of One Species Limit the Population Size of Others?

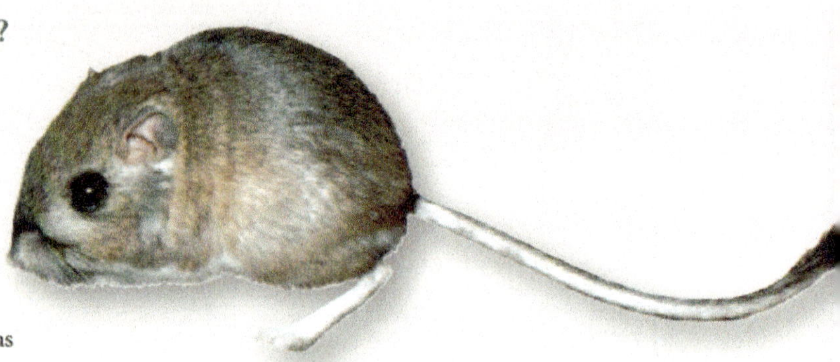

Implicit in Darwin's theory of evolution is the idea that species in nature compete for limiting resources. Does this really happen? Some of the best evidence of competition between species comes from experimental field studies, studies conducted not in the laboratory but out in natural populations. By setting up experiments in which two species occur either alone or together, scientists can determine whether the presence of one species has a negative impact on the size of the population of the other species. The experiment discussed here concerns a variety of seed-eating rodents that occur in North American deserts. In 1988, researchers set up a series of 50-meter × 50-meter enclosures to investigate the effect of kangaroo rats on smaller seed-eating rodents. Kangaroo rats were removed from half of the enclosures, but not from the other enclosures. The walls of all the enclosures had holes that allowed rodents to come and go, but in plots without kangaroo rats the holes were too small to allow the kangaroo rats to enter.

The graph to the right displays data collected over the course of the next three years as researchers monitored the number of the smaller rodents present in the enclosures. To estimate the population sizes, researchers determined how many small rodents could be captured in a fixed interval. Data were collected for each enclosure immediately after the kangaroo rats were removed in 1988, and at three-month intervals thereafter. The graph presents the relative population size—that is, the total number of captures averaged over the number of enclosures (an **average** is the numerical mean value, calculated by adding a list of values and then dividing this sum by the number of items in the list. For example, if a total of 30 rats were captured from 3 enclosures, the average would be 10 rats). As you can see, the two kinds of enclosures do not contain the same number of small rodents.

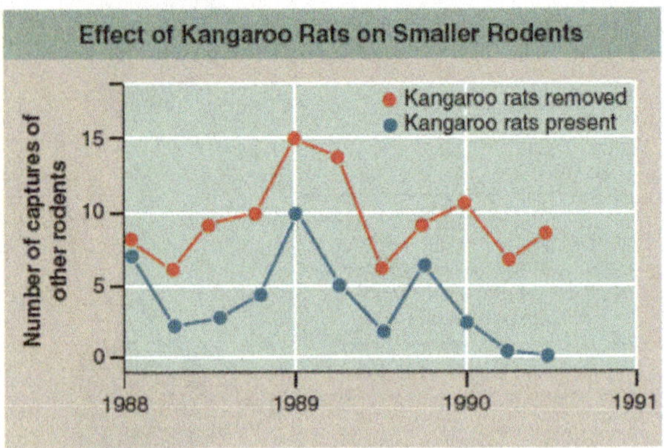

1. **Applying Concepts**
   a. Variable. In the graph, what is the dependent variable?
   b. Relative Magnitude. Which of the two kinds of enclosures maintains the highest population of small rodents? Does it have kangaroo rats or have they been removed?
2. **Interpreting Data**
   a. What is the average number of small rodents in each of the two plots immediately after kangaroo rats were removed? After one year? After two?
   b. At what point is the difference between the two kinds of enclosures the greatest?
3. **Making Inferences**
   a. What precisely is the observed impact of kangaroo rats on the population size of small rodents?
   b. Examine the magnitude of the difference between the number of small rodents in the two plots. Is there a trend?
4. **Drawing Conclusions**
   Do these results support the hypothesis that kangaroo rats compete with other small rodents to limit their population sizes?
5. **Further Analysis**
   a. Can you think of any cause other than competition that would explain these results? Suggest an experiment that could potentially eliminate or confirm this alternative.
   b. Do the populations of the two kinds of enclosures change in synchrony (that is, grow and shrink at the same times) over the course of a year? If so, why might this happen? How would you test this hypothesis?

Chapter Review

Biology and the Living World

1.1 The Diversity of Life

- Biology is the study of life. All living organisms share common characteristics, but they are also diverse and are categorized into six groups called kingdoms. The six kingdoms are Bacteria, Archaea, Protista, Fungi, Plantae, and Animalia (**figure 1.1**).

1.2 Properties of Life

- All living organisms share five basic properties: cellular organization, metabolism, homeostasis, growth and reproduction, and heredity. Cellular organization indicates that all living organisms are composed of cells. Metabolism means that all living organisms, like the kingfisher shown here from **figure 1.3**, use energy. Homeostasis is the process whereby all living organisms maintain stable internal conditions. The properties of growth and reproduction indicate that all living organisms grow in size and reproduce. The property of heredity describes how all living organisms possess genetic information in DNA that determines how each organism looks and functions, and this information is passed on to future generations.

1.3 Organization of Life

- Living organisms exhibit increasing levels of complexity within their cells a(cellular level), within their bodies (organismal level), and within ecosystems (populational level) (**figure 1.4**).
- Novel properties that appear in each level of the hierarchy of life are called emergent properties.

1.4 Biological Themes

- Five themes emerge from the study of biology: evolution, the flow of energy, cooperation, structure determines function, and homeostasis. These themes are used to examine the similarities and differences among organisms (**table 1.1**).

The Scientific Process

1.5 How Scientists Think

- Scientists use reasoning. Deductive reasoning is the process of using general principles to explain individual observations. Inductive reasoning is the process of using specific observations to formulate general principles (**figure 1.5**).

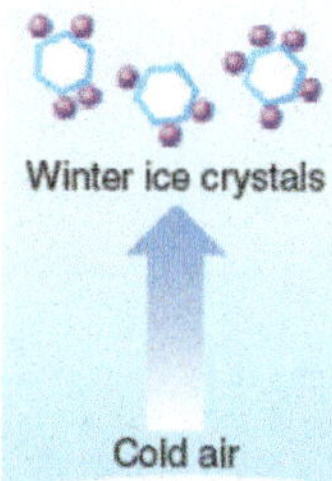

1.6 Science in Action: A Case Study

- Scientists observed a thinning of the ozone layer over Antarctica. Their scientific investigation of the "ozone hole" revealed that industrially produced CFCs were responsible for the thinning of the ozone layer in the earth's atmosphere (**figure 1.6**).

1.7 Stages of a Scientific Investigation

- Scientists use observations to formulate hypotheses. Hypotheses are possible explanations that are used to form predictions. These predictions are tested experimentally. Some hypotheses are rejected based on experimental results, while others are tentatively accepted.
- Scientific investigations often use a series of stages, called the scientific process, to study a scientific question. These stages are observations, forming hypotheses, making predictions, testing, establishing controls, and drawing conclusions (**figure 1.7**).
- The discovery of the hole in the ozone required careful observations of data collected from the atmosphere. Scientists proposed a hypothesis to explain what caused a decrease in the levels of ozone over the Antarctic. They then formed predictions and tested the hypothesis against controls.

1.8 Theory and Certainty

- Hypotheses that hold up to testing over time are combined into statements called theories. Theories carry a higher degree of certainty, although no theory in science is absolute.
- Science can only study what can be tested experimentally. A hypothesis can only be established through science if it can be tested and potentially disproven.

Core Ideas of Biology

1.9 Four Theories Unify Biology as a Science

- There are four unifying theories in biology: cell theory, gene theory, the theory of heredity, and the theory of evolution.
- The cell theory states that all living organisms are composed of cells, which grow and reproduce to form other cells (**figure 1.10**).
- The gene theory states that long molecules of DNA carry instructions for producing cellular components. These instructions are encoded in the nucleotide sequences in the strands of DNA, like this section of DNA from **figure 1.11**. The nucleotides are organized into discrete units called genes, and the genes determine how an organism looks and functions (**figure 1.12**).

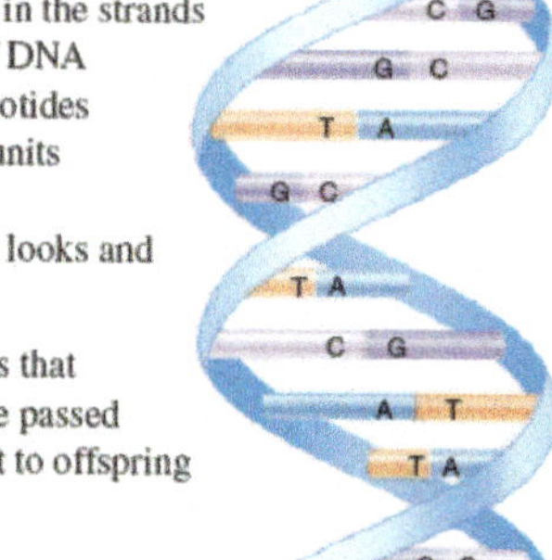

- The theory of heredity states that the genes of an organism are passed as discrete units from parent to offspring (**figure 1.13**).
- Organisms are organized into kingdoms based on similar characteristics. The kingdoms are further organized into three major groups called domains based on their cellular characteristics. The three domains are Bacteria, Archaea, and Eukarya (**figure 1.14**).
- The theory of evolution states that modifications in genes that are passed from parent to offspring result in changes in future generations. These changes lead to greater diversity among organisms over time (**figure 1.15**).

Test Your Understanding

1. Biologists categorize all living things based on related characteristics into large groups, called
   a. kingdoms. c. populations.
   b. species. d. ecosystems.
2. Living things can be distinguished from nonliving things because they have
   a. complexity. c. cellular organization.
   b. movement. d. response to a stimulus.
3. Living things are organized. Choose the answer that illustrates this organization and that is arranged from smallest to largest.
   a. cell, atom, molecule, tissue, organelle, organ, organ system, organism, population, species, community, ecosystem
   b. atom, molecule, organelle, cell, tissue, organ, organ system, organism, population, species, community, ecosystem
   c. atom, molecule, organelle, cell, tissue, organ, organ system, organism, community, population, species, ecosystem
   d. atom, molecule, cell wall, cell, organ, organelle, organism, species, population, community, ecosystem
4. At each level in the hierarchy of living things, properties occur that were not present at the simpler levels. These properties are referred to as
   a. novelistic properties. c. incremental properties.
   b. complex properties. d. emergent properties.
5. The five general biological themes include
   a. evolution, energy flow, competition, structure determines function, and homeostasis.
   b. evolution, energy flow, cooperation, structure determines function, and homeostasis.
   c. evolution, growth, competition, structure determines function, and homeostasis.
   d. evolution, growth, cooperation, structure determines function, and homeostasis.
6. When you are trying to understand something new, you begin by observation, and then put the observations together in a logical fashion to form a general principle. This method is called
   a. inductive reasoning. c. theory production.
   b. rule enhancement. d. deductive reasoning.
7. When trying to figure out explanations for observations, you usually construct a series of possible hypotheses. Then you make predictions of what will happen if each hypothesis is true, and
   a. test each hypothesis, using appropriate controls, to determine which hypothesis is true.
   b. test each hypothesis, using appropriate controls, to rule out as many as possible.
   c. use logic to determine which hypothesis is most likely true.
   d. reject those that seem unlikely.
8. Which of the following statements is correct regarding a hypothesis?
   a. After sufficient testing, you can conclude that it is true.
   b. If it explains the observations, it doesn't need to be tested.
   c. After sufficient testing, you can accept it as probable, being aware that it may be revised or rejected in the future.
   d. You never have any degree of certainty that it is true; there are too many variables.
9. Cell theory states that
   a. all organisms have cell walls and all cell walls come from other cells.
   b. all cellular organisms undergo sexual reproduction.
   c. all living organisms use cells for energy, either their own or they ingest cells of other organisms.
   d. all living organisms consist of cells, and all cells come from other cells.
10. The gene theory states that all the information that specifies what a cell is and what it does
    a. is different for each cell type in the organism.
    b. is passed down, unchanged, from parents to offspring.
    c. is contained in a long molecule called DNA.
    d. All of the above.

Apply Your Understanding

1. For over two centuries global temperatures have been warming, and over this same period of time the number of pirate ship attacks has steadily decreased. Does this graph support the conclusion that the number of pirate attacks has decreased because of warmer temperatures? Explain.

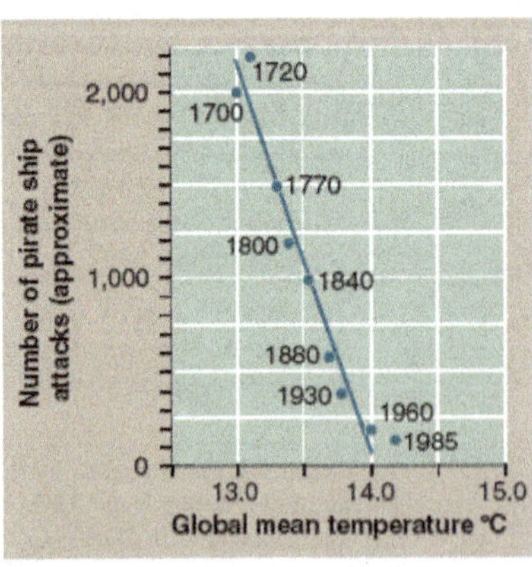

2. **Figure 1.5** You notice that on cloudy days people often carry umbrellas, folded or in a case. You also note that when umbrellas are open there are many car accidents. You conclude that open umbrellas cause car accidents. Referring back to figure 1.5, explain the type of reasoning used to reach this conclusion, and why it can sometimes be a problem.

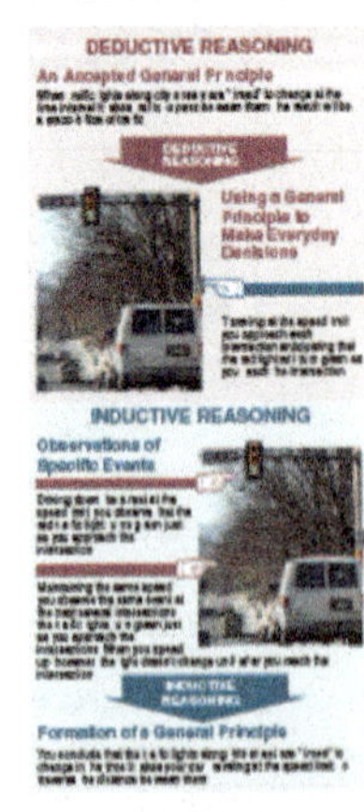

Synthesize What You Have Learned

1. You are the biologist in a group of scientists who have traveled to a distant star system and landed on a planet. You see an astounding array of shapes and forms. You have three days to take samples of living things before returning to earth. How do you decide what is alive?
2. St. John's wort is an herb that has been used for hundreds of years as a remedy for mild depression. How might a modern-day scientist research its effectiveness?

2

The Chemistry of Life

Learning Objectives

Some Simple Chemistry

2.1 Atoms

1. Define matter.
2. Describe the basic structure of an atom in terms of three subatomic particles.
3. Differentiate mass from weight.
4. Identify which of the three subatomic particles determines the chemical behavior of atoms, and explain why.
5. Explain how electrons carry energy.
6. Differentiate between an electron shell and an electron orbital.

2.2 Ions and Isotopes

1. Differentiate between an ion and an isotope.
2. Describe the process of radioactive decay.
3. Describe ^{14}C radioisotopic dating, and explain why it cannot be used to date dinosaur fossils.

2.3 Molecules

1. Define molecule.
2. List the three principal kinds of chemical bonds.
3. Explain why ionic bonds promote crystal formation, but covalent bonds do not.
4. Distinguish between polar and nonpolar covalent bonds.
5. Explain why hydrogen bonds cannot form stable molecules by themselves, whereas covalent bonds can.
6. Distinguish between chemical bonds and van der Waals forces.

Water: Cradle of Life

2.4 Hydrogen Bonds Give Water Unique Properties

1. List and describe the five general properties of water.
2. Explain why ice floats.
3. Explain why sweating cools you off.
4. Explain how an insect can walk on water, whereas you cannot.
5. Explain why table salt will dissolve in water, and vegetable oil will not.

2.5 Water Ionizes

1. Define pH.
2. Predict the change in hydrogen ion concentration represented by a difference of 1 on the pH scale.
3. Distinguish an acid from a base.
4. Explain how a buffer maintains a constant pH.
5. Name the key buffer in human blood and describe how it works to keep blood pH constant.

Today's Biology: Acid Rain

Inquiry & Analysis: Using Radioactive Decay to Date the Iceman

These trees have been seriously damaged by acid rain. The death of this forest must have seemed a calamity to the animals that lived there. A porcupine knows no chemistry, has no way to comprehend what has happened, or why. Later in this chapter, you will explore what causes acid rain and snow, and how the acid has killed forests like this one. A famous conservation saying is that "you cannot save what you don't understand." In order to understand acid rain, you must first come to understand some simpler things, the nuts and bolts that underlie what happens in nature. All living things—in fact, everything you can see in the picture above—are made of tiny particles called atoms, linked together in assemblies called molecules. This is where we will have to start, if we want to understand things like what happened to this forest. Then, with molecules under our belt, we will need to get more specific and consider the nature of rain. What are rain and snow made of? Water. We will need to take a very careful look at water. When we do, we will see that when some chemicals are added to water, a chemically active mixture called an acid results. Acid rain is water containing such chemicals. Understanding this gives us the mental tool we need to attack the problem of what happened to this forest and determine how to stop it. In just this way, chemistry underlies much of what you will learn in biology.

Some Simple Chemistry

2.1 Atoms

Biology is the science of life, and all life, in fact even all nonlife, is made of substances. **Chemistry** is the study of the properties of these substances. So, while it may seem tedious or unrelated to examine chemistry in a biology text, it is essential. Organisms are chemical machines (figure 2.1), and to understand them we must learn a little chemistry.

Any substance in the universe that has mass and occupies space is defined as **matter.** All matter is composed of extremely small particles called **atoms.** An atom is the smallest particle into which a substance can be divided and still retain its chemical properties.

Figure 2.1 Replacing electrolytes.

During extreme exercise, athletes will often consume drinks that contain "electrolytes," chemicals such as calcium, potassium, and sodium that play an important role in muscle contraction. Electrolytes can also be depleted in other types of dehydration.

Every atom has the same basic structure you see in figure 2.2. At the center of every atom is a small, very dense nucleus formed of two types of subatomic particles, **protons** (illustrated by purple balls) and **neutrons** (the pink balls). Whizzing around the core is an orbiting cloud of a third kind of subatomic particle, the **electron** (depicted by yellow balls on concentric rings). Neutrons have no electrical charge, whereas protons have a positive charge and electrons have a negative one. In each atom, there is an orbiting electron for every proton in the nucleus. The electron's negative charge balances the proton's positive charge. The atom is said to be electrically neutral.

An atom is typically described by the number of protons in its nucleus or by the overall mass of the atom. The terms *mass* and *weight* are often used interchangeably, but they have slightly different meanings. Mass refers to the amount of a substance, whereas weight refers to the force gravity exerts on a substance. Hence, an object has the same mass whether it is on the earth or the moon, but its weight will be greater on the earth, because the earth's gravitational force is greater than the moon's. For example, an astronaut weighing 180 pounds on earth will weigh about 30 pounds on the moon. He didn't lose any significant mass during his flight to the moon, there is just less gravitational pull on his mass.

The number of protons in the nucleus of an atom is called the **atomic number.** For example, the atomic number of carbon is 6 because it has six protons. Atoms with the same atomic number (that is, the same number of protons) have the same chemical properties and are said to belong to the same **element.** Formally speaking, an element is any substance that cannot be broken down into any other substance by ordinary chemical means.

Neutrons are similar to protons in mass, and the number of protons and neutrons in the nucleus of an atom is called the **mass number.** A carbon atom that has six protons and six neutrons has a mass number of 12. An electron's contribution to the overall mass of an atom is negligible. The atomic numbers and mass numbers of some of the most common elements on earth are shown in table 2.1.

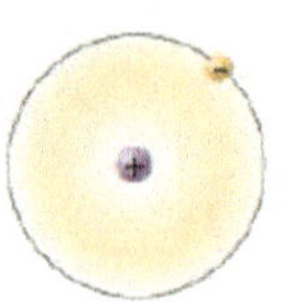

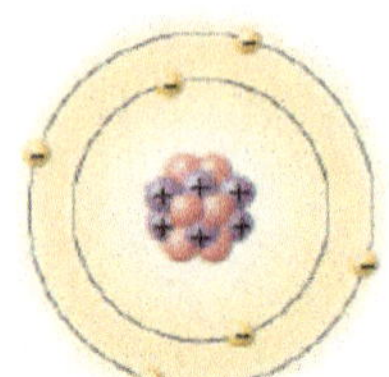

Figure 2.2 Basic structure of atoms.

All atoms have a nucleus consisting of protons and neutrons, except hydrogen, the smallest atom, which has only one proton and no neutrons in its nucleus. Carbon, for example, has six protons and six neutrons in its nucleus. Electrons spin around the nucleus in orbitals a far distance away from the nucleus. The electrons determine how atoms react with each other.

TABLE 2.1 ELEMENTS COMMON IN LIVING ORGANISMS

Element	Symbol	Atomic Number	Mass Number
Hydrogen	H	1	1.008
Carbon	C	6	12.011
Nitrogen	N	7	14.007
Oxygen	O	8	15.999
Sodium	Na	11	22.989
Phosphorus	P	15	30.974
Sulfur	S	16	32.064
Chlorine	Cl	17	35.453
Potassium	K	19	39.098
Calcium	Ca	20	40.080
Iron	Fe	26	55.847

Electrons Determine What Atoms Are Like

Electrons have very little mass (only about 1/1,840 the mass of a proton). Of all the mass contributing to your weight, the portion that is contributed by electrons is less than the mass of your eyelashes. And yet electrons determine the chemical behavior of atoms because they are the parts of atoms that come close enough to each other in nature to interact. Almost all the volume of an atom is empty space. Protons and neutrons lie at the core of this space, whereas orbiting electrons are very far from the nucleus. If the nucleus of an atom were the size of an apple, the orbit of the nearest electron would be more than a mile out!

Electrons Carry Energy

Because electrons are negatively charged, they are attracted to the positively charged nucleus, but they also repel the negative charges of each other. It takes work to keep them in orbit, just as it takes work to hold an apple in your hand when gravity is pulling the apple down toward the ground. The apple in your hand is said to possess **energy,** the ability to do work, because of its position—if you were to release it, the apple would fall. Similarly, electrons have energy of position, called *potential energy*. It takes work to oppose the attraction of the nucleus, so moving the electron farther out from the nucleus, as shown by the set of arrows on the right side of figure 2.3, requires an input of energy and results in an electron with greater potential energy. Moving an electron in toward the nucleus has the opposite effect (the set of arrows on the left side); energy is released, and the electron has less potential energy. Consider again an apple held in your hand. If you carry the apple up to a second-story window, it has a greater potential energy when you drop it, compared to when it is dropped at ground level. Similarly, if you lower the apple until it is 6 inches from the ground, it has less potential energy. Cells use the potential energy of atoms to drive chemical reactions, as we will discuss in chapter 5.

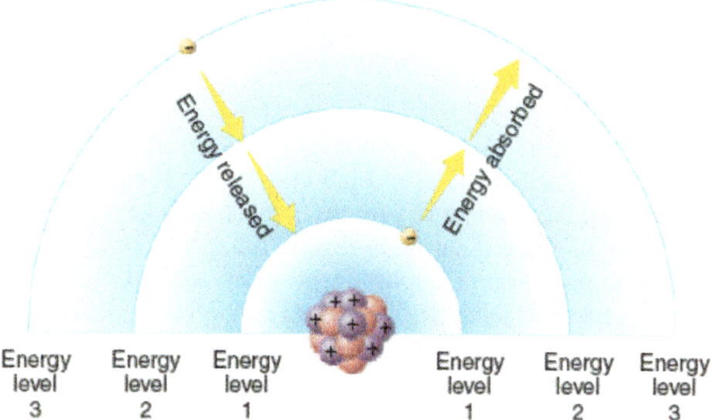

Figure 2.3 The electrons of atoms possess potential energy.
Electrons that circulate rapidly around the nucleus contain energy, and depending on their distance from the nucleus, they may contain more or less energy. Energy level 1 is the lowest potential energy level because it is closest to the nucleus. When an electron absorbs energy, it moves from level 1 to the next higher energy level (level 2). When an electron loses energy, it falls to a lower energy level closer to the nucleus.

While the energy levels of an atom are often visualized as well-defined circular orbits around a central nucleus as was shown in figure 2.2, such a simple picture is not accurate. These energy levels, called *electron shells,* often consist of complex three-dimensional shapes, and the exact location of an individual electron at any given time is impossible to specify. However, some locations are more probable than others, and it is often possible to say where an electron is *most likely* to be located. The volume of space around a nucleus where an electron is most likely to be found is called the **orbital** of that electron.

Each electron shell has a specific number of orbitals, and each orbital can hold up to two electrons. The first shell in any atom contains one orbital. Helium, shown in figure 2.4*a*, has one electron shell with one orbital that corresponds to the lowest energy level. The orbital contains two electrons, shown above and below the nucleus. In atoms with more than one electron shell, the second shell contains four orbitals and holds up to eight electrons. Nitrogen, shown in figure 2.4*b*, has two electron shells; the first one is completely filled with two electrons, but three of the four orbitals in the second electron shell are not filled because nitrogen's second shell contains only five electrons (openings in orbitals are indicated with dotted circles). In atoms with more than two electron shells, subsequent shells also contain up to four orbitals and a maximum of eight electrons. Atoms with unfilled electron orbitals tend to be more reactive because they lose, gain, or share electrons in order to fill their outermost electron shell. Losing, gaining, or sharing electrons is the basis for chemical reactions in which chemical bonds form between atoms. Chemical bonds will be discussed later in this chapter.

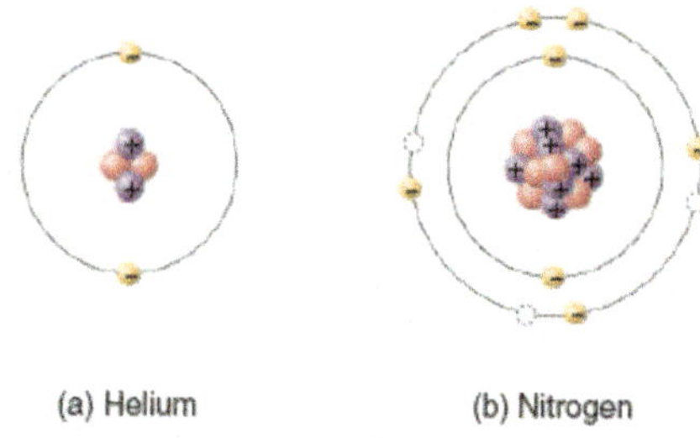

Figure 2.4 Electrons in electron shells.
(*a*) An atom of helium has two protons, two neutrons, and two electrons. The electrons fill the one orbital in its one electron shell, the lowest energy level. (*b*) An atom of nitrogen has seven protons, seven neutrons, and seven electrons. Two electrons fill the orbital in the innermost electron shell, and five electrons occupy orbitals in the second electron shell (the second energy level). The orbitals in the second electron shell can hold up to eight electrons; therefore there are three vacancies in the outer electron shell of a nitrogen atom.

Key Learning Outcome 2.1 Atoms, the smallest particles into which a substance can be divided, are composed of electrons orbiting a nucleus that contains protons and neutrons. Electrons determine the chemical behavior of atoms.

2.2 Ions and Isotopes

Ions

Sometimes an atom may gain or lose an electron from its outer shell. Atoms in which the number of electrons does not equal the number of protons because they have gained or lost one or more electrons are called **ions.** All ions are electrically charged. For example, an atom of sodium (on the left in figure 2.5) becomes a positively charged ion, called a *cation* (on the right), when it loses an electron, such that one proton in the nucleus is left with an unbalanced charge (11 positively charged protons and only 10 negatively charged electrons). Negatively charged ions, called *anions*, also form when an atom gains an electron from another atom.

Isotopes

The number of neutrons in an atom of a particular element can vary without changing the chemical properties of the element. Atoms that have the same number of protons but different numbers of neutrons are called **isotopes.** Isotopes of an atom have the same atomic number but differ in their mass number. Most elements in nature exist as mixtures of different isotopes. For example, there are three isotopes of the element carbon, all of which possess six protons (the purple balls in figure 2.6). The most common isotope of carbon (99% of all carbon) has six neutrons (the pink balls). Because its mass number is 12 (six protons plus six neutrons), it is referred to as carbon-12. The isotope carbon-14 (on the right) is rare (1 in 1 trillion atoms of carbon) and unstable, such that its nucleus tends to break up into particles with lower atomic numbers, a process called **radioactive decay.** Radioactive isotopes are used in medicine and in dating fossils.

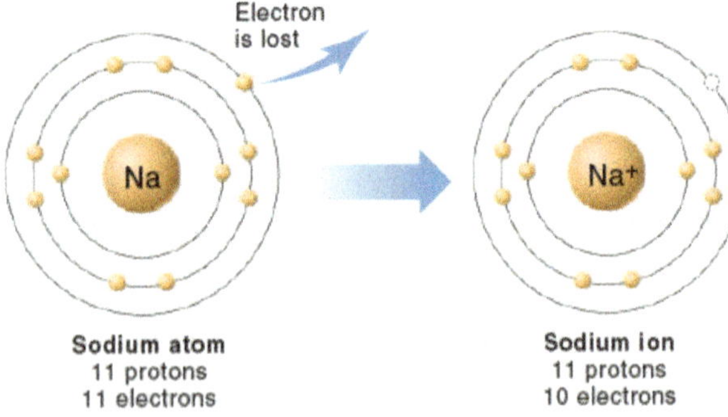

Figure 2.5 Making a sodium ion.
An electrically neutral sodium atom has 11 protons and 11 electrons. Sodium ions bear one positive charge when they ionize and lose one electron. Sodium ions have 11 protons and only 10 electrons.

Medical Uses of Radioactive Isotopes

When most people hear the word "radioactive" they picture atomic bombs exploding into mushroom clouds and the devastation that results. While it is true that the radiation emitted from radioactive isotopes can damage cells of the human body, it is also true that isotopes can be used in many medical procedures. Short-lived isotopes, those that decay fairly rapidly and produce harmless products, are commonly used as tracers in the body. A **tracer** is a radioactive substance that is taken up and used by the body. Emissions from the radioactive isotope tracer are detected using special laboratory equipment, and can reveal key diagnostic information about the functioning of the body. For example, PET and PET/CT (positron emission tomography/computerized tomography) imaging procedures can be used to identify a cancerous area in the body. First, a radioactive tracer is injected into the body. This tracer is taken up by all cells, but it is taken up in larger amounts in cells with higher metabolic activities, such as cancer cells. Images are then taken of the body, and areas emitting greater amounts of the tracer can be seen. For example, in the images in figure 2.7, the radioactive-emitting cancer site appears as a black area on the left and a yellow glowing area on the right. There are many other uses of radioactive isotopes in medicine, both in detection and treatment of disorders.

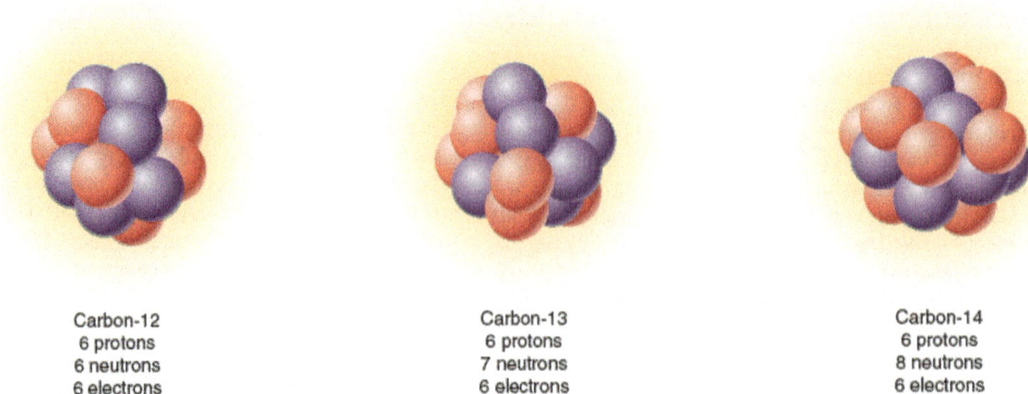

Figure 2.6 Isotopes of the element carbon.
The three most abundant isotopes of carbon are carbon-12, carbon-13, and carbon-14. The yellow "clouds" in the diagrams represent the orbiting electrons, whose numbers are the same for all three isotopes. Protons are shown in purple, and neutrons are shown in pink.

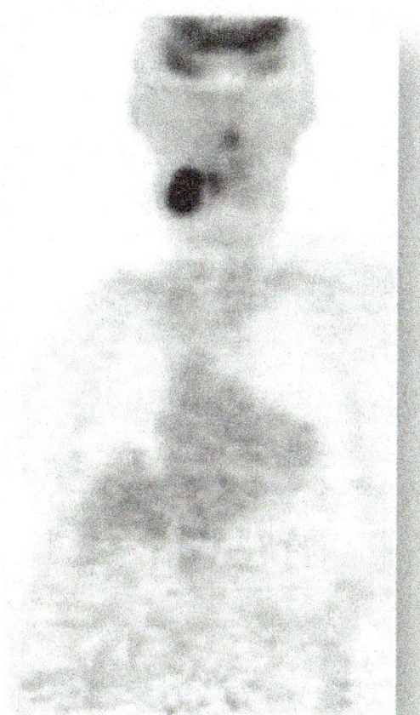

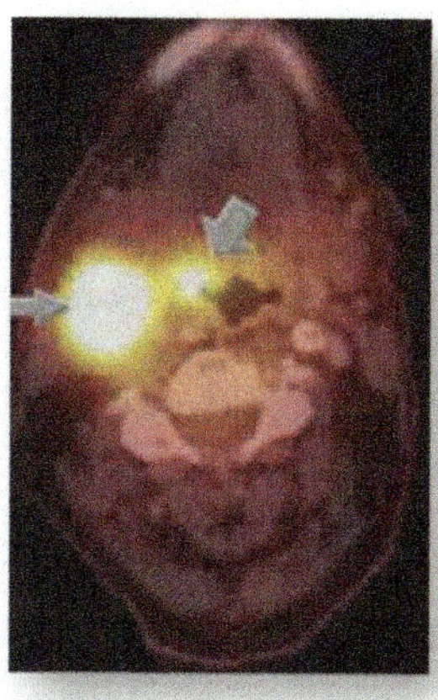

Figure 2.7 Using a radioactive tracer to identify cancer.

In certain medical imaging procedures, the patient ingests or is injected intravenously with a radioactive tracer that is absorbed in greater amounts by cancer cells. The tracer emits radioactivity that is detected using PET and PET/CT equipment. A cancerous area in the neck is seen in these two images, as a dark black area on the left and a bright yellow glowing area on the right.

Dating Fossils

Fossils are created when the remains, footprints, or other traces of organisms become buried in sand or sediment. Over time, the calcium in bone and other hard tissues becomes mineralized as the sediment is converted to rock. A fossil is any record of prehistoric life—generally taken to mean older than 10,000 years. By dating the rocks in which fossils occur, biologists can get a very good idea of how old the fossils are. Rocks are usually dated by measuring the degree of radioactive decay of certain radioactive isotopes among rock-forming minerals. A radioactive isotope is one whose nucleus is unstable and eventually flies apart, creating more stable atoms of another element. Because the rate of decay of a radioactive element (the percent of isotopes that undergo decay in a minute) is constant, scientists can use the amount of radioactive decay to date fossils. The older the fossil, the greater the fraction of its radioactive isotopes that have decayed.

A widely employed method of dating fossils less than 50,000 years old is the carbon-14 (^{14}C) **radioisotopic dating** method illustrated in figure 2.8. Most carbon atoms have a mass number of 12 (^{12}C). However, a tiny but fixed proportion of the carbon atoms in the atmosphere consists of carbon atoms with a mass number of 14 (^{14}C). This isotope of carbon is created by the bombardment of nitrogen-14 atoms with cosmic rays. This proportion of ^{14}C (designated *A* in the figure) is captured by plants in photosynthesis, and is the proportion present in the carbon molecules of the animal's body that eats the plants, in this case a rabbit. After the plant or animal dies, it no longer accumulates carbon, and the ^{14}C present at the time of death gradually decays over time back to nitrogen-14 (^{14}N). The amount of ^{14}C (*A*) decreases while the amount of ^{12}C stays the same. Scientists can determine how long ago an organism died by measuring the ratio of ^{14}C to ^{12}C

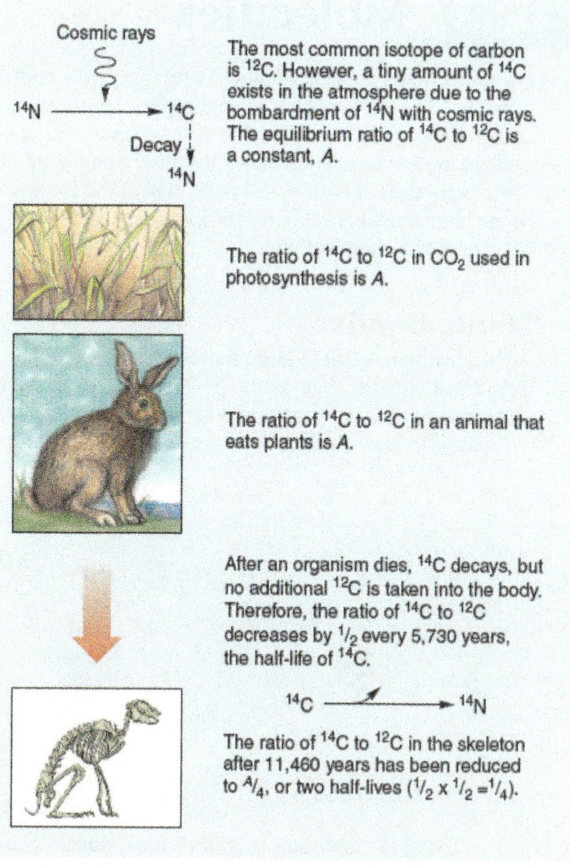

Figure 2.8 Radioactive isotope dating.

This diagram illustrates radioactive dating using carbon-14, a short-lived isotope.

in its remains or in the fossilized rock. Over time, the ratio of ^{14}C to ^{12}C decreases. It takes 5,730 years for half of the ^{14}C (1/2*A* or *A*/2) present in the sample to be converted to ^{14}N by this process. This length of time is called the **half-life** of the isotope. Because the half-life of an isotope is a constant that never changes, the extent of radioactive decay allows you to date a sample. Thus a sample that had a quarter of its original proportion of ^{14}C remaining (1/4*A* or *A*/4) would be approximately 11,460 years old (two half-lives—5,730 years for half of the ^{14}C to decay to a level of *A*/2 and another 5,730 years for the remaining ^{14}C to decay to a level of *A*/4).

For fossils older than 50,000 years, there is too little ^{14}C remaining to measure precisely, and scientists instead examine the decay of potassium-40 (^{40}K) into argon-40 (^{40}Ar), which has a half-life of 1.3 billion years.

Key Learning Outcome 2.2 When an atom gains or loses one or more electrons, it is called an ion. Isotopes of an element differ in the number of neutrons they contain, but all have the same chemical properties.

2.3 Molecules

A **molecule** is a group of atoms held together by energy. The energy acts as "glue," ensuring that the various atoms stick to one another. The energy or force holding two atoms together is called a **chemical bond.** Chemical bonds determine the shapes of the large biological molecules that will be discussed in chapter 3. There are three principal kinds of chemical bonds: ionic bonds, where the force is generated by the attraction of oppositely charged ions; covalent bonds, where the force results from the sharing of electrons; and hydrogen bonds, where the force is generated by the attraction of opposite partial electrical charges. Another type of chemical attraction called van der Waals forces will be discussed later, but keep in mind that this type of interaction is not considered a chemical bond.

Ionic Bonds

Chemical bonds called **ionic bonds** form when atoms are attracted to each other by opposite electrical charges. Just as the positive pole of a magnet is attracted to the negative pole of another, so an atom can form a strong link with another atom if they have opposite electrical charges. Because an atom with an electrical charge is an ion, these bonds are called ionic bonds.

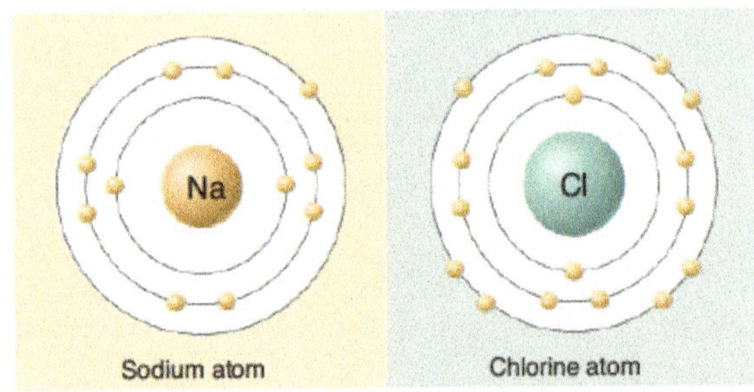

Everyday table salt is built of ionic bonds. The sodium and chlorine atoms of table salt are ions. The sodium you see in the yellow panels gives up the sole electron in its outermost shell (the shell underneath has eight) and chlorine, in the light green panels, gains an electron to complete its outermost shell. Recall from section 2.1 that an atom is more stable when its outermost electron shell is filled (with two electrons in the innermost shell or eight electrons in shells that are farther out from the nucleus).

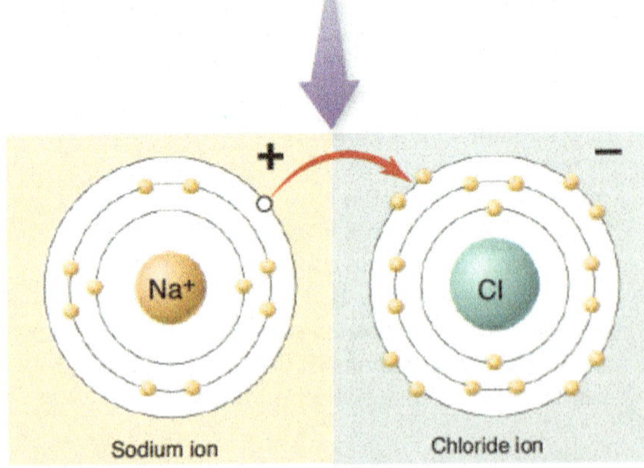

To achieve this stability, an atom will give up or accept electrons from another atom. As a result of this electron hopping, sodium atoms in table salt are positive sodium ions and chlorine atoms are negative chloride ions.

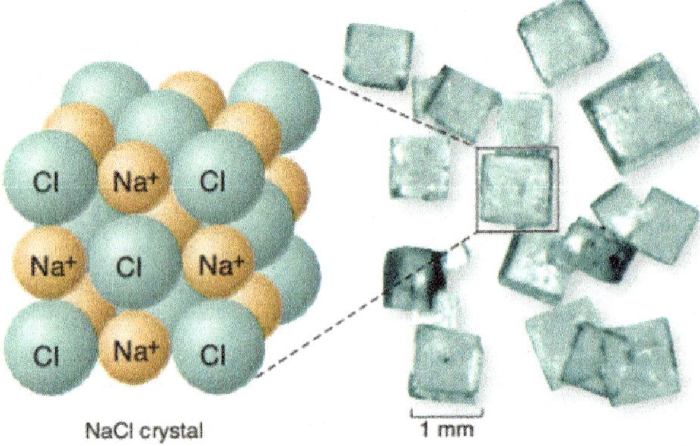

Because each ion is attracted electrically to all surrounding ions of opposite charge, this causes the formation of an elaborate matrix of sodium and chloride ionic bonds—a crystal. The sodium chloride crystal shown above reveals an organized structure of alternating sodium (yellow) and chloride (light green) ions. That is why table salt is composed of tiny crystals and is not a powder.

The two key properties of ionic bonds that make them form crystals are that they are strong (although not as strong as covalent bonds) and that they are *not* directional. A charged atom is attracted to the electrical field contributed by all nearby atoms of opposite charge. Ionic bonds do not play an important part in most biological molecules because of this lack of directionality. Complex, stable shapes require the more specific associations made possible by directional bonds.

Covalent Bonds

Strong chemical bonds called **covalent bonds** form between two atoms when they share electrons. Most of the atoms in your body are linked to other atoms by covalent bonds. Why do atoms in molecules share electrons? Remember, all atoms seek to fill up their outermost shell of orbiting electrons, which in all atoms (except tiny hydrogen and helium) takes eight electrons.

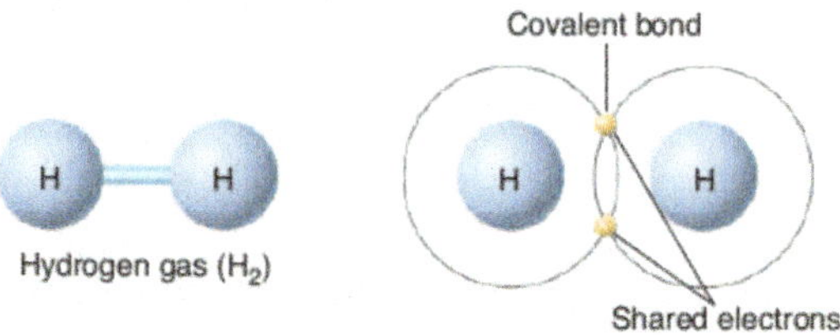

A covalent bond is formed when electrons are shared between atoms. The atoms sharing the electrons may be of the same element or different elements. Some atoms, like hydrogen (H), can form only one covalent bond, because hydrogen needs only one more electron to fill its outermost shell.

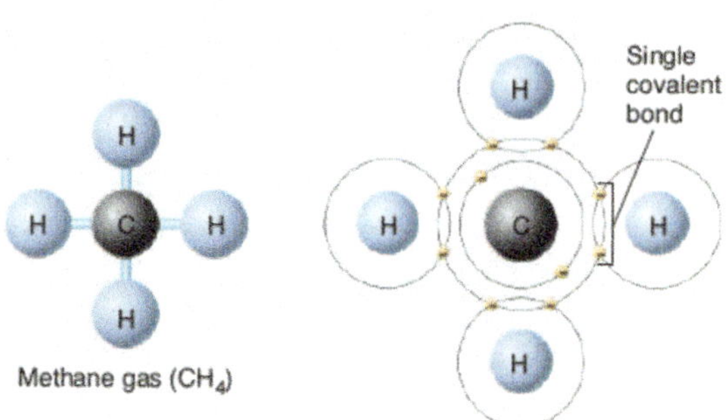

Other atoms, such as carbon (C), nitrogen (N), or oxygen (O), can form more than one covalent bond, depending on the space available in their outermost electron shells. The carbon atom has four electrons in its outermost shell, and carbon can form as many as four covalent bonds in its attempt to fully populate its outermost shell of electrons. Because there are many ways four covalent bonds can form, carbon atoms participate in many different kinds of molecules.

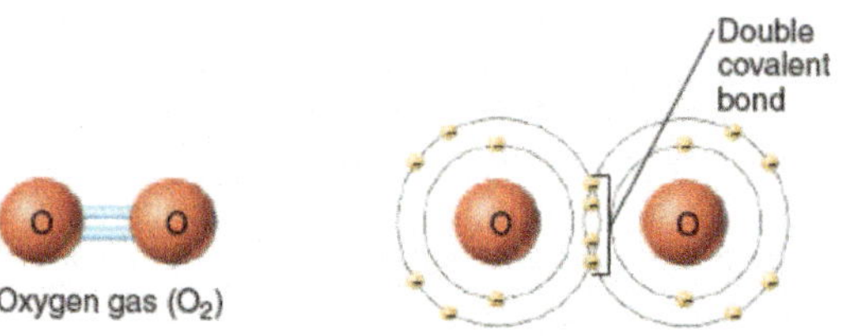

Most covalent bonds are *single bonds,* which involve the sharing of two electrons, but *double bonds* (in which four electrons are shared) are also common. *Triple bonds* (in which six electrons are shared) are much less frequent in nature, but are found in some common compounds, like nitrogen gas (N_2).

Energy is often released when covalent bonds are broken. The *Hindenberg* dirigible was filled with hydrogen gas when it exploded and burned in 1937; the energy of the inferno came from the breaking of H_2 covalent bonds.

Polar and Nonpolar Covalent Bonds When a covalent bond forms between two atoms, one nucleus may be much better at attracting the shared electrons than the other, an aspect of the atom called its *electronegativity.* In water, for example, the shared electrons are much more strongly attracted to the oxygen atom than to the hydrogen atoms; oxygen has a higher electronegativity. When this happens, shared electrons spend more time in the vicinity of the oxygen atom, which as a result becomes somewhat negative in charge; they spend less time in the vicinity of the hydrogens, and these become somewhat positive in charge. The charges are not full electrical charges like in ions but rather tiny *partial charges* (see next page), signified by the Greek letter delta (δ). What you end up with is a sort of molecular magnet, with positive and negative ends, or "poles." Molecules like this are said to be **polar molecules,** and the bonds between the atoms are called *polar covalent bonds.* Molecules that don't exhibit a large difference in electronegativities of its atoms, like the carbon-hydrogen bonds of methane, are called **nonpolar molecules** and contain *nonpolar covalent bonds.*

The two key properties of covalent bonds that make them ideal for their molecule-building role in living systems are that (1) they are strong, involving the sharing of lots of energy; and (2) they are very directional—bonds form between two specific atoms, rather than a generalized attraction of one atom for its neighbors.

Hydrogen Bonds

Polar molecules like water are attracted to one another, a special type of weak chemical bond called a **hydrogen bond.** Hydrogen bonds occur when the positive end of one polar molecule is attracted to the negative end of another, like two magnets drawn to each other.

In a hydrogen bond, an electropositive hydrogen from one polar molecule is attracted to an electronegative atom, often oxygen (O) or nitrogen (N), from another polar molecule.

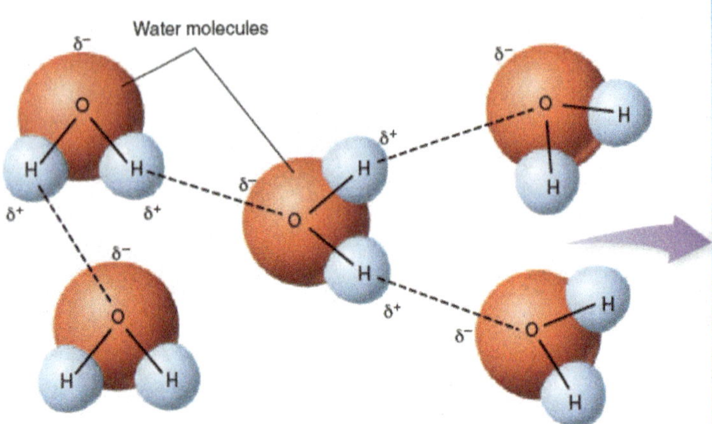

Because the oxygen atoms in water molecules are more electronegative than the hydrogen atoms, water molecules are polar. Water molecules form strong hydrogen bonds with each other, giving liquid water many unique properties. Each oxygen has a partial negative charge (δ^-), and each hydrogen has a partial positive charge (δ^+). Hydrogen bonds (shown as dashed lines) form between the positive end of one polar molecule and the negative end of another polar molecule. This attraction of partial charges attracts water molecules to one another.

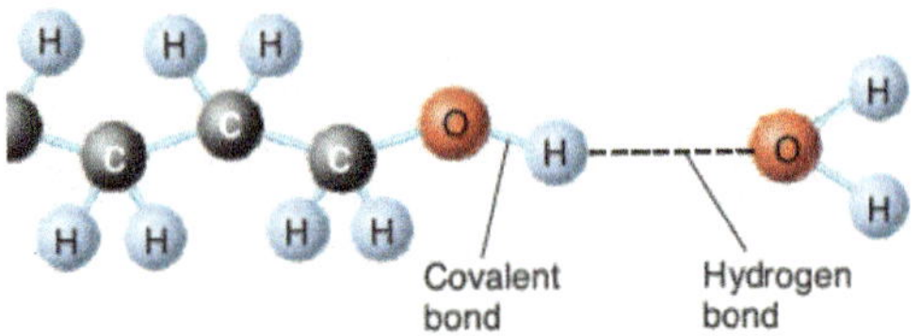

Two key properties of hydrogen bonds cause them to play an important role in the molecules found in organisms. First, they are weak and so are not effective over long distances like more powerful covalent and ionic bonds. Hydrogen bonds are too weak to actually form stable molecules by themselves. Instead, they act like Velcro, forming a tight bond by the additive effects of many weak interactions. Second, hydrogen bonds are highly directional. In chapter 3, we will discuss the role of hydrogen bonding in maintaining the structures of large biological molecules such as proteins and DNA.

Van der Waals Forces

Another important kind of weak chemical attraction is a nondirectional attractive force called *van der Waals forces* (or van der Waals interactions). These chemical forces come into play only when two atoms are very close to one another. The attraction is very weak, and disappears if the atoms move even a little apart. It becomes significant when numerous atoms in one molecule simultaneously come close to numerous atoms of another molecule—that is, when the shapes of the molecules match precisely. For example, this interaction is important when antibodies in your blood recognize the shape of an invading virus as foreign.

Key Learning Outcome 2.3 Molecules are atoms linked together by chemical bonds. Ionic bonds, covalent bonds, and hydrogen bonds are the three principal types of bonds, and van der Waals forces are weaker interactions.

2.4 Hydrogen Bonds Give Water Unique Properties

Three-fourths of the earth's surface is covered by liquid water. About two-thirds of your body is water, and you cannot exist long without it. All other organisms also require water. It is no accident that tropical rain forests are bursting with life, whereas dry deserts seem almost lifeless except after rain. The chemistry of life, then, is water chemistry.

Water has a simple atomic structure, an oxygen atom linked to two hydrogen atoms by single covalent bonds. The chemical formula for water is thus H_2O. It is because the oxygen atom attracts the shared electrons more strongly than the hydrogen atoms that water is a *polar molecule* and so can form *hydrogen bonds*. Water's ability to form hydrogen bonds is responsible for much of the organization of living chemistry, from membrane structure to how proteins fold.

The weak hydrogen bonds that form between a hydrogen atom of one water molecule and the oxygen atom of another produce a lattice of hydrogen bonds within liquid water. Each of these bonds is individually very weak and short-lived—a single bond lasts only 1/100,000,000,000 of a second. However, like the grains of sand on a beach, the cumulative effect of large numbers of these bonds is enormous and is responsible for many of the important physical properties of water (table 2.2).

TABLE 2.2 THE PROPERTIES OF WATER

Property	Explanation
Heat storage	Hydrogen bonds require considerable heat before they break, minimizing temperature changes.
Ice formation	Water molecules in an ice crystal are spaced relatively far apart because of hydrogen bonding.
High heat of vaporization	Many hydrogen bonds must be broken for water to evaporate.
Cohesion	Hydrogen bonds hold molecules of water together.
High polarity	Water molecules are attracted to ions and polar compounds.

Heat Storage

The temperature of any substance is a measure of how rapidly its individual molecules are moving. Because of the many hydrogen bonds that water molecules form with one another, a large input of thermal energy is required to disrupt the organization of liquid water and raise its temperature. Because of this, water heats up more slowly than almost any other compound and holds its temperature longer. That is a major reason why your body is able to maintain a relatively constant internal temperature.

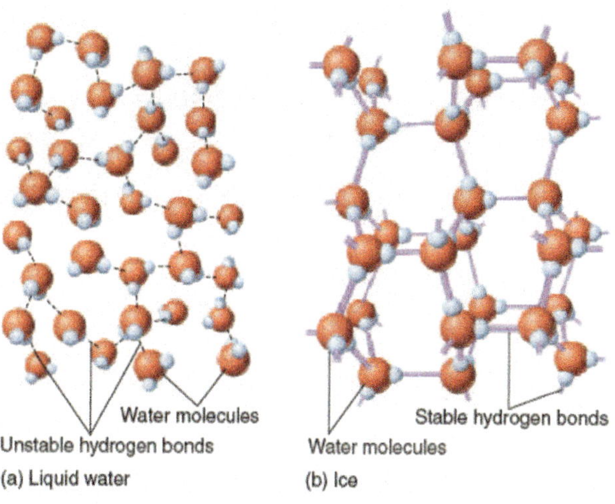

Figure 2.9 Ice formation.
When water (*a*) cools below 0°C, it forms a regular crystal structure (*b*) that floats. The individual water molecules are spaced apart and held in position by hydrogen bonds.

Ice Formation

If the temperature is low enough, very few hydrogen bonds break in water. Instead, the lattice of these bonds assumes a crystal-like structure, forming a solid we call ice. Interestingly, ice is less dense than water—that is why icebergs and ice cubes float. Why is ice less dense? This is best understood by comparing the molecular structures of water and ice that you see in figure 2.9. At temperatures above freezing (0°C or 32°F), water molecules in figure 2.9*a* move around each other with hydrogen bonds breaking and forming. As temperatures drop, the movement of water molecules decreases, allowing hydrogen bonds to stabilize, holding individual molecules farther apart, as in figure 2.9*b*, making the ice structure less dense.

High Heat of Vaporization

If the temperature is high enough, many hydrogen bonds break in water, with the result that the liquid is changed into vapor. A considerable amount of heat energy is required to do this—every gram of water that evaporates from your skin removes 2,452 joules of heat from your body, which is equal to the energy released by lowering the temperature of 586 grams of water 1°C. That is why sweating cools you off; as the sweat evaporates (vaporizes) it takes energy with it, in the form of heat, cooling the body.

(a)

(b)

Figure 2.10 Cohesion.

(a) Cohesion allows water molecules to stick together and form droplets. *(b)* Surface tension is a property derived from cohesion—that is, water has a "strong" surface due to the force of its hydrogen bonds. Some insects, such as this water strider, literally walk on water.

Cohesion

Because water molecules are very polar, they are attracted to other polar molecules—hydrogen bonds bind polar molecules to each other. When the other polar molecule is another water molecule, the attraction is called **cohesion.** The surface tension of water is created by cohesion. Surface tension is the force that causes water to bead, like on the spider web in figure 2.10, or supports the weight of the water strider. When the other polar molecule is a different substance, the attraction is called **adhesion.** Capillary action—such as water moving up into a paper towel—is created by adhesion. Water clings to any substance, such as paper fibers, with which it can form hydrogen bonds. Adhesion is why things get "wet" when they are dipped in water and why waxy substances do not—they are composed of nonpolar molecules that don't form hydrogen bonds with water molecules.

High Polarity

Water molecules in solution always tend to form the maximum number of hydrogen bonds possible. Polar molecules form hydrogen bonds and are attracted to water molecules. Polar molecules are called **hydrophilic** (from the Greek *hydros,* water, and *philic,* loving), or water-loving, molecules. Water molecules gather closely around any molecule that exhibits an electrical charge, whether a full charge (ion) or partial charge (polar molecule). When a salt crystal dissolves in water as you see happening in figure 2.11, what really happens is that individual ions break off from the crystal and become surrounded by water molecules. The blue hydrogen atoms of water are attracted to the negative charge of the chloride ions and the red oxygen atoms are attracted to the positive charge of the sodium ions. Water molecules orient around each ion like a swarm of bees attracted to honey, and this shell of water molecules, called a *hydration shell,* prevents the ions from reassociating with the crystal. Similar shells of water form around all polar molecules, and polar molecules that dissolve in water in this way are said to be **soluble** in water.

Nonpolar molecules like oil do not form hydrogen bonds and are not water-soluble. When nonpolar molecules are placed in water, the water molecules shy away, instead forming hydrogen bonds with other water molecules. The nonpolar molecules are forced into association with one another, crowded together to minimize their disruption of the hydrogen bonding of water. It seems almost as if the nonpolar compounds shrink from contact with water, and for this reason they are called **hydrophobic** (from the Greek *hydros,* water, and *phobos,* fearing). Many biological structures are shaped by such hydrophobic forces, as will be discussed in chapter 3.

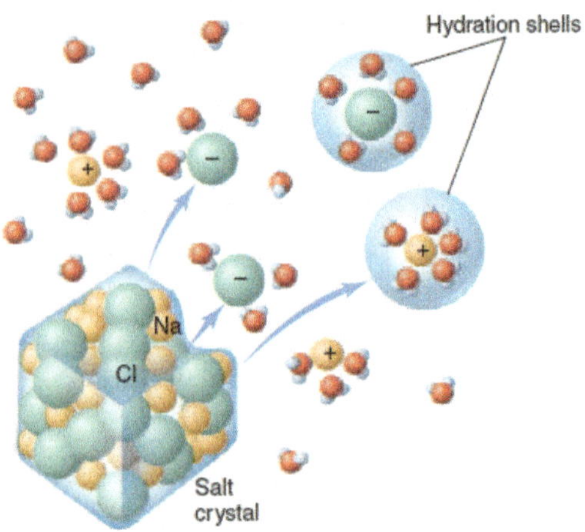

Figure 2.11 How salt dissolves in water.

Salt is soluble in water because the partial charges on water molecules are attracted to the charged sodium and chloride ions. The water molecules surround the ions, forming what are called hydration shells. When all of the ions have been separated from the crystal, the salt is said to be dissolved.

Key Learning Outcome 2.4 Water molecules form a network of hydrogen bonds in liquid and dissolve other polar molecules. Many of the key properties of water arise because it takes considerable energy to break liquid water's many hydrogen bonds.

2.5 Water Ionizes

The covalent bonds within a water molecule sometimes break spontaneously. When it happens, one of the protons (hydrogen atom nuclei) dissociates from the molecule. Because the dissociated proton lacks the negatively charged electron it was sharing in the covalent bond with oxygen, its own positive charge is no longer counterbalanced, and it becomes a positively charged ion, **hydrogen ion** (H^+). The rest of the dissociated water molecule, which has retained the shared electron from the covalent bond, is negatively charged and forms a **hydroxide ion** (OH^-). This process of spontaneous ion formation is called **ionization.** It can be represented by a simple chemical equation, in which the chemical formulas for water and the two ions are written down, with an arrow showing the direction of the dissociation:

$$\underset{\text{water}}{H_2O} \longleftrightarrow \underset{\text{hydroxide ion}}{OH^-} + \underset{\text{hydrogen ion}}{H^+}$$

Because covalent bonds are strong, spontaneous ionization is not common. In a liter of water, only roughly 1 molecule out of each 550 million is ionized at any instant in time, corresponding to 1/10,000,000 (that is, 10^{-7}) of a mole of hydrogen ions. (A mole is a measurement of weight. One mole of any object is the weight of 6.022×10^{23} units of that object.) The concentration of H^+ in water can be written more easily by simply counting the number of decimal places after the digit "1" in the denominator:

$$[H^+] = \frac{1}{10{,}000{,}000}$$

pH

A more convenient way to express the hydrogen ion concentration of a solution is to use the pH scale (figure 2.12). This scale defines pH as the negative logarithm of the hydrogen ion concentration in the solution:

$$pH = -\log [H^+]$$

Since the logarithm of the hydrogen ion concentration is simply the exponent of the molar concentration of H^+, the pH equals the exponent times –1. Thus, pure water, with an $[H^+]$ of 10^{-7} mole/liter, has a pH of 7. Recall that for every hydrogen ion formed when water dissociates, a hydroxide ion is also formed, meaning that the dissociation of water produces H^+ and OH^- in equal amounts. Therefore, a pH value of 7 indicates neutrality—a balance between H^+ and OH^-—on the pH scale.

Note that the pH scale is *logarithmic,* which means that a difference of 1 on the pH scale represents a 10-fold change in hydrogen ion concentration. This means that a solution with a pH of 4 has *10 times* the concentration of H^+ present in one with a pH of 5.

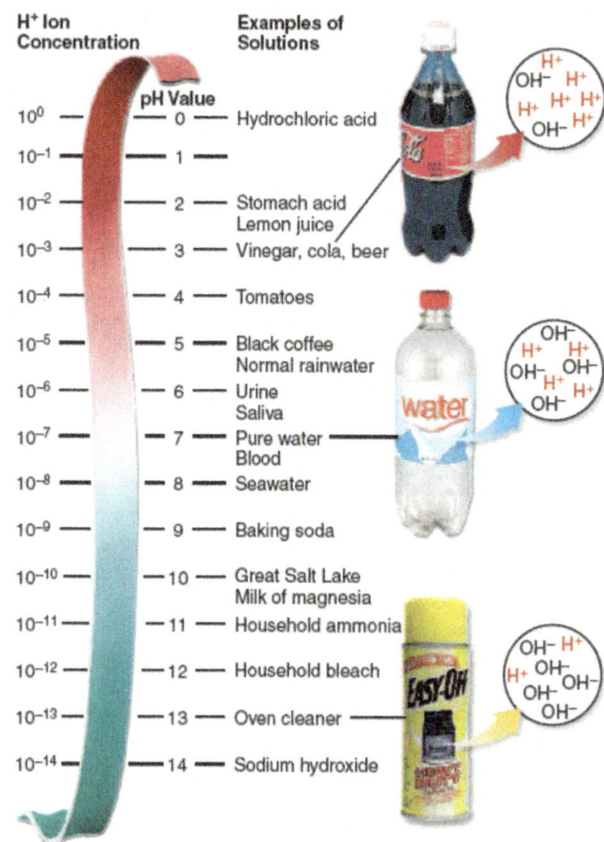

Figure 2.12 The pH scale.
A fluid is assigned a value according to the number of hydrogen ions present in a liter of that fluid. The scale is logarithmic, so that a change of only 1 means a 10-fold change in the concentration of hydrogen ions; thus lemon juice with a pH of 2 is 100 times more acidic than tomatoes with a pH of 4, and seawater is 10 times more basic than pure water.

Acids Any substance that dissociates in water to increase the concentration of H^+ is called an **acid.** Acidic solutions have pH values below 7. The stronger an acid is, the more H^+ it produces and the lower its pH. For example, hydrochloric acid (HCl), which is abundant in your stomach, ionizes completely in water. This means that a dilution of 10^{-1} mole/liter of HCl will dissociate to form 10^{-1} mole/liter of H^+, giving the solution a pH of 1. The pH of champagne, which bubbles because of the carbonic acid dissolved in it, is about 3.

Bases A substance that combines with H^+ when dissolved in water is called a **base.** By combining with H^+, a base lowers the H^+ concentration in the solution. Basic (or alkaline) solutions, therefore, have pH values above 7. Very strong bases, such as sodium hydroxide (NaOH), have pH values of 12 or more.

Today's *Biology*

Acid Rain

As you study biology, you will learn that hydrogen ions play many roles in the chemistry of life. When conditions become overly acidic—too many hydrogen ions—serious damage to organisms often results. One important example of this is acid precipitation, more informally called **acid rain.** Acid precipitation is just what it sounds like, the presence of acid in rain or snow. Where does the acid come from? Tall smokestacks from coal-burning power plants send smoke high into the atmosphere through these stacks, each of which is over 65 meters tall. The smoke the stacks belch out contains high concentrations of sulfur dioxide (SO_2), because the coal that the plants burn is rich in sulfur. The sulfur-rich smoke is dispersed and diluted by winds and air currents. Since the 1950s, such tall stacks have become popular in the United States and Europe—there are now over 800 of them in the United States alone.

In the 1970s, 20 years after the stacks were introduced, ecologists began to report evidence that the tall stacks were not eliminating the problems associated with the sulfur, just exporting the ill effects elsewhere. The lakes and forests of the Northeast suffered drastic drops in biodiversity, forests dying and lakes becoming devoid of life. It turned out that the SO_2 introduced into the upper atmosphere by high smokestacks combines with water vapor to produce sulfuric acid (H_2SO_4). When this water later falls back to earth as rain or snow, it carries the sulfuric acid with it. When schoolchildren measured the pH of natural rainwater as part of a nationwide project in 1989, rain and snow in the Northeast often had a pH as low as 2 or 3—more acidic than vinegar.

After accumulating in soils for over 50 years, the effects of acid rain are now only too evident. The impact of acid rain on forests first became evident in the Northeast. Some 15% of the lakes in New England have become chronically acidic and are dying biologically as their pH levels fall to below 5.0. Many of the forests of the northeastern United States and Canada have also been seriously damaged. The trees in this photo and on the first page of this chapter show the ill effects of acid precipitation. In the last decades, acid added to forest soils has caused the loss from these soils of over half the essential plant nutrients calcium and magnesium. Researchers blame excess acids for dissolving Ca^{++} and Mg^{++} ions into drainage waters much faster than weathering rocks can replenish them. Without them, trees stop growing and die.

Now, some 30 years later, acid rain effects are becoming apparent in the Southeast as well. Researchers suggest the reason for the delay is that southern soils are generally thicker than northern ones and thus able to sponge up far more acid. But now that southern forest soils are becoming saturated, they too are beginning to die. In a third of the southeastern streams studied, fish are declining or already gone.

The solution is straightforward: Capture and remove the emissions instead of releasing them into the atmosphere. Progressively tougher pollution laws over the past three decades have reduced U.S. emissions of sulfur dioxide by about 40% from its 1973 peak of 28.8 metric tons a year. Despite this significant progress, much remains to be done. Unless levels are cut further, researchers predict forests may not recover for centuries.

An informed public will be essential. While textbook treatments have in the past tended to minimize the impact of this issue on students ("the vast majority of North American forests are not suffering substantially from acid precipitation"), it is important that we face the issue squarely and support continued efforts to address this serious problem.

Buffers

The pH inside almost all living cells, and in the fluid surrounding cells in multicellular organisms, is fairly close to 7. The many proteins that govern metabolism are all extremely sensitive to pH, and slight alterations in pH can cause the molecules to take on different shapes that disrupt their activities. For this reason, it is important that a cell maintain a constant pH level. The pH of your blood, for example, is 7.4, and you would survive only a few minutes if it were to fall to 7.0 or rise to 7.8.

Yet the chemical reactions of life constantly produce acids and bases within cells. Furthermore, many animals eat substances that are acidic or basic; Coca-Cola, for example, is acidic, and egg white is basic. What keeps an organism's pH constant? Cells contain chemical substances called buffers that minimize changes in concentrations of H^+ and OH^-.

A **buffer** is a substance that takes up or releases hydrogen ions into solution as the hydrogen ion concentration of the solution changes. Hydrogen ions are donated to the solution when their concentration falls and taken from the solution when their concentration rises. The graph in figure 2.13 shows how buffers work. The blue line indicates changes in pH. As a base is added to the solution, the H^+ concentration falls and the pH should rise sharply, but by contributing H^+ to the solution the buffer works to keep the pH within a range, called the buffering range (the darker blue bar). Only when the buffering capacity is exceeded does the pH begin to rise. What sort of substance will act in this way? Within organisms, most buffers consist of pairs of substances, one an acid and the other a base.

The key buffer in human blood is an acid-base pair consisting of *carbonic acid* (acid) and *bicarbonate* (base). These two substances interact in a pair of reversible reactions. First, carbon dioxide (CO_2) and H_2O join to form carbonic acid (H_2CO_3) (step 2 in figure 2.14), which in a second reaction dissociates to yield bicarbonate ion (HCO_3^-) and H^+ (step 3). If some acid or other substance adds H^+ to the blood, the HCO_3^- acts as a base and removes the excess H^+ by forming H_2CO_3. Similarly, if a basic substance removes H^+ from the blood, H_2CO_3 dissociates, releasing more H^+ into the blood. The forward and reverse reactions that interconvert H_2CO_3 and HCO_3^- thus stabilize the blood's pH.

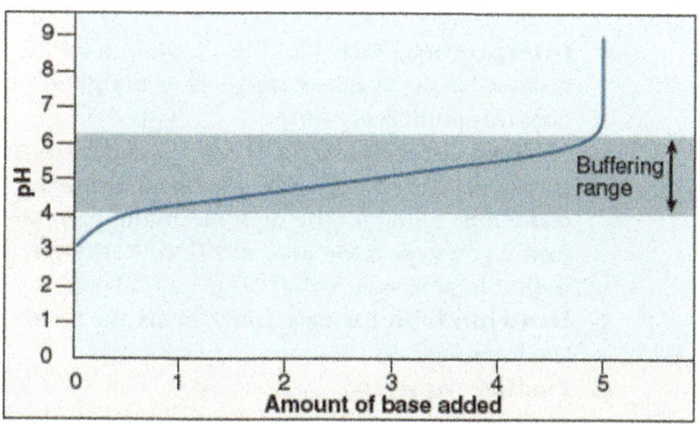

Figure 2.13 Buffers minimize changes in pH.
Adding a base to a solution neutralizes some of the acid present and so raises the pH. Thus, as the curve moves to the right, reflecting more and more base, it also rises to higher pH values. What a buffer does is to make the curve rise or fall very slowly over a portion of the pH scale, called the "buffering range" of that buffer.

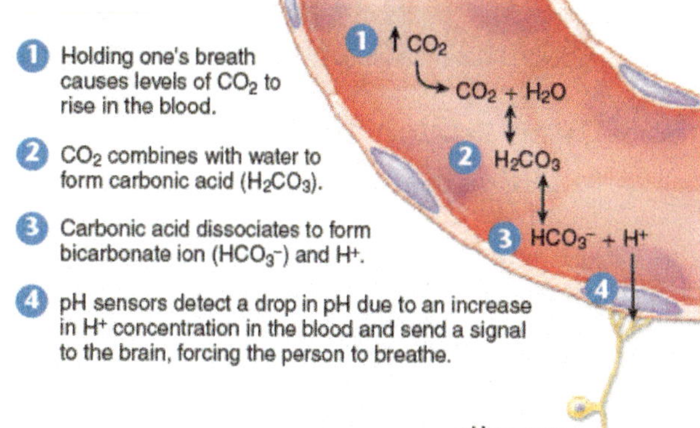

Figure 2.14 Holding your breath.
CO_2 accumulates in the blood when a person holds his/her breath. CO_2 combines with water, forming carbonic acid. Carbonic acid dissociates into bicarbonate and H^+, which acts to lower the pH. The drop in pH is detected by sensors that stimulate the brain, causing the person to breathe.

For example, when you breathe in, your body takes up oxygen from the air and when you breathe out, your body releases carbon dioxide. When you hold your breath, CO_2 accumulates in your blood and drives the chemical reactions in figure 2.14, producing carbonic acid. Can you hold your breath indefinitely? No, but not for the reason you might think. It is not lack of oxygen that forces you to breathe, but too much carbon dioxide. If you try and hold your breath for very long, CO_2 accumulates in the blood, as shown in step 1 in figure 2.14, triggering the formation of carbonic acid (step 2) that dissociates into bicarbonate ion and H^+ (step 3). This increase in H^+ causes the blood to become more acidic. If the pH in the blood drops too low, pH sensors that are located in some of the major blood vessels of the body detect the change (step 4) and send signals to the brain. These signals, along with other sensory processes, stimulate the area of the brain that controls respiration, causing it to increase the rate of breathing. Hyperventilating, breathing very quickly, has the opposite effect, lowering the levels of CO_2 in the blood. That is why you are told to breathe into a paper bag when you are hyperventilating, to increase your intake of CO_2.

Key Learning Outcome 2.5 A tiny fraction of water molecules spontaneously ionize at any moment, forming H^+ and OH^-. The pH of a solution is a measure of its H^+ concentration. Low pH values indicate high H^+ concentrations (acidic solutions), and high pH values indicate low H^+ concentrations (basic solutions).

INQUIRY & ANALYSIS

Using Radioactive Decay to Date the Iceman

In the fall of 1991, sticking out of the melting snow on the crest of a high pass near the mountainous border between Italy and Austria, two Austrian hikers found a corpse. Right away it was clear the body was very old, frozen in an icy trench where he had sought shelter long ago and only now released as the ice melted. In the years since this startling find, scientists have learned a great deal about the dead man, whom they named Ötzi. They know his age, his health, the shoes and clothing he wore, what he ate, and that he died from an arrow that ripped through his back. Its tip is still embedded in the back of his left shoulder. From the distribution of chemicals in his teeth and bones, we know he lived his life within 60 kilometers of where he died.

How long ago did this Iceman die? Scientists answered this key question by measuring the degree of decay of the short-lived carbon isotope ^{14}C in Ötzi's body. This procedure is discussed earlier in this chapter (see figure 2.8). The graph to the right displays the radioactive decay curve of the carbon isotope carbon-14 (^{14}C); it takes 5,730 years for half of the ^{14}C present in a sample to decay to nitrogen-14 (^{14}N). When Ötzi's carbon isotopes were analyzed, researchers determined that the ratio of ^{14}C to ^{12}C (a **ratio** is the size of one variable relative to another), also written as the fraction $^{14}C/^{12}C$, in Ötzi's body is 0.435 of the fraction found in tissues of a person who has recently died.

1. **Applying Concepts**
   a. Variable. In the graph, what is the dependent variable?
   b. Proportion. What proportion (a **proportion** is the size of a variable relative to the whole) of the ^{14}C present in Ötzi's body when he died is still there today?

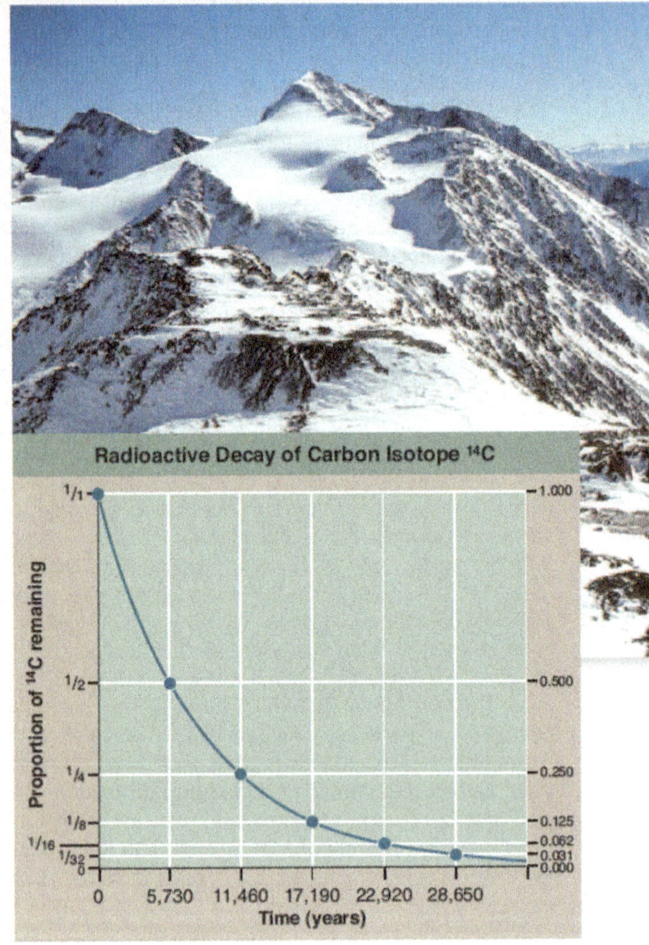

2. **Interpreting Data** Plot this proportion on the ^{14}C radioactive decay curve above. How many half-lives does this point represent?
3. **Making Inferences** If Ötzi were indeed a recent corpse, made to look old by the harsh weather conditions found on the high mountain pass, what would you expect the ratio of ^{14}C to ^{12}C to be, relative to that in your own body?
4. **Drawing Conclusions** How old are the remains of the Iceman Ötzi?
5. **Further Analysis**
   a. The radioactive iodine isotope ^{131}I decays at a half-life of eight days. Plotted on the graph above, would its radioactive decay curve be above or below that of ^{14}C?
   b. Scientists often employ the radioactive decay of isotope potassium-40 (^{40}K) into argon-40 (^{40}Ar) to date old material. ^{40}K has a half-life of 1.3 billion years. Would it be a better or poorer isotope than ^{14}C to use in dating Ötzi?

Chapter Review

Some Simple Chemistry

2.1 Atoms

- An atom is the smallest particle that retains the chemical properties of its substance. Atoms, like the carbon atom shown here from **figure 2.2**, contain a core nucleus of protons and neutrons; electrons spin around the nucleus. The number of electrons equals the number of protons in an atom.
- The number of protons in an atom is called its atomic number. The mass that is contributed by the protons and neutrons is called the atom's mass number. All atoms that have the same atomic number are said to be the same element.
- Protons are positively charged particles and neutron particles carry no charge. Electrons are negatively charged particles that orbit around the nucleus at different energy levels. Electrons determine the chemical behavior of an atom because they are the subatomic particles that interact with other atoms.
- It takes energy to hold the electrons in their orbits; this energy of position is called potential energy. The amount of potential energy of an electron is based on its distance from the nucleus (**figure 2.3**).
- Most electron shells hold up to eight electrons and atoms will undergo chemical reactions in order to fill the outermost electron shell, either by gaining, losing, or sharing electrons (**figure 2.4**).

2.2 Ions and Isotopes

- Ions are atoms that have either gained one or more electrons (negative ions called anions) or lost one or more electrons (positive ions called cations) (**figure 2.5**).
- Isotopes are atoms that have the same number of protons but differing numbers of neutrons (**figure 2.6**). Isotopes tend to be unstable and break up into other elements through a process called radioactive decay. Some isotopes have applications in medicine (**figure 2.7**) and dating fossils (**figure 2.8**).

2.3 Molecules

- Molecules form when atoms are held together with energy. The force holding atoms together is called a chemical bond. There are three main types of chemical bonds.
- Ionic bonds form when ions of opposite charge are attracted to each other. Table salt is formed by ionic bonds between positive sodium ions and negative chloride ions (**integrated art, page 38**).
- Covalent bonds form when two atoms share electrons, attempting to fill empty electron orbitals (**integrated art, page 39**). Covalent bonds are stronger when more electrons are shared.

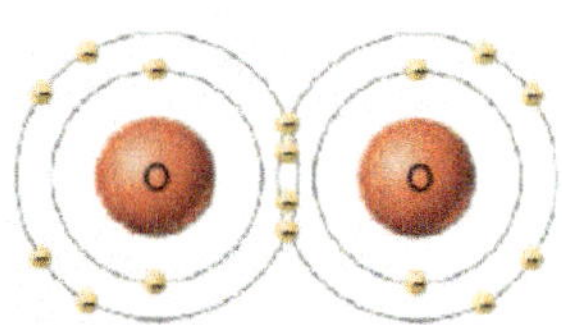

- The atoms in a polar molecule are held together by covalent bonds in which the shared electrons are unevenly distributed around their nuclei, giving the molecule a slightly positive end and a slightly negative end. Hydrogen bonds form when the positive end of one polar molecule is attracted to the negative end of another (**integrated art, page 40**).

Water: Cradle of Life

2.4 Hydrogen Bonds Give Water Unique Properties

- Water molecules are polar molecules that form hydrogen bonds with each other and with other polar molecules. Many of the physical properties of water are attributed to hydrogen bonding.
- Water molecules held together through hydrogen bonding are more difficult to separate, and as a result, a significant amount of heat energy is needed to pull the molecules apart. For this reason, water heats up slowly and holds it temperature longer.
- The hydrogen bonds that hold water molecules together become more stable at lower temperatures and as a result, they lock water molecules into place in solid crystal structures called ice, such as that shown here from **figure 2.9**.
- In order for water to vaporize into a gas, a significant input of heat energy is needed to break the hydrogen bonds. This high heat of vaporization is a property of water used by our bodies in thermoregulation.
- Because water molecules are polar molecules, they will form hydrogen bonds with other polar molecules. If the other polar molecules are water molecules, the process is called cohesion (**figure 2.10**). If the other polar molecules are some other substance, the process is called adhesion.
- When water molecules form hydrogen bonds with other polar molecules, water molecules will tend to surround other polar molecules, forming a barrier around them called a hydration shell. Polar molecules are said to be hydrophilic and are water-soluble (**figure 2.11**). Nonpolar molecules do not form hydrogen bonds and will cluster together when placed in water. They are said to be hydrophobic and are water-insoluble.

2.5 Water Ionizes

- Water molecules dissociate forming negatively charged hydroxide ions (OH^-) and positively charged hydrogen ions (H^+). This property of water is significant because the concentration of hydrogen ions in a solution determines its pH.
- A solution with a higher hydrogen ion concentration is an acid with specific chemical properties, and a solution with a lower hydrogen ion concentration is a base with different chemical properties (**figure 2.12**).
- Substances called buffers control changes in pH by taking up or releasing H^+ into the solution and control pH within a range called the buffering range (**figure 2.13**).
- A buffer that functions in the human body is an acid-base pair consisting of carbonic acid and bicarbonate. This buffering process involves a series of two reversible reactions, one that generates hydrogen ions, reducing the pH, and the reverse reaction that takes up hydrogen ions from solution, increasing pH (**figure 2.14**).
- This carbonic acid-bicarbonate buffering system works in the blood to regulate blood pH (**figure 2.14**).

Test Your Understanding

1. The smallest particle into which a substance can be divided and still retain all of its chemical properties is
   a. matter.
   b. an atom.
   c. a molecule.
   d. mass.
2. An atom that has gained or lost one or more electrons is
   a. an isotope.
   b. a neutron.
   c. an ion.
   d. radioactive.
3. Atoms are held together by a force called a bond. The three types of bonds are
   a. positive, negative, and neutral.
   b. hydrophobic, hydrophilic, and van der Waals interactions.
   c. magnetic, electric, and radioactive.
   d. ionic, covalent, and hydrogen.
4. Carbon has four electrons in its outer electron shell, therefore
   a. it has a completely filled outer electron shell.
   b. it can form four single covalent bonds.
   c. it does not react with any other atom.
   d. it has a positive charge.
5. The partial separation of charge in the water molecule
   a. results from the electrons' greater attraction to the oxygen atom.
   b. means the molecule has a positive end and a negative end.
   c. indicates that the water molecule is a polar molecule.
   d. All of the above.
6. Water has some very unusual properties. These properties occur because of the
   a. hydrogen bonds between the individual water molecules.
   b. covalent bonds between the individual water molecules.
   c. hydrogen bonds within each individual water molecule.
   d. ionic bonds between the individual water molecules.
7. Which of the following properties are somehow related to the need for significant heat energy to break hydrogen bonds?
   a. cohesion and adhesion
   b. hydrophobic and hydrophilic
   c. heat storage and heat of vaporization
   d. ice formation and high polarity
8. The attraction of water molecules to other water molecules is called
   a. cohesion.
   b. capillary action.
   c. solubility.
   d. adhesion.
9. Water sometimes ionizes, a single molecule breaking apart into a hydrogen ion and a hydroxide ion. Other materials may dissociate in water, resulting in either (1) an increase of hydrogen ions or (2) a decrease of hydrogen ions in the solution. We call the results
   a. (1) acids and (2) bases.
   b. (1) bases and (2) acids.
   c. (1) neutral solutions and (2) neutronic solutions.
   d. (1) hydrogen solutions and (2) hydroxide solutions.
10. Which of the following is not true about buffers?
   a. A buffer takes up H^+ from the solution.
   b. A buffer keeps the pH relatively constant.
   c. A buffer stops water from ionizing.
   d. A buffer releases H^+ into the solution.

Apply Your Understanding

1. **Figure 2.8** Based on the figure, explain why it is difficult to use carbon-14 dating on 100-million-year-old dinosaurs.

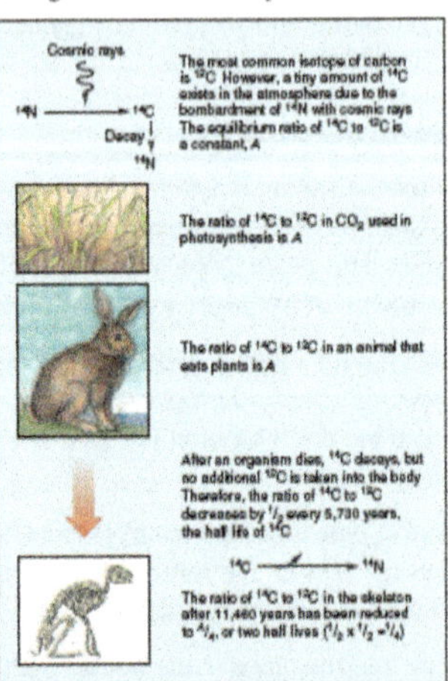

2. This figure shows an oxygen atom forming covalent bonds with two hydrogen atoms. A carbon atom, like oxygen, has two electrons in its innermost shell, but only four electrons in its outermost shell. Using this water molecule and page 39 as guides, draw a diagram showing how carbon forms covalent bonds with two oxygen atoms in a carbon dioxide (CO_2) molecule.

Oxygen
8+
8n
Hydrogen
Hydrogen
Electrons from hydrogen

3. **Figure 2.10** This insect is walking on water. Why doesn't it sink?

Synthesize What You Have Learned

1. Imagine you were to find a fossilized animal bone containing one-eighth the amount of carbon-14 present in earth's atmosphere today. How long ago did the animal die?
2. Explain why it is possible for a bacterium to break the triple covalent bond of N_2 gas, whereas you and other animals cannot.
3. You are on a 10-day backpacking trip with a small group of friends. Yvonne has washed out a set of water bottles with bleach. Before she can rinse them, Carlos, hot and thirsty, picks one up and drinks it, drops it, and begins to choke. What is Carlos' problem, and what can you do for him?

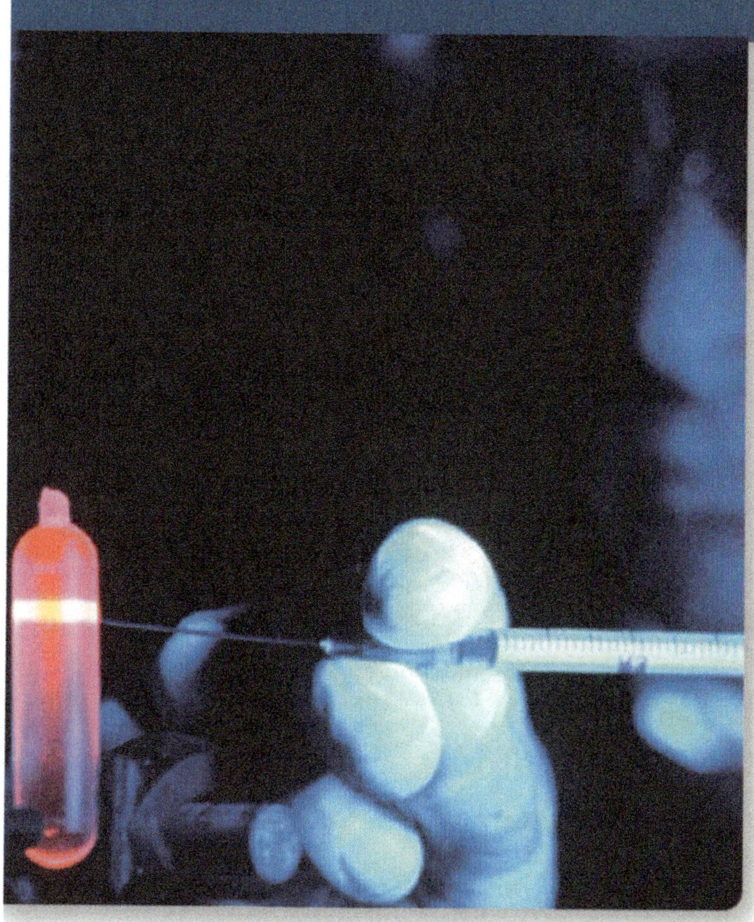

3

Molecules of Life

Learning Objectives

Forming Macromolecules

3.1 Polymers Are Built of Monomers

1. Distinguish between a polymer and a monomer.
2. Distinguish between organic and inorganic molecules.
3. Name the four different kinds of biological macromolecules.
4. Contrast hydrolysis with dehydration synthesis.

Types of Macromolecules

3.2 Proteins

1. List six different functional groupings of proteins.
2. Diagram the structure of an amino acid and the formation of a peptide bond between two amino acids.
3. Distinguish among primary, secondary, tertiary, and quaternary structure.
4. Explain how the polar nature of water influences protein folding.
5. Explain how a chaperone protein works.
6. Define prion.

3.3 Nucleic Acids

1. Name the three parts of a nucleotide.
2. Identify the five kinds of nucleotides.
3. State the two major chemical differences between DNA and RNA.
4. Identify what two base pairings are possible in a DNA molecule, and explain why the other four potential base pairings do not usually occur.

A Closer Look: Discovering the Structure of DNA

3.4 Carbohydrates

1. Define carbohydrate.
2. Distinguish among monosaccharide, disaccharide, and polysaccharide, and name one example of each.
3. Distinguish between starch and glycogen.
4. Explain why you cannot digest cellulose, but a termite can.

3.5 Lipids

1. Define lipid.
2. Diagram the basic structure of a fat.
3. Distinguish between saturated and unsaturated fats, and explain why one is a solid, the other a liquid.
4. Explain how trans-fats may contribute to bad health.
5. Explain why phospholipids are polar while triglycerides are not.
6. Describe the normal function of cholesterol in biological membranes.
7. Explain why excessive levels of cholesterol are dangerous to health.

Biology and Staying Healthy: Anabolic Steroids in Sports

Author's Corner: My Battle with Cholesterol

Inquiry & Analysis: How Does pH Affect a Protein's Function?

Using a syringe, this researcher is gently removing a glowing band of DNA, to be used in an experiment studying heredity. DNA, the carrier of an organism's genes, is one of several kinds of very large molecules found in all organisms. A molecule is a collection of tiny atoms linked together. Atoms are the basic chemical elements. Only a few are found in any significant numbers in living things. An essential atom for life is carbon, which can assemble into DNA and other very large molecules. Interacting with water, these long carbon chains twist about each other, or fold up into compact masses. Much of the chemistry that goes on in organisms, determining what each individual is like, depends on the actions of large folded molecules called proteins. By promoting particular chemical reactions, proteins trigger the production of structural materials like carbohydrates and energy storage molecules like lipids. Because DNA encodes the information needed to assemble each protein present in an organism, it is the library of life.

Forming Macromolecules

3.1 Polymers Are Built of Monomers

The bodies of organisms contain thousands of different kinds of molecules and atoms. Organisms obtain many of these molecules from their surroundings and from what they consume. You might be familiar with some of the substances listed on nutritional labels, such as the one shown in figure 3.1. But what do the words on these labels mean? Some of them are names of minerals (atoms, see chapter 2), such as calcium and iron (also discussed in chapters 22, 24, 33, and others). Others are vitamins, which are discussed in later chapters (see chapter 25). Still others are the subject of this chapter: large molecules that make up the bodies of organisms and are found in our food, such as proteins, carbohydrates (including sugars), and lipids (including, fats, trans fats, saturated fats, and cholesterol). These molecules, called **organic molecules,** are formed by living organisms and consist of a carbon-based core with special groups attached. These groups of atoms have special chemical properties and are referred to as *functional groups.* Functional groups tend to act as units during chemical reactions and to confer specific chemical properties on the molecules that possess them. Five principal functional groups are listed in figure 3.2; the last column indicates the types of organic molecules that contain these functional groups.

The bodies of organisms contain thousands of different kinds of organic molecules, but much of the body is made of just four kinds: *proteins, nucleic acids, carbohydrates,* and *lipids.* Called **macromolecules** because they can be very large, these four are the building materials of cells, the "bricks and mortar" that make up the bodies of cells and the machinery that runs within them.

The body's macromolecules are assembled by sticking smaller bits, called **monomers,** together, much as a train is built by linking railcars together. A molecule built up of long chains of similar subunits is called a **polymer.** Table 3.1 lists the monomers in the first column that make up the polymers that are the basis for many cellular structures.

Nutrition Facts

Serving Size 2 tbsp (33g)
(makes 3.5 cups popped)
Servings Per Bag about 3
Servings Per Box about 9

Amount Per Serving	2 tbsp (33g) Unpopped	Per 1 cup Popped
Calories	170	35
Calories from Fat	90	20
	% Daily Value**	
Total Fat 10g*	15%	3%
Saturated Fat 2g	10%	0%
Trans Fat 3.5g		
Cholesterol 0mg	0%	0%
Sodium 440mg	18%	4%
Total Carbohydrate 19g	6%	1%
Dietary Fiber 3g	12%	4%
Sugars 0g		
Protein 3g		
Iron	6%	0%

Figure 3.1 What's in a nutritional label?
Fats, cholesterol, carbohydrates, sugars, and proteins are just some of the molecules found in popcorn and discussed in this chapter.

Group	Structural Formula	Ball-and-Stick Model	Found In
Hydroxyl	—OH	O H	Carbohydrates
Carbonyl	>C=O	C O	Lipids
Carboxyl	—C(=O)OH	C O O H	Proteins
Amino	—N(H)H	H N H	Proteins
Phosphate	—O—P(O⁻)(=O)—O⁻	O⁻ O P O⁻ O	DNA, ATP

Figure 3.2 Five principal functional groups.
These functional groups can be transferred from one molecule to another and are common in organic molecules.

TABLE 3.1 MACROMOLECULES	
Monomer	**Polymer**
Amino Acid	Protein
Alanine	Ala Val Ser Val Ala
Nucleotide	Nucleic acid (DNA)
Monosaccharide	Carbohydrate (starch)
Fatty acid	Lipid (fat molecule)

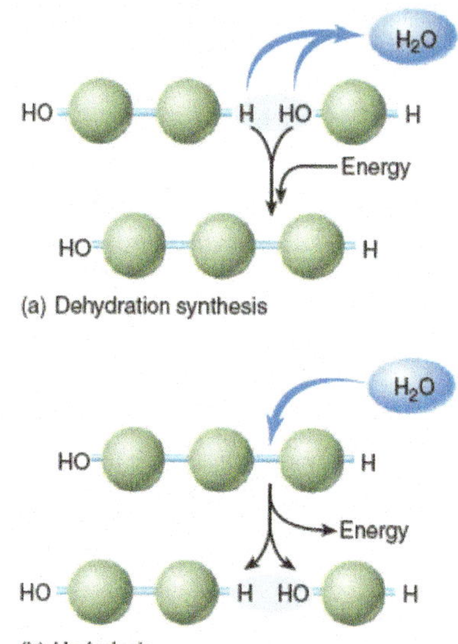

Figure 3.3 Dehydration and hydrolysis.
(*a*) Biological molecules are formed by linking subunits. The covalent bond between subunits is formed in dehydration synthesis, a process during which a water molecule is eliminated. (*b*) Breaking such a bond requires the addition of a water molecule, a reaction called hydrolysis.

Making (and Breaking) Macromolecules

The four different kinds of macromolecules (proteins, nucleic acids, carbohydrates, and lipids) are all put together in the same way: a covalent bond is formed between two subunits in which a hydroxyl group (OH) is removed from one subunit and a hydrogen (H) is removed from the other. This process (illustrated in figure 3.3*a*) is called **dehydration synthesis** because, in effect, the removal of the OH and H groups (highlighted by the blue oval) constitutes removal of a molecule of water—the word *dehydration* means "taking away water." This process requires the help of a special class of proteins called **enzymes** to facilitate the positioning of the molecules so that the correct chemical bonds are stressed and broken. The process of tearing down a molecule, such as the protein or fat contained in food that is consumed, is essentially the reverse of dehydration synthesis: Instead of removing a water molecule, one is added. When a water molecule comes in, as shown in figure 3.3*b*, a hydrogen becomes attached to one subunit and a hydroxyl to another, and the covalent bond is broken. The breaking up of a polymer in this way is called **hydrolysis.**

Key Learning Outcome 3.1 Macromolecules are formed by linking subunits together into long chains, removing a water molecule as each link is formed. In contrast, macromolecules are broken down into their subunits by hydrolysis reactions, the addition of water molecules.

Types of Macromolecules

3.2 Proteins

Complex macromolecules called **proteins** are a major group of biological macromolecules within the bodies of organisms. Perhaps the most important proteins are *enzymes,* which have the key role in cells of lowering the energy required to initiate particular chemical reactions. Other proteins play structural roles. Cartilage, bones, and tendons all contain a structural protein called collagen. Keratin, another structural protein, forms hair, the horns of a rhinoceros, and feathers. Still other proteins act as chemical messengers within the brain and throughout the body. Figure 3.4 presents an overview of the wide-ranging functions of proteins.

Amino Acids

Despite their diverse functions, all proteins have the same basic structure: a long polymer chain made of subunits called amino acids. **Amino acids** are small molecules with a simple basic structure: a central carbon atom to which an amino group (—NH_2), a carboxyl group (—COOH), a hydrogen atom (H), and a functional group, designated "R," are bonded.

There are 20 common kinds of amino acids that differ from one another by the identity of their functional R group. The 20 amino acids are classified into four general groups, with representative amino acids shown in figure 3.5 (their R groups are highlighted in white). Six of the amino acids are nonpolar, differing chiefly in size—the most bulky contain ring structures (like phenylalanine in the upper left), and amino acids containing them are called *aromatic.* Another six are polar but uncharged (like asparagine in the upper right), and these differ from one another in the strength of their polarity. Five more are polar and are capable of ionizing to a charged form (like aspartic acid in the lower left). The remaining three possess special chemical groups (like the white highlighted area of proline in the lower right) that are important in forming links between protein chains or in forming kinks in their shapes. The polarity of the R groups is important to the proper folding of the protein into its functional shape, which is discussed later.

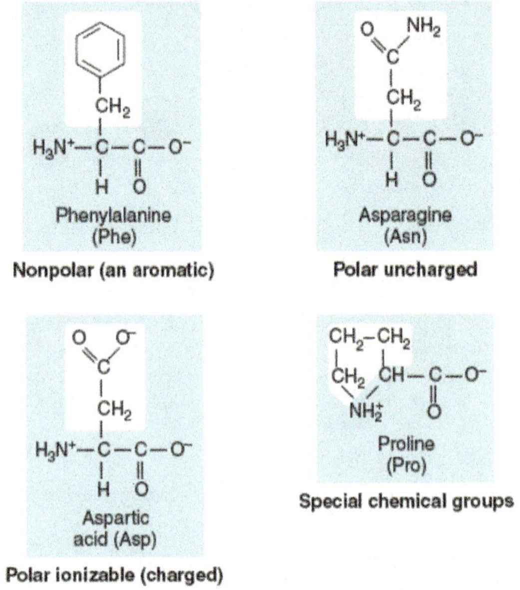

Phenylalanine (Phe)
Nonpolar (an aromatic)

Asparagine (Asn)
Polar uncharged

Aspartic acid (Asp)
Polar ionizable (charged)

Proline (Pro)
Special chemical groups

Figure 3.5 Examples of amino acids.
There are four general groups of amino acids that differ in their functional groups (highlighted in *white*).

Linking Amino Acids

An individual protein is made by linking specific amino acids together in a particular order, just as a word is made by linking specific letters of the alphabet together in a particular order. The covalent bond linking two amino acids together is called a **peptide bond** and forms by dehydration synthesis. Recall from section 3.1 that in dehydration synthesis, water is formed as a by-product of the reaction. You can see in figure 3.6 that a water molecule is released as the peptide bond forms. Long chains of amino acids linked by peptide bonds are called **polypeptides.** Functional polypeptides are more commonly called proteins.

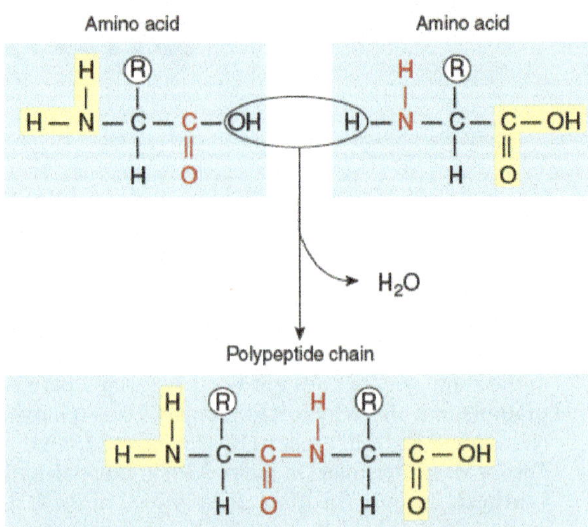

Figure 3.6 The formation of a peptide bond.
Every amino acid has the same basic structure, with an amino group (-NH_2) at one end and a carboxyl group (-COOH) at the other. The only variable is the functional, or "R," group. Amino acids are linked by dehydration synthesis to form peptide bonds. Chains of amino acids linked in this way are called polypeptides and are the basic structural components of proteins.

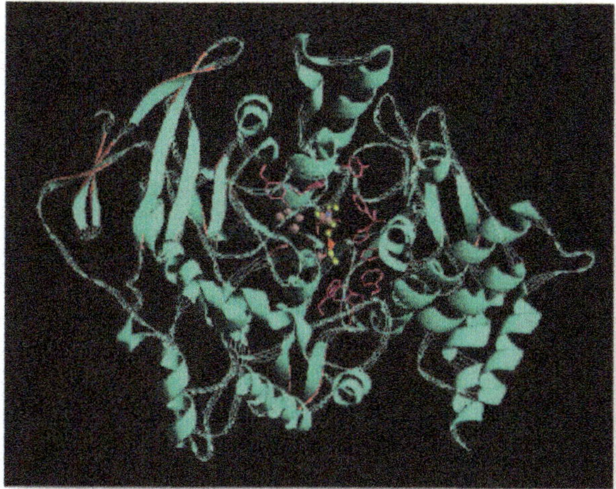

(a) **Enzymes:** Globular proteins called enzymes play a key role in many chemical reactions.

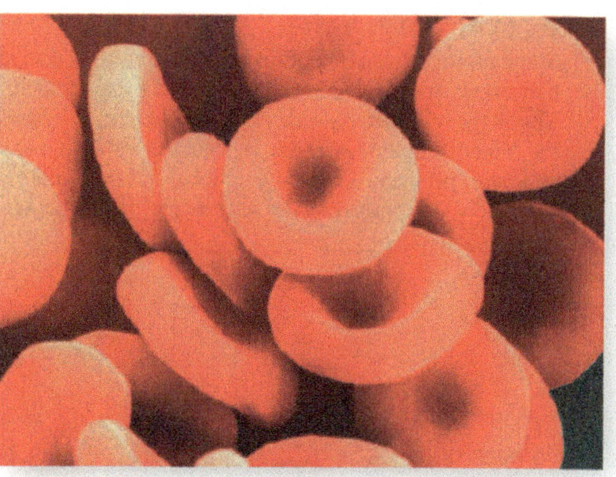

(b) **Transport proteins:** Red blood cells contain the protein hemoglobin, which transports oxygen and carbon dioxide in the body.

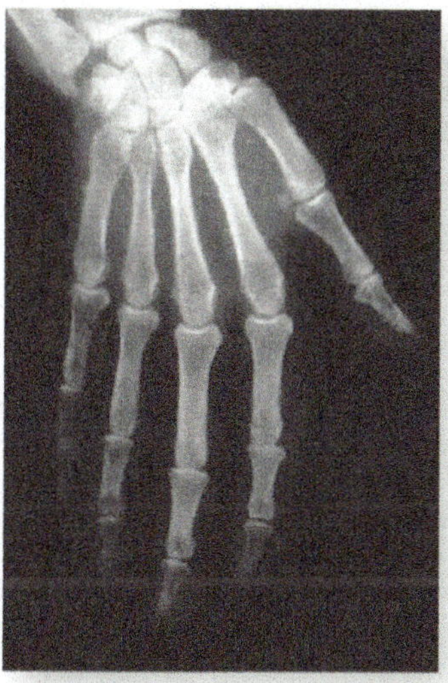

(c) **Structural proteins (collagen):** Collagen is present in bones, tendons, and cartilage.

(d) **Structural proteins (keratin):** Keratin forms hair, nails, feathers, and components of horns.

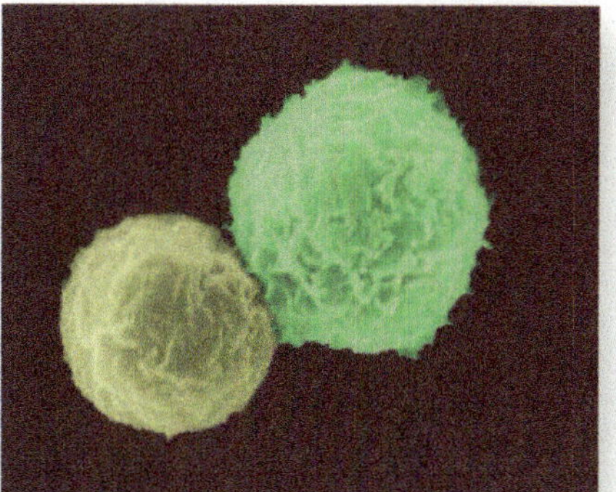
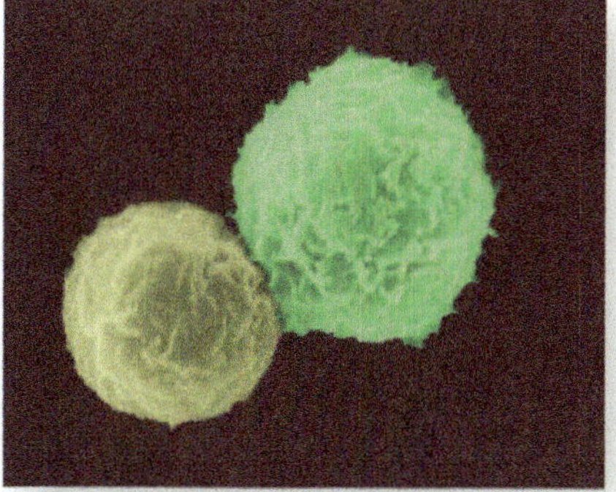

(e) **Defensive proteins:** White blood cells destroy cells without the proper identity proteins and make antibody proteins that attack invaders.

(f) **Contractile proteins:** Proteins called actin and myosin are present in muscles.

Figure 3.4 Some of the different types of proteins.

Protein Structure

Some proteins form long, thin fibers, whereas others are globular, their strands coiled up and folded back on themselves. The shape of a protein is very important because it determines the protein's function. There are four general levels of protein structure: primary, secondary, tertiary, and quaternary; all are ultimately determined by the sequence of amino acids.

Primary Structure The sequence of amino acids of a polypeptide chain is termed the polypeptide's **primary structure.** The amino acids are linked together by peptide bonds, forming long chains like a "beaded strand." The primary structure of a protein, the sequence of its amino acids, determines all other levels of protein structure. Because amino acids can be assembled in any sequence, a great diversity of proteins is possible.

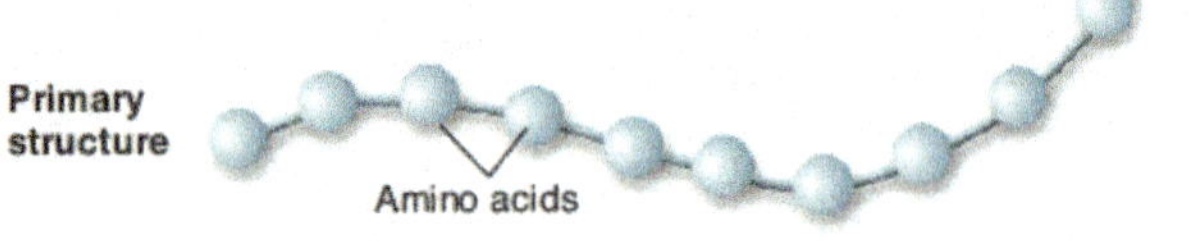

Primary structure

Amino acids

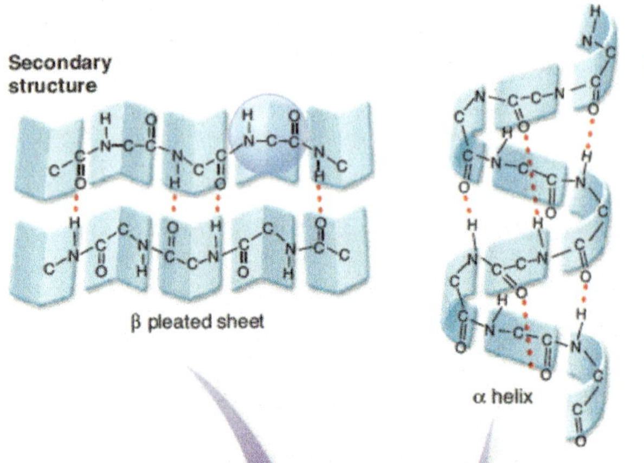

Secondary structure

β pleated sheet

α helix

Secondary Structure Hydrogen bonds forming between different parts of the polypeptide chain stabilize the folding of the polypeptide. As you can see, these stabilizing hydrogen bonds, indicated by red dotted lines, do not involve the R groups themselves, but rather the polypeptide backbone. This initial folding is called the **secondary structure** of a protein. Hydrogen bonding within this secondary structure can fold the polypeptide into coils, called α-helices, and sheets, called β-pleated sheets.

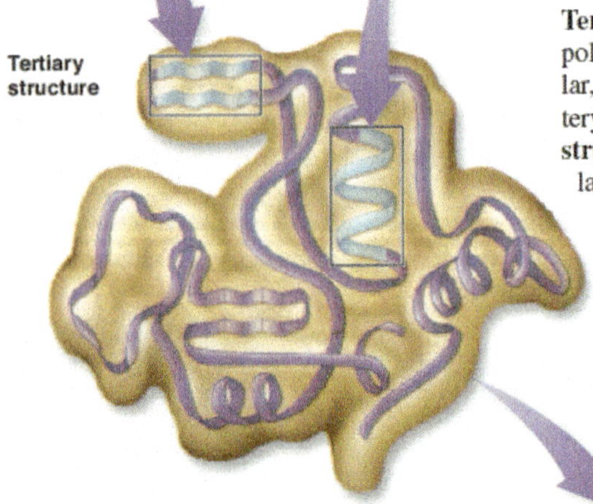

Tertiary structure

Tertiary Structure Because some of the amino acids are nonpolar, a polypeptide chain folds up in water, which is very polar, pushing nonpolar amino acid functional groups from the watery environment. The final three-dimensional shape, or **tertiary structure,** of the protein, folded and twisted in the case of a globular molecule, is determined by exactly where in a polypeptide chain the nonpolar amino acids occur.

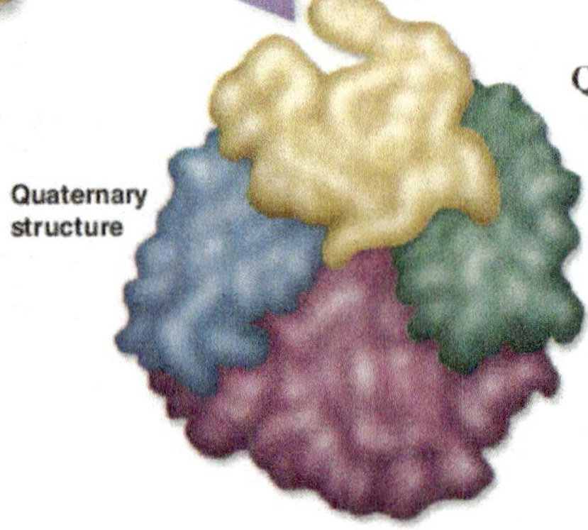

Quaternary structure

Quaternary Structure When a protein is composed of more than one polypeptide chain, the spatial arrangement of the several component chains is called the **quaternary structure** of the protein. For example, four subunits make up the quaternary structure of the protein hemoglobin.

How Proteins Fold into Their Functional Shape

The polar nature of the watery environment in the cell influences how the polypeptide folds into the functional protein. A protein is folded in such a way that allows it to carry out its function.

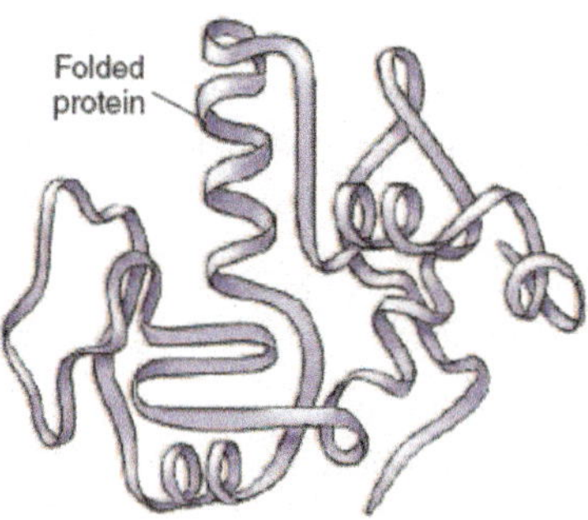

If the polar nature of the protein's environment changes by either increasing temperature or lowering pH, both of which alter hydrogen bonding, the protein may unfold, as in the lower right of the figure. When this happens the protein is said to be **denatured.**

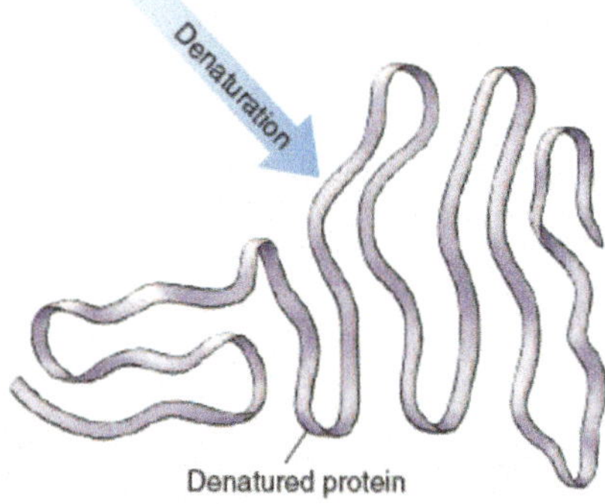

When the polar nature of the solvent is reestablished, some proteins may spontaneously refold. When proteins are denatured, they usually lose their ability to function properly. That is the rationale behind traditional methods of salt-curing and pickling food. Prior to the ready availability of refrigerators and freezers, the only practical way to keep microorganisms from growing in food was to keep the food in a solution containing a high concentration of salt or vinegar, which denatured proteins in the microorganisms and kept them from growing on the food.

Protein Structure Determines Function

The structure of a protein determines its function, and because the primary structure of a protein, its sequence of amino acids, determines how the protein folds into its functional shape, a change in the identity of even one amino acid can have profound effects on a protein's ability to function properly.

Enzymes are globular proteins that have three-dimensional shapes. For enzymes to function properly, they need to fold correctly. Enzymes have grooves or depressions that precisely fit a particular sugar or other chemical (like the red molecule binding to the enzyme to the right); once in the groove, the chemical is encouraged to undergo a reaction—often, one of its chemical bonds is stressed as the chemical is bent by the enzyme, like a foot in a flexing shoe. This process of enhancing chemical reactions is called **catalysis,** and proteins are the catalytic agents of cells, determining what chemical processes take place and where and when.

Active-site cleft

Many structural proteins form long cables that have architectural roles in cells, providing strength and determining shape. As you will discover in chapter 4, cells contain a network of protein cables that maintain the shape of the cell and function in transporting materials throughout the cell (figure 3.7). Contractile proteins function in muscle contraction, which is the shortening of a muscle. A muscle shortens when two proteins that are anchored on opposite ends of a muscle fiber slide past each other, bringing the ends of the fiber closer together (discussed in more detail in chapter 22).

Figure 3.7 Protein structure determines function. Fluorescently-labeled structural proteins within a cell.

Chaperone Proteins

How does a protein fold into a specific shape? As just discussed, nonpolar amino acids play a key role. Until recently, investigators thought that newly made proteins fold spontaneously as hydrophobic interactions with water shove nonpolar amino acids into the protein interior. We now know this is too simple a view. Proteins can fold in so many different ways that trial and error would simply take too long. In addition, as the open chain folds its way toward its final form, nonpolar "sticky" interior portions are exposed during intermediate stages. If these intermediate forms are placed in a test tube in the same protein environment that occurs in a cell, they stick to other unwanted protein partners, forming a gluey mess.

How do cells avoid this? A vital clue came in studies of unusual mutations (changes in DNA) that prevented viruses from replicating in bacterial cells—it turned out the virus proteins could not fold properly! Further study revealed that normal cells contain special proteins called **chaperone proteins** that help new proteins fold correctly. When the bacterial gene encoding its chaperone protein is disabled by mutation, the bacteria die, clogged with lumps of incorrectly folded proteins. Fully 30% of the bacteria's proteins fail to fold into the right shape.

Molecular biologists have now identified more than 17 kinds of proteins that act as molecular chaperones. Many are heat shock proteins, produced in greater amounts if a cell is exposed to elevated temperature; high temperatures cause proteins to unfold, and heat shock chaperone proteins help the cell's proteins refold.

To understand how a chaperone works, examine figure 3.8 closely. The misfolded protein enters inside the chaperone. There, in a way not clearly understood, the visiting protein is induced to unfold, and then refold again, before it leaves. You can see in the third panel of the diagram the protein has unfolded into a long polypeptide chain. In the fourth panel, the polypeptide chain has then refolded into a different shape. The chaperone protein has in this way "rescued" a protein that was caught in a wrongly folded state, and given it another chance to fold correctly. To demonstrate this rescue capability, investigators "fed" a deliberately misfolded protein malate dehydrogenase to chaperone proteins; the malate dehydrogenase was rescued, refolding to its active shape.

Protein Folding and Disease

There are tantalizing suggestions that chaperone protein deficiencies may play a role in Alzheimer's disease. By failing to facilitate the intricate folding of key proteins, the deficiency leads to the amyloid protein clumping in brain cells characteristic of the disease. Mad cow disease and the similar human disorder called variant Creutzfeldt-Jacob disease are both caused by misfolded brain proteins called **prions.** The misfolded prions induce other brain prion proteins to misfold in turn, creating a chain reaction of misfolding that kills ever more brain cells, leading to progressive loss of brain function and eventual death.

> **Key Learning Outcome 3.2** **Proteins are made up of chains of amino acids that fold into complex shapes. The sequence of its amino acids determines a protein's function. Chaperone proteins help newly produced proteins to fold properly.**

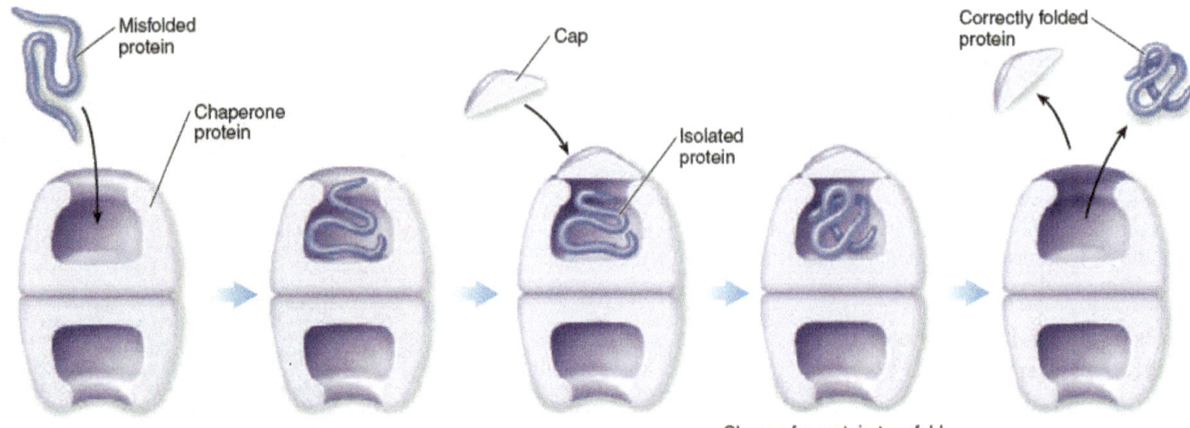

Figure 3.8 How one type of chaperone protein works.
This barrel-shaped chaperone protein is a heat shock protein, produced in elevated amounts at high temperatures. An incorrectly folded protein enters one chamber of the barrel, and a cap seals the chamber and confines the protein. The isolated protein is now prevented from aggregating with other misfolded proteins, and it has a chance to refold properly. After a short time, the protein is ejected, folded or unfolded, and the cycle can repeat itself.

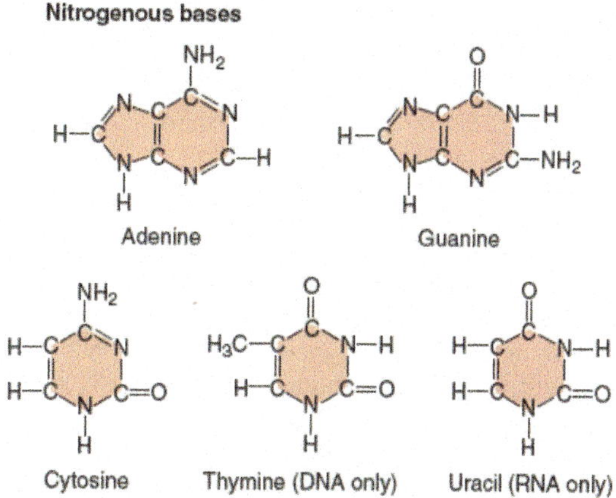

Figure 3.9 The structure of a nucleotide.
(*a*) Nucleotides are composed of three parts: a five-carbon sugar, a phosphate group, and an organic nitrogenous base. (*b*) The nitrogenous base can be one of five.

3.3 Nucleic Acids

Very long polymers called **nucleic acids** serve as the information storage devices of cells, just as CDs or hard drives store the information that computers use. Nucleic acids are long polymers of repeating subunits called **nucleotides.** Each nucleotide is a complex organic molecule composed of three parts shown in figure 3.9*a:* a five-carbon sugar (in blue), a phosphate group (in yellow, PO_4), and an organic nitrogen-containing base (in orange). In the formation of a nucleic acid, the individual sugars with their attached nitrogenous bases are linked in a line by the phosphate groups in very long **polynucleotide chains** (shown to the right).

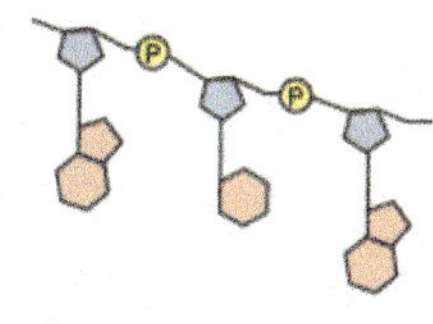

How does the long, chainlike structure of a nucleic acid permit it to store the information necessary to specify what a human being is like? If nucleic acids were simply a monotonous repeating polymer, it could not encode the message of life. Imagine trying to write a story using only the letter *E* and no spaces or punctuation. All you could ever say is "EEEEEEE. . . ." You need more than one letter to communicate—the English alphabet uses 26 letters. Nucleic acids can encode information because they contain more than one kind of nucleotide. There are five different kinds of nucleotides: two larger ones that contain the nitrogenous bases adenine and guanine (shown in the top row of figure 3.9*b*), and three smaller ones that contain the nitrogenous bases cytosine, thymine, and uracil (in the bottom row). Nucleic acids encode information by varying the identity of the nucleotide at each position in the polymer.

DNA and RNA

Nucleic acids come in two varieties, **deoxyribonucleic acid (DNA)** and **ribonucleic acid (RNA),** both polymers of nucleotides with some differences. RNA is similar to DNA, but with two major chemical differences. First, RNA molecules contain the sugar ribose, in which the 2' carbon (this is the carbon labeled 2' in figure 3.9*a*) is bonded to a hydroxyl group (—OH). In DNA, this hydroxyl group is replaced with a hydrogen atom. Second, RNA molecules do not contain the thymine nucleotide; they contain uracil instead. Structurally, RNA is also different. RNA is a long, single strand of nucleotides and is used by cells in making proteins using genetic instructions encoded within DNA. The sequence of nucleotides in DNA determines the order of amino acids in the primary structure of the protein. DNA consists of *two* polynucleotide chains wound around each other in a **double helix,** like strands of a pearl necklace twisted together. You can see this difference in structure by comparing the blue double-stranded DNA molecule in figure 3.10 with the green single-stranded RNA molecule.

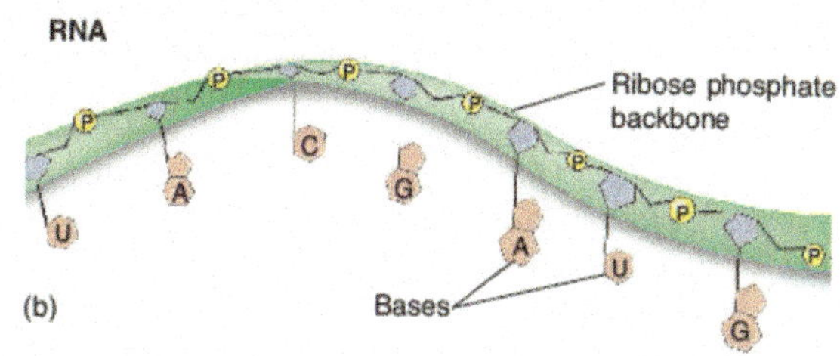

Figure 3.10 How DNA structure differs from RNA.
(*a*) DNA contains two polynucleotide strands wrapped around each other, while (*b*) RNA is single-stranded.

The Double Helix

Why is DNA a *double* helix? When scientists looked carefully at the structure of the DNA double helix, they found that the bases of each chain point inward toward the other (like the DNA strands shown in figure 3.11). The bases of the two chains are linked in the middle of the molecule by hydrogen bonds (the dotted lines between the two strands), like two columns of people holding hands across. The key to understanding why DNA is a double helix is revealed by looking at the bases: *only two base pairs are possible.* Because the distance between the two strands is consistent, this suggests that two big bases cannot pair together—the combination is simply too bulky to fit; similarly, two little ones cannot pair, as they pinch the helix inward too much. To form a double helix, it is necessary to pair a big base with a little one. *In every DNA double helix, adenine (A) pairs with thymine (T) and guanine (G) pairs with cytosine (C).* The reason A doesn't pair with C and G doesn't pair with T is that these base pairs cannot form proper hydrogen bonds—the electron-sharing atoms are not aligned with each other.

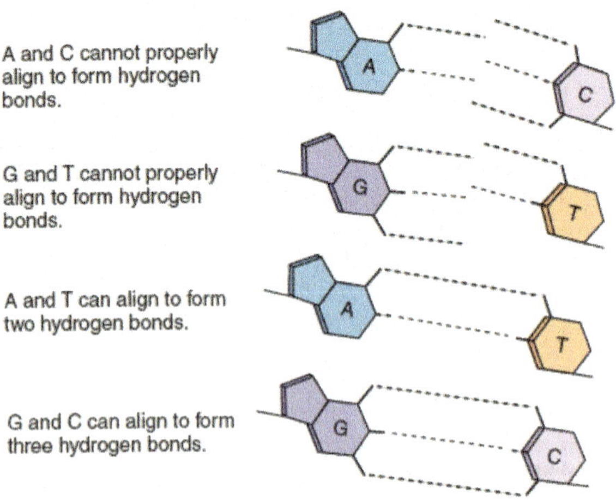

The simple A–T, G–C base pairs within the DNA double helix allow the cell to copy the information in a very simple way. It just unzips the helix and adds the complementary bases to each new strand! That is the great advantage of a double helix—it actually contains two copies of the information, one the mirror image of the other. If the sequence of one chain is ATTGCAT, the sequence of its partner in the double helix *must* be TAACGTA. The fidelity with which hereditary information is passed from one generation to the next is a direct result of this simple double-entry bookkeeping, which makes accurate copying of the genetic message possible.

Key Learning Outcome 3.3 Nucleic acids like DNA are composed of long chains of nucleotides. The sequence of the nucleotides specifies the amino acid sequence of proteins.

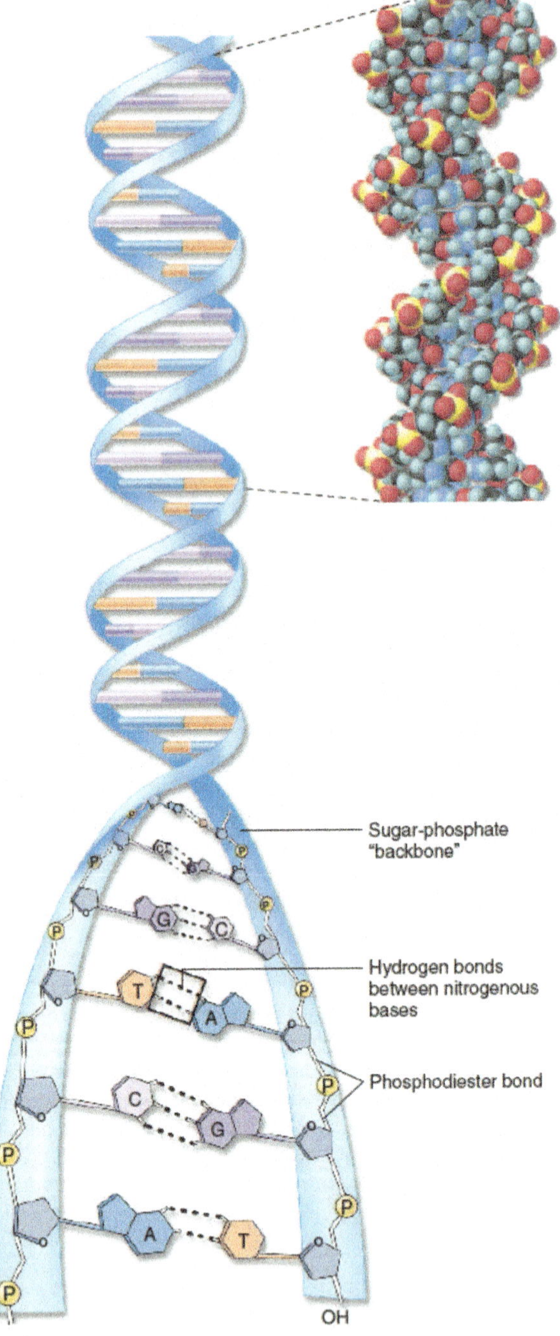

Figure 3.11 The DNA double helix.
The DNA molecule is composed of two polynucleotide chains twisted together to form a double helix. The two chains of the double helix are joined by hydrogen bonds between the A–T and G–C base pairs. The section of DNA on the upper right is a space-filling model of DNA, where atoms are indicated by colored balls.

A Closer Look

Discovering the Structure of DNA

By the middle of the last century, biologists were increasingly sure that DNA was the molecule that stored the hereditary information, but investigators were puzzled over how such a seemingly simple molecule could carry out such a complex function.

A key observation was made by chemist Erwin Chargaff shortly after the end of the Second World War. He noted that in DNA molecules, the amount of adenine, A, always equals the amount of thymine, T, and the amount of guanine, G, always equals the amount of cytosine, C. This observation (A=T, G=C), known as Chargaff's rule, strongly suggested that DNA has a regular structure, but did not reveal what it was.

The significance of the regularities pointed out by Chargaff were not immediately obvious, but they became clear when a young British chemist, Rosalind Franklin, carried out an X-ray diffraction analysis of DNA. In X-ray diffraction, a molecule is bombarded with a beam of X rays. When individual rays encounter atoms, their path is bent or diffracted, and the diffraction pattern is recorded on photographic film. The pattern that resulted using DNA resembled the ripples created by tossing a rock into a smooth lake. When carefully analyzed, a molecule's pattern can reveal information about the three-dimensional structure of the molecule.

X-ray diffraction works best on substances that can be prepared as perfectly regular crystalline arrays. However at the time of Franklin's analysis, it was impossible to obtain true crystals of natural DNA, so she had to use DNA in the form of fibers. Franklin worked in the same laboratory as Oxford biochemist Maurice Wilkins, who was able to prepare more uniformly oriented DNA fibers than anyone had previously. Using these fibers, Franklin succeeded in obtaining crude diffraction information on natural DNA. The diffraction patterns she obtained seemed to suggest that the DNA molecule had the shape of a coiled spring or corkscrew, a form called a helix.

Learning informally of Franklin's results before they were published in 1953, James Watson and Francis Crick, two young investigators at Cambridge University, quickly worked out a likely structure for the DNA molecule, which we now know was substantially correct. The key to their understanding the structure of DNA was Watson and Crick's insight that each DNA molecule is actually made up of two chains of nucleotides that are intertwined—a double helix.

In Watson and Crick's historic 1953 model (in the photograph, Watson is peering at the model as Crick points), each DNA molecule is composed of two complementary polynucleotide strands that form a double helix, with the bases extending into the interior of the helix. An analogy that is often made is to a spiral staircase where the two strands of the double helix are the handrails on the staircase.

What holds the two strands together? Watson and Crick proposed that the bases from opposite strands can form hydrogen bonds with each other to join the two complementary strands. Although each individual base pair is of low energy, the sum of many base pairs has enough energy that the molecule is very stable. To return to our spiral staircase analogy, where the backbone is the handrails, the base pairs are the stairs themselves.

Because of differences in size and position of particular atoms, only two hydrogen bonding pairs are possible in such a double helix: adenine (A) can form hydrogen bonds with thymine (T), and guanine (G) can form hydrogen bonds with cytosine (C). The Watson-Crick model thus, in a very direct and simple way, explained what had until then been one of the great mysteries of DNA, Chargaff's observation that adenine and thymine always occur in the same proportions in any DNA molecule, as do guanine and cytosine.

At the heart of the Watson-Crick model of DNA is a seemingly simple concept with some profound implications. Because only two base pairs are possible, if we know the sequence of one strand, we automatically know the sequence of the other strand; wherever there is an A in one strand, there must be a T in the other, and wherever there is a G in one strand, there must be a C in the other. This concept of a mirror-image relationship is called *complementarity*. You see the importance: If more than two base pairs were possible, then the sequence of one DNA strand would not allow us to know for sure the sequence of the other. It is this fundamental insight that has made Watson and Crick's discovery one of the most profound of the 20th century.

Watson and Crick continued on in the area of DNA research, but Franklin's career was cut short by her untimely death due to cancer at the age of 37.

3.4 Carbohydrates

Polymers called **carbohydrates** make up the structural framework of cells and play a critical role in energy storage. A carbohydrate is any molecule that contains carbon, hydrogen, and oxygen in the ratio 1:2:1. Some carbohydrates are small monomers or dimers and are called **simple carbohydrates.** Others are long polymers and are called **complex carbohydrates.** Because they contain many carbon-hydrogen (C–H) bonds, carbohydrates are well-suited for energy storage. Such C–H bonds are the ones most often broken by organisms to obtain energy. Table 3.2 on the facing page shows some examples of carbohydrates.

Simple Carbohydrates

The simplest carbohydrates are the *simple sugars* or **monosaccharides** (from the Greek *monos,* single, and *saccharon,* sweet). These molecules consist of one subunit. For example, glucose, the sugar that carries energy to the cells of your body, is made of six carbons and has the chemical formula $C_6H_{12}O_6$. A molecule of glucose is pictured in several ways in figure 3.12. The long chain of carbon atoms at the top of the figure is its formal chemical structure. When placed in water, the chain folds into the ring structure shown on the lower right. The individual atoms are depicted in the "3-D" space-filling model you see in the lower left. Another type of simple carbohydrate is a **disaccharide,** which forms when two monosaccharides link together through a dehydration reaction. In figure 3.13 you can see how the disaccharide sucrose (table sugar) is made by linking two six-carbon sugars together, a glucose (orange) and a fructose (green).

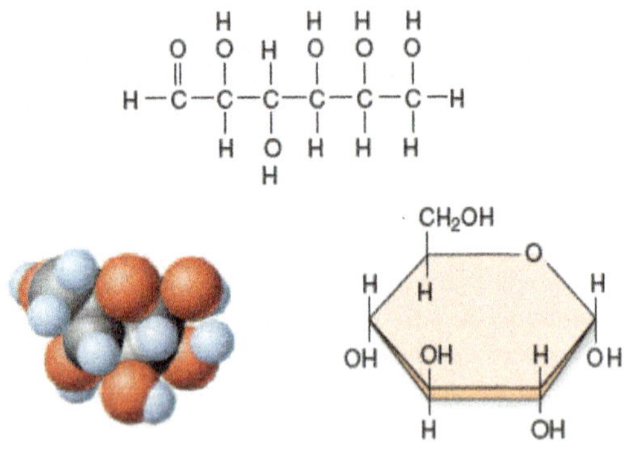

Figure 3.12 The structure of glucose.
Glucose is a monosaccharide and consists of a linear six-carbon molecule that forms a ring when added to water. This illustration shows three ways glucose can be represented diagrammatically.

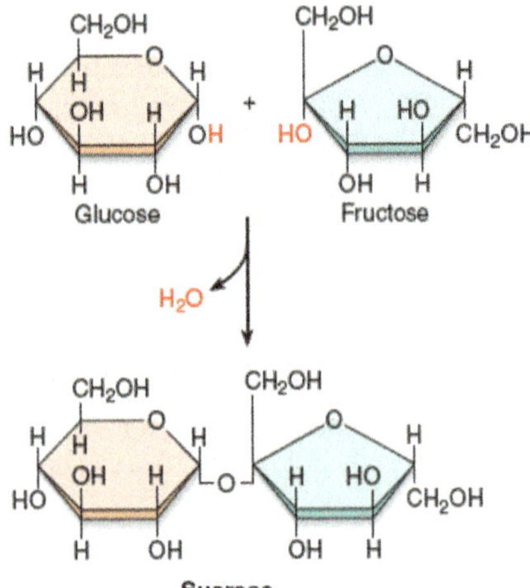

Figure 3.13 Formation of sucrose.
The disaccharide sucrose is formed from glucose and fructose in a dehydration reaction.

Complex Carbohydrates

Organisms store their metabolic energy by converting sugars, which are soluble, into insoluble forms that can be deposited in specific storage areas in the body. This trick is achieved by linking the sugars together into long polymer chains called **polysaccharides.** Plants and animals store energy in polysaccharides formed from glucose. The glucose polysaccharide that plants use to store energy is called **starch**—that is why potatoes are referred to as "starchy" food. In animals, energy is stored in **glycogen,** a highly insoluble macromolecule formed of glucose polysaccharides that are very long and highly branched. Plants and animals also use glucose chains as building materials, linking the subunits together in different orientations not recognized by most enzymes. These structural polysaccharides are **chitin,** in animals, and **cellulose,** in plants. The cellulose deposited in the cell walls of plant cells, like the cellulose strand shown in figure 3.14, cannot be digested by humans and makes up the fiber in our diets.

Key Learning Outcome 3.4 Carbohydrates are molecules made of C, H, and O atoms. As sugars they store energy in C–H bonds, and as long polysaccharide chains they can provide structural support.

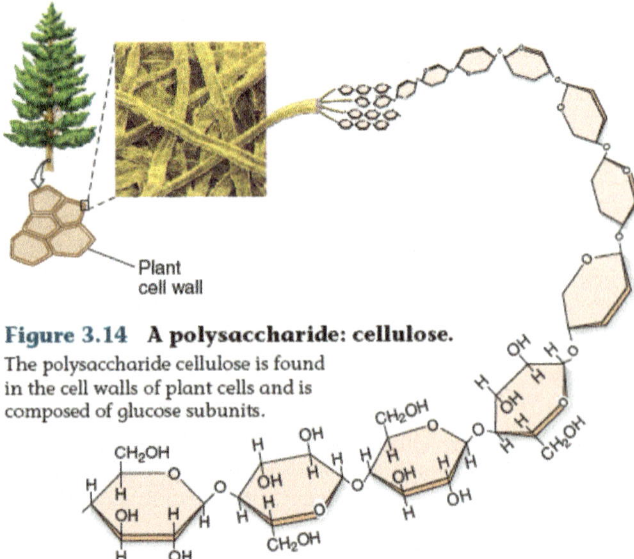

Figure 3.14 A polysaccharide: cellulose.
The polysaccharide cellulose is found in the cell walls of plant cells and is composed of glucose subunits.

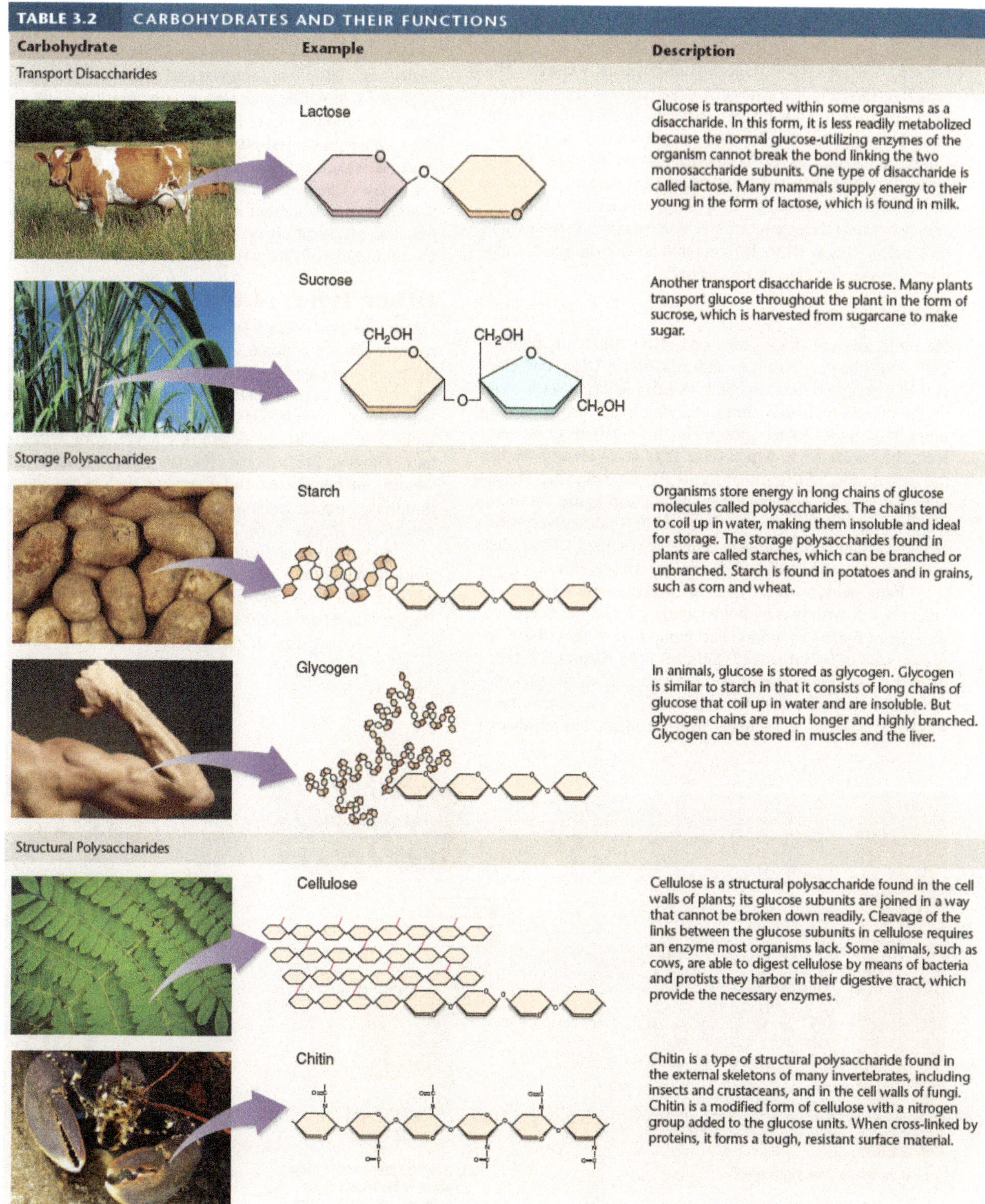

TABLE 3.2 CARBOHYDRATES AND THEIR FUNCTIONS

Carbohydrate	Example	Description
Transport Disaccharides		
	Lactose	Glucose is transported within some organisms as a disaccharide. In this form, it is less readily metabolized because the normal glucose-utilizing enzymes of the organism cannot break the bond linking the two monosaccharide subunits. One type of disaccharide is called lactose. Many mammals supply energy to their young in the form of lactose, which is found in milk.
	Sucrose	Another transport disaccharide is sucrose. Many plants transport glucose throughout the plant in the form of sucrose, which is harvested from sugarcane to make sugar.
Storage Polysaccharides		
	Starch	Organisms store energy in long chains of glucose molecules called polysaccharides. The chains tend to coil up in water, making them insoluble and ideal for storage. The storage polysaccharides found in plants are called starches, which can be branched or unbranched. Starch is found in potatoes and in grains, such as corn and wheat.
	Glycogen	In animals, glucose is stored as glycogen. Glycogen is similar to starch in that it consists of long chains of glucose that coil up in water and are insoluble. But glycogen chains are much longer and highly branched. Glycogen can be stored in muscles and the liver.
Structural Polysaccharides		
	Cellulose	Cellulose is a structural polysaccharide found in the cell walls of plants; its glucose subunits are joined in a way that cannot be broken down readily. Cleavage of the links between the glucose subunits in cellulose requires an enzyme most organisms lack. Some animals, such as cows, are able to digest cellulose by means of bacteria and protists they harbor in their digestive tract, which provide the necessary enzymes.
	Chitin	Chitin is a type of structural polysaccharide found in the external skeletons of many invertebrates, including insects and crustaceans, and in the cell walls of fungi. Chitin is a modified form of cellulose with a nitrogen group added to the glucose units. When cross-linked by proteins, it forms a tough, resistant surface material.

3.5 Lipids

For long-term energy storage, organisms usually convert glucose into fats, another kind of storage molecule that contains more energy-rich C–H bonds than carbohydrates. Fats and all other biological molecules that are not soluble in water but soluble in oil are called **lipids.** Lipids are insoluble in water not because they are long chains like starches but rather because they are nonpolar. In water, fat molecules cluster together because they cannot form hydrogen bonds with water molecules. This is why oil forms into a layer on top of water when the two substances are mixed.

Fats

Fat molecules are lipids composed of two kinds of subunits: fatty acids (the gray boxed structures in figure 3.15*a*) and glycerol (the orange boxed structure). A **fatty acid** is a long chain of carbon and hydrogen atoms (called a hydrocarbon) ending in a carboxyl (—COOH) group. The three carbons of glycerol form the backbone to which three fatty acids are attached in the dehydration reaction that forms the fat molecule. That is why the carboxyl groups of the fatty acids in figure 3.15*a* are not apparent, because they are involved in bonds with glycerol. Because it contains three fatty acids, the resulting fat molecule is sometimes called a *triacylglycerol,* or **triglyceride.**

Fatty acids with all internal carbon atoms forming covalent bonds with two hydrogen atoms contain the maximum number of hydrogen atoms. Fats composed of these fatty acids are said to be **saturated** (figure 3.15*b*). Saturated fats are solid at room temperature. On the other hand, fats composed of fatty acids that have double bonds between one or more pairs of carbon atoms contain fewer than the maximum number of hydrogen atoms and are called **unsaturated** (figure 3.15*c*). The double bonds create kinks in the fatty acid tails, which usually makes the unsaturated fats liquid at room temperature. Many plant fats are unsaturated and occur as oils. Animal fats, in contrast, are often saturated and occur as hard fats. In some cases, unsaturated fats in food products may be artificially *hydrogenated* (industrial addition of hydrogens) to make them more saturated, extending the shelf life of these products. In some cases, the hydrogenation creates *trans fats,* a type of unsaturated fat in which some of the double bonds are less kinked than those in naturally occurring unsaturated fats. Eating trans fats and saturated fats may increase the risk of heart disease.

Other Types of Lipids

Organisms also contain other types of lipids that play many roles in cells in addition to energy storage. The male and female sex hormones testosterone and estradiol are lipids called *steroids.* Unlike the structure of fats shown in figure 3.15, the structure of steroids consists of multiple ring structures and looks somewhat like a section of chicken wire. Other important biological lipids include phospholipids, cholesterol (also a steroid), rubber, waxes, and pigments, such as the chlorophyll that makes plants green and the retinal that your eyes use to detect light (figure 3.16).

If you look at figure 3.17, you will see two types of lipids that play key roles in the membranes that encase the cells of your body: *phospholipid* molecules and *cholesterol.* Phospholipids are modified triacylglycerol molecules, where one of the carbons in the glycerol backbone bonds to a

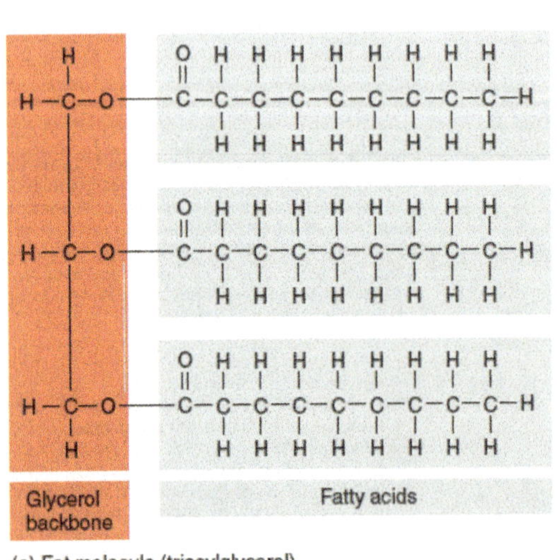

(a) Fat molecule (triacylglycerol)

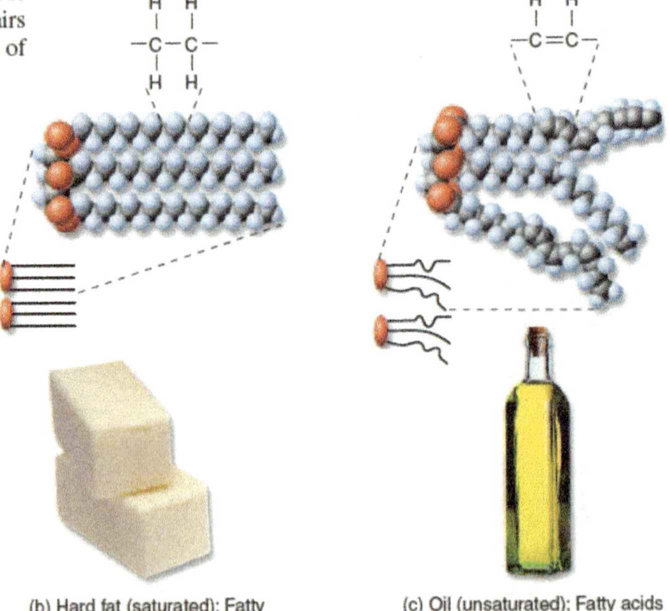

(b) Hard fat (saturated): Fatty acids with single bonds between all carbon pairs

(c) Oil (unsaturated): Fatty acids that contain double bonds between one or more pairs of carbon atoms

Figure 3.15 Saturated and unsaturated fats.
(*a*) Fat molecules each contain a three-carbon glycerol to which is attached three fatty acid tails. (*b*) Most animal fats are "saturated" (every carbon atom carries the maximum load of hydrogens). Their fatty acid chains fit closely together, and these triacylglycerols form immobile arrays called hard fats. (*c*) Most plant fats are unsaturated, which prevents close association between triacylglycerols and produces oils.

Biology and *Staying Healthy*

Anabolic Steroids in Sports

Among the most notorious of lipids in recent years has been the class of synthetic hormones known as anabolic steroids. Since the 1950s some athletes have been taking these chemicals to build muscle and so boost athletic performance. Both because of the intrinsic unfairness of this and because of health risks, the use of anabolic steroids has been banned in sports for decades. Controversy over their use in professional baseball has recently returned anabolic steroids to the nation's front pages.

Home run slugger Barry Bonds was involved in a steroid controversy in 2006.

Anabolic steroids were developed in the 1930s to treat hypogonadism, a condition in which the male testes do not produce sufficient amounts of the hormone testosterone for normal growth and sexual development. Scientists soon discovered that by slightly altering the chemical structure of testosterone, they could produce synthetic versions that facilitated the growth of skeletal muscle in laboratory animals. The word "anabolic" means growing or building. Further tweaking reduced the added impact of these new chemicals on sexual development. More than 100 different anabolic steroids have been developed, most of which have to be injected to be effective. All require a prescription to be used legally in the United States, and all are banned in professional, college, and high school sports.

Another way to increase the body's level of testosterone is to use a chemical that is not itself anabolic but one that the body converts to testosterone. One such chemical is 4-androstenedione, more commonly called "andro." It was first developed in the 1970s by East German scientists to try to enhance their athletes' Olympic performances. Because andro does not have the same side effects as anabolic steroids, it was legally available until 2004. It was used by Mark McGwire, but it is now banned in all sports, and possession of andro is a federal crime.

Anabolic steroids work by signalling muscle cells to make more protein. They bind to special "androgenic receptor" proteins within the cells of muscle tissue. Like jabbing these proteins with a poker, the binding prods the receptors into action, causing them to activate genes on the cell's chromosomes that produce muscle tissue proteins, triggering an increase in protein synthesis. At the same time, the anabolic steroid molecules bind to so-called "cortisol receptor" proteins in the cell, preventing these receptors from doing their job of causing protein breakdown, the muscle cell's way of suppressing inflammation and promoting the use of proteins for fuel during exercise. By increasing protein production and inhibiting the breakdown of proteins in muscle cells after workouts, anabolic steroids significantly increase the mass of an athlete's muscle tissue.

If the only effect of anabolic steroids on your body was to enhance your athletic performance by increasing your muscle mass, using them would still be wrong, for one very simple and important reason: fairness. To gain advantage in competition by concealed use of anabolic steroids—"doping"—is simply cheating. That is why these drugs are banned in sports.

The use of anabolic steroids by athletes and others is not only wrong, but also illegal, because increased muscle mass is not the only effect of using these chemicals. Among adolescents, anabolic steroids can also lead to premature termination of the adolescent growth spurt, so that for the rest of their lives, users remain shorter than they would have been without the drugs. Adolescents and adults are also affected by steroids in the following ways. Anabolic steroids can lead to potentially fatal liver cysts and liver cancer (the liver is the organ of the body that attempts to detoxify the blood), cholesterol changes and hypertension (both of which can promote heart attack and stroke), and acne. Other signs of steroid use in men include reduced size of testicles, balding, and development of breasts. In women, signs include the growth of facial hair, lowering of the voice, and cessation of menstruation.

In the fall of 2003, athletic organizations learned that some athletes were using a new performance-enhancing anabolic steroid undetectable by standard antidoping tests, tetrahydrogestrinone (THG). The use of THG was only discovered because an anonymous coach sent a spent syringe to U.S. antidoping officials. THG's chemical structure is similar to gestrinone, a drug used to treat a form of pelvic inflammation, and can be made from it by simply adding four hydrogen atoms, an easy chemical task. THG tends to break down when prepared for analysis by standard means, which explains why antidoping tests had failed to detect it. New urine tests for THG that were developed in 2004 have been used to catch several well-known sports figures. Olympic athlete Marion Jones and baseball sluggers Rafael Palmeiro, Barry Bonds, and Mark McGwire have all been involved in steroid use.

Author's Corner

My Battle with Cholesterol

By a cruel twist of fate I was born loving steak: a 1 1/2 inch Porterhouse is my idea of culinary perfection. I love French fries, too, and more than anything else, Big Macs. What is cruel about my love affair with Big Macs is that a dozen years ago my doctors informed me I have high levels of cholesterol in my blood, well over the upper recommended level of 200. Because all this cholesterol in my blood tends to deposit itself along the insides of arteries, this puts me at high risk of heart attack if I don't do something about it.

Thus began my decade-long battle with cholesterol. My old friends the Big Mac and French fries pretty much became history, and steak a much more casual acquaintance. My wife and three daughters, knowing my weakness of character when it comes to things bovine, started to do more of the grocery shopping, buying lots of chicken.

Didn't do any good. After two years of Big Mac-less days and steakless nights, my cholesterol levels rose higher than before. When measured in October of 1999, my total cholesterol was 274, the highest it had ever been. When you consider that every 1% increase in level over 200 increases my risk of heart disease 2%, this report was frightening.

I called my brothers to report the bad news, only to find that they too wage the same battle. The problem, it appears, is that I and my brothers inherited a defective steak-hating gene. We suffer from hypercholesterolemia. This mouthful of a word simply means "inherited high cholesterol." People inheriting a copy of the defective gene from one of their parents have elevated levels of cholesterol in their blood serum that dietary restriction (taking away my steaks) cannot reduce.

My frustrating on-going battle with cholesterol is not unique. Over half a million Americans suffer from hypercholesterolemia, the most frequent of all gene disorders. None of my hundreds of thousands of brothers and sisters in this war, fellow victims of this genetic quirk, are able to reduce cholesterol levels by diet and exercise alone. They, and I, must rely on modern medicine to defeat our Mendelian enemy.

Even a quick look at the biology of cholesterol suggests a likely line of attack. Cholesterol is a special kind of fat that your body uses to control the flexibility of its cell membranes and to insulate nerves. Your liver makes about 80% of the cholesterol in your body, up to 800 mg a day. You take in the rest when you eat steak and other fatty foods.

You can see why my attempting to lower total cholesterol by reducing dietary input of fat (steaks, peanut butter, fried food, chocolate, ice cream—all the good stuff) didn't get the job done. Dietary cholesterol is only 20% of my body's total. Most is manufactured by my liver, and because of hypercholesterolemia, my liver is churning out cholesterol twice as fast as a normal liver does. To solve the problem, I need to put the brakes on my liver's cholesterol-making frenzy.

In the 1980s researchers determined that the rate-limiting step in the liver's manufacture of cholesterol occurs early in the process, when a six-carbon molecule called mevalonate is converted to something called hydroxy methyl glutaryl CoA (HMG-CoA, for short). This reaction is carried out by an enzyme called HMG-CoA reductase. If we want to slow cholesterol production, that's our target.

In the last 20 years a series of drugs called statins have been developed that reduce levels of cholesterol by inhibiting HMG-CoA reductase. Among them are fluvastatin (Lescol), lovastatin (Mevacor), simvastatin (Zocor), and pravastatin (Pravachol).

Atorvastatin (Lipitor), introduced in 1997, is a particularly potent inhibitor of HMG-CoA reductase with fewer side effects than other statins. Lipitor is a blockbuster drug. It generated $4.4 billion in sales its first year, and became the top-selling global drug by 2005, with sales in excess of $10 billion.

It is upon Lipitor that I have banked my future. I started taking it in 1999, a little white pill every morning before breakfast. It is likely I shall be taking it the rest of my life.

How well did it work? Very well indeed. Within three months of starting daily Lipitor tablets, my cholesterol level has fallen to below 200. In the years since then, it has never risen above 200. Not once. When last measured at the start of 2010, my cholesterol level was 158.

In *The Naming of Cats,* the T.S. Eliot poem on which the Broadway musical *Cats* was based, "a cat must have three different names." I think of my anticholesterol drug as being like T.S. Eliot's cat. Like his cat, my drug has a "sensible everyday name", Lipitor, a name that anyone can use. Then Eliot says a cat needs a second name, "a name that's particular," "a name that never belonged to more than one cat." Atorvastatin is such a name, one that drug manufacturers and scientists use.

According to T.S. Eliot, every cat also needs a third very private name, a "deep and inscrutable singular name." For Lipitor, this name, its chemical soul, is [R-(R*,R*)]-2-(4-fluorophenyl)-β,δ-dihydroxy-5-(1-methylethyl)-3-phenyl-4-[(phenylamino)carbonyl]-1H-pyrrole-1-heptanoic acid.

To someone looking at a cholesterol level 116 points below 274, that's beautiful.

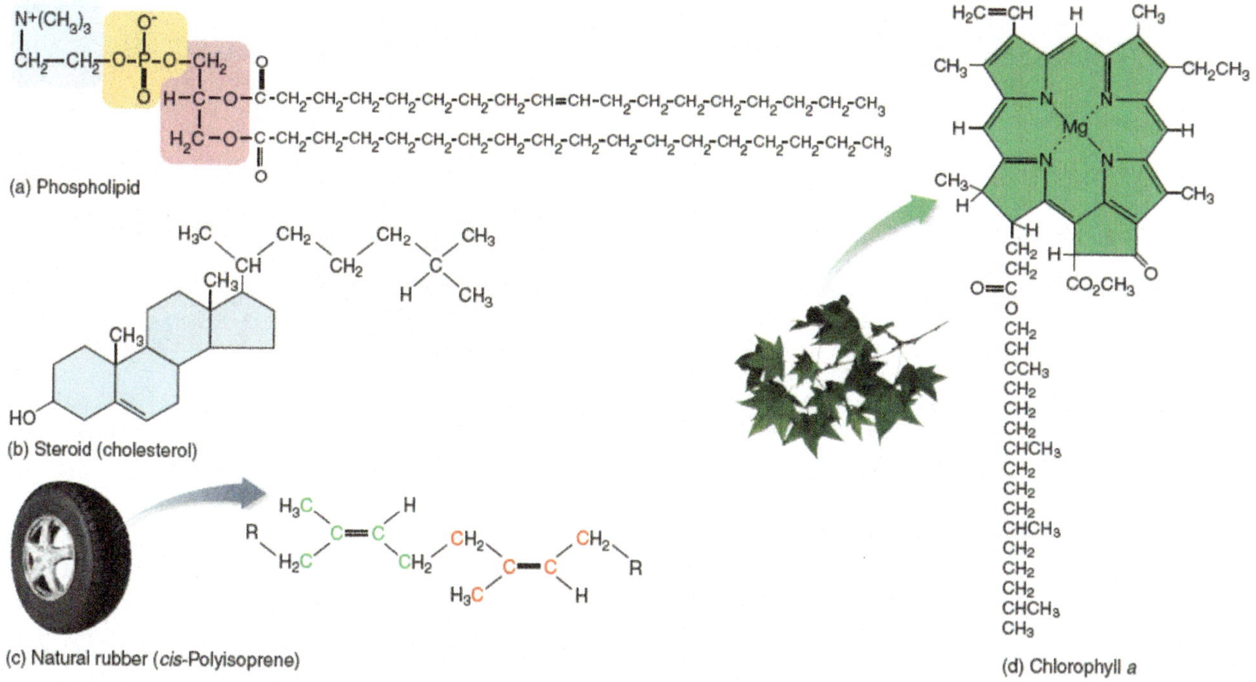

(a) Phospholipid

(b) Steroid (cholesterol)

(c) Natural rubber (*cis*-Polyisoprene)

(d) Chlorophyll *a*

Figure 3.16 Types of lipid molecules.

(*a*) A phospholipid is similar to a fat molecule with the exception that one of the fatty acid tails is replaced with a polar group linked to the glycerol by a phosphate group. (*b*) Steroids like cholesterol are lipids with complex ring structures. (*c*) Natural rubber is a linear polymer of a 5-carbon unit called isoprene (two isoprene units are shown here; one unit has its carbons in *green* and the other in *red*). Rubber, harvested from the rubber tree, is used in many products such as the tires on your car. (*d*) Chlorophyll *a* has a multiringed area and a long hydrocarbon tail. Chlorophyll is the primary pigment used in photosynthesis, and it is what makes leaves green, as you will see in chapter 6.

phosphate group rather than to a third fatty acid (compare figures 3.15*a* and 3.16*a*). This gives the molecule an overall shape of a polar head and two nonpolar tails. In water, the nonpolar ends of the phospholipids group together through hydrophobic interactions, leaving the polar heads to interact with water molecules. These interactions result in the formation of two layers of molecules with the nonpolar tails pointed inside—a lipid bilayer. All biological membranes, those that surround the cell and those found inside the cell, have this arrangement. Most animal cell membranes also contain the steroid cholesterol, a lipid composed of four carbon rings (see figure 3.16*b*). Cholesterol helps membranes stay flexible. However, excess saturated fat intake can cause plugs of cholesterol to form in the blood vessels, which may lead to blockage, high blood pressure, stroke, or heart attack. Membranes are discussed in more detail in chapter 4.

Key Learning Outcome 3.5 **Lipids are not water-soluble. Fats contain chains of fatty acid subunits that store energy. Other lipids include phospholipids, steroids, rubber, and pigment molecules.**

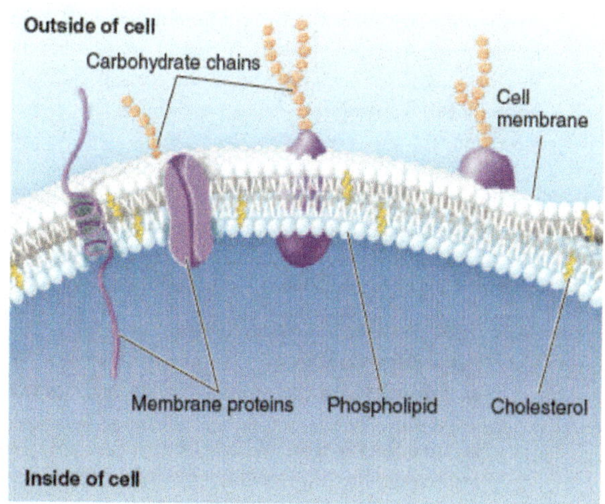

Figure 3.17 Lipids are a key component of biological membranes.

Lipids are one of the most common molecules in the human body, because the membranes of all the body's 10 trillion cells are composed largely of lipids called phospholipids. Membranes also contain cholesterol, another type of lipid.

INQUIRY & ANALYSIS

How Does pH Affect a Protein's Function?

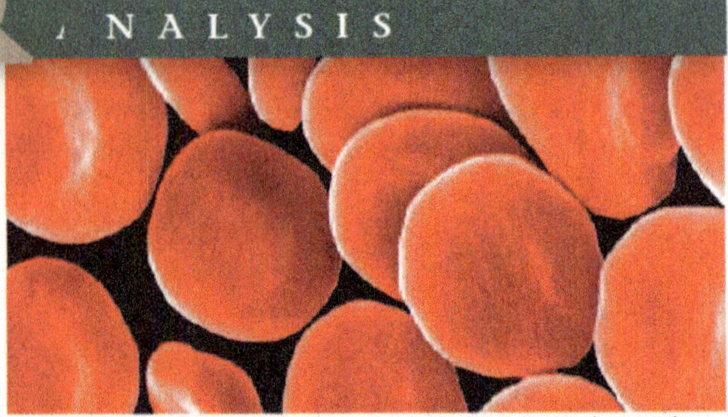

The red blood cells you see to the right carry oxygen to all parts of your body. These cells are red because they are chock full of a large iron-rich protein called *hemoglobin.* The iron atoms in each hemoglobin molecule provide a place for oxygen gas molecules to stick to the protein. When oxygen levels are highest (in the lungs), oxygen atoms bind to hemoglobin tightly, and a large percent of the hemoglobin molecules in a cell possess bound oxygen atoms. When oxygen levels are lower (in the tissues of the body), hemoglobin doesn't bind oxygen atoms as tightly, and as a consequence hemoglobin releases its oxygen to the tissues. What causes this difference between lungs and tissues in how hemoglobin loads and unloads oxygen? Oxygen concentration is not the only factor that might be responsible. Blood pH, for example, also differs between lungs and body tissues (**pH** is a measure of how many H^+ ions a solution contains). Tissues are slightly more acid (that is, they have more H^+ ions and a lower pH) because their metabolic activities release CO_2 into the blood, which you will recall from chapter 2, quickly becomes converted to carbonic acid.

The graph to the right displays so-called "oxygen loading curves" that reveal the effectiveness with which hemoglobin binds oxygen. The more effective the binding, the less oxygen required before hemoglobin becomes fully loaded, and the farther to the left a loading curve is shifted. To assess the impact of pH on this process, O_2 loading curves were carried out at three different blood pH values. In the graph, oxygen levels in the blood are presented on the *x* axis, and for each data point the corresponding % hemoglobin saturation (a %, or **percent**, is the numerator [top part] of a fraction whose denominator [bottom part] is 100—in this case, a measure of the fraction of the hemoglobin that is bound to oxygen) is presented on the *y* axis. The oxygen-loading curve was repeated at pH values of 7.6, 7.4, and 7.2, corresponding to the blood pH that might be expected in resting, exercising, and very active muscle tissue, respectively.

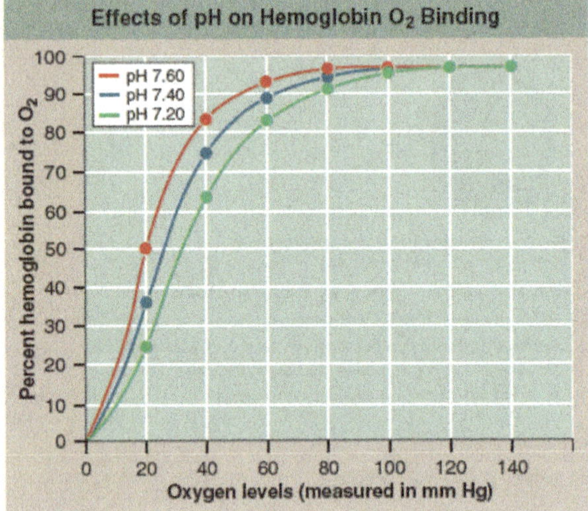

1. **Applying Concepts**
   a. Variable. In the graph, what is the dependent variable?
   b. Concentration. Which of the three pH values represents the highest concentration of hydrogen ions? (The **concentration** of a substance is the amount of that substance present in a given volume.) Is this value more acidic or more basic than the other two?
2. **Interpreting Data**
   a. What is the percent hemoglobin bound to O_2 for each of the three pH concentrations at saturation? At an oxygen level of 20 mm Hg? At 40 mm Hg? at 60 mm Hg?
   b. What general statement can be made regarding the effect of the oxygen levels in the blood (as measured by partial pressure of oxygen, measured in mm Hg) on the binding of oxygen to hemoglobin?
   c. Are there any significant differences in the hemoglobin saturation values for the three pHs at high oxygen levels?
3. **Making Inferences** At an oxygen level of 40 mm Hg, would hemoglobin bind oxygen more tightly at a pH of 7.8 or 7.0?
4. **Drawing Conclusions** How does pH affect the release of oxygen from hemoglobin?
5. **Further Analysis**
   a. Carbon dioxide acts to lower the pH of the blood. Predict what would happen to hemoglobin loading when carbon dioxide enters the blood as the blood circulates through the tissues of the body.
   b. In the lungs, oxygen levels are high and carbon dioxide leaves the blood and is exhaled, leading to lower carbon dioxide levels in the blood. Predict what happens to hemoglobin's oxygen loading under these conditions.

Chapter Review

Forming Macromolecules

3.1 Polymers Are Built of Monomers

- Living organisms produce organic macromolecules, which are large carbon-based molecules. The chemical properties of these molecules are due to unique functional groups that are attached to the carbon core (**figure 3.2**).
- Large molecules called macromolecules are formed by dehydration reactions, like the reaction shown here from **figure 3.3*a***, in which molecular subunits called monomers are linked together through covalent bonding. The reaction is called a dehydration reaction because a water molecule is a product of the reaction.

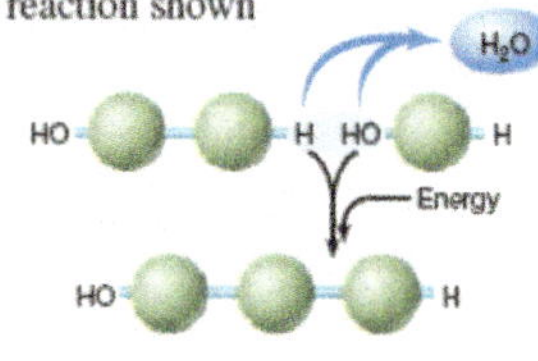

- The breakdown of a macromolecule involves a hydrolysis reaction, where a water molecule is broken down into H^+ and OH^-. These ions disrupt the covalent bond that holds two monomers together, causing the bond to break (**figure 3.3*b***).
- Amino acids are the subunits that link together to form polypeptides. Nucleotide monomers link together to form nucleic acids. Monosaccharide monomers link together to form carbohydrates. Fatty acids are the monomers that link together to form a type of lipid called fats (**table 3.1**).

Types of Macromolecules

3.2 Proteins

- Proteins are macromolecules that carry out many functions in the cell (**figure 3.4**). They are produced by the linking together of amino acid subunits that form a chain called a polypeptide.
- There are 20 different amino acids found in proteins. All amino acids have the same basic core structure. They differ in the type of functional group attached to the core. The functional groups are referred to as R groups. Some functional groups are polar, some are nonpolar, and still others give the amino acid unique chemical properties (**figure 3.5**).
- Amino acids are linked together with covalent bonds referred to as peptide bonds (**figure 3.6**).
- The sequence of amino acids within the polypeptide is the primary structure of the protein (**integrated art, page 54**). The chain of amino acids can twist into a secondary structure, where hydrogen bonding holds portions of the polypeptide in a coiled shape, called an α-helix, or in sheets called β-pleated sheets. Further bending and folding of the polypeptide results in its tertiary structure. When two or more polypeptides are present in a protein, the interaction of these polypeptide subunits is its quaternary structure.
- Changes in environmental conditions that disrupt hydrogen bonding can cause a protein to unfold, a process called denaturation (**integrated art, page 55**). Globular proteins cannot function if they are denatured. Some denatured proteins can refold back into their functional shapes. The structure of the protein determines its function (**figure 3.7**).
- Chaperone proteins aid in the folding of the amino acid chain into the correct shape (**figure 3.8**). Some diseases may be caused by misfolded, and therefore nonfunctional, proteins.

3.3 Nucleic Acids

- Nucleic acids, such as DNA and RNA, are long chains of nucleotides. Nucleotides contain three parts: a five-carbon sugar, a phosphate group, and a nitrogenous base (**figure 3.9**). DNA and RNA function as information storage molecules in the cell, storing the information needed to build proteins. The information is stored as different sequences of nucleotides.
- DNA and RNA differ chemically in that the sugar found in DNA is deoxyribose and in RNA is ribose. The nitrogenous bases in DNA are cytosine, adenine, guanine, and thymine, and the same are found in RNA except for thymine, which is substituted in RNA with uracil.
- DNA and RNA also differ structurally. DNA contains two strands of nucleotides wound around each other, called a double helix. RNA, as shown here from **figure 3.10**, is a single strand of nucleotides.

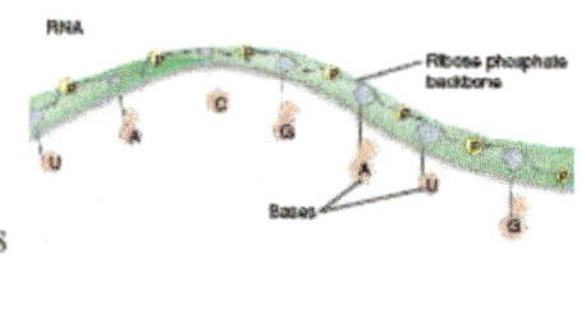

- The two nucleotide strands of the DNA double helix are held together through hydrogen bonding between nitrogenous bases: adenine (A) pairs with thymine (T) and cytosine (C) pairs with guanine (G) (**figure 3.11**).

3.4 Carbohydrates

- Carbohydrates, also referred to as sugars, are macromolecules that serve two primary functions in the cell: structural framework and energy storage.
- Carbohydrates that consist of only one or two monomers are called simple carbohydrates (**figures 3.12 and 3.13**). Carbohydrates that consist of long chains of monomers are called complex carbohydrates or polysaccharides (**figure 3.14**).

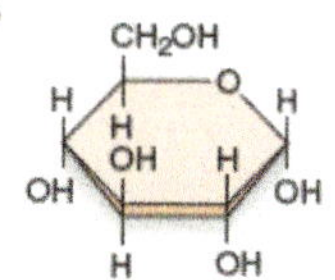

- Polysaccharides such as starch and glycogen provide a means of storing energy in the cell. They are broken down in the cells when energy is needed. Carbohydrates such as cellulose and chitin provide structural integrity (**table 3.2**).

3.5 Lipids

- Lipids are large nonpolar molecules that are insoluble in water. Fats that function in long-term energy storage include saturated fats, as shown here from **figure 3.15**, and unsaturated fats. Saturated fats are solid at room temperature and are found in animals, while unsaturated fats are liquid (oils) at room temperature and found in plants.

- Other lipids include the steroids (including the sex steroids and cholesterol), rubber, and pigments such as chlorophyll. All have very different structures and very different functions (**figure 3.16**).
- All cells are encased in a plasma membrane, called the lipid bilayer, that is composed of two layers of modified fats called phospholipids (**figure 3.17**).

Test Your Understanding

1. The four kinds of organic macromolecules are
   a. hydroxyls, carboxyls, aminos, and phosphates.
   b. proteins, carbohydrates, lipids, and nucleic acids.
   c. DNA, RNA, simple sugars, and amino acids.
   d. carbon, hydrogen, oxygen, and nitrogen.
2. Organic molecules are made up of monomers. Which of the following is *not* considered a monomer of organic molecules?
   a. amino acids c. polypeptides
   b. simple sugars d. nucleotides
3. Your body is filled with many types of proteins. Each type has a distinctive sequence of amino acids that determines both its specialized _____ and its specialized _____.
   a. number, weight c. structure, function
   b. length, mass d. charge, pH
4. A peptide bond forms
   a. by the removal of a water molecule.
   b. by a dehydration reaction.
   c. between two amino acids.
   d. All of the above.
5. Nucleic acids
   a. are the energy source for our bodies.
   b. act on other molecules, breaking them apart or building new ones to help us function.
   c. are only found in a few, specialized locations within the body.
   d. are information storage devices found in body cells.
6. The two strands of a DNA molecule are held together through hydrogen bonds between nucleotide bases. Which of the following best describes this base pairing in DNA?
   a. Adenine forms hydrogen bonds with thymine.
   b. Adenine forms hydrogen bonds with cytosine.
   c. Cytosine forms hydrogen bonds with thymine.
   d. Guanine forms hydrogen bonds with adenine.
7. Carbohydrates are used for
   a. structure and energy. c. fat storage and hair.
   b. information storage. d. hormones and enzymes.
8. Which of the following carbohydrates is *not* found in plants?
   a. glycogen c. starch
   b. cellulose d. All are found in plants.
9. A characteristic common to fat molecules is
   a. that they contain long chains of C–H bonds.
   b. that they are insoluble in water.
   c. that they have a glycerol backbone.
   d. All of these are characteristics of fats.
10. Lipids are used for
    a. motion and defense.
    b. information storage.
    c. energy storage and for some hormones.
    d. enzymes and for some hormones.

Apply Your Understanding

1. **Figures 3.6 and 3.13** The molecule below, on the left, is a peptide made from monomers of amino acids. The molecule below, on the right, is a disaccharide made from monomers of simple sugars. Both molecules were synthesized using a common chemical reaction. What is the chemical reaction that formed these molecules and what is the common by-product of both these reactions?

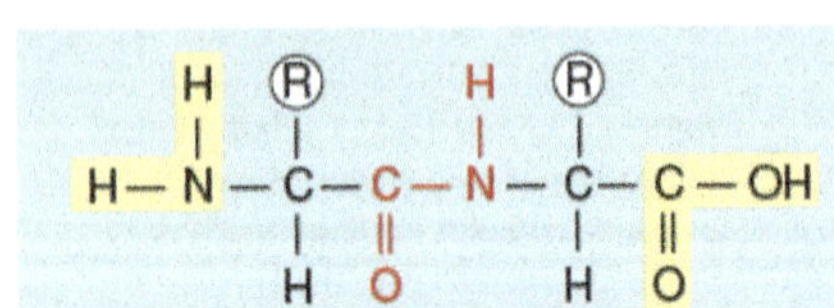

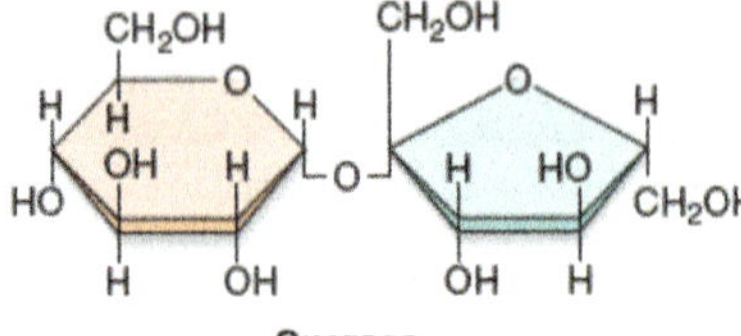

2. **Figure 3.15** The following are two lipid molecules. The lipid on the left is a saturated fat and the one on the right is an unsaturated fat. What is the difference in the chemical structure of their fatty acid tails and how does this affect their physical properties?

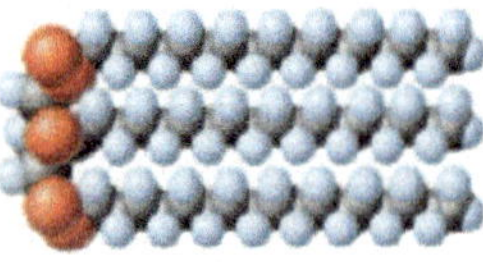

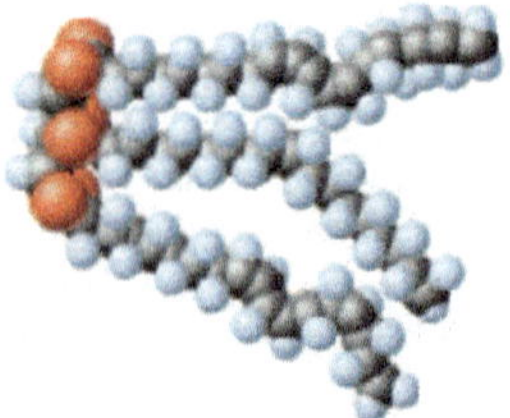

Synthesize What You Have Learned

1. How many molecules of water are used up in the breakdown of a polypeptide that is 15 amino acids in length?
2. The amylase enzyme present in the cells of your body can break the bonds between the glucose monomers in starch but it cannot break the bonds between the glucose monomers in cellulose. You learned that enzymes are very specific in what molecules they bind to. After examining the chemical structures of starch and cellulose in table 3.2, explain why the same enzyme that breaks down starch cannot break down cellulose.

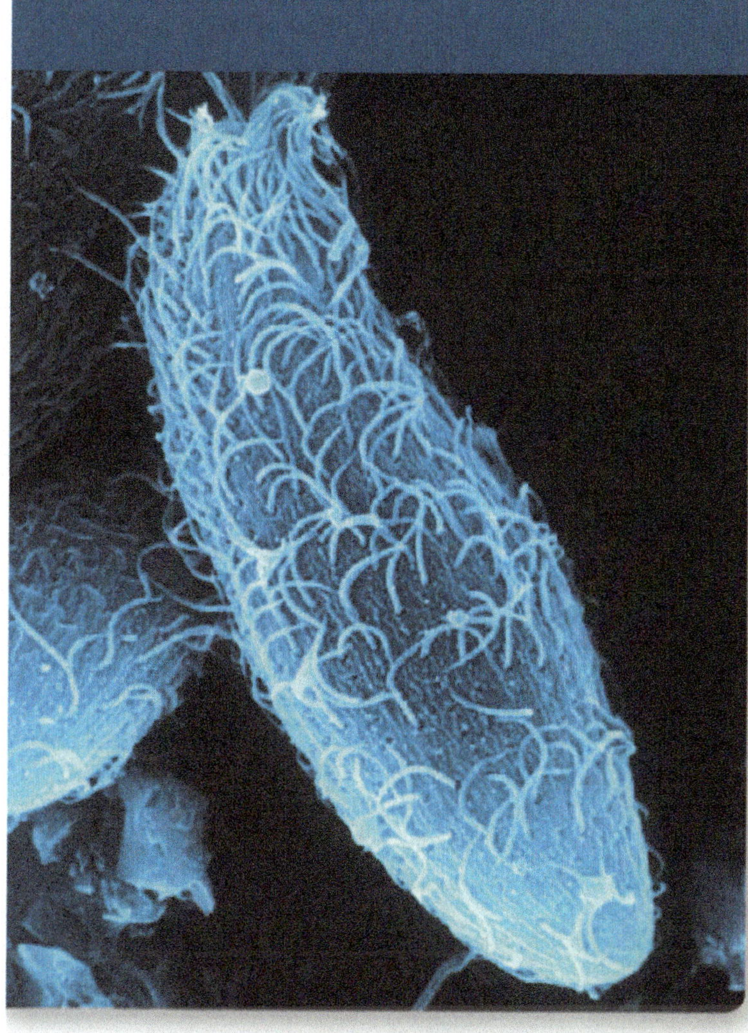

4
Cells

Learning Objectives

This alarming-looking creature is the single-celled organism *Dileptus,* magnified a thousand times. Too small to see with the unaided eye, *Dileptus* is one of hundreds of inhabitants of a drop of pond water. Everything that a living organism does to survive and prosper, *Dileptus* must do with only the equipment this tiny cell provides. Just as you move about using legs to walk, so *Dileptus* uses the hairlike projections (called cilia) that cover its surface to propel itself through the water. Just as your brain is the control center of your body, so the compartment called the nucleus, deep within the interior of *Dileptus,* controls the many activities of this complex and very active cell. *Dileptus* has no mouth, but it takes in food particles and other molecules through its surface. This versatile protist is capable of leading a complex life because its interior is subdivided into compartments, in each of which it carries out different activities. Functional specialization is the hallmark of this cell's interior, a powerful approach to cellular organization that is shared by all eukaryotes.

The World of Cells

4.1 Cells

Hold your finger up and look at it closely. What do you see? Skin. It looks solid and smooth, creased with lines and flexible to the touch. But if you were able to remove a bit and examine it under a microscope, it would look very different—a sheet of tiny, irregularly shaped bodies crammed together like tiles on a floor. Figure 4.1 takes you on a journey into your fingertip. The crammed bodies you see in panels 3 and 4 are skin cells, laid out like a tiled floor. As your journey inward continues, you travel inside one of the cells and see organelles, structures in the cell that perform specific functions. Proceeding even farther inward, you encounter the molecules of which the structures are made, and finally the atoms shown in panels 8 and 9. While some organisms are composed of a single cell, your body is composed of many cells. A single human being has as many cells as the stars in a galaxy, between 10 and 100 trillion (depending on how big you are). All cells, however, are small. In this chapter we look more closely at cells and learn something of their internal structure and how they communicate with their environment.

The Cell Theory

Because cells are so small, no one observed them until microscopes were invented in the mid-seventeenth century. Robert Hooke first described cells in 1665, when he used a microscope he had built to examine a thin slice of nonliving plant tissue called cork. Hooke observed a honeycomb of tiny, empty (because the cells were dead) compartments. He called the compartments in the cork *cellulae* (Latin, small rooms), and the term has come down to us as **cells.** For another century and a half, however, biologists failed to recognize the importance

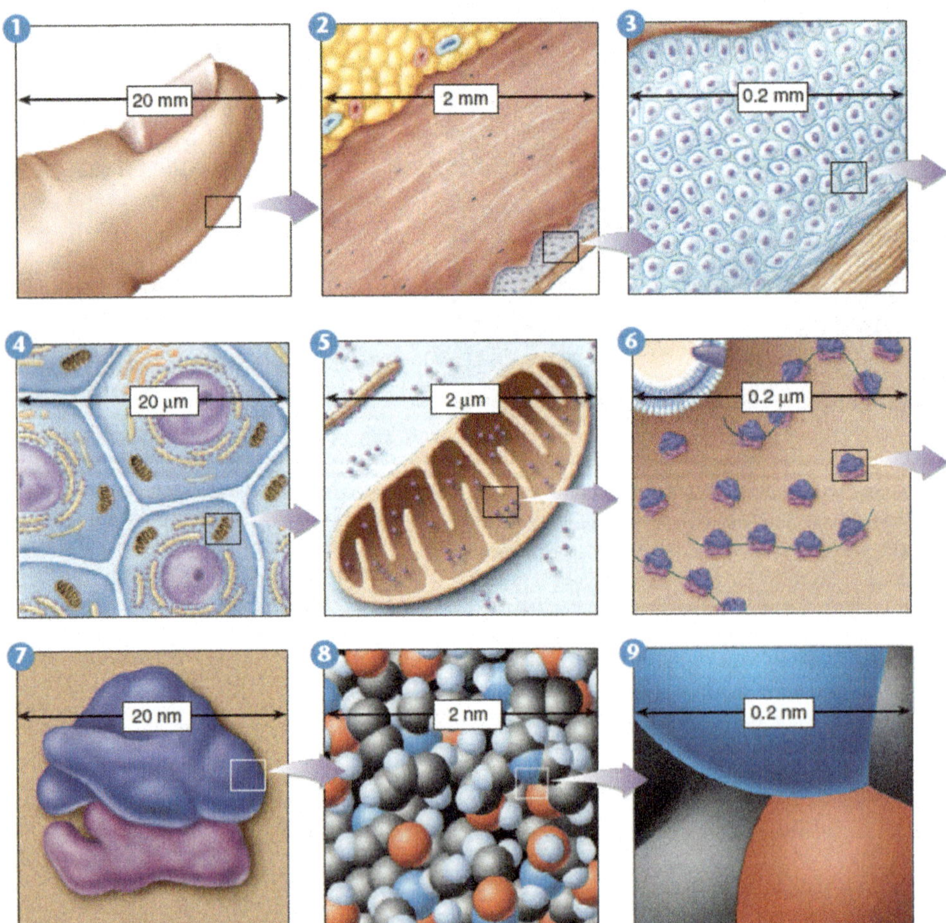

Figure 4.1 The size of cells and their contents.
This diagram shows the size of human skin cells, organelles, and molecules. In general, the diameter of a human skin cell 4 is a little less than 20 micrometers (µm), of a mitochondrion 5 is 2 µm, of a ribosome 7 is 20 nanometers (nm), of a protein molecule 8 is 2 nm, and of an atom 9 is 0.2 nm.

of cells. In 1838, botanist Matthias Schleiden made a careful study of plant tissues and developed the first statement of the cell theory. He stated that all plants "are aggregates of fully individualized, independent, separate beings, namely the cells themselves." In 1839, Theodor Schwann reported that all animal tissues also consist of individual cells.

The idea that all organisms are composed of cells is called the **cell theory.** In its modern form, the cell theory includes three principles:

1. All organisms are composed of one or more cells, within which the processes of life occur.
2. Cells are the smallest living things. Nothing smaller than a cell is considered alive.
3. Cells arise only by division of a previously existing cell. Although life likely evolved spontaneously in the environment of the early earth, biologists have concluded that no additional cells are originating spontaneously at present. Rather, life on earth represents a continuous line of descent from those early cells.

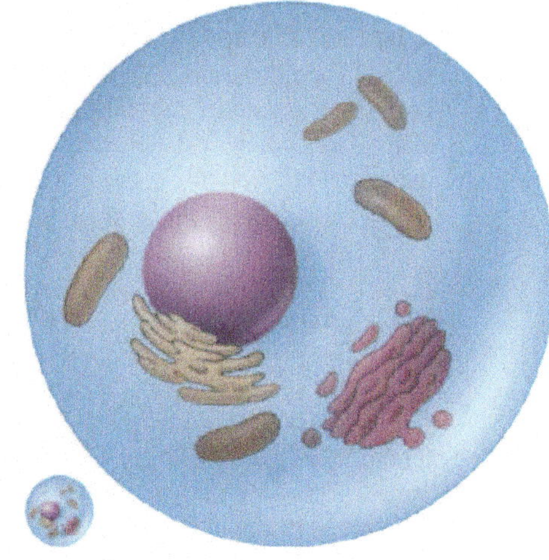

Cell radius (r)	1 unit	10 units
Surface area ($4\pi r^2$)	12.57 units^2	1,257 units^2
Volume ($\frac{4}{3}\pi r^3$)	4.189 units^3	4,189 units^3

Figure 4.2 Surface-to-volume ratio.
As a cell gets larger, its volume increases at a faster rate than its surface area. If the cell radius increases by 10 times, the surface area increases by 100 times, but the volume increases by 1,000 times. A cell's surface area must be large enough to meet the needs of its volume.

Most Cells Are Very Small

Most cells are relatively small, but not all are the same size. The cells of your body are typically from 5 to 20 micrometers (a micrometer, μm, is one-millionth of a meter) in diameter, too small to see with the naked eye. Bacteria cells are even smaller than yours, only a few micrometers thick. However, there are some cells that are larger; individual marine alga cells, for example, can be up to 5 centimeters long—as long as your little finger.

Why Aren't Cells Larger?

Why are most cells so tiny? Most cells are small because larger cells do not function as efficiently. In the center of every cell is a command center that must issue orders to all parts of the cell, directing the synthesis of certain enzymes, the entry of ions and molecules from the exterior, and the assembly of new cell parts. These orders must pass from the core to all parts of the cell, and it takes them a very long time to reach the periphery of a large cell. For this reason, an organism made up of relatively small cells is at an advantage over one composed of larger cells.

Another reason cells are not larger is the advantage of having a greater **surface-to-volume ratio.** As cell size increases, volume grows much more rapidly than surface area. For a round cell, surface area increases as the square of the diameter, whereas volume increases as the cube. To visualize this, consider the two single cells in figure 4.2. The large cell to the right is 10 times bigger than the small cell, but while its surface area is 100 times greater (10^2), its volume is 1,000 times (10^3) the volume of the smaller cell. A cell's surface provides the interior's only opportunity to interact with the environment, with substances passing into and out of the cell across its surface. Large cells have far less surface for each unit of volume than do small ones.

Some larger cells, however, function quite efficiently in part because they have structural features that increase surface area. For example, cells in the nervous system called neurons are long, slender cells, some extending more than a meter in length. These cells efficiently interact with their environment because although they are long, they are thin, some less than 1 micrometer in diameter, and so their interior regions are not far from the surface at any given point.

Another structural feature that increases the surface area of a cell are small "fingerlike" projections called microvilli. The cells that line the small intestines of the human digestive system are covered with microvilli that dramatically increase the surface area of the cells.

With few exceptions however, cells don't usually grow much larger than 50 micrometers. For organisms to get much larger, they are usually composed of many cells. By grouping together many smaller cells, these multicellular organisms vastly increase their total surface-to-volume ratio.

An Overview of Cell Structure

All cells are surrounded by a delicate membrane, called a *plasma membrane,* that controls the permeability of the cell to water and dissolved substances. A semifluid matrix called *cytoplasm* fills the interior of the cell. It used to be thought that the cytoplasm was uniform, like Jell-O™, but we now know that it is highly organized. Your cells, for example, have an internal framework that both gives the cell its shape and positions components and materials within its interior. In the following sections, we explore the membranes that encase all living cells and then examine in detail their interiors.

Visualizing Cells

How many cells are big enough to see with the unaided eye? Other than egg cells, not many are (figure 4.3). Most are less than 50 micrometers in diameter, far smaller than the period at the end of this sentence.

The Resolution Problem How do we study cells if they are too small to see? The key is to understand why we can't see them. The reason we can't see such small objects is the limited resolution of the human eye. **Resolution** is defined as the minimum distance two points can be apart and still be distinguished as two separated points. On the visibility scale in figure 4.3 below, you can see that the limit of resolution of the human eye (the blue bar at the bottom) is about 100 micrometers. This limit occurs because when two objects are closer together than about 100 micrometers, the light reflected from each strikes the same "detector" cell at the rear of the eye. Only when the objects are farther apart than 100 micrometers will the light from each strike different cells, allowing your eye to resolve them as two objects rather than one.

Microscopes One way to increase resolution is to increase magnification so that small objects appear larger. Robert Hooke and Anton van Leeuwenhoek used glass lenses to magnify small cells and cause them to appear larger than the 100-micrometer limit imposed by the human eye. The glass lens adds additional focusing power. Because the glass lens makes the object appear closer, the image on the back of the eye is bigger than it would be without the lens.

Modern *light microscopes* use two magnifying lenses (and a variety of correcting lenses) to achieve very high magnification and clarity. The first lens focuses the image of the object on the second lens, which magnifies it again and focuses it on the back of the eye. Microscopes that magnify in stages using several lenses are called **compound microscopes.** They can resolve structures that are separated by more than 200 nanometers (nm). The six entries in the upper portion of table 4.1 are images viewed through various types of light microscopes.

Increasing Resolution Light microscopes, even compound ones, are not powerful enough to resolve many structures within cells. For example, a membrane is only 5 nanometers thick. Why not just add another magnifying stage to the microscope and so increase its resolving power? Because when two objects are closer than a few hundred nanometers, the light beams reflecting from the two images start to overlap. The only way two light beams can get closer together and still be resolved is if their wavelengths are shorter.

One way to avoid overlap is by using a beam of electrons rather than a beam of light. Electrons have a much shorter wavelength, and a microscope employing electron beams has 1,000 times the resolving power of a light microscope. A **transmission electron microscope (TEM),** so called because the electrons used to visualize the specimens are transmitted through the material, is capable of resolving objects only 0.2 nanometer apart—just twice the diameter of a hydrogen atom! The entry on the left under electron microscopes in table 4.1 is an example of an image captured using a TEM.

A second kind of electron microscope, the **scanning electron microscope (SEM),** beams the electrons onto the surface of the specimen. The electrons reflected back from the surface of the specimen, together with other electrons that the specimen itself emits as a result of the bombardment, are amplified and transmitted to a screen, where the image can be viewed and photographed. Scanning electron microscopy yields striking three-dimensional images and has improved our understanding of many biological and physical phenomena. The entry on the right in table 4.1 under electron microscopes is an SEM image.

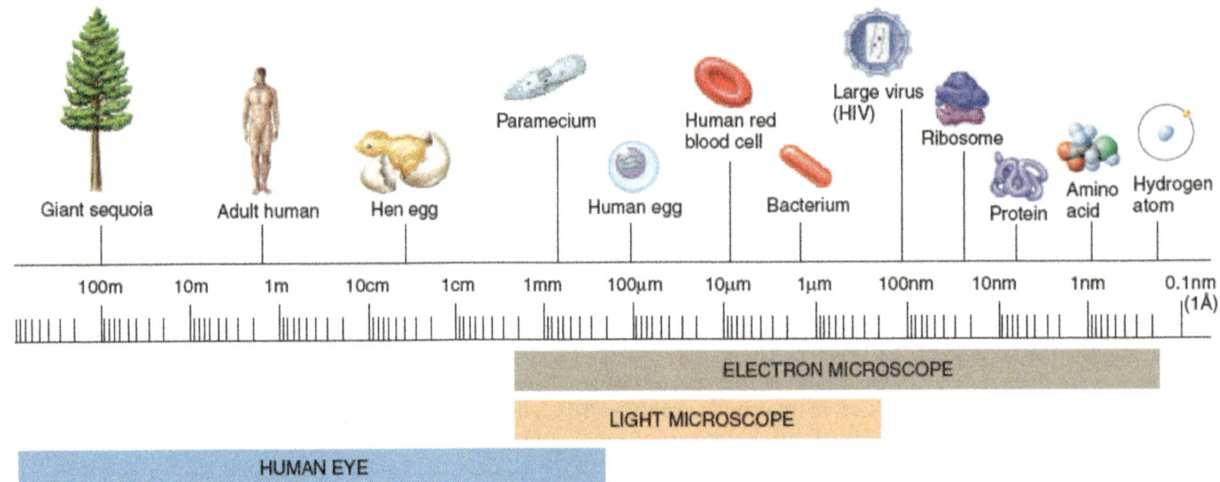

Figure 4.3 A scale of visibility.
Most cells are microscopic in size, although vertebrate eggs are typically large enough to be seen with the unaided eye. Prokaryotic cells are generally 1 to 2 micrometers (µm) across.

TABLE 4.1 TYPES OF MICROSCOPES

Light Microscopes

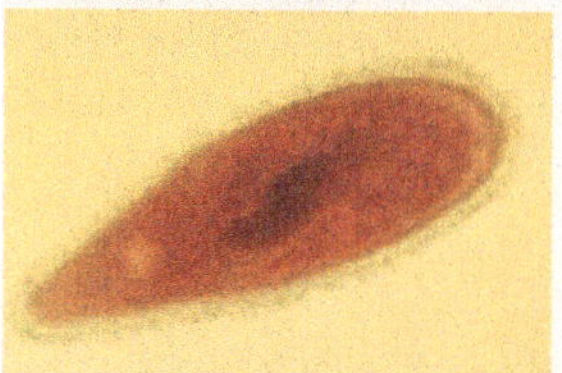

Bright-field microscope: Light is simply transmitted through a specimen in culture, giving little contrast. Staining specimens improves contrast but requires that cells be fixed (not alive), which can cause distortion or alteration of components.

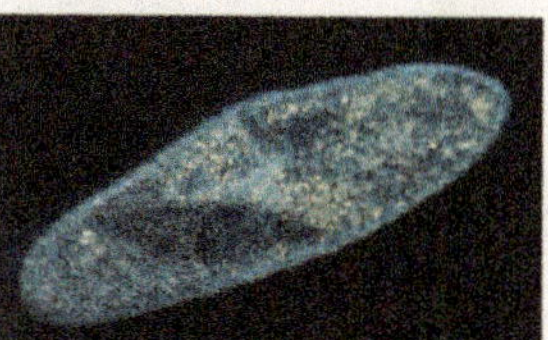

Dark-field microscope: Light is directed at an angle toward the specimen; a condenser lens transmits only light reflected off the specimen. The field is dark, and the specimen is light against this dark background.

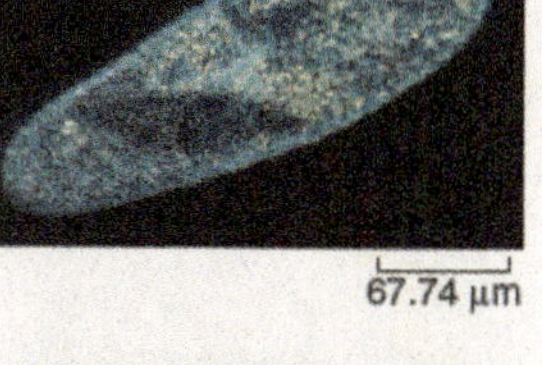

Phase-contrast microscope: Components of the microscope bring light waves out of phase, which produces differences in contrast and brightness when the light waves recombine.

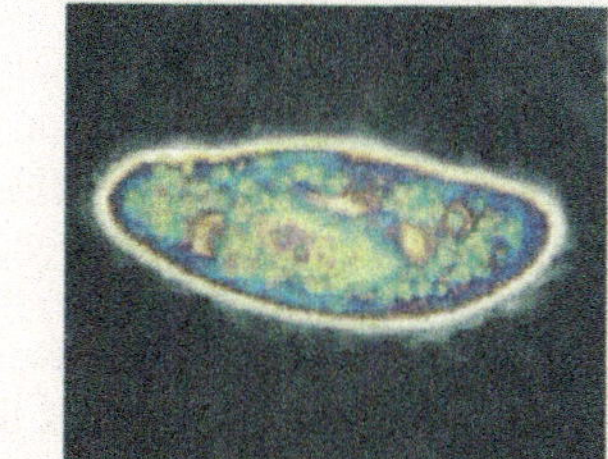

Differential-interference-contrast microscope: Out-of-phase light waves that produce differences in contrast are combined with two beams of light travelling close together, which creates even more contrast, especially at the edges of structures.

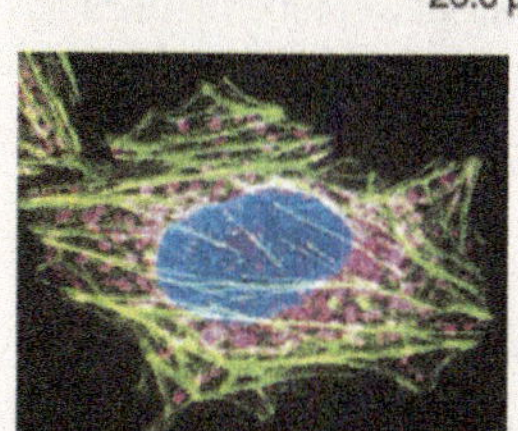

Fluorescence microscope: A set of filters transmits only light that is emitted by fluorescently stained molecules or tissues.

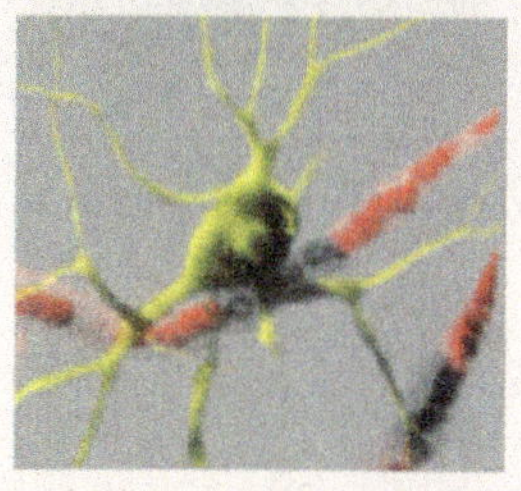

Confocal microscope: Light from a laser is focused to a point and scanned across the specimen in two directions. Clear images of one plane of the specimen are produced, while other planes of the specimen are excluded and do not blur the image. Fluorescent dyes and false coloring enhance the image.

Electron Microscopes

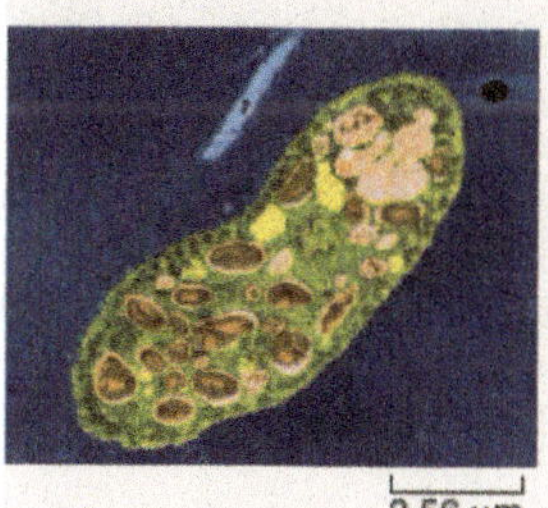

Transmission electron microscope: A beam of electrons is passed through the specimen. Electrons that pass through are used to form an image. Areas of the specimen that scatter electrons appear dark. False coloring has been added to this image.

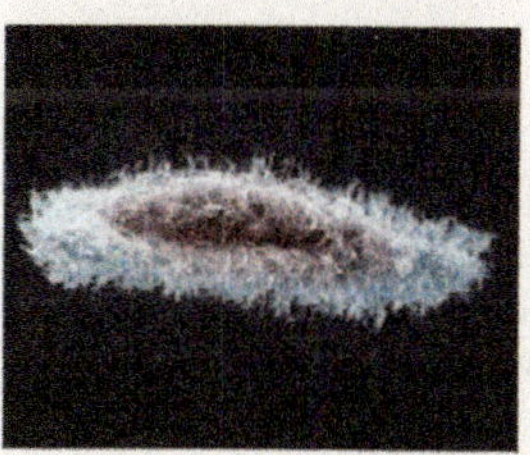

Scanning electron microscope: An electron beam is scanned across the surface of the specimen, and electrons are knocked off the surface. Thus, the surface topography of the specimen determines the contrast and the content of the image. False coloring enhances the image.

Visualizing Cell Structure by Staining Specific Molecules

A powerful tool for the analysis of cell structure has been the use of stains that bind to specific molecular targets. This approach has been used in the analysis of tissue samples, or histology, for many years and has been improved dramatically with the use of antibodies that bind to very specific molecular structures. This process, called immunocytochemistry, uses antibodies generated in animals such as rabbits or mice. When these animals are injected with specific proteins, they will produce antibodies that specifically bind to the injected protein, which can be purified from their blood. These purified antibodies can then be chemically bonded to enzymes, stains, or fluorescent molecules that glow when exposed to specific wavelengths of light. When cells are washed in a solution containing the antibodies, the antibodies bind to cellular structures that contain the target molecule and can be seen with light microscopy. The image produced using fluorescence microscopy in table 4.1 shows the cytoskeleton made of cablelike structures inside the cell. This approach has been used extensively in the analysis of cell structure and function.

> **Key Learning Outcome 4.1** **All living things are composed of one or more cells, each a small volume of cytoplasm surrounded by a plasma membrane. Most cells and their components can only be viewed using microscopes.**

4.2 The Plasma Membrane

Encasing all living cells is a delicate sheet of molecules called the **plasma membrane.** It would take more than 10,000 of these molecular sheets, which are about 5 nanometers thick, piled on top of one another to equal the thickness of this sheet of paper. However, the sheets are not simple in structure, like a soap bubble's skin. Rather, they are made up of a diverse collection of proteins floating within a lipid framework like small boats bobbing on the surface of a pond. Regardless of the kind of cell they enclose, all plasma membranes have the same basic structure of proteins embedded in a sheet of lipids, called the **fluid mosaic model.**

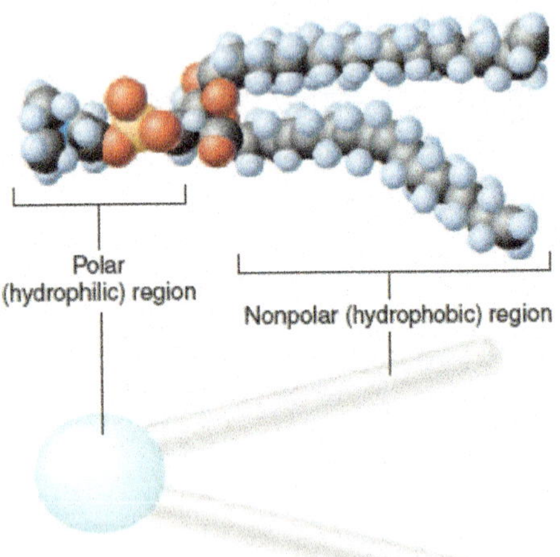

The lipid layer that forms the foundation of a plasma membrane is composed of modified fat molecules called **phospholipids.** A phospholipid molecule can be thought of as a polar head with two nonpolar tails attached to it, as shown above. The head of a phospholipid molecule has a phosphate chemical group linked to it, making it extremely polar (and thus water-soluble). The other end of the phospholipid molecule is composed of two long fatty acid chains. Recall from chapter 3 that fatty acids are long chains of carbon atoms with attached hydrogen atoms. The carbon atoms are the gray spheres you see above. The fatty acid tails are strongly nonpolar and thus water-insoluble. The phospholipid is often depicted diagrammatically as a ball with two tails.

Imagine what happens when a collection of phospholipid molecules is placed in water. A structure called a **lipid bilayer** forms spontaneously. How can this happen? The long nonpolar tails of the phospholipid molecules are pushed away by the water molecules that surround them, shouldered aside as the water molecules seek partners that

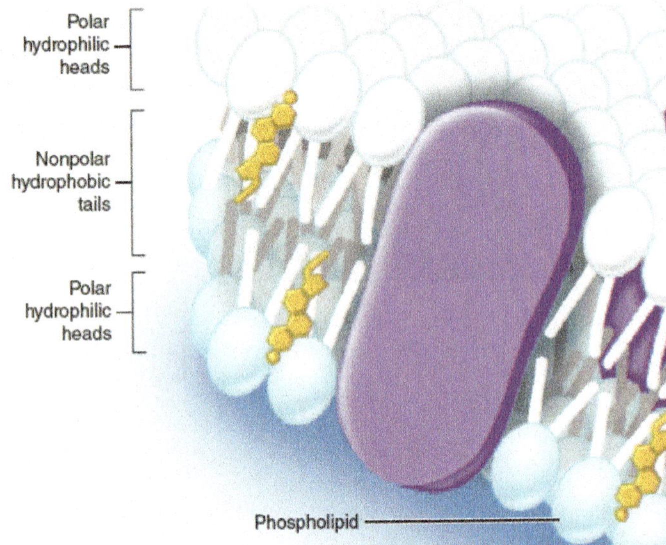

can form hydrogen bonds. After much shoving and jostling, every phospholipid molecule ends up with its polar head facing water and its nonpolar tail facing away from water. The phospholipid molecules form a *double* layer, called a bilayer. As you can see in the figure above, the watery environments inside and outside the plasma membrane push the nonpolar tails to the interior of the bilayer. Because there are two layers with the tails facing each other, no tails are ever in contact with water. Thus, the interior of a lipid bilayer is completely nonpolar, and it repels any water-soluble molecules that attempt to pass through it, just as a layer of oil stops the passage of a drop of water (that's why ducks do not get wet).

Cholesterol, another nonpolar lipid molecule, resides in the interior portion of the bilayer. Cholesterol is a multi-ringed molecule that affects the fluid nature of the membrane. Although cholesterol is important in maintaining the integrity of the plasma membrane, it can accumulate in blood vessels, forming plaques that lead to cardiovascular disease.

Proteins Within the Membrane

The second major component of every biological membrane is a collection of **membrane proteins** that float within the lipid bilayer. Membrane proteins function as channels, receptors, and cell surface markers. As you can see here in the figure, some proteins pass through the lipid bilayer, providing channels through which molecules and information pass. While some membrane proteins are fixed into position, others move about freely.

Many membrane proteins project up from the surface of the plasma membrane like buoys, often with carbohydrate chains or lipids attached to their tips like flags. These **cell surface proteins** act as markers to identify particular types of cells, or as beacons to bind specific hormones or proteins to the cell.

Protein channels that extend all the way across the bilayer provide passageways for ions and polar molecules like water so they can pass into and out of the cell. How do these **transmembrane proteins** manage to span the membrane, rather than just floating on the surface in the way that a drop of water floats on oil? The part of the protein that actually traverses the lipid bilayer is a specially constructed spiral helix of nonpolar amino acids—the red coiled areas of the transmembrane protein you see in the inset. Water responds to these nonpolar amino acids much as it does to nonpolar lipid chains, and as a result the helical spiral is held within the lipid interior of the bilayer, anchored there by the strong tendency of water to avoid contact with these nonpolar amino acids.

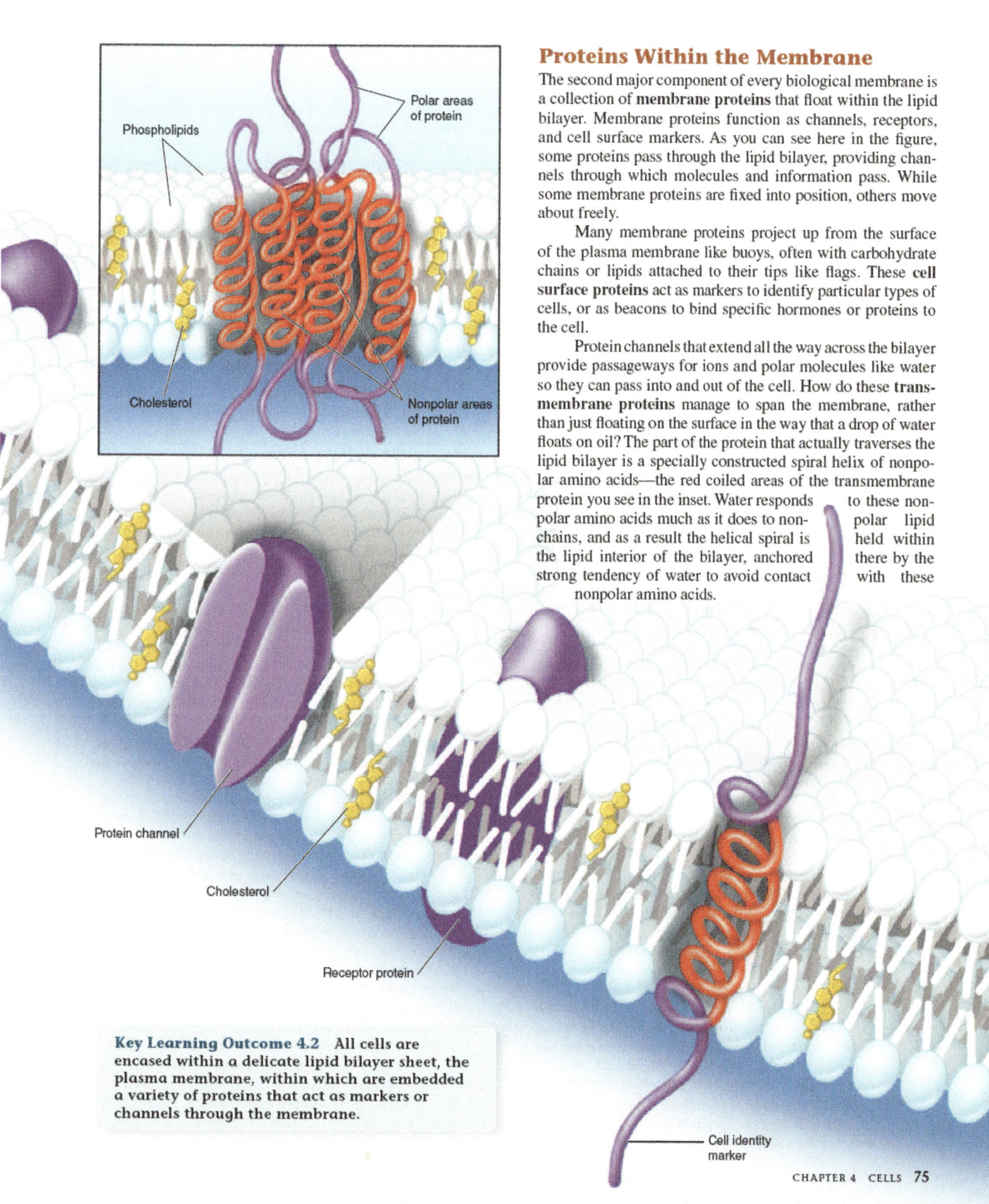

Key Learning Outcome 4.2 All cells are encased within a delicate lipid bilayer sheet, the plasma membrane, within which are embedded a variety of proteins that act as markers or channels through the membrane.

A Closer Look

How Water Crosses the Plasma Membrane

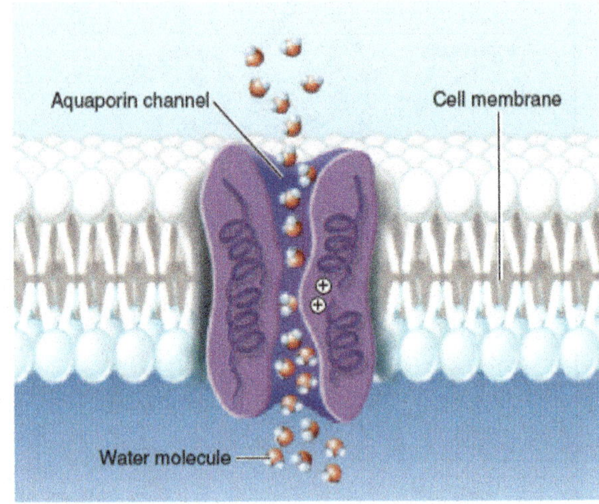

One of the enduring mysteries of cellular biology has been the free movement of water into and out of cells. As early as the middle of the nineteenth century, biologists understood that there must be a way for water to pass across the plasma membrane. With the understanding in the middle of the 1950s that the plasma membrane was composed of a lipid bilayer, the problem posed by the free movement of water became even more puzzling. How could very polar water molecules traverse the very nonpolar environment found in the lipid core of the bilayer?

While some proposed that water leaked into cells through tiny imperfections in the bilayer, or through gaps that open up in the bilayer when the hydrocarbon tails flex and bend, these hypotheses were soon rejected, as they failed to explain how cell membranes managed to prevent the diffusion of protons (hydrogen ions), which are smaller than water. This ability is crucial to the life of a cell, as the difference in proton concentration between the inside and outside of cell organelles is the basis of energy metabolism, as described in chapters 6 and 7.

Clearly the answer to this puzzle lay with the proteins associated with the lipid bilayer. For 30 years researchers searched for the protein machinery that prevented ions from passing across the membrane while allowing water molecules to pass freely. With the acceptance of the fluid mosaic model of membrane structure proposed in 1972, the search focused on proteins that bridge the bilayer. In the mid-1980s, a Johns Hopkins University researcher named Peter Agre, studying red blood cell proteins, identified a previously unknown protein. "No one had seen it before, but we found that it was the fifth most abundant protein in the cell," said Agre. "That's like coming across a big town that's not on the map. It gets your attention."

Agre determined the amino acid sequence of the mysterious protein, and saw that it had long nonpolar segments that would allow it to repeatedly traverse the lipid bilayer, just as a cellular water channel would have to do. Perhaps this was the protein so many had sought.

To test this hypothesis, Agre carried out a simple experiment. He compared normal red blood cells that contained the protein with mutant red blood cells that lacked it. When he placed the cells in distilled lab water, cells with the protein in their membrane absorbed water and began to swell, while cells lacking the protein did nothing—they failed to absorb water and did not swell.

To be sure that some other undiscovered membrane protein could not be the cause of the water movement across the red blood cell plasma membrane, Agre repeated the experiment with liposomes, which are artificial cells made of pure lipid bilayer with no proteins—in essence, soap bubbles. As you might expect, water could not cross into liposomes, and he found they did not swell when immersed in distilled lab water. However, the liposomes did become permeable to water if Agre planted the protein in their bilayer.

As a final test, Agre knew that mercury ions poison cells by keeping them from taking up and releasing water, and he showed that water transport through liposomes with his protein in their membranes was prevented by mercury too. Agre concluded that the protein he had discovered was indeed the water channel, and named the protein *aquaporin*, "water pore." Peter Agre was awarded the Nobel Prize in Chemistry in 2003 for his discovery.

Working with other research teams on this exciting discovery, Agre reported in 2000 the results of X-ray diffraction studies that revealed the three-dimensional structure of aquaporin in atomic detail. Now it was possible to see in detail how the water channel functions.

The aquaporin channel Agre described is illustrated above. Individual water molecules pass through the narrow open channel single file, snaking their way along by orienting themselves in the local electrical field formed by the atoms of the channel wall. And now we see the key feature that allows the channel to exclude protons while passing larger water molecules. Look in the center of the pore. A cluster of positively-charged amino acids line the pore there. While this has no impact on the passage of water molecules because of their partial negative charges, the positive charges repel protons, which are also positively charged. The positive charges in the interior of the aquaporin channel act as a filter to prevent protons leaking through the passage.

In the last 10 years researchers have identified aquaporins in many kinds of bacteria, plants, and animals. There are at least 11 kinds in the human body alone. One kind, called AQP1, acts in your kidney to recover water that would otherwise be lost in your urine. Over 24 hours, AQP1 channels in your kidneys recover about 120 liters of water!

Kinds of Cells

4.3 Prokaryotic Cells

There are two major kinds of cells: prokaryotes and eukaryotes. **Prokaryotes** have a relatively uniform cytoplasm that is not subdivided by interior membranes into separate compartments. They do not, for example, have special membrane-bounded compartments, called *organelles,* or a *nucleus* (a membrane-bounded compartment that holds the hereditary information). As discussed in chapter 1, figure 1.1, the two main groups of prokaryotes are *bacteria* and *archaea;* all other organisms are eukaryotes.

Prokaryotes are the simplest cellular organisms. Over 5,000 species are recognized, but doubtless many times that number actually exist and have not yet been described. Although these species are diverse in form, their organization is fundamentally similar: They are single-celled organisms; the cells are small (typically about 1 to 10 micrometers thick); the cells are enclosed by a plasma membrane; and there are no distinct interior compartments (figure 4.4). Outside of almost all bacteria and archaea is a *cell wall,* composed of different molecules in different groups (see table 15.1 and section 16.3). In some bacteria another layer called the *capsule* encloses the cell wall. Archaea are an extremely diverse group that inhabit diverse environments (figure 4.5*a*). Bacteria are abundant and play critical roles in many biological processes. Bacteria assume many shapes, like the sausage or spiral shapes shown in figure 4.5*b, c*. They can also adhere in chains and masses like the spherical cells in figure 4.5*d,* but in these cases the individual cells remain functionally separate from one another.

The interior of a prokaryotic cell has little or no structural support (the cell wall supports the cell's shape), but scattered throughout the cytoplasm are small structures called *ribosomes.* Ribosomes are the sites where proteins are made, but they are not considered organelles because they lack a membrane boundary. Prokaryotic DNA is found in a region of the cytoplasm called the *nucleoid region,* which is not enclosed within an internal membrane. Some prokaryotes use a **flagellum** (plural, **flagella**) to move. Flagella are long, threadlike structures, made of protein fibers that project from the surface of a cell. They are used in locomotion and feeding. There may be none, one, or more per cell depending on the species. Bacteria can swim at speeds of up to 20 cell diameters per second, rotating their flagella like screws. **Pili** (singular, **pilus**) are short flagella (only several micrometers long, and about 7.5 to 10 nanometers thick) that occur on the cells of some prokaryotes. Pili help the prokaryotic cells attach to appropriate substrates and aid in the exchange of genetic information between cells.

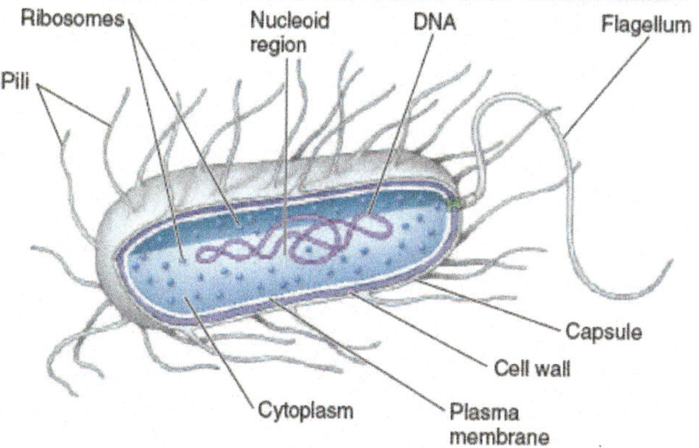

Figure 4.4 Organization of a prokaryotic cell.
Prokaryotic cells lack internal compartments. Not all prokaryotic cells have a flagellum or a capsule like the one illustrated here, but all have a nucleoid region, ribosomes, a plasma membrane, cytoplasm, and a cell wall.

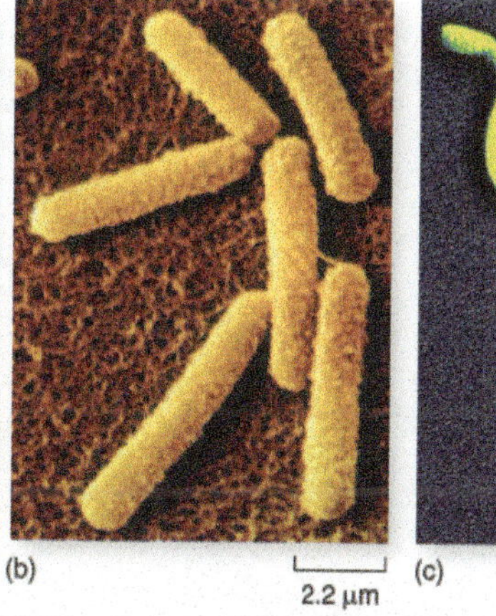

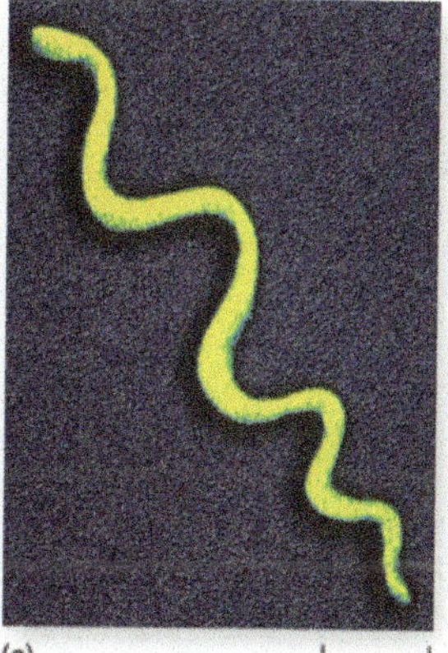

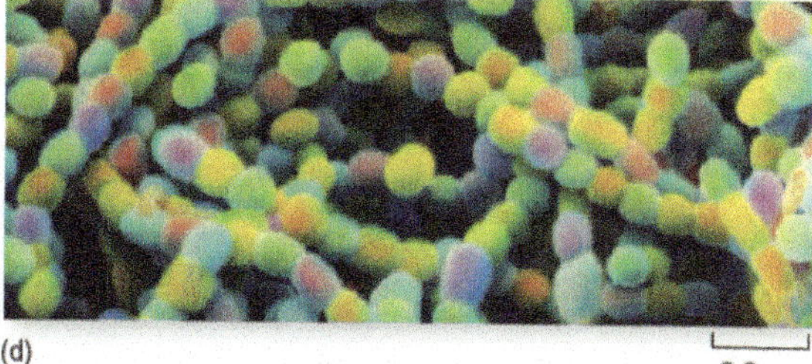

Figure 4.5 Diversity of prokaryotes.
(*a*) *Methanococcus* is only able to survive in environments with no oxygen. (*b*) *Bacillus* is a rod-shaped bacterium. (*c*) *Treponema* is a coil-shaped bacterium; rotation of internal filaments produces a corkscrew movement. (*d*) *Streptomyces* is a more or less spherical bacterium in which the individuals adhere in chains.

Key Learning Outcome 4.3 Prokaryotic cells lack a nucleus and do not have an extensive system of interior membranes.

4.4 Eukaryotic Cells

For the first 1 billion years of life on earth, all organisms were prokaryotes, cells with very simple interiors. Then, about 1.5 billion years ago, a new kind of cell appeared for the first time, the eukaryotic cell. Eukaryotic cells are much larger and profoundly different from prokaryotic cells, with a complex interior organization. All cells alive today except bacteria and archaea are of this new kind.

Figures 4.6 and 4.7 present cross-sectional diagrams of idealized animal and plant cells. As you can see, the interior of a eukaryotic cell is much more complex than the prokaryotic cell you encountered in figure 4.4. The **plasma membrane** ❶ encases a semifluid matrix called the **cytoplasm** ❷, which contains within it the nucleus and various cell structures called organelles. An **organelle** is a specialized structure within which particular cell processes occur. Each organelle, such as a **mitochondrion** ❸, has a specific function in the eukaryotic cell. The organelles are anchored at specific locations in the cytoplasm by an interior scaffold of protein fibers, the **cytoskeleton** ❹.

One of the organelles is very visible when these cells are examined with a microscope, filling the center of the cell like the pit of a peach. Seeing it, the English botanist Robert Brown in 1831 called it the **nucleus** ❺ (plural, *nuclei*), from the Latin word for "kernel." Inside the nucleus, the DNA is wound tightly around proteins and packaged into compact units called chromosomes. It is the nucleus that gives **eukaryotes** their name, from the Greek words *eu*, true, and *karyon*, nut; by way of contrast, the earlier-evolving bacteria and archaea are called prokaryotes ("before the nut").

If you examine the organelles in figures 4.6 and 4.7, you can see that most of them form separate compartments within the cytoplasm, bounded by their own membranes. *The hallmark of the eukaryotic cell is this compartmentalization.* This internal compartmentalization is achieved by an extensive **endomembrane system** ❻ that weaves through the cell interior, providing extensive surface area for many membrane-associated cell processes to occur.

Vesicles ❼ (small membrane-bounded sacs that store and transport materials) form in the cell either by budding off of the endomembrane system or by the incorporation of lipids and protein in the cytoplasm. These many closed-off compartments allow different processes to proceed simultaneously without interfering with one another, just as rooms do in a house. Thus the

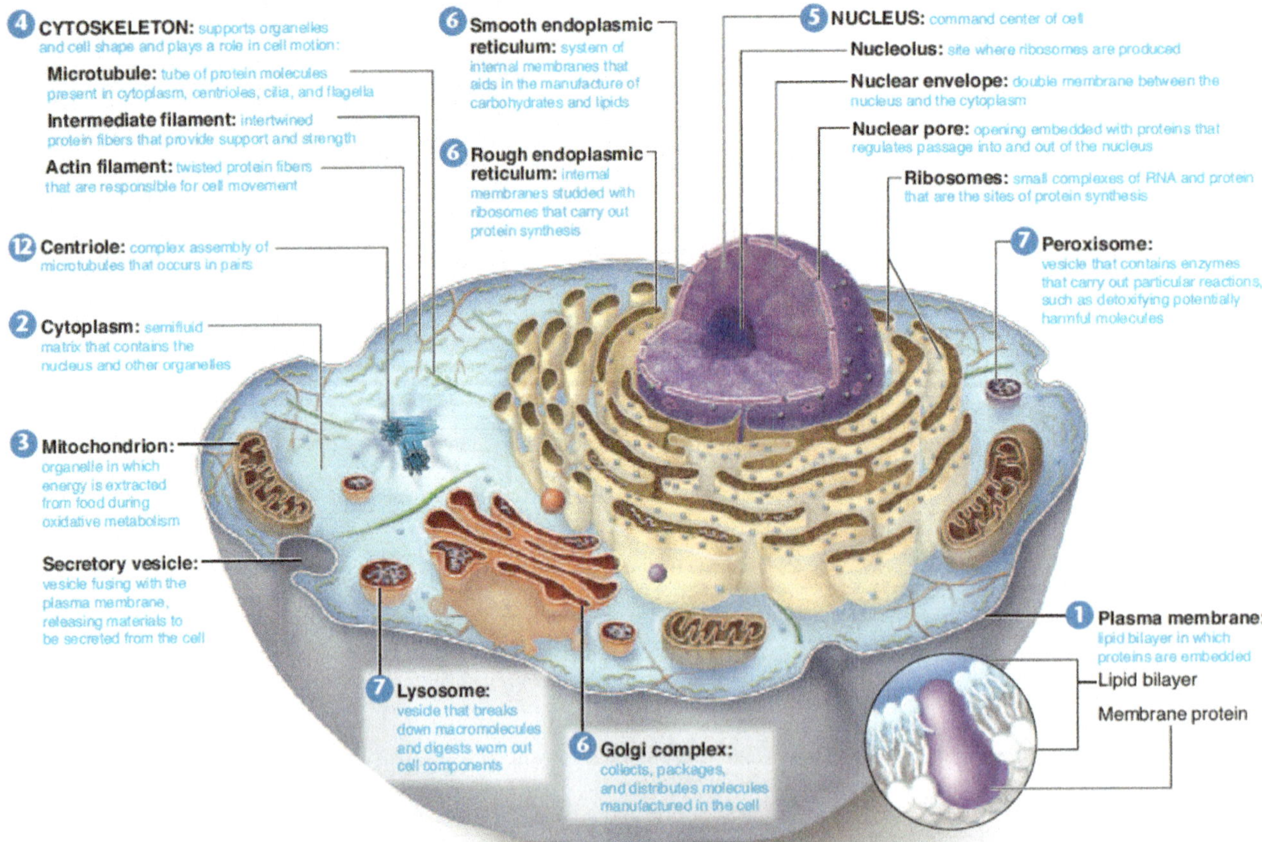

Figure 4.6 Structure of an animal cell.

In this generalized diagram of an animal cell, the plasma membrane encases the cell, which contains the cytoskeleton and various cell organelles and interior structures suspended in a semifluid matrix called the cytoplasm. Some kinds of animal cells possess fingerlike projections called microvilli. Other types of eukaryotic cells, for example many protist cells, may possess flagella, which aid in movement, or cilia, which can have many different functions.

organelles called *lysosomes* are recycling centers. Their very acid interiors break down old organelles, and the component molecules are recycled. This acid would be very destructive if released into the cytoplasm. Similarly, chemical isolation is essential to the function of the organelles called *peroxisomes*. Toxic chemicals are degraded and food molecules are processed within peroxisomes by enzymes that act by removing electrons and associated hydrogen atoms. If not isolated within the peroxisomes, these enzymes would tend to short-circuit chemical reactions occurring in the cytoplasm, which often involves adding hydrogen atoms to molecules.

If you compare figure 4.6 with figure 4.7, you will see the same set of organelles, with a few interesting exceptions. For example, the cells of plants, fungi, and many protists have strong thick exterior **cell walls** ⑧ composed of cellulose or chitin fibers, while the cells of animals lack cell walls. All plants and many kinds of protists have **chloroplasts** ⑨, within which photosynthesis occurs. No animal or fungal cells contain chloroplasts. Plant cells also contain a large **central vacuole** ⑩ that stores water, and cytoplasmic connections through openings in the cell wall called **plasmodesmata** ⑪. **Centrioles** ⑫ are present in animal cells but absent in plant and fungal cells. Some kinds of animal cells possess fingerlike projections called *microvilli*. Many animal and protist cells possess *flagella*, which aid in movement, or *cilia*, which have many different functions. Flagella occur in sperm of a few plant species, but are otherwise absent in plant and fungal cells.

We will now journey into the interior of a typical eukaryotic cell and explore it in more detail, using diagrams with the particular organelle you are examining highlighted. While the various organelles are color-coded for easier identification, remember that most are actually colorless.

Key Learning Outcome 4.4 **Eukaryotic cells have a system of interior membranes and organelles that subdivide the interior into functional compartments.**

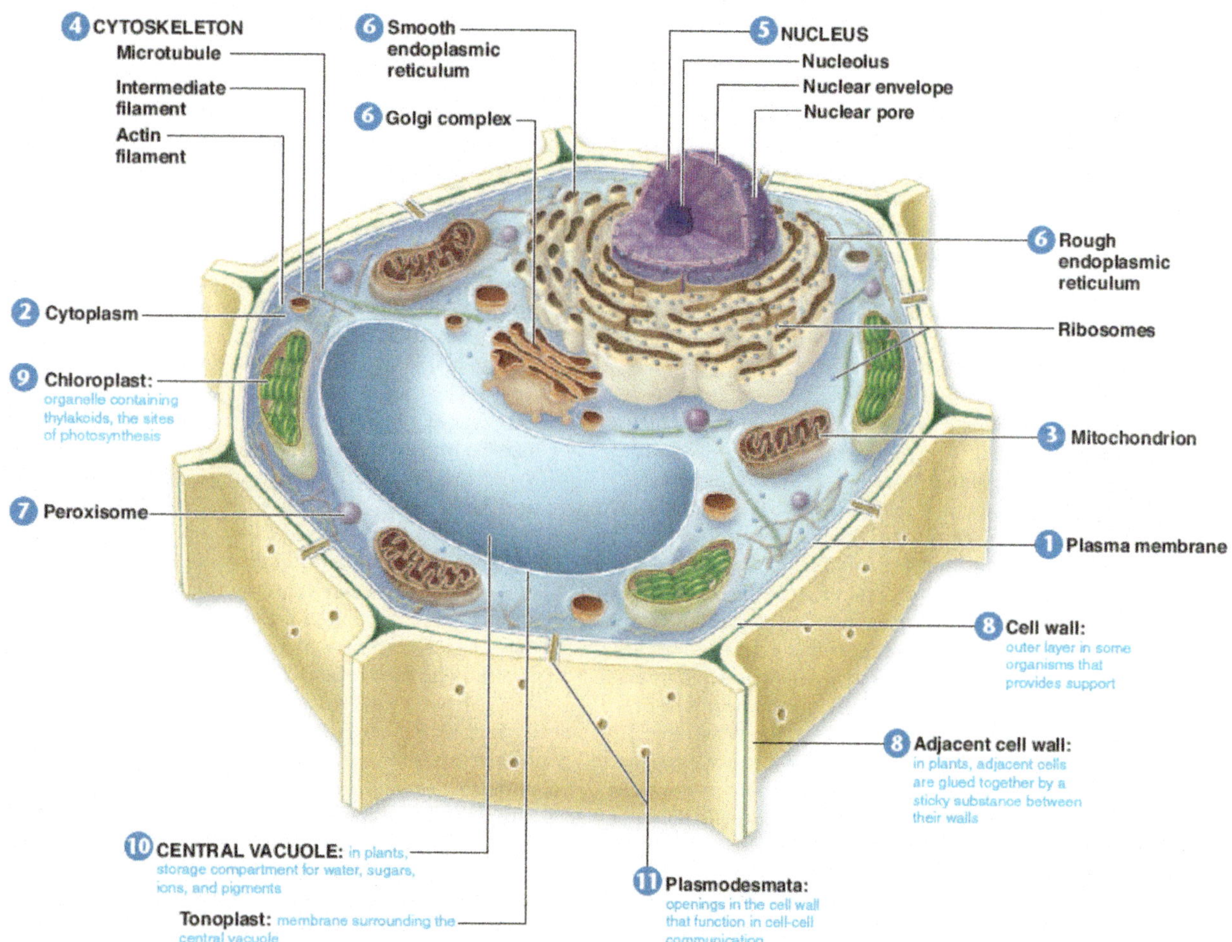

Figure 4.7 Structure of a plant cell.

Most mature plant cells contain large central vacuoles that occupy a major portion of the internal volume of the cell and organelles called chloroplasts, within which photosynthesis takes place. The cells of plants, fungi, and some protists have cell walls, although the composition of the walls varies among the groups. Plant cells have cytoplasmic connections through openings in the cell wall called plasmodesmata. Flagella occur in sperm of a few plant species, but are otherwise absent in plant and fungal cells. Centrioles are also absent in plant and fungal cells

Today's *Biology*

Membrane Defects Can Cause Disease

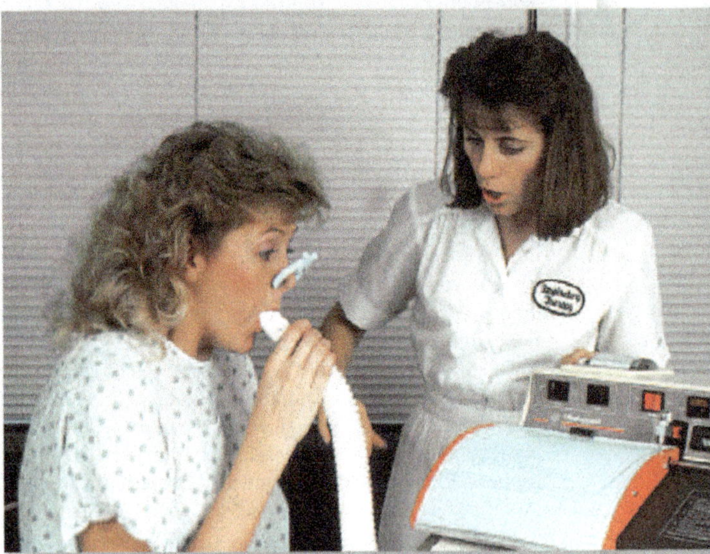

The year 1993 marked an important milestone in the treatment of human disease. That year the first attempt was made to cure **cystic fibrosis (CF)**, a deadly genetic disorder, by transferring healthy genes into sick individuals. Cystic fibrosis is a fatal disease in which the body cells of affected individuals secrete a thick mucus that clogs the airways of the lungs. The cystic fibrosis patient in the photograph is breathing into a Vitalograph, a device that measures lung function. These same secretions block the ducts of the pancreas and liver so that the few patients who do not die of lung disease die of liver failure. Cystic fibrosis is usually thought of as a children's disease because until recently few affected individuals lived long enough to become adults. Even today half die before their mid-twenties. There is no known cure.

Cystic fibrosis results from a defect in a single gene that is passed down from parent to child. It is the most common fatal genetic disease of Caucasians. One in 20 individuals possesses at least one copy of the defective gene. Most of these individuals are not afflicted with the disease; only those children who inherit a copy of the defective gene from each parent succumb to cystic fibrosis—about 1 in 2,500 infants.

Cystic fibrosis has proven difficult to study. Many organs are affected, and until recently it was impossible to identify the nature of the defective gene responsible for the disease. In 1985 the first clear clue was obtained. An investigator, Paul Quinton, seized on a commonly observed characteristic of cystic fibrosis patients, that their sweat is abnormally salty, and performed the following experiment. He isolated a sweat duct from a small piece of skin and placed it in a solution of salt (NaCl) that was three times as concentrated as the NaCl inside the duct. He then monitored the movement of ions. Diffusion tends to drive both the sodium (Na^+) and the chloride (Cl^-) ions into the duct because of the higher outer ion concentrations. In skin isolated from normal individuals, Na^+ and Cl^- both entered the duct, as expected. In skin isolated from cystic fibrosis individuals, however, only Na^+ entered the duct—no Cl^- entered. For the first time, the molecular nature of cystic fibrosis became clear. Water accompanies chloride, and was not entering the ducts because chloride was not, creating thick mucus. Cystic fibrosis is a defect in a plasma membrane protein called CFTR (*cystic fibrosis transmembrane conductance regulator*) that normally regulates passage of Cl^- into and out of the body's cells.

The defective *cf* gene was isolated in 1987, and its position on a particular human chromosome (chromosome 7) was pinpointed in 1989. Interestingly, many cystic fibrosis patients produce a CFTR protein with a normal amino acid sequence. The *cf* mutation in these cases appears to interfere with how the CFTR protein folds, preventing it from folding into a functional shape.

Soon after the *cf* gene was isolated, experiments were begun to see if it would be possible to cure cystic fibrosis by gene therapy—that is, by transferring healthy *cf* genes into the cells with defective ones. In 1990 a working *cf* gene was successfully transferred into human lung cells growing in tissue culture, using adenovirus, a cold virus, to carry the gene into the cells. The CFTR-defective cells were "cured," becoming able to transport chloride ions across their plasma membranes. Then in 1991 a team of researchers successfully transferred a normal human *cf* gene into the lung cells of a living animal—a rat. The *cf* gene was first inserted into the adenovirus genome because adenovirus is a cold virus and easily infects lung cells. The treated virus was then inhaled by the rat. Carried piggyback, the *cf* gene entered the rat lung cells and began producing the normal human CFTR protein within these cells!

These results were very encouraging, and at first the future for all cystic fibrosis patients seemed bright. Clinical tests using adenovirus to introduce healthy *cf* genes into cystic fibrosis patients were begun with much fanfare in 1993.

They were not successful. As described further in chapter 13, there were insurmountable problems with the adenovirus being used to transport the *cf* gene into cystic fibrosis patients. The difficult and frustrating challenge that cystic fibrosis researchers had faced was not over. Research into clinical problems is often a time-consuming and frustrating enterprise, never more so than in this case. Recently, as chapter 13 recounts, new ways of introducing the healthy *cf* gene have been tried with better results. The long, slow journey toward a cure has taught us not to leap to the assumption that a cure is now at hand, but the steady persistence of researchers has taken us a long way, and again the future for cystic fibrosis patients seems bright.

4.5 The Nucleus: The Cell's Control Center

Comparing the animal and plant cells on the previous pages, you cannot help but notice that many parts of the cells are remarkably similar. In paramecia, petunias, and primates, cell organelles look similar and carry out similar functions (see table 4.2 on pages 88–89).

If you were to journey far into the interior of one of your cells, you would eventually reach the center of the cell. There you would find, cradled within a network of fine filaments like a ball in a basket, the **nucleus** (figure 4.8). The nucleus is the command and control center of the cell, directing all of its activities. It is also the genetic library where the hereditary information is stored.

Nuclear Membrane

The surface of the nucleus is bounded by a special kind of membrane called the **nuclear envelope.** The nuclear envelope is actually *two* membranes, one outside the other, like a sweater over a shirt. The nuclear envelope acts as a barrier between the nucleus and the cytoplasm, but substances need to pass through the envelope. The exchange of materials occurs through openings scattered over the surface of this envelope. Called **nuclear pores,** these openings form when the two membrane layers of the nuclear envelope pinch together. A nuclear pore is not an empty opening, however; rather, it has many proteins embedded within it that permit proteins and RNA to pass into and out of the nucleus.

Chromosomes

In both prokaryotes and eukaryotes, all hereditary information specifying cell structure and function is encoded in DNA. However, unlike the circular prokaryotic DNA, the DNA of eukaryotes is divided into several segments and associated with protein, forming **chromosomes.** The proteins in the chromosome permit the DNA to wind tightly and condense during cell division. Under a light microscope, these condensed chromosomes are readily seen in dividing cells as densely staining rods. After cell division, eukaryotic chromosomes uncoil and fully extend into threadlike strands called **chromatin** that can no longer be distinguished individually with a light microscope within the nucleoplasm. Once uncoiled, the chromatin is available for protein synthesis. RNA copies of genes are made from the DNA in the nucleus. The RNA molecules leave the nucleus through the nuclear pores and enter the cytoplasm where proteins are synthesized.

Nucleolus

To make its many proteins, the cell employs a special structure called a **ribosome,** a kind of platform on which the proteins are built. Ribosomes read the RNA copy of a gene and use that information to direct the construction of a protein. Ribosomes are made up of several special forms of RNA called *ribosomal RNA,* or *rRNA,* bound up within a complex of several dozen different proteins.

You will notice in figure 4.8 that one region of the nucleus appears darker than the rest; this darker region is called the **nucleolus.** There a cluster of several hundred genes encode rRNA where the ribosome subunits assemble. These subunits leave the nucleus through the nuclear pores and enter the cytoplasm, where final assembly of ribosomes takes place.

Key Learning Outcome 4.5 The nucleus is the command center of the cell, issuing instructions that control cell activities. It also stores the cell's hereditary information.

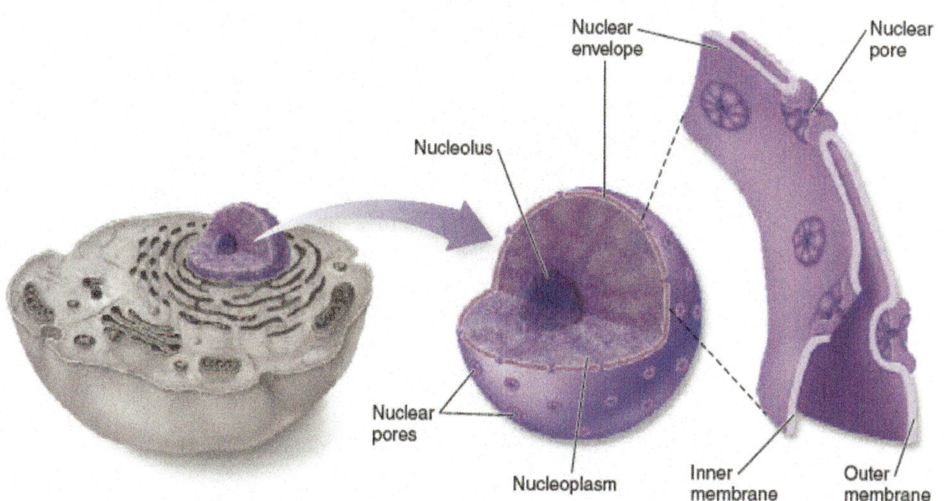

Figure 4.8
The nucleus.
The nucleus is composed of a double membrane, called a nuclear envelope, enclosing a fluid-filled interior containing the chromosomes. In cross section, the individual nuclear pores are seen to extend through the two membrane layers of the envelope. The pore is lined with protein, which acts to control access through the pore.

4.6 The Endomembrane System

Surrounding the nucleus within the interior of the eukaryotic cell is a tightly packed mass of membranes. They fill the cell, dividing it into compartments, channeling the transport of molecules through the interior of the cell and providing the surfaces on which enzymes act. The system of internal compartments created by these membranes in eukaryotic cells constitutes the most fundamental distinction between the cells of eukaryotes and prokaryotes.

Ribosomes
Rough endoplasmic reticulum
Smooth endoplasmic reticulum

Endoplasmic reticulum

Endoplasmic Reticulum: The Transportation System

The extensive system of internal membranes is called the **endoplasmic reticulum,** often abbreviated **ER.** The ER creates a series of channels and interconnections, and it also isolates some spaces as membrane-enclosed sacs called **vesicles.** The surface of the ER is the place where the cell makes proteins intended for export (such as enzymes secreted from the cell surface). The surface of those regions of the ER devoted to the synthesis of transported proteins is heavily studded with ribosomes and appears pebbly, like the surface of sandpaper, when seen through an electron microscope. For this reason, these regions are called **rough ER.** Regions in which ER-bound ribosomes are relatively scarce are correspondingly called **smooth ER.** The surface of the smooth ER is embedded with enzymes that aid in the manufacture of carbohydrates and lipids.

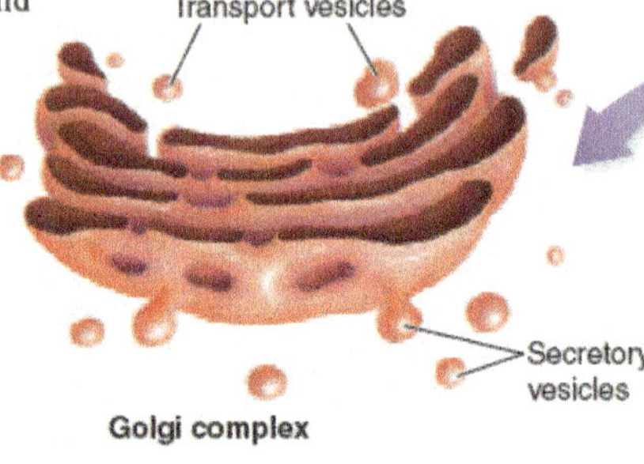

Golgi complex

The Golgi Complex: The Delivery System

As new molecules are made on the surface of the ER, they are passed from the ER to flattened stacks of membranes called **Golgi bodies.** The number of Golgi bodies a cell contains ranges from 1 or a few in protists, to 20 or more in animal cells, and several hundred in certain plant cells. Golgi bodies function in the collection, packaging, and distribution of molecules manufactured in the cell. Scattered throughout the cytoplasm, Golgi bodies are collectively referred to as the **Golgi complex.**

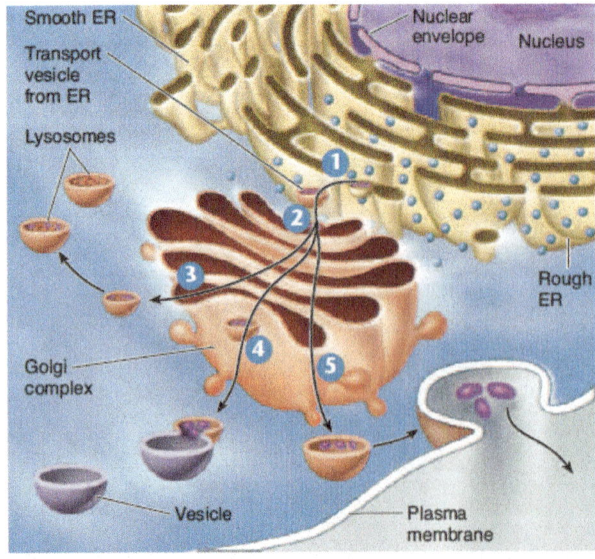

How the endomembrane system works

The rough ER, smooth ER, and Golgi work together as a transport system in the cell. Proteins and lipids that are manufactured on the ER membranes are transported throughout the channels of the ER and are packaged into transport vesicles that bud off from the ER ❶. The vesicles fuse with the membrane of the Golgi bodies, dumping their contents into the Golgi ❷. Within the Golgi bodies the molecules may take one of many paths, indicated by the branching arrows in the figure. Many of these molecules become tagged with carbohydrates. The molecules collect at the ends of the membranous folds of the Golgi bodies. Vesicles that pinch off from the ends carry the molecules to the different compartments of the cell ❸ and ❹, or to the inner surface of the plasma membrane, where molecules to be secreted are released to the outside ❺.

Lysosomes: Recycling Centers

Other organelles called **lysosomes** arise from the Golgi complex (the light orange vesicle budding at 3) and contain a concentrated mix of the powerful enzymes manufactured in the rough ER that break down macromolecules. Lysosomes are also the recycling centers of the cell, digesting worn-out cell components to make way for newly formed ones while recycling the proteins and other materials of the old parts. Large organelles called mitochondria are replaced in some human tissues every 10 days, with lysosomes digesting the old ones as the new ones are produced. In addition to breaking down organelles and other structures within cells, lysosomes also eliminate particles (including other cells) that the cell has engulfed.

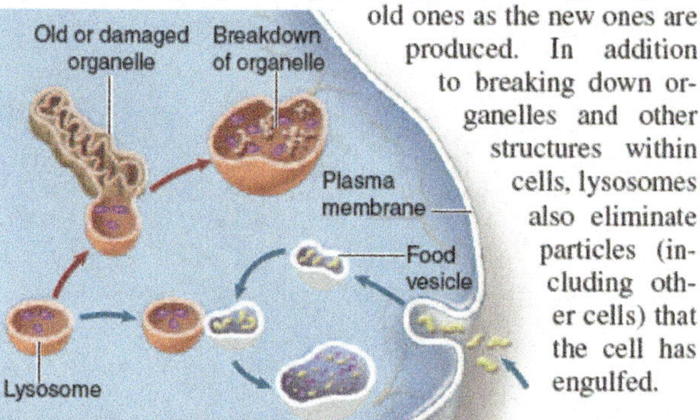

Peroxisomes: Chemical Specialty Shops

The interior of the eukaryotic cell contains a variety of membrane-bounded spherical organelles derived from the ER that carry out particular chemical functions. Almost all eukaryotic cells, for example, contain peroxisomes. Peroxisomes are spherical organelles that may contain a large crystal structure composed of protein (a peroxisome containing a crystal has been colored purple in the micrograph below). Peroxisomes contain two sets of enzymes. One set found in plant seeds converts fats to carbohydrates, and the other set found in all eukaryotes detoxifies various potentially harmful molecules—strong oxidants—that form in cells. They do this by using molecular oxygen to remove hydrogen atoms from specific molecules. These chemical reactions would be very destructive to the cell if not confined to the peroxisomes.

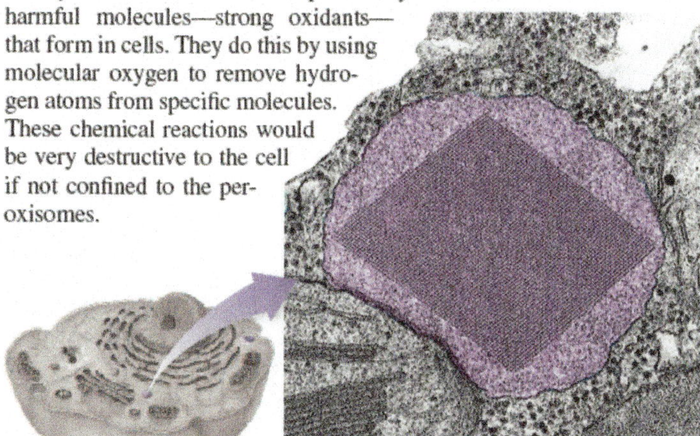

The Endomembrane System and Your Health

Several serious human diseases result when the endomembrane system does not work as it should. One of the most serious of these involves a breakdown in the ability of the Golgi complex to properly address proteins destined for delivery to the cell's lysosomes. In a normal cell, the address label is a modified sugar called mannose-6-phosphate, which the Golgi complex attaches to lysosome-bound proteins with a special enzyme. In individuals lacking this enzyme, the proteins that should have gone to lysosomes lack the mannose-6-phosphate address tag, and instead are delivered to the plasma membrane and secreted from the cell. Because almost all the recycling enzymes normally found in lysosomes are missing, the lysosomes swell with undegraded materials. Viewed under a microscope, the lysosomes appear to contain numerous inclusions where the undegraded material has clumped together, giving this condition its name, *inclusion-cell disease.* The swollen lysosomes ultimately cause serious damage to the cells of a developing human embryo, leading to facial and skeletal abnormalities and mental retardation.

About 40 different human diseases are caused by failure of the endomembrane system to deliver particular enzymes to the lysosomes. Called *lysosome storage disorders,* these disorders follow the pattern you have seen for inclusion-cell disease. For lack of the necessary recycling enzyme, a particular cell material accumulates undegraded in the lysosomes, eventually leading to cell damage and death. Most of these disorders are fatal in early childhood.

In the lysosome storage disorder called *Pompe's disease,* lysosomes lack an enzyme necessary to digest glycogen. Glycogen, which the body uses as a ready source of energy, is stored in muscle and liver cells. When energy is required, lysosomes digest the glycogen, chopping off sugar units that the cell can use to produce energy (we explore how this happens in chapter 7). Pompe's disease lysosomes, because they lack the necessary enzyme, do not degrade glycogen. Instead, harmful amounts of glycogen accumulate in the muscle and liver cells.

In the lysosome storage disease called *Tay-Sachs disease,* lysosomes are missing the enzyme necessary to degrade specific glycolipids called gangliosides that are abundant in the plasma membranes of brain cells. In Tay-Sachs individuals, the glycolipids accumulate in brain cells, which swell and eventually burst, releasing oxidative enzymes that kill the brain cells. The affected individual begins to exhibit rapid mental deterioration at about six to eight months of age. Within a year after birth, affected children are blind. Paralysis and death follow within five years.

Key Learning Outcome 4.6 An extensive system of interior membranes organizes the interior of the cell into functional compartments that manufacture and deliver proteins and carry out a variety of specialized chemical processes.

4.7 Organelles That Contain DNA

Eukaryotic cells contain several kinds of complex, cell-like organelles that contain their own DNA and appear to have been derived from ancient bacteria assimilated by ancestral eukaryotes in the distant past. The two principal kinds are mitochondria (which occur in the cells of all but a very few eukaryotes) and chloroplasts (which do not occur in animal or fungi cells—they occur only in algae and plants).

Mitochondria: Powerhouses of the Cell

Eukaryotic organisms extract energy from organic molecules ("food") in a complex series of chemical reactions called **oxidative metabolism,** which takes place only in their mitochondria. **Mitochondria** (singular, **mitochondrion**) are sausage-shaped organelles about the size of a bacterial cell. Mitochondria are bounded by two membranes. The outer membrane, shown partially cut away in figure 4.9, is smooth and apparently derives from the plasma membrane of the host cell that first took up the bacterium long ago. The inner membrane, apparently the plasma membrane of the bacterium that gave rise to the mitochondrion, is bent into numerous folds called **cristae** (singular, **crista**) that resemble the folded plasma membranes in various groups of bacteria. The cutaway view of the figure shows how the cristae partition the mitochondrion into two compartments, an inner **matrix** and an outer compartment, called the *intermembrane space*. As you will learn in chapter 7, this architecture is critical to successfully carrying out oxidative metabolism.

During the 1.5 billion years in which mitochondria have existed in eukaryotic cells, most of their genes have been transferred to the chromosomes of the host cells. But mitochondria still have some of their original genes, contained in a circular, closed, naked molecule of DNA (called *mitochondrial DNA*, or *mtDNA*) that closely resembles the circular DNA molecule of a bacterium. On this mtDNA are several genes that produce some of the proteins essential for oxidative metabolism. In both mitochondria and bacteria, the circular DNA molecule is replicated during the process of division. When a mitochondrion divides, it copies its DNA located in the matrix and splits into two by simple fission, dividing much as bacteria do.

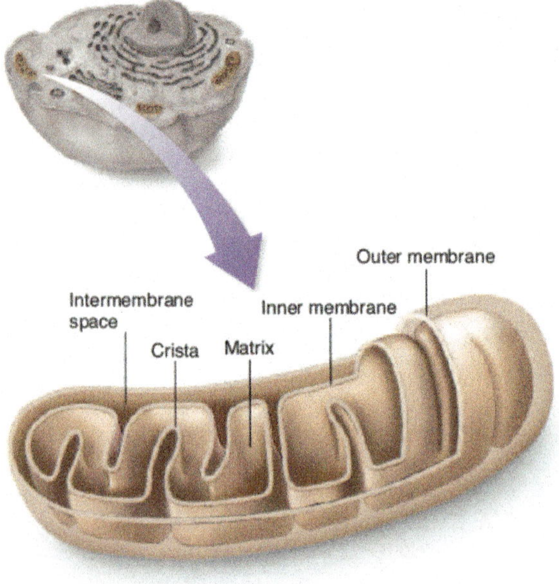

(a)

(b)

Figure 4.9 Mitochondria.
The mitochondria of a cell are sausage-shaped organelles within which oxidative metabolism takes place, and energy is extracted from food using oxygen. (*a*) A mitochondrion has a double membrane. The inner membrane is shaped into folds called cristae. The space within the cristae is called the matrix. The cristae greatly increase the surface area for oxidative metabolism. (*b*) Micrograph of two mitochondria, one in cross section, the other cut lengthwise.

Chloroplasts: Energy-Capturing Centers

All photosynthesis in plants and algae takes place within another bacteria-like organelle, the **chloroplast** (figure 4.10). There is strong evidence that chloroplasts, like mitochondria, were derived from bacteria by symbiosis. A chloroplast is bounded, like a mitochondrion, by two membranes, the inner derived from the original bacterium and the outer resembling the host cell's ER. Chloroplasts are larger than mitochondria, and have a more complex organization. Inside the chloroplast, another series of membranes are fused to form stacks of closed vesicles called **thylakoids,** the green disklike structures visible in the interior of the chloroplast in figure 4.10. The light-dependent reactions of photosynthesis take place within the thylakoids. The thylakoids are stacked on top of one another to form a column called a **granum** (plural, **grana**). The interior of a chloroplast is bathed with a semiliquid substance called the **stroma.**

Like mitochondria, chloroplasts have a circular DNA molecule. On this DNA are located many of the genes coding for the proteins necessary to carry out photosynthesis. Plant cells can contain from one to several hundred chloroplasts, depending on the species. Neither mitochondria nor chloroplasts can be grown in a cell-free culture; they are totally dependent on the cells within which they occur.

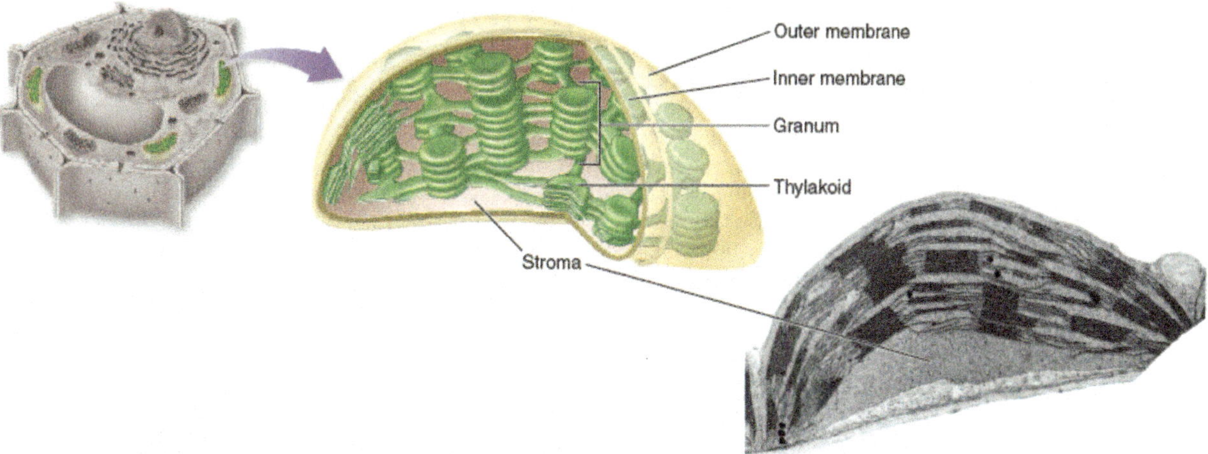

Figure 4.10 A chloroplast.

Bacteria-like organelles called chloroplasts are the sites of photosynthesis in photosynthetic eukaryotes. Like mitochondria, they have a complex system of internal membranes on which chemical reactions take place. The internal membranes of a chloroplast are fused to form stacks of closed vesicles called thylakoids. Photosynthesis occurs within these thylakoids. Thylakoids are stacked one on top of the other in columns called grana. The interior of the chloroplast is bathed in a semiliquid substance called the stroma.

Endosymbiosis

Symbiosis is a close relationship between organisms of different species that live together. The theory of **endosymbiosis** proposes that some of today's eukaryotic organelles evolved by a symbiosis in which one cell of a prokaryotic species was engulfed by and lived inside the cell of another species of prokaryote that was a precursor to eukaryotes. Figure 4.11 shows how this is thought to have occurred. Many cells take up food or other substances through endocytosis, a process whereby the plasma membrane of a cell wraps around the substance, enclosing it within a vesicle inside the cell. Oftentimes, the contents in the vesicle are broken down with digestive enzymes. According to the endosymbiont theory this did not occur; instead, the engulfed prokaryotes provided their hosts with certain advantages associated with their special metabolic abilities. Two key eukaryotic organelles just described are believed to be the descendants of these endosymbiotic prokaryotes: mitochondria, which are thought to have originated as bacteria capable of carrying out oxidative metabolism; and chloroplasts, which apparently arose from photosynthetic bacteria.

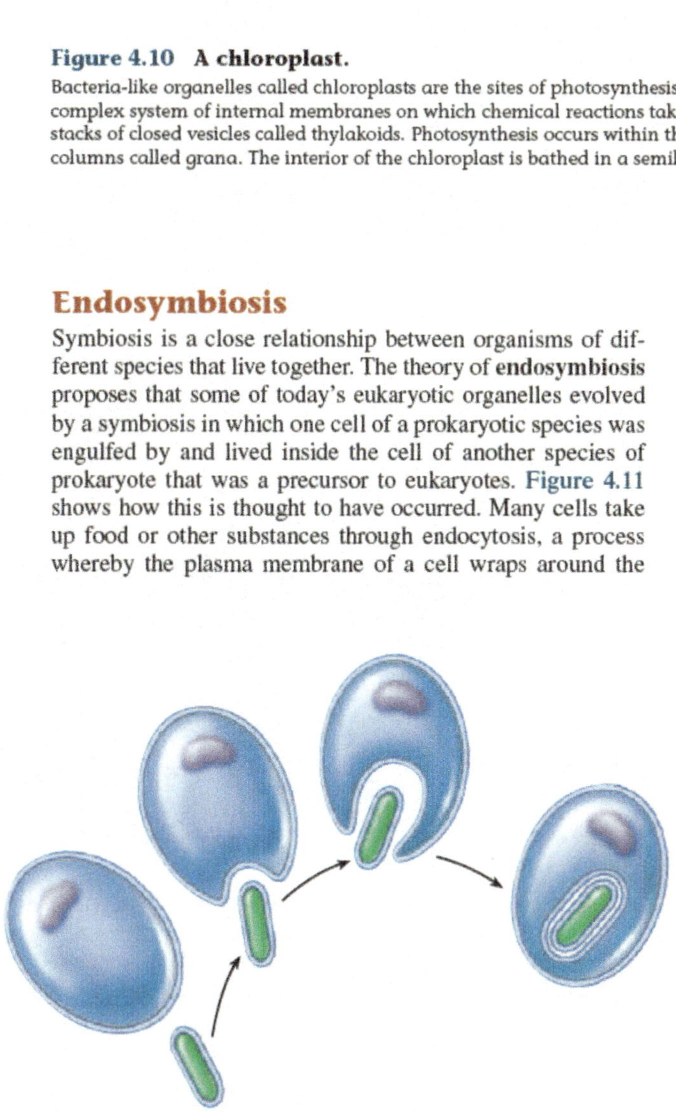

The endosymbiont theory is supported by a wealth of evidence. Both mitochondria and chloroplasts are surrounded by two membranes; the inner membrane probably evolved from the plasma membrane of the engulfed bacterium, while the outer membrane is probably derived from the plasma membrane or endoplasmic reticulum of the host cell. Mitochondria are about the same size as most bacteria, and the cristae formed by their inner membranes resemble the folded membranes in various groups of bacteria. Mitochondrial ribosomes are also similar to bacterial ribosomes in size and structure. Both mitochondria and chloroplasts contain circular molecules of DNA similar to those in bacteria. Finally, mitochondria divide by simple fission, splitting in two just as bacterial cells do, and they apparently replicate and partition their DNA in much the same way as bacteria do.

Figure 4.11 Endosymbiosis.

This figure shows how a double membrane may have been created during the symbiotic origin of mitochondria or chloroplasts.

Key Learning Outcome 4.7 **Eukaryotic cells contain complex organelles that have their own DNA and are thought to have arisen by endosymbiosis from ancient bacteria.**

4.8 The Cytoskeleton: Interior Framework of the Cell

If you were to shrink down and enter into the interior of a eukaryotic cell, your view would be similar to what you see in the illustration shown here: A dense network of protein fibers called the **cytoskeleton** provides a framework that supports the shape of the cell. The cytoskeleton also anchors organelles like the mitochondria to fixed locations within the cell interior. The protein fibers of the cytoskeleton are a dynamic system, constantly being formed and disassembled. There are three different kinds of protein fibers that make up the cytoskeleton, shown as enlargements below and in table 4.2.

Microfilaments (Actin Filaments) Actin filaments are long fibers about 7 nanometers in diameter. Each filament is composed of two protein chains loosely twined together like two strands of pearls. Each "pearl," or subunit, on the chains is the globular protein **actin.** Actin filaments are found throughout the cell but are most highly concentrated just inside the plasma membrane. Actin filaments are responsible for cellular movements such as contraction, crawling, "pinching" during division, and formation of cellular extensions.

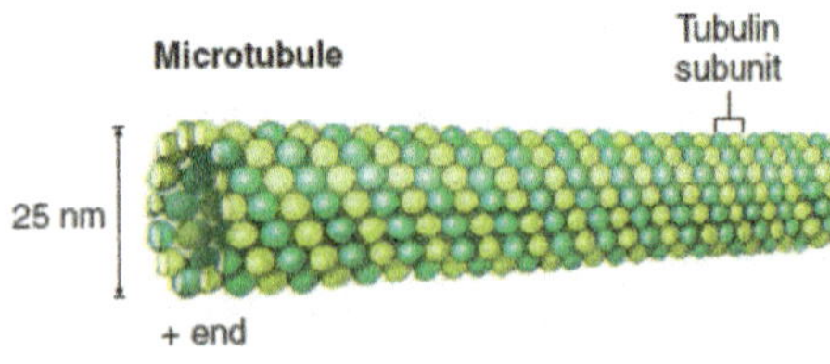

Microtubules Microtubules are hollow tubes about 25 nanometers in diameter composed of *tubulin* protein subunits arranged side by side to form a tube. In many cells, microtubules form from nucleation centers near the center of the cell and radiate toward the periphery. The ends of the microtubule are designated as "+" (away from the nucleation center) or "–" (toward the nucleation center). Microtubules are comparatively stiff cytoskeletal elements that serve to organize metabolism and intracellular transport in the nondividing cell and to stabilize cell structure. They are also responsible for the movement of chromosomes in mitosis.

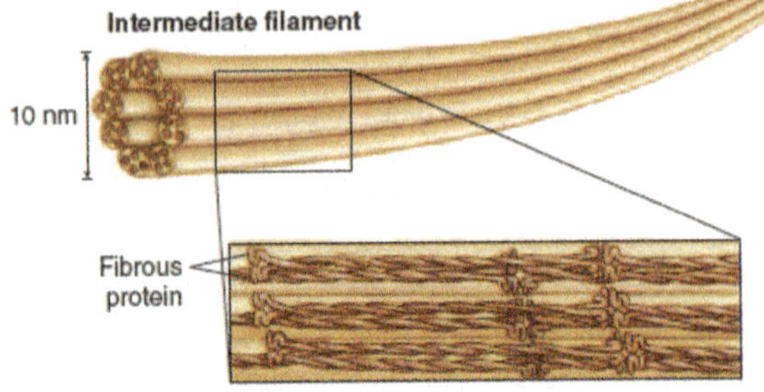

Intermediate Filaments Intermediate filaments are composed of overlapping staggered tetramers of protein. These tetramers are then bundled into cables. This molecular arrangement allows for a ropelike structure that imparts tremendous mechanical strength to the cell. Intermediate filaments are characteristically 8 to 10 nanometers in diameter, intermediate in size between actin filaments and microtubules (which is why they are called *intermediate* filaments). Once formed, intermediate filaments are stable and usually do not break down. They provide structural reinforcement to the cell and organelles.

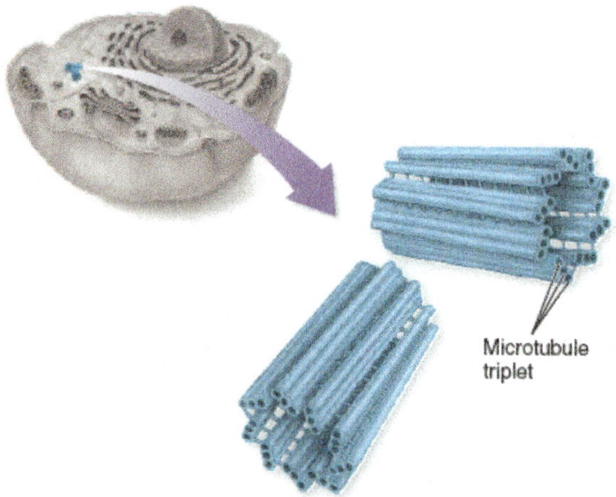

Figure 4.12 Centrioles.
Centrioles anchor and assemble microtubules. Centrioles usually occur in pairs and are composed of nine triplets of microtubules.

The cytoskeleton plays a major role in determining the shape of animal cells, which lack rigid cell walls. Because filaments can form and dissolve readily, the shape of an animal cell can change rapidly. If you examine the surface of an animal cell with a microscope, you will often find it alive with motion, projections shooting out from the surface and then retracting, only to shoot out elsewhere moments later.

The cytoskeleton is not only responsible for the cell's shape, but it also provides a scaffold both for ribosomes to carry out protein synthesis and for enzymes to be localized within defined areas of the cytoplasm. By anchoring particular enzymes near one another, the cytoskeleton participates with organelles in organizing the cell's activities.

Centrioles

Complex structures called **centrioles** assemble microtubules from tubulin subunits in the cells of animals and most protists. Centrioles occur in pairs within the cytoplasm, usually located at right angles to one another as you can see in figure 4.12. They are usually found near the nuclear envelope and are among the most structurally complex microtubular assemblies of the cell. In cells that contain flagella or cilia, each flagellum or cilium is anchored by a form of centriole called a *basal body*. Most animal and protist cells have both centrioles and basal bodies; higher plants and fungi lack them, instead organizing microtubules without such structures. Although they lack a membrane, centrioles resemble spirochete bacteria in many other respects. Some biologists believe that centrioles, like mitochondria and chloroplasts, originated as symbiotic bacteria.

Vacuoles: Storage Compartments

Within the interiors of plant and many protist cells, the cytoskeleton positions not only organelles, but also storage compartments that are membrane-bounded, called **vacuoles.** The center of the plant cell shown above in figure 4.13, as in all plant cells, contains a large, apparently empty space, called the *central vacuole*. This vacuole is not really empty; it contains large amounts of water and other materials, such as sugars, ions, and pigments. The central vacuole functions as a storage center for these important substances.

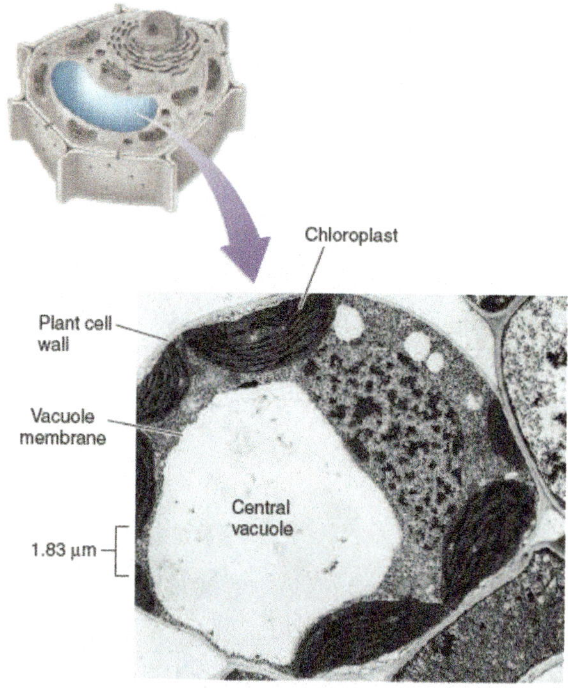

Figure 4.13 A plant central vacuole.
A plant's central vacuole stores dissolved substances and can increase in size to increase the surface area of a plant cell.

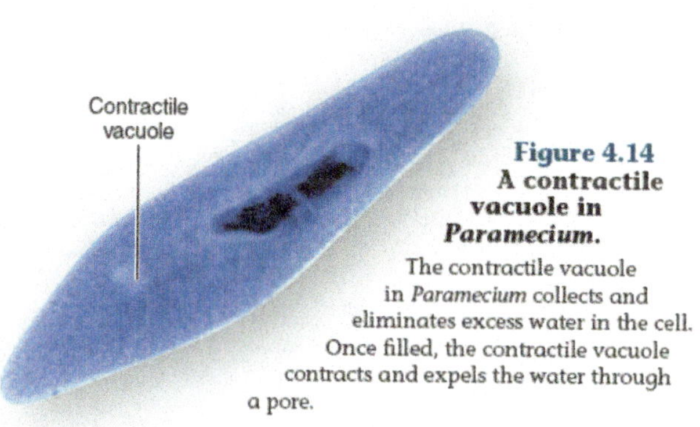

Figure 4.14 A contractile vacuole in *Paramecium*.
The contractile vacuole in *Paramecium* collects and eliminates excess water in the cell. Once filled, the contractile vacuole contracts and expels the water through a pore.

Certain protists, like *Paramecium* in figure 4.14, contain a **contractile vacuole** near the cell surface that accumulates excess water. This vacuole is bounded by actin filaments and has a small pore that opens to the outside of the cell. By rhythmic ATP-powered contractions, it pumps accumulated water out through the pore.

TABLE 4.2 EUKARYOTIC CELL STRUCTURES AND THEIR FUNCTIONS

Structure	Description
Plasma Membrane 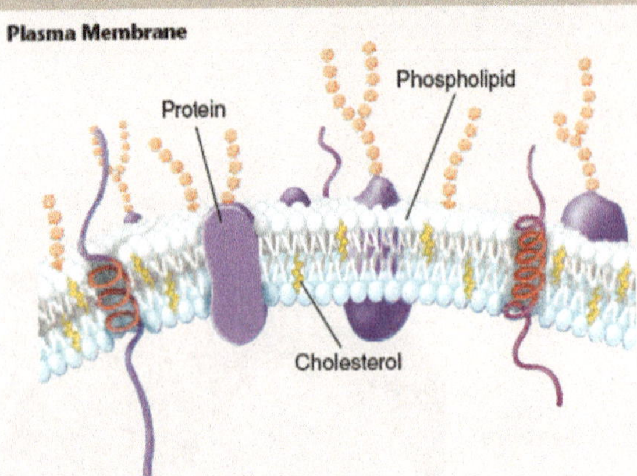	The plasma membrane is a phospholipid bilayer embedded with proteins that encloses a cell and separates its contents from its surroundings. The bilayer results from the tail-to-tail packing of the phospholipid molecules that make up the membrane. The proteins embedded in the lipid bilayer are in large part responsible for a cell's ability to interact with its environment. Transport proteins provide channels through which molecules and ions enter and leave the cell across the plasma membrane. Receptor proteins induce changes within the cell when they come in contact with specific molecules in the environment, such as hormones, or with molecules on the surface of neighboring cells.
Nucleus 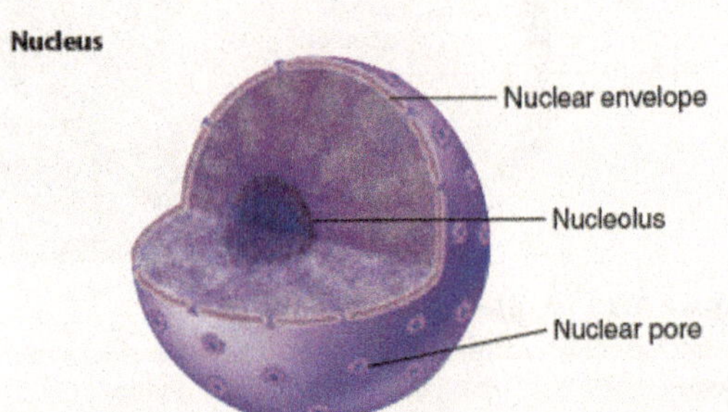	Every cell contains DNA, the hereditary material. The DNA of eukaryotes is isolated within the nucleus, a spherical organelle surrounded by a double membrane structure called the nuclear envelope. This envelope is studded with pores that control traffic into and out of the nucleus. The DNA contains the genes that code for the proteins synthesized by the cell. Stabilized by proteins, it forms chromatin, the major component of the nucleus. When the cell prepares to divide, the chromatin of the nucleus condenses into threadlike chromosomes.
Endoplasmic Reticulum 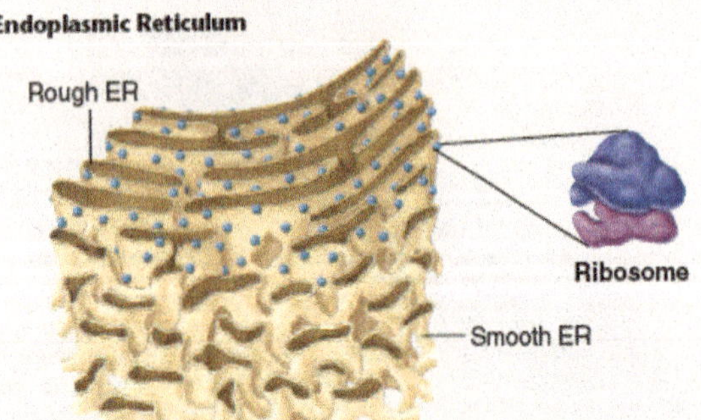	The hallmark of the eukaryotic cell is compartmentalization, achieved by an extensive endomembrane system that weaves through the cell interior. The membrane network is called the endoplasmic reticulum, or ER. The ER begins at the nuclear envelope and extends out into the cytoplasm, its sheets of membrane weaving through the cell interior. Rough ER contains numerous ribosomes that give it a bumpy appearance. These ribosomes manufacture proteins destined for the ER or for other parts of the cell. ER without these attached ribosomes is called smooth ER, which often functions to detoxify harmful substances or to aid in the synthesis of lipids. Sugar side chains are added to molecules as they pass through the ER. Delivery of molecules to other parts of the cell is via vesicles that pinch off from the borders of the rough ER.
Golgi Complex 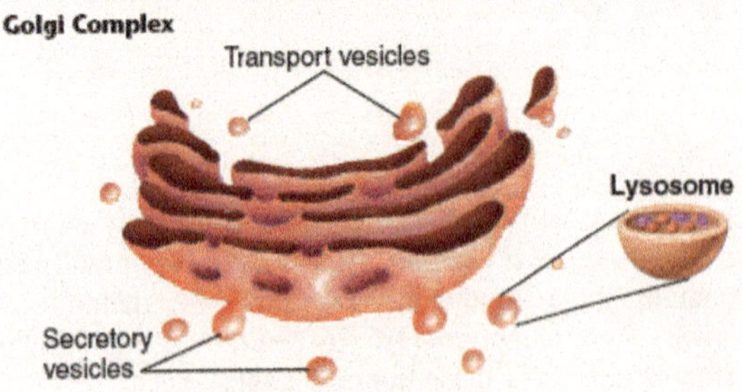	At various locations within the cytoplasm flattened stacks of membranes occur. Animal cells may contain 20, plant cells several hundred. Collectively, they are referred to as the Golgi complex. Molecules manufactured in the ER pass to the Golgi complex within vesicles. The Golgi sorts and packages these molecules and also synthesizes carbohydrates. The Golgi adds sugar side chains to molecules as they pass through the stacks of membranes. The Golgi then directs the molecules to lysosomes, secretory vesicles, or the plasma membrane.

Structure	Description
Mitochondrion 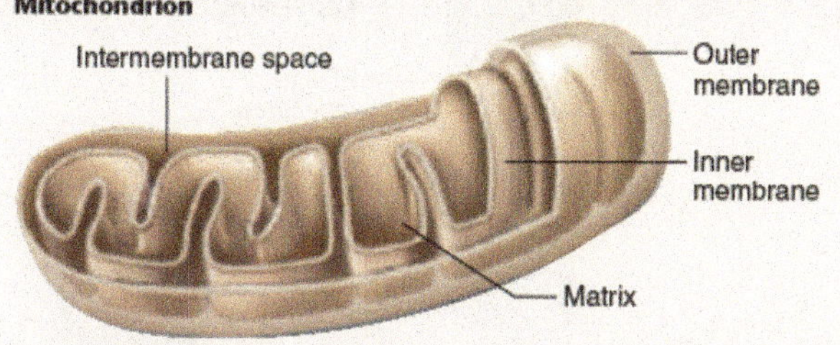	Mitochondria are bacteria-like organelles that are responsible for extracting most of the energy a cell derives from the food molecules it consumes. Two membranes encase each mitochondrion, separated by an intermembrane space. The key energy-harvesting chemical reactions occur within the interior matrix. The energy is used to pump protons from the matrix into the intermembrane space; their return across this membrane drives the synthesis of ATP, the energy currency of the cell.
Chloroplast 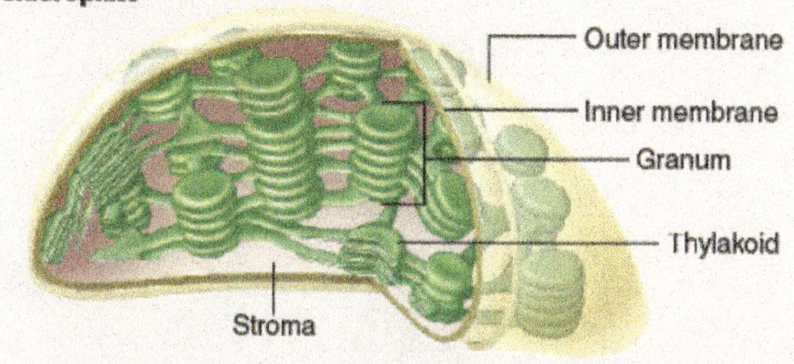	The green color of plants and algae results from cell organelles called chloroplasts rich in the photosynthetic green pigment chlorophyll. Photosynthesis is the sunlight-powered process that converts CO_2 in the air to the organic molecules of which all living things are composed. Chloroplasts, like mitochondria, are composed of two membranes separated by an intermembrane space. In a chloroplast, the inner membrane pinches into a series of sacs called thylakoids, which pile up in columns called grana. The chlorophyll-facilitated light reactions of photosynthesis take place within the thylakoids. These are suspended in a semiliquid substance called the stroma.
Cytoskeleton 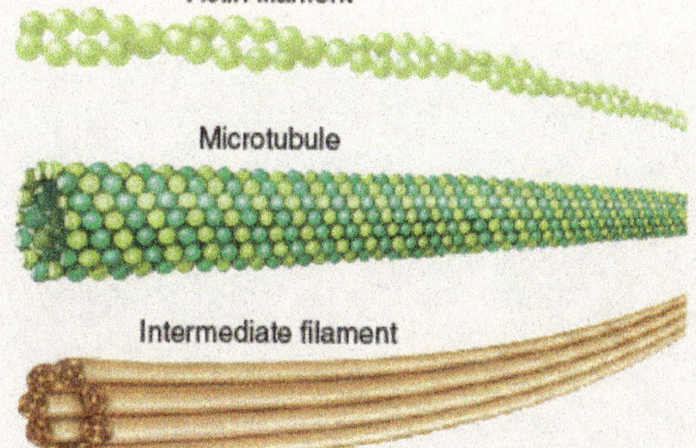	The cytoplasm of all eukaryotic cells is crisscrossed by a network of protein fibers called the cytoskeleton that supports the shape of the cell and anchors organelles to fixed locations. The cytoskeleton is a dynamic structure, composed of three kinds of fibers. Long actin filaments are responsible for cellular movements such as contraction, crawling, and the "pinching" that occurs as cells divide. Hollow microtubule tubes, constantly forming and disassembling, facilitate cellular movements and are responsible for moving materials within the cell. Special motor proteins move organelles around the cell on microtubular "tracks." Durable intermediate filaments provide the cell with structural stability.
Centrioles 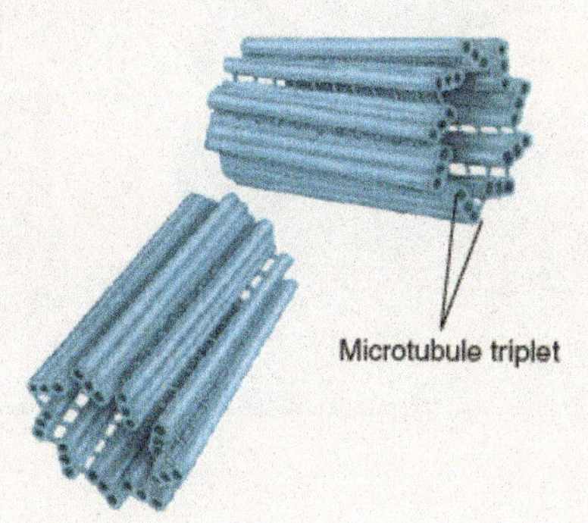	Centrioles are barrel-shaped organelles found in the cells of animals and most protists. They occur in pairs, usually located at right angles to each other near the nucleus. Centrioles help assemble the animal cell's microtubules, playing a key role in producing the microtubules that move chromosomes during cell division. Centrioles are also involved in the formation of cilia and flagella, which are composed of sets of microtubules. Cells of plants and fungi lack centrioles, and cell biologists are still in the process of characterizing their microtubule-organizing centers.

Cell Movement

Essentially, all cell motion is tied to the movement of actin filaments, microtubules, or both. Intermediate filaments act as intracellular tendons, preventing excessive stretching of cells, and actin filaments play a major role in determining the shape of cells. Because actin filaments can form and dissolve so readily, they enable some cells to change shape quickly. If you look at the surfaces of such cells under a microscope, you will find them moving and changing shape.

Some Cells Crawl It is the arrangement of actin filaments within the cell cytoplasm that allows cells to "crawl," literally! Crawling is a significant cellular phenomenon, essential to inflammation, clotting, wound healing, and the spread of cancer. White blood cells in particular exhibit this ability. Produced in the bone marrow, these cells are released into the circulatory system and then eventually crawl out of capillaries and into the tissues to destroy potential pathogens. The crawling mechanism is an exquisite example of cellular coordination.

Cytoskeletal fibers play a role in other types of cell movement. For example, during animal cell reproduction (see chapters 8 and 9), chromosomes move to opposite sides of a dividing cell because they are attached to shortening microtubules. The cell then pinches in two when a belt of actin filaments contracts like a purse string. Muscle cells also use actin filaments to contract their cytoskeletons. The fluttering of an eyelash, the flight of an eagle, and the awkward crawling of a baby all depend on these cytoskeletal movements within muscle cells.

Swimming with Flagella and Cilia Some eukaryotic cells contain **flagella** (singular, **flagellum**), fine, long, threadlike organelles protruding from the cell surface. The cutaway view in figure 4.15 shows how a flagellum arises from a microtubular structure called a **basal body,** with groups of microtubules arranged in rows of three, as you can see in the cross-sectional view. Some of these microtubules extend up into the flagellum, which consists of a circle of nine microtubule pairs surrounding two central ones (again seen in the cross-sectional view). This **9 + 2 arrangement** is a fundamental feature of eukaryotes and apparently evolved early in their history. Even in cells that lack

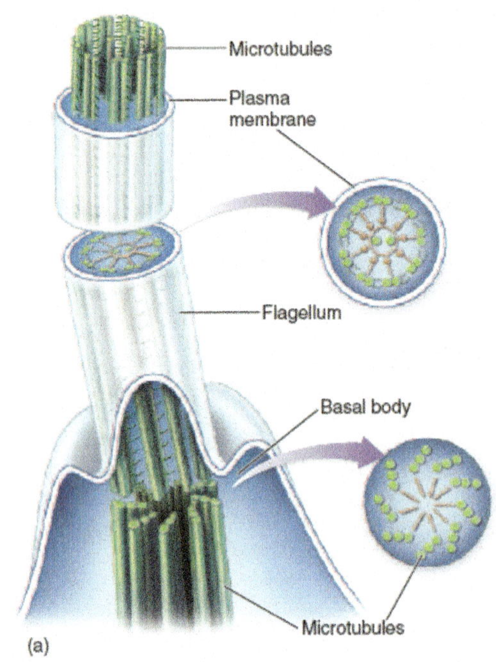

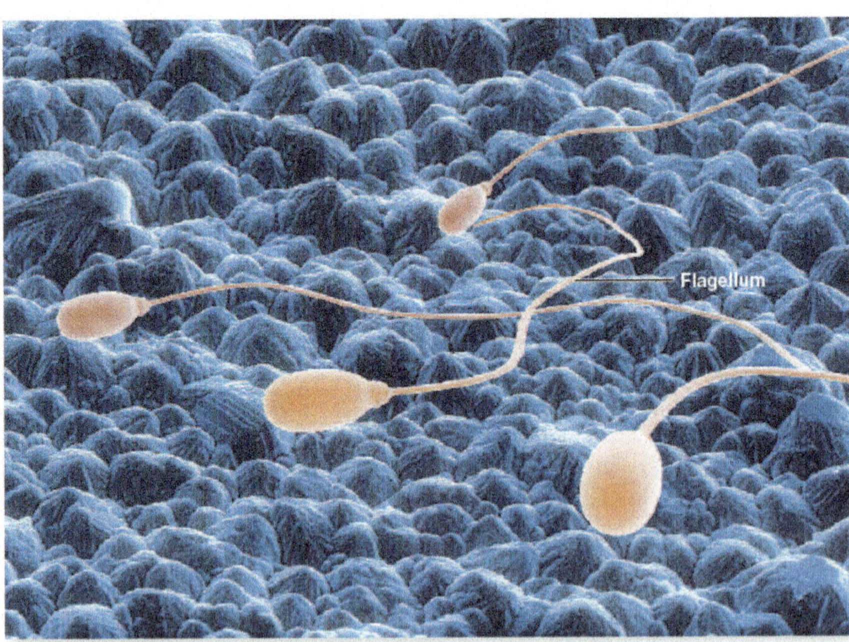

(b) Human sperm cells

Figure 4.15 Flagella.
(*a*) A eukaryotic flagellum springs directly from a basal body and is composed of a ring of nine pairs of microtubules with two microtubules in its core. (*b*) Human sperm cells have one flagellum.

flagella, derived structures with the same 9 + 2 arrangement often occur, like in the sensory hairs of the human ear. In humans, we find a single long flagellum on each sperm cell that propels the cell in a swimming motion. If flagella are numerous and organized in dense rows, they are called **cilia.** Cilia do not differ from flagella in their structure, but cilia are usually short. The *Paramecium* in figure 4.16 is covered with cilia, giving it a furry appearance. In humans, dense mats of cilia project from cells that line our breathing tube, the trachea, to move mucus and dust particles out of the respiratory tract into the throat (where we can expel these unneeded contaminants by spitting or swallowing). Eukaryotic flagella serve a similar function as flagella found in prokaryotes, discussed in section 4.3, but they are very different structurally.

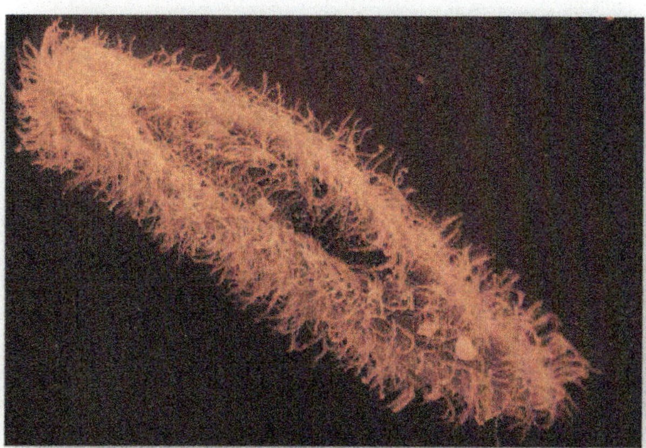

Figure 4.16 Cilia.
The surface of this *Paramecium* is covered with a dense forest of cilia.

Moving Materials Within the Cell

All eukaryotic cells must move materials from one place to another in the cytoplasm. Most cells use the endomembrane system as an intracellular highway. As you saw on page 82, the Golgi complex packages materials into vesicles that come from the channels of the endoplasmic reticulum to the far reaches of the cell. However, this highway is only effective over short distances. When a cell has to transport materials through long extensions like the axon of a nerve cell, the endomembrane system highways are too slow. For these situations, eukaryotic cells have developed high-speed locomotives that run along microtubular tracks. Lysosomes move along such microtubular tracks to reach a food vacuole, and mitochondria travel down them to the far-away tips of long axons.

For long-distance intracellular transport, four components are required, illustrated in figure 4.17: a vesicle or organelle (the light tan structure) that is to be transported; a motor molecule, in this case dynein, that provides the energy-driven motion; connector molecules that connect the vesicle to the motor molecule (in the figure, they are the dynactin complex and other associated proteins); and microtubules (the green tube) on which the vesicle will ride like a train on a rail, pulled by a locomotive. As nature's tiniest motors, these motor proteins literally pull the transport vesicles along the microtubular tracks.

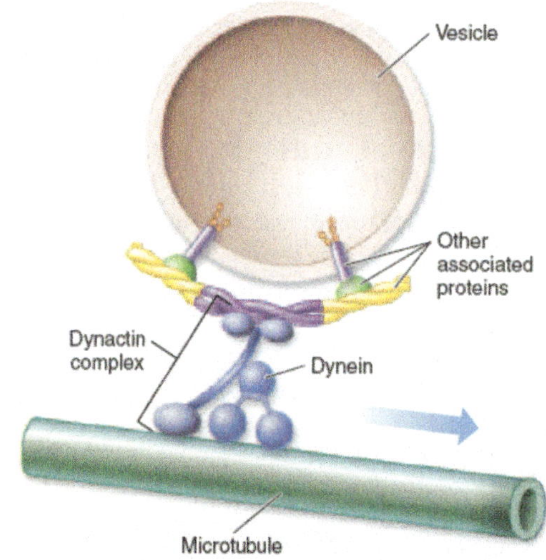

Figure 4.17 Molecular motors.
Vesicles that are transported within cells are attached with connector molecules, such as the dynactin complex shown here, to motor molecules, like dynein, which move along microtubules.

How does such a tiny motor work? A motor protein uses ATP to power its movement. Scientists have proposed that they use a type of stepping action. Also, the *direction* of movement is different for different motor proteins. The **dynein** motor protein shown in figure 4.17 moves inward toward the cell's center, dragging the vesicle with it as it travels along toward the "–" end of a microtubule. Another motor protein, **kinesin,** directs movement in the opposite direction, toward the "+" end of a microtubule, which is toward the periphery of the cell. The destination of a particular transport vesicle and its contents is thus determined by the nature of the linking protein embedded within the vesicle's membrane. Like possessing a ticket to one of two destinations, if a vesicle links to kinesin, it moves outward; if it links to dynein, it moves inward.

Key Learning Outcome 4.8 **The cytoskeleton is a latticework of protein fibers that determines a cell's shape and anchors organelles to particular locations within the cytoplasm. Cells can move by changing their shape and move substances around the cell using molecular motors.**

4.9 Outside the Plasma Membrane

Cell Walls Offer Protection and Support

Plants, fungi, and many protists cells share a characteristic with bacteria that is not shared with animal cells—that is, they have **cell walls,** which protect and support their cells. Eukaryotic cell walls are chemically and structurally different from bacterial cell walls. In plants, cell walls are composed of fibers of the polysaccharide cellulose, while in fungi they are composed of chitin. The **primary walls** of plant cells are laid down when the cell is still growing. These are the thinner, outer walls of the cells shown below in figure 4.18. Between the walls of adjacent cells is a sticky substance called the **middle lamella,** which glues the cells together. Some plant cells produce strong **secondary walls,** which are deposited inside the primary walls. As you can see in the photo, the secondary cell walls are very thick compared to the primary walls and therefore are not deposited until the cell has finished increasing in size.

An Extracellular Matrix Surrounds Animal Cells

As we discussed, many types of eukaryotic cells possess a cell wall exterior to the plasma membrane. The wall acts to protect the cell, maintain its shape, and prevent excessive water uptake. Animal cells are the great exception, lacking the cell walls that encase the cells of plants, fungi, and most protists. Animal cells secrete an elaborate mixture of **glycoproteins** (proteins with short chains of sugars attached to them) into the space around them, forming the **extracellular matrix (ECM),** which performs a function different than cell walls.

The fibrous protein collagen, the same protein in cartilage and ligaments, is abundant in the ECM. Figure 4.19 shows how these fibers of collagen and another fibrous protein, elastin, are embedded within a complex web of other glycoproteins called proteoglycans, which form a protective layer over the cell surface.

The ECM is attached to the plasma membrane by a third kind of glycoprotein, **fibronectin.** As you can see in the figure, fibronectin molecules bind not only to ECM glycoproteins but also to proteins called **integrins,** which are an integral part of the plasma membrane. Integrins extend into the cytoplasm, where they are attached to the microfilaments of the cytoskeleton. Linking ECM and cytoskeleton, integrins allow the ECM to influence cell behavior in important ways, altering gene expression and cell migration patterns by a combination of mechanical and chemical signaling pathways. In this way, the ECM can help coordinate the behavior of all the cells in a particular tissue.

Key Learning Outcome 4.9 **Plant and protist cells encase themselves within a strong cell wall. In animal cells, which lack a cell wall, the cytoskeleton is linked by integrin proteins to a web of glycoproteins called the extracellular matrix.**

Cell 1
Primary walls
Secondary wall
Middle lamella
Cell 2

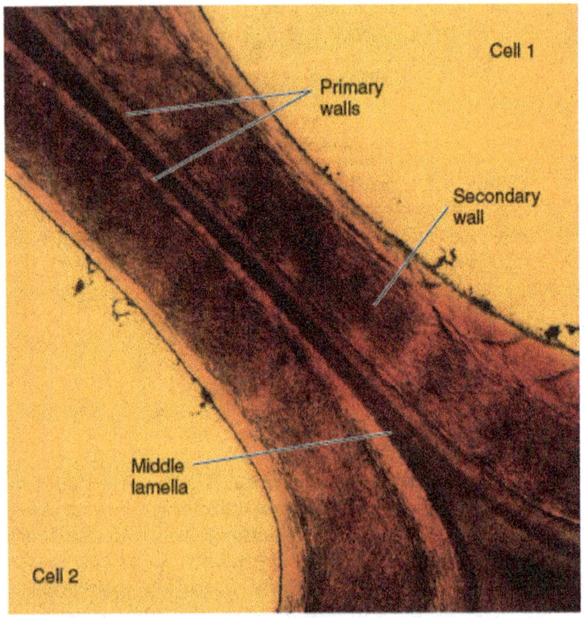

Figure 4.18 Cell walls in plants.

Plant cell walls are thick, strong, and rigid. Primary cell walls are laid down when the cell is young. Thicker secondary cell walls may be added later when the cell is fully grown. The middle lamella lies between the walls of adjacent cells and glues the cells together.

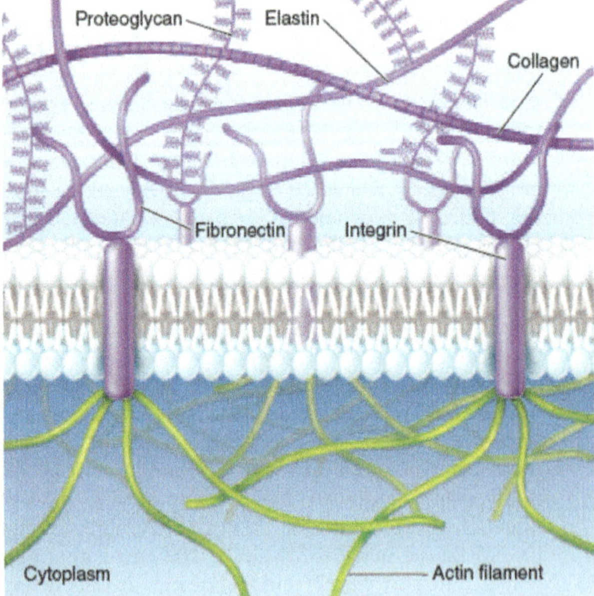

Figure 4.19 The extracellular matrix.

Animal cells are surrounded by an extracellular matrix (ECM) composed of various glycoproteins. The ECM carries out a variety of functions that influence cell behavior, including cell migration, gene expression, and the coordination of signaling between cells.

4.10 Diffusion

For cells to survive, food particles, water, and other materials must pass into the cell, and waste materials must be eliminated. All of this moving back and forth across the cell's plasma membrane occurs in one of three ways: (1) water and other substances diffuse through the membrane; (2) proteins in the membrane act as doors that admit certain molecules only; or (3) food particles and sometimes liquids are engulfed by the membrane folding around them. First we will examine diffusion.

Diffusion

Most molecules are in constant motion. How a molecule moves—just where it goes—is totally random, like shaking marbles in a cup. So, if two kinds of molecules are added together, they soon mix. The random motion of molecules always tends to produce uniform mixtures. To see how this works, visualize the simple experiment shown in the Key Biological Process illustration below, in which a lump of sugar is dropped into a beaker of water. The lump slowly breaks apart into individual sugar molecules, which move about randomly until eventually the sugar molecules become evenly distributed throughout the water in the beaker (panel 4 below) as individual molecules take long random journeys. This process of random molecular mixing is called **diffusion.**

Selective Permeability

Diffusion becomes important to the life of a cell because of the chemical nature of biological membranes. As figure 4.20 illustrates, the nonpolar nature of the lipid bilayer determines which substances can pass and which ones cannot. Oxygen gas and carbon dioxide are not repelled by the bilayer and can freely cross, and so can small nonpolar fats and lipids. However, sugars like glucose cannot, and neither can proteins. In fact, no polar molecule can freely travel across a biological membrane because of the barrier to diffusion imposed by the lipid bilayer. This is true of charged ions like Na^+, Cl^-, and H^+ and is also true of water molecules, which are very polar. It was once thought that water somehow "leaked" across the plasma membrane, slipping through gaps that open up when the hydrocarbon tails of the bilayer flex and bend, but biologists now discount this, as discussed on page 76. Later in this chapter we will further explore the movement of water molecules across plasma membranes.

Because the plasma membrane admits some substances and not others, it is said to be **selectively permeable.** The selective permeability of a cell's plasma membrane is arguably its most important property. What a cell is and how it functions are largely determined by which molecules it admits and which it refuses. As you proceed through this chapter, you will encounter a variety of ways in which particular cell types exercise control over their admission policy.

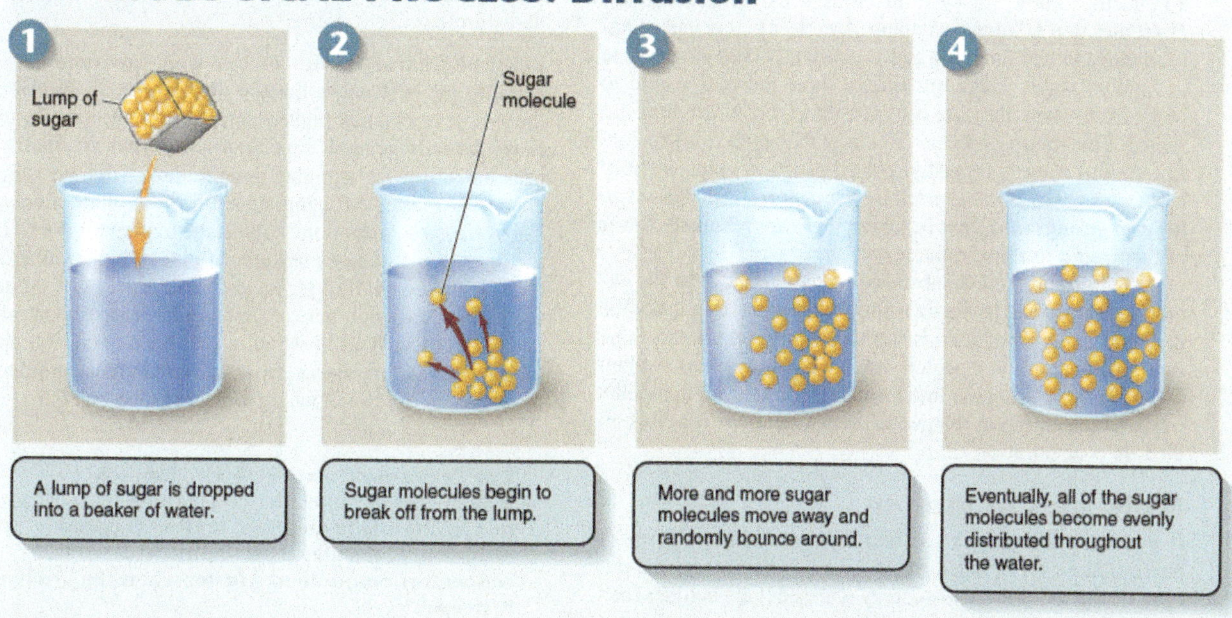

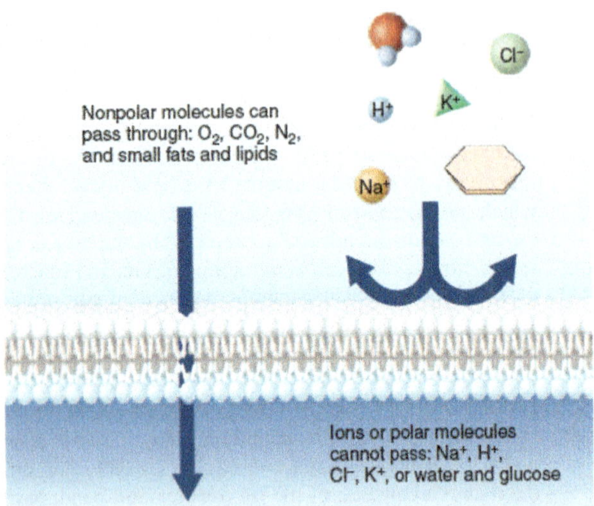

Figure 4.20 Membranes are selectively permeable.
Nonpolar molecules, such as those on the *left*, can pass through the membrane, but polar molecules and ions (on the *right*) cannot.

Concentration Gradients

The number of molecules present per unit volume on one side of a membrane is its *concentration* there. As a direct consequence of a cell membrane's selective permeability (figure 4.20), polar and charged molecules and ions often have a different concentration on one side of a membrane than on the other. This difference in concentration levels is called a **concentration gradient.**

When a substance moves from regions where its concentration is high to regions where its concentration is lower, it is said to move *down* its concentration gradient. How does a molecule "know" in what direction to move? It doesn't—molecules don't "know" anything. A molecule is equally likely to move in any direction and is constantly changing course in random ways. There are simply more molecules able to move from where they are common than from where they are scarce. This mixing process is simply the result of diffusion. In fact, diffusion is formally defined as *the net movement of molecules down a concentration gradient toward regions of lower concentration* (that is, where there are relatively fewer of them) *as a result of random motion.*

Importantly, each substance tends to diffuse in the direction established by its own concentration gradient, not the gradients established by other substances present in the same fluid. Thus the rate at which oxygen diffuses across a plant cell's plasma membrane into the cell is affected by that cell's oxygen concentration relative to the air, but not by its carbon dioxide concentration.

Molecules in Motion

If you think about it, a concentration gradient is a form of energy. Just as a boulder perched on a hilltop stores the energy it took to lift it there—energy released if the boulder rolls downhill, so a concentration gradient across a membrane can drive the movement of molecules across the membrane in the direction of lower concentration.

How fast molecules diffuse across a membrane (what physiologists call the rate of diffusion) is determined by two characteristics of the cell, and also by the physical characteristics of the environment in which the cell finds itself.

The Steepness of the Concentration Gradient Diffusion is most rapid when gradients are steepest. You can see why this must be so: Many more molecules randomly move away from a region of high concentration than into it from a region of low concentration. As this process continues, the number of molecules in the region of high concentration continually decreases, and the rate of diffusion slows down as fewer molecules leave and more arrive. Eventually, the same number is arriving as leaving. At this point, diffusion slows to a halt. Molecules are still moving, but the relative concentrations of the two regions no longer change—what a physiologist calls a state of dynamic equilibrium.

The Area of the Membrane Available for Diffusion Some gases like oxygen and a variety of small lipid-soluble molecules diffuse readily across the lipid bilayer of biological membranes, and the rate of their diffusion into or out of a cell is most rapid when the proportion of membrane surface occupied by the lipid bilayer is greatest. Membranes with larger portions of embedded proteins will exhibit lower rates of diffusion of gases and lipids moving through the bilayer, as the area available for their diffusion is less. Similarly, the rate of diffusion of charged ions like Na^+ and polar molecules like sugars and amino acids is greatest in membranes with the greatest number of protein channels with passages through which these molecules can pass. Because these channels are often quite specific, allowing only a particular ion or molecule to pass, the rate of diffusion of a substance is only affected by the number of channels actually available to it, rather than by the overall proportion of the membrane surface taken up by protein.

Physical Characteristics of the Cell Environment Temperature has a strong influence on the rate of diffusion, for the simple reason that higher temperatures cause molecules to move faster. In general, the rate of diffusion is greater for cells of organisms living at higher temperatures. High pressure also encourages faster diffusion, as molecules collide more often. This effect becomes quite important for organisms living in the deep ocean, where pressures are much higher than on the earth's surface. A third physical characteristic that influences the rate of diffusion across some cells in an important way is the electrical field in which a cell finds itself. An electric gradient across a nerve cell membrane can have an important influence on the rate at which ions diffuse in and out.

Key Learning Outcome 4.10 **Random movements of molecules cause them to mix uniformly in solution, a process called diffusion. Molecules tend to diffuse down their concentration gradients, faster when the gradient is steeper.**

4.11 Facilitated Diffusion

The selective permeability of biological membranes is perhaps their most important property. Ions and polar molecules can only cross the lipid core of membranes by passing through protein channels that bridge the bilayer.

Open Channels

The simplest of these channels are so-called *open channels*, shaped like tubes and functioning as open doors. As long as a molecule fits the channel, it is free to pass through in either direction, like a marble through a donut. Diffusion tends to equalize the concentration of such molecules on both sides of the membrane, with the molecules moving toward the side where they are scarcest. Many of the cell's water and ion channels are open channels, simple open pores that span the membrane. The open channels are selective, as only ions and molecules that precisely fit the pore can diffuse through it, in either direction. Often the pore has a "gate," a door that must be opened before an ion can pass through. Open ion channels with gates that swing open or shut in response to electrical charge play an essential role in signaling by the nervous system.

Carrier Proteins

There are limits to the specificity of open channels, as many different polar molecules are roughly the same size, shape, and charge. To increase the selectivity of membrane transport, cells employ a more complex channel that requires the diffusing molecule to bind to the surface of a "carrier" protein. Such hand-in-glove binding can be very specific. Once having bound their cargo, a carrier protein then physically carries the diffusing molecule across the membrane.

Each carrier protein binds with only a certain molecule, such as a particular sugar, amino acid, or ion, physically binding them on one side of the membrane and releasing them on the other. The direction of the molecule's net movement depends only on its concentration gradient across the membrane. If the concentration is greater outside the cell, the molecule is more likely to bind to the carrier on the extracellular side of the membrane, as in panel 1 of the Key Biological Process illustration below, and be released on the cytoplasmic side, as in panel 3. The net movement always occurs from high concentration to low, just as it does in simple diffusion, but the process is facilitated by the carriers. For this reason, this mechanism of transport is given a special name, **facilitated diffusion.**

A characteristic feature of transport by carrier proteins is that its rate can be saturated. If the concentration of a substance is progressively increased, the rate of transport of the substance increases up to a certain point and then levels off. There are a limited number of carrier proteins in the membrane, and when the concentration of the transported substance is raised high enough, all the carriers will be in use. The transport system is then said to be "saturated."

Key Learning Outcome 4.11 **Facilitated diffusion is the selective transport of substances across a membrane using a protein channel or carrier in the direction of lower concentration.**

KEY BIOLOGICAL PROCESS: Facilitated Diffusion

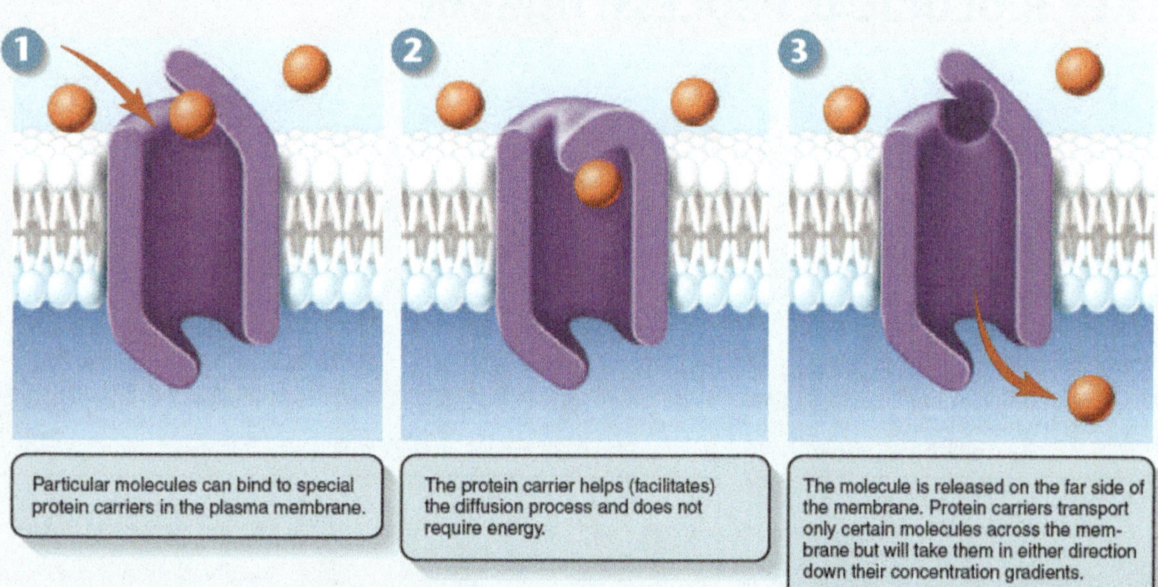

4.12 Osmosis

Diffusion allows molecules like oxygen, carbon dioxide, and nonpolar lipids to cross the plasma membrane. The movement of water molecules is not blocked because there are many small channels, called aquaporins, that allow water to pass freely through the membrane (see page 76).

Because water is so important, biologically, the diffusion of water molecules from an area of high concentration to an area of lower concentration is given a specific name, called **osmosis.** However, the number of water molecules that are free to diffuse across the membrane is dependent upon the concentration of other substances in solution. To understand how water moves into and out of a cell, let's focus on the water molecules already present inside a cell. What are they doing? Many of them are interacting with the sugars, proteins, and other polar molecules inside. Remember, water is very polar itself and readily interacts with other polar molecules. These "social" water molecules are not randomly moving about as they were outside; instead, they remain clustered around the polar molecules they are interacting with. As a result, while water molecules keep coming into the cell by random motion, they don't randomly come out again. Because more water molecules come in than go out, there is a net movement of water into the cell. The experiment presented in the Key Biological Process illustration below shows what happens. The right side of the beaker represents the inside of a cell and the left side is a watery environment. When the polar molecule urea is present in the cell, as in panel 2 below, water molecules cluster around each urea molecule and are no longer able to pass through the membrane to the "outside." In effect, the polar solute has reduced the number of free water molecules. Because the "outside" of the cell (on the left) has more unbound water molecules, water moves by diffusion into the cell (from the left to the right).

The concentration of all particles dissolved in a solution (called **solutes**) is called the osmotic concentration of the solution. If two solutions have unequal osmotic concentrations, the solution with the higher solute concentration, like the right side of the beaker below, is said to be **hypertonic** (Greek *hyper,* more than), and the solution with the lower concentration, like the left side of the beaker, is **hypotonic** (Greek *hypo,* less than). If the osmotic concentrations of the two solutions are equal, the solutions are **isotonic** (Greek *iso,* the same).

In cells, the plasma membrane separates two aqueous solutions, one inside the cell (the cytoplasm) and one outside (the extracellular fluid). The direction of the net diffusion of water across this membrane is determined by the osmotic concentrations of the solutions on either side. For example, if the cytoplasm of a cell was hypotonic to the extracellular fluid, water would diffuse out of the cell, toward the solution with the higher concentration of solutes (and, therefore, the lower concentration of unbound water molecules). This loss of water from the cytoplasm would cause the cell to shrink until the osmotic concentrations of the cytoplasm and the extracellular fluid become equal.

Osmotic Pressure

What would happen if the cell's cytoplasm were hypertonic to the extracellular fluid? In this situation, water would diffuse into the cell from the extracellular fluid, causing the cell to swell. The pressure of the cytoplasm pushing out against the cell membrane, called **hydrostatic pressure,** would increase. On the other hand, the **osmotic pressure,** defined as

KEY BIOLOGICAL PROCESS: Osmosis

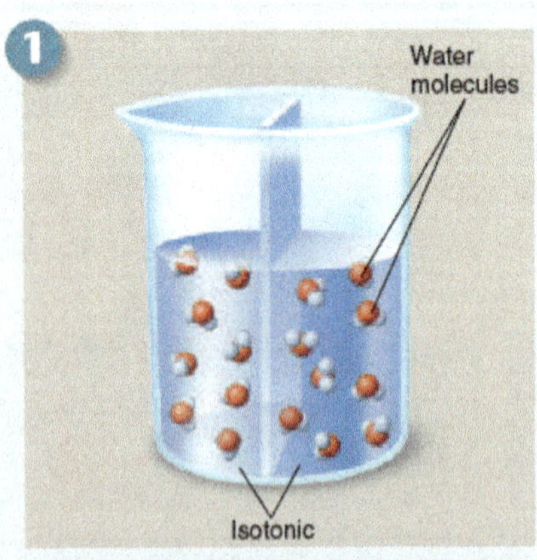

Diffusion causes water molecules to distribute themselves equally on both sides of a semipermeable membrane.

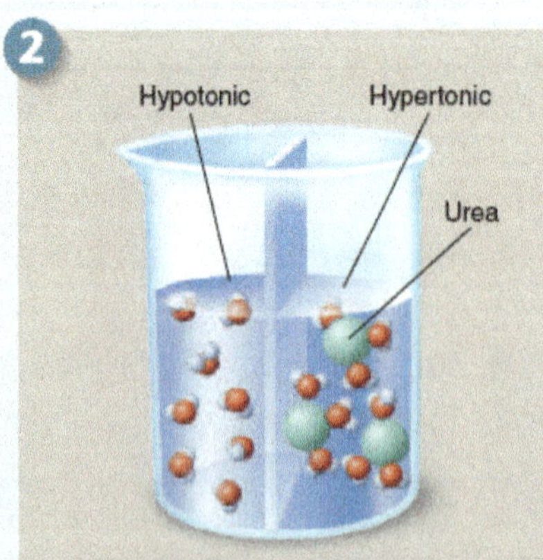

Addition of solute molecules that cannot cross the membrane reduces the number of free water molecules on that side, as they bind to the solute.

Diffusion then causes free water molecules to move from the side where their concentration is higher to the solute side, where their concentration is lower.

the pressure that must be applied to stop the osmotic movement of water across a membrane, would also be at work. If the membrane were strong enough, the cell would reach an equilibrium, at which the osmotic pressure, which tends to drive water into the cell, is exactly counterbalanced by the hydrostatic pressure, which tends to drive water back out of the cell. However, a plasma membrane by itself cannot withstand large internal pressures, and an isolated cell under such conditions would burst like an overinflated balloon. Accordingly, it is important for animal cells to maintain isotonic conditions.

Figure 4.21 illustrates how solutes create osmotic pressure. Look first at the red blood cells at the top of the figure. On the left, in a hypertonic solution like ocean water, there is a net movement of water molecules out of the red blood cell toward the higher concentration of solutes outside, causing the cell to shrivel. In an isotonic solution (in the middle), the concentration of solutes on either side of the red blood cell membrane is the same. Osmosis still occurs, but water diffuses into and out of the cell at the same rate, and the cell doesn't change size. In a hypotonic solution on the right, the concentration of solutes is higher within the cell than outside, so the net movement of water is into the cell. This is the situation utilized by Peter Agre in the experiment described on page 76 in which he demonstrated that aquaporin was a functioning water channel. A red blood cell is an enclosed structure, so as water entered a cell placed in a hypotonic solution (in his case, pure water), pressure is applied to the cell membrane causing the cell to swell, becoming spherical. This swelling can continue until the cell membrane can stretch no more and ruptures.

Now look at the plant cells at the bottom of figure 4.21. In these cells, unlike animal cells, the hydrostatic pressure generated by osmosis is counterbalanced by osmotic pressure, the force required to stop the flow of water into the cell. Plant cells have strong cell walls that can apply adequate osmotic pressure to keep the cell from rupturing.

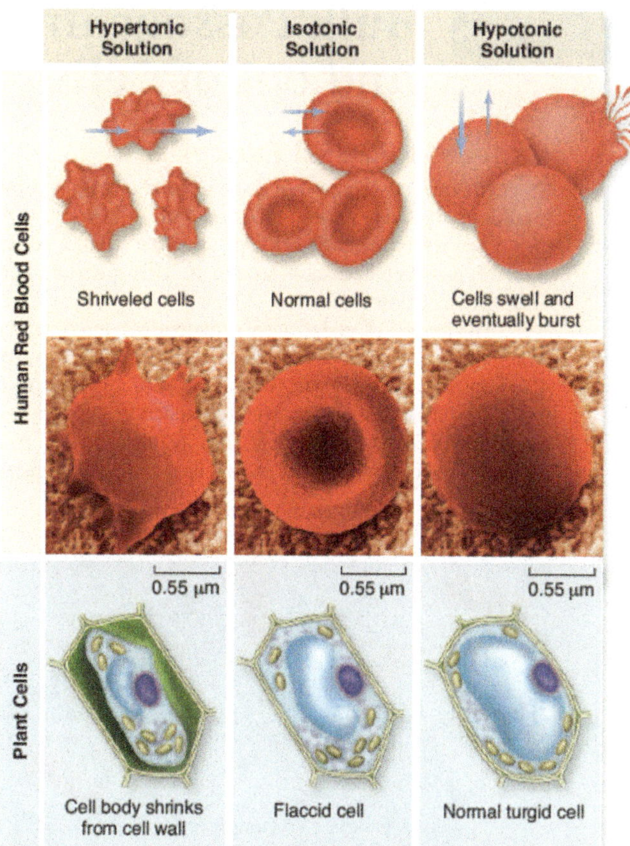

Figure 4.21 How solutes create osmotic pressure.
A cell is an enclosed structure, and so as water enters the cell from a hypotonic solution, pressure is applied to the plasma membrane until the cell ruptures. This hydrostatic pressure is counterbalanced by osmotic pressure, the force required to stop the flow of water into the cell. Plant cells have strong cell walls that can apply adequate osmotic pressure to keep the cell from rupturing.

Maintaining Osmotic Balance

Organisms have developed many solutions to the osmotic dilemma posed by being hypertonic to their environment.

Extrusion Some single-celled eukaryotes like the protist *Paramecium* use organelles called contractile vacuoles to remove water. Each vacuole collects water from various parts of the cytoplasm and transports it to the central part of the vacuole, near the cell surface. The vacuole possesses a small pore that opens to the outside of the cell. By contracting rhythmically, the vacuole pumps water out through the pore that is continuously seeping into the cell by osmosis.

Isosmotic Solutions Some organisms that live in the ocean adjust their internal concentration of solutes to match that of the surrounding seawater. Isotonic with respect to their environment, there is no net flow of water into or out of these cells. Many terrestrial animals solve the problem in a similar way, by circulating a fluid through their bodies that bathes cells in an isotonic solution. The blood in your body, for example, contains a high concentration of the protein albumin, which elevates the solute concentration of the blood to match your cells. Sharks maintain a high concentration of urea in their blood and body fluids, keeping their cells isosmotic with respect to the sea water in which they swim.

Turgor Most plant cells are hypertonic to their immediate environment, containing a high concentration of solutes in their central vacuoles. The resulting internal hydrostatic pressure, known as **turgor pressure,** presses the plasma membrane firmly against the interior of the cell wall as you saw in figure 4.7, making the cell rigid. Most green plants depend on turgor pressure to maintain their shape, and wilt when they lack sufficient water.

Key Learning Outcome 4.12 Water molecules associated with polar solutes are not free to diffuse, so there is a net movement of water across a membrane toward the side with less "free" water. Osmosis is the diffusion of water, but not solutes, across a membrane. Cells must maintain an osmotic balance to function properly.

4.13 Bulk Passage into and out of Cells

Endocytosis and Exocytosis

The cells of many eukaryotes take in food and liquids by extending their plasma membranes outward toward food particles. The membrane engulfs the particle and forms a vesicle—a membrane-bounded sac—around it. This process is called **endocytosis** (figure 4.22).

The reverse of endocytosis is **exocytosis**, the discharge of material from vesicles at the cell surface. The vesicle in figure 4.23 contains a substance to be discharged, or released, from the cell. The purple particles remain suspended in the vesicle as it fuses with the plasma membrane. The membrane that forms the vesicle is made of phospholipids, and as it comes in contact with the plasma membrane, the phospholipids of both membranes interact, forming a pore through which the contents leave the vesicle to the outside. In plant cells, exocytosis is an important means of exporting the materials needed to construct the cell wall through the plasma membrane. Among protists, the discharge of a contractile vacuole is a form of exocytosis. In animal cells, exocytosis provides a mechanism for secreting many hormones, neurotransmitters, digestive enzymes, and other substances.

Phagocytosis and Pinocytosis

If the material the cell takes in is particulate (made up of discrete particles), such as an organism, like the red bacterium in figure 4.22*a*, or some other fragment of organic matter, the process is called **phagocytosis** (Greek *phagein*, to eat, and *cytos*, cell). If the material the cell takes in is liquid or substances dissolved in a liquid like the small particles in figure 4.22*b*, it

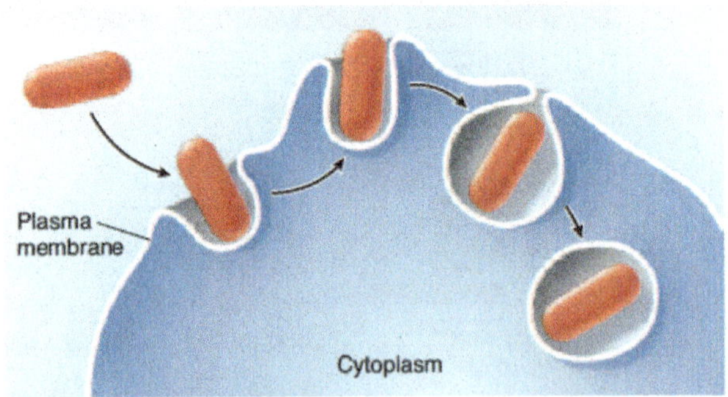

(a) Phagocytosis

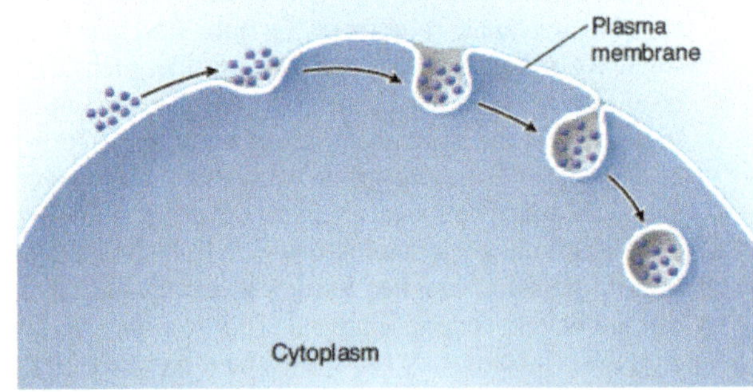

(b) Pinocytosis

Figure 4.22 Endocytosis.
Endocytosis is the process of engulfing material by folding the plasma membrane around it, forming a vesicle. (*a*) When the material is an organism or some other relatively large fragment of organic matter, the process is called phagocytosis. (*b*) When the material is a liquid, the process is called pinocytosis.

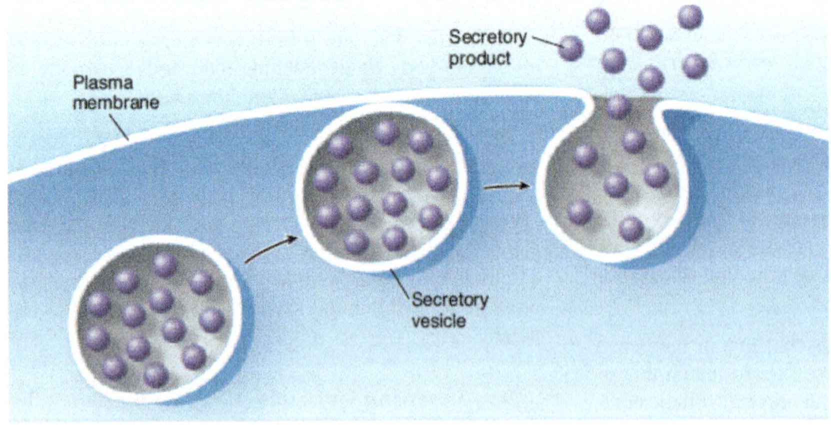

(a)

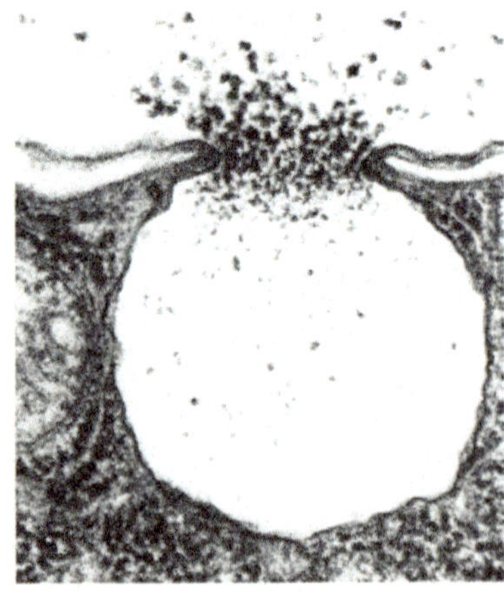

(b)

Figure 4.23 Exocytosis.
Exocytosis is the discharge of material from vesicles at the cell surface. (*a*) Proteins and other molecules are secreted from cells in small pockets called secretory vesicles, whose membranes fuse with the plasma membrane, thereby allowing the secretory vesicles to release their contents to the cell surface. (*b*) In the photomicrograph, you can see exocytosis taking place explosively.

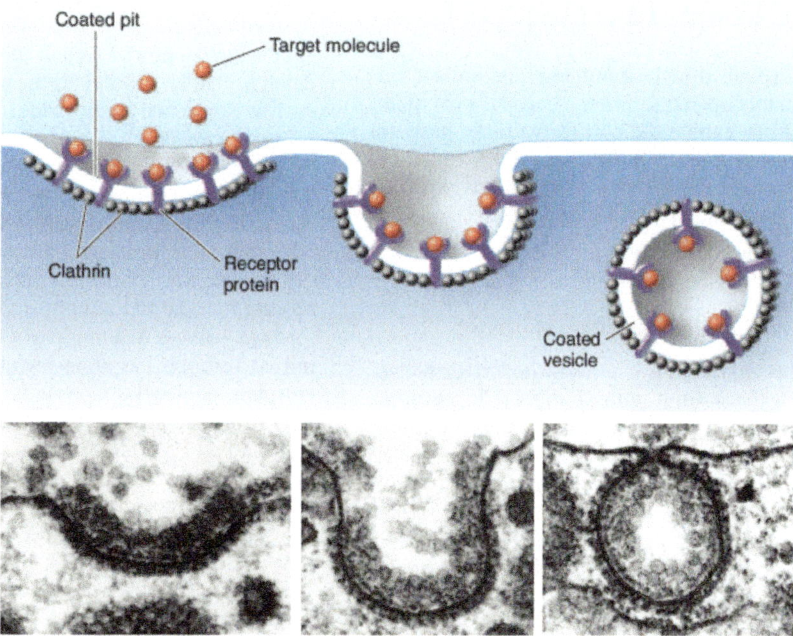

Figure 4.24 Receptor-mediated endocytosis.
Cells that undergo receptor-mediated endocytosis have pits coated with the protein clathrin that initiate endocytosis when target molecules bind to receptor proteins in the plasma membrane. In the photomicrographs, a coated pit appears in the plasma membrane of a developing egg cell, covered with a layer of proteins (80,000×). When an appropriate collection of molecules gathers in the coated pit, the pit deepens, and eventually seals off to form a vesicle.

is called **pinocytosis** (Greek *pinein,* to drink). Pinocytosis is common among animal cells. Mammalian egg cells, for example, "nurse" from surrounding cells; the nearby cells secrete nutrients that the maturing egg cell takes up by pinocytosis. Virtually all eukaryotic cells constantly carry out these kinds of endocytosis, trapping particles and extracellular fluid in vesicles and ingesting them. Endocytosis rates vary from one cell type to another. They can be surprisingly high: Some types of white blood cells ingest 25% of their cell volume each hour!

Receptor-Mediated Endocytosis

Specific molecules are often transported into eukaryotic cells through **receptor-mediated endocytosis,** illustrated in figure 4.24. Molecules to be transported into the cell, the red balls in the figure, first bind to specific receptors in the plasma membrane. The transport process is specific to only molecules that have a shape that fits snugly into the receptor. The plasma membrane of a particular kind of cell contains a characteristic battery of receptor types, each for a different kind of molecule.

The portion of the receptor molecule inside the membrane is trapped in an indented pit coated with the protein clathrin, visible in the photos as well as in the drawing above. The pits act like molecular mousetraps, closing over to form an internal vesicle when the right molecule enters the pit. The trigger that releases the trap is the binding of the properly fitted target molecule to a receptor embedded in the membrane of the pit. When binding occurs, the cell reacts by initiating endocytosis. The process is highly specific and very fast.

One type of molecule that is taken up by receptor-mediated endocytosis is called low-density lipoprotein (LDL). The LDL molecules bring cholesterol into the cell where it can be incorporated into membranes. Cholesterol plays a key role in determining the stiffness of the cell's membrane. In the human genetic disease called hypercholesterolemia, the receptors lack tails and so are never caught in the clathrin-coated pits and, thus, are never taken up by the cells. The cholesterol stays in the bloodstream of affected individuals, coating their arteries and leading to heart attacks.

It is important to understand that receptor-mediated endocytosis in itself does not bring substances directly into the cytoplasm of a cell. The material taken in is still separated from the cytoplasm by the membrane of the vesicle. Other processes break down or release the contents.

Key Learning Outcome 4.13 The plasma membrane can engulf materials by endocytosis, folding the membrane around the material to encase it within a vesicle. Exocytosis is essentially this process in reverse, expelling substances using vesicles. Receptor-mediated endocytosis brings in only selected substances.

4.14 Active Transport

Other channels through the plasma membrane are closed doors. These channels open only when energy is provided. They are designed to enable the cell to maintain high or low concentrations of certain molecules, much more or less than exists outside the cell. Like motor-driven turnstiles, the channels operate to move a certain substance *up* its concentration gradient. The operation of these one-way, energy-requiring channels results in **active transport,** the movement of molecules across a membrane to a region of higher concentration by the expenditure of energy.

The Sodium-Potassium Pump The most important active transport channel is the **sodium-potassium (Na^+-K^+) pump,** which expends metabolic energy to actively pump sodium ions (Na^+) in one direction, out of cells, and potassium ions (K^+) in one direction, into cells. More than one-third of all the energy expended by your body's cells is spent driving Na^+-K^+ pump channels. This energy is derived from *adenosine triphosphate (ATP),* a molecule we will learn more about in chapter 5. The transportation of two different ions in opposite directions happens because energy causes a change in the shape of the membrane protein carrier. The Key Biological Process illustration below walks you through one cycle of the pump. Each channel can move over 300 sodium ions per second when working full tilt. As a result of all this pumping, there are far fewer sodium ions in the cell. This concentration gradient, paid for by the expenditure of considerable metabolic energy in the form of ATP molecules, is exploited by your cells in many ways. Two of the most important are (1) the conduction of signals along nerve cells (discussed in detail in chapter 28) and (2) the pulling of valuable molecules such as sugars and amino acids into the cell *against* their concentration gradient!

We will focus for a moment on this second process. The plasma membranes of many cells are studded with facilitated diffusion channels, which offer a path for sodium ions that have been pumped out by the Na^+-K^+ pump to diffuse back in. There is a catch, however; these channels require that the sodium ions have a partner in order to pass through—like a dancing party where only couples are admitted through the door—which is why these are called **coupled channels.** Coupled channels won't let sodium ions across unless another molecule tags along, crossing hand in hand with the sodium ion. In some cases the partner molecule is a sugar (see the last entry in table 4.3), in others an amino acid or other molecule. Because the concentration gradient for sodium is so large, many sodium ions are trying to get back in, and this diffusion pressure drags in the partner molecules as well, even if they are already in high concentration within the cell. In this way, sugars and other actively transported molecules enter the cell—via special coupled channels.

Key Learning Outcome 4.14 **Active transport is energy-driven transport across a membrane toward a region of higher concentration.**

KEY BIOLOGICAL PROCESS: Sodium-Potassium Pump

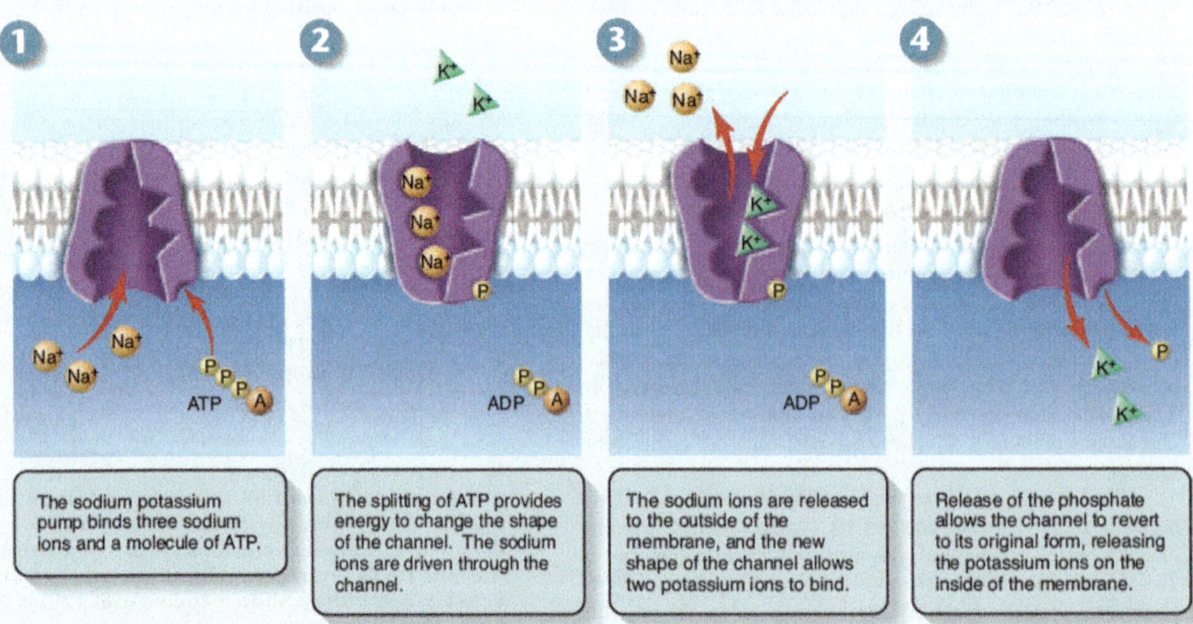

TABLE 4.3 MECHANISMS FOR TRANSPORT ACROSS CELL MEMBRANES

Process	Passage Through Membrane	How it Works	Example
Passive Processes			
Diffusion			
Direct		Random molecular motion produces net migration of molecules toward region of lower concentration.	Movement of oxygen into cells
Protein channel		Polar molecules pass through a protein channel.	Movement of ions in or out of cell
Facilitated Diffusion			
Protein carrier		Molecule binds to carrier protein in membrane and is transported across; net movement is toward region of lower concentration.	Movement of glucose into cells
Osmosis			
Aquaporins		Diffusion of water across selectively permeable membrane.	Movement of water into cells placed in a hypotonic solution
Active Processes			
Endocytosis			
Membrane vesicle			
Phagocytosis		Particle is engulfed by membrane, which folds around it and forms a vesicle.	Ingestion of bacteria by white blood cells
Pinocytosis		Fluid droplets are engulfed by membrane, which forms vesicles around them.	"Nursing" of human egg cells
Receptor-mediated endocytosis		Endocytosis is triggered by the binding of a target molecule to a specific receptor.	Cholesterol uptake
Exocytosis			
Membrane vesicle		Vesicles fuse with plasma membrane and eject contents.	Secretion of mucus
Active Transport			
Protein carrier			
Na^+-K^+ pump		Carrier expends energy to transport a substance across a membrane against its concentration gradient.	Movement of Na^+ and K^+ against their concentration gradients
Coupled transport		Molecules are transported across a membrane against their concentration gradients by the cotransport of another substance down its concentration gradient.	Coupled uptake of glucose into cells against its concentration gradient

INQUIRY & ANALYSIS

Why Does a Cell's Disposal of Damaged Proteins Consume Energy?

Much of modern biology is devoted to learning how cells build things—how the information encoded in DNA is used by cells to manufacture the proteins that make us what we are. The Nobel Prize in Chemistry was awarded in 2004 to researchers for their discovery of how the opposite, less glamorous process works: how cells break down and recycle proteins that are damaged or have outlived their usefulness.

It turns out that a cell's recycling of proteins is much more than just "taking out the trash." Particular proteins are removed, often quite quickly, and cells use such targeted removals to control a lot of their activities, timing when a cell carries out particular functions, when it divides, and even when it dies. Of the 25,000 genes in your DNA, about 1,000 take part in this protein recycling system.

Our understanding of how this system works begins with a puzzle first noted in the 1950s. Most enzymes that break down proteins, including those that digest food, do not need energy to work. But a cell's recycling of its own proteins does consume energy. Researchers had no idea why energy was needed.

The answer to this puzzle came from an unexpected direction. In 1975 scientists discovered a small protein in calves' brains consisting of just 76 amino acids. Soon they realized that exactly the same protein is found in all eukaryotes, from yeasts to humans. They called this ubiquitous ("found everywhere") protein *ubiquitin*.
In the early 1980s researchers worked out that ubiquitin was a label that the cell attaches to proteins to mark them for destruction, a sort of molecular "kiss of death." The process of attaching ubiquitin takes energy, solving the puzzle of why protein recycling requires energy. The tagged proteins are taken to a barrel-shaped chamber in the cell's cytoplasm called a *proteasome,* which slices the proteins into bits that are then recycled by the cell into new protein.

The graph above displays the sort of protein recycling experiment that revealed ubiquitin's key role. The experiment monitors levels of a particular protein involved in cell division (the "target" protein) within human cells growing in culture in a laboratory flask. Two cultures are monitored in side-by-side experiments: In the culture indicated by red dots, cells contain functional copies of the ubiquitin gene (ubi^+); in the culture indicated by blue dots the ubiquitin gene has been deleted from the DNA (ubi^-). After 20 minutes, energy in the form of ATP is made available to the growing cells, which until then had been energy-starved.

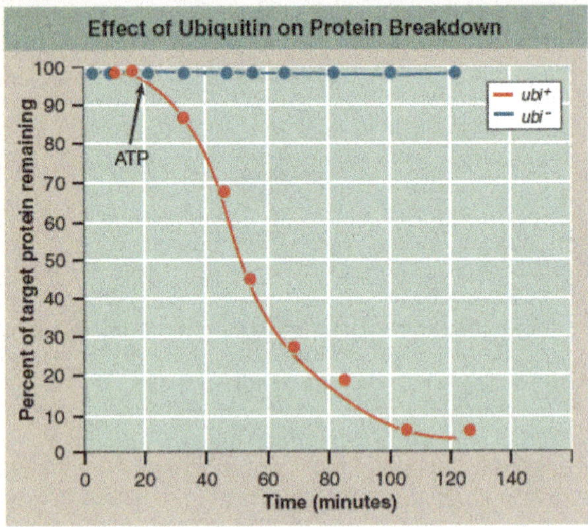

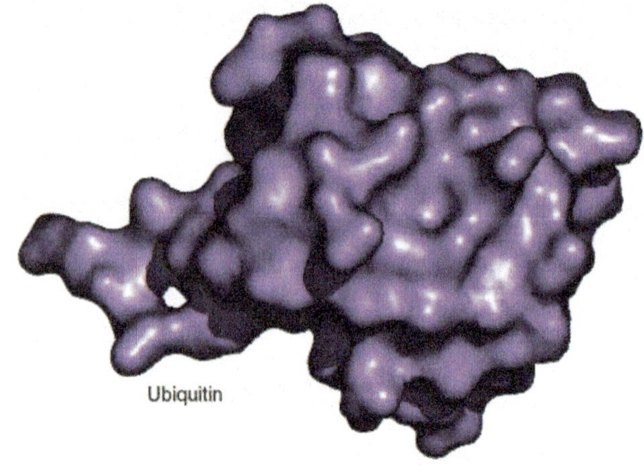

1. **Applying Concepts**
   a. Variable. In the graph, what is the dependent variable?
   b. Concentration. After 100 minutes, which of the two cultures represents the higher concentration of target protein?
2. **Interpreting Data** Does the addition of ATP affect the level of target protein in either culture? Which one?
3. **Making Inferences** How does this culture differ from the other? Why might ATP stimulate removal of target protein from this culture, but not the other?
4. **Drawing Conclusions** Using the information in the graph, suggest why the functioning of ubiquitin requires ATP energy for the effective removal of the target protein.

Chapter Review

The World of Cells

4.1 Cells

- Cells are the smallest living structure. They consist of cytoplasm enclosed in a plasma membrane. Organisms can be composed of a single cell or multiple cells.
- Materials pass into and out of cells across the plasma membrane. A smaller cell has a larger surface-to-volume ratio, which increases the area through which materials may pass (**figure 4.2**). Because cells are so small (**figure 4.3**), a microscope is needed to view and study them (**table 4.1**).

4.2 The Plasma Membrane

- The plasma membrane that encloses all cells consists of a double layer of lipids, called the lipid bilayer, in which proteins are embedded. The structure of the plasma membrane is called the fluid mosaic model.
- The lipid bilayer (**integrated art, pages 74-75**) is made up of special lipid molecules called phospholipids, which have a polar (water-soluble) end and a nonpolar (water-insoluble) end. The bilayer forms because the nonpolar ends move away from the watery surroundings, forming the two layers. There are many different types of proteins associated with the plasma membrane.

Kinds of Cells

4.3 Prokaryotic Cells

- Prokaryotic cells are simple unicellular organisms that lack nuclei or other internal organelles and are usually encased in a rigid cell wall (**figure 4.4**). They vary in shape and may contain external structures (**figure 4.5**).

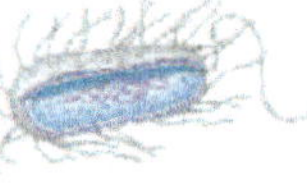

4.4 Eukaryotic Cells

- Eukaryotic cells are larger and more structurally complex than prokaryotic cells. They contain nuclei, organelles, and internal membrane systems (**figures 4.6 and 4.7**).

Tour of a Eukaryotic Cell

4.5 The Nucleus: The Cell's Control Center

- The nucleus is the command and control center of the cell. It contains the cell's DNA, which encodes the hereditary information that runs the cell (**figure 4.8**). The nucleolus is a darker area inside the nucleus where ribosomal RNA is made.

4.6 The Endomembrane System

- The endomembrane system (**integrated art, pages 82-83**) is a collection of interior membranes that organize and divide the cell's interior into functional areas. The endoplasmic reticulum is a transport system that modifies and moves proteins and other molecules produced in the ER to the Golgi complex. There are two types of ER in the cell, rough and smooth. The Golgi complex is a delivery system that carries molecules to the surface of the cell where they are released to the outside.

4.7 Organelles That Contain DNA

- The mitochondrion (**figure 4.9**) is called the powerhouse of the cell because it is the site of oxidative metabolism, an energy-extracting process. Chloroplasts are the site of photosynthesis and are present in plants and algal cells (**figure 4.10**). Mitochondria and chloroplasts are cell-like organelles that appear to be ancient bacteria that formed endosymbiotic relationships with early eukaryotic cells (**figure 4.11**).

4.8 The Cytoskeleton: Interior Framework of the Cell

- The interior of the cell contains a network of protein fibers that make up the cytoskeleton (**integrated art, page 86**). The cytoskeleton supports the shape of the cell and anchors organelles in place.
- Centrioles assemble microtubules (**figure 4.12**). Vacuoles are storage compartments (**figures 4.13 and 4.14**).
- Cilia and flagella propel the cell through its environment (**figures 4.15 and 4.16**). Motor proteins move materials throughout the cell (**figure 4.17**).

4.9 Outside the Plasma Membrane

- The cells of plants, fungi, and many protists have cell walls that serve a similar function as prokaryotic cell walls, but are composed of different molecules (**figure 4.18**). Animal cells lack cell walls but contain an outer layer of glycoproteins, called the extracellular matrix (**figure 4.19**).

Transport Across Plasma Membranes

4.10 Diffusion

- Materials pass into and out of the cell passively through diffusion (**Key Biological Process, page 93**). Diffusion is the movement of molecules from an area of high concentration to an area of low concentration, down their concentration gradients.

4.11 Facilitated Diffusion

- In facilitated diffusion (**Key Biological Process, page 95**) a substance travels down its concentration gradient, but must bind to a protein carrier in order to pass across the membrane.

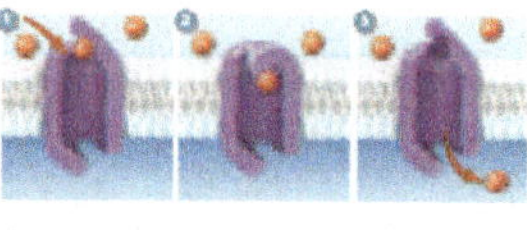

4.12 Osmosis

- Osmosis (**Key Biological Process, page 96**) is the movement of water into and out of the cell, driven by differing concentrations of solute. Water molecules move to areas of higher solute concentrations (**figure 4.21**).

4.13 Bulk Passage into and out of Cells

- Larger structures or larger quantities of material move into and out of the cell through endocytosis and exocytosis, respectively (**figures 4.22 and 4.23**). Receptor-mediated endocytosis is a selective transport process, bringing in only those substances that are able to bind to specific receptors (**figure 4.24**).

4.14 Active Transport

- Active transport involves the input of energy to transport a substance against (or up) its concentration gradient. Examples include the sodium-potassium pump (**Key Biological Process, page 100**) and coupled channels, where one substance travels down its concentration gradient but is cotransported with another substance against its concentration gradient.

Test Your Understanding

1. Cell theory includes the principle that
   a. cells are the smallest living things; nothing smaller than a cell is considered alive.
   b. all cells are surrounded by cell walls that protect them.
   c. all organisms are made up of many cells arranged in specialized, functional groups.
   d. all cells contain membrane-bounded structures called organelles.
2. The plasma membrane is
   a. a carbohydrate layer that surrounds groups of cells to protect them.
   b. a double lipid layer with proteins inserted in it, which surrounds every cell individually.
   c. a thin sheet of structural proteins that encloses cytoplasm.
   d. composed of proteins that form a protective barrier.
3. Organisms that have cells with a relatively uniform cytoplasm and no organelles are called _____, and organisms whose cells have organelles and a nucleus are called _____.
   a. cellulose, nuclear
   b. eukaryotes, prokaryotes
   c. flagellated, streptococcal
   d. prokaryotes, eukaryotes
4. Within the nucleus of a cell you can find
   a. a nucleolus.
   b. a cytoskeleton.
   c. mitochondria.
   d. All of these.
5. The endomembrane system within a cell includes the
   a. cytoskeleton and the ribosomes.
   b. prokaryotes and the eukaryotes.
   c. endoplasmic reticulum and the Golgi complex.
   d. mitochondria and the chloroplasts.
6. It was once thought that only the nucleus of each cell contained DNA. We now know that DNA is also found in the
   a. cytoskeleton and the ribosomes.
   b. prokaryotes and the eukaryotes.
   c. endoplasmic reticulum and the Golgi bodies.
   d. mitochondria and the chloroplasts.
7. Which of the following statements is true?
   a. All cells have a cell wall for protection and structure.
   b. Eukaryotic cells in plants and fungi, and all prokaryotes, have a cell wall.
   c. There is a second membrane composed of structural carbohydrates surrounding all cells.
   d. Prokaryotes and all cells of eukaryotic animals have a cell wall.
8. If you put a drop of food coloring into a glass of water, the drop of color will
   a. fall to the bottom of the glass and sit there unless you stir the water; this is because of hydrogen bonding.
   b. float on the top of the water, like oil, unless you stir the water; this is because of surface tension.
   c. instantly disperse throughout the water; this is because of osmosis.
   d. slowly disperse throughout the water; this is because of diffusion.
9. When large molecules, such as food particles, need to get into a cell, they cannot easily pass through the plasma membrane, and so they move across the membrane through the processes of
   a. diffusion and osmosis.
   b. endocytosis and phagocytosis.
   c. exocytosis and pinocytosis.
   d. facilitated diffusion and active transport.
10. Active transport of specific molecules involves
    a. facilitated diffusion.
    b. endocytosis and phagocytosis.
    c. energy and specialized pumps or channels.
    d. permeability and a concentration gradient.

Apply Your Understanding

1. **Figure 4.3** The first microscope was used in about 1590. Electron microscopes came into common use about 70 years ago. Just over 100 years ago most physicians did not wash up between patients, even when someone had just died, or was very sick. Explain why it took so long to convince doctors to wash their hands.
2. In the lungs, there are steep concentration gradients for oxygen and carbon dioxide molecules such that large numbers of these molecules move across the plasma membrane of the cells that line the lungs. These molecules pass through the plasma membranes by simple diffusion. This process is fast and efficient. Would this process be just as efficient if the oxygen and carbon dioxide molecules passed through the membranes by facilitated diffusion? Why or why not?

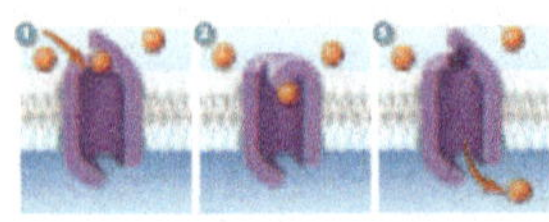

Synthesize What You Have Learned

1. You are using a computer program to design a new single-celled organism. Discuss why a flat, platelike cell will be more efficient in transporting materials than a spherical, ball-like cell of the same volume.
2. Antibiotics are medicines that target bacterial infections in vertebrates. How can the antibiotic penicillin kill all the bacterial cells and not harm vertebrate cells? Hint: What part of the bacterial cell might the antibiotic be targeting?
3. Compare the cellular organelles and other structures to the parts of a city—for example, the nucleus is city hall and the DNA is all the city's laws and instructions.
4. A ribosome contains two subunits. The ribosome subunits are assembled within the nucleus, but ribosomes act in the cytoplasm. How do you imagine the subunits get out of the nucleus and into the cytoplasm?

5
Energy and Life

Learning Objectives

Cells and Energy

5.1 The Flow of Energy in Living Things
1. Define energy.
2. Differentiate between kinetic and potential energy.
3. Define thermodynamics.
4. Define chemical reaction.

5.2 The Laws of Thermodynamics
1. State the first law of thermodynamics.
2. Defend the proposition that heat is kinetic energy.
3. Describe how heat can be harnessed to do work.
4. State the second law of thermodynamics.
5. Define entropy.

Cell Chemistry

5.3 Chemical Reactions
1. Differentiate between reactant and product; between endergonic and exergonic.
2. Define activation energy.
3. Describe the effect of catalysis on activation energy.

Enzymes

5.4 How Enzymes Work
1. Describe the biological function of enzymes.
2. Differentiate between active site and binding site.
3. Diagram the three stages of the Key Biological Process of enzyme catalysis.
4. Distinguish between chemical reaction and biochemical pathway.
5. Describe the effects of temperature on an enzyme-catalyzed reaction, and explain the reason for its influence.
6. Describe the effect of pH on enzyme-catalyzed reactions, and explain the reason for its influence.

5.5 How Cells Regulate Enzymes
1. Describe how repressors interact with allosteric sites of enzymes, and the consequences of the interaction.
2. Define feedback inhibition.
3. Explain how you could determine whether an enzyme inhibitor was competitive or noncompetitive.

How Cells Use Energy

5.6 ATP: The Energy Currency of the Cell
1. Diagram the chemical structure of an ATP molecule.
2. Explain how the chain of three phosphate groups stores potential energy, and describe how organisms use ATP to power endergonic reactions.
3. Define coupled reaction.

Inquiry & Analysis: Do Enzymes Physically Attach to Their Substrates?

All of life is driven by energy. It took energy for these mice to climb up the wheat stalks. Their mousely activities while perched on the stalks—looking for danger, generating body heat, wiggling their whiskers—take energy. The energy comes from the wheat kernels and other foods these mice eat. By breaking the chemical bonds of carbohydrates and other molecules in the wheat kernels, and transferring the energy of these bonds to those of a "molecular currency" called ATP, the mice are able to capture the chemical energy in their food and put it to work. The cells of the mice perform this feat with the aid of enzymes, which are macromolecules with highly specialized shapes. Each enzyme's shape has a surface cavity called an active site, into which some specific chemical in the cell fits precisely, like a foot fits into a shoe of the proper size. When the chemical nudges in, the enzyme responds by bending, stressing particular covalent bonds in the chemical and so triggering a specific chemical reaction. The chemistry of life is enzyme chemistry.

Cells and Energy

5.1 The Flow of Energy in Living Things

We are about to begin our discussion of energy and cellular chemistry. Although these subjects may seem difficult at first, remember that all life is driven by energy. The concepts and processes discussed in the next three chapters are key to life. We are chemical machines, powered by chemical energy, and for the same reason that a successful race car driver must learn how the engine of a car works, we must look at cell chemistry. Indeed, if we are to understand ourselves, we must "look under the hood" at the chemical machinery of our cells and see how it operates.

As described in chapter 2, **energy** is defined as the ability to do work. It can be considered to exist in two states: kinetic energy and potential energy. **Kinetic energy** is the energy of motion. Objects that are not in the process of moving but have the capacity to do so are said to possess **potential energy,** or stored energy. The difference in the two states of energy is being experienced by the young man in figure 5.1. A boulder perched on a hilltop (figure 5.1*a*) has potential energy; after the man pushes the boulder and it begins to roll downhill (figure 5.1*b*), some of the boulder's potential energy is converted into kinetic energy. All of the work carried out by living organisms also involves the transformation of potential energy to kinetic energy.

Energy exists in many forms: mechanical energy, heat, sound, electric current, light, or radioactive radiation. Because it can exist in so many forms, there are many ways to measure energy. The most convenient is in terms of heat, because all other forms of energy can be converted to heat. Thus the study of energy is called **thermodynamics,** meaning "heat changes."

Energy flows into the biological world from the sun, which shines a constant beam of light on the earth. It is estimated that the sun provides the earth with more than 13×10^{23} calories per year, or 40 million billion calories per second! Plants, algae, and certain kinds of bacteria capture a fraction of this energy through photosynthesis. In photosynthesis, energy garnered from sunlight is used to combine small molecules (water and carbon dioxide) into more complex molecules (sugars). These complex sugar molecules have potential energy due to the arrangement of their atoms. This potential energy, in the form of chemical energy, does the work in cells. Recall from chapter 2 that an atom consists of a central nucleus surrounded by one or more orbiting electrons, and a covalent bond forms when two atomic nuclei share electrons. Breaking such a bond requires energy to pull the nuclei apart. Indeed, the strength of a covalent bond is measured by the amount of energy required to break it. For example, it takes 98.8 kcal to break 1 mole (6.023×10^{23}) of carbon–hydrogen (C–H) bonds.

All the chemical activities within cells can be viewed as a series of chemical reactions between molecules. A **chemical reaction** is the making or breaking of chemical bonds—gluing atoms together to form new molecules or tearing molecules apart and sometimes sticking the pieces onto other molecules.

Key Learning Outcome 5.1 **Energy is the capacity to do work, either actively (kinetic energy) or stored for later use (potential energy). Chemical reactions occur when the covalent bonds linking atoms together are formed or broken.**

(a) Potential energy

(b) Kinetic energy

Figure 5.1 Potential and kinetic energy.
Objects that have the capacity to move but are not moving have potential energy, while objects that are in motion have kinetic energy. (*a*) The energy required to move the ball up the hill is stored as potential energy. (*b*) This stored energy is released as kinetic energy as the ball rolls down the hill.

5.2 The Laws of Thermodynamics

Running, thinking, singing, reading these words—all activities of living organisms involve changes in energy. A set of universal laws we call the laws of thermodynamics govern these and all other energy changes in the universe.

The First Law of Thermodynamics

The first of these universal laws, the **first law of thermodynamics,** concerns the amount of energy in the universe. It states that energy can change from one state to another (from potential to kinetic, for example) but it can never be destroyed, nor can new energy be made. The total amount of energy in the universe remains constant.

A lion eating a giraffe is in the process of acquiring energy. Rather than creating new energy or capturing the energy in sunlight, the lion is merely transferring some of the potential energy stored in the giraffe's tissues to its own body (just as the giraffe obtained the potential energy stored in the plants it ate while it was alive). Within any living organism, this chemical potential energy can be shifted to other molecules and stored in chemical bonds, or it can be converted into kinetic energy, or into other forms of energy such as light or electrical energy. During each conversion, some of the energy dissipates into the environment as **heat energy,** a measure of the random motions of molecules (and, hence, a measure of one form of kinetic energy). Energy continuously flows through the biological world in one direction, with new energy from the sun constantly entering the system to replace the energy dissipated as heat.

Heat can be harnessed to do work only when there is a heat gradient—that is, a temperature difference between two areas. This is how a steam engine functions. In old steam locomotives like you see in figure 5.2, heat was used to move the wheels. First, a boiler (not shown) heats up water to create steam. The steam is then pumped into the cylinder of the steam engine, where it moves the piston to the right. The moving of this piston then does the work of the steam engine by moving a lever that turns the wheel. Cells are too small to maintain significant internal temperature differences, so heat energy is incapable of doing the work of cells. Thus, although the total amount of energy in the universe remains constant, the energy available to do useful work in a cell decreases, as progressively more of it dissipates as heat.

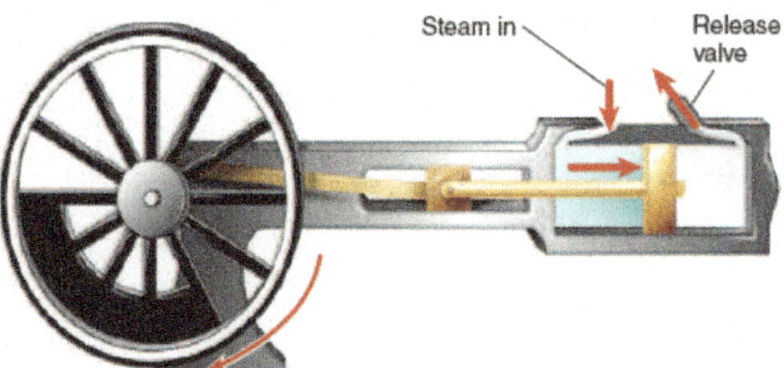

Figure 5.2 A steam engine.
In a steam engine, heat is used to produce steam. The expanding steam pushes against a piston that causes the wheel to turn.

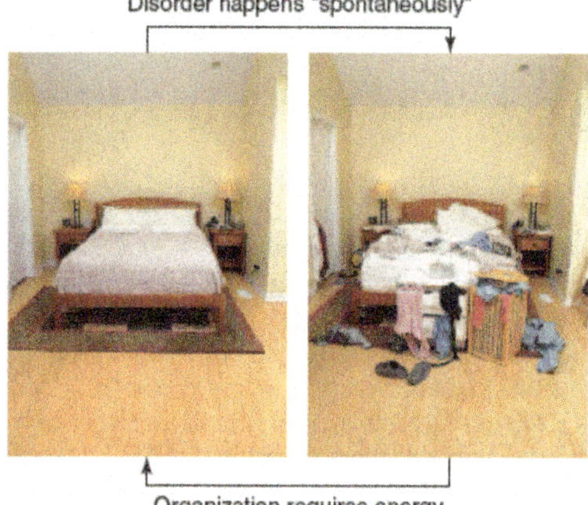

Figure 5.3 Entropy in action.
As time elapses, a teenager's room becomes more disorganized. It takes energy to clean it up.

The Second Law of Thermodynamics

The **second law of thermodynamics** concerns this transformation of potential energy into heat, or random molecular motion. It states that the disorder in a closed system like the universe is continuously increasing. Put simply, disorder is more likely than order. For example, it is much more likely that a column of bricks will tumble over than that a pile of bricks will arrange themselves spontaneously to form a column. In general, energy transformations proceed spontaneously to convert matter from a more ordered, less stable form, to a less ordered, more stable form. Without an input of energy from the teenager (or a parent), the ordered room in figure 5.3 falls into disorder.

Entropy

Entropy is a measure of the degree of disorder of a system, so the second law of thermodynamics can also be stated simply as "entropy increases." When the universe formed 10 to 20 billion years ago, it held all the potential energy it will ever have. It has become progressively more disordered ever since, with every energy exchange increasing the entropy of the universe.

Key Learning Outcome 5.2 The first law of thermodynamics states that energy cannot be created or destroyed; it can only undergo conversion from one form to another. The second law states that disorder (entropy) in the universe tends to increase. Life converts energy from the sun to other forms of energy that drive life processes; the energy is never lost, but as it is used, more and more of it is converted to heat.

Cell Chemistry

5.3 Chemical Reactions

In a chemical reaction, the original molecules before the chemical reaction occurs are called **reactants,** or sometimes **substrates,** whereas the molecules that result after the reaction has taken place are called the **products** of the reaction. Not all chemical reactions are equally likely to occur. Just as a boulder is more likely to roll downhill than uphill, so a reaction is more likely to occur if it releases energy than if it needs to have energy supplied. Consider how the chemical reaction proceeds in figure 5.4 ❶. Like when rolling a boulder uphill, energy needs to be supplied. This is because the product of the reaction contains more energy than the reactant. This type of chemical reaction, called **endergonic,** does not occur spontaneously. By contrast, an **exergonic** reaction, shown in ❷, tends to occur spontaneously because the product has less energy than the reactant, like a boulder that has rolled downhill.

Activation Energy

If all chemical reactions that release energy tend to occur spontaneously, it is fair to ask, "Why haven't all exergonic reactions occurred already?" Clearly they have not. If you ignite gasoline, it burns with a release of energy. So why doesn't all the gasoline in all the automobiles in the world just burn up right now? It doesn't because the burning of gasoline, and almost all other chemical reactions, requires an input of energy to get it started—a kick in the pants such as a match or spark plug. Even in exergonic reactions where the product contains or stores less energy than the reactants, it is first necessary to break existing chemical bonds in the reactants, and this takes energy. The extra energy required to destabilize existing chemical bonds and so initiate a chemical reaction is called **activation energy,** indicated by brackets in figure 5.4 ❷ and ❸. You must first nudge a boulder out of the hole it sits in before it can roll downhill. Activation energy is simply a chemical nudge.

Catalysis

One way to make an exergonic reaction more likely to happen is to lower the necessary activation energy. Like digging away the ground below your boulder, lowering activation energy reduces the nudge needed to get things started. The process of lowering the activation energy of a reaction is called **catalysis.** Catalysis cannot make an endergonic reaction occur spontaneously—you cannot avoid the need to supply energy—but it can make a reaction, endergonic or exergonic, proceed much faster. Compare the activation energy levels (the red arched arrows) in ❷ and ❸ below: The catalyzed reaction has a lower barrier to overcome.

Key Learning Outcome 5.3 Endergonic reactions require an input of energy. Exergonic reactions release energy. Activation energy that initiates chemical reactions can be lowered by catalysis.

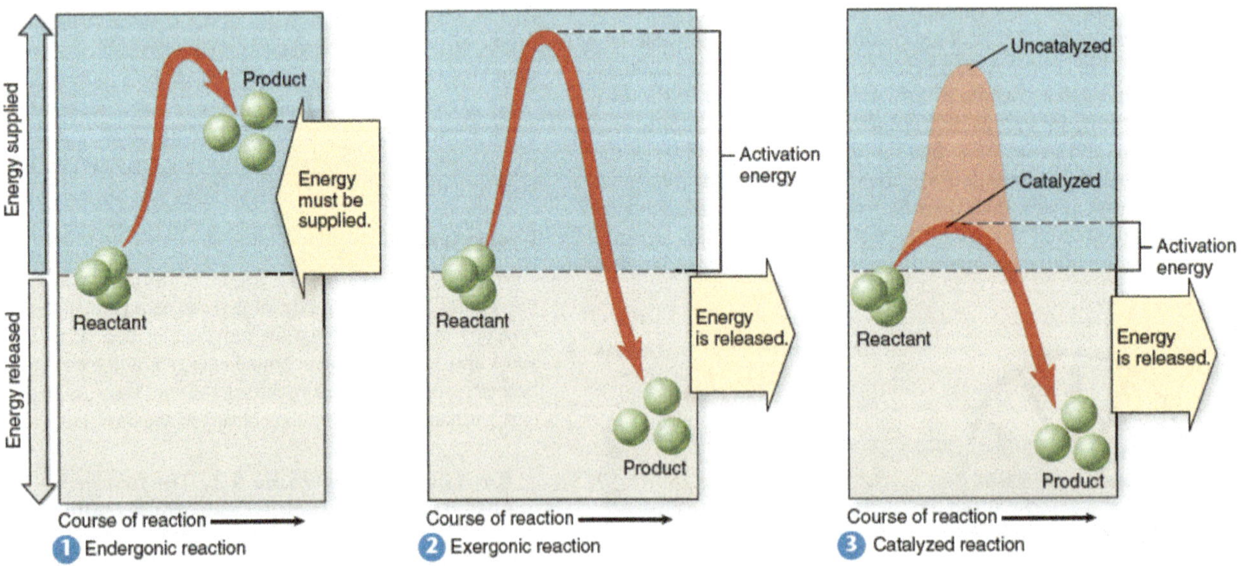

Figure 5.4 Chemical reactions and catalysis.
❶ The products of endergonic reactions contain more energy than the reactants. ❷ The products of exergonic reactions contain less energy than the reactants, but exergonic reactions do not necessarily proceed rapidly because it takes energy to get them going. The "hill" in this energy diagram represents energy that must be supplied to destabilize existing chemical bonds. ❸ Catalyzed reactions occur faster because the amount of activation energy required to initiate the reaction—the height of the energy hill that must be overcome—is lowered, and the reaction proceeds to its end faster.

5.4 How Enzymes Work

Enzymes, which can be made of proteins or nucleic acids, are the catalysts used by cells to touch off particular chemical reactions. By controlling which enzymes are present, and when they are active, cells are able to control what happens within themselves, just as a conductor controls the music an orchestra produces by dictating which instruments play when.

An enzyme works by binding to a specific molecule and stressing the bonds of that molecule in such a way as to make a particular reaction more likely. The key to this activity is the shape of the enzyme. An enzyme is specific for a particular reactant, or substrate, because the enzyme surface provides a mold that very closely fits the shape of the desired reactant. For example, the blue-colored lysozyme enzyme in figure 5.5 is contoured to fit a specific sugar molecule (the yellow reactant). Other molecules that fit less perfectly simply don't adhere to the enzyme's surface. The site on the enzyme surface where the reactant fits is called the **active site** (panel 1 below). The site on the reactant that binds to an enzyme is called the **binding site.** Enzymes are not rigid. The binding of the reactant induces the enzyme to change its shape slightly. In figure 5.5*b* and in panel 2 of the Key Biological Process illustration below, the edges of the enzyme now hug the reactant(s), leading to an "induced fit" between the enzyme and its reactant, like a hand wrapping around a baseball.

An enzyme lowers the activation energy of a particular reaction. In the case of lysozyme, an enzyme found in human tears, the enzyme has an antibacterial function, encouraging the breaking of a particular chemical bond in molecules that make up the cell wall of bacteria (figure 5.5). The enzyme weakens the bond by drawing away some of its electrons. Alternatively, an enzyme may encourage the formation of a link between two reactants, like the blue and red colored molecules in panel 2 below by holding them near each other. Regardless of the type of reaction, the enzyme is not affected by the chemical reaction and is available to be used again.

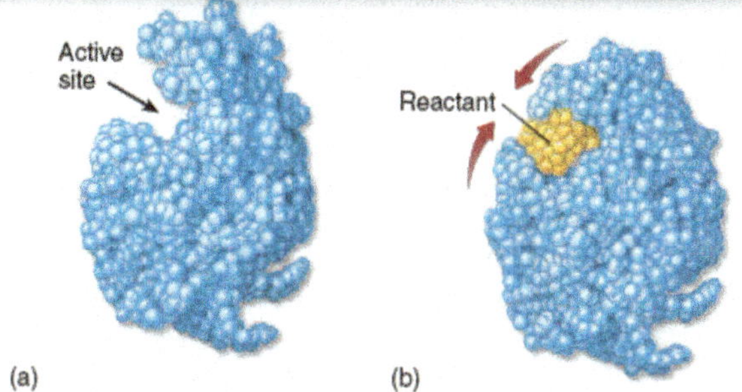

Figure 5.5 Enzyme shape determines its activity.
(*a*) A groove runs through the lysozyme enzyme (*blue* in this diagram) that fits the shape of the reactant (in this case, a chain of sugars). (*b*) When such a chain of sugars, indicated in *yellow,* slides into the groove, it induces the protein to change its shape slightly and embrace the substrate more intimately. This induced fit causes a chemical bond between two sugar molecules within the chain to break.

KEY BIOLOGICAL PROCESS: How Enzymes Work

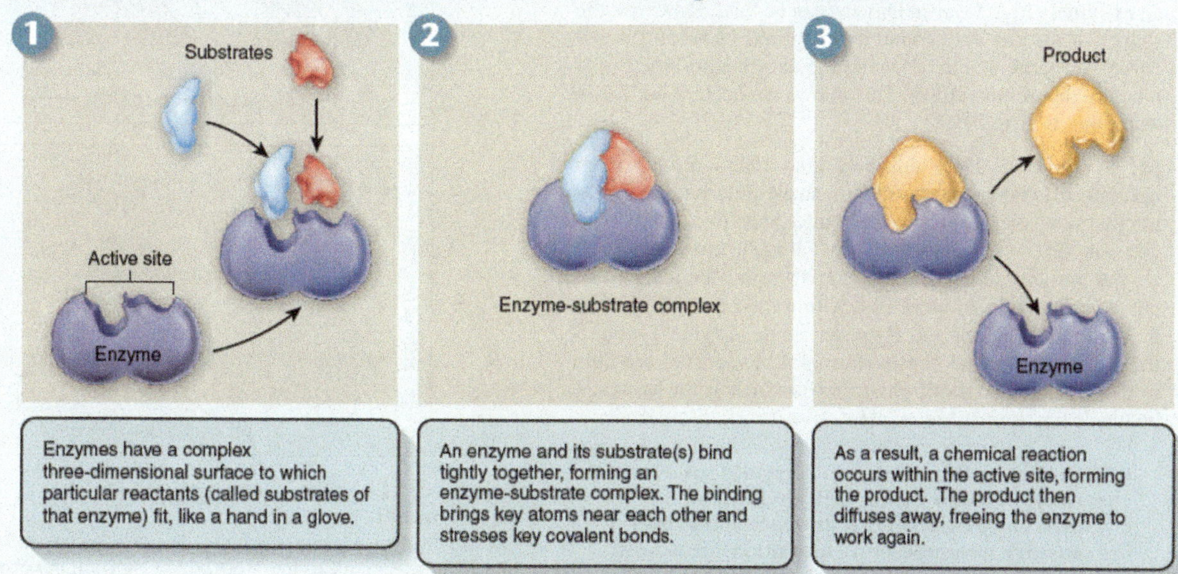

Biochemical Pathways

Every organism contains thousands of different kinds of enzymes that together catalyze a bewildering variety of reactions. Often several of these reactions occur in a fixed sequence called a **biochemical pathway,** the product of one reaction becoming the substrate for the next. You can see in the biochemical pathway shown in figure 5.6 how the initial substrate is altered by enzyme 1 so that it now fits into the active site of another enzyme, becoming the substrate for enzyme 2, and so on until the final product is produced. Because these reactions occur in sequence, the enzymes involved are often positioned near each other in the cell. For example, the enzymes involved in this biochemical pathway are all embedded in a membrane near each other. Many biochemical pathways occur in membranes, although enzyme assemblies also occur in organelles and within the cytoplasm. The close proximity of the enzymes allows the reactions of the biochemical pathway to proceed faster. Biochemical pathways are the organizational units of metabolism. We will discuss them more in chapters 6 and 7.

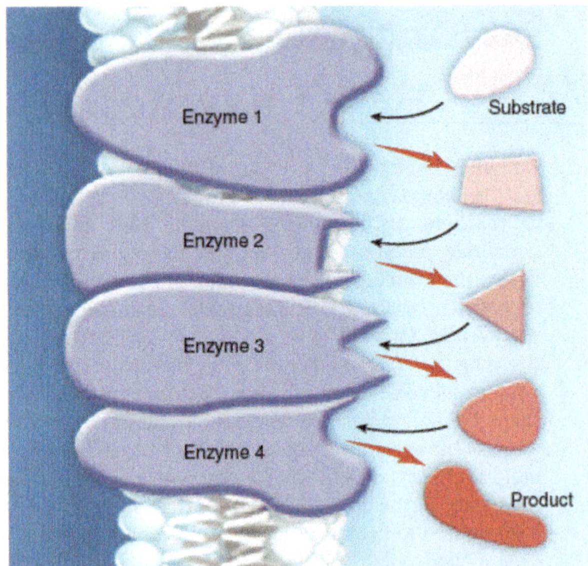

Figure 5.6 A biochemical pathway.
The original substrate is acted on by enzyme 1, changing the substrate to a new form recognized by enzyme 2. Each enzyme in the pathway acts on the product of the previous stage.

Factors Affecting Enzyme Activity

Temperature and pH can have a major influence on the action of enzymes. Enzyme activity is affected by any change in condition that alters the enzyme's three-dimensional shape.

Temperature When the temperature increases, the bonds that determine enzyme shape are too weak to hold the enzyme's peptide chains in the proper position, and the enzyme denatures. As a result, enzymes function best within an optimum temperature range, which is relatively narrow for most human enzymes. In the human body, enzymes work best at temperatures near the normal body temperature of 37°C, as shown by the brown curve in figure 5.7*a*. Also notice that the rates of enzyme reactions tend to drop quickly at higher temperatures, when the enzyme begins to unfold. This is why an extremely high fever in humans can be fatal. However, the shapes of the enzymes found in hotsprings bacteria (the red curve) are more stable, allowing the enzymes to function at much higher temperatures. This allows the bacteria to live in water that is near 70°C.

pH In addition, most enzymes also function within an optimal pH range, because the shape-determining polar interactions of enzymes are quite sensitive to hydrogen ion (H^+) concentration. Most human enzymes, such as the protein-degrading enzyme trypsin (the dark blue curve in figure 5.7*b*) work best within the range of pH 6 to 8. Blood has a pH of 7.4. However, some enzymes, such as the digestive enzyme pepsin (the light blue curve) are able to function in very acidic environments such as the stomach, but can't function at higher pHs.

Key Learning Outcome 5.4 Enzymes catalyze chemical reactions within cells and can be organized into biochemical pathways. Enzymes are sensitive to temperature and pH because both of these variables influence enzyme shape.

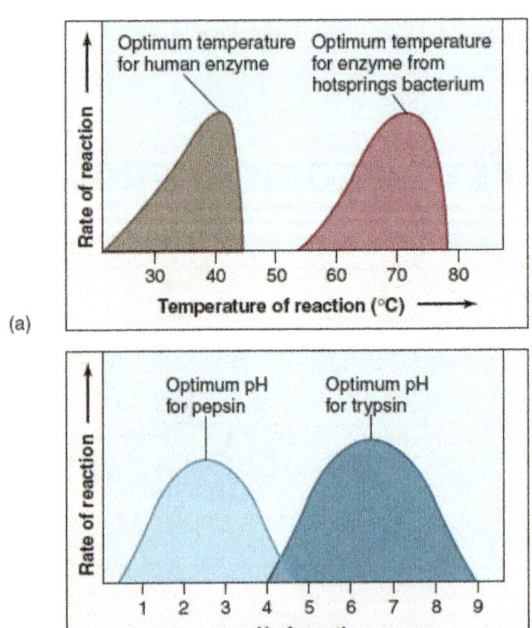

Figure 5.7 Enzymes are sensitive to their environment.
The activity of an enzyme is influenced by both (*a*) temperature and (*b*) pH. Most human enzymes work best at temperatures of about 40°C and within a pH range of 6 to 8.

KEY BIOLOGICAL PROCESS: Allosteric Enzyme Regulation

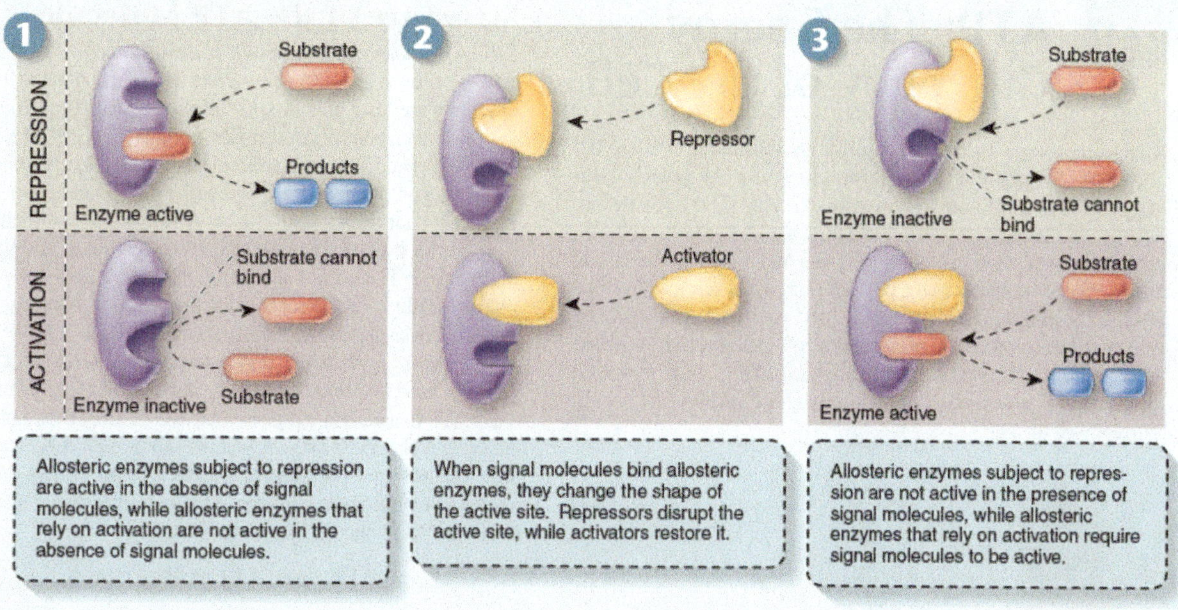

5.5 How Cells Regulate Enzymes

Because an enzyme must have a precise shape to work correctly, it is possible for the cell to control when an enzyme is active by altering its shape. Many enzymes have shapes that can be altered by the binding of "signal" molecules to their surfaces. Such enzymes are called *allosteric* (Latin, other shape). Enzymes can be inhibited or activated by the binding of signal molecules. For example, the upper tan panels in the Key Biological Process illustration above show an enzyme that is inhibited. The binding of a signal molecule, called a **repressor** (panel 2), alters the shape of the enzyme's active site such that it cannot bind the substrate. In other cases, the enzyme may not be able to bind the reactants *unless* the signal molecule is bound to the enzyme. The lower set of panels shows a signal molecule serving as an **activator.** The red substrate cannot bind to the enzyme's active site unless the activator (the yellow molecule) is in place, altering the shape of the active site. The site where the signal molecule binds to the enzyme surface is called the **allosteric site.**

Enzymes are often regulated by a mechanism called **feedback inhibition,** where the product of the reaction acts as the repressor. Feedback inhibition can occur in two ways: *competitive inhibitors* and *noncompetitive inhibitors*. The blue molecule in figure 5.8*a* functions as a competitive inhibitor, blocking the active site so that the substrate cannot bind. The yellow molecule in figure 5.8*b* functions as a noncompetitive inhibitor. It binds to an allosteric site, changing the shape of the enzyme such that it is unable to bind to the substrate.

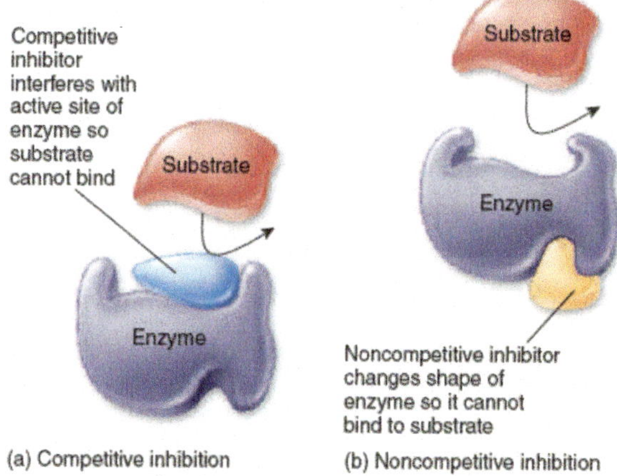

Figure 5.8 How enzymes can be inhibited.
(*a*) In competitive inhibition, the inhibitor interferes with the active site of the enzyme. (*b*) In noncompetitive inhibition, the inhibitor binds to the enzyme at a place away from the active site, effecting a conformational change in the enzyme so that it can no longer bind to its substrate. In feedback inhibition, the inhibitor molecule is the product of the reaction.

Many drugs and antibiotics work by inhibiting enzymes. Statin drugs like Lipitor lower cholesterol by inhibiting a key enzyme cells used to make cholesterol. The antibiotic penicillin inhibits an enzyme bacteria used in making cell walls. Because humans lack this enzyme, we are not harmed by the drug.

Key Learning Outcome 5.5 **An enzyme's activity can be affected by signal molecules that bind to it, changing its shape.**

How Cells Use Energy

5.6 ATP: The Energy Currency of the Cell

Cells use energy to do all those things that require work, but how does the cell use energy from the sun or the potential energy stored in molecules to power its activities? The sun's radiant energy and the energy stored in molecules are energy sources, but like money that is invested in stocks and bonds or real estate, these energy sources cannot be used directly to run a cell. To be useful, the energy from the sun or food molecules must first be converted to a source of energy that a cell can use, like someone converting stocks and bonds to ready cash. The "cash" molecule in the body is **adenosine triphosphate (ATP).** ATP is the energy currency of the cell.

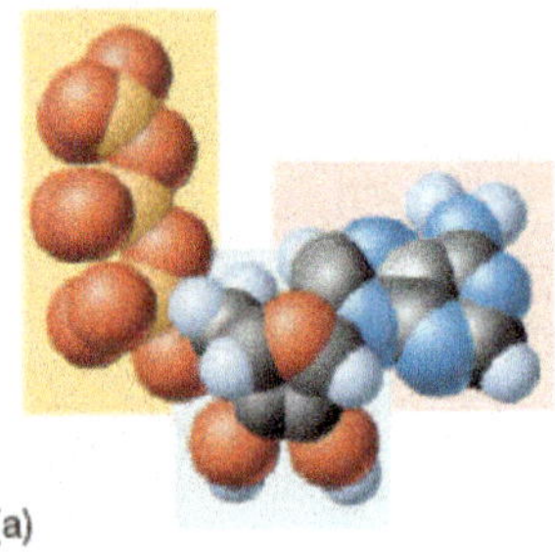

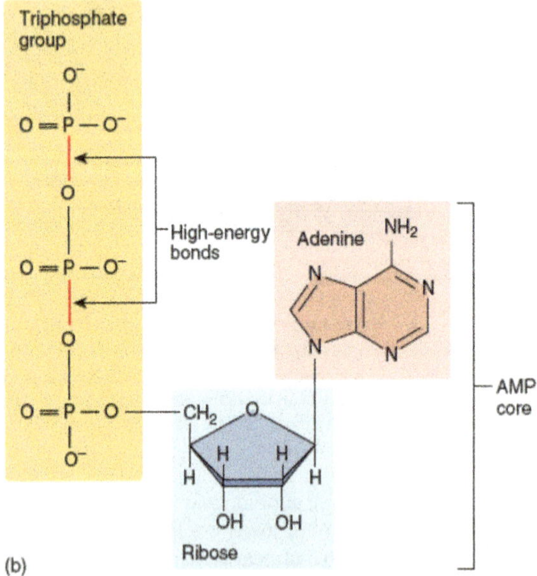

Figure 5.9 The parts of an ATP molecule.
The model (*a*) and structural diagram (*b*) both show that ATP consists of three phosphate groups attached to a ribose (five-carbon sugar) molecule. The ribose molecule is also attached to an adenine molecule (also one of the nitrogenous bases of DNA and RNA). When the endmost phosphate group is split off from the ATP molecule, considerable energy is released.

Structure of the ATP Molecule

Each ATP molecule is composed of the three parts shown in figure 5.9: (1) a sugar (colored blue) serves as the backbone to which the other two parts are attached, (2) adenine (colored peach) is one of the four nitrogenous bases in DNA and RNA, and (3) a chain of three phosphates (colored yellow) contain high-energy bonds.

As you can see in the figure, the phosphates carry negative electrical charges, and so it takes considerable chemical energy to hold the line of three phosphates next to one another at the end of ATP. Like a coiled spring, the phosphates are poised to push apart. It is for this reason that the chemical bonds linking the phosphates are such chemically reactive bonds.

When the endmost phosphate is broken off an ATP molecule, a sizable packet of energy is released. The reaction converts ATP to adenosine diphosphate, ADP. The second phosphate group can also be removed, yielding additional energy and leaving adenosine monophosphate (AMP). Most energy exchanges in cells involve cleavage of only the outermost bond, converting ATP into ADP and P_i, inorganic phosphate:

$$\text{ATP} \longleftrightarrow \text{ADP} + \text{P}_i + \text{energy}$$

Exergonic reactions require activation energy, and endergonic reactions require the input of even more energy, and so these reactions in the cell are usually coupled with the breaking of the phosphate bond in ATP, called *coupled reactions*. Because almost all chemical reactions in cells require less energy than is released by this reaction, ATP is able to power many of the cell's activities, producing heat as a by-product. Table 5.1 introduces you to some of the key cellular activities powered by the breakdown of ATP. ATP is continually recycled from ADP and P_i through the ATP-ADP cycle.

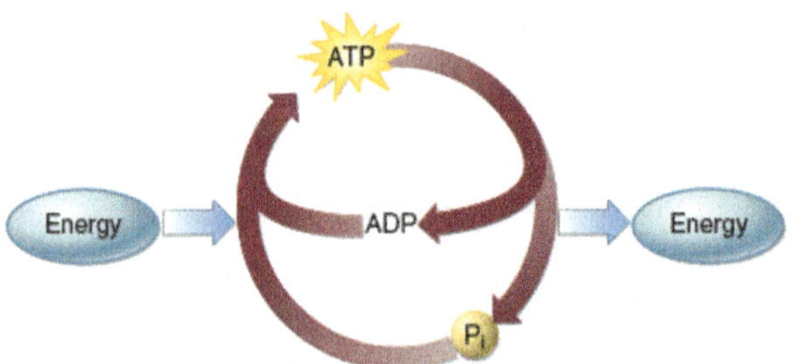

Cells use two different but complementary processes to convert energy from the sun and potential energy found in food molecules into ATP. Some cells convert energy from the sun into molecules of ATP through the process of **photosynthesis,** the subject of chapter 6. This ATP is then used to manufacture sugar molecules, converting the energy from ATP into potential energy stored in the bonds that hold the atoms together. All cells convert the potential energy found in food molecules into ATP through **cellular respiration,** the subject of chapter 7.

Key Learning Outcome 5.6 Cells use the energy in ATP molecules to drive chemical reactions.

TABLE 5.1 HOW CELLS USE ATP ENERGY TO POWER CELLULAR WORK

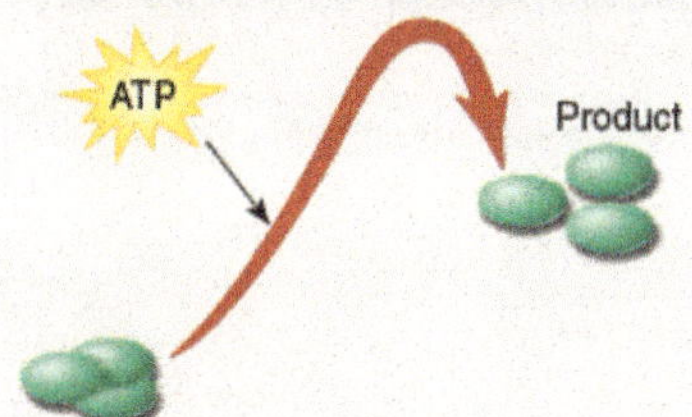

Biosynthesis

Cells use the energy released from the exergonic hydrolysis of ATP to drive endergonic reactions like those of protein synthesis, an approach called energy coupling.

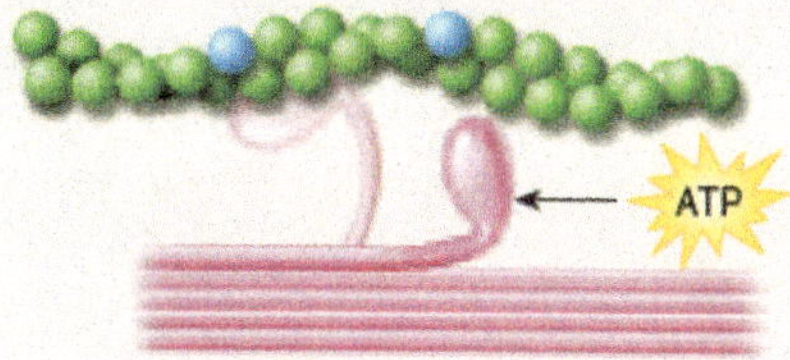

Contraction

In muscle cells, filaments of protein repeatedly slide past each other to achieve contraction of the cell. An input of ATP is required for the filaments to reset and slide again.

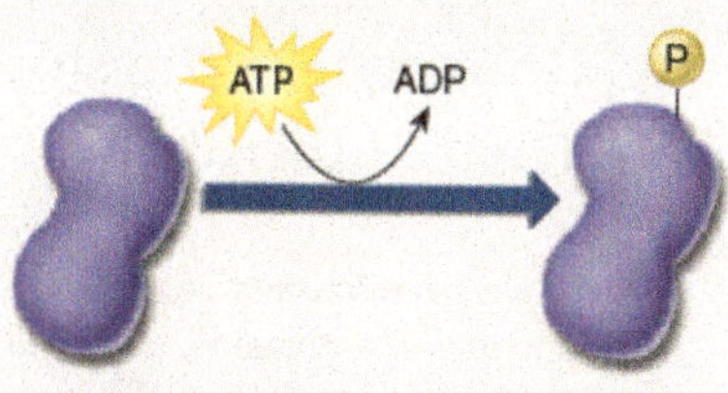

Chemical Activation

Proteins can become activated when a high-energy phosphate from ATP attaches to the protein, activating it. Other types of molecules can also become phosphorylated by transfer of a phosphate from ATP.

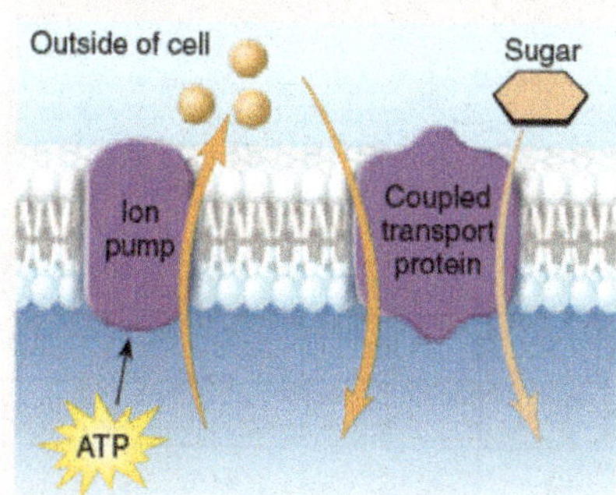

Importing Metabolites

Metabolite molecules such as amino acids and sugars can be transported into cells against their concentration gradients by coupling the intake of the metabolite to the inward movement of an ion moving down its concentration gradient, this ion gradient being established using ATP.

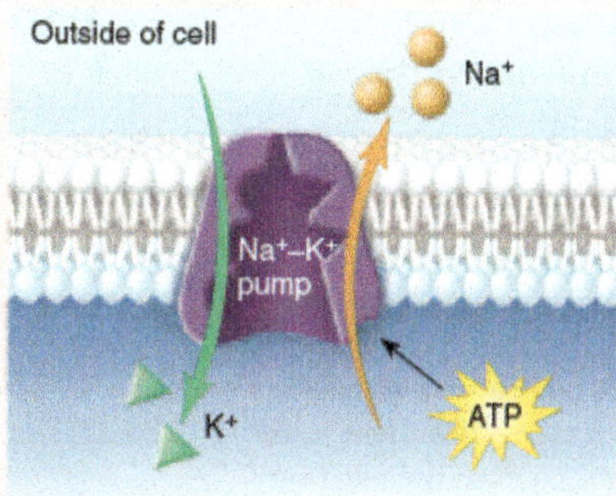

Active Transport: Na^+–K^+ Pump

Most animal cells maintain a low internal concentration of Na^+ relative to their surroundings, and a high internal concentration of K^+. This is achieved using a protein called the sodium-potassium pump, which actively pumps Na^+ out of the cell and K^+ in, using energy from ATP.

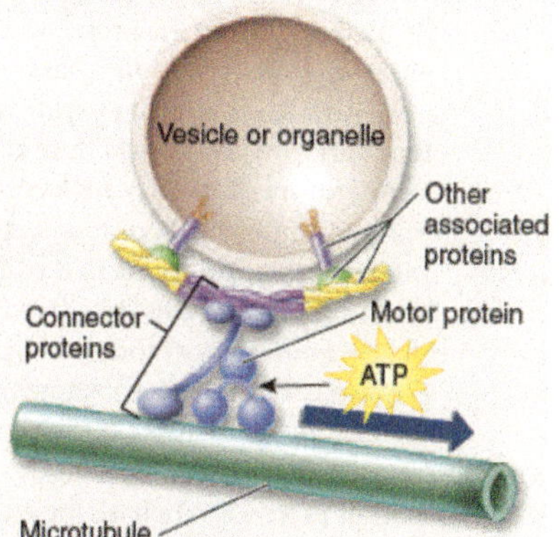

Cytoplasmic Transport

Within a cell's cytoplasm, vesicles or organelles can be dragged along microtubular tracks using molecular motor proteins, which are attached to the vesicle or organelle with connector proteins. The motor proteins use ATP to power their movement.

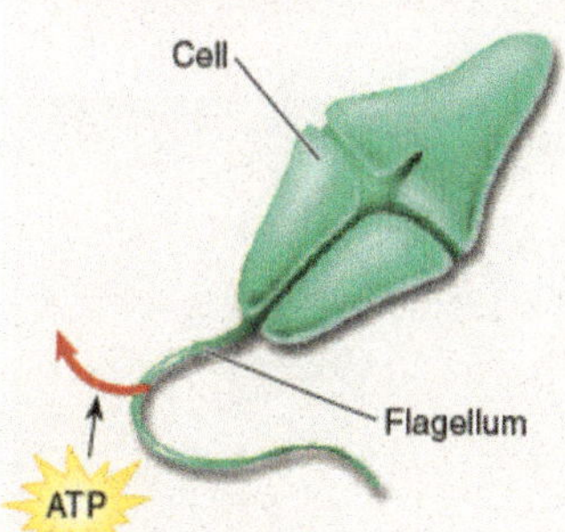

Flagellar Movements

Microtubules within flagella slide past each other to produce flagellar movements. ATP powers the sliding of the microtubules.

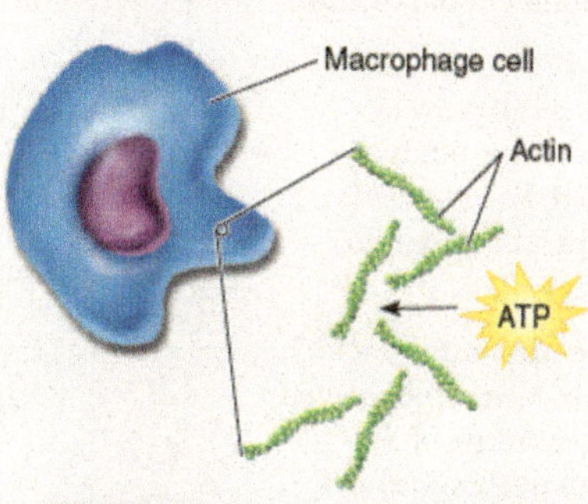

Cell Crawling

Actin filaments in a cell's cytoskeleton continually assemble and disassemble to achieve changes in cell shape and to allow cells to crawl over substrates or engulf materials. The dynamic character of actin is controlled by ATP molecules bound to actin filaments.

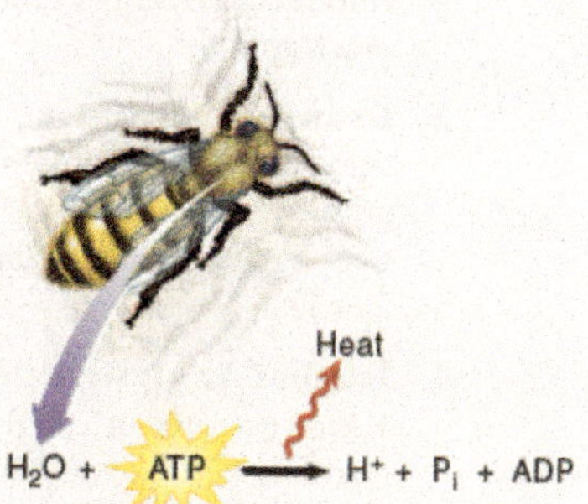

Heat Production

The hydrolysis of the ATP molecule releases heat. Reactions that hydrolyze ATP often take place in mitochondria or in contracting muscle cells and may be coupled to other reactions. The heat generated by these reactions can be used to maintain an organism's temperature.

Do Enzymes Physically Attach to Their Substrates?

When scientists first began to examine the chemical activities of organisms, no one knew that biochemical reactions were catalyzed by enzymes. The first enzyme was discovered in 1833 by French chemist Anselme Payen. He was studying how beer is made from barley: First barley is pressed and gently heated so its starches break down into simple 2-sugar units; then yeasts convert these units into ethanol. Payen found that the initial breakdown requires a chemical factor that is not alive, and that does not seem to be used up during the process—a catalyst. He called this first enzyme *diastase* (we call it *amylase* today).

Did this catalyst operate at a distance, increasing reaction rate all around it, much as raising the temperature of nearby molecules might do? Or did it operate in physical contact, actually attaching to the molecules whose reaction it catalyzed (its "substrate")?

The answer was discovered in 1903 by French chemist Victor Henri. He saw that the hypothesis that an enzyme physically binds to its substrate makes a clear and testable prediction: In a solution of substrate and enzyme, there must be a maximum reaction rate, faster than which the reaction cannot proceed. When all the enzyme molecules are working full tilt, the reaction simply cannot go any faster, no matter how much more substrate you add to the solution. To test this prediction, Henri carried out the experiment whose results you see in the graph, measuring the reaction rate (V) of diastase at different substrate concentrations (S).

1. **Making Inferences** As S increases, does V increase? If so, in what manner—steadily, or by smaller and smaller amounts? Is there a maximum reaction rate?
2. **Drawing Conclusions** Does this result provide support for the hypothesis that an enzyme binds physically to its substrate? Explain. If the hypothesis were incorrect, what would you expect the graph to look like?
3. **Further Analysis** If the smaller amounts by which V increases are strictly the result of fewer unoccupied enzymes being available at higher values of S, then the curve in Henri's experiment should show a pure exponential decline in V—mathematically, meaning a reciprocal plot ($1/V$ versus $1/S$) should be a straight line. If some other factor is also at work that reacts differently to substrate concentration, then the reciprocal plot would curve upward or downward. Fill in the reciprocal values in the table to the right, and then plot the values on the lower graph ($1/S$ on the x axis and $1/V$ on the y axis). Is a reciprocal plot of Henri's data a straight line?

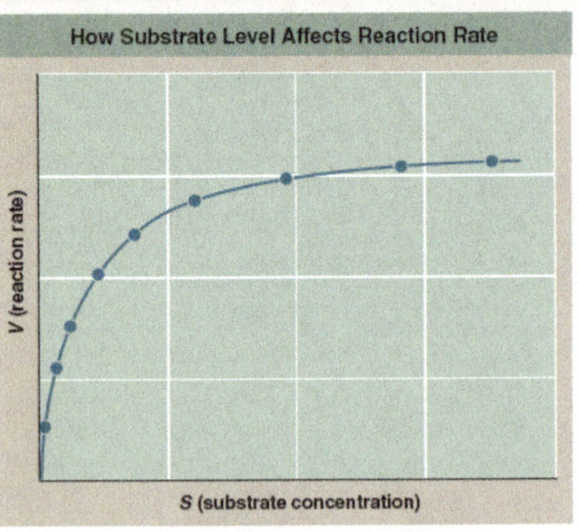

Trial	S	1/S	V	1/V
1	5	0.200	7.7	0.130
2	10		15.4	
3	25		23.1	
4	50		30.8	
5	75		38.5	
6	125		40.7	
7	200		46.2	
8	275		47.7	
9	350		48.5	

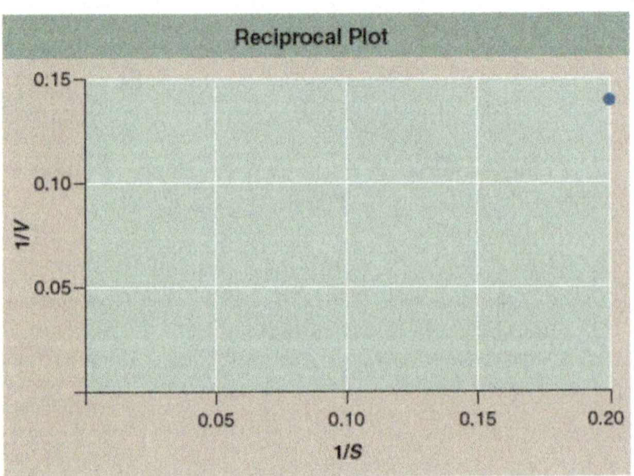

Chapter Review

Cells and Energy

5.1 The Flow of Energy in Living Things

- Energy is the ability to do work. Energy exists in two states: kinetic energy and potential energy.
- Kinetic energy is the energy of motion. Potential energy is stored energy, which exists in objects that aren't in motion but have the capacity to move, like a ball poised at the top of a hill shown here from **figure 5.1.** All of the work carried out by living things involves the transformation of potential energy into kinetic energy.
- Energy flows from the sun to the earth, where it is trapped by photosynthetic organisms and stored in carbohydrates as potential energy. This energy is transferred during chemical reactions.

5.2 The Laws of Thermodynamics

- The laws of thermodynamics describe changes in energy in our universe. The first law of thermodynamics explains that energy can never be created or destroyed, only changed from one state to another. The total amount of energy in the universe remains constant.
- Energy exists in different forms in the universe, such as light, electrical, or heat energy. This energy, such as heat energy, can be harnessed to do work (**figure 5.2**).
- The second law of thermodynamics explains that the conversion of potential energy into random molecular motion is constantly increasing. This conversion of energy progresses from an ordered but less stable form to a disordered but stable form. Entropy, which is a measure of disorder in a system, is constantly increasing such that disorder is more likely than order. Energy must be used to maintain order (**figure 5.3**).

Cell Chemistry

5.3 Chemical Reactions

- Chemical reactions involve the breaking or formation of covalent bonds. The starting molecules are called the reactants, and the molecules produced by the reaction are called the products. Chemical reactions in which the products contain more potential energy than the reactants are called endergonic reactions. Chemical reactions that release energy are called exergonic reactions, as shown here from **figure 5.4**, and are more likely to occur.
- All chemical reactions require an input of energy. The energy required to start a reaction is called activation energy. A chemical reaction proceeds faster when its activation energy is lowered, and this occurs through a process called catalysis.

Enzymes

5.4 How Enzymes Work

- Enzymes are macromolecules that lower the activation energy of chemical reactions in the cell. Enzymes are catalysts.
- An enzyme, like the lysozyme shown here from **figure 5.5**, binds the reactant, or substrate. The substrate binds to the enzyme's active site. The enzyme forms around the reactant and acts to stress covalent bonds or bring atoms into closer proximity. The actions of the enzyme increase the likelihood that chemical bonds will break or form. An enzyme lowers the activation energy of the reaction. The enzyme is not affected by the reaction and can be used over and over again (**Key Biological Process, page 109**).
- Sometimes enzymes work in a series of reactions called a biochemical pathway. The product of one reaction becomes the substrate for the next reaction. The enzymes that are involved are usually located near each other in the cell (**figure 5.6**).
- In the cell, chemical reactions are regulated by controlling which enzymes are active. Other factors, such as temperature and pH, can also affect enzyme function, and so most enzymes have an optimal temperature and pH range (**figure 5.7**).
- Higher temperatures can disrupt the bonds that hold the enzyme in its proper shape, decreasing its ability to catalyze a chemical reaction. The bonds that hold the enzyme's shape are also affected by hydrogen ion concentrations, and so increasing or decreasing the pH can disrupt the enzyme's function.

5.5 How Cells Regulate Enzymes

- An enzyme can be inhibited or activated in the cell as a means of regulation by temporarily altering the enzyme's shape (**Key Biological Process, page 111**). An enzyme can be inhibited when a molecule, called a repressor, binds to the enzyme, altering the shape of the active site so that it cannot bind the substrate. Some enzymes need to be activated, or turned on, in order to bind to their substrate. A molecule called an activator binds to the enzyme, changing the shape of the active site so that it is able to bind the substrate. Enzymes that are controlled in this way are allosteric enzymes.
- A repressor molecule can bind to the active site of the enzyme, blocking it. This is called a competitive inhibition. In noncompetitive inhibition, the repressor binds to a different site on the enzyme, altering the shape of the active site so it cannot bind its substrate (**figure 5.8**). Enzymes are often regulated by feedback inhibition, a process where the product of the reaction functions as a repressor, shutting down its own synthesis.

How Cells Use Energy

5.6 ATP: The Energy Currency of the Cell

- Cells require energy to do work in the form of ATP (**table 5.1**). ATP contains a sugar, an adenine, and a chain of three phosphates, as shown here from **figure 5.9**. The three phosphates are held together with high-energy bonds. When the endmost phosphate bond breaks, considerable energy is released. A cell uses this energy to drive reactions in the cell by coupling the breakdown of ATP with other chemical reactions in the cell.

Test Your Understanding

1. The ability to do work is the definition for
   a. thermodynamics. c. energy.
   b. radiation. d. entropy.
2. The first law of thermodynamics
   a. says that energy recycles constantly, as organisms use and reuse it.
   b. says that entropy, or disorder, continually increases in a closed system.
   c. is a formula for measuring entropy.
   d. says that energy can change forms, but cannot be made or destroyed.
3. The second law of thermodynamics
   a. says that energy recycles constantly, as organisms use and reuse it.
   b. says that entropy, or disorder, continually increases in a closed system.
   c. is a formula for measuring entropy.
   d. says that energy can change forms, but cannot be made or destroyed.
4. Chemical reactions that occur spontaneously are called
   a. exergonic and release energy.
   b. exergonic and their products contain more energy.
   c. endergonic and release energy.
   d. endergonic and their products contain more energy.
5. The catalysts that help an organism carry out needed chemical reactions are called
   a. hormones. c. reactants.
   b. enzymes. d. substrates.
6. Factors that affect the activity of an enzyme molecule include
   a. peptides and energy.
   b. thermodynamics.
   c. temperature and pH.
   d. entropy.
7. In order for an enzyme to work properly
   a. it must have a particular shape.
   b. the temperature must be within certain limits.
   c. the pH must be within certain limits.
   d. All of the above.
8. In competitive inhibition
   a. an enzyme molecule has to compete with other enzyme molecules for the necessary substrate.
   b. an enzyme molecule has to compete with other enzyme molecules for the necessary energy.
   c. an inhibitor molecule competes with the substrate for the active site on the enzyme.
   d. two different products compete for the same binding site on the enzyme.
9. Which of the following is not a component of ATP?
   a. active site c. adenine
   b. ribose d. phosphate groups
10. Endergonic reactions can occur in the cell because they are coupled with
    a. the breaking of phosphate bonds in ATP.
    b. uncatalyzed reactions.
    c. activators.
    d. all of the above.

Apply Your Understanding

1. **Figure 5.7** Examine the graphs shown here. Describe what happens to a human enzyme when the body's temperature is raised to 50°C. Looking at part (b), what happens to trypsin's ability to function as the surrounding concentration of H^+ ions increases? How do you think the enzyme pepsin would respond to a change in pH from 4 to 3?

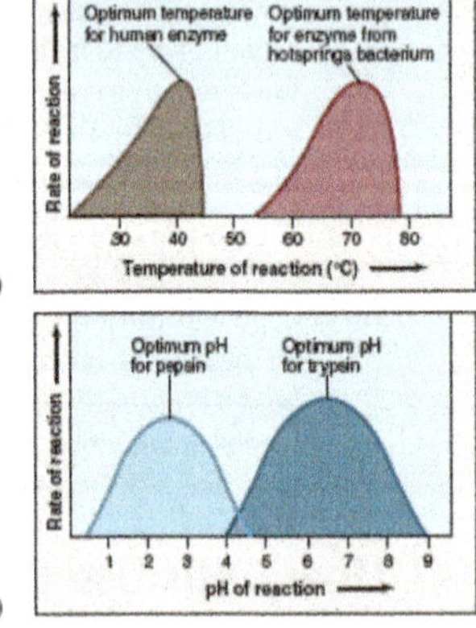

2. **Table 5.1** ATP forms primarily from the breakdown of glucose. If your blood glucose level were to drop precipitously, what sorts of problems would you expect this to cause in your body?

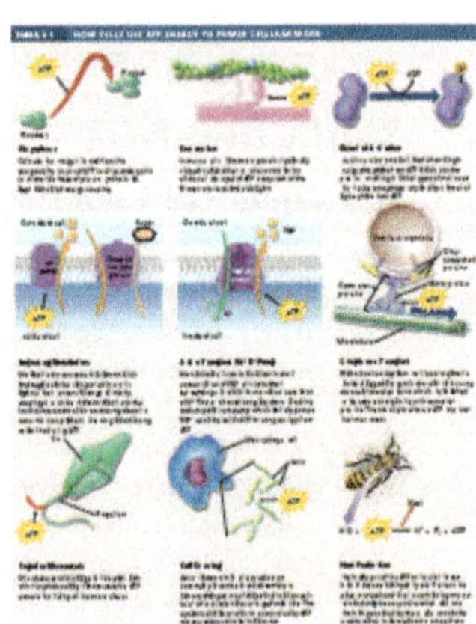

Synthesize What You Have Learned

1. Photosynthetic organisms, such as plants, algae, and bacteria, capture the sun's energy and transform it into sugar molecules that other organisms like us can harvest for energy. Explain where the sun's energy is stored in these sugar molecules, and how it is harvested by animals that eat the sugar.
2. A restriction endonuclease is an enzyme that cuts DNA at a specific, unique sequence, like GAATTC. How does a particular restriction enzyme "know" when it has found its target sequence?
3. When a baseball thrown by a pitcher encounters the swinging bat of a slugger, what happens to the ball's kinetic energy? What happens to the bat's kinetic energy?

6

Photosynthesis: Acquiring Energy from the Sun

Learning Objectives

Photosynthesis

6.1 An Overview of Photosynthesis

1. Define photosynthesis and name the three basic kinds of organisms that carry out photosynthesis.
2. Explain why no photosynthesis occurs within an oak tree's stem.
3. Name the three layers of a leaf through which light must pass to reach chloroplasts.
4. Explain why the plasma membrane of mesophyll cells does not absorb light, while chloroplasts within these cells do.
5. Diagram the structure of a chloroplast, and state where within it chlorophyll is found.
6. Describe how a photosystem captures photons.
7. Contrast the light-dependent and light-independent reactions of photosynthesis.

6.2 How Plants Capture Energy from Sunlight

1. Describe what a photon is made of, and state in what way a photon's energy is related to its wavelength.
2. Identify what color(s) of light are *not* absorbed by the pigment chlorophyll.
3. Explain why leaves change color in autumn.

6.3 Organizing Pigments into Photosystems

1. List and describe the five stages of the light-dependent reactions.
2. Differentiate reaction center chlorophyll molecules from other photosystem chlorophyll molecules.

6.4 How Photosystems Convert Light to Chemical Energy

1. Describe the function of the electron transport system.
2. Define chemiosmosis, and state the function of ATP synthase.
3. Differentiate between photosystems I and II.

6.5 Building New Molecules

1. Describe the function of the Calvin cycle, and explain why it requires NADPH as well as ATP.

Photorespiration

6.6 Photorespiration: Putting the Brakes on Photosynthesis

1. Contrast C_3 and C_4 photosynthesis.
2. Differentiate between C_4 photosynthesis and CAM photosynthesis.

Today's Biology: Cold-Tolerant C_4 Photosynthesis

Inquiry & Analysis: Does Iron Limit the Growth of Ocean Phytoplankton?

In this forest glade, you can literally see the pulse of life flowing through the organisms of an ecosystem. Sunlight beams down, a stream of energy in the form of packets of light called photons. Everywhere the light falls, there are plants—trees and shrubs and flowers and grasses, all with green leaves intercepting the energy as it rains down. In the cells of each leaf are organelles called chloroplasts that contain light-gathering pigments in their membranes. These pigments, notably the pigment chlorophyll, which makes leaves green, absorb photons of light and use the energy to strip electrons from water molecules. The chloroplasts use these electrons to reduce CO_2—that is, to add hydrogens (a hydrogen atom, you will recall, is just a proton with an associated electron)—and so make organic molecules. This process of capturing the sun's energy to build molecules is called photosynthesis—literally, using "light to build." In this chapter, we will delve into photosynthesis, tracing how light energy is captured, converted to chemical energy, and put to work assembling organic molecules. In the roots and other tissues of the plants, the opposite process is taking place. Organic molecules are being broken down in the process of cellular respiration to provide energy to power growth and cellular activities. These reactions, which take place largely in another kind of organelle called a mitochondrion, are the subject of the following chapter. Together, chloroplasts and mitochondria carry out a flow of energy driven by the power of sunlight.

6.1 An Overview of Photosynthesis

Life is powered by sunshine. All of the energy used by almost all living cells comes ultimately from the sun, captured by plants, algae, and some bacteria through the process of **photosynthesis.** Every oxygen atom in the air we breathe was once part of a water molecule, liberated by photosynthesis as you will discover in this chapter. Life as we know it is only possible because our earth is awash in energy streaming inward from the sun. Each day, the radiant energy that reaches the earth is equal to that of about 1 million Hiroshima-sized atomic bombs. About 1% of it is captured by photosynthesis and provides the energy that drives almost all life on earth. Use the arrows on this page and the next three pages to follow the path of energy from the sun through photosynthesis.

Trees Many kinds of organisms carry out photosynthesis, not only the many kinds of plants that make our world green, but also bacteria and algae. Photosynthesis is somewhat different in bacteria, but we will focus our attention on photosynthesis in plants, starting with this maple tree crowned with green leaves. Later we will look at the grass growing beneath the maple tree—it turns out that grasses and other related plants sometimes take a different approach to photosynthesis depending on the conditions.

Leaves To learn how this maple tree captures energy from sunlight, follow the light. It comes beaming in from the sun, down through earth's atmosphere, bathing the top of the tree in light. What part of the maple tree is actually being struck by this light? The green leaves are. Each branch at the top of the tree ends in a spread of these leaves, each leaf flat and thin like the page of a book. Within these green leaves is where photosynthesis occurs. No photosynthesis occurs within this tree's stem, covered with bark, and none in the roots, covered with soil—no light reaches these parts of the plant. The tree has a very efficient internal plumbing system that transports the products of photosynthesis to the stem, roots, and other parts of the plant so that they too may benefit from the capture of the sun's energy.

The Leaf Surface Now follow the light as it passes into a leaf. The beam of light first encounters a waxy protective layer called the cuticle. The cuticle acts a bit like a layer of clear fingernail polish, providing a thin, watertight and surprisingly strong layer of protection. Light passes right through this transparent wax, and then proceeds to pass right on through a layer of cells immediately beneath the cuticle called the epidermis. Only one cell thick, this epidermis acts as the "skin" of the leaf, providing more protection from damage and, very importantly, controlling how gases and water enter and leave the leaf. Very little of the light has been absorbed by the leaf at this point—neither the cuticle nor the epidermis absorb much.

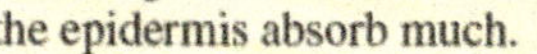

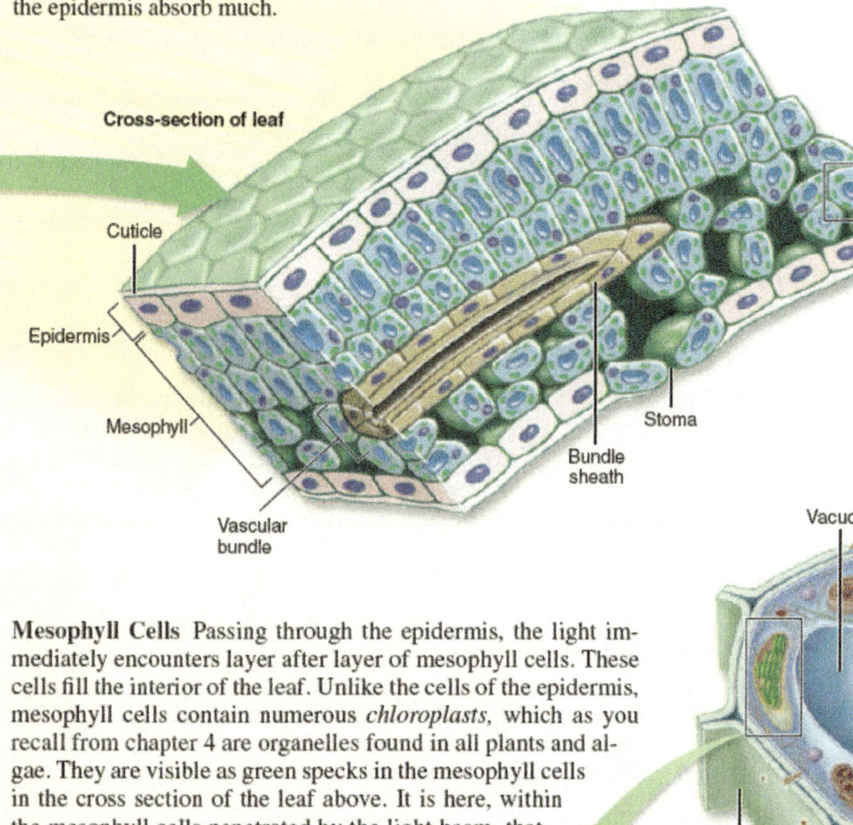

Mesophyll Cells Passing through the epidermis, the light immediately encounters layer after layer of mesophyll cells. These cells fill the interior of the leaf. Unlike the cells of the epidermis, mesophyll cells contain numerous *chloroplasts,* which as you recall from chapter 4 are organelles found in all plants and algae. They are visible as green specks in the mesophyll cells in the cross section of the leaf above. It is here, within the mesophyll cells penetrated by the light beam, that photosynthesis occurs.

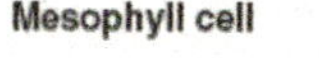

Mesophyll cell

Thylakoid
Inner membrane
Outer membrane
Granum
Stroma

Chloroplast

Chloroplasts Light penetrates into mesophyll cells. The cell walls of the mesophyll cells don't absorb it, nor does the plasma membrane or nucleus or mitochondria. Why not? Because these elements of the mesophyll cell contain few if any molecules that absorb visible light. If chloroplasts were not also present in these cells, most of this light would pass right through, just as it passed through the epidermis. But chloroplasts are present, lots of them. One chloroplast is highlighted by a box in the mesophyll cell above. Light passes into the cell to the chloroplast, and when it reaches the chloroplast, it passes through the outer and inner membranes to reach the thylakoid structures within the chloroplast, clearly seen as the green disks in the cutaway chloroplast shown here.

Inside the Chloroplast

All the important events of photosynthesis happen inside the chloroplast. The journey of light into the chloroplasts ends when the light beam encounters a series of internal membranes within the chloroplast organized into flattened sacs called *thylakoids.* Often, numerous thylakoids are stacked on top of one another in columns called *grana*. In the drawing below, the grana look not unlike piles of dishes. While each thylakoid is a separate compartment that functions more-or-less independently, the membranes of the individual thylakoids are all connected, part of a single continuous membrane system. Occupying much of the interior of the chloroplast, this thylakoid membrane system is submerged within a semiliquid substance called *stroma,* which fills the interior of the chloroplast in much the same way that cytoplasm fills the interior of a cell. Suspended within the stroma are many enzymes and other proteins, including the enzymes that act later in photosynthesis to assemble organic molecules from carbon dioxide (CO_2) in reactions that do not require light and which are discussed later.

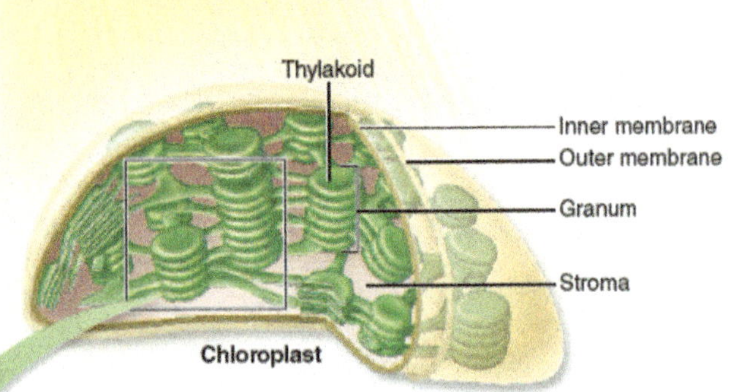

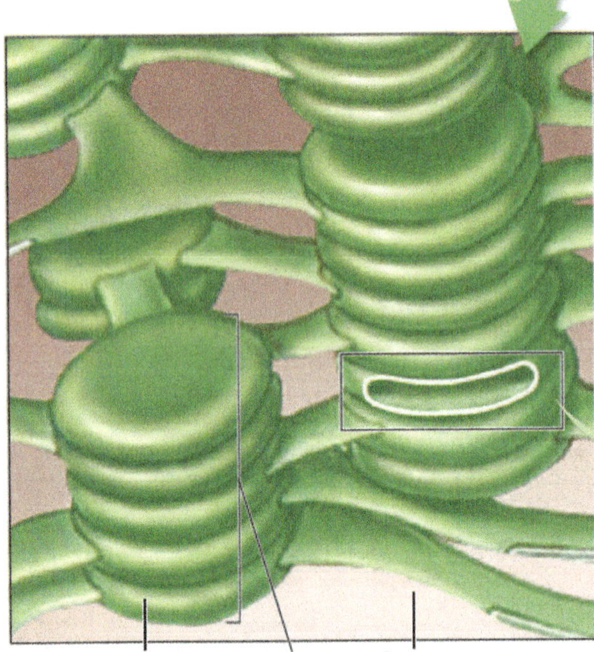

Penetrating the Thylakoid Surface The first key event of photosynthesis occurs when a beam of sunlight strikes the surface membrane of a thylakoid. Embedded within this membrane, like icebergs on an ocean, are clusters of light-absorbing pigments. A pigment molecule is a molecule that absorbs light energy. The primary pigment molecule in most photosystems is **chlorophyll,** an organic molecule that absorbs red and blue light, but does not absorb green wavelengths. The green light is instead reflected, giving the thylakoid and the chloroplast that contains it an intense green color. Plants are green because they are rich in green chloroplasts. Except for some alternative pigments also present in thylakoids, which we will discuss later, no other parts of the plant absorb visible light with such intensity.

Striking the Photosystem Within each pigment cluster, the chlorophyll molecules are arranged in a network called a *photosystem.* The light-absorbing chlorophyll molecules of a photosystem act together as an antenna to capture photons (units of light energy). A lattice of structural proteins, indicated by the purple element inserted into the thylakoid membrane in the diagram on the facing page, anchors each of the chlorophyll molecules of a photosystem into a precise position, such that every chlorophyll molecule is touching several others. Wherever a photon of light strikes the photosystem, some chlorophyll molecule will be in position to receive it.

Energy Absorption When a photon of sunlight strikes any chlorophyll molecule in the photosystem, the chlorophyll molecule it hits absorbs that photon's energy. The energy becomes part of the chlorophyll molecule, boosting some of its electrons to higher energy levels. Possessing these more energetic electrons, the chlorophyll molecule is said to now be "excited." With this key event, the biological world has captured energy from the sun.

Excitation of the Photosystem The excitation that the absorption of light creates is then passed from the chlorophyll molecule that was hit to another, and then to another, like a hot potato being passed down a line of people. This shuttling of excitation is not a chemical reaction, in which an electron physically passes between atoms. Rather, it is energy that passes from one chlorophyll molecule to its neighbor. A crude analogy to this form of energy transfer is the initial "break" in a game of pool. If the cue ball squarely hits the point of the triangular array of 15 billiard balls, the two balls at the far corners of the triangle fly off, and none of the central balls move at all. The kinetic energy is transferred through the central balls to the most distant ones. In much the same way, the photon's excitation energy moves through the photosystem from one chlorophyll to the next.

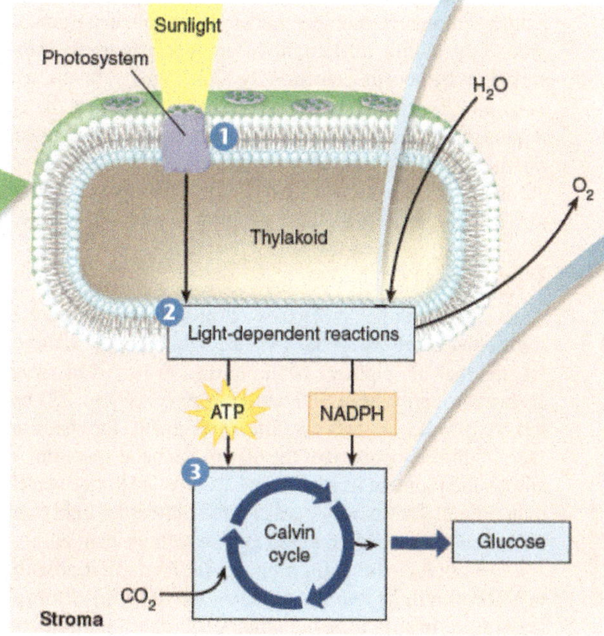

Energy Capture As the energy shuttles from one chlorophyll molecule to another within the photosystem network, it eventually arrives at a key chlorophyll molecule, the only one that is touching a membrane-bound protein. Like shaking a marble in a box with a walnut-sized hole in it, the excitation energy will find its way to this special chlorophyll just as sure as the marble will eventually find its way to and through the hole in the box. The special chlorophyll then transfers an excited (high-energy) electron to the acceptor molecule it is touching.

The Light-Dependent Reactions Like a baton being passed from one runner to another in a relay race, the electron is then passed from that acceptor protein to a series of other proteins in the membrane that put the energy of the electron to work making ATP and NADPH. In a way you will explore later in this chapter, the energy is used to power the movement of protons across the thylakoid membrane to make ATP and another key molecule, NADPH. So far, photosynthesis has consisted of two stages, indicated by numbers in the diagram to the lower left: ❶ capturing energy from sunlight—accomplished by the photosystem; and ❷ using the energy to make ATP and NADPH. These first two stages of photosynthesis take place only in the presence of light, and together are traditionally called the **light-dependent reactions.** ATP and NADPH are important energy-rich chemicals, and after this, the rest of photosynthesis becomes a chemical process.

The Light-Independent Reactions The ATP and NADPH molecules generated by the light-dependent reactions described above are then used to power a series of chemical reactions in the stroma of the chloroplast, each catalyzed by an enzyme present there. Acting together like the many stages of a manufacturing assembly line, these reactions accomplish the synthesis of carbohydrates from CO_2 in the air ❸. This third stage of photosynthesis, the formation of organic molecules like glucose from atmospheric CO_2, is called the **Calvin cycle,** but is also referred to as the **light-independent reactions** because it doesn't require light directly. We will examine the Calvin cycle in detail later in this chapter.

This completes our brief overview of photosynthesis. In the rest of the chapter we will revisit each stage and consider its elements in more detail. For now, the overall process may be summarized by the following simple equation:

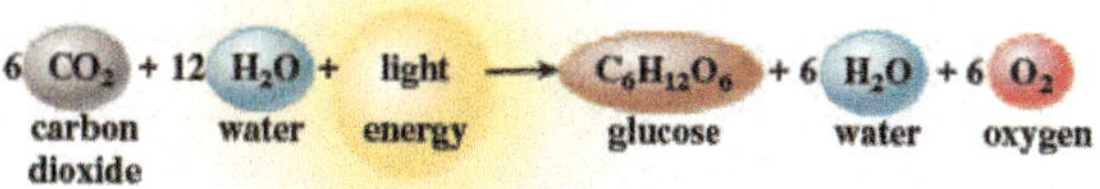

Key Learning Outcome 6.1 **Photosynthesis uses energy from sunlight to power the synthesis of organic molecules from CO_2 in the air. In plants, photosynthesis takes place in specialized compartments within chloroplasts.**

6.2 How Plants Capture Energy from Sunlight

Where is the energy in light? What is there about sunlight that a plant can use to create chemical bonds? The revolution in physics in the twentieth century taught us that light actually consists of tiny packets of energy called **photons,** which have properties both of particles and of waves. When light shines on your hand, your skin is being bombarded by a stream of these photons smashing onto its surface.

Sunlight contains photons of many energy levels, only some of which we "see." We call the full range of these photons the **electromagnetic spectrum.** As you can see in figure 6.1, some of the photons in sunlight have shorter wavelengths (toward the left side of the spectrum) and carry a great deal of energy—for example, gamma rays and ultraviolet (UV) light. Others such as radio waves carry very little energy and have longer wavelengths (hundreds to thousands of meters long). Our eyes perceive photons carrying intermediate amounts of energy as **visible light,** because the retinal pigment molecules in our eyes, which are different from chlorophyll, absorb only those photons of intermediate wavelength. Plants are even more picky, absorbing mainly blue and red light and reflecting back what is left of the visible light. To understand why plants are green, look at the green tree in figure 6.2. The full spectrum of visible light shines on the leaves of this tree, and only the green wavelengths of light are not absorbed. They are reflected off the leaf, which is why our eyes perceive leaves as green.

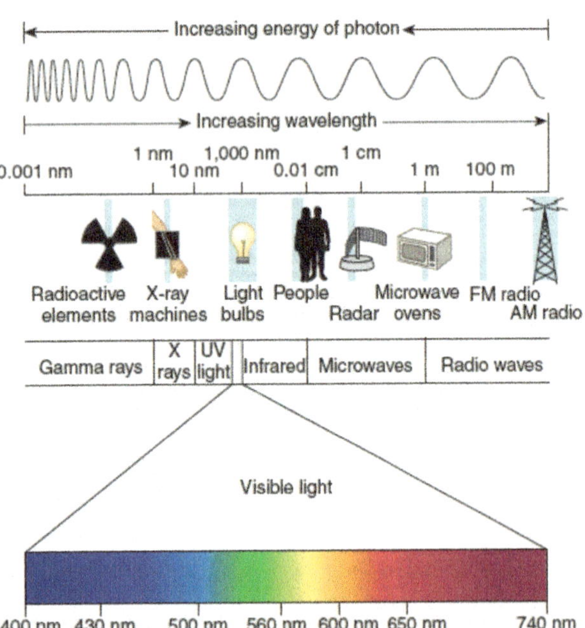

Figure 6.1 Photons of different energy: the electromagnetic spectrum.
Light is composed of packets of energy called photons. Some of the photons in light carry more energy than others. Light, a form of electromagnetic energy, is conveniently thought of as a wave. The shorter the wavelength of light, the greater the energy of its photons. Visible light represents only a small part of the electromagnetic spectrum, that with wavelengths between about 400 and 740 nanometers.

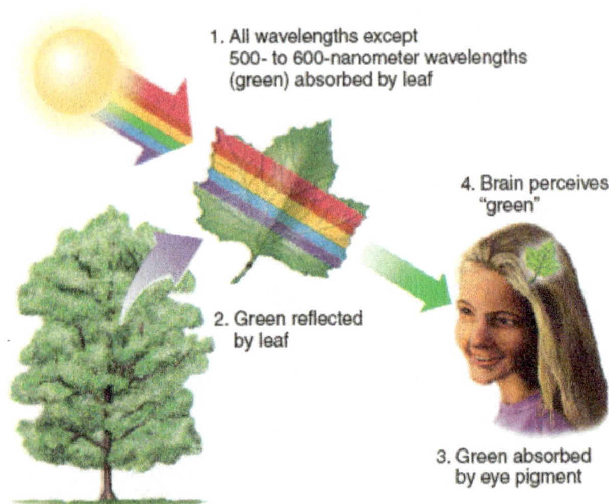

Figure 6.2 Why are plants green?
A leaf containing chlorophyll absorbs a broad range of photons—all the colors in the spectrum except for the photons around 500 to 600 nanometers. The leaf reflects these colors. These reflected wavelengths are absorbed by the visual pigments in our eyes, and our brains perceive the reflected wavelengths as "green."

How can a leaf or a human eye choose which photons to absorb? The answer to this important question has to do with the nature of atoms. Remember that electrons spin in particular orbits around the atomic nucleus, at different energy levels. Atoms absorb light by boosting electrons to higher energy levels, using the energy in the photon to power the move. Boosting the electron requires just the right amount of energy, no more and no less, just as when climbing a ladder you must raise your foot just so far to climb a rung. A particular kind of atom absorbs only certain photons of light, those with the appropriate amount of energy.

Pigments

As mentioned earlier, molecules that absorb light energy are called **pigments.** When we speak of visible light, we refer to those wavelengths that the pigment within human eyes, called *retinal,* can absorb—roughly from 380 nanometers (violet) to 750 nanometers (red). Other animals use different pigments for vision and thus "see" a different portion of the electromagnetic spectrum. For example, the pigment in insect eyes absorbs at shorter wavelengths than retinal. That is why bees can see ultraviolet light, which we cannot see, but are blind to red light, which we can see.

As noted, the main pigment in plants that absorbs light is chlorophyll. Its two forms, chlorophyll *a* and chlorophyll *b,* are similar in structure, but slight differences in their chemical "side groups" produce slight differences in their absorption spectra. An absorption spectrum is a graph indicating how effectively a pigment absorbs different wavelengths of visible light. For example, chlorophyll molecules will absorb photons

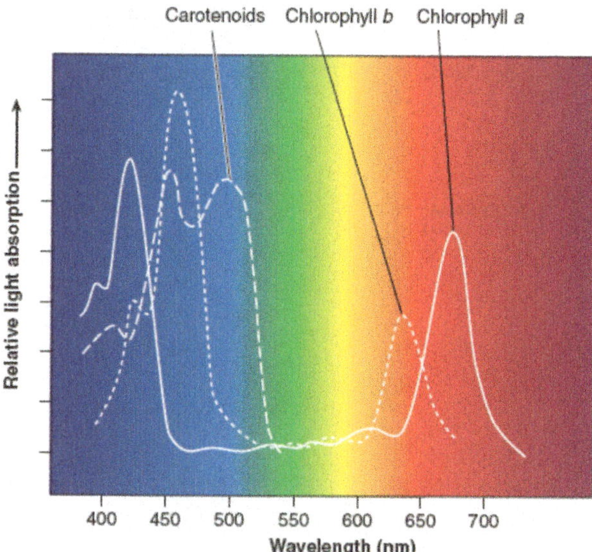

Figure 6.3 Absorption spectra of chlorophylls and carotenoids.

The peaks represent wavelengths of sunlight strongly absorbed by the two common forms of photosynthetic pigment, chlorophyll *a* and chlorophyll *b*, and by accessory carotenoid pigments. Chlorophylls absorb predominantly violet-blue and red light, in two narrow bands of the spectrum, while they reflect the green light in the middle of the spectrum. Carotenoids absorb mostly blue and green light and reflect orange and yellow light.

at the ends of the visible spectrum, the peaks you see in figure 6.3. While chlorophyll absorbs fewer kinds of photons than our visual pigment retinal, it is much more efficient at capturing them. Chlorophyll molecules capture photons with a metal ion (magnesium) that lies at the center of a complex carbon ring. Photons excite electrons of the magnesium ion, which are then channeled away by the carbon atoms.

While chlorophyll is the primary pigment involved in photosynthesis, plants also contain other pigments called *accessory pigments* that absorb light of wavelengths not captured by chlorophyll. **Carotenoids** are a group of accessory pigments that capture violet to blue-green light. As you can see in figure 6.3, these wavelengths of light are not efficiently absorbed by chlorophyll.

Accessory pigments give color to flowers, fruits, and vegetables but are also present in leaves, their presence usually masked by chlorophyll. During the warm months, when plants are actively producing food through photosynthesis, their cells are filled with chlorophyll-containing chloroplasts that cause the leaves to appear green, like the oak leaves on the left side of figure 6.4. In the fall, the days become shorter and cooler and for many species, leaves stop their food-making processes. Their chlorophyll molecules break down and are not replaced. When this happens, the colors reflected by accessory pigments become visible. The leaves turn colors of yellow, orange, and red, like the oak leaves on the right.

Key Learning Outcome 6.2 Plants use pigments like chlorophyll to capture photons of blue and red light, reflecting photons of green wavelengths.

Figure 6.4 Fall colors are produced by pigments such as carotenoids.

During the spring and summer, chlorophyll masks the presence in leaves of other pigments called carotenoids. Cool temperatures in the fall cause leaves of deciduous trees to cease manufacturing chlorophyll. With chlorophyll no longer present to reflect green light, the orange and yellow light reflected by carotenoids gives bright colors to the autumn leaves.

6.3 Organizing Pigments into Photosystems

The light-dependent reactions of photosynthesis occur on membranes. In most photosynthetic bacteria, the proteins involved in the light-dependent reactions are embedded within the plasma membrane. In algae, intracellular membranes contain the proteins that drive the light-dependent reactions. In plants, photosynthesis occurs in specialized organelles called chloroplasts. The chlorophyll molecules and proteins involved in the light-dependent reactions are embedded in the thylakoid membranes inside the chloroplasts. A portion of a thylakoid membrane is enlarged in figure 6.5. The chlorophyll molecules can be seen as the green spheres embedded along with accessory pigment molecules within a matrix of proteins (the purple area) within the thylakoid membrane. This complex of protein and pigment makes up the **photosystem.**

The light-dependent reactions take place in five stages, illustrated in figure 6.6. Each stage will be discussed in detail later in this chapter:

1. **Capturing light.** In stage ❶, a photon of light of the appropriate wavelength is captured by a pigment molecule, and the excitation energy is passed from one chlorophyll molecule to another.
2. **Exciting an electron.** In stage ❷, the excitation energy is funneled to a key chlorophyll *a* molecule called the **reaction center.** The excitation energy causes the transfer of an excited electron from the reaction center to another molecule that is an electron acceptor. The reaction center replaces this "lost" electron with an electron from the breakdown of a water molecule. Oxygen is produced as a by-product of this reaction.
3. **Electron transport.** In stage ❸, the excited electron is then shuttled along a series of electron-carrier molecules embedded in the membrane. This is called the **electron transport system (ETS).** As the electron passes along the electron transport system, the energy from the electron is "siphoned" out in small amounts. This energy is used to pump hydrogen ions (protons) across the membrane, indicated by the blue arrow, eventually building up a high concentration of protons inside the thylakoid.
4. **Making ATP.** In stage ❹, the high concentration of protons can be used as an energy source to make ATP. Protons are only able to move back across the membrane via special channels, the protons flooding through them like water through a dam. The kinetic energy that is released by the movement of protons is transferred to potential energy in the building of ATP molecules from ADP. This process, called **chemiosmosis,** makes the ATP that will be used in the Calvin cycle to make carbohydrates.

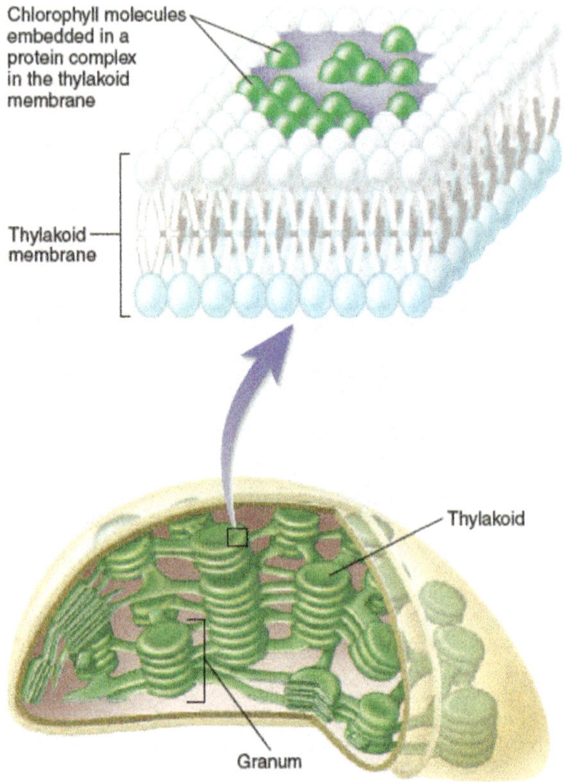

Figure 6.5 Chlorophyll embedded in a membrane. Chlorophyll molecules are embedded in a network of proteins that hold the pigment molecules in place. The proteins are embedded within the membranes of thylakoids.

5. **Making NADPH.** The electron leaves the electron transport system and enters another photosystem where it is "reenergized" by the absorption of another photon of light. In ❺, this energized electron enters another electron transport system, where it is again shuttled along a series of electron-carrier molecules. The result of this electron transport system is not the synthesis of ATP, but rather the formation of NADPH. The electron is transferred to a molecule, $NADP^+$, and a hydrogen ion that forms NADPH. This molecule is important in the synthesis of carbohydrates in the Calvin cycle.

Architecture of a Photosystem

In all but the most primitive bacteria, light is captured by photosystems. Like a magnifying glass focusing light on a precise point, a photosystem channels the excitation energy gathered by any one of its pigment molecules to a specific chlorophyll *a* molecule, the reaction center chlorophyll. For example, in figure 6.7, a chlorophyll molecule on the outer edge of the photosystem is excited by the photon, and this energy passes from one chlorophyll molecule to another, indicated by the yellow zig-zag arrow, until it reaches the reaction center molecule. This molecule then passes the energy, in the form of an excited electron, out of the photosystem to drive the synthesis of ATP and organic molecules.

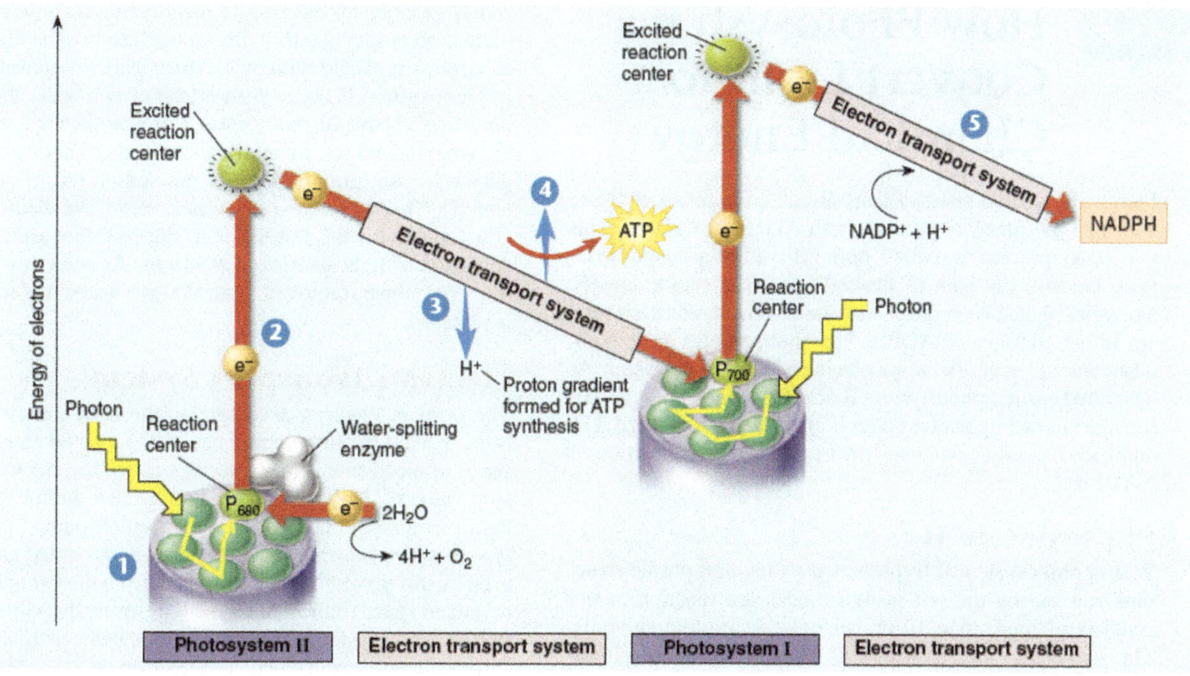

Figure 6.6 Plants use two photosystems.
In stage ❶, a photon excites pigment molecules in photosystem II. In stage ❷, a high-energy electron from photosystem II is transferred to the electron transport system. In stage ❸, the excited electron is used to pump a proton across the membrane. In stage ❹, the concentration gradient of protons is used to produce a molecule of ATP. In stage ❺, the ejected electron then passes to photosystem I, which uses it, with a photon of light energy, to drive the formation of NADPH.

Using Two Photosystems

Plants and algae use two photosystems, photosystems I and II, indicated by the two purple cylinders in figure 6.6. Photosystem II captures the energy that is used to produce the ATP needed to build sugar molecules. The light energy that it captures is used to transfer the energy of a photon of light ❶ to an excited electron ❷; the energy of this electron is then used by the electron transport system ❸ to produce ATP ❹.

Photosystem I powers the production of the hydrogen atoms needed to build sugars and other organic molecules from CO_2 (which has no hydrogen atoms). Photosystem I is used to energize an electron that, carried by a hydrogen ion (a proton), forms NADPH from $NADP^+$ ❺. NADPH shuttles hydrogens to the Calvin cycle where sugars are made.

The photosystems are not numbered in the order in which they are used. Photosystem II actually acts first in the series, and photosystem I acts second. The confusion arises because the photosystems were named in the order in which they were discovered, and photosystem I was discovered before photosystem II.

Key Learning Outcome 6.3 Photon energy is captured by pigments that employ it to excite electrons that are channeled away to do the chemical work of producing ATP and NADPH.

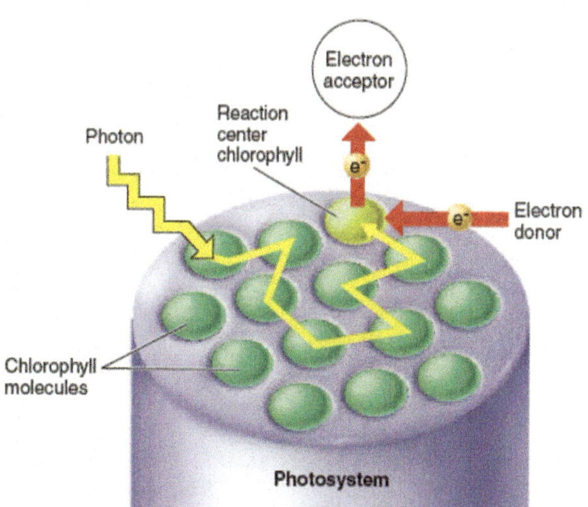

Figure 6.7 How a photosystem works.
When light of the proper wavelength strikes any pigment molecule within a photosystem, the light is absorbed and its excitation energy is then transferred from one molecule to another within the cluster of pigment molecules until it encounters the reaction center, which exports the energy as high-energy electrons to an acceptor molecule.

6.4 How Photosystems Convert Light to Chemical Energy

Plants use the two photosystems discussed in series, first one and then the other, to produce both ATP and NADPH. This two-stage process is called **noncyclic photophosphorylation,** because the path of the electrons is not a circle—the electrons ejected from the photosystems do not return to them, but rather end up in NADPH. The photosystems are replenished instead with electrons obtained by splitting water. As described earlier, photosystem II acts first. High-energy electrons generated by photosystem II are used to synthesize ATP and then passed to photosystem I to drive the production of NADPH.

Photosystem II

Within photosystem II (represented by the first purple structure you see on the left in figure 6.8), the reaction center consists of more than 10 transmembrane protein subunits. The *antenna complex,* which is the portion of the photosystem that contains all the pigment molecules, consists of some 250 molecules of chlorophyll *a* and accessory pigments bound to several protein chains. The antenna complex captures energy from a photon and funnels it to a reaction center chlorophyll. You can also see the antenna complex in the photosystem illustrated in figure 6.7. The reaction center gives up an excited electron to a primary electron acceptor in the electron transport system. The path of the excited electron is indicated with the red arrow. After the reaction center gives up an electron to the electron transport system, there is an empty electron orbital that needs to be filled. This electron is replaced with an electron from a water molecule. In photosystem II the oxygen atoms of two water molecules bind to a cluster of manganese atoms embedded within an enzyme and bound to the reaction center (notice the light gray water-splitting enzyme at the bottom left of photosystem II). This enzyme splits water, removing electrons one at a time to fill the holes left in the reaction center by the departure of light-energized electrons. As soon as four electrons have been removed from the two water molecules, O_2 is released.

Electron Transport System

The primary electron acceptor for the light-energized electrons leaving photosystem II passes the excited electron to a series of electron-carrier molecules called the *electron transport system.* These proteins are embedded within the thylakoid membrane; one of them is a "proton pump" protein, a type of active transport channel. The energy of the electron is used by this protein to pump a proton from the stroma into the thylakoid space (indicated by the blue arrow through the electron transport system). A nearby protein in the membrane then carries the now energy-depleted electron on to photosystem I.

Making ATP: Chemiosmosis

Before progressing onto photosystem I, let's see what happens with the protons that were pumped into the thylakoid by the electron transport system. Each thylakoid is a closed compartment into which protons are pumped. The thylakoid membrane is impermeable to protons, so protons build up inside the thylakoid space, creating a very large concentra-

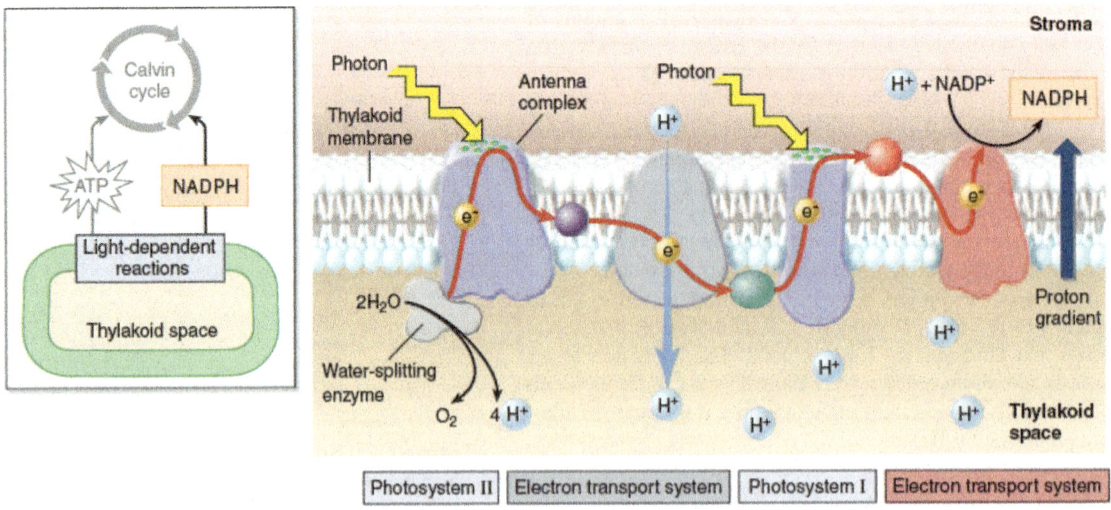

Figure 6.8 The photosynthetic electron transport system.

tion gradient. As you may recall from chapter 4, molecules in solution diffuse from areas of higher concentration to areas of lower concentration. Here, protons diffuse back out of the thylakoid space, down their concentration gradient, passing through special protein channels called ATP synthases. *ATP synthase* is an enzyme that can use the concentration gradient of protons to drive the synthesis of ATP from ADP. ATP synthase channels protrude like knobs on the external surface of the thylakoid membrane (figure 6.9). As protons pass out of the thylakoid through the ATP synthase channels, ADP is phosphorylated to ATP and released into the stroma (the fluid matrix inside the chloroplast). Because the chemical formation of ATP is driven by a diffusion process similar to osmosis, this type of ATP formation is called *chemiosmosis*.

Photosystem I

Now, with ATP formed, let's return our attention to the right half of figure 6.8, with photosystem I accepting an electron from the electron transport system. The reaction center of photosystem I is a membrane complex consisting of at least 13 protein subunits. Energy is fed to it by an antenna complex consisting of 130 chlorophyll *a* and accessory pigment molecules. The electron arriving from the first electron transport system has by no means lost all of its light-excited energy; almost half remains. Thus, the absorption of another photon of light energy by photosystem I boosts the electron leaving its reaction center to a very high energy level.

Making NADPH

Like photosystem II, photosystem I passes electrons to an electron transport system. When two of these electrons reach the end of this electron transport system, they are then donated to a molecule of $NADP^+$ to form NADPH (one electron is transferred with a proton as a hydrogen atom). This reaction, which takes place on the stromal side of the thylakoid (as shown in figure 6.8), involves an $NADP^+$, two electrons, and a proton. Because the reaction occurs on the stromal side of the membrane and involves the uptake of a proton in forming NADPH, it contributes further to the proton concentration gradient established during photosynthetic electron transport.

Products of the Light-Dependent Reactions

The light-dependent reactions can be seen more as a stepping stone, rather than an end point of photosynthesis. All of the products of the light-dependent reactions are either waste products, such as oxygen, or are ultimately used elsewhere in the cell. The ATP and NADPH produced in the light-dependent reactions end up being passed on to the Calvin cycle in the stroma of the chloroplast. The stroma contains the enzymes that catalyze the light-independent reactions, in which ATP is used to power chemical reactions that build carbohydrates. NADPH is used as the source of "reducing power," providing the hydrogens and electrons used in building carbohydrates. The next section discusses the Calvin cycle of photosynthesis.

Key Learning Outcome 6.4 The light-dependent reactions of photosynthesis produce the ATP and NADPH needed to build organic molecules, and release O_2 as a by-product of stripping hydrogen atoms and their associated electrons from water molecules.

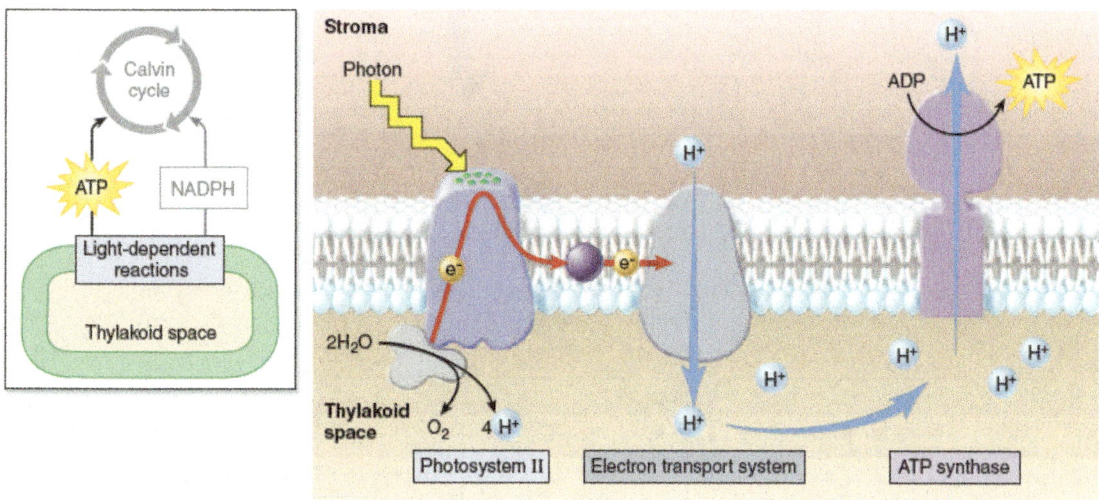

Figure 6.9 Chemiosmosis in a chloroplast.

The energy of the electron absorbed by photosystem II powers the pumping of protons into the thylakoid space. These protons then pass back out through ATP synthase channels, their movement powering the production of ATP.

6.5 Building New Molecules

The Calvin Cycle

Stated very simply, photosynthesis is a way of making organic molecules from carbon dioxide (CO_2). To build organic molecules, cells use raw materials provided by the light-dependent reactions:

1. **Energy.** ATP (provided by the ETS of photosystem II) drives the endergonic reactions.
2. **Reducing power.** NADPH (provided by the ETS of photosystem I) provides a source of hydrogens and the energetic electrons needed to bind them to carbon atoms. A molecule that accepts an electron is said to be reduced, as will be discussed in detail in chapter 7.

The actual assembly of new molecules employs a complex battery of enzymes in what is called the **Calvin cycle,** or **C_3 photosynthesis** (C_3 because the first molecule produced in the process is a three-carbon molecule). The Calvin cycle takes place in the stroma of the chloroplasts. The NADPH and the ATP that were generated by the light-dependent reactions are used in the Calvin cycle to build carbohydrate molecules. In the Key Biological Process illustration below, the number of carbon atoms at each stage is indicated by the number of balls. It takes six turns of the cycle to make one six-carbon molecule of glucose. The process takes place in three stages, highlighted in the three panels of the Key Biological Process illustration below. These three stages are also indicated by different-colored pie-shaped pieces in the more detailed look at the Calvin cycle provided in figure 6.10. Both figures indicate that three turns of the cycle are needed to produce one molecule of glyceraldehyde 3-phosphate. In any *one* turn of the cycle, a carbon atom from a carbon dioxide molecule

KEY BIOLOGICAL PROCESS: The Calvin Cycle

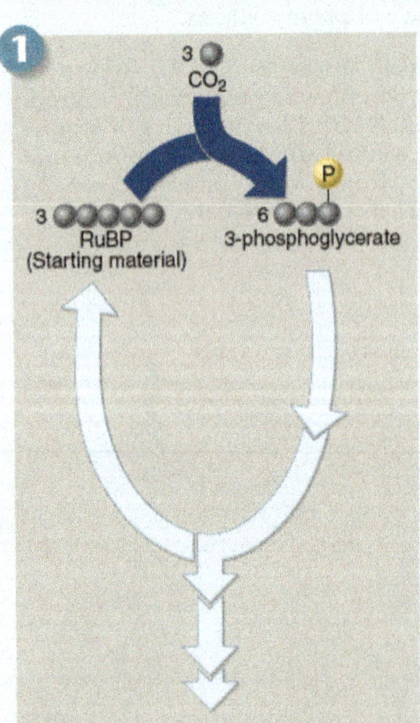

The Calvin cycle begins when a carbon atom from a CO_2 molecule is added to a five-carbon molecule (the starting material). The resulting six-carbon molecule is unstable and immediately splits into three-carbon molecules. (Three "turns" of the cycle are indicated here with three molecules of CO_2 entering the cycle.)

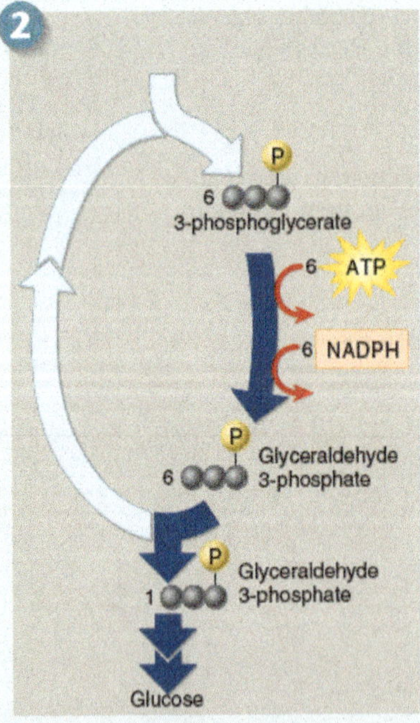

Then, through a series of reactions, energy from ATP and hydrogens from NADPH (the products of the light-dependent reactions) are added to the three-carbon molecules. The now-reduced three-carbon molecules either combine to make glucose or are used to make other molecules.

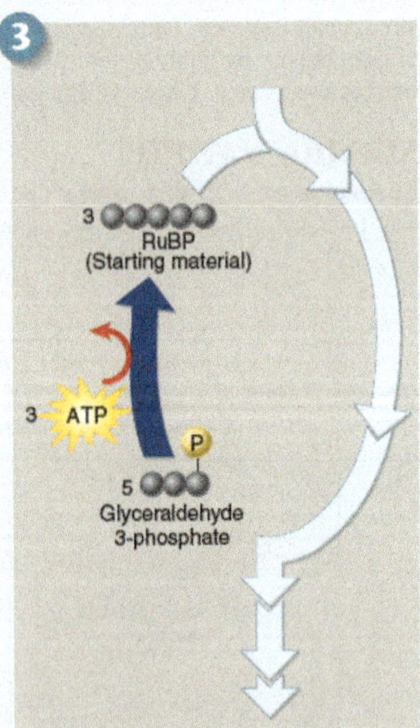

Most of the reduced three-carbon molecules are used to regenerate the five-carbon starting material, thus completing the cycle.

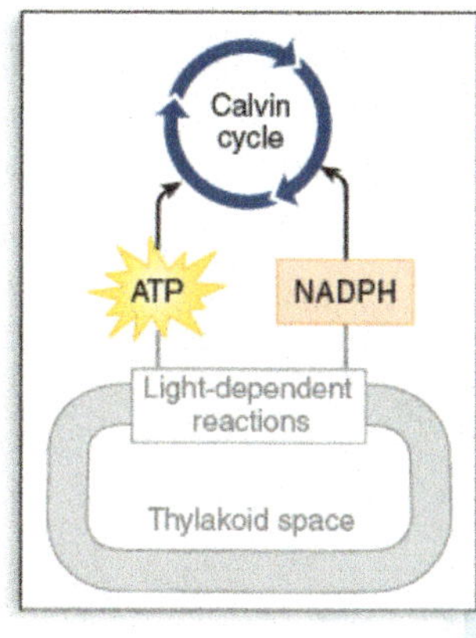

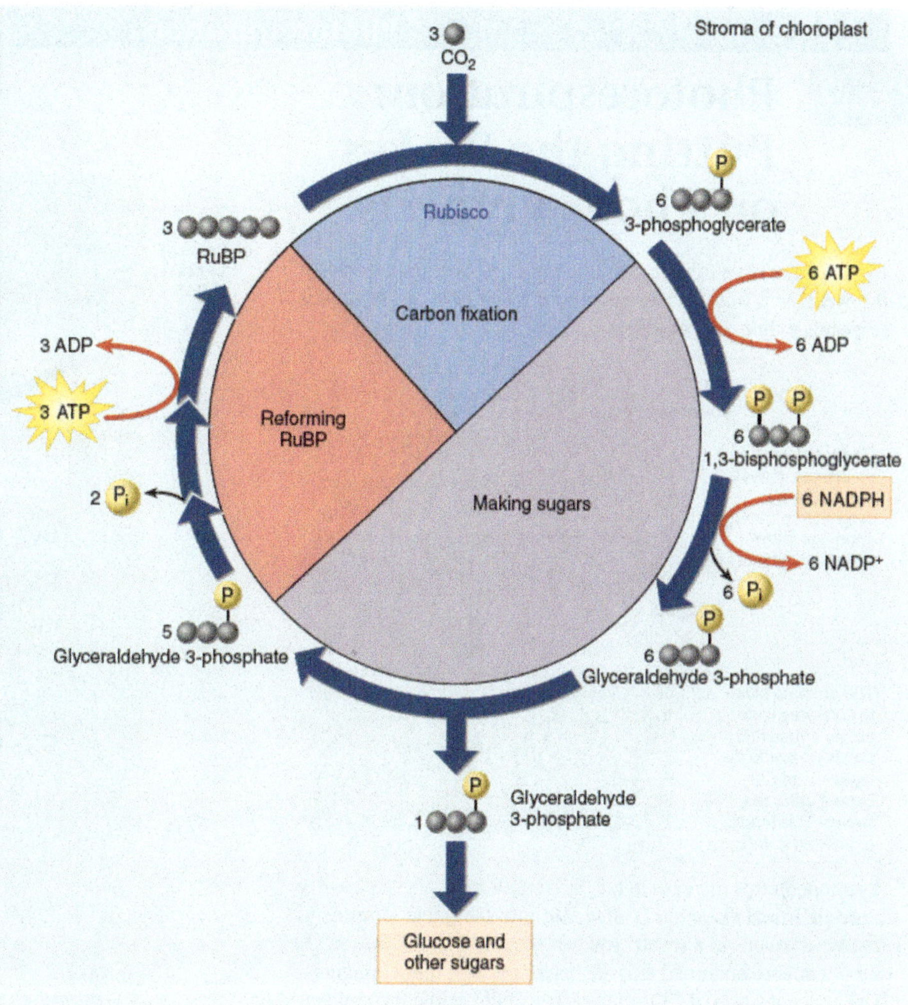

Figure 6.10
Reactions of the Calvin cycle.
For every three molecules of CO_2 that enter the cycle, one molecule of the three-carbon compound glyceraldehyde 3-phosphate (G3P) is produced. Notice that the process requires energy stored in ATP and NADPH, which are generated by the light-dependent reactions. This process occurs in the stroma of the chloroplast. The large 16-subunit enzyme that catalyzes the reaction, RuBP carboxylase, or **rubisco**, is the most abundant protein in chloroplasts and is thought to be the most abundant protein on earth.

is first added to a five-carbon sugar, producing two three-carbon sugars. This process, highlighted by the dark blue arrow in panel 1 of the Key Biological Process illustration and the blue pie-shaped area in figure 6.10, is called *carbon fixation* because it attaches a carbon atom that was in a gas to an organic molecule.

Then, in a long series of reactions, the carbons are shuffled about. Eventually some of the resulting molecules are channeled off to make sugars (shown by the dark blue arrows in panel 2 of the Key Biological Process illustration and at the bottom of the cycle within the purple colored area in figure 6.10). Other molecules are used to re-form the original five-carbon sugar (the dark blue arrow in panel 3 of the Key Biological Process illustration and the light-red-colored area in figure 6.10), which is then available to restart the cycle. The cycle has to "turn" six times in order to form a new glucose molecule, because each turn of the cycle adds only one carbon atom from CO_2, and glucose is a six-carbon sugar.

Recycling ADP and $NADP^+$

The products of the light-dependent reactions, ATP and NADPH, feed into the light-independent reactions of the Calvin cycle to make sugar molecules. To keep photosynthesis moving along, the cells must continually supply the light-dependent reactions with more ADP and $NADP^+$. This is accomplished by recycling these products from the Calvin cycle. After the phosphate bonds are broken in ATP, ADP is available for chemiosmosis. After the hydrogens and electrons are stripped from NADPH, $NADP^+$ is available to cycle back to the electron transport system of photosystem I.

Key Learning Outcome 6.5 In a series of reactions that do not directly require light, cells use ATP and NADPH provided by photosystems II and I to assemble new organic molecules.

6.6 Photorespiration: Putting the Brakes on Photosynthesis

Many plants have trouble carrying out C_3 photosynthesis when the weather is hot. A cross section of a leaf here shows how it responds to hot, arid weather:

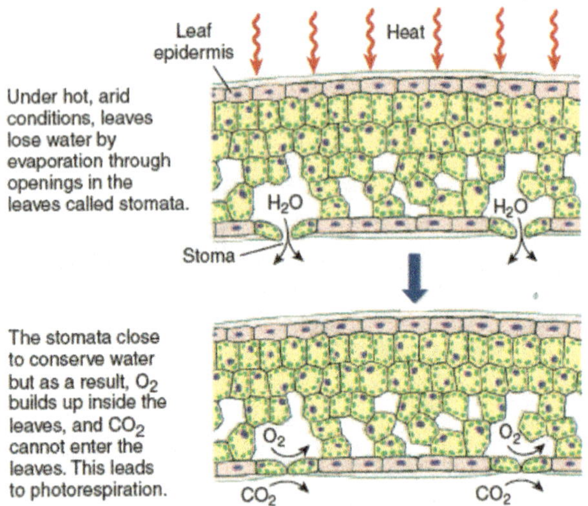

As temperatures increase in hot, arid conditions, plants partially close their leaf openings, called **stomata** (singular, **stoma**), to conserve water. As a result, you can see above that CO_2 and O_2 are not able to enter and exit the leaves through these openings. The concentration of CO_2 in the leaves falls, while the concentration of O_2 in the leaves rises. Under these conditions rubisco, the enzyme that carries out the first step of the Calvin cycle, engages in **photorespiration,** where the enzyme incorporates O_2, not CO_2, into the cycle and when this occurs, CO_2 is ultimately released as a by-product. Photorespiration thus short-circuits the successful performance of the Calvin cycle.

C_4 Photosynthesis

Some plants are able to adapt to climates with higher temperatures by performing **C_4 photosynthesis.** In this process, plants such as sugarcane, corn, and many grasses are able to fix carbon using different types of cells and chemical reactions within their leaves, thereby avoiding a reduction in photosynthesis due to higher temperatures.

A cross section of a leaf from a C_4 plant is shown in figure 6.11. Examining it, you can see how these plants solve the problem of photorespiration. In the enlargement, you see two cell types: The green cell is a mesophyll cell and the tan cell is a bundle-sheath cell. In the mesophyll cell, CO_2 combines with a three-carbon molecule instead of RuBP as it did in figure 6.10, producing a four-carbon molecule, oxaloacetate (hence the name, C_4 photosynthesis), rather than the three-carbon molecule phosphoglycerate you saw in figure 6.10. C_4 plants carry out this process in the mesophyll cells of their leaves, using a different enzyme. The oxaloacetate is then converted to malate, which is transferred to the bundle-sheath cells of the leaf. In the tan bundle-sheath cell, malate is broken down to regenerate CO_2, which enters the Calvin cycle you are familiar with from figure 6.10, and sugars are synthesized. Why go to all this trouble? Because the bundle-sheath cells are impermeable to CO_2 and so the concentration of CO_2 increases within them, so much that the rate of photorespiration is substantially lowered.

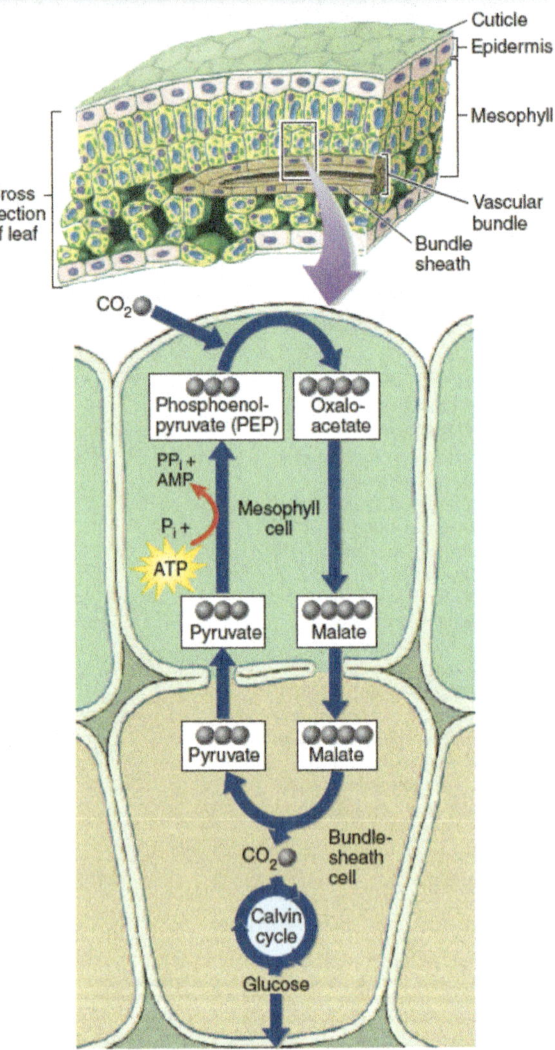

Figure 6.11 Carbon fixation in C_4 plants.
This process is called the C_4 pathway because the first molecule formed in the pathway is a four-carbon sugar, oxaloacetate. This molecule is converted into malate that is transported into bundle-sheath cells. Once there, malate undergoes a chemical reaction producing carbon dioxide. The carbon dioxide is trapped in the bundle-sheath cell, where it enters the Calvin cycle.

Today's *Biology*

Cold-Tolerant C_4 Photosynthesis

Corn (*Zea mays*), one of humanity's most important agricultural crops, is highly productive when grown at warm temperatures. However, its commercial use in northern areas is severely limited by its much poorer performance at low temperatures. Much of corn's high productivity results from its use of the C_4 photosynthetic pathway, which has the highest efficiency of photosynthesis known. However, much of this efficiency is lost below 20° C. At 5° C, 80% of photosynthesis is lost.

In C_4 species like corn, sugarcane, sorghum, and switchgrass, sensitivity to low temperatures appears to depend on the sensitivity of key C_4 photosynthetic enzymes, particularly the Calvin cycle enzyme catalyzing the final stage illustrated in figure 6.11. This enzyme has the imposing name pyruvate orthophosphate dikinase and is abbreviated PPDK. PPDK, which appears to be the rate-limiting step in corn C_4 photosynthesis, is very sensitive to low temperature, with little activity remaining when temperatures fall below 10° C.

One relative of corn recently has been shown to be strikingly different. Chinese silver grass (*Miscanthus giganteus*) is a perennial grass that uses the same C_4 pathway as corn. However, in marked contrast to corn, it produces efficiently at temperatures as low as 5° C. With its greater tolerance of low temperatures, this species thrives at chilling temperatures, with individual stalks growing as high as 13 feet! Similar temperatures severely limit C_4 photosynthesis in its relative.

What is the cause of *Miscanthus*'s tolerance of cold? At low temperatures, when amounts of PPDK fall in corn, PPDK activity actually rises in *Miscanthus*. Researchers are currently examining the *Miscanthus* PPDK gene to better understand the cold-tolerance it confers. If these early results are confirmed, genetic engineers can explore the possibility of replacing the corn PPDK gene with the *Miscanthus* version, in the hope of greatly extending the northern range of corn, a key agricultural crop.

A second strategy to decrease photorespiration is used by many succulent (water-storing) plants such as cacti and pineapples. This mode of initial carbon fixation is called **crassulacean acid metabolism (CAM)** after the plant family Crassulaceae in which it was first discovered. In these plants, the stomata open during the night when it's cooler, and close during the day. CAM plants initially fix CO_2 into organic compounds at night, using the C_4 pathway. These organic compounds accumulate at night and are subsequently broken down during the following day, releasing CO_2. These high levels of CO_2 drive the Calvin cycle and decrease photorespiration. To understand how photosynthesis differs in CAM plants and C_4 plants, examine figure 6.12. In C_4 plants (on the left), the C_4 pathway occurs in mesophyll cells, while the Calvin cycle occurs in bundle-sheath cells. In CAM plants (on the right), the C_4 pathway and the Calvin cycle occur in the same cell, a mesophyll cell, but they occur at different times of the day, the C_4 cycle at night and the Calvin cycle during the day.

Key Learning Outcome 6.6 **Photorespiration occurs due to a buildup of oxygen within photosynthetic cells. C_4 plants get around photorespiration by synthesizing sugars in bundle-sheath cells, and CAM plants delay the light-independent reactions until night, when stomata are open.**

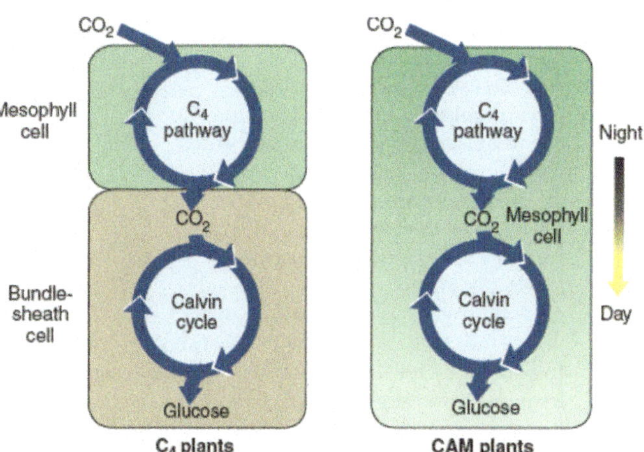

Figure 6.12 Comparing carbon fixation in C_4 and CAM plants.

Both C_4 and CAM plants utilize the C_4 and C_3 pathways. In C_4 plants, the pathways are separated spatially; the C_4 pathway takes place in the mesophyll cells and the C_3 pathway (the Calvin cycle) in the bundle-sheath cells. In CAM plants, the two pathways occur in mesophyll cells but are separated temporally; the C_4 pathway is utilized at night and the C_3 pathway during the day.

INQUIRY & ANALYSIS

Does Iron Limit the Growth of Ocean Phytoplankton?

Phytoplankton are microscopic organisms that live in the oceans, carrying out much of the earth's photosynthesis. The photo below is of *Chaetoceros*, a phytoplankton. Decades ago, scientists noticed "dead zones" in the ocean where little photosynthesis occurred. Looking more closely, they found that phytoplankton collected from these waters are not able to efficiently fix CO_2 into carbohydrates. In an attempt to understand why not, the scientists hypothesized that lack of iron (needed by the ETS) was the problem, and predicted that fertilizing these ocean waters with iron could trigger an explosively rapid growth of phytoplankton.

To test this idea, they carried out a field experiment, seeding large areas of phytoplankton-poor ocean waters with iron crystals to see if this triggered phytoplankton growth. Other similarly phytoplankton-poor areas of ocean were not seeded with iron and served as controls.

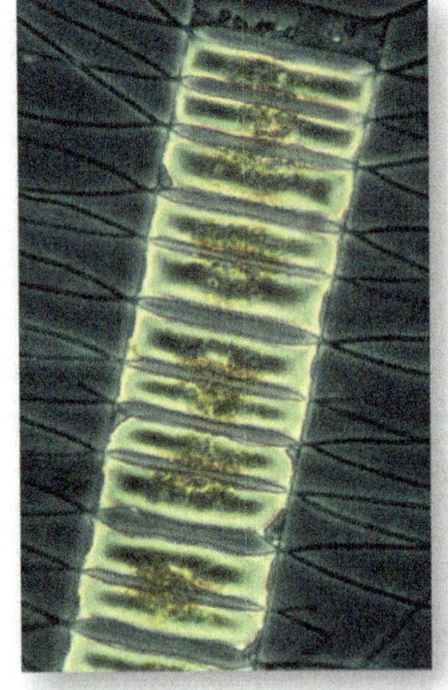

In one such experiment, the results of which are presented in the graph to the right, a 72-km² grid of phytoplankton-deficient ocean water was seeded with iron crystals and a tracer substance in three successive treatments, indicated with arrows on the *x* axis of the graph (on days 0, 3, and 7). The multiple seedings were carried out to reduce the effect of the iron crystals dissipating over time. A smaller control grid, 24 km², was seeded with just the tracer substance.

To assess the numbers of phytoplankton organisms carrying out photosynthesis in the ocean water, investigators did not actually count organisms. Instead, they estimated the amount of chlorophyll *a* in water samples as an easier-to-measure index. An **index** is a parameter that accurately reflects the quantity of another less-easily-measured parameter. In this instance, the level of chlorophyll *a*, easily measured by monitoring the wavelengths of light absorbed by a liquid sample, is a suitable index of phytoplankton, as this pigment is found nowhere else in the ocean other than within phytoplankton.

Chlorophyll *a* measurements were made periodically on both test and control grids for 14 days. The results are plotted on the graph. Red points indicate chlorophyll *a* concentrations in iron-seeded waters; blue points indicate chlorophyll *a* levels in the control grid waters that were not seeded.

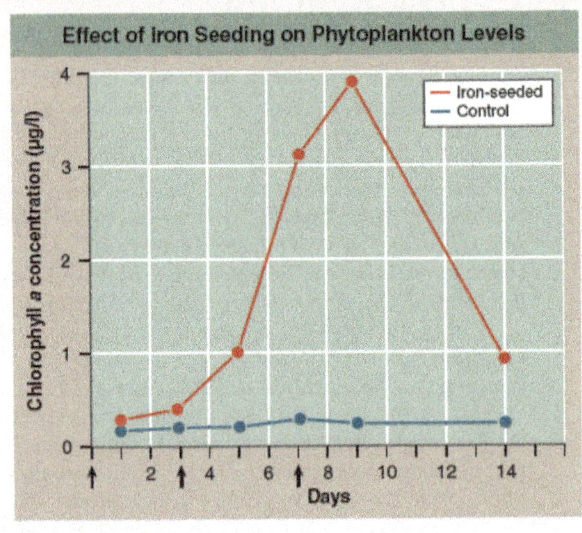

1. **Applying Concepts**
   a. **Variable.** In the graph above, which is the dependent variable?
   b. **Index.** What does the increase in levels of chlorophyll *a* say about numbers of phytoplankton?
   c. **Control.** What substance is lacking in the waters sampled in the blue-dot plots ?
2. **Interpreting Data**
   a. What happened to the levels of chlorophyll *a* in the test areas of the ocean (red dots)?
   b. What happened to the levels of chlorophyll *a* in the control areas (blue dots)?
   c. Comparing the red line to the blue line, about how many times more numerous are phytoplankton in iron-seeded waters on the three days of seeding?
3. **Making Inferences**
   a. What general statement can be made regarding the effect of seeding phytoplankton-poor regions of the ocean with iron?
   b. Why did chlorophyll *a* levels drop by day 14?
4. **Drawing Conclusions** Do these results support the claim that lack of iron is limiting the growth of phytoplankton, and thus of photosynthesis, in certain areas of the oceans?
5. **Further Analysis** Based on this experiment, what would be a potential drawback of using this method of seeding with iron to increase levels of ocean photosynthesis?

Chapter Review

Photosynthesis

6.1 An Overview of Photosynthesis

- Photosynthesis is a biochemical process whereby energy from the sun is captured and used to build carbohydrates from CO_2 gas and water (**integrated art, pages 118-119**).
- Photosynthesis consists of a series of chemical reactions that occurs in two stages: The light-dependent reactions that produce ATP and NADPH occur on the thylakoid membranes of chloroplasts in plants, while the light-independent reactions (the Calvin cycle) that synthesize carbohydrates occur in the stroma (**integrated art, pages 120-121**).

6.2 How Plants Capture Energy from Sunlight

- Pigments are molecules that capture light energy. Energy present in visible light is captured by the pigment chlorophyll and other accessory pigments present in chloroplasts (**figure 6.1**).
- Plants appear green because of their chlorophyll pigments (**figure 6.2**). Chlorophyll absorbs wavelengths in the far ends of the visual spectrum (the blue and red wavelengths) and reflect the green wavelengths, which is why leaves appear green.
- Accessory pigments, such as carotenoids, capture energy from different areas of the spectrum than chlorophyll and give different colors to flowers, fruits, and other parts of plants that are not green (**figure 6.4**).

6.3 Organizing Pigments into Photosystems

- The light-dependent reactions occur on the thylakoid membranes of chloroplasts in plants. The chlorophyll molecules and other pigments involved in photosynthesis are embedded in a complex of proteins within the membrane called a photosystem (**figure 6.5**).
- Light energy is captured by the photosystem, where it excites an electron that is passed to an electron transport system. There, it is used to generate ATP and NADPH, both of which power the Calvin cycle. Plants utilize two photosystems that occur in series (**figure 6.6**). Photosystem II leads to the formation of ATP, and photosystem I leads to the formation of NADPH.
- The energy from a photon of light is absorbed by a chlorophyll molecule and is transferred between chlorophyll molecules in the photosystem, as shown here from **figure 6.7**. Once the energy is passed to the reaction center, it excites an electron, which is transferred to the electron transport system.

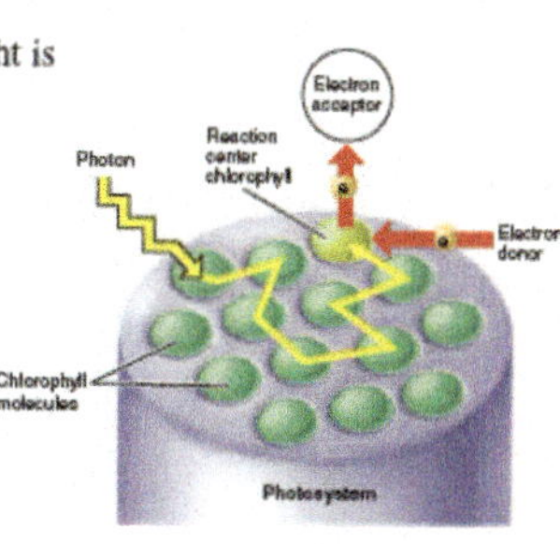

6.4 How Photosystems Convert Light to Chemical Energy

- The excited electron that leaves the reaction center of photosystem II and enters the electron transport system is replenished with an electron from the breakdown of a water molecule.
- The excited electron is passed from one protein to another in the electron transport system, where energy from the electron is used to operate a proton pump that pumps hydrogen ions across the membrane against a concentration gradient (**figure 6.8**).
- The hydrogen ion concentration gradient is used as a source of energy to generate molecules of ATP. This energy is used to drive H^+ back across the membrane through a specialized channel protein called ATP synthase, which catalyzes the formation of ATP, as shown here from **figure 6.9**.

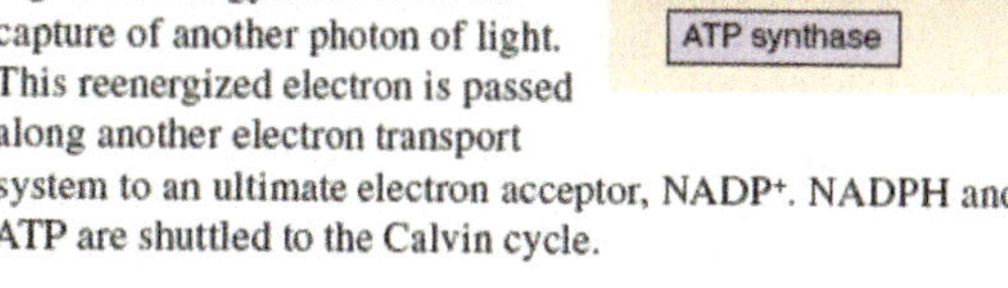

- After the electron passes along the first electron transport system, it is then transferred to a second photosystem, photosystem I, where it gets an energy boost from the capture of another photon of light. This reenergized electron is passed along another electron transport system to an ultimate electron acceptor, $NADP^+$. NADPH and ATP are shuttled to the Calvin cycle.

6.5 Building New Molecules

- The Calvin cycle is carried out by a series of enzymes that use the energy from ATP and electrons and hydrogen ions from NADPH to build molecules of carbohydrates by reducing CO_2 (**Key Biological Process, page 128** and **figure 6.10**).
- During the Calvin cycle, ADP and $NADP^+$ are recycled as by-products and feed back into the light-dependent reactions.

Photorespiration

6.6 Photorespiration: Putting the Brakes on Photosynthesis

- In hot, dry weather, plants will close the stomata in their leaves to conserve water. As a result, the levels of O_2 increase in the leaves, and CO_2 levels drop, as shown here from **page 130**. Under these conditions, the Calvin cycle, also called C_3 photosynthesis, is disrupted. When there is a higher internal concentration of oxygen, O_2 rather than CO_2 enters the Calvin cycle in a process called photorespiration. In this case, the first enzyme in the Calvin cycle, rubisco, binds oxygen instead of carbon dioxide.

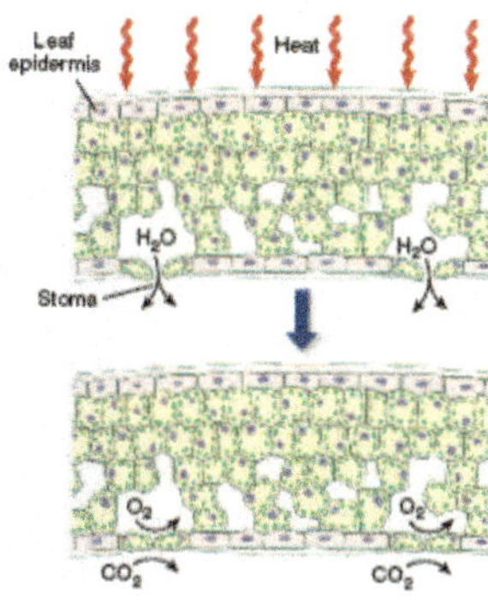

- C_4 plants reduce the effects of photorespiration by modifying the carbon-fixation step, splitting it into two steps that take place in different cells. The C_4 pathway produces malate in mesophyll cells (**figure 6.11**).
- In CAM plants, carbon dioxide is processed into intermediate organic molecules through the C_4 pathway during the night when stomata are open (**figure 6.12**).

Test Your Understanding

1. The energy that is used by almost all living things on our planet comes from the sun. It is captured by plants, algae, and some bacteria through the process of
   a. thylakoid. c. photosynthesis.
   b. chloroplasts. d. the Calvin cycle.
2. Plants capture sunlight
   a. through photorespiration.
   b. with molecules called pigments that absorb photons and use their energy.
   c. with the light-independent reactions.
   d. with the electron transport system.
3. Visible light occupies what part of the electromagnetic spectrum?
   a. the entire spectrum
   b. the upper half of the spectrum (with longer wavelengths)
   c. a small portion in the middle of the spectrum
   d. the lower half of the spectrum (with shorter wavelengths)
4. The colors of light that are absorbed by chlorophyll are
   a. red and blue.
   b. green and yellow.
   c. infrared and ultraviolet.
   d. All colors are equally effective.
5. Once a plant has initially captured the energy of a photon,
   a. a series of reactions occurs in thylakoid membranes of the cell.
   b. the energy is transferred through several steps into a molecule of ATP.
   c. a water molecule is broken down, releasing oxygen.
   d. All of the above.
6. Plants use two photosystems to capture energy used to produce ATP and NADPH. The electrons used in these photosystems
   a. recycle through the system constantly, with energy added from the photons.
   b. recycle through the system several times and then are lost due to entropy.
   c. only go through the system once; they are obtained by splitting a water molecule.
   d. only go through the system once; they are obtained from the photon.
7. During photosynthesis, ATP molecules are generated by
   a. the Calvin cycle.
   b. chemiosmosis.
   c. the splitting of a water molecule.
   d. photons of light being absorbed by chlorophyll molecules.
8. NADPH is recycled during photosynthesis. It is produced during the ________ and used in the__________.
   a. electron transport system of photosystem I, Calvin cycle
   b. process of chemiosmosis, Calvin cycle
   c. electron transport system of photosystem II, electron transport system of photosystem I
   d. light-independent reactions, light-dependent reactions
9. The overall purpose of the Calvin cycle is to
   a. generate molecules of ATP.
   b. generate NADPH.
   c. build sugar molecules.
   d. produce oxygen.
10. Many plants cannot carry out the typical C_3 photosynthesis in hot weather, so some plants
    a. use the ATP cycle.
    b. use C_4 photosynthesis or CAM.
    c. shut down photosynthesis completely.
    d. All of these are true for different plants.

Apply Your Understanding

1. **Figure 6.3** Why do most leaves have more than one type of pigment?

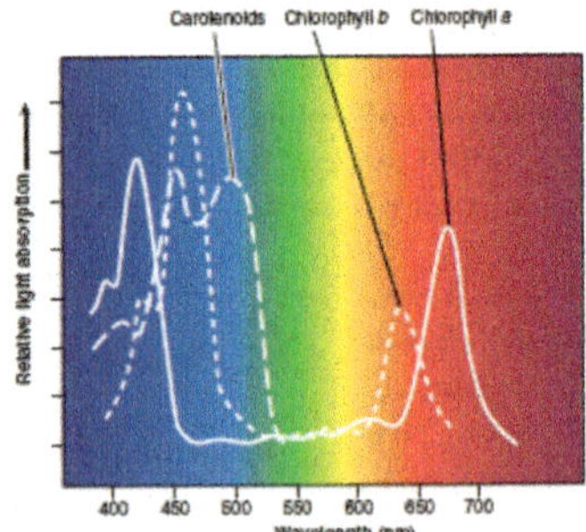

2. **Figure 6.9** Could a plant cell produce ATP through chemiosmosis if the thylakoid membrane was "leaky" with regard to protons? Explain.

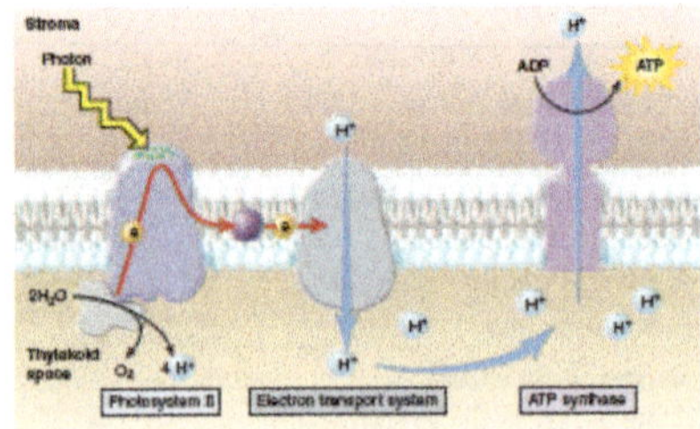

Synthesize What You Have Learned

1. To reduce six molecules of carbon dioxide to one molecule of glucose via photosynthesis, how many molecules of NADPH and ATP are required?
2. In theory, a plant kept in total darkness could still manufacture glucose, if it were supplied with which molecules?
3. If you were going to design a plant that would survive in the deserts of Arizona and New Mexico, how would you balance its need for CO_2 with its need to avoid water loss in the hot summer temperatures?

7

How Cells Harvest Energy from Food

Learning Objectives

An Overview of Cellular Respiration

7.1 Where Is the Energy in Food?

1. Distinguish between oxidation and reduction.
2. Define cellular respiration.
3. Write a chemical equation for the oxidation of glucose.
4. Identify where in a cell glycolysis takes place.
5. State which of the two stages of cellular respiration is anaerobic and which is aerobic.

Respiration Without Oxygen: Glycolysis

7.2 Using Coupled Reactions to Make ATP

1. Define a coupled reaction, and explain how coupled reactions are used to extract energy from glucose.
2. Define substrate-level phosphorylation.
3. State how many molecules of ATP are made from a glucose molecule by glycolysis in the absence of oxygen.
4. Explain why the presence of oxygen would affect the yield of ATP molecules in glycolysis.

Respiration With Oxygen: The Krebs Cycle

7.3 Harvesting Electrons from Chemical Bonds

1. Name and describe the enzyme that removes CO_2 from pyruvate.
2. State where in the cell the Krebs cycle takes place.
3. Identify the substrate for the nine-reaction Krebs cycle, and the products.

A Closer Look: Metabolic Efficiency and the Length of Food Chains

7.4 Using the Electrons to Make ATP

1. Identify the components of the electron transport chain.
2. Describe the journey of an electron through the chain, and identify its final destination.
3. Describe the location and function of ATP synthase.
4. Calculate how many ATP molecules a cell can harvest from a glucose molecule in the presence of oxygen and in its absence.

Harvesting Electrons Without Oxygen: Fermentation

7.5 Cells Can Metabolize Food Without Oxygen

1. Define fermentation.
2. Distinguish between ethanol and lactic acid fermentation.

A Closer Look: Beer and Wine—Products of Fermentation

Other Sources of Energy

7.6 Glucose Is Not the Only Food Molecule

1. Describe how cells garner energy from proteins and from fats.

Biology and Staying Healthy: Fad Diets and Impossible Dreams

Inquiry & Analysis: How Do Swimming Fish Avoid Low Blood pH?

Animals such as this chipmunk depend on the energy stored in the chemical bonds of the food they eat to power their life processes. Their lives are driven by energy. All the activities this chipmunk carries out—climbing trees, chewing on acorns, seeing and smelling and hearing its surroundings, thinking the thoughts that chipmunks think—use energy. But unlike the oak tree that produces the nuts on which this chipmunk is dining, no part of the chipmunk is green. It cannot carry out photosynthesis like an oak tree, and so cannot harvest energy from the sun as the tree does. Instead, it must get its energy secondhand, by consuming organic molecules manufactured by plants. The chemical energy that the oak tree invested in making its molecules is harvested by the chipmunk in a process called cellular respiration. The same processes are used by all animals to harvest energy from molecules—and by plants too. There is no sunlight under the soil where the oak tree's roots penetrate, and like the cells of the chipmunk, these plant root cells obtain the energy to fuel their lives from cellular respiration. In this chapter, we examine cellular respiration up close. As you will see, cellular respiration and photosynthesis have much in common.

An Overview of Cellular Respiration

7.1 Where Is the Energy in Food?

In both plants and animals, and in fact in almost all organisms, the energy for living is obtained by breaking down the organic molecules originally produced in plants. The ATP energy and reducing power invested in building the organic molecules are retrieved by stripping away the energetic electrons and using them to make ATP. When electrons are stripped away from chemical bonds, the food molecules are being oxidized (remember, oxidation is the loss of electrons). The oxidation of foodstuffs to obtain energy is called **cellular respiration.** Do not confuse the term cellular respiration with the breathing of oxygen gas that your lungs carry out, which is called simply respiration.

The cells of plants fuel their activities with sugars and other molecules that they produce through photosynthesis and breakdown in cellular respiration. Nonphotosynthetic organisms eat plants, extracting energy from plant tissue in cellular respiration. Other animals, like the lion gnawing with such relish on a giraffe leg in figure 7.1, eat these animals.

Eukaryotes produce the majority of their ATP by harvesting electrons from chemical bonds of the food molecule glucose. The electrons are transferred along an electron transport chain (similar to the electron transport system in photosynthesis), and eventually donated to oxygen gas. Chemically, there is little difference between this oxidation of carbohydrates in a cell and the burning of wood in a fireplace. In both instances, the reactants are carbohydrates and oxygen, and the products are carbon dioxide, water, and energy:

$$C_6H_{12}O_6 + 6\,O_2 \longrightarrow 6\,CO_2 + 6\,H_2O + \text{energy (heat or ATP)}$$

In many of the reactions of photosynthesis and cellular respiration, electrons pass from one atom or molecule to another. When an atom or molecule loses an electron, it is said to be *oxidized,* and the process by which this occurs is called **oxidation.** The name reflects the fact that in biological systems, oxygen, which attracts electrons strongly, is the most common electron acceptor. Conversely, when an atom or molecule gains an electron, it is said to be *reduced,* and the process is called **reduction.** Oxidation and reduction always take place together, because every electron that is lost by an atom through oxidation is gained by some other atom through reduction. Therefore, chemical reactions of this sort are called **oxidation-reduction (redox) reactions.** In redox reactions, energy follows the electron, as shown in figure 7.2.

Cellular respiration is carried out in two stages, illustrated in figure 7.3. The first stage uses coupled reactions to make ATP. This stage, *glycolysis,* takes place in the cell's cytoplasm. Importantly, it is anaerobic (that is, it does not require oxygen). This ancient energy-extracting process is thought to have evolved over 2 billion years ago, when there was no oxygen in the earth's atmosphere.

Figure 7.1 Lion at lunch.
Energy that this lion extracts from its meal of giraffe will be used to power its roar, fuel its running, and build a bigger lion.

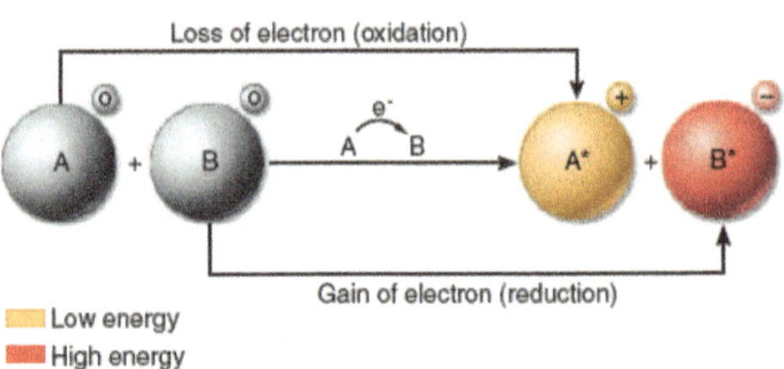

Figure 7.2 Redox reactions.
Oxidation is the loss of an electron; reduction is the gain of an electron. Here the charges of molecules A and B are shown in small circles to the upper right of each molecule. Molecule A loses energy as it loses an electron, while molecule B gains energy as it gains an electron.

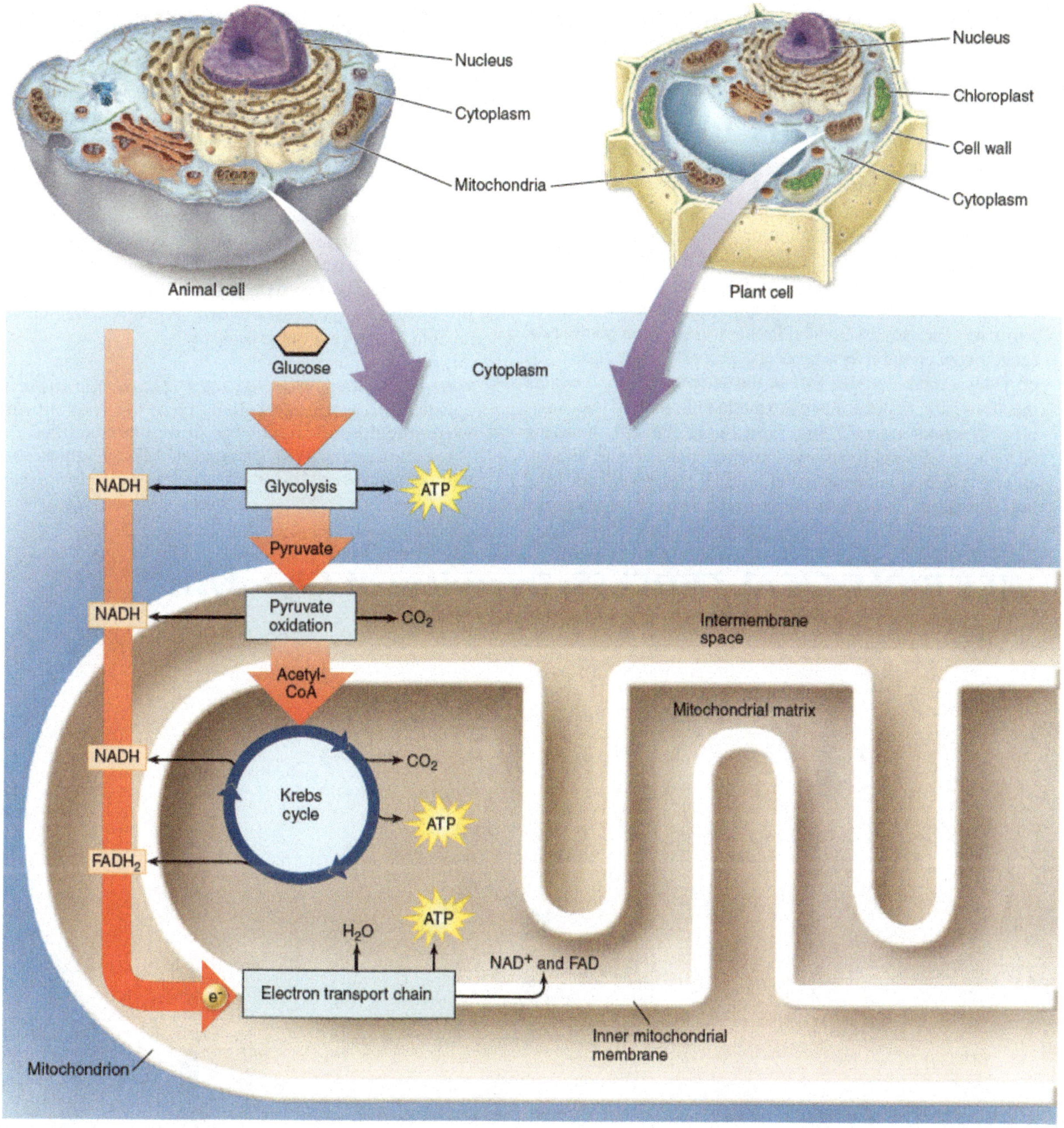

Figure 7.3 An overview of cellular respiration.

The second stage is aerobic (requires oxygen) and takes place within the mitochondrion. The focal point of this stage is the *Krebs cycle,* a cycle of chemical reactions that harvests electrons from C—H chemical bonds and passes the energy-rich electrons to carrier molecules, NADH and $FADH_2$. These molecules deliver the electrons to the electron transport chain, which uses their energy to power the production of ATP. This harvesting of electrons, a form of *oxidation,* is far more powerful than glycolysis at recovering energy from food molecules, and is how the bulk of the energy used by eukaryotic cells is extracted from food molecules.

Key Learning Outcome 7.1 **Cellular respiration is the dismantling of food molecules to obtain energy. In aerobic respiration, the cell harvests energy from glucose molecules in two stages, glycolysis and oxidation.**

Respiration Without Oxygen: Glycolysis

7.2 Using Coupled Reactions to Make ATP

The first stage in cellular respiration, called **glycolysis,** is a series of sequential biochemical reactions, a *biochemical pathway.* In 10 enzyme-catalyzed reactions, the six-carbon sugar glucose is cleaved into two three-carbon molecules called pyruvate. The Key Biological Process illustration below presents a conceptual overview of the process, while figure 7.4 provides a more detailed look at the series of 10 biochemical reactions. Where is the energy extracted? In each of two "coupled" reactions (steps 7 and 10 in figure 7.4), the breaking of a chemical bond in an exergonic reaction releases enough energy to drive the formation of an ATP molecule from ADP (an endergonic reaction). This transfer of a high-energy phosphate group from a substrate to ADP is called **su' s'ra'e-level phosphorylation.** In the process, electrons and hydrogen atoms are extracted and donated to a carrier molecule called NAD^+. The NAD^+ carries the electrons as NADH to join the other electrons extracted during oxidative respiration, discussed in the following section. Only a small number of ATP molecules are made in glycolysis itself, two for each molecule of glucose, but in the absence of oxygen this is the only way organisms can get energy from food.

Glycolysis is thought to have been one of the earliest of all biochemical processes to evolve. Every living creature is capable of carrying out glycolysis.

Key Learning Outcome 7.2 **In the first stage of cellular respiration, called glycolysis, cells shuffle chemical bonds in glucose so that two coupled reactions can occur, producing ATP by substrate-level phosphorylation.**

KEY BIOLOGICAL PROCESS: Overview of Glycolysis

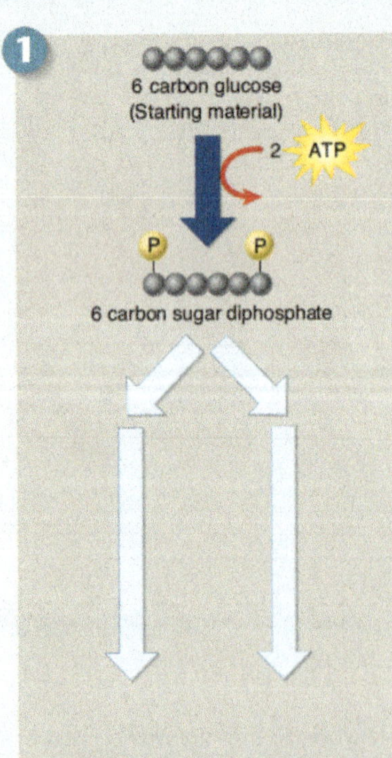

Priming reactions. Glycolysis begins with the addition of energy. Two high energy phosphates from two molecules of ATP are added to the six carbon molecule glucose, producing a six carbon molecule with two phosphates.

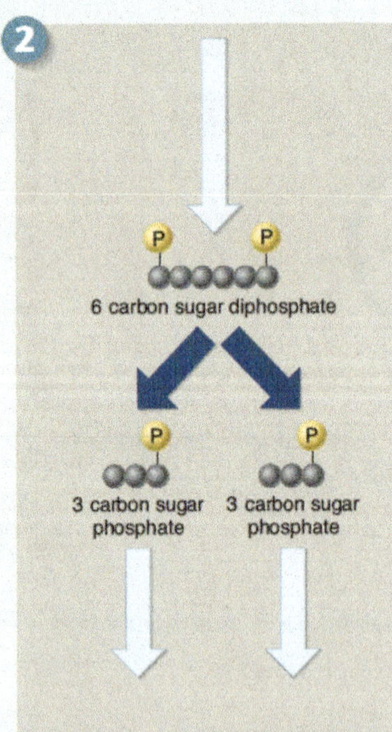

Cleavage reactions. Then, the phosphorylated six carbon molecule is split in two, forming two three carbon sugar phosphates.

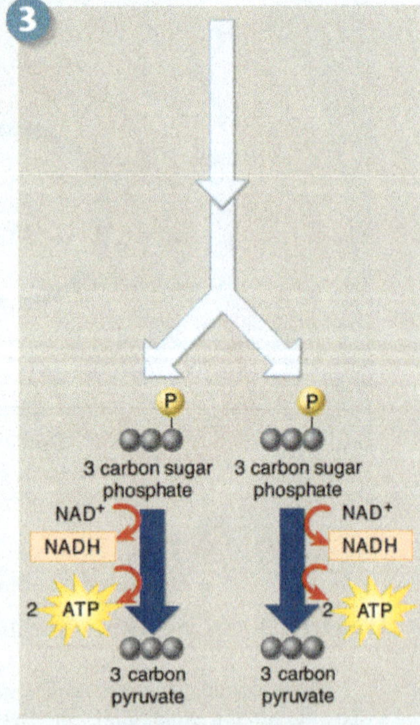

Energy harvesting reactions. Finally, in a series of reactions, each of the two three carbon sugar phosphates is converted to pyruvate. In the process, an energy rich hydrogen is harvested as NADH, and two ATP molecules are formed for each pyruvate.

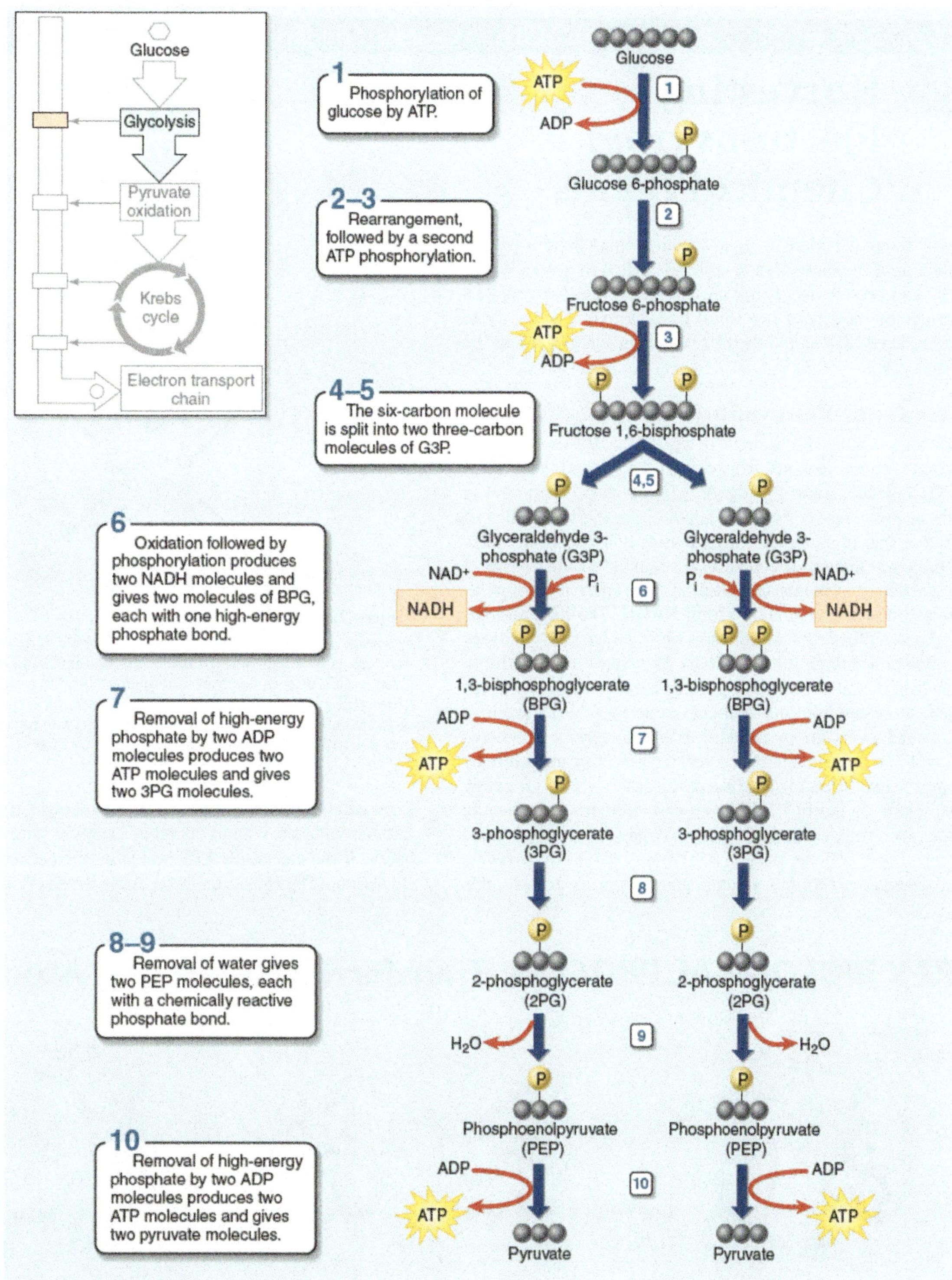

Figure 7.4 The reactions of glycolysis.
The process of glycolysis involves 10 enzyme-catalyzed reactions.

Respiration With Oxygen: The Krebs Cycle

7.3 Harvesting Electrons from Chemical Bonds

The first step of oxidative respiration in the mitochondrion is the oxidation of the three-carbon molecule called pyruvate, which is the end product of glycolysis. The cell harvests pyruvate's considerable energy in two steps: first, by oxidizing pyruvate to form acetyl-CoA, and then by oxidizing acetyl-CoA in the Krebs cycle.

Step One: Producing Acetyl-CoA

Pyruvate is oxidized in a single reaction that cleaves off one of pyruvate's three carbons. This carbon then departs as part of a CO_2 molecule, shown in figure 7.5 coming off the pathway with the green arrow. Pyruvate dehydrogenase, the complex of enzymes that removes CO_2 from pyruvate, is one of the largest enzymes known. It contains 60 subunits! In the course of the reaction, a hydrogen and electrons are removed from pyruvate and donated to NAD^+ to form NADH. The Key Biological Process illustration below shows how an enzyme catalyzes this reaction, bringing the substrate (pyruvate) into proximity with NAD^+. Cells use NAD^+ to carry hydrogen atoms and energetic electrons from one molecule to another. NAD^+ oxidizes energy-rich molecules by acquiring their hydrogens (this proceeds $1 \longrightarrow 2 \longrightarrow 3$ in the figure) and then reduces other molecules by giving the hydrogens to them (this proceeds $3 \longrightarrow 2 \longrightarrow 1$). Now focus again on figure 7.5. The two-carbon fragment (called an acetyl group) that remains after removing CO_2 from pyruvate is joined to a cofactor called coenzyme A (CoA) by pyruvate dehydrogenase, forming a compound known as **acetyl-CoA.**

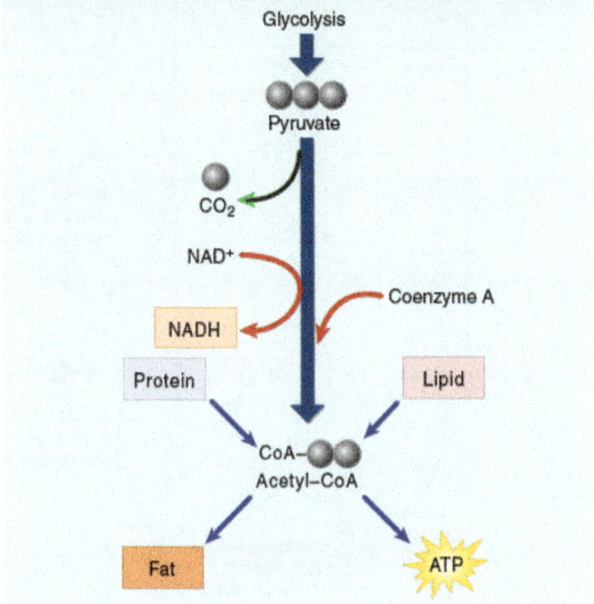

Figure 7.5 Producing acetyl-CoA.
Pyruvate, the three-carbon product of glycolysis, is oxidized to the two-carbon molecule acetyl-CoA, in the process losing one carbon atom as CO_2 and an electron (donated to NAD^+ to form NADH). Almost all the molecules you use as foodstuffs are converted to acetyl-CoA; the acetyl-CoA is then channeled into fat synthesis or into ATP production, depending on your body's needs.

If the cell has plentiful supplies of ATP, acetyl-CoA is funneled into fat synthesis, with its energetic electrons preserved for later needs. If the cell needs ATP now, the fragment is directed instead into ATP production through the Krebs cycle.

KEY BIOLOGICAL PROCESS: Transferring Hydrogen Atoms

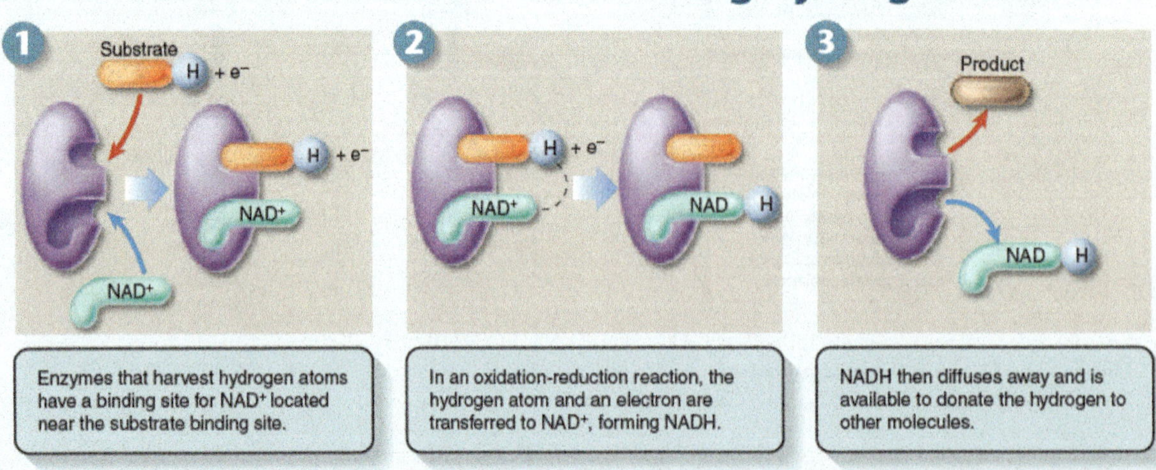

A **Closer** *Look*

Metabolic Efficiency and the Length of Food Chains

In the earth's ecosystems, the organisms that carry out photosynthesis are often consumed as food by other organisms. We call these "organism-eaters" *heterotrophs.* Humans are heterotrophs, as no human photosynthesizes.

It is thought that the first heterotrophs were ancient bacteria living in a world where photosynthesis had not yet introduced much oxygen into the oceans or atmosphere. The only mechanism they possessed to harvest chemical energy from their food was glycolysis. Neither oxygen-generating photosynthesis nor the oxidative stage of cellular respiration had evolved yet. It has been estimated that a heterotroph limited to glycolysis, as these ancient bacteria were, captures only 3.5% of the energy in the food it consumes. Hence, if such a heterotroph preserves 3.5% of the energy in the photosynthesizers it consumes, then any other heterotrophs that consume the first heterotroph will capture through glycolysis 3.5% of the energy in it, or 0.12% of the energy available in the original photosynthetic organisms. A very large base of photosynthesizers would thus be needed to support a small number of heterotrophs.

When organisms became able to extract energy from organic molecules by oxidative cellular respiration, which we discuss on the next page, this constraint became far less severe, because the efficiency of oxidative respiration is estimated to be about 32%. This increased efficiency results in the transmission of much more energy from one trophic level to another than does glycolysis. (A *trophic level* is a step in the movement of energy through an ecosystem.) The efficiency of oxidative cellular respiration has made possible the evolution of food chains, in which photosynthesizers are consumed by heterotrophs, which are consumed by other heterotrophs, and so on. You will read more about food chains in chapter 36.

Even with this very efficient oxidative metabolism, approximately two-thirds of the available energy is lost at each trophic level, and that puts a limit on how long a food chain can be. Most food chains, like the East African grassland ecosystem illustrated here, involve only three or rarely four trophic levels. Too much energy is lost at each transfer to allow chains to be much longer than that. For example, it would be impossible for a large human population to subsist by eating lions captured from the grasslands of East Africa; the amount of grass available there would not support enough zebras and other herbivores to maintain the number of lions needed to feed the human population. Thus, the ecological complexity of our world is fixed in a fundamental way by the chemistry of oxidative cellular respiration.

Photosynthesizers. The grass under this yellow fever tree grows actively during the hot, rainy season, capturing the energy of the sun and storing it in molecules of glucose, which are then converted into starch and stored in the grass.

Herbivores. These zebras consume the grass and transfer some of its stored energy into their own bodies.

Carnivores. The lion feeds on zebras and other animals, capturing part of their stored energy and storing it in its own body.

Scavengers. This hyena and the vultures occupy the same stage in the food chain as the lion. They also consume the body of the dead zebra, after it has been abandoned by the lion.

Refuse utilizers. These butterflies, mostly *Precis octavia,* are feeding on the material left in the hyena's dung after the food the hyena consumed had passed through its digestive tract.

A food chain in the savannas, or open grasslands, of East Africa.

At each of these levels in the food chain, only about a third or less of the energy present is used by the recipient.

Step Two: The Krebs Cycle

The next stage in oxidative respiration is called the **Krebs cycle,** named after the man who discovered it. The Krebs cycle (not to be confused with the Calvin cycle in photosynthesis) takes place within the mitochondrion. While a complex process, its nine reactions can be broken down into three stages, as indicated by the overview presented in the Key Biological Process illustration below:

Stage 1. Acetyl-CoA joins the cycle, binding to a four-carbon molecule and producing a six-carbon molecule.
Stage 2. Two carbons are removed as CO_2, their electrons donated to NAD^+, and a four-carbon molecule is left. A molecule of ATP is also produced.
Stage 3. More electrons are extracted, forming NADH and $FADH_2$; the four-carbon starting material is regenerated.

To examine the Krebs cycle in more detail, follow along the series of individual reactions illustrated in figure 7.6. The cycle starts when the two-carbon acetyl-CoA fragment produced from pyruvate is stuck onto a four-carbon sugar called oxaloacetate. Then, in rapid-fire order, a series of eight additional reactions occur (steps 2 through 9). When it is all over, two carbon atoms have been expelled as CO_2, one ATP molecule has been made in a coupled reaction, eight more energetic electrons have been harvested and taken away as NADH or on other carriers, such as $FADH_2$, which serves the same function as NADH, and we are left with the same four-carbon sugar we started with. The process of reactions is a cycle—that is, a circle of reactions. In each turn of the cycle, a new acetyl group replaces the two CO_2 molecules lost, and more electrons are extracted. Note that a single glucose molecule produces *two* turns of the cycle, one for each of the two pyruvate molecules generated by glycolysis.

In the process of cellular respiration, glucose is entirely consumed. The six-carbon glucose molecule is first cleaved into a pair of three-carbon pyruvate molecules during glycolysis. One of the carbons of each pyruvate is then lost as CO_2 in the conversion of pyruvate to acetyl-CoA, and the other two carbons are lost as CO_2 during the oxidations of the Krebs cycle. All that is left to mark the passing of the glucose molecule into six CO_2 molecules is its energy, preserved in four ATP molecules and electrons carried by 10 NADH and two $FADH_2$ carriers.

Key Learning Outcome 7.3 The end product of glycolysis, pyruvate, is oxidized to the two-carbon acetyl-CoA, yielding a pair of electrons plus CO_2. Acetyl-CoA then enters the Krebs cycle, yielding ATP, many energized electrons, and two CO_2 molecules.

KEY BIOLOGICAL PROCESS: Overview of the Krebs Cycle

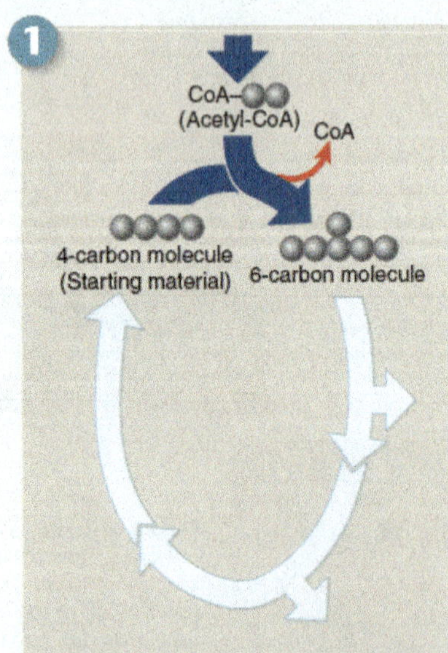

The Krebs cycle begins when a two-carbon fragment is transferred from acetyl-CoA to a four-carbon molecule (the starting material).

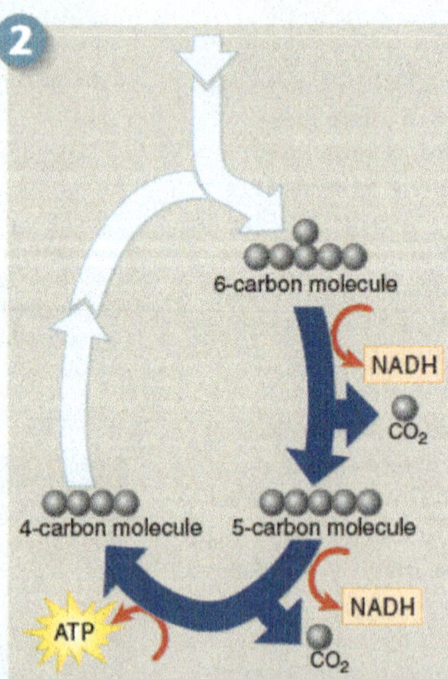

Then, the resulting six-carbon molecule is oxidized (a hydrogen removed to form NADH) and decarboxylated (a carbon removed to form CO_2). Next, the five-carbon molecule is oxidized and decarboxylated again, and a coupled reaction generates ATP.

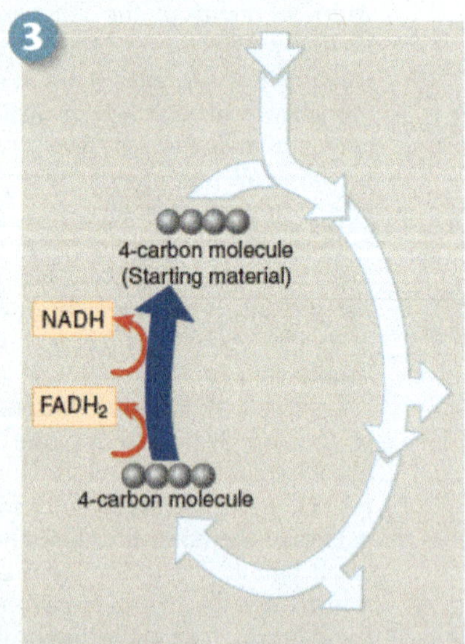

Finally, the resulting four-carbon molecule is further oxidized (hydrogens removed to form $FADH_2$ and NADH). This regenerates the four-carbon starting material, completing the cycle.

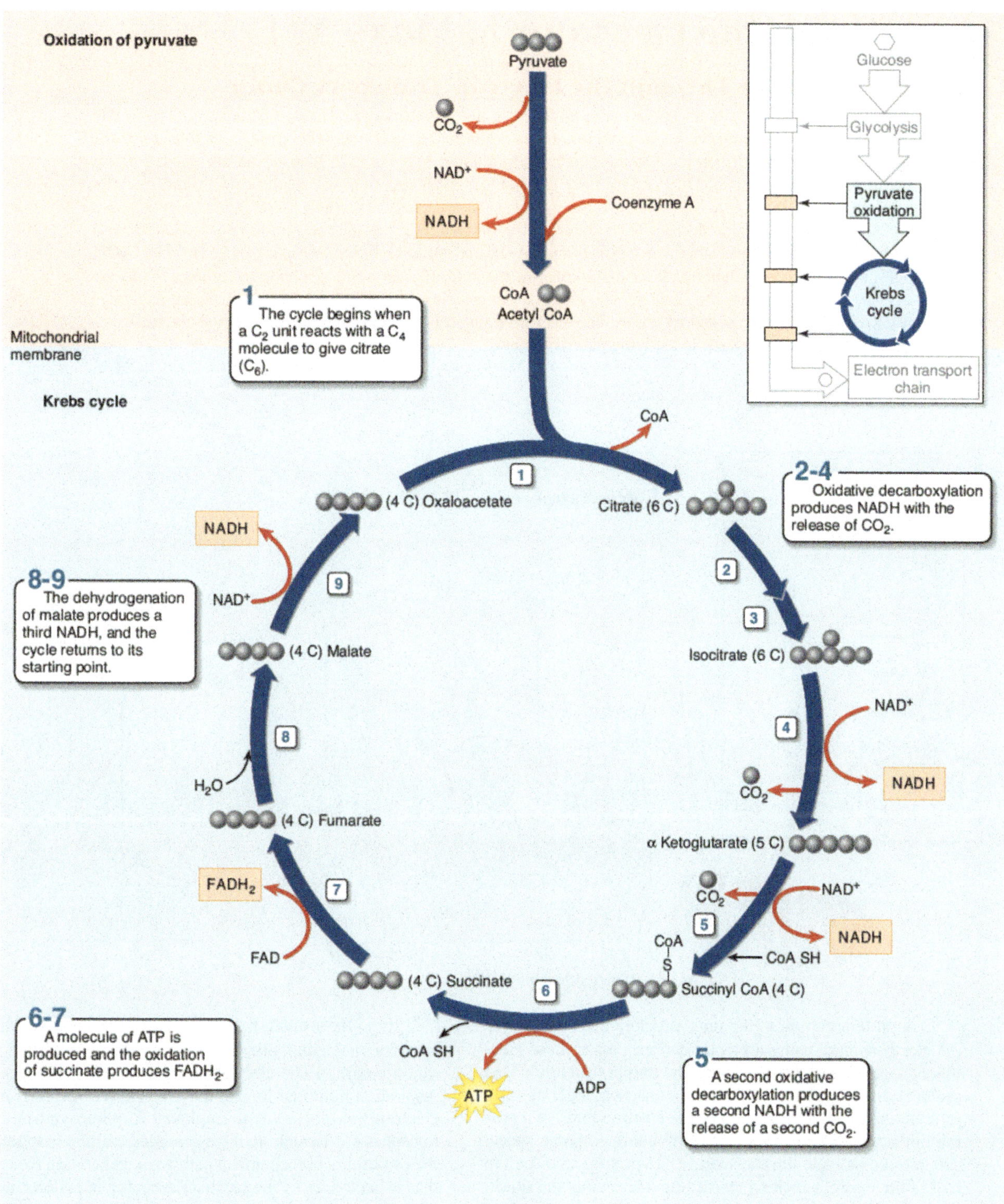

Figure 7.6 The Krebs cycle.
This series of nine enzyme-catalyzed reactions takes place within the mitochondrion.

7.4 Using the Electrons to Make ATP

Moving Electrons Through the Electron Transport Chain

In eukaryotes, aerobic respiration takes place within the mitochondria present in virtually all cells. The internal compartment, or **matrix,** of a mitochondrion contains the enzymes that carry out the reactions of the Krebs cycle. As described earlier, the electrons harvested by oxidative respiration are passed along the electron transport chain, and the energy they release transports protons out of the matrix and into the **intermembrane space.**

The NADH and $FADH_2$ molecules formed during the first stages of aerobic respiration each contain electrons and hydrogens that were gained when NAD^+ and FAD were reduced (refer back to figure 7.3). The NADH and $FADH_2$ molecules carry their electrons to the inner mitochondrial membrane (an enlarged area of the membrane is shown below), where they transfer the electrons to a series of membrane-associated molecules collectively called the **electron transport chain.** The electron transport chain works much as does the electron transport system you encountered in studying photosynthesis.

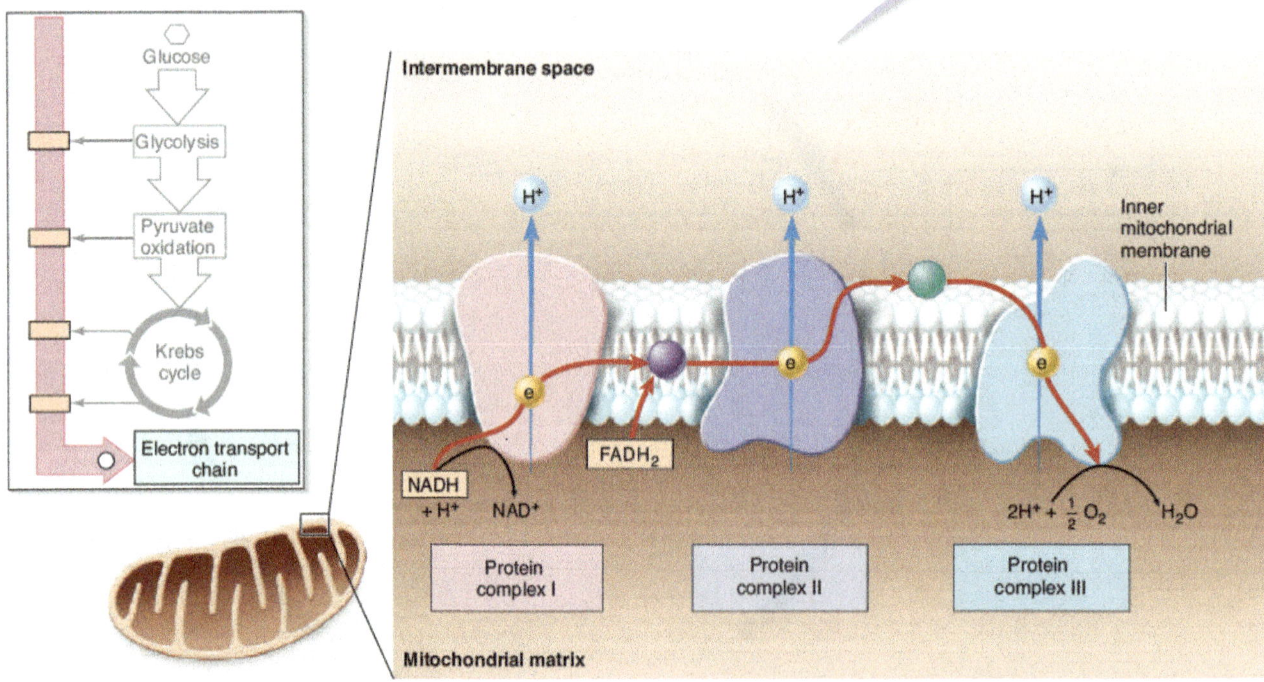

A protein complex (the pink structure above) receives the electrons and, using a mobile carrier, passes these electrons to a second protein complex (the purple structure). This protein complex, along with others in the chain, operates as a proton pump, using the energy of the electron to drive a proton out across the membrane into the intermembrane space. The arrows indicate the transport of the protons into the top half of the figure, which represents the intermembrane space.

The electron is then carried by another carrier to a third protein complex (the light blue structure). This complex uses electrons such as this one to link oxygen atoms with hydrogen ions to form molecules of water.

It is the availability of a plentiful supply of electron acceptor molecules like oxygen that makes oxidative respiration possible. The electron transport chain used in aerobic respiration is similar to, and may well have evolved from, the electron transport system employed in photosynthesis. Photosynthesis is thought to have preceded cellular respiration in the evolution of biochemical pathways, generating the oxygen that is necessary as the electron acceptor in cellular respiration. Natural selection didn't start from scratch and design a new biochemical pathway for cellular respiration; instead, it built on the photosynthetic pathway that already existed, and uses many of the same reactions.

Producing ATP: Chemiosmosis

As the proton concentration in the intermembrane space rises above that in the matrix, the concentration gradient induces the protons to reenter the matrix by diffusion through a special proton channel called **ATP synthase.** ATP synthase channels are embedded in the inner mitochondrial membrane, as shown in the figure to the left. As the protons pass through, these channels synthesize ATP from ADP and P_i within the matrix. The ATP is then transported by facilitated diffusion out of the mitochondrion and into the cell's cytoplasm. This ATP synthesizing process is the same chemiosmosis process that you encountered in studying photosynthesis in chapter 6.

Although we have discussed electron transport and chemiosmosis as separate processes, in a cell they are integrated, as shown below, left. The electron transport chain uses two electrons harvested in glycolysis, two harvested in pyruvate oxidation, and eight harvested in aerobic respiration (red arrows) to pump a large number of protons out across the inner mitochondrial membrane (shown in the upper right). Their subsequent reentry back into the mitochondrial matrix drives the synthesis of 34 ATP molecules by chemiosmosis (shown in the lower right). Two additional ATPs were harvested by a coupled reaction in glycolysis, and two more in the Krebs cycle. As two ATPs must be expended to transport NADH into the mitochondria by active transport, the grand total of ATPs harvested is thus 36 molecules.

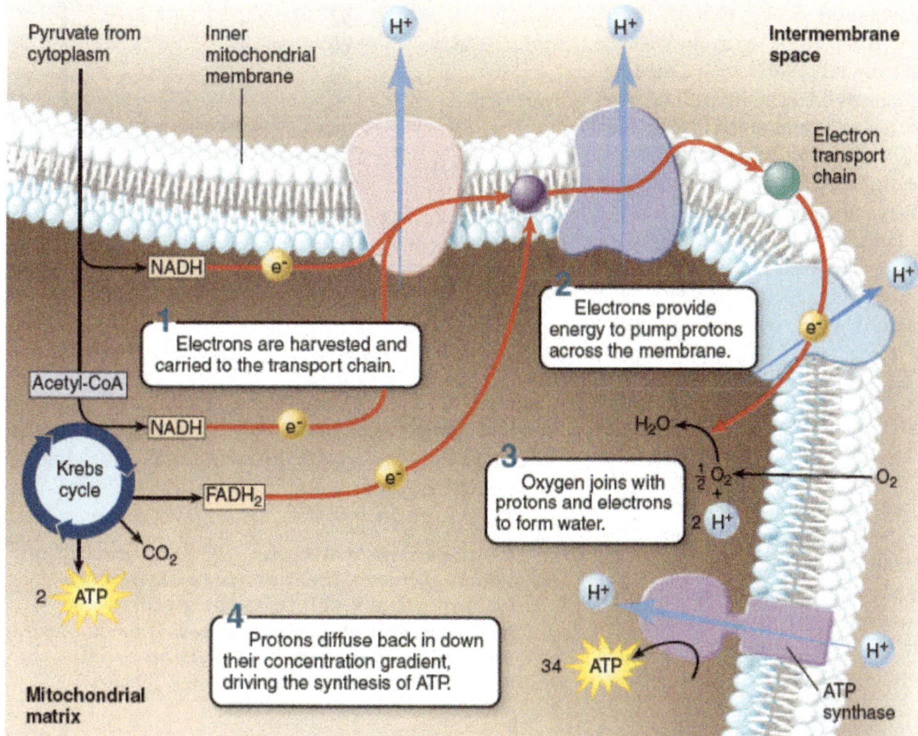

Key Learning Outcome 7.4 The electrons harvested by oxidizing food molecules are used to power proton pumps that chemiosmotically drive the production of ATP.

Harvesting Electrons Without Oxygen: Fermentation

7.5 Cells Can Metabolize Food Without Oxygen

Fermentation

In the absence of oxygen, aerobic metabolism cannot occur, and cells must rely exclusively on glycolysis to produce ATP. Under these conditions, the hydrogen atoms generated by glycolysis are donated to organic molecules in a process called **fermentation,** a process that recycles NAD^+, the electron acceptor required for glycolysis to proceed.

Bacteria carry out more than a dozen kinds of fermentations, all using some form of organic molecule to accept the hydrogen atom from NADH and thus recycle NAD^+:

Organic molecule + NADH	⟷	Reduced organic molecule + NAD^+

Often the reduced organic compound is an organic acid—such as acetic acid, butyric acid, propionic acid, or lactic acid—or an alcohol.

Ethanol Fermentation Eukaryotic cells are capable of only a few types of fermentation. In one type, which occurs in single-celled fungi called yeast, the molecule that accepts hydrogen from NADH is pyruvate, the end product of glycolysis itself. Yeast enzymes remove a CO_2 group from pyruvate through decarboxylation, producing a two-carbon molecule called acetaldehyde. The CO_2 released causes bread made with yeast to rise, while bread made without yeast (unleavened bread) does not. The acetaldehyde accepts a hydrogen atom from NADH, producing NAD^+ and ethanol (figure 7.7). This particular type of fermentation is of great interest to humans because it is the source of the ethanol in wine and beer. Ethanol is a by-product of fermentation that is actually toxic to yeast; as it approaches a concentration of about 12%, it begins to kill the yeast. That explains why naturally fermented wine contains only about 12% ethanol.

Lactic Acid Fermentation Most animal cells regenerate NAD^+ without decarboxylation. Muscle cells, for example, use an enzyme called lactate dehydrogenase to transfer a hydrogen atom from NADH back to the pyruvate that is produced by glycolysis. This reaction converts pyruvate into lactic acid and regenerates NAD^+ from NADH (figure 7.7). It therefore closes the metabolic circle, allowing glycolysis to continue as long as glucose is available. Circulating blood removes excess lactate (the ionized form of lactic acid) from muscles. It was once thought that during strenuous exercise, when the removal of lactic acid cannot keep pace with its production, the accumulation of lactic acid induces muscle fatigue. However, as you will learn in chapter 22, muscle fatigue actually has a quite different cause, involving the leakage of calcium ions within muscles.

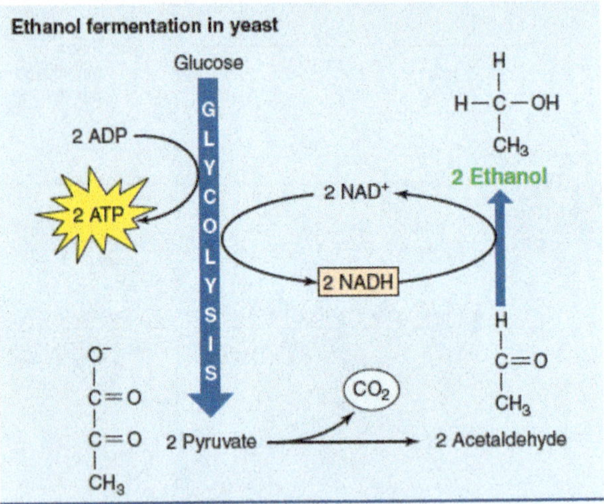

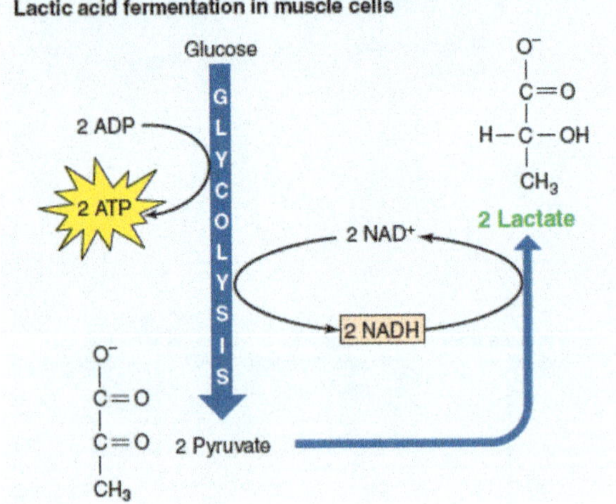

Figure 7.7 Fermentation.
Yeasts carry out the conversion of pyruvate to ethanol. Muscle cells convert pyruvate into lactate, which is less toxic than ethanol. In both cases, NAD^+ is regenerated to allow glycolysis to continue.

Key Learning Outcome 7.5 In fermentation, which occurs in the absence of oxygen, the electrons that result from the glycolytic breakdown of glucose are donated to an organic molecule, regenerating NAD^+ from NADH.

A Closer Look

Beer and Wine—Products of Fermentation

Beer fermentors.

Alcoholic fermentation, the anaerobic conversion of sugars into alcohol discussed on the facing page, predates human history. Like many other natural processes, humans seem to have first learned to control this process by stumbling across its benefit—the pleasures of beer and wine. Artifacts from the third millennium B.C. contain wine residue, and we know from the pollen record that domesticated grapes first became abundant at that time. The production of beer started even earlier, back in the fifth millennium B.C., making beer possibly one of the oldest beverages produced by humans.

The process of making beer is called brewing, and involves the fermentation of cereal grains. Grains are rich in starches, which you will recall from chapter 3 are long chains of glucose molecules linked together. First, the cereal grains are crushed and soaked in warm water, creating an extract called *mash*. Enzymes called amylases are added to the mash to break down the starch within the grains into free sugars. The mash is filtered, yielding a darker, sugary liquid called wort. The wort is boiled in large tanks, like the ones shown in the photo here, called fermentors. This helps break down the starch and kills any bacteria or other microorganisms that might be present before they begin to feast on all the sugar. Hops, another plant product, is added during the boiling stage. Hops gives the solution a bitter taste that complements the sweet flavor of the sugar in the wort.

Now we are ready to make beer. The temperature of the wort is brought down and yeast is added. This begins the fermentation process. It can be done in either of two ways, yielding the two basic kinds of beer.

Lager. The yeast *Saccharomyces uvarum* is a bottom-fermenting yeast that settles on the bottom of the vat during fermentation. This yeast produces a lighter, pale beer called lager. Bottom fermentation is the most widespread method of brewing. Bottom fermentation occurs at low temperatures (5–8°C). After fermentation is complete, the beer is cooled further to 0°C, allowing for the beer to mature before the yeast is filtered out.

Ale. The yeast *S. cerevisiae*, often called "brewer's yeast," is a top-fermenting yeast. It rises to the top during fermentation and produces a darker, more aromatic beer called ale (also called porter or stout). In top fermentation, less yeast is used. The temperature is higher, around 15–25°C, and so fermentation occurs more quickly but less sugar is converted into alcohol, giving the beer a sweeter taste.

Most yeasts used in beer brewing are alcohol-tolerant only up to about 5% alcohol—alcohol levels above that kill the yeast. That is why commercial beer is typically 5% alcohol.

Carbon dioxide is also a product of fermentation, and gives beer the bubbles that form the head of the beer. However, because only a little CO_2 forms naturally, the last step in most brewing involves artificial carbonation, with CO_2 gas being injected into the beer.

Wine. Wine fermentation is similar to beer fermentation in that yeasts are used to ferment sugars into alcohol. Wine fermentation, however, uses grapes. Rather than the starches found in cereal grains, grapes are fruits rich in the sugar sucrose, a disaccharide combination of glucose and fructose. Grapes are crushed and placed into barrels. There is no need for boiling to break down the starch, and grapes have their own amylase enzymes to cleave sucrose into free glucose and fructose sugars. To start the fermentation, yeast is added directly to the crush. *Saccharomyces cerevisiae* and *S. bayanus* are common wine yeasts. They differ from the yeast used in beer fermentation in that they have higher alcohol tolerances—up to 12% alcohol (some wine yeasts have even higher alcohol tolerances).

A common misconception about wine is that the color of the wine results from the color of the grape juice used, red wine using juice from red grapes and white wines using juice from white grapes. This is not true. The juice of all grapes is similar in color, usually light. The color of red wine is produced by leaving the skins of red or black grapes in the crush during the fermentation process. White wines can be made from any color of grapes, and are white simply because the skins that contribute red coloring are removed before fermentation.

White wines are typically fermented at low temperatures (8–19°C). Red wines are fermented at higher temperatures of 25–32°C with yeasts that have a higher heat tolerance. During wine fermentation, the carbon dioxide is vented out, leaving no carbonation in the wine. Champagnes have some natural carbonation that results from a two-step fermentation process: In the first fermentation step, the carbon dioxide is allowed to escape; in a second fermentation step, the container is sealed, trapping the carbon dioxide. But, like beer brewing, the natural carbonation is often supplemented with artificial carbonation.

Other Sources of Energy

7.6 Glucose Is Not the Only Food Molecule

We have considered in detail the fate of a molecule of glucose, a simple sugar, in cellular respiration. But how much of what you eat is sugar? As a more realistic example of the food you eat, consider the fate of a fast-food hamburger. The hamburger you eat is composed of carbohydrates, fats, protein, and many other molecules. This diverse collection of complex molecules is broken down by the process of digestion in your stomach and intestines into simpler molecules. Carbohydrates are broken down into simple sugars, fats into fatty acids, and proteins into amino acids. These breakdown reactions produce little or no energy themselves, but prepare the way for cellular respiration—that is, glycolysis and oxidative metabolism. Nucleic acids are also present in the food you eat and are broken down during digestion, but these macromolecules store little energy that the body actually uses.

We have seen what happens to the glucose. What happens to the amino acids and fatty acids? These subunits undergo chemical modifications that convert them into products that feed into cellular respiration.

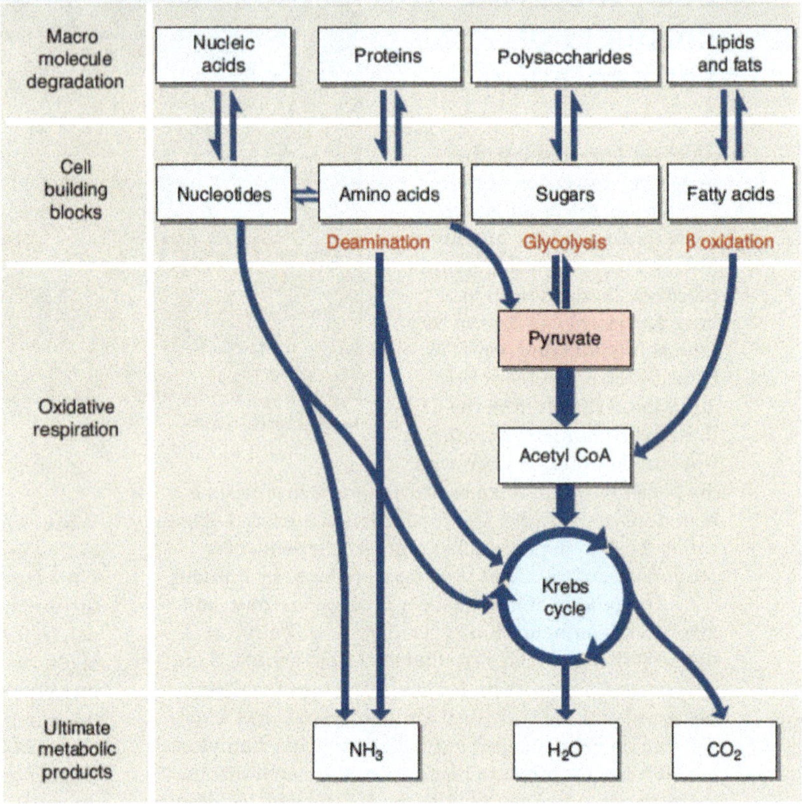

Figure 7.8 How cells obtain energy from foods.
Most organisms extract energy from organic molecules by oxidizing them. The first stage of this process, breaking down macromolecules into their subunits, yields little energy. The second stage, cellular respiration, extracts energy, primarily in the form of high-energy electrons. The subunit of many carbohydrates, glucose, readily enters glycolysis and passes through the biochemical pathways of oxidative respiration. However, the subunits of other macromolecules must be converted into products that can enter the biochemical pathways found in oxidative respiration.

Cellular Respiration of Protein

Proteins (the second category in figure 7.8) are first broken down into their individual amino acids. A series of *deamination* reactions removes the nitrogen side groups (called amino groups) and converts the rest of the amino acid into a molecule that takes part in the Krebs cycle. For example, alanine is converted into pyruvate, glutamate into α-ketoglutarate, and aspartate into oxaloacetate. The reactions of the Krebs cycle then extract the high-energy electrons from these molecules and put them to work making ATP.

Cellular Respiration of Fat

Lipids and fats (the fourth category in figure 7.8) are first broken down into fatty acids. A fatty acid typically has a long tail of sixteen or more $—CH_2$ links, and the many hydrogen atoms in these long tails provide a rich harvest of energy. Enzymes in the matrix of the mitochondrion first remove one two-carbon acetyl group from the end of a fatty acid tail, and then another, and then another, in effect chewing down the length of the tail in two-carbon bites. Eventually the entire fatty acid tail is converted into acetyl groups. Each acetyl group then combines with coenzyme A to form acetyl-CoA, which feeds into the Krebs cycle. This process is known as *β-oxidation*.

Thus, in addition to the carbohydrates, the proteins and fats in the hamburger also become important sources of energy.

Key Learning Outcome 7.6 Cells garner energy from proteins and fats, which are broken down into products that feed into cellular respiration.

Biology and *Staying Healthy*

Fad Diets and Impossible Dreams

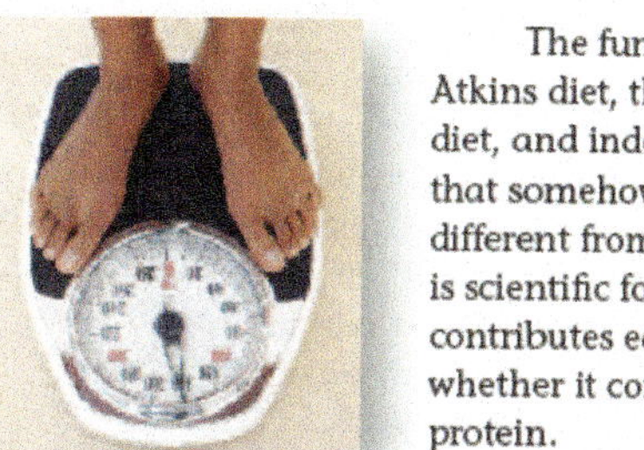

Most Americans put on weight in middle age, slowly adding 30 or more pounds. They did not ask for that weight, do not want it, and are constantly looking for a way to get rid of it. It is not a lonely search—it seems like everyone past the flush of youth is trying to lose weight. Many have been seduced by fad diets, investing hope only to harvest frustration. The much discussed Atkins diet is the fad diet most have tried—*Dr. Atkins' Diet Revolution* is one of the 10 best-selling books in history, and was (and is) prominently displayed in bookstores. The reason this diet doesn't deliver on its promise of pain-free weight loss is well understood by science, but not by the general public. Only hope and hype make it a perpetual best seller.

The secret of the Atkins diet, stated simply, is to avoid carbohydrates. Atkins's basic proposition is that your body, if it does not detect blood glucose (from metabolizing carbohydrates), will think it is starving and start to burn body fat, even if there is lots of fat already circulating in your bloodstream. You may eat all the fat and protein you want, all the steak and eggs and butter and cheese, and you will still burn fat and lose weight—just don't eat any carbohydrates, any bread or pasta or potatoes or fruit or candy. Despite the title of Atkins's book, this diet is hardly revolutionary. A basic low-carbohydrate diet was first promoted over a century ago in the 1860s by William Banting, an English casket maker, in his best-selling book *Letter on Corpulence*. Books promoting low-carbohydrate diets have continued to be best sellers ever since.

Those who try the Atkins diet often lose 10 pounds in two to three weeks. In three months it is all back, and then some. So what happened? Where did the pounds go, and why did they come back? The temporary weight loss turns out to have a simple explanation. Carbohydrates act as water sponges in your body, and so forcing your body to become depleted of carbohydrates causes your body to lose water. The 10 pounds lost on this diet was not fat weight but water weight, quickly regained with the first starchy foods eaten.

The Atkins' diet is the sort of diet the American Heart Association tells us to avoid (all those saturated fats and cholesterol), and it is difficult to stay on. If you do hang in there, you will lose weight, simply because you eat less. Other popular diets these days, *The Zone* diet of Dr. Barry Sears and *The South Beach Diet* of Dr. Arthur Agatston, are also low-carbohydrate diets, although not as extreme as the Atkins diet. Like the Atkins diet, they work not for the bizarre reasons claimed by their promoters, but simply because they are low-calorie diets.

There are two basic laws that no diet can successfully violate:

1. All calories are equal.
2. (calories in) – (calories out) = fat.

The fundamental fallacy of the Atkins diet, the Zone diet, the South Beach diet, and indeed of all fad diets, is the idea that somehow carbohydrate calories are different from fat and protein calories. This is scientific foolishness. Every calorie you eat contributes equally to your eventual weight, whether it comes from carbohydrate, fat, or protein.

To the extent these diets work at all, they do so because they obey the second law. By reducing calories in, they reduce fat. If that were all there was to it, we should all go out and buy a diet book. Unfortunately, losing weight isn't that simple, as anyone who has seriously tried already knows. The problem is that your body will not cooperate.

If you try to lose weight by exercising and eating less, your body will attempt to compensate by metabolizing more efficiently. It has a fixed weight, what obesity researchers call a "set point," a weight to which it will keep trying to return. A few years ago, a group of researchers at Rockefeller University in New York, in a landmark study, found that if you lose weight, your metabolism slows down and becomes more efficient, burning fewer calories to do the same work—your body will do everything it can to gain the weight back! Similarly, if you gain weight, your metabolism speeds up. In this way your body uses its own natural weight control system to keep your weight at its set point. No wonder it's so hard to lose weight!

Clearly our bodies don't keep us at one weight all our adult lives. It turns out your body adjusts its fat thermostat—its set point—depending on your age, food intake and amount of physical activity. Adjustments are slow, however, and it seems to be a great deal easier to move the body's set point up than to move it down. Apparently higher levels of fat reduce the body's sensitivity to the leptin hormone that governs how efficiently we burn fat. That is why you can gain weight, despite your set point resisting the gain—your body still issues leptin alarm calls to speed metabolism, but your brain doesn't respond with as much sensitivity as it used to. Thus the fatter you get, the less effective your weight control system becomes.

This doesn't mean that we should give up and learn to love our fat. Rather, now that we are beginning to understand the biology of weight gain, we must accept the hard fact that we cannot beat the requirements of the two diet laws. The real trick is not to give up. Eat less and exercise more, and keep at it. In one year, or two, or three, your body will readjust its set point to reflect the new reality you have imposed by constant struggle. There simply isn't any easy way to lose weight.

How Do Swimming Fish Avoid Low Blood pH?

Animals that live in oxygen-poor environments, like worms living in the oxygen-free mud at the bottom of lakes, are not able to obtain the energy required for muscle movement from the Krebs cycle. Their cells lack the oxygen needed to accept the electrons stripped from food molecules. Instead, these animals rely on glycolysis to obtain ATP, donating the electron to pyruvate, forming lactic acid. While much less efficient than the Krebs cycle, glycolysis does not require oxygen. Even when oxygen is plentiful, the muscles of an active animal may use up oxygen more quickly than it can be supplied by the bloodstream and so be forced to temporarily rely on glycolysis to generate the ATP for continued contraction.

This presents a particular problem for fish. Fish blood is much lower in carbon dioxide than yours is, and as a consequence, the amount of sodium bicarbonate acting as a buffer in fish blood is also quite low. Now imagine you are a trout, and need to suddenly swim very fast to catch a mayfly for dinner. The vigorous swimming will cause your muscles to release large amounts of lactic acid into your poorly-buffered blood; this could severely disturb the blood's acid-base balance and so impede contraction of your swimming muscles before the prey is captured.

The graph to the right presents the results of an experiment designed to explore how a trout solves this dilemma. In the experiment, the trout was made to swim vigorously for 15 minutes in a laboratory tank, and then allowed a day's recovery. The lactic acid concentration in its blood was monitored periodically during swimming and recovery phases.

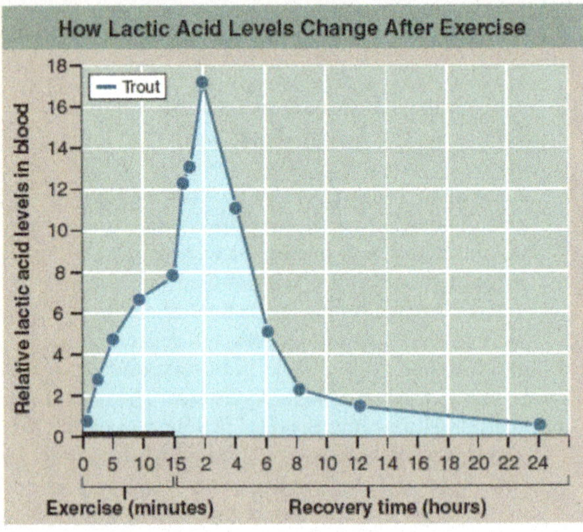

1. **Applying Concepts**
   a. Variable. What is the dependent variable?
   b. Recording Data. Lactic acid levels are presented for both swimming and recovery periods. In what time units are the swimming data presented? The recovery data?
2. **Interpreting Data**
   a. What is the effect of exercise on the level of lactic acid in the trout's blood?
   b. Does the level of lactic acid change after exercise stops? How?
3. **Making Inferences** About how much of the total lactic acid created by vigorous swimming is released after this exercise stops? [Hint: Notice the *x* axis scale changes from minutes to hours, so replot all points to minutes and compare areas under curve.]
4. **Drawing Conclusions** Is this result consistent with the hypothesis that fish maintain blood pH levels by delaying the release of lactic acid from muscles? Why might this be beneficial to the fish?

Chapter Review

An Overview of Cellular Respiration

7.1 Where Is the Energy in Food?

- Nonphotosynthetic organisms acquire energy from the breakdown of food, either by eating plants that store the food or by eating animals that have eaten plants (**figure 7.1**). Energy stored in carbohydrate molecules is extracted through the process of cellular respiration and is stored in the cell as ATP.
- Cellular respiration is carried out in two stages: glycolysis occurring in the cytoplasm and oxidation occurring in the mitochondria (**figure 7.3**).

Respiration Without Oxygen: Glycolysis

7.2 Using Coupled Reactions to Make ATP

- Glycolysis is an energy-extracting process. It is a series of 10 chemical reactions in which glucose is broken down into two three-carbon pyruvate molecules. The energy is extracted from glucose by two exergonic reactions that are coupled with an endergonic reaction that leads to the formation of ATP. This is called substrate-level phosphorylation.
- The initial steps in glycolysis, called the priming reactions, require the input of energy from ATP. After priming, the glucose molecule splits into two smaller molecules in the cleavage reactions (**Key Biological Process, page 138**). The energy harvesting reactions that follow produce ATP, but very little net ATP is produced (**figure 7.4**).
- Electrons extracted from glucose are donated to a carrier molecule, NAD^+, which becomes NADH. NADH carries electrons and hydrogen atoms to be used in a later stage of oxidative respiration.

Respiration With Oxygen: The Krebs Cycle

7.3 Harvesting Electrons from Chemical Bonds

- The two molecules of pyruvate formed in glycolysis are passed into the mitochondrion where they are converted into two molecules of acetyl-coenzyme A (**figure 7.5**). What the cell does with acetyl-CoA depends on the needs of the cell. If the cell has enough ATP, acetyl-CoA is used in synthesizing fat molecules. If the cell needs energy, acetyl-CoA is directed to the Krebs cycle.
- The formation of NADH is an enzyme catalyzed reaction (**Key Biological Process, page 140**). The enzyme brings the substrate and NAD^+ into close proximity. Through a redox reaction, a hydrogen atom and an electron are transferred to NAD^+, reducing it to NADH. NADH then carries the electrons and hydrogen to a later step in oxidative respiration.

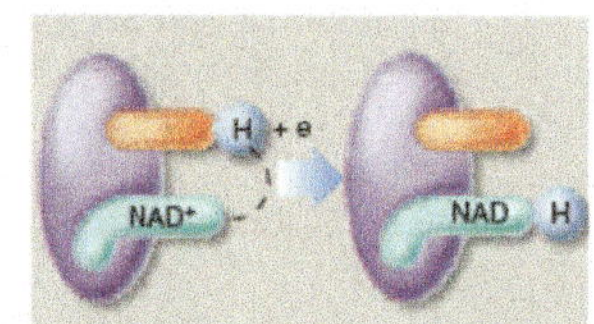

- Acetyl-CoA enters a series of chemical reactions called the Krebs cycle, where one molecule of ATP is produced in a coupled reaction. Also, energy is harvested in the form of electrons that are transferred to molecules of NAD^+ and FAD to produce NADH and $FADH_2$, respectively (**Key Biological Process, page 142** and **figure 7.6**).
- The Krebs cycle makes two turns for every molecule of glucose that is oxidized.

7.4 Using the Electrons to Make ATP

- The molecules of NADH and $FADH_2$ that were produced during glycolysis and the Krebs cycle carry electrons to the inner mitochondrial membrane. Here they give up electrons to the electron transport chain. The electrons, along with their energy, are passed along the electron transport chain. The energy from the electrons drives proton pumps that pump H^+ across the inner membrane from the matrix to the intermembrane space, creating an H^+ concentration gradient (**integrated art, page 144**).
- When the electrons reach the end of the electron transport chain, they bind with oxygen and hydrogen to form water molecules.
- ATP is produced in the mitochondrion through chemiosmosis. The H^+ concentration gradient in the intermembrane space drives H^+ back across the membrane through ATP synthase channels (**integrated art, top of page 145**). The energy from the movement of H^+ through the channel is transferred to the chemical bonds in ATP. Thus, the energy stored in the glucose molecule is harvested through glycolysis and the Krebs cycle with the formation of NADH and $FADH_2$ and ultimately the formation of ATP. The energy carried by NADH and $FADH_2$ is transferred to the electron transport chain and is stored in ATP (**integrated art, bottom of page 145**).

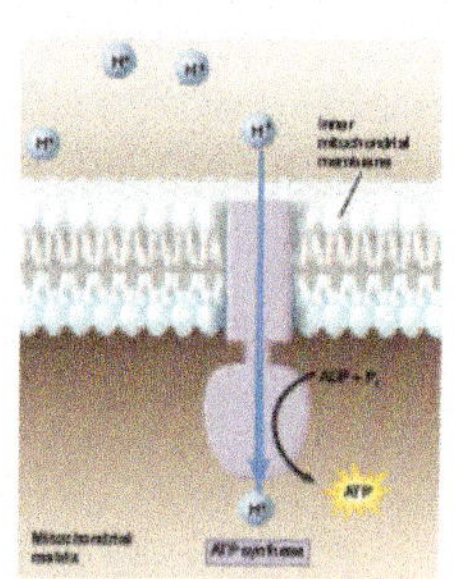

Harvesting Electrons Without Oxygen: Fermentation

7.5 Cells Can Metabolize Food Without Oxygen

- In the absence of oxygen, other molecules can be used as electron acceptors. When the electron acceptor is an organic molecule, the process is called fermentation (**figure 7.7**). Depending on what type of organic molecule accepts the electrons, either ethanol or lactic acid, in the form of lactate, is formed.

Other Sources of Energy

7.6 Glucose Is Not the Only Food Molecule

- Food sources other than glucose are also used in oxidative respiration. Macromolecules, such as proteins, lipids, and nucleic acids, are broken down into intermediate products that enter cellular respiration in different reaction steps, as shown here from **figure 7.8**.

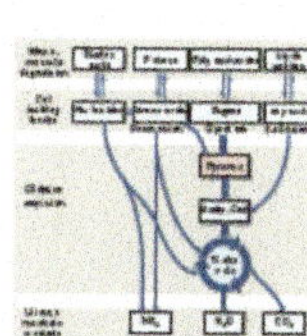

Test Your Understanding

1. In animals, the energy for life is obtained by cellular respiration. This involves
   a. breaking down the organic molecules that were consumed.
   b. capturing photons from plants.
   c. obtaining ATP from plants.
   d. breaking down CO_2 that was produced by plants.
2. During glycolysis, ATP forms by
   a. the breakdown of pyruvate.
   b. chemiosmosis.
   c. substrate-level phosphorylation.
   d. NAD^+.
3. Which of the following processes can occur in the absence of oxygen?
   a. the Krebs cycle
   b. glycolysis
   c. chemiosmosis
   d. All of the above.
4. Every living creature on this planet is capable of carrying out the rather inefficient biochemical process of glycolysis, which
   a. makes glucose, using the energy from ATP.
   b. makes ATP by splitting glucose and capturing the energy.
   c. phosphorylates ATP to make ADP.
   d. makes glucose, using oxygen and carbon dioxide and water.
5. The electrons generated from the Krebs cycle are transferred to ______________, which then carries them to ________________.
   a. NAD^+, oxygen
   b. NAD^+, the electron transport chain
   c. NADH, oxygen
   d. NADH, the electron transport chain
6. After glycolysis, the pyruvate molecules go to the
   a. nucleus of the cell and provide energy.
   b. membranes of the cell and are broken down in the presence of CO_2 to make more ATP.
   c. mitochondria of the cell and are broken down in the presence of O_2 to make more ATP.
   d. Golgi bodies and are packaged and stored until needed.
7. The vast majority of the ATP molecules produced within a cell are produced
   a. during pyruvate oxidation.
   b. during glycolysis.
   c. during the Krebs cycle.
   d. during the electron transport chain.
8. NAD^+ is recycled during
   a. glycolysis.
   b. fermentation.
   c. the Krebs cycle.
   d. the formation of acetyl-CoA.
9. The final electron acceptor in lactic acid fermentation is
   a. pyruvate.
   b. NAD^+.
   c. lactic acid.
   d. O_2.
10. Cells can extract energy from foodstuffs other than glucose because
    a. proteins, fatty acids, and nucleic acids get converted to glucose and then enter oxidative respiration.
    b. each type of macromolecule has its own oxidative respiration pathway.
    c. each type of macromolecule is broken down into its subunits, which enter the oxidative respiration pathway.
    d. they can all enter the glycolytic pathway.

Apply Your Understanding

1. Consider the structure of a mitochondrion, shown here in a cutaway view. If you were able to poke a hole in this mitochondrion, do you think it still would be able to perform oxidative respiration? Explain.

2. **Figure 7.8** Your friend Yevgeny wants to go on a low-carbohydrate diet so that he can lose some of the "baby fat" he's still carrying. He asks your advice; what do you tell him?

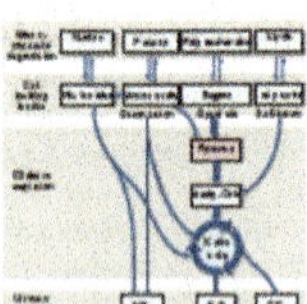

Synthesize What You Have Learned

1. The electron carrier cytochrome *c* is one of many different cytochromes, but unlike the others, the sequence of *cytochrome c* is nearly identical in all species. Why do you suppose this is so? Among humans, no genetic disorder affecting cytochrome *c* has ever been reported. Why do you suppose this is so?
2. How much less ATP would be generated in the cells of a person who consumed a diet of pyruvate instead of glucose (use one molecule of each for your calculation)?
3. Which of the following food molecules would generate the most ATP molecules, assuming that glycolysis, the Krebs cycle, and the electron transport chain were all functioning and that the foods were consumed in equal amounts: carbohydrates, proteins, or fats? Explain your answer.
4. Soft drinks are artificially carbonated, which is what causes them to fizz. Beer and sparkling wines are naturally carbonated. How does this natural carbonation occur?

Unit Three | The Continuity of Life

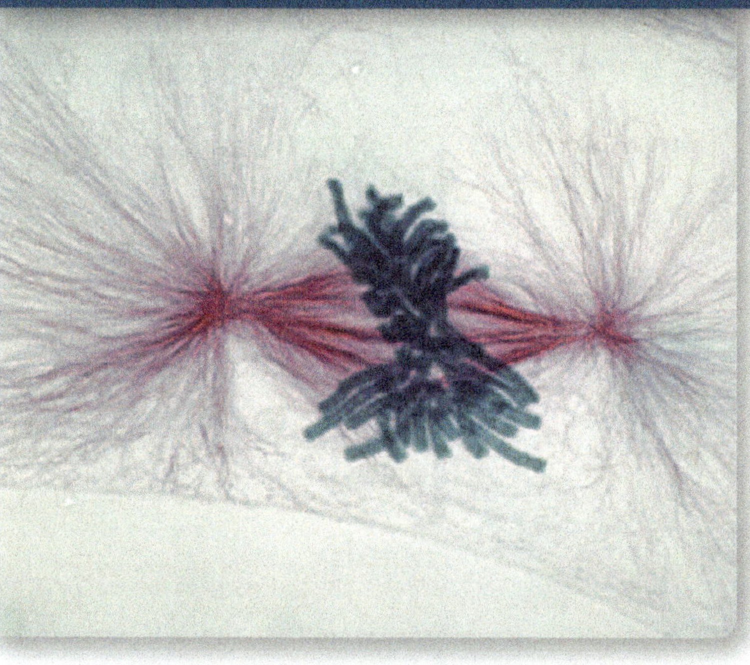

8 Mitosis

Learning Objectives

Cell Division

8.1 Prokaryotes Have a Simple Cell Cycle
1. Define binary fission.
2. Describe a prokaryotic chromosome.
3. Diagram the prokaryotic cell cycle, identifying DNA replication, DNA partitioning, and cell fission.

8.2 Eukaryotes Have a Complex Cell Cycle
1. List and describe the three phases of interphase.
2. Distinguish between M phase and C phase.

8.3 Chromosomes
1. Distinguish between homologue and sister chromatid.
2. State the total number of chromosomes in a human body cell and the number of DNA nucleotides in a typical human chromosome.
3. Diagram a nucleosome and discuss its function.

8.4 Cell Division
1. Name and describe the four stages of mitosis.
2. Distinguish between kinetochore and centromere.
3. Distinguish karyokinesis from cytokinesis.

8.5 Controlling the Cell Cycle
1. Locate the three points on the cell cycle where checkpoints occur.
2. Name and describe the three checkpoints.
3. Describe the Hayflick limit.
4. Discuss the role of telomeres in limiting cell proliferation.

Cancer and the Cell Cycle

8.6 What Is Cancer?
1. Distinguish between tumors and metastases.
2. Explain how mutation is linked to cancer.
3. Name and describe the two classes of growth factor genes usually involved in cancer.

8.7 Cancer and Control of the Cell Cycle
1. Describe the role of p53 in preventing cancer.

Biology and Staying Healthy: Curing Cancer

Inquiry & Analysis: Why Do Human Cells Age?

The cell you see above is a dividing cell of the Oregon newt *Taricha granulosa,* a kind of salamander. The micrograph (that is, a photo taken through a microscope) captures the cell in the metaphase stage of mitosis, when all the blue-stained chromosomes are lined up on the metaphase plate. Soon the red-stained spindle fibers will draw duplicates of the homologous chromosomes to opposite poles of the cell. When cell division is complete, two daughter cells will result, each containing the same amount of DNA as the parent cell. Different types of cells divide at different rates. Some human cells divide frequently, particularly those subjected to a lot of wear and tear. The epithelial cells of your skin divide so often that your skin replaces itself every two weeks. The lining of your stomach is replaced every few days! Nerve cells, on the other hand, can live for 100 years without dividing. Cells use a battery of genes to regulate when and how frequently they divide. If some of these genes become disabled, a cell may begin to divide ceaselessly, a condition we call cancer. Exposure to DNA-damaging chemicals such as those in cigarette smoke greatly increases the chance of this sort of event occurring in the tissues exposed to the smoke, which is why smokers will more likely get lung cancer than colon cancer.

Cell Division

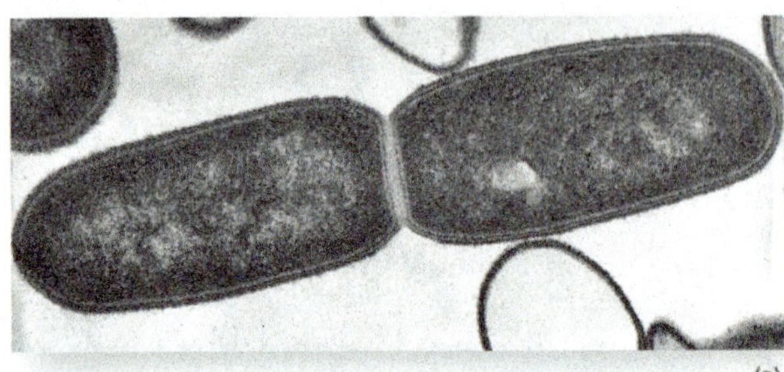

(a)

8.1 Prokaryotes Have a Simple Cell Cycle

All species reproduce, passing their hereditary information on to their offspring. In this chapter, we begin our consideration of heredity with a look at how cells reproduce. Cell division in prokaryotes takes place in two stages, which together make up a simple cell cycle. First the DNA is copied, and then the cell splits in half by a process called **binary fission.** The cell in figure 8.1*a* is undergoing binary fission.

In prokaryotes, the hereditary information—that is, the genes that specify the prokaryote—is encoded in a single circle of DNA, called a prokaryotic chromosome. Before the cell itself divides, the DNA circle makes a copy of itself, a process called *replication.* Starting at one point, the origin of replication (the point where the two strands of DNA are connected at the top of figure 8.1*b*), the double helix of DNA begins to unzip, exposing the two strands. The enlargement on the right of figure 8.1*b* shows how the DNA replicates. The purple strand is from the original DNA and the red strand is the newly formed DNA. The new double helix is formed from each naked strand by placing on each exposed nucleotide its complementary nucleotide (that is, A with T, G with C, as discussed in chapter 3). DNA replication is discussed in more detail in chapter 11. When the unzipping has gone all the way around the circle, the cell possesses two copies of its hereditary information.

When the DNA has been copied, the cell grows, resulting in elongation. The newly replicated DNA molecules are partitioned toward each end of the cell. This partitioning process involves DNA sequences near the origin of replication, and results in these sequences being attached to the membrane. When the cell reaches an appropriate size, the prokaryotic cell begins to split into two equal halves. New plasma membrane and cell wall are added at a point between where the two DNA copies are partitioned, indicated by the green divider in figure 8.1*b*. As the growing plasma membrane pushes inward, the cell is constricted in two, eventually forming two *daughter cells.* Each contains one prokaryotic chromosome and is a complete living cell in its own right.

Key Learning Outcome 8.1 Prokaryotes divide by binary fission after the DNA has replicated.

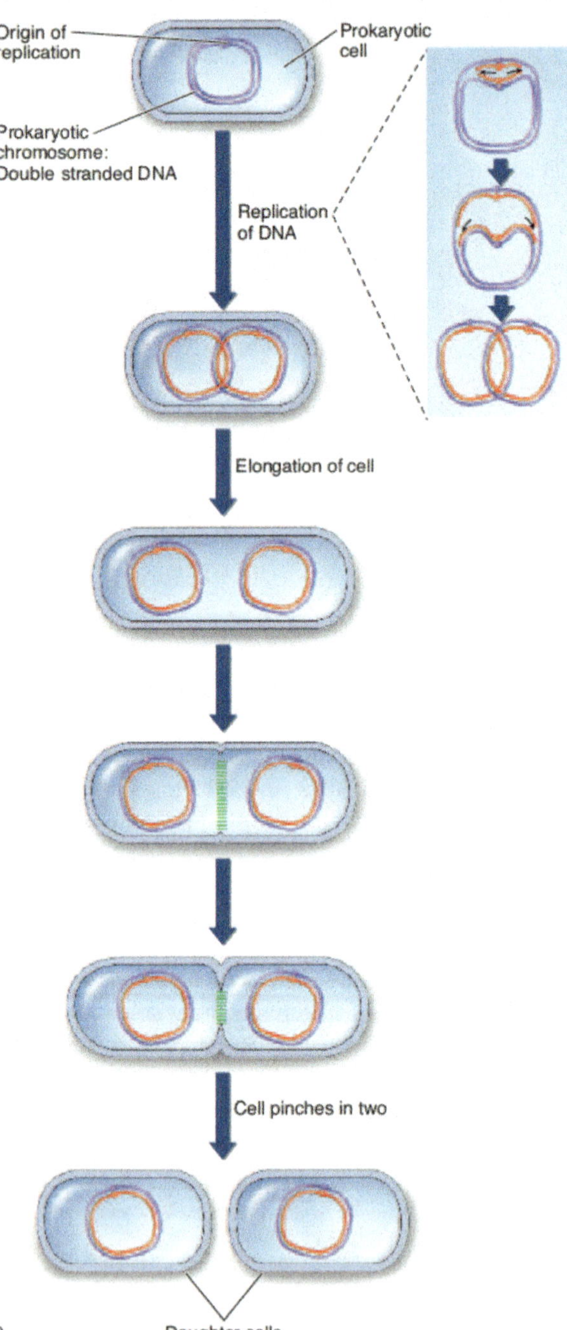

Figure 8.1 Cell division in prokaryotes.
(*a*) Prokaryotes divide by a process of binary fission. Here, a cell has divided in two and is about to be pinched apart by the growing plasma membrane. (*b*) Before the cell splits, the circular DNA molecule of a prokaryote initiates replication at a single site, called the origin of replication, moving out in both directions. When the two moving replication points meet on the far side of the molecule, its replication is complete. The cell then undergoes binary fission, where the cell divides into two daughter cells.

8.2 Eukaryotes Have a Complex Cell Cycle

The evolution of the eukaryotes introduced several additional factors into the process of cell division. Eukaryotic cells are much larger than prokaryotic cells, and they contain much more DNA. Eukaryotic DNA is contained in a number of linear chromosomes, whose organization is much more complex than that of the single, circular DNA molecules in prokaryotes. A eukaryotic **chromosome** is a single, long DNA molecule wound tightly around proteins, called *histones,* into a compact shape.

Cell division in eukaryotes is more complex than in prokaryotes, both because eukaryotes contain far more DNA and because it is packaged differently. The cells of eukaryotic organisms either undergo mitosis or meiosis to divide up the DNA. **Mitosis** is the mechanism of cell division that occurs in an organism's nonreproductive cells, or *somatic cells.* An alternate process, called **meiosis,** divides the DNA in cells that participate in sexual reproduction, or *germ cells.* Meiosis results in the production of gametes, such as sperm and eggs, and is discussed in chapter 9.

The events that prepare the eukaryotic cell for division and the division process itself constitute a **complex cell cycle.** The Key Biological Process illustration below walks you through the phases of the cell cycle:

Interphase. This is the first phase of the cell cycle, step 1 in the figure below, and is usually considered a resting phase, but the cell is far from resting. Interphase is itself made up of three phases:

G_1 phase. This "first gap" phase is the cell's primary growth phase. For most organisms, this phase occupies the major portion of the cell's life span.

S phase. In this "synthesis" phase, the DNA replicates, producing two copies of each chromosome.

G_2 phase. Cell division preparation continues in the "second gap" phase with the replication of mitochondria, chromosome condensation, and the synthesis of microtubules.

M phase. In mitosis, a microtubular apparatus binds to the chromosomes and moves them apart, shown in steps 2–5.

C phase. In cytokinesis, the cytoplasm divides, creating two daughter cells, shown in step 6.

Human cells growing in culture typically have a 22-hour cell cycle. Most cell types take about 80 minutes in this 22 hours to complete cell division: prophase—23 minutes, metaphase—29 minutes, anaphase—10 minutes, telophase—14 minutes, and cytokinesis—4 minutes. The proportion of the cell cycle spent in any one phase of mitosis varies considerably in different tissues.

Key Learning Outcome 8.2 **Eukaryotic cells divide by separating duplicate copies of their chromosomes into daughter cells.**

KEY BIOLOGICAL PROCESS: The Cell Cycle

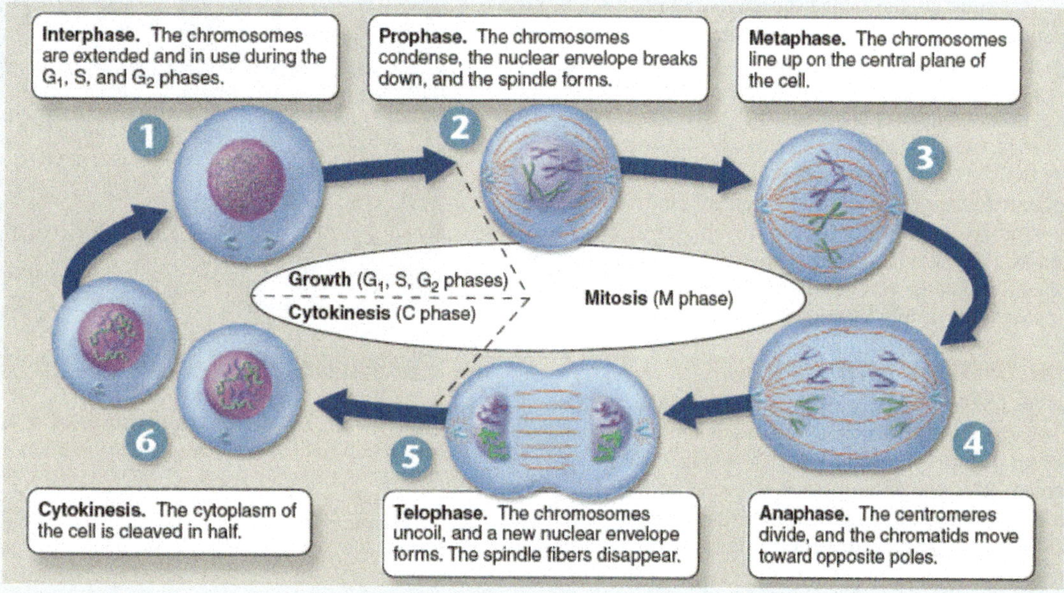

8.3 Chromosomes

Chromosomes were first observed by the German embryologist Walther Fleming in 1879 while he was examining the rapidly dividing cells of salamander larvae. When Fleming looked at the cells through what would now be a rather primitive light microscope, he saw minute threads within their nuclei that appeared to be dividing lengthwise. Fleming called their division *mitosis,* based on the Greek word *mitos,* meaning "thread."

Chromosome Number

Since their initial discovery, chromosomes have been found in the cells of all eukaryotes examined. Their number may vary enormously from one species to another. A few kinds of organisms—such as the Australian ant *Myrmecia* spp.; the plant *Haplopappus gracilis,* a relative of the sunflower that grows in North American deserts; and the fungus *Penicillium*—have only 1 pair of chromosomes, while some ferns have more than 500 pairs. Most eukaryotes have between 10 and 50 chromosomes in their body cells.

Homologous Chromosomes

Chromosomes exist in somatic cells as pairs, called **homologous chromosomes,** or **homologues.** Homologues carry information about the same traits at the same locations on each chromosome but the information can vary between homologues, which will be discussed in chapter 10. Cells that have two of each type of chromosome are called **diploid cells.** One chromosome of each pair is inherited from the mother (colored green in figure 8.2) and the other from the father (colored purple). Before cell division, each homologous chromosome replicates, resulting in two identical copies, called **sister chromatids.** You see in figure 8.2 that the sister chromatids remain joined together after replication at a special linkage site called the **centromere,** the knoblike structure in the middle of each chromosome. Human body cells have a total of 46 chromosomes, which are actually 23 pairs of homologous chromosomes. In their duplicated state, before mitosis, there are still only 23 pairs of chromosomes, but each chromosome has duplicated and consists of two sister chromatids, for a total of 92 chromatids. The duplicated sister chromatids can make it confusing to count the number of chromosomes in an organism, but keep in mind that the number of centromeres doesn't increase with replication, and so you can always determine the number of chromosomes simply by counting the centromeres.

The Human Karyotype

The 46 human chromosomes can be paired as homologues by comparing size, shape, location of centromeres, and so on. This arrangement of chromosomes is called a *karyotype.* An example of a human karyotype is shown in figure 8.3. You can see how the different sizes and shapes of chromosomes allow scientists to pair together the ones that are homologous.

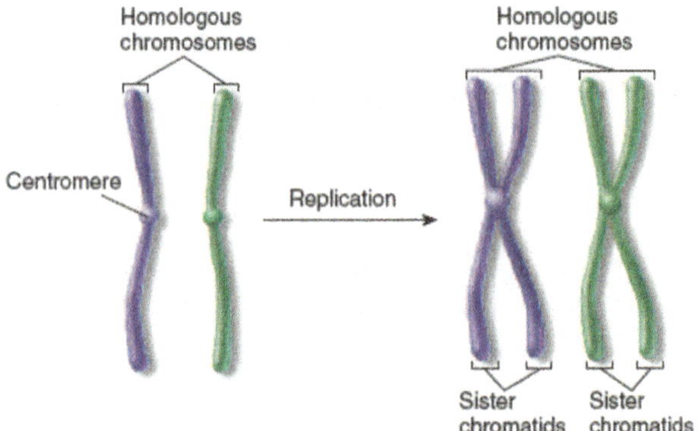

Figure 8.2 The difference between homologous chromosomes and sister chromatids.
Homologous chromosomes are a pair of the same chromosome—say, chromosome number 16. Sister chromatids are the two replicas of a single chromosome held together by the centromere after DNA replication. A duplicated chromosome looks somewhat like an X.

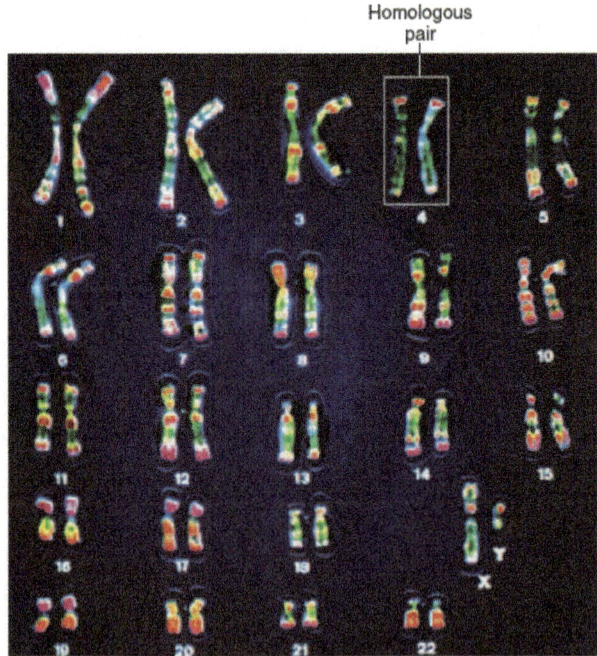

Figure 8.3 The 46 chromosomes of a human.
In this presentation, photographs of the individual chromosomes of a human male have been cut out and paired with their homologues, creating an organized display called a *karyotype.* The chromosomes are in a duplicated state, and the sister chromatids can actually be seen in many of the homologous pairs.

For example, chromosome 1 is much larger than chromosome 14, and its centromere is more centrally located on the chromosome. Each chromosome contains thousands of genes that play important roles in determining how a person's body develops and functions. For this reason, possession of all the chromosomes is essential to survival. Humans missing even one chromosome, a condition called monosomy, do not usually survive embryonic development. Nor does the human embryo develop properly with an extra copy of any one chromosome, a condition called trisomy. For all but a few of the smallest chromosomes, trisomy is fatal; even in those cases, serious problems result. We will revisit this issue of differences in chromosome number in chapter 10.

Chromosome Structure

Chromosomes are composed of **chromatin,** a complex of DNA and protein; most are about 40% DNA and 60% protein. A significant amount of RNA is also associated with chromosomes because chromosomes are the sites of RNA synthesis. The DNA of a chromosome is one very long, double-stranded fiber that extends unbroken through the entire length of the chromosome. A typical human chromosome contains about 140 million (1.4×10^8) nucleotides in its DNA. Furthermore, if the strand of DNA from a single chromosome were laid out in a straight line, it would be about 5 centimeters (2 inches) long. The amount of information in one human chromosome would fill about 2,000 printed books of 1,000 pages each! Fitting such a strand into a nucleus is like cramming a string the length of a football field into a baseball—and that's only 1 of 46 chromosomes! In the cell, however, the DNA is coiled, allowing it to fit into a much smaller space than would otherwise be possible.

Chromosome Coiling

The DNA of eukaryotes is divided into several chromosomes, although the chromosomes you see in figure 8.3 hardly look like long, double-stranded molecules of DNA. These chromosomes, duplicated as sister chromatids, are formed by winding and twisting the long DNA strands into a much more compact structure. Winding up DNA presents an interesting challenge. Because the phosphate groups of DNA molecules have negative charges, it is impossible to just tightly wind up DNA because all the negative charges would simply repel one another. As you can see in figure 8.4, the DNA helix wraps around proteins with positive charges called **histones.** The positive charges of the histones counteract the negative charges of the DNA, so that the complex has no net charge. Every 200 nucleotides, the DNA duplex is coiled around a core of eight histone proteins, forming a complex known as a **nucleosome.** The nucleosomes, which resemble beads on a string in figure 8.4, are further coiled into a solenoid. This solenoid is then organized into looped domains. The final organization of the chromosome is not known, but it appears to involve further radial looping into rosettes around a preexisting scaffolding of protein. This complex of DNA and histone proteins, coiled tightly, forms a compact chromosome.

Key Learning Outcome 8.3 All eukaryotic cells store their hereditary information in chromosomes, but different kinds of organisms use very different numbers of chromosomes to store this information. Coiling of the DNA into chromosomes allows it to fit in the nucleus.

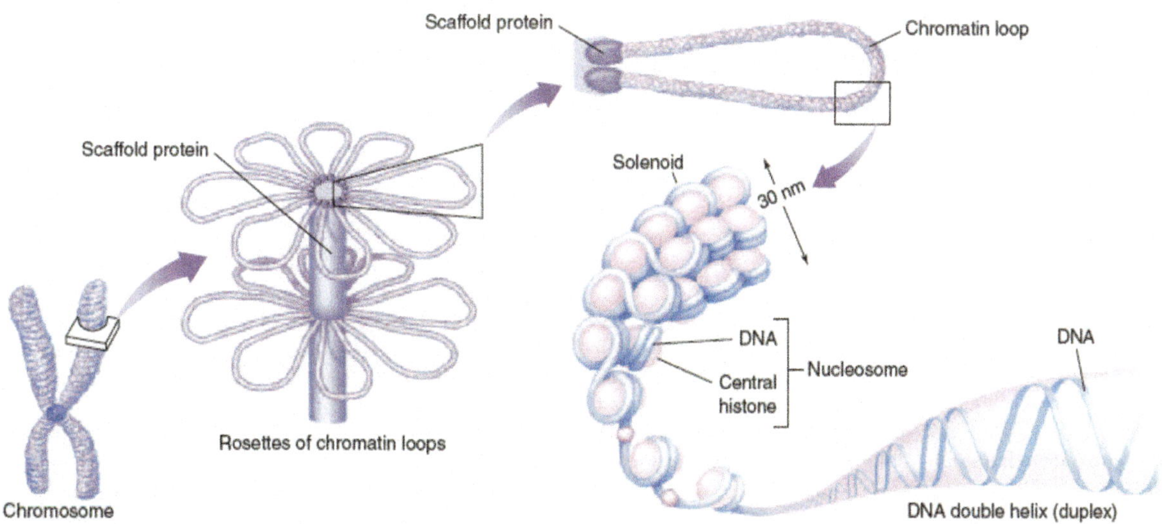

Figure 8.4 Levels of eukaryotic chromosomal organization.

Compact, rod-shaped chromosomes are in fact highly wound-up molecules of DNA. The arrangement illustrated here is one of many possibilities.

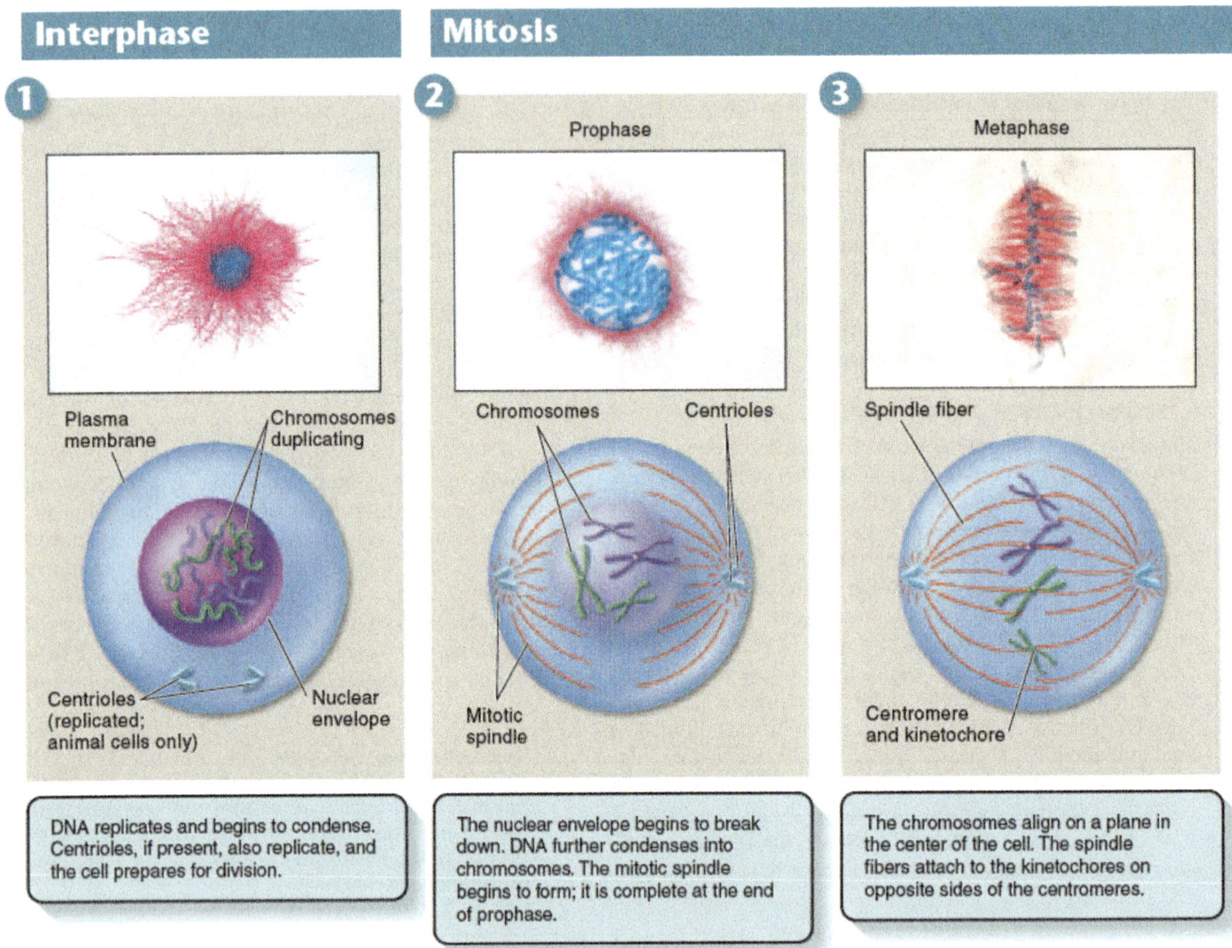

Figure 8.5 How cell division works.
Cell division in eukaryotes begins in interphase, carries through the four stages of mitosis, and ends with cytokinesis. Several features of the spindle illustrated in the drawings above appear in dividing animal cells but not in plant cells, and cannot be seen in the photographs, which are of the African blood lily *Haemanthus katharinae*. (In these exceptional photographs, the chromosomes are stained *blue* and microtubules stained *red*.)

8.4 Cell Division

Interphase

When cell division begins in interphase, chromosomes first replicate, and then begin to wind up tightly, a process called **condensation.** Sister chromatids are held together by a complex of proteins called *cohesin*. Chromosomes are not usually visible during interphase, but to clarify what is happening, they are shown in panel 1 of figure 8.5 as if they were.

Mitosis

Interphase is not a phase of mitosis, but it sets the stage for cell division. It is followed by nuclear division, called *mitosis*. Although the process of mitosis is continuous, with the stages flowing smoothly one into another, for ease of study, mitosis is traditionally subdivided into four stages: prophase, metaphase, anaphase, and telophase. We will be referring to the panels in figure 8.5 in the following descriptions.

Prophase: Mitosis Begins In **prophase,** the individual condensed chromosomes, the blue structures in the photo of panel 2, first become visible with a light microscope. As the replicated chromosomes condense, the nucleolus disappears and the cell dismantles the nuclear envelope and begins to assemble the apparatus it will use to pull the replicated sister chromatids to opposite ends ("poles") of the cell. In the center of an animal cell, the centrioles have replicated, and the two pairs of centrioles move apart toward opposite poles of the cell, forming between them as they move apart a network of protein cables called the **spindle.** In panel 2, the centrioles are positioned at the poles; the red structures in the drawing and photo are the protein cables that make up the spindle. Each cable is called a *spindle fiber* and is made of microtubules, which are long, hollow tubes of protein. Plant

Cytokinesis

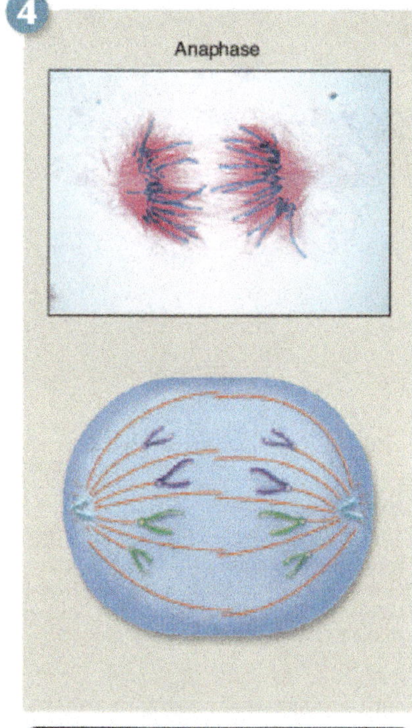

The centromeres replicate. The sister chromatids separate and move to opposite poles.

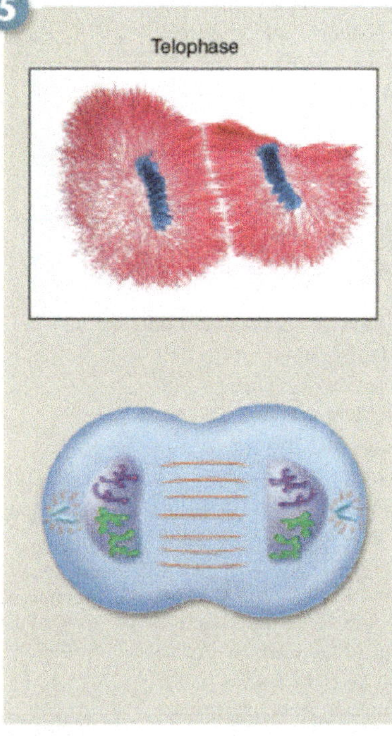

The nuclear envelope reappears. The chromosomes decondense. As telophase progresses, cytokinesis also occurs.

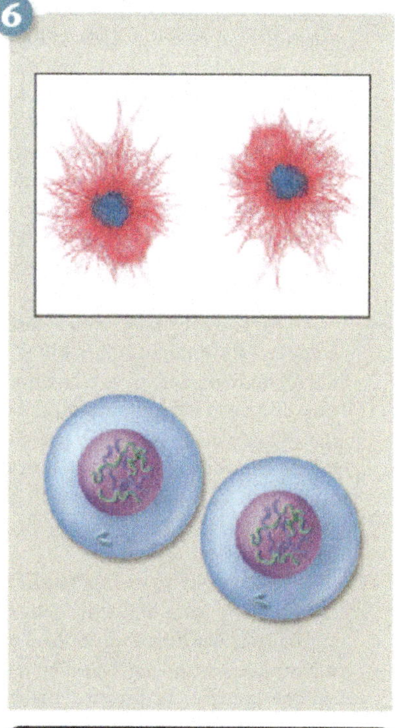

In cytokinesis two daughter cells form. Each cell is a replicate of the parent cell and is diploid.

cells lack centrioles and instead brace the ends of the spindle toward the poles.

As condensation of the chromosomes continues, a second group of microtubules extends out from the poles toward the centromeres of the chromosomes. Each set of microtubules continues to grow longer until it makes contact with a disk of protein, called a *kinetochore,* associated with each side of the centromere. When the process is complete, one sister chromatid of each pair is attached by microtubules to one pole and the other sister chromatid to the other pole.

Metaphase: Alignment of the Chromosomes The second phase of mitosis, **metaphase,** begins when the chromosomes, each consisting of a pair of sister chromatids, align in the center of the cell along an imaginary plane that divides the cell in half, referred to as the equatorial plane. Panel 3 shows the chromosomes beginning to align along the equatorial plane. Microtubules attached to the kinetochores of the centromeres are fully extended back toward the opposite poles of the cell.

Anaphase: Separation of the Chromatids In **anaphase,** enzymes cleave the cohesin link holding sister chromatids together, the kinetochores split, and the sister chromatids are freed from each other. Cell division is now simply a matter of reeling in the microtubules, dragging to the poles the sister chromatids, now referred to as daughter chromosomes. In panel 4 you see the daughter chromosomes being pulled by their centromeres, the arms of the chromosomes dangling behind. The ends of the microtubules are dismantled, one bit after another, making the tubes shorter and shorter and so drawing the chromosome attached to the far end closer and closer to the opposite poles of the cell. When they finally arrive, each pole has one complete set of chromosomes.

Telophase: Re-formation of the Nuclei The only tasks that remain in **telophase** are the dismantling of the stage and the removal of the props. The mitotic spindle is disassembled, and a nuclear envelope forms around each set of chromosomes while they begin to uncoil, as shown in panel 5, and the nucleolus reappears.

Cytokinesis

At the end of telophase, mitosis is complete. The cell has divided its replicated chromosomes into two nuclei, which are positioned at opposite ends of the cell. Mitosis is also referred to as **karyokinesis.** You may recall from chapter 4 that the nucleus is also referred to as *karyon* (Latin for "kernel"); therefore, karyokinesis is the division of the nucleus. Toward the end of mitosis, **cytokinesis,** the division of the cytoplasm, occurs, and the cell is cleaved into roughly equal halves. Cytoplasmic organelles have already been replicated and resorted to the areas that will separate and become the daughter cells. Cytokinesis, shown in panel 6 of figure 8.5, signals the end of cell division.

In animal cells, which lack cell walls, cytokinesis is achieved by pinching the cell in two with a contracting belt of actin filaments. As contraction proceeds, a *cleavage furrow* becomes evident around the cell's circumference, where the cytoplasm is being progressively pinched inward by the decreasing diameter of the actin belt. In figure 8.6*a* you see an animal cell pinching in half during cytokinesis. Imagine the cleavage furrow deepening further, until the cell is literally pinched in two.

Plant cells have rigid walls that are far too strong to be deformed by actin filament contraction. A different approach to cytokinesis has therefore evolved in plants. Plant cells assemble membrane components in their interior, at right angles to the mitotic spindle. In figure 8.6*b*, you can see how membrane is deposited between the daughter cells by vesicles that fuse together. This expanding partition, called a *cell plate*, grows outward until it reaches the interior surface of the plasma membrane and fuses with it, at which point it has effectively divided the cell in two. Cellulose is then laid down over the new membranes, forming the cell walls of the two new cells.

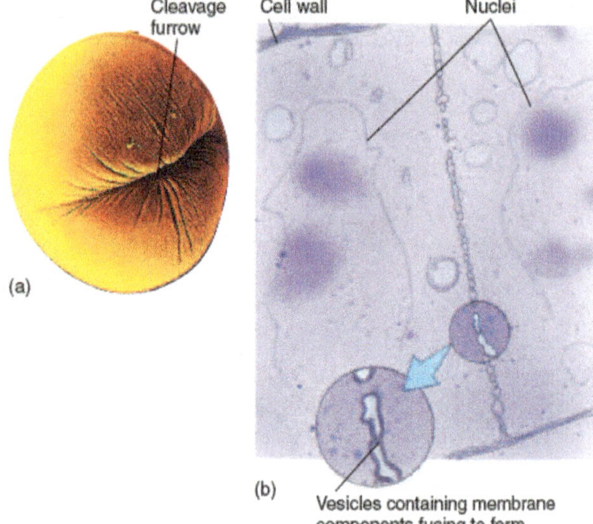

Figure 8.6 Cytokinesis.
The division of cytoplasm that occurs after mitosis is called cytokinesis and cleaves the cell into roughly equal halves. (*a*) In an animal cell, such as this sea urchin egg, a cleavage furrow forms around the dividing cell. (*b*) In this dividing plant cell, a cell plate is forming between the two newly forming daughter cells.

Cell Death

Despite the ability to divide, no cell lives forever. The ravages of living slowly tear away at a cell's machinery. To some degree, damaged parts can be replaced, but no replacement process is perfect. And sometimes the environment intervenes. If food supplies are cut off, for example, animal cells cannot obtain the energy necessary to maintain their lysosome membranes. The cells die, digested from within by their own enzymes.

During fetal development, many cells are programmed to die. In human embryos, hands and feet appear first as "paddles," but the skin cells between bones die on schedule to form the separated toes and fingers. Figure 8.7 shows a developing human hand looking like a paddle. The cells in the tissue between the bones will later die, leaving behind a set of fingers. In ducks, this cell death is not part of the developmental program, which is why ducks have webbed feet and you don't.

Human cells appear to be programmed to undergo only so many cell divisions and then die, following a plan written into the genes. In tissue culture, cell lines divide about 50 times, and then the entire population of cells dies off. Even if some of the cells are frozen for years, when they are thawed they simply resume where they left off and die on schedule. Only cancer cells appear to thwart these instructions, dividing endlessly. All other cells in your body contain a hidden clock that keeps time by counting cell divisions, and when the alarm goes off the cells die.

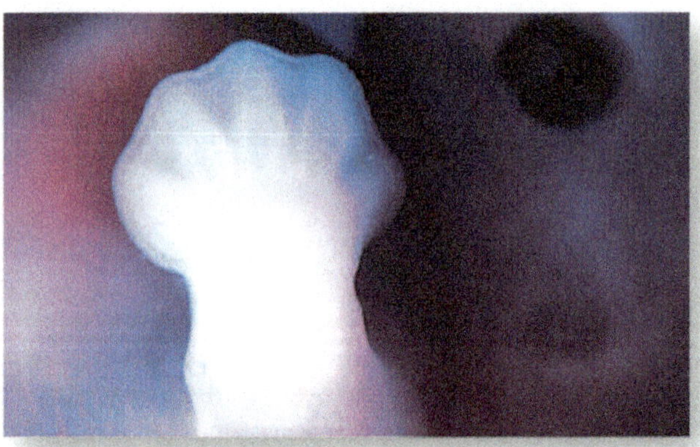

Figure 8.7 Programmed cell death.
In the human embryo, programmed cell death results in the formation of fingers and toes from paddlelike hands and feet.

Key Learning Outcome 8.4 The eukaryotic cell cycle starts in interphase with the condensation of replicated chromosomes; in mitosis, these chromosomes are drawn by microtubules to opposite ends of the cell; in cytokinesis, the cell is split into two daughter cells.

8.5 Controlling the Cell Cycle

The events of the cell cycle are coordinated in much the same way in all eukaryotes. The control system that human cells use first evolved among the protists over a billion years ago; today, it operates in essentially the same way in fungi as it does in humans.

Check Points

The goal of controlling any cyclic process is to adjust the duration of the cycle to allow sufficient time for all events to occur. In principle, a variety of methods can achieve this goal. For example, an internal clock can be employed to allow adequate time for each phase of the cycle to be completed. This is how many organisms control their daily activity cycles. The disadvantage of using such a clock to control the cell cycle is that it is not very flexible. One way to achieve a more flexible and sensitive regulation of a cycle is simply to let the completion of each phase of the cycle trigger the beginning of the next phase, as a runner passing a baton starts the next leg in a relay race. Until recently, biologists thought this type of mechanism controlled the cell division cycle. However, we now know that eukaryotic cells employ a separate, centralized controller to regulate the process: At critical points in the cell cycle, further progress depends upon a central set of "go/no-go" switches that are regulated by feedback from the cell.

This mechanism is the same one engineers use to control many processes. For example, the furnace that heats a home in the winter typically goes through a daily heating cycle. When the daily cycle reaches the morning "turn on" checkpoint, sensors report whether the house temperature is below the set point (for example, 70°F). If it is, the thermostat triggers the furnace, which warms the house. If the house is already at least that warm, the thermostat does not start the furnace. Similarly, the cell cycle has key checkpoints where feedback signals from the cell about its size and the condition of its chromosomes can either trigger subsequent phases of the cycle or delay them to allow more time for the current phase to be completed.

Three principal checkpoints control the cell cycle in eukaryotes (figure 8.8):

1. **Cell growth is assessed at the G_1 checkpoint.** Located near the end of G_1 and just before entry into S phase, the G_1 checkpoint makes the key decision of whether the cell should divide, delay division, or enter a resting stage. In yeasts, where researchers first studied this checkpoint, it is called START. If conditions are favorable for division, the cell begins to copy its DNA, initiating S phase. The G_1 checkpoint is where the more complex eukaryotes typically arrest the cell cycle if environmental conditions make cell division impossible or if the cell passes into an extended resting period called G_0 (figure 8.9).

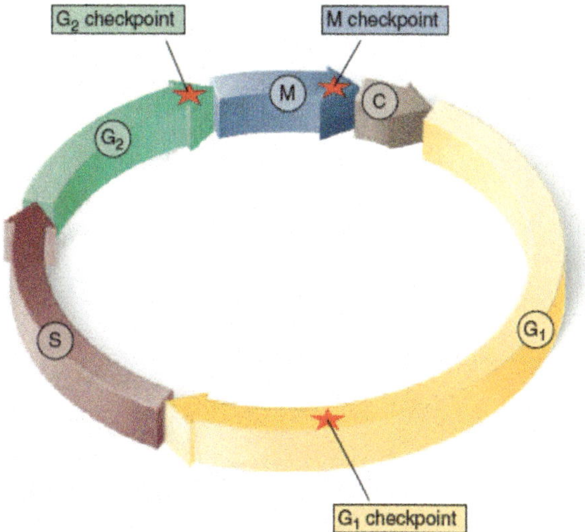

Figure 8.8 Control of the cell cycle.
Cells use a centralized control system to check whether proper conditions have been achieved before passing three key checkpoints in the cell cycle.

Figure 8.9 The G_1 checkpoint.
Feedback from the cell determines whether the cell cycle will proceed to the S phase, pause, or withdraw into G_0 for an extended rest period.

2. **DNA replication is assessed at the G_2 checkpoint.** The second checkpoint, the G_2 checkpoint, triggers the start of M phase. If this checkpoint is passed, the cell initiates the many molecular processes that signal the beginning of mitosis.
3. **Mitosis is assessed at the M checkpoint.** The third checkpoint, the M checkpoint, occurs at metaphase and triggers the exit from mitosis and cytokinesis and the beginning of G_1.

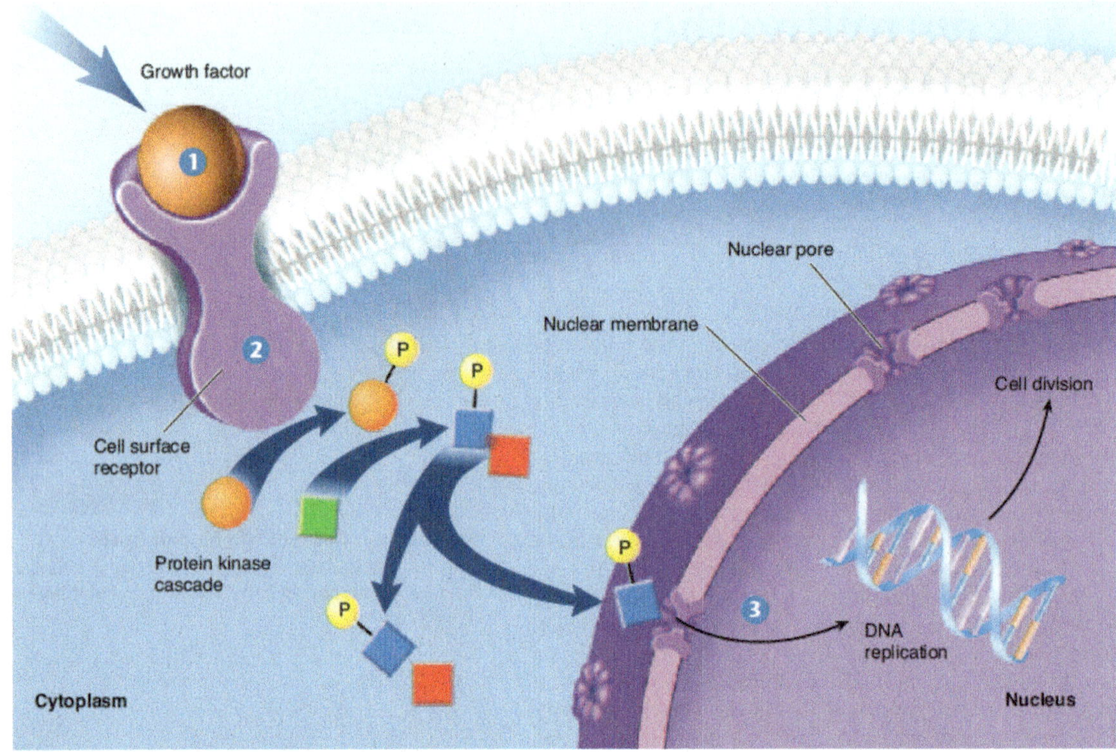

Figure 8.10 The cell proliferation-signaling pathway.
Binding of a growth factor sets in motion a cascading intracellular signaling pathway, which activates proteins in the nucleus that trigger cell division.

Growth Factors Trigger Cell Division

Cell division is initiated by small proteins called **growth factors.** Growth factors work by binding to the plasma membrane and triggering intracellular signaling systems. Fibroblasts are cells that form connective tissue in the body and possess numerous receptors on their plasma membranes for one of the first growth factors to be identified: platelet-derived growth factor (PDGF). When PDGF binds to a membrane receptor, it initiates an amplifying chain of internal cell signals that stimulates cell division.

PDGF was discovered when investigators found that fibroblasts would grow and divide in tissue culture only if the growth medium contained blood serum (the liquid that remains after blood clots); blood plasma (blood from which the cells have been removed without clotting) would not work. The researchers hypothesized that platelets in the blood clots were releasing into the serum one or more factors required for fibroblast growth. Eventually, they isolated such a factor and named it PDGF.

Growth factors such as PDGF override cellular controls that otherwise inhibit cell division. When a tissue is injured, a blood clot forms and the release of PDGF triggers neighboring cells to divide, helping to heal the wound. Only a tiny amount of PDGF (approximately 10^{-10} M) is required to stimulate cell division.

Characteristics of Growth Factors. Over 50 different proteins that function as growth factors have been isolated, and more undoubtedly exist. A specific cell surface receptor "recognizes" each growth factor, its shape fitting that growth factor precisely. Figure 8.10 shows what happens when the growth factor ❶ binds with its receptor ❷. The receptor is activated and reacts by triggering a series of events within the cell, indicated by the arrows, that end with the replication of DNA and cell division ❸. The cellular selectivity of a particular growth factor depends upon which target cells bear its unique receptor. Some growth factors, like PDGF and epidermal growth factor (EGF), affect a broad range of cell types, while others affect only specific types. For example, nerve growth factor (NGF) promotes the growth of certain classes of neurons, and erythropoietin triggers cell division in red blood cell precursors. Most animal cells need a combination of several different growth factors to overcome the various controls that inhibit cell division.

The G_0 Phase. If cells are deprived of appropriate growth factors, they stop at the G_1 checkpoint of the cell cycle. With their growth and division arrested, they remain in the G_0 phase, as mentioned earlier. This nongrowing state is distinct from the interphase stages of the cell cycle, G_1, S, and G_2.

It is the ability to enter G_0 that accounts for the incredible diversity seen in the length of the cell cycle among different tissues. Epithelial cells lining the gut divide more than twice a day, constantly renewing the lining of the digestive tract. By contrast, liver cells divide only once every year or two, spending most of their time in G_0 phase. Mature neurons and muscle cells usually never leave G_0.

Aging and the Cell Cycle

All humans die. However, while each of us knows we shall someday die, few of us can escape wishing we could delay the process. Some succeed. The oldest documented living person, Jeanne Calment of France, reached the age of 122 years in 1997. The tantalizing possibility of long life that she represents is one reason why there is such interest in the aging process; if we knew enough about it perhaps we could slow it. A wide variety of theories have been advanced to explain why we age. In recent years, scientists have come a long way toward unraveling the puzzle.

The first clue was the discovery that cells appear to die on schedule, as if following a script. In a famous experiment carried out in 1961, geneticist Leonard Hayflick demonstrated that fibroblast cells growing in tissue culture will divide only a certain number of times. As you can see in figure 8.11, after about 50 population doublings, cell division stops, the cell cycle blocked just before DNA replication. If a cell sample is frozen after the cell has undergone 20 doublings, when thawed the cell resumes growth for 30 more doublings, then stops.

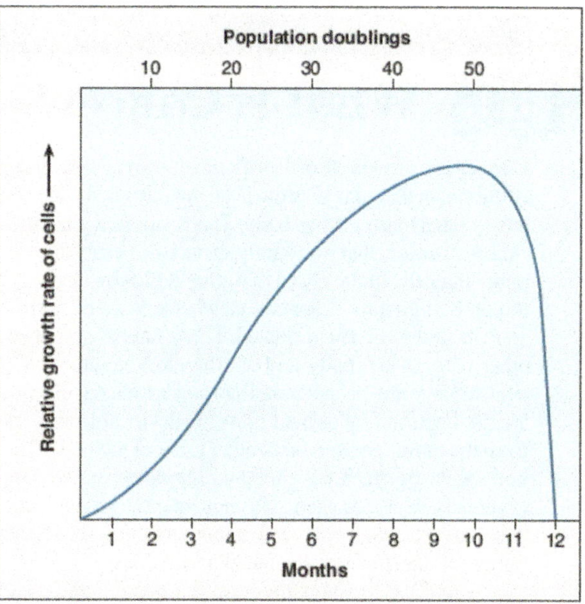

Figure 8.11 The Hayflick limit.
Normal human fibroblast (connective tissue) cells stop growing in culture after about 40 doublings, and within 10 more doublings all the cells are dead (*blue line*). When genetic engineers induce fibroblast cells to express the enzyme telomerase, proliferation continues long after 40 doublings.

An explanation of the "Hayflick limit" was suggested in 1978 when Elizabeth Blackburn of the University of California, San Francisco, first glimpsed an extra length of DNA at the ends of chromosomes. These telomeric regions, about 5,000 nucleotides long, are each composed of several thousand repeats of the sequence TTAGGG. Blackburn found the telomeric region to be substantially shorter in body tissue chromosomes than in those of germ-line cells, the egg and sperm. She speculated that in body cells a portion of the telomere cap was lost by a chromosome during each cycle of DNA replication.

Blackburn was right. The cell machinery that replicates the DNA of each chromosome sits on the last 100 units of DNA at the chromosome's tip, and so cannot copy that bit. So each time the cell divides, its chromosomes get a little shorter. Eventually, after some 50 replication cycles, the protective telomeric cap is used up, and the cell line then enters senescence, no longer able to proliferate.

How do sperm and egg cells avoid this trap, dividing continuously for decades? Blackburn and collaborator Jack Szostak proposed that cells must possess a special enzyme that lengthens telomeres. In 1984 Blackburn's graduate student Carol Greider found the enzyme, dubbed "telomerase." Using it, egg and sperm maintain their chromosomes at a constant length of 5,000 nucleotide units. In body cells, by contrast, the telomerase gene is silent. For their discoveries, Blackburn, Greider, and Szostak received the 2009 Nobel Prize in Physiology or Medicine.

Later research has provided direct evidence for a causal relation between telomeric shortening and cell senescence. Using genetic engineering, teams of researchers from California and Texas in 1998 transferred into human body cell cultures a DNA fragment that unleashes each cell's telomerase gene. The result was unequivocal. New telomeric caps were added to the chromosomes of the cells, and the cells with the artificially elongated telomeres did not senesce at the Hayflick limit, continuing to divide in a healthy and vigorous manner for more than 20 additional generations.

This research shows clearly that loss of telomere DNA eventually restrains the ability of human cells to proliferate. And yet every human cell possesses a copy of the telomerase gene that, if expressed, would rebuild the telomere. Why do our cells accept aging if they need not? The answer, it seems, is to avoid cancer. By limiting the number of divisions allotted to human cell lines, the body ensures that no cell can continue to divide indefinitely. Suppression of the telomerase gene is, in a very real sense, cancer suppression. When scientists examine cancer cells, they commonly find their telomerase genes have been activated and are maintaining telomeres at full length. Thus telomere shortening is a tumor-suppressing mechanism, one of your body's key safeguards against cancer.

Key Learning Outcome 8.5 The complex cell cycle of eukaryotes is controlled with feedback at three checkpoints by protein signals called growth factors that initiate cell division. Telomeres play a key role in limiting cell proliferation.

Cancer and the Cell Cycle

8.6 What Is Cancer?

Cancer is a growth disorder of cells. It starts when an apparently normal cell begins to grow in an uncontrolled way, spreading out to other parts of the body. The result is a cluster of cells, called a **tumor,** that constantly expands in size. The cluster of pink lung cells in the photo in figure 8.12 have begun to form a tumor. Benign tumors are completely enclosed by normal tissue and are said to be encapsulated. These tumors do not spread to other parts of the body and are therefore noninvasive. Malignant tumors are invasive and not encapsulated. Because they are not enclosed by normal tissue, cells are able to break away from the tumor and spread to other areas of the body. The tumor you see in figure 8.12, called a *carcinoma,* grows larger and eventually begins to shed cells that enter the bloodstream. Cells that leave a tumor and spread throughout the body, forming new tumors at distant sites, are called **metastases.**

Cancer is perhaps the most devastating and deadly disease. Most of us have had family or friends affected by the disease. In 2010, about 1.5 million American men and women were diagnosed with cancer. One in every two Americans born will be diagnosed with some form of cancer during their lifetime; nearly one in four are projected to die from cancer.

In the U.S., the three deadliest human cancers are lung cancer, cancer of the colon and rectum, and breast cancer. Lung cancer, responsible for the most cancer deaths, is largely preventable; most cases result from smoking cigarettes. Colorectal cancers appear to be fostered by the high-meat diets so favored in the United States. The cause of breast cancer is still a mystery.

Not surprisingly, researchers are expending a great deal of effort to learn the cause of cancer. Scientists have made considerable progress in the last 30 years using molecular biological techniques, and the rough outlines of understanding are now emerging. We now know that cancer is a gene disorder of somatic tissue, in which damaged genes fail to properly control cell growth and division. The cell division cycle is regulated by a sophisticated group of proteins called growth factors. Cancer results from the damage of these genes encoding these proteins. Damage to DNA, such as damage to these genes, is called **mutation.**

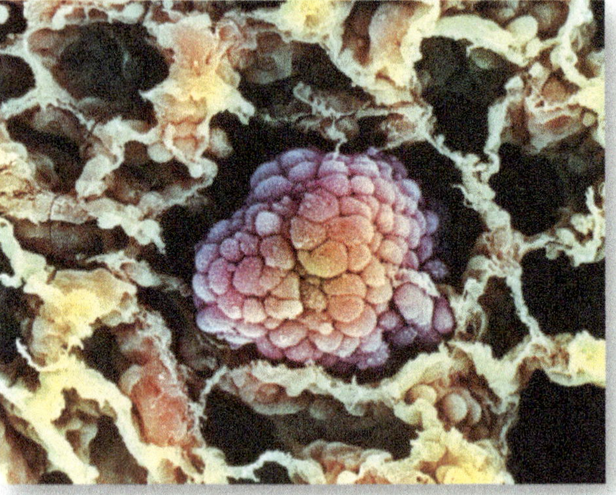

Figure 8.12 Lung cancer cells (300×).
These cells are from a tumor located in the alveolus (air sac) of a human lung.

There are two general classes of growth factor genes that are usually involved in cancer: proto-oncogenes and tumor-suppressor genes. Genes known as **proto-oncogenes** encode proteins that stimulate cell division. Mutations to these genes can cause cells to divide excessively. Mutated proto-oncogenes become cancer-causing genes called **oncogenes.**

The second class of cancer-causing genes is called **tumor-suppressor genes.** Cell division is normally turned off in healthy cells by proteins encoded by tumor-suppressor genes. Mutations to these genes essentially "release the brakes," allowing the cell containing the mutated gene to divide uncontrolled.

Key Learning Outcome 8.6 Cancer is unrestrained cell growth and division caused by damage to genes regulating the cell division cycle.

8.7 Cancer and Control of the Cell Cycle

Cancer can be caused by chemicals like those in cigarette smoke, by environmental factors such as UV rays that damage DNA, or in some instances by viruses that circumvent the cell's normal growth and division controls. Whatever the immediate cause, however, all cancers are characterized by unrestrained cell growth and division. The cell cycle never stops in a cancerous line of cells.

Cancer results from damaged genes failing to control cell division. Researchers have identified several of these genes. One particular gene seems to be a key regulator of the cell cycle. Officially dubbed *p53* (researchers italicize the gene symbol to differentiate it from the protein), this gene plays a key role in the G_1 checkpoint of cell division. Figure 8.13 illustrates how the product of this gene, the p53 protein, monitors the integrity of DNA, checking that it has been successfully replicated and is undamaged. If the p53 protein detects damaged DNA, as it does in the upper panel, it halts cell division and stimulates the activity of special enzymes to repair the damage. Once the DNA has been repaired, p53 allows cell division to continue, indicated by the upper path of arrows. In cases where the DNA cannot be repaired, p53 then directs the cell to kill itself, activating an apoptosis (cell suicide) program, indicated by the lower path of arrows.

By halting division in damaged cells, *p53* prevents the formation of tumors (even though its activities are not limited to cancer prevention). Scientists have found that *p53* is itself damaged beyond use in most human cancers they have examined. It is precisely because *p53* is nonfunctional that these cancer cells are able to repeatedly undergo cell division without being halted at the G_1 checkpoint. The lower panel of figure 8.13 shows what happens when p53 doesn't function properly. The abnormal p53 does not stop cell division and the damaged strand is replicated, which results in damaged cells. As more and more damage occurs to these cells, they become cancerous. To test this, scientists administered healthy p53 protein to rapidly dividing cancer cells in a petri dish: The cells soon ceased dividing and died. Scientists have further reported that cigarette smoke causes mutations in the *p53* gene, reinforcing the strong link between smoking and cancer that you will encounter in chapter 24, section 24.6.

In about 50% of cancers, the *p53* cancer defense malfunctions because the *p53* gene itself has been damaged by chemicals or radiation, so that the protein the gene encodes no longer functions properly. In the other 50%, however, the defects lie in other genes. In many instances, the DNA is damaged at a site suppressing the production of a small molecule called MDM2 that is a potent natural inhibitor of the p53 protein. When the *MDM2* gene becomes overactive as a result of mutation, its protein product suppresses p53's activity.

A very promising approach to cancer prevention involves this second kind of p53 malfunction. Researchers studying the MDM2 protein found a relatively deep and well-defined pocket on its surface that proved to be the site that makes contact with the p53 protein. Perhaps, they hypothesized, a small molecule could be found that would fit into the site, and in doing so prevent p53 from binding. Such a molecule might prevent 50% of cancers! Searching for the key to fit this lock, they identified a family of synthetic chemicals they called "nutlins." When tumor cells that had functional *p53* genes were treated with nutlin, levels of p53 protein in the treated cells went up, and the tumor cells were killed, while normal cells treated in the same way were not killed. Nutlin is one of a host of new cancer therapies under active investigation.

Key Learning Outcome 8.7 Mutations disabling key elements of the G_1 checkpoint are associated with many cancers.

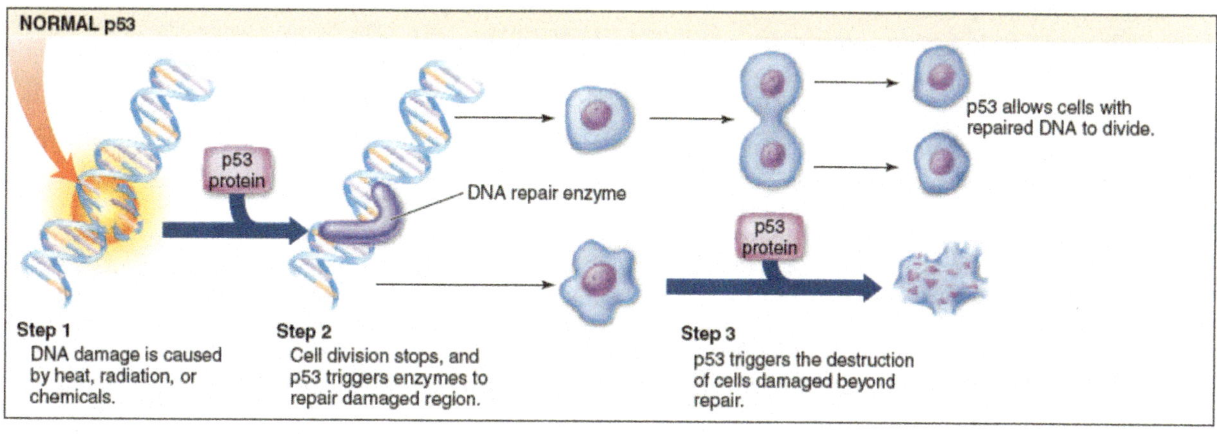

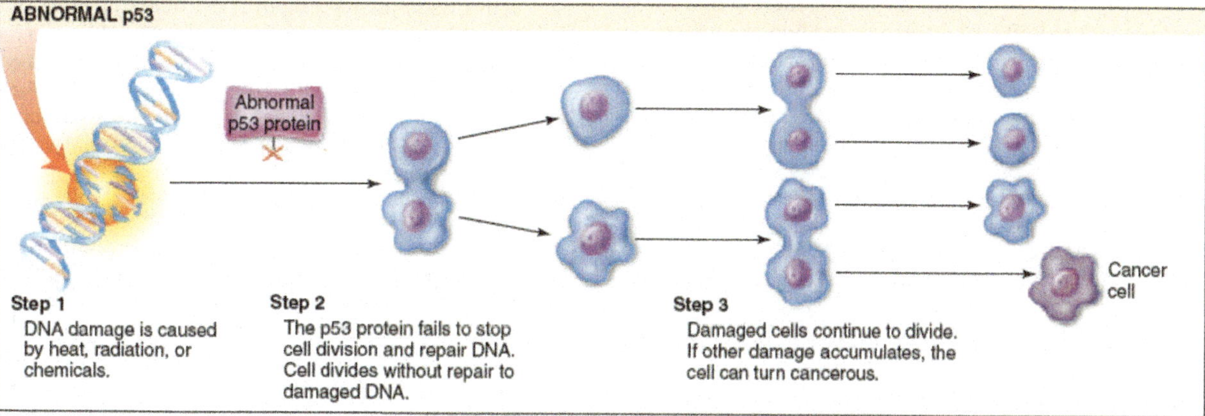

Figure 8.13 Cell division and p53 protein.
Normal p53 protein monitors DNA, destroying cells with irreparable damage to their DNA. Abnormal p53 protein fails to stop cell division and repair DNA. As damaged cells proliferate, cancer develops.

Biology and *Staying Healthy*

Curing Cancer

Half of all Americans will face cancer at some point in their lives. Potential cancer therapies are being developed on many fronts. Some act to prevent the start of cancer within cells. Others act outside cancer cells, preventing tumors from growing and spreading. The figure on the right indicates targeted areas for the development of cancer treatments. The following discussion will examine each of these areas.

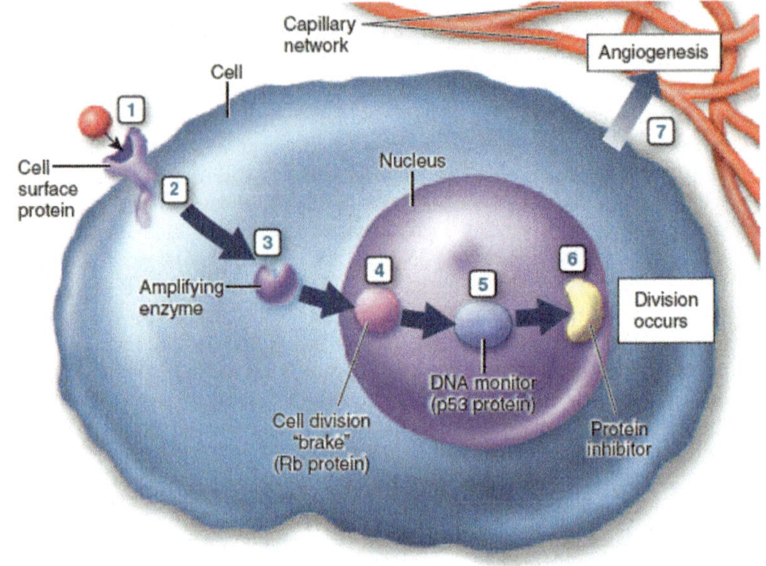

Seven different stages in the cancer process.

(1) On the cell surface, a growth factor's signal to divide is increased. (2) Just inside the cell, a protein relay switch that passes on the divide signal gets stuck in the "ON" position. (3) In the cytoplasm, enzymes that amplify the signal are amplified even more. In the nucleus, (4) a "brake" preventing DNA replication is inoperable, (5) proteins that check for damage in the DNA are inactivated, and (6) other proteins that inhibit the elongation of chromosome tips are destroyed. (7) The new tumor promotes angiogenesis, the formation of new blood vessels that promote growth.

Preventing the Start of Cancer

Many promising cancer therapies act within potential cancer cells, focusing on different stages of the cell's "Shall I divide?" decision-making process.

1 **Receiving the Signal to Divide** The first step in the decision process is receiving a "divide" signal, usually a small protein called a growth factor released from a neighboring cell. The growth factor, the red ball at #1 in the figure, is received by a protein receptor on the cell surface. Like banging on a door, its arrival signals that it's time to divide. Mutations that increase the number of receptors on the cell surface amplify the division signal and so lead to cancer. Over 20% of breast cancer tumors prove to overproduce a protein called HER2 associated with the receptor for epidermal growth factor (EGF).

Therapies directed at this stage of the decision process utilize the human immune system to attack cancer cells. Special protein molecules called *monoclonal antibodies*, created by genetic engineering, are the therapeutic agents. These monoclonal antibodies are designed to seek out and stick to HER2. Like waving a red flag, the presence of the monoclonal antibody calls down attack by the immune system on the HER2 cell. Because breast cancer cells overproduce HER2, they are killed preferentially. The biotechnology research company Genentech's recently approved monoclonal antibody, called herceptin, has given promising results in clinical tests.

Up to 70% of colon, prostate, lung, and head/neck cancers have excess copies of a related receptor, epidermal growth factor 1 (HER1). The monoclonal antibody C225, directed against HER1, has succeeded in shrinking 22% of advanced, previously incurable colon cancers in early clinical trials. Apparently blocking HER1 interferes with the ability of tumor cells to recover from chemotherapy or radiation.

2 **Passing the Signal via a Relay Switch** The second step in the decision process is the passage of the signal into the cell's interior, the cytoplasm. This is carried out in normal cells by a protein called Ras that acts as a relay switch, #2 in the figure. When growth factor binds to a receptor like EGF, the adjacent Ras protein acts like it has been "goosed," contorting into a new shape. This new shape is chemically active, and initiates a chain of reactions that passes the "divide" signal inward toward the nucleus. Mutated forms of the Ras protein behave like a relay switch stuck in the "ON" position, continually instructing the cell to divide when it should not. Thirty percent of all cancers have a mutant form of Ras. So far, no effective therapies have been developed targeting this step.

3 **Amplifying the Signal** The third step in the decision process is the amplification of the signal within the cytoplasm. Just as a TV signal needs to be amplified in order to be received at a distance, so a "divide" signal must be amplified if it is to reach the nucleus at the interior of the cell, a very long journey at a molecular scale. To get a signal all the way into the nucleus, the cell employs a sort of pony express. The "ponies" in this case are enzymes called *tyrosine kinases*, #3 in the figure. These enzymes add

phosphate groups to proteins, but only at a particular amino acid, tyrosine. No other enzymes in the cell do this, so the tyrosine kinases form an elite core of signal carriers not confused by the myriad of other molecular activities going on around them.

Cells use an ingenious trick to amplify the signal as it moves toward the nucleus. Ras, when "ON," activates the initial protein kinase. This protein kinase activates other protein kinases that in their turn activate still others. The trick is that once a protein kinase enzyme is activated, it goes to work like a demon, activating hoards of others every second! And each and every one it activates behaves the same way too, activating still more, in a cascade of ever-widening effect. At each stage of the relay, the signal is amplified a thousandfold.

Mutations stimulating any of the protein kinases can dangerously increase the already amplified signal and lead to cancer. Some 15 of the cell's 32 internal tyrosine kinases have been implicated in cancer. Five percent of all cancers, for example, have a mutant hyperactive form of the protein kinase Src. The trouble begins when a mutation causes one of the tyrosine kinases to become locked into the "ON" position, sort of like a stuck doorbell that keeps ringing and ringing.

To cure the cancer, you have to find a way to shut the bell off. Each of the signal carriers presents a different problem, as you must quiet it without knocking out all the other signal pathways the cell needs. The cancer therapy drug Gleevec, a monoclonal antibody, has just the right shape to fit into a groove on the surface of the tyrosine kinase called "abl." Mutations locking abl "ON" are responsible for chronic myelogenous leukemia, a lethal form of white blood cell cancer. Gleevec totally disables abl. In clinical trials, blood counts revert to normal in more than 90% of cases.

4 **Releasing the Brake** The fourth step in the decision process is the removal of the "brake" the cell uses to restrain cell division. In healthy cells this brake, a tumor-suppressor protein called Rb, blocks the activity of a protein called E2F, #4 in the figure. When free, E2F enables the cell to copy its DNA. Normal cell division is triggered to begin when Rb is inhibited, unleashing E2F. Mutations that destroy Rb release E2F from its control completely, leading to ceaseless cell division. Forty percent of all cancers have a defective form of Rb.

Therapies directed at this stage of the decision process are only now being attempted. They focus on drugs able to inhibit E2F, which should halt the growth of tumors arising from inactive Rb. Experiments in mice in which the E2F genes have been destroyed provide a model system to study such drugs, which are being actively investigated.

5 **Checking That Everything Is Ready** The fifth step in the decision process is the mechanism used by the cell to ensure that its DNA is undamaged and ready to divide. This job is carried out in healthy cells by the tumor-suppressor protein p53, which inspects the integrity of the DNA, #5 in the figure. When it detects damaged or foreign DNA, p53 stops cell division and activates the cell's DNA repair systems. If the damage doesn't get repaired in a reasonable time, p53 pulls the plug, triggering events that kill the cell. In this way, mutations such as those that cause cancer are either repaired or the cells containing them eliminated. If p53 is itself destroyed by mutation, future damage accumulates unrepaired. Among this damage are mutations that lead to cancer. Fifty percent of all cancers have a disabled p53. Fully 70% to 80% of lung cancers have a mutant inactive p53—the chemical benzo[a]pyrene in cigarette smoke is a potent mutagen of p53.

6 **Stepping on the Gas** Cell division starts with replication of the DNA. In healthy cells, another tumor suppressor "keeps the gas tank nearly empty" for the DNA replication process by inhibiting production of an enzyme called *telomerase*. Without this enzyme, a cell's chromosomes lose material from their tips, called *telomeres*. Every time a chromosome is copied, more tip material is lost. After some 30 divisions, so much is lost that copying is no longer possible. Cells in the tissues of an adult human have typically undergone 25 or more divisions. Cancer can't get very far with only the five remaining cell divisions, so inhibiting telomerase is a very effective natural brake on the cancer process, #6 in the figure. It is thought that almost all cancers involve a mutation that destroys the telomerase inhibitor, releasing this brake and making cancer possible. It should be possible to block cancer by reapplying this inhibition. Cancer therapies that inhibit telomerase are just beginning clinical trials.

Preventing the Spread of Cancer

7 **Stopping Tumor Growth** Once a cell begins cancerous growth, it forms an expanding tumor. As the tumor grows ever-larger, it requires an increasing supply of food and nutrients, obtained from the body's blood supply. To facilitate this necessary grocery shopping, tumors leak out substances into the surrounding tissues that encourage the formation of small blood vessels, a process called angiogenesis, #7 in the figure. Chemicals that inhibit this process are called *angiogenesis inhibitors*. Two such natural angiogenesis inhibitors, angiostatin and endostatin, caused tumors to regress to microscopic size in mice, but initial human trials were disappointing.

Laboratory drugs are more promising. A monoclonal antibody drug called Avastin, targeted against a blood vessel growth-promoting substance called vascular endothelial growth factor (VEGF), destroys the ability of VEGF to carry out its blood-vessel-forming job. Given to hundreds of advanced colon cancer patients in 2003 as part of a large clinical trial, Avastin improved colon cancer patients's chance of survival by 50% over chemotherapy.

INQUIRY & ANALYSIS

Why Do Human Cells Age?

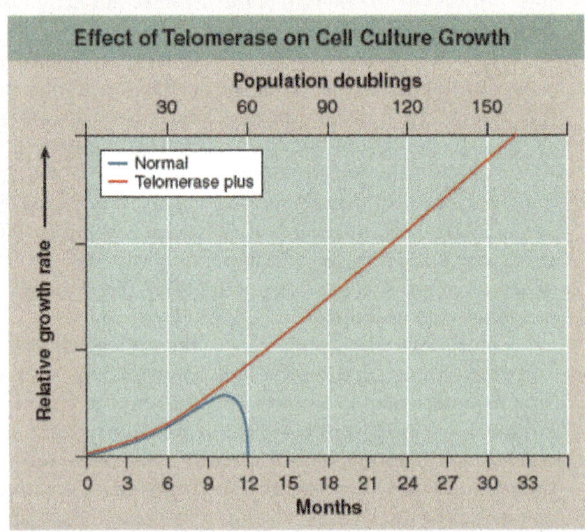

Human cells appear to have built-in life spans. As you learned in this chapter, cell biologist Leonard Hayflick reported in 1961 the startling result that skin cells growing in tissue culture, such as those growing in culture flasks in the photo below, will divide only a certain number of times. After about 50 population doublings, cell division stops (a **doubling** is a round of cell division producing two daughter cells for each dividing cell; for example, going from a population of 30 cells to 60 cells). If a cell sample is taken after 20 doublings and frozen, when thawed it resumes growth for 30 more doublings, and then stops. An explanation of the "Hayflick limit" was suggested in 1978 when researchers first glimpsed an extra length of DNA at the end of chromosomes. Dubbed telomeres, these lengths proved to be composed of the simple DNA sequence TTAGGG, repeated nearly a thousand times. Importantly, telomeres were found to be substantially shorter in the cells of older body tissues. This led

TTAGGG TTAGGG TTAGGG TTAGGG TTAGGG--------

to the hypothesis that a run of some 16 TTAGGGs was where the DNA replicating enzyme, called *polymerase*, first sat down on the DNA (16 TTAGGGs being the size of the enzyme's "footprint"), and because of being its docking spot, the polymerase was unable to copy that bit. Thus a 100-base portion of the telomere was lost by a chromosome during each doubling as DNA replicated. Eventually, after some 50 doubling cycles, each with a round of DNA replication, the telomere would be used up and there would be no place for the DNA replication enzyme to sit. The cell line would then enter senescence, no longer able to proliferate.

This hypothesis was tested in 1998. Using genetic engineering, researchers transferred into newly established human cell cultures a gene that leads to expression of an enzyme called *telomerase* that all cells possess but no body cell uses. This enzyme adds TTAGGG sequences back to the end of telomeres, in effect rebuilding the lost portions of the telomere. Laboratory cultures of cell lines with (telomerase plus) and without (normal) this gene were then monitored for many generations. The graph above displays the results.

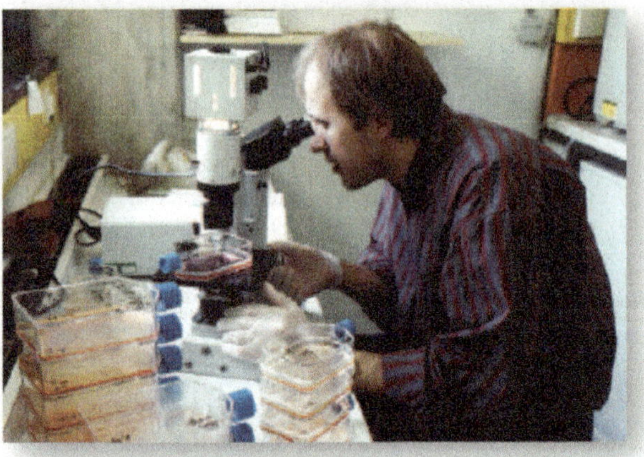

1. **Applying Concepts**
   a. Variable. In the graph, what is the dependent variable?
   b. Comparing Continuous Processes. How do normal skin cells (blue line) differ in their growth history from telomerase plus cells with the telomerase gene (red line)?
2. **Interpreting Data**
   a. After how many doublings do the normal cells cease to divide? Is this consistent with the telomerase hypothesis?
   b. After how many doublings do the telomerase plus cells cease to divide in this experiment?
3. **Making Inferences** After nine population doublings, would the rate of cell division be different between the two cultures? After 15? Why?
4. **Drawing Conclusions** How does the addition of the telomerase gene affect the senescence of skin cells growing in culture? Does this result confirm the telomerase hypothesis this experiment had set out to test?
5. **Further Analysis**
   a. Cancer cells are thought to possess mutations disabling the cell's ability to keep the telomerase gene shut off. How would you test this hypothesis?
   b. Sperm-producing cells continue to divide throughout a male's adult life. How might this be possible? How would you test this idea?

Chapter Review

Cell Division

8.1 Prokaryotes Have a Simple Cell Cycle

- Prokaryotic cells divide in a two-step process: DNA replication and binary fission. The DNA is present as a single loop called a chromosome. The DNA begins replication at a site called the origin of replication. The DNA double strand unzips, and new strands form along the original strands, producing two circular chromosomes that separate to the ends of the cell. New plasma membrane and cell wall is added down the middle of the cell, as shown here from **figure 8.1**, splitting the cell in two. This cell division, called binary fission, produces two daughter cells that are genetically identical to the parent cell.

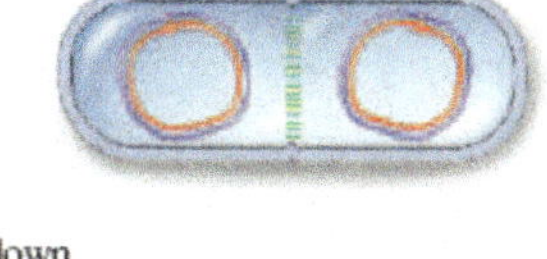

8.2 Eukaryotes Have a Complex Cell Cycle

- Cell division in eukaryotes is more complex than in prokaryotes. The eukaryotic cell cycle occurs in several phases: interphase, M phase, and C phase. Interphase is considered a resting phase, but during interphase the cell is growing and preparing for cell division. Interphase is further broken down into its own phases. The G_1 phase is the growing phase and takes up the major portion of the cell's life cycle. The S phase is the synthesis phase and is when the DNA is replicated. The G_2 phase involves the final preparations for cell division.
- During the M and C phases, the chromosomes are distributed to opposite sides of the cell, which then divides its cytoplasm into two separate daughter cells (**Key Biological Process, page 155**).

8.3 Chromosomes

- Eukaryotic DNA is organized into chromosomes, and two chromosomes that carry copies of the same genes, like the two shown here from **figure 8.2**, are called homologous chromosomes. Before cells divide, the DNA replicates forming two identical copies of each chromosome, called sister chromatids. Sister chromatids are held together at the centromere region. Human somatic cells have 46 chromosomes (**figure 8.3**).

- The DNA in a chromosome is one long double-stranded fiber called chromatin. After the DNA is replicated, it begins to coil up in a process called condensation. The DNA wraps around histone proteins with each DNA-histone complex forming a nucleosome. The string of nucleosome then folds and loops on itself forming a compact chromosome (**figure 8.4**).

8.4 Cell Division

- Interphase (**figure 8.5, panel 1**) begins the cell cycle, followed by mitosis, which consists of four phases: prophase, metaphase, anaphase, and telophase.
- Prophase (**figure 8.5, panel 2**) signals the beginning of mitosis. The DNA that was replicated during interphase condenses into chromosomes. The sister chromatids stay attached at the centromeres. The nucleolus and nuclear envelope disappear. Centrioles migrate to opposite sides of the cell and begin forming the spindle. Microtubules that form the spindle extend from the poles and attach to the kinetochores, which are proteins located at the centromere areas of the chromosomes.
- Metaphase (**figure 8.5, panel 3**) involves the alignment of sister chromatids along the equatorial plane. Microtubules connect the kinetochores of sister chromatids to each of the poles.
- During anaphase (**figure 8.5, panel 4**), the kinetochores split, and enzymes break down the cohesin, freeing the sister chromatids. The microtubules shorten, pulling the sister chromatids apart and toward opposite poles.
- Telophase (**figure 8.5, panel 5**) signals the completion of nuclear division. The microtubule spindle is dismantled, the chromosomes begin to uncoil; the nuclear envelope and nucleolus re-form.
- Following mitosis, the cell separates into two daughter cells in a process called cytokinesis (**figure 8.6**). Cytokinesis in animal cells involves a pinching in of the cell around its equatorial plane until the cell eventually splits into two cells. Cytokinesis in plant cells involves the assembly of cell membranes and cell walls between the two poles, eventually forming two separate cells.
- Many cells are programmed to die, either as part of development (**figure 8.7**) or after a set number of cell divisions (usually about 50 divisions). Only cancer cells appear to divide endlessly.

8.5 Controlling the Cell Cycle

- The cell cycle is controlled at three checkpoints. The G_1 and G_2 checkpoints occur during interphase, and the third occurs during mitosis (**figure 8.8**). At the G_1 checkpoint the cell either initiates division or enters a period of rest called G_0 (**figure 8.9**). If cell division is initiated at the G_1 checkpoint, DNA replication begins and is checked at the G_2 checkpoint. Mitosis is initiated if the DNA has been accurately replicated.
- Cell division is initiated by proteins called growth factors (**figure 8.10**). After about 50 cell doublings, regions at the ends of chromosomes called telomeres become too short to allow cell division; this acts to suppress cancer (**figure 8.11**).

Cancer and the Cell Cycle

8.6 What Is Cancer?

- Cancer is a growth disorder of cells, where there is a loss of control over cell division. Cells begin to divide in an uncontrolled way, forming a mass of cells called a tumor (**figure 8.12**). A tumor in which cells break away from the mass and spread to other tissues is called metastasis. Cancer results when genes that encode proteins that control the cell cycle, such as proto-oncogenes and tumor-suppressor genes, are damaged.

8.7 Cancer and Control of the Cell Cycle

- The *p53* gene plays a key role in the G_1 checkpoint, checking the condition of the DNA. If the DNA is damaged, the p53 protein will stop cell division so the damaged DNA can be repaired. If the DNA cannot be repaired, the p53 protein will trigger the destruction of the cell. When the *p53* gene is damaged by mutations, the DNA is not checked, and cells with damaged DNA can divide. Further mutations to the DNA can accumulate in the cells and result in cancerous cells (**figure 8.13**).

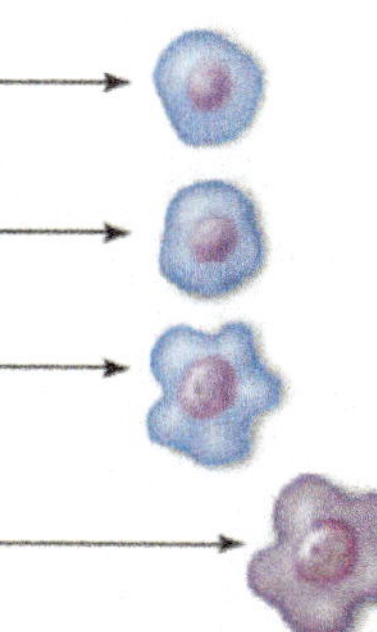

Test Your Understanding

1. Prokaryotes reproduce new cells by
   a. copying DNA then undergoing binary fission.
   b. splitting in half.
   c. undergoing mitosis.
   d. copying DNA then undergoing the M phase.
2. The eukaryotic cell cycle is different from prokaryotic cell division in all the following ways *except*
   a. the amount of DNA present in the cells.
   b. how the DNA is packaged.
   c. in the production of genetically identical daughter cells.
   d. the involvement of microtubules.
3. In eukaryotes, the genetic material is found in chromosomes
   a. and the more complex the organism, the more pairs of chromosomes it has.
   b. and many organisms have only one chromosome.
   c. and most eukaryotes have between 10 and 50 pairs of chromosomes.
   d. and most eukaryotes have between 2 and 10 pairs of chromosomes.
4. Homologous chromosomes
   a. are also referred to as sister chromatids.
   b. are genetically identical.
   c. carry information about the same traits located in the same places on the chromosomes.
   d. are connected to each other at their centromeres.
5. Eukaryotic chromosomes are composed of
   a. DNA
   b. proteins.
   c. histones.
   d. All of the above.
6. In mitosis, when the duplicated chromosomes line up in the center of the cell, that stage is called
   a. prophase.
   b. metaphase.
   c. anaphase.
   d. telophase.
7. The division of the cytoplasm in the eukaryotic cell cycle is called
   a. interphase.
   b. mitosis.
   c. cytokinesis.
   d. binary fission.
8. The cell cycle is controlled by
   a. a series of checkpoints.
   b. telomerase.
   c. the G_0 phase.
   d. All of the above.
9. When cell division becomes unregulated, and a cluster of cells begins to grow without regard for the normal controls, that is called
   a. a mutation.
   b. cancer.
   c. metastases.
   d. oncogenes.
10. The normal function of the *p53* gene in the cell is
    a. as a tumor-suppressor gene.
    b. to monitor the DNA for damage.
    c. to trigger the destruction of cells not capable of DNA repair.
    d. All of the above.

Apply Your Understanding

1. **Figure 8.3** This karyotype shows a complete set of human chromosomes of an individual. At what stage of the cell cycle are such photos taken? Explain.

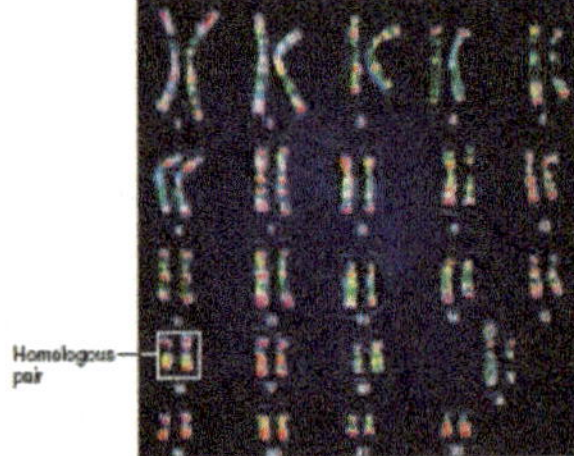

2. **Figure 8.4** During interphase the DNA is not visible through a microscope. Why isn't the DNA visible during interphase, and why would you expect this to be the case?

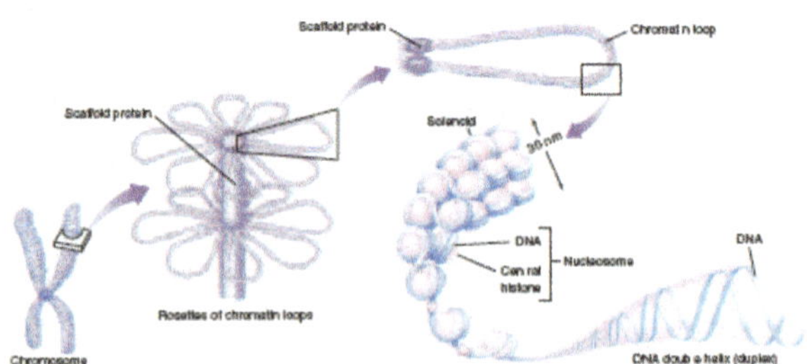

Synthesize What You Have Learned

1. Why does the DNA in a cell need to change periodically from a long, double-helix chromatin molecule into a tightly wound-up chromosome? What does it do in one configuration that it cannot do in the other?
2. Despite all we know about cancer today, some types of cancers are still increasing in frequency. Lung cancer in women is one of those. What reason(s) might there be for this increasing problem? Can you suggest a solution?

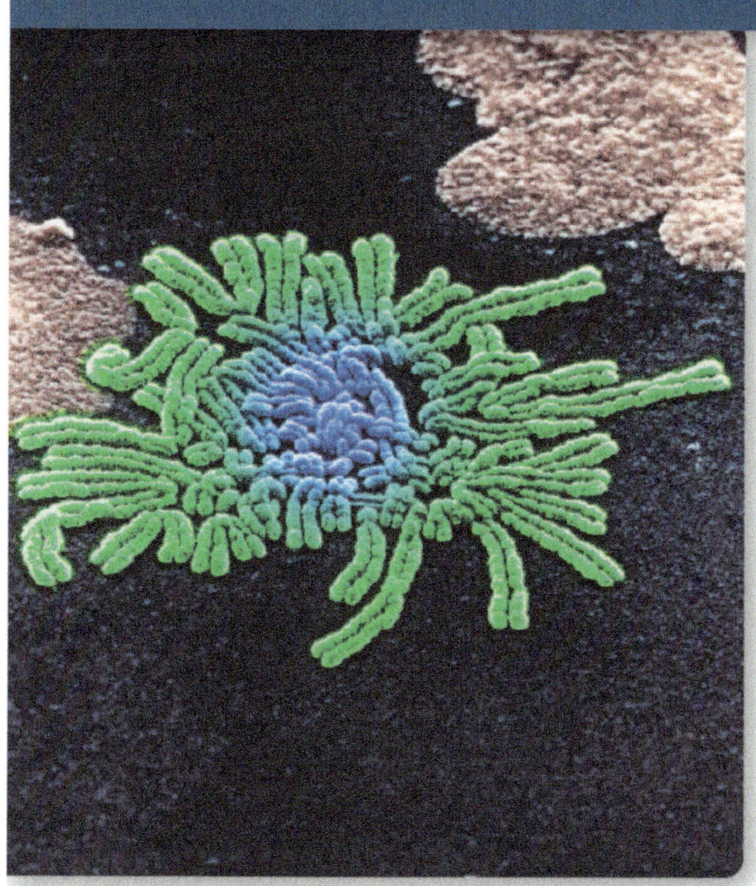

9

Meiosis

Learning Objectives

Humans, like most animals and plants, reproduce sexually. That is how you came into being: Your father contributed a sperm cell that united with an egg cell from your mother to form a cell called a zygote, containing both sets of chromosomes. Dividing repeatedly by mitosis, this zygote cell eventually gave rise to your adult body, made up of an astonishing number of cells—some 10 to 100 trillion. The sperm and egg that joined to form your initial cell were the products of a special form of cell division called meiosis, the subject of this chapter. Far more intricate than mitosis, the details of meiosis are not as well understood. The basic process, however, is clear. A cell dividing by meiosis goes through two nuclear divisions, replicating the DNA before the first division but not between the two divisions. In the photo above, chromosomes are lining up and getting ready to be pulled to opposite ends of the cell by microtubules too tiny to be visible to our eyes. When the two meiotic divisions are all over, there will be four cells, each with only half as much DNA as the initial cell. Confused? So were biologists when they first discovered meiosis. Hopefully, this chapter will make things clearer. It is important that you understand meiosis clearly, because meiosis and sexual reproduction play key roles in generating the tremendous genetic diversity that is the raw material of evolution.

Meiosis

9.1 Discovery of Meiosis

Only a few years after Walther Fleming's discovery of chromosomes in 1879, Belgian cytologist Pierre-Joseph van Beneden was surprised to find different numbers of chromosomes in different types of cells in the roundworm *Ascaris.* Specifically, he observed that the **gametes** (eggs and sperm) each contained two chromosomes, whereas the somatic (nonreproductive) cells of embryos and mature individuals each contained four.

Fertilization

From his observations, van Beneden proposed in 1887 that an egg and a sperm, each containing half the complement of chromosomes found in other cells, fuse to produce a single cell called a **zygote.** The zygote, like all of the somatic cells ultimately derived from it, contains two copies of each chromosome. The fusion of gametes to form a new cell is called **fertilization,** or **syngamy.**

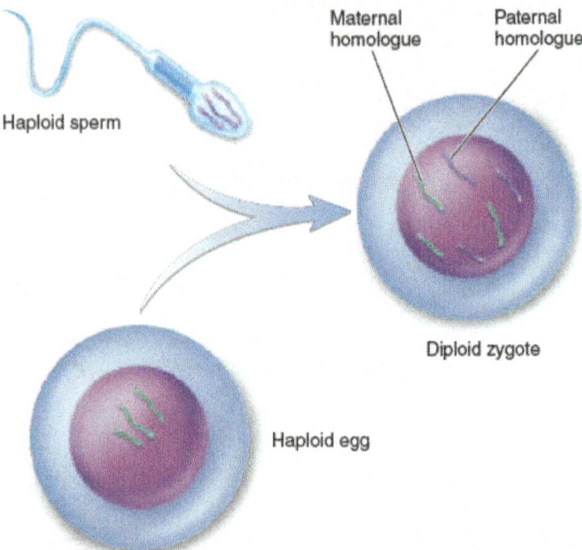

Figure 9.1 Diploid cells carry chromosomes from two parents.
A diploid cell contains two versions of each chromosome, a maternal homologue contributed by the mother's haploid egg, and a paternal homologue contributed by the father's haploid sperm.

Meiosis

It was clear even to early investigators that gamete formation must involve some mechanism that reduces the number of chromosomes to half the number found in other cells. If it did not, the chromosome number would double with each fertilization, and after only a few generations, the number of chromosomes in each cell would become impossibly large. For example, in just 10 generations, the 46 chromosomes present in human cells would increase to over 47,000 (46×2^{10}) chromosomes.

The number of chromosomes does not explode in this way because of a special reduction division that occurs during gamete formation, producing cells with half the normal number of chromosomes. The subsequent fusion of two of these cells ensures a consistent chromosome number from one generation to the next. This reduction division process, known as **meiosis,** is the subject of this chapter.

The Sexual Life Cycle

Meiosis and fertilization together constitute a cycle of reproduction. Two sets of chromosomes are present in the somatic cells of adult individuals, making them **diploid** cells (Greek, *di,* two), but only one set is present in the gametes, which are thus **haploid** (Greek, *haploos,* one). Figure 9.1 shows how two haploid cells, a sperm cell containing three chromosomes contributed by the father and an egg cell containing three chromosomes contributed by the mother, fuse to form a diploid zygote with six chromosomes. Reproduction that involves this alternation of meiosis and fertilization is called **sexual reproduction.** Some organisms however, reproduce by mitotic division and don't involve the fusion of gametes. Reproduction in these organisms is referred to as **asexual reproduction.** Binary fission of prokaryotes shown in chapter 8 is an example of asexual reproduction. Some organisms are able to reproduce both asexually and sexually. The strawberry plant pictured in figure 9.2 reproduces sexually by fertilization that occurs in its flowers. Strawberries also reproduce asexually by sending out runners, stems that grow along the ground and produce new roots and shoots that give rise to genetically identical plants.

Figure 9.2 Sexual and asexual reproduction.
Reproduction in an organism is not always either sexual or asexual. The strawberry plant reproduces both asexually (runners) and sexually (flowers).

Key Learning Outcome 9.1 Meiosis is a process of cell division in which the number of chromosomes in certain cells is halved during gamete formation.

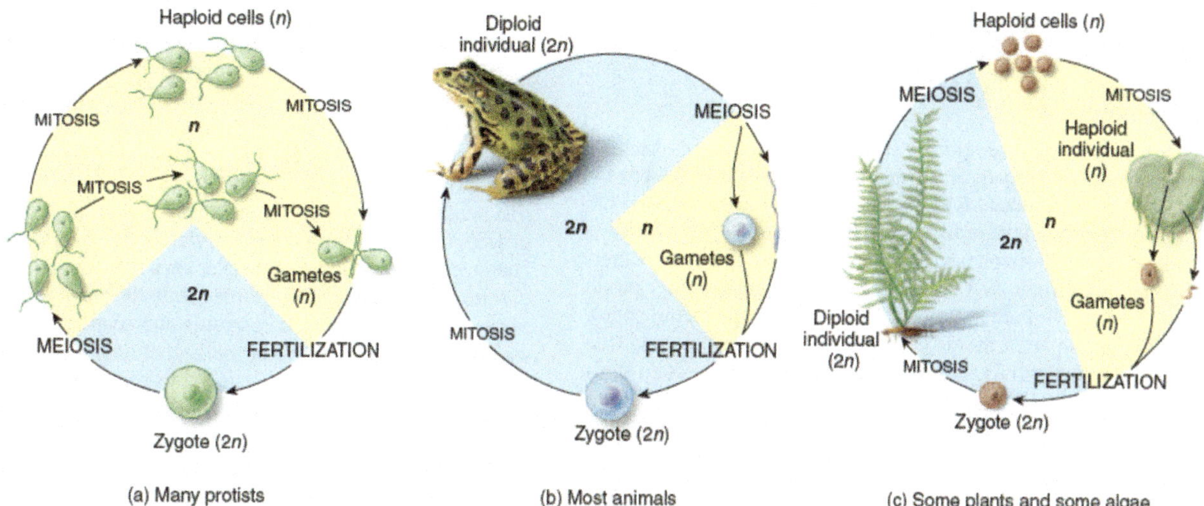

Figure 9.3 Three types of sexual life cycles.
In sexual reproduction, haploid cells or organisms alternate with diploid cells or organisms.

9.2 The Sexual Life Cycle

Somatic Tissues

The life cycles of all sexually reproducing organisms follow the same basic pattern of alternation between diploid chromosome numbers (the blue areas of the life cycles illustrated in figure 9.3) and haploid ones (the yellow areas). In most animals, fertilization results in the formation of a diploid zygote, shown in figure 9.3*b*, that begins to divide by mitosis. This single diploid cell eventually gives rise to all of the cells in the adult frog shown in the figure. These cells are called **somatic** cells, from the Latin word for "body." Each is genetically identical to the zygote.

In unicellular eukaryotic organisms like the protists shown in figure 9.3*a*, individual haploid cells function as gametes, fusing with other gamete cells. In plants like the fern shown in figure 9.3*c*, the haploid cells that meiosis produces divide by mitosis, forming a multicellular haploid phase, the heart-shaped structure in the figure. Some cells of this haploid phase eventually differentiate into eggs or sperm, which fuse to form a diploid zygote.

Germ-Line Tissues

In animals, the cells that will eventually undergo meiosis to produce gametes are set aside from somatic cells early in the course of development. These cells are often referred to as **germ-line** cells. Both the somatic cells and the gamete-producing germ-line cells are diploid, as indicated by blue arrows in figure 9.4. Somatic cells undergo mitosis to form genetically identical, diploid daughter cells. The germ-line cells undergo meiosis, indicated by the yellow arrows, producing haploid gametes.

> **Key Learning Outcome 9.2** In the sexual life cycle, there is an alternation of diploid and haploid phases.

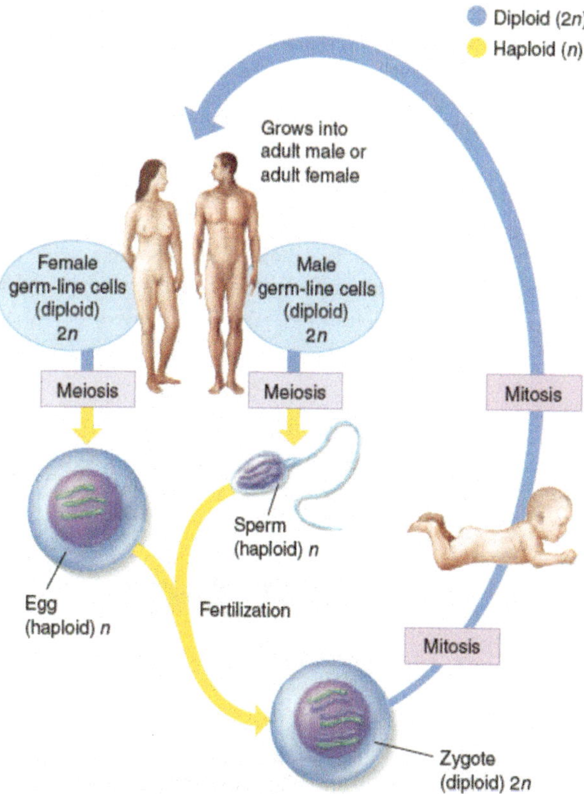

Figure 9.4 The sexual life cycle in animals.
In animals, the completion of meiosis is followed soon by fertilization. Thus, the vast majority of the life cycle is spent in the diploid stage. In this text, *n* stands for haploid and 2*n* stands for diploid. Germ-line cells are set aside early in development and undergo meiosis to form haploid gametes (eggs or sperm). The rest of the body cells are called somatic cells.

9.3 The Stages of Meiosis

Now, let's look more closely at the process of meiosis. Meiosis consists of two rounds of cell division, called meiosis I and meiosis II, which produce four haploid cells. Just as in mitosis, the chromosomes have replicated before meiosis begins, during a period called interphase. The first of the two divisions of meiosis, called **meiosis I** (meiosis I is shown in the outer circle of the Key Biological Process illustration on the facing page), serves to separate the two versions of each chromosome (the homologous chromosomes or homologues); the second division, **meiosis II** (the inner circle), serves to separate the two replicas of each version, called *sister chromatids.* Thus when meiosis is complete, what started out as one diploid cell ends up as four haploid cells. Because there was one replication of DNA but *two* cell divisions, the process reduces the number of chromosomes by half.

Meiosis I

Meiosis I is traditionally divided into four stages:

1. **Prophase I.** The two versions of each chromosome (the two homologues) pair up and exchange segments.
2. **Metaphase I.** The chromosomes align on a central plane.
3. **Anaphase I.** One homologue with its two sister chromatids still attached moves to a pole of the cell, and the other homologue moves to the opposite pole.
4. **Telophase I.** Individual chromosomes gather together at each of the two poles.

In **prophase I,** individual chromosomes first become visible, as viewed with a light microscope, as their DNA coils more and more tightly. Because the chromosomes (DNA) have replicated before the onset of meiosis, each of the threadlike chromosomes actually consists of two sister chromatids associated along their lengths (held together by cohesin proteins in a process called *sister chromatid cohesion*) and joined at their centromeres, just as in mitosis. However, now meiosis begins to differ from mitosis. During prophase I, the two homologous chromosomes line up side by side, physically touching one another, as you see in figure 9.5. It is at this point that a process called **crossing over** is initiated, in which DNA is exchanged between the two nonsister chromatids of homologous chromosomes. The chromosomes actually break in the same place on both nonsister chromatids and sections of chromosomes are swapped between the homologous chromosomes, producing a hybrid chromosome that is part maternal chromosome (the green sections) and part paternal chromosome (the purple sections). Two elements hold the homologous chromosomes together: (1) cohesion between sister chromatids; and (2) crossovers between nonsister chromatids (homologues). Late in prophase, the nuclear envelope disperses.

In **metaphase I,** the spindle apparatus forms, but because homologues are held close together by crossovers, spindle fibers can attach to only the outward-facing kinetochore of each centromere. For each pair of homologues, the orientation on the metaphase plate is random; which homologue is oriented toward which pole is a matter of chance. Like shuffling a deck of cards, many combinations are possible—in fact, 2 raised to a power equal to the number of chromosome pairs. For example, in a hypothetical cell that has three chromosome pairs, there are eight possible orientations (2^3). Each orientation results in gametes with different combinations of parental chromosomes. This process is called **independent assortment.** The chromosomes in figure 9.6 line up along the metaphase plate, but whether the maternal chromosome (the green chromosomes) is on the right or left of the plate is completely random.

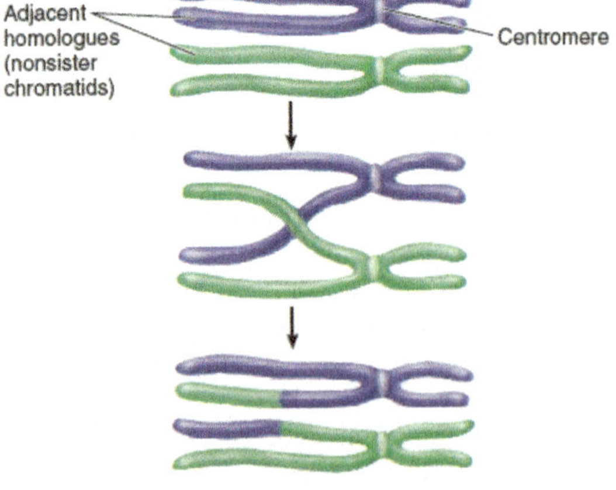

Figure 9.5 Crossing over.
In crossing over, the two homologues of each chromosome exchange portions. During the crossing over process, nonsister chromatids that are next to each other exchange chromosome arms or segments.

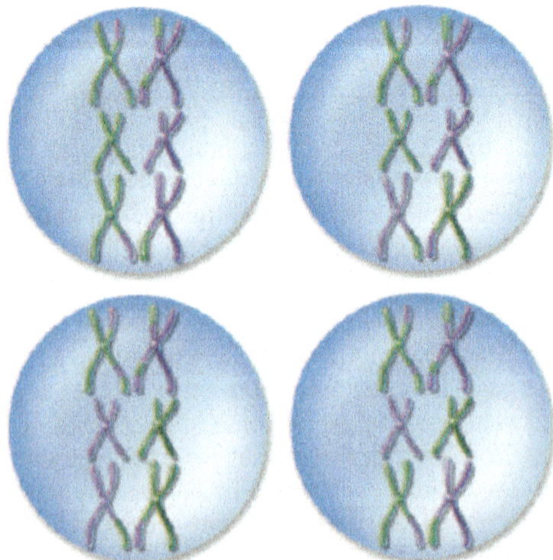

Figure 9.6 Independent assortment.
Independent assortment occurs because the orientation of chromosomes on the metaphase plate is random. Shown here are four possible orientations of chromosomes in a hypothetical cell. Each of the many possible orientations results in gametes with different combinations of parental chromosomes.

KEY BIOLOGICAL PROCESS: Meiosis

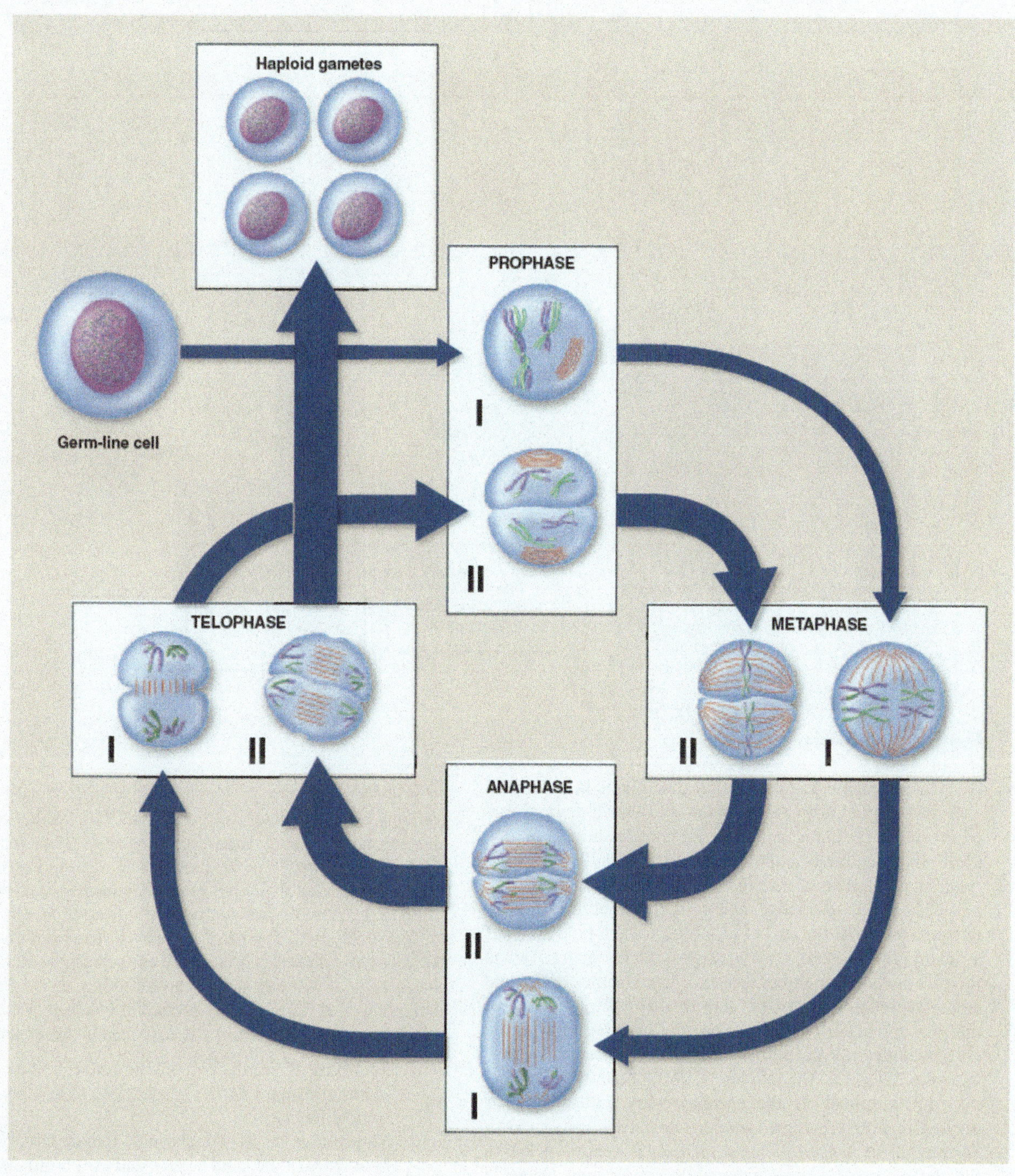

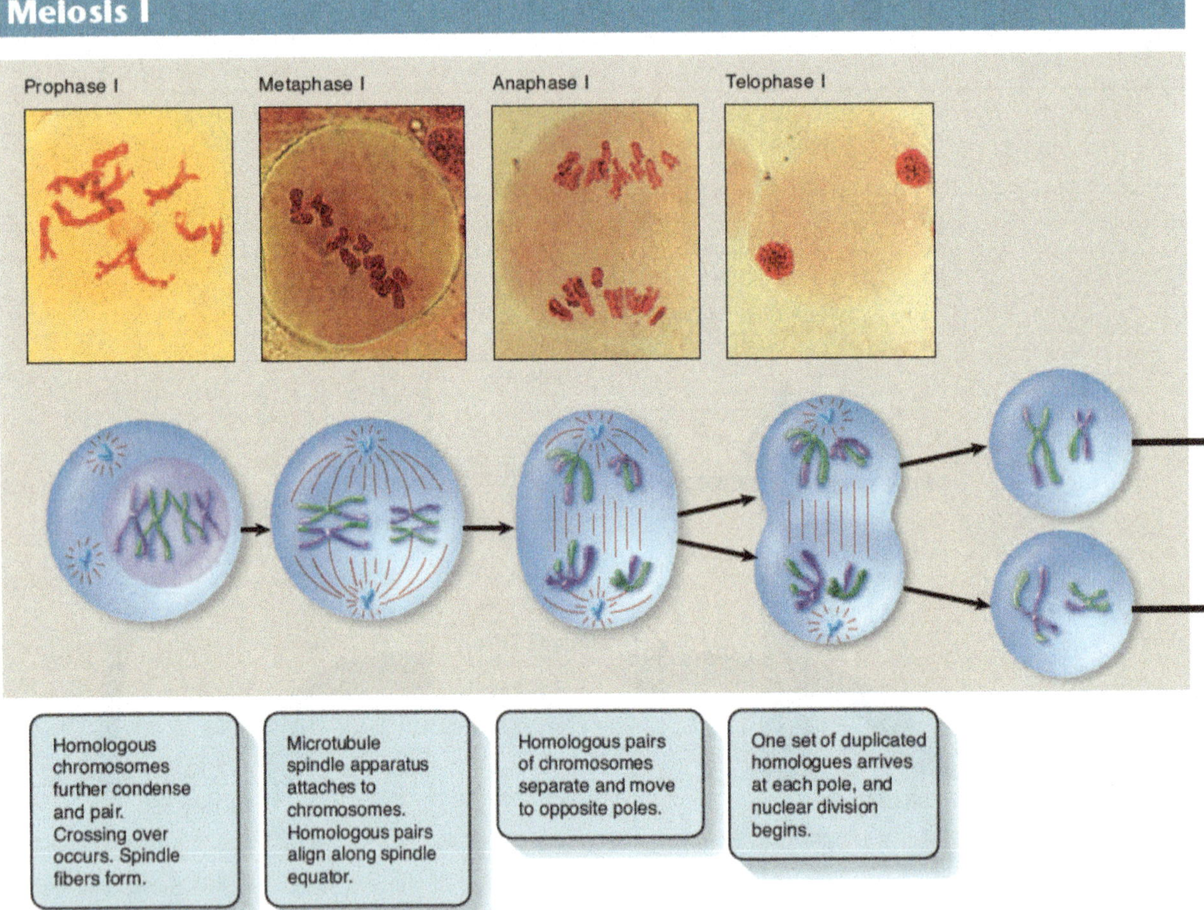

Figure 9.7 Meiosis.

In **anaphase I,** the spindle attachment is complete, and homologues are pulled apart and move toward opposite poles. Sister chromatids are not separated at this stage. Because the orientation along the spindle equator is random, the chromosome that a pole receives from each pair of homologues is also random with respect to all chromosome pairs. At the end of anaphase I, each pole has half as many chromosomes as were present in the cell when meiosis began. Remember that the chromosomes replicated and thus contained two sister chromatids before the start of meiosis, but sister chromatids are not counted as separate chromosomes. As in mitosis, count the number of centromeres to determine the number of chromosomes.

In **telophase I,** the chromosomes gather at their respective poles to form two chromosome clusters. After an interval of variable length, meiosis II occurs in which the sister chromatids are separated as in mitosis. Meiosis can be thought of as two consecutive cycles, as shown in the Key Biological Process illustration on the previous page. The outer cycle contains the phases of meiosis I and the inner cycle contains the phases of meiosis II, discussed next.

Meiosis II

After a brief interphase, in which no DNA synthesis occurs, the second meiotic division begins. Meiosis II is simply a mitotic division involving the products of meiosis I, except that the sister chromatids are not genetically identical, as they are in mitosis, because of crossing over. You can see this by looking at figure 9.7, where some of the arms of the sister chromatids contain two different colors. At the end of anaphase I, each pole has a haploid complement of chromosomes, each of which is still composed of two sister chromatids attached at the centromere. Like meiosis I, meiosis II is divided into four stages:

1. **Prophase II.** At the two poles of the cell, the clusters of chromosomes enter a brief prophase II, where a new spindle forms.
2. **Metaphase II.** In metaphase II, spindle fibers bind to both sides of the centromeres and the chromosomes line up along a central plane.
3. **Anaphase II.** The spindle fibers shorten, splitting the centromeres and moving the sister chromatids to opposite poles.
4. **Telophase II.** Finally, the nuclear envelope re-forms around the four sets of daughter chromosomes.

Meiosis II

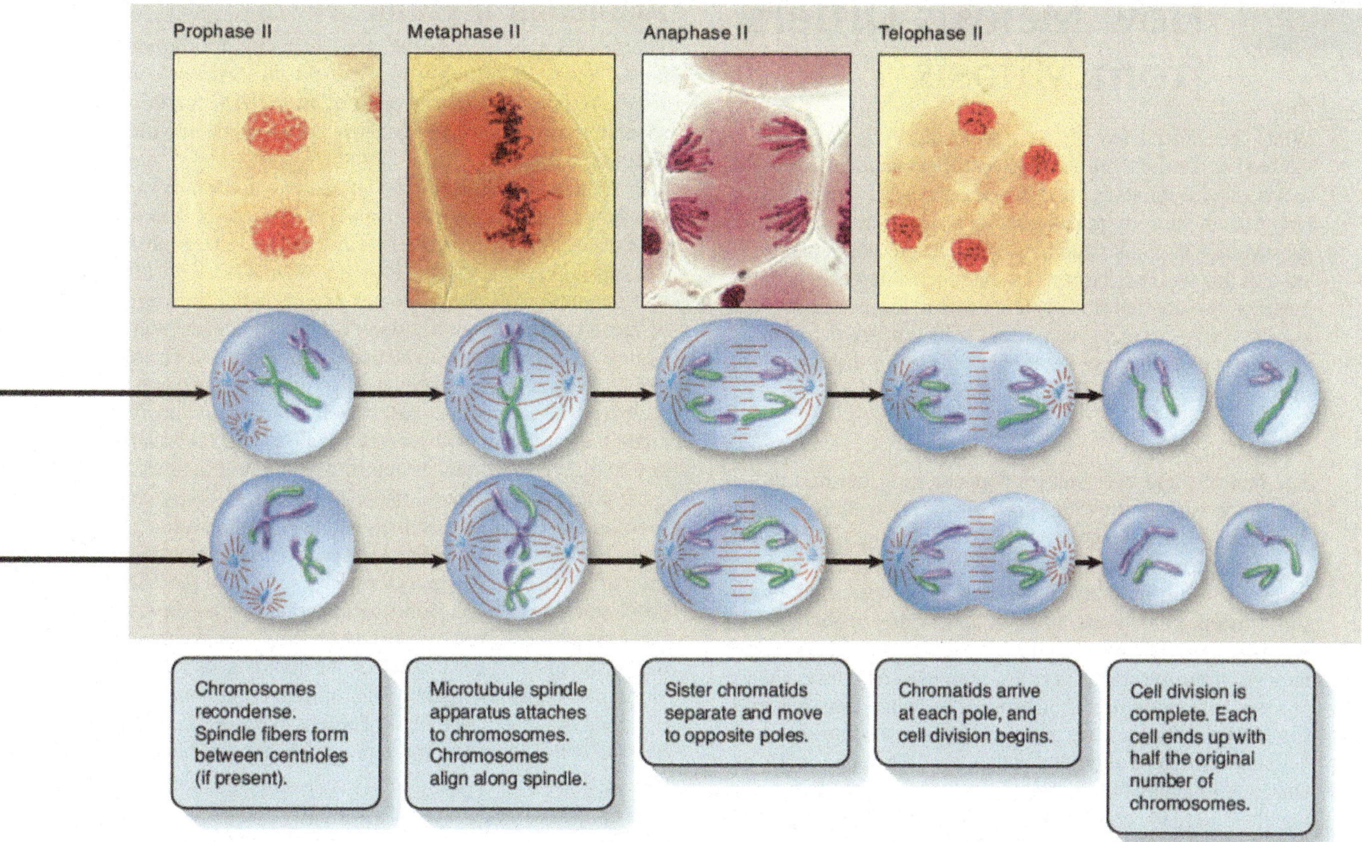

The main outcome of the four stages of meiosis II—prophase II, metaphase II, anaphase II, and telophase II—is to separate the sister chromatids. The final result of this division is four cells containing haploid sets of chromosomes. No two are alike because of the crossing over in prophase I. The nuclei are then reorganized, and nuclear envelopes form around each haploid set of chromosomes. The cells that contain these haploid nuclei may develop directly into gametes, as they do in most animals. Alternatively, they may themselves divide mitotically, as they do in plants, fungi, and many protists, eventually producing greater numbers of gametes or, as in the case of some plants and insects, adult haploid individuals.

The Important Role of Crossing Over

If you think about it, the key to meiosis is that the sister chromatids of each chromosome are not separated from each other in the first division. Why not? What prevents microtubules from attaching to them and pulling them to opposite poles of the cell, just as eventually happens later in the second meiotic division? The answer is the crossing over that occurred early in the first division. By exchanging segments, the two homologues are tied together by strands of DNA. It is because microtubules can gain access to only one side of each homologue that they cannot pull the two sister chromatids apart! Imagine two people dancing closely—you can tie a rope to the back of each person's belt, but you cannot tie a second rope to their belt buckles because the two dancers are facing each other and are very close. In just the same way, microtubules cannot attach to the inner sides of the homologues because crossing over holds the homologous chromosomes together like dancing partners.

Key Learning Outcome 9.3 During meiosis I, homologous chromosomes move to opposite poles of the cell. At the end of meiosis II, each of the four haploid cells contains one copy of every chromosome in the set, rather than two. Because of crossing over, no two cells are the same.

Comparing Meiosis and Mitosis

9.4 How Meiosis Differs from Mitosis

While the general features of meiosis that you have just reviewed are similar in all eukaryotes, the detailed mechanism of meiosis varies somewhat in different organisms. This is particularly true of chromosomal separation mechanisms, which differ substantially in protists and fungi from the process in plants and animals that we describe here. Despite such differences in detail, however, two consistent features are seen in the meiotic processes of every eukaryote: synapsis and reduction division. Indeed, these two unique features are the key differences that distinguish meiosis from mitosis, which you studied in chapter 8.

Synapsis

The first of these two features happens early during the first nuclear division. Following chromosome replication, homologous chromosomes or homologues *pair all along their lengths,* with sister chromatids being held together, as mentioned earlier, by cohesin proteins. While homologues are thus physically joined, *genetic exchange occurs at one or more points between them.* The process of forming these complexes of homologous chromosomes is called **synapsis,** and the exchange process between paired homologues is called crossing over. Figure 9.8*a* shows how the homologous chromosomes are held together close enough that they are able to physically exchange segments of their DNA. Chromosomes are then drawn together along the equatorial plane of the dividing cell; subsequently, homologues are pulled apart by microtubules toward opposite poles of the cell. When this process is complete, the cluster of chromosomes at each pole contains one of the two homologues of each chromosome. Each pole is haploid, containing half the number of chromosomes present in the original diploid cell. Sister chromatids do not separate from each other in the first nuclear division, so each homologue is still composed of two chromatids joined at the centromere, and still considered one chromosome.

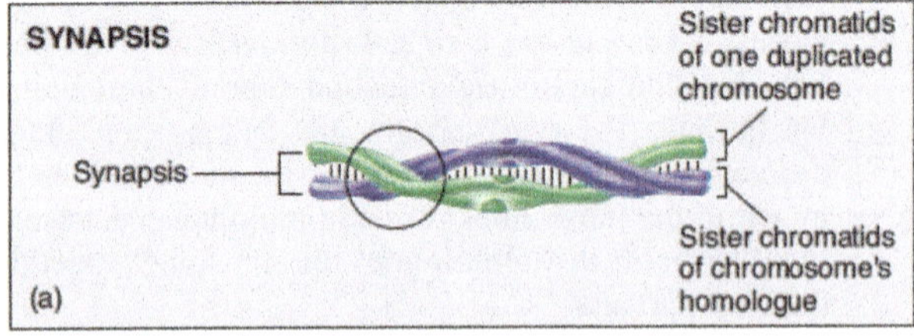

Reduction Division

The second unique feature of meiosis is that *the chromosome homologues do not replicate between the two nuclear divisions,* so that chromosome assortment in the second division separates sister chromatids of each chromosome into different daughter cells.

In most respects, the second meiotic division is identical to a normal mitotic division. However, because of the crossing over that occurred during the first division, the sister chromatids in meiosis II are not identical to each other. Also, there are only half the number of chromosomes in each cell at the beginning of meiosis II because only one of the homologues is present. Figure 9.8*b* shows how reduction division occurs. The diploid cell contains four chromosomes (two homologous pairs). After meiosis I, the cells contain just two chromosomes (remember to count the number of *centromeres,* because sister chromatids are not considered separate chromosomes). During meiosis II, the sister chromatids separate, but each gamete still only contains two chromosomes, half as many of the germ-line cell.

Because mitosis and meiosis use similar terminology, it is easy to confuse the two processes. Figure 9.9 compares the two processes side-by-side. Both processes start with a diploid cell, but during meiosis, you can see that crossing over occurs and that during the first division in meiosis, the homologous pairs line up along the metaphase plate, while in mitosis, centromeres line up along the metaphase plate. These two differences result in haploid cells in meiosis and diploid cells in mitosis.

Key Learning Outcome 9.4 In meiosis, homologous chromosomes become intimately associated and do not replicate between the two nuclear divisions.

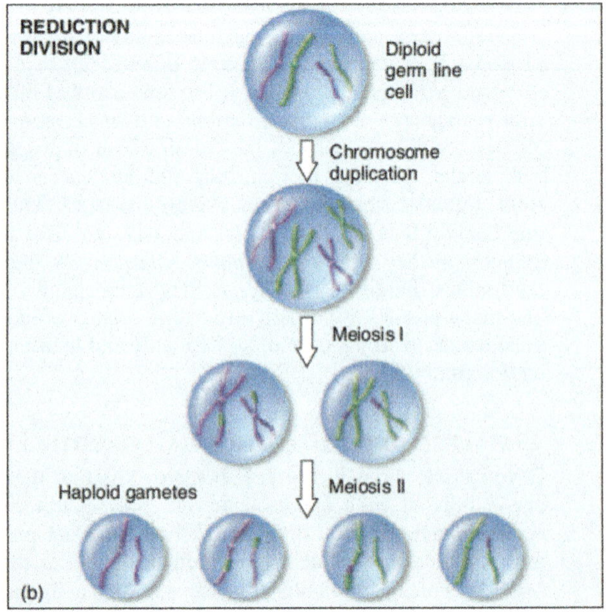

Figure 9.8 Unique features of meiosis.
(*a*) Synapsis draws homologous chromosomes together, all along their lengths, creating a situation (indicated by the circle) where two homologues can physically exchange portions of arms, a process called crossing over. (*b*) Reduction division, omitting a chromosome duplication before meiosis II, produces haploid gametes, thus ensuring that the chromosome number remains the same as that of the parents, following fertilization.

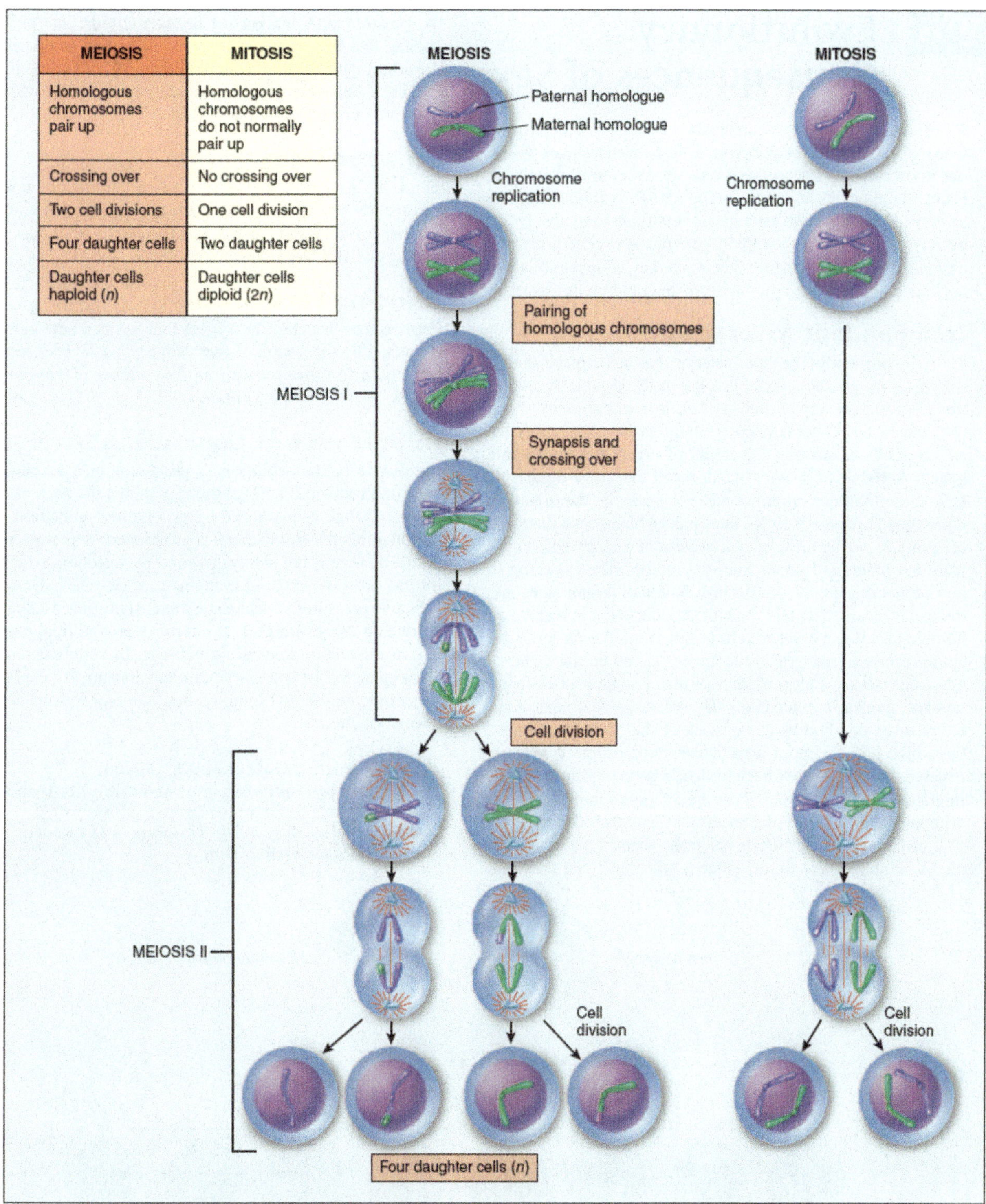

MEIOSIS	MITOSIS
Homologous chromosomes pair up	Homologous chromosomes do not normally pair up
Crossing over	No crossing over
Two cell divisions	One cell division
Four daughter cells	Two daughter cells
Daughter cells haploid (*n*)	Daughter cells diploid (2*n*)

Figure 9.9 A comparison of meiosis and mitosis.

Meiosis differs from mitosis in several key ways, highlighted by the orange boxes. Meiosis involves two nuclear divisions with no DNA replication between them. It thus produces four daughter cells, each with half the original number of chromosomes. Also, crossing over occurs in prophase I of meiosis. Mitosis involves a single nuclear division after DNA replication. Thus, it produces two daughter cells, each containing the original number of chromosomes, which are genetically identical to those in the parent cell.

9.5 Evolutionary Consequences of Sex

As you can now appreciate, meiosis is a lot more complicated than mitosis. Why has evolution gone to so much trouble? While our knowledge of how meiosis and sex evolved is sketchy, it is abundantly clear that meiosis and sexual reproduction have an enormous impact on how species continue to evolve today because of their ability to rapidly generate new genetic combinations. Three mechanisms each make key contributions: independent assortment, crossing over, and random fertilization.

Independent Assortment

The reassortment of genetic material that takes place during meiosis is the principal factor that has made possible the evolution of eukaryotic organisms, in all their bewildering diversity, over the past 1.5 billion years. Sexual reproduction represents an enormous advance in the ability of organisms to generate genetic variability. To understand, recall that most organisms have more than one chromosome. For example, the organism represented in figure 9.10 has three pairs of chromosomes, each offspring receiving three homologues from each parent, purple from the father and green from the mother. The offspring in turn produces gametes, but the distribution of homologues into the gametes is completely random. A gamete could receive all homologues that are paternal in origin, as on the far left; or it could receive all maternal homologues, as on the far right, or any combination. Independent assortment alone leads to eight possible gamete combinations. In human beings, each gamete receives one homologue of each of the 23 chromosomes, but which homologue of a particular chromosome it receives is determined randomly. Each of the 23 pairs of chromosomes migrates independently, so there are 2^{23} (more than 8 million) different possible kinds of gametes that can be produced.

To make this point to his class, one professor offers an "A" course grade to any student who can write down all the possible combinations of heads and tails (an "either/or" choice, like that of a chromosome migrating to one pole or the other) with flipping a coin 23 times (like 23 chromosomes moving independently). No student has ever won an "A," as there are over 8 million possibilities.

Crossing Over

The DNA exchange that occurs when the arms of nonsister chromatids cross over adds even more recombination. The number of possible genetic combinations that can occur among gametes is virtually unlimited.

Random Fertilization

Furthermore, because the zygote that forms a new individual is created by the fusion of two gametes, each produced independently, fertilization squares the number of possible outcomes ($2^{23} \times 2^{23}$ = 70 trillion).

Importance of Generating Diversity

Paradoxically, the evolutionary process is both revolutionary and conservative. It is revolutionary in that the pace of evolutionary change is quickened by genetic recombination, much of which results from sexual reproduction. It is conservative in that change is not always favored by selection, which may instead preserve existing combinations of genes. These conservative pressures appear to be greatest in some asexually reproducing organisms that do not move around freely and that live in especially demanding habitats. In vertebrates, on the other hand, the evolutionary premium appears to have been on versatility, and sexual reproduction is the predominant mode of reproduction.

Key Learning Outcome 9.5 Sexual reproduction increases genetic variability through independent assortment in metaphase I of meiosis, crossing over in prophase I of meiosis, and random fertilization.

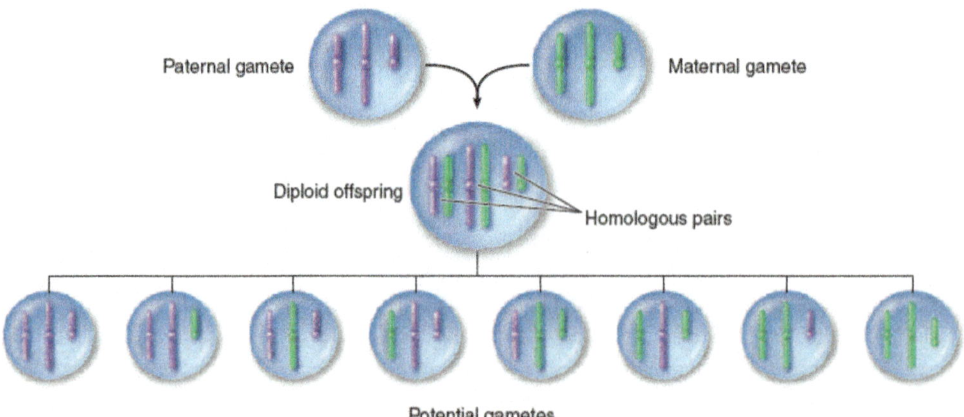

Figure 9.10 Independent assortment increases genetic variability.
Independent assortment contributes new gene combinations to the next generation because the orientation of chromosomes on the metaphase plate is random. In the cell shown here with three chromosome pairs, there are eight different gametes that can result, each with different combinations of parental chromosomes.

A Closer Look

Why Sex?

Not all reproduction is sexual. In **asexual reproduction,** an individual inherits all of its chromosomes from a single parent and is, therefore, genetically identical to its parent. Prokaryotic cells reproduce asexually, undergoing binary fission to produce two daughter cells containing the same genetic information.

Most protists reproduce asexually except under conditions of stress; then they switch to sexual reproduction. Among plants and fungi, asexual reproduction is common.

In animals, asexual reproduction often involves the budding off of a localized mass of cells, which grows by mitosis to form a new individual.

Even when meiosis and the production of gametes occur, there may still be reproduction without sex. The development of an adult from an unfertilized egg, called **parthenogenesis,** is a common form of reproduction in arthropods. Among bees, for example, fertilized eggs develop into diploid females, but unfertilized eggs develop into haploid males. Parthenogenesis even occurs among the vertebrates. Some lizards, fishes, and amphibians are capable of reproducing in this way; their unfertilized eggs undergo a mitotic nuclear division without cell cleavage to produce a diploid cell, which then develops into an adult. In some plants, such as hawkweeds, dandelions, and blackberries, a process similar to parthenogenesis called *apomixis* can occur.

If reproduction can occur without sex, why does sex occur at all? This question has generated considerable discussion, particularly among evolutionary biologists. Sex is of great evolutionary advantage for populations or species, which benefit from the variability generated in meiosis by random orientation of chromosomes and by crossing over. However, evolution occurs because of changes at the level of *individual* survival and reproduction, rather than at the population level, and no obvious advantage accrues to the progeny of an individual that engages in sexual reproduction. In fact, recombination is a destructive as well as a constructive process in evolution. The segregation of chromosomes during meiosis tends to disrupt advantageous combinations of genes more often than it creates new, better adapted combinations; as a result, some of the diverse progeny produced by sexual reproduction will not be as well adapted as their parents were. In fact, the more complex the adaptation of an individual organism, the less likely that recombination will improve it, and the more likely that recombination will disrupt it. It is, therefore, a puzzle to know what a well-adapted individual gains from participating in sexual reproduction, as *all* of its progeny could maintain its successful gene combinations if that individual simply reproduced asexually.

The DNA Repair Hypothesis. Several geneticists have suggested that sex occurs because only a diploid cell can effectively repair certain kinds of chromosome damage, particularly double-strand breaks in DNA. Both radiation and chemical events within cells can induce such breaks. As organisms became larger and longer-lived, it must have become increasingly important for them to be able to repair such damage. Synapsis, which in early stages of meiosis precisely aligns pairs of homologous chromosomes, may well have evolved originally as a mechanism for repairing double-strand damage to DNA. The undamaged homologous chromosome could be used as a template to repair the damaged chromosome. A transient diploid phase would have provided an opportunity for such repair. In yeast, mutations that inactivate the repair system for double-strand breaks of the chromosomes also prevent crossing over, suggesting a common mechanism for both synapsis and repair processes.

Muller's Ratchet. The geneticist Herman Muller pointed out in 1965 that asexual populations incorporate a kind of mutational ratchet mechanism—once harmful mutations arise, asexual populations have no way of eliminating them, and they accumulate over time, like turning a ratchet. Sexual populations, on the other hand, can employ recombination to generate individuals carrying fewer mutations, which selection can then favor. Sex may just be a way to keep the mutational load down.

The Red Queen Hypothesis. One evolutionary advantage of sex may be that it allows populations to "store" forms of a trait that are currently bad but have promise for reuse at some time in the future. Because populations are constrained by a changing physical and biological environment, selection is constantly acting against such traits. But in sexual species, selection can never get rid of those variants sheltered by more dominant forms of the trait.

The evolution of most sexual species, most of the time, thus manages to keep pace with ever-changing physical and biological constraints. This "treadmill evolution" is sometimes called the "Red Queen hypothesis," after the Queen of Hearts in Lewis Carroll's *Through the Looking Glass,* who tells Alice, "Now, here, you see, it takes all the running you can do, to keep in the same place."

INQUIRY & ANALYSIS

Are New Microtubules Made When the Spindle Forms?

During interphase, before the beginning of meiosis, relatively few long microtubules extend from the centrosome (a zone around the centrioles of animal cells where microtubules are organized) to the cell periphery. Like most microtubules, these are refreshed at a low rate with resynthesis. Late in prophase, however, a dramatic change is seen—the centrosome divides into two, and a large increase is seen in the number of microtubules radiating from each of the two daughter centrosomes. The two clusters of new microtubules are easily seen as the green fibers connecting to the two sets of purple chromosomes in the micrograph of early prophase below (a **micrograph** is a photo taken through a microscope). This burst of microtubule assembly marks the beginning of the formation of the spindle characteristic of metaphase. When it first became known to cell biologists, they asked whether these were existing microtubules being repositioned in the spindle, or newly synthesized microtubules only produced just before metaphase begins.

The graph to the upper right displays the results of an experiment designed to answer this question. Mammalian cells in culture (cells **in culture** are growing in the laboratory on artificial medium) were injected with microtubule subunits (tubulin) to which a fluorescent dye had been attached (a **fluorescent dye** is one that glows when exposed to ultraviolet or short-wavelength visual light). After the fluorescent subunits had become incorporated into the cells's microtubules, all the fluorescence in a small region of a cell was bleached by an intense laser beam, destroying the microtubules there. Any subsequent rebuilding of microtubules in the bleached region would have to employ the fluorescent subunits present in the cell, causing recovery of fluorescence in the bleached region. The graph reports this recovery as a function of time, for interphase and metaphase cells. The dotted line represents the time for 50% recovery of fluorescence ($t_{1/2}$) (that is, $t_{1/2}$ is the time required for half of the microtubules in the region to be resynthesized).

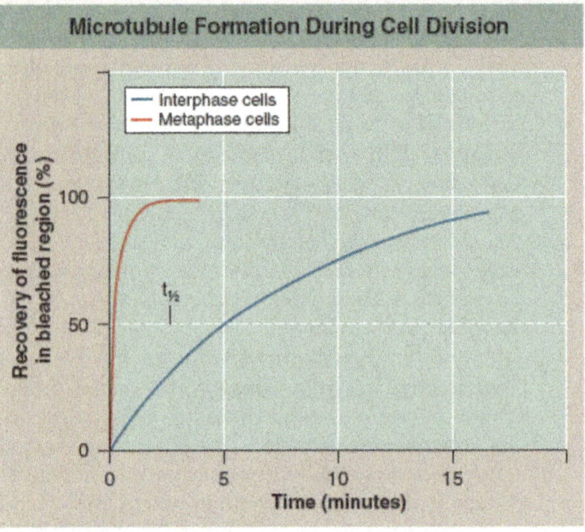

1. **Applying Concepts**
   a. Variable. In the graph, what is the dependent variable?
   b. $t_{1/2}$. Are new microtubules synthesized during interphase? What is the $t_{1/2}$ of this replacement synthesis? Are new microtubules synthesized during metaphase? What is the $t_{1/2}$ of this replacement synthesis?
2. **Interpreting Data** Is there a difference in the rate at which microtubules are synthesized during interphase and metaphase? How big is the difference? What might account for it?
3. **Making Inferences**
   a. What general statement can be made regarding the relative rates of microtubule production before and during meiosis?
   b. Is there any difference in the final amount of microtubule synthesis that would occur if this experiment were to be continued for an additional 15 minutes?
4. **Drawing Conclusions** When are the microtubules of the spindle assembled?
5. **Further Analysis** The spindle breaks down after cell division is completed. Design an experiment to test whether the tubulin subunits of the spindle microtubules are recycled into other cell components, or destroyed, after meiosis.

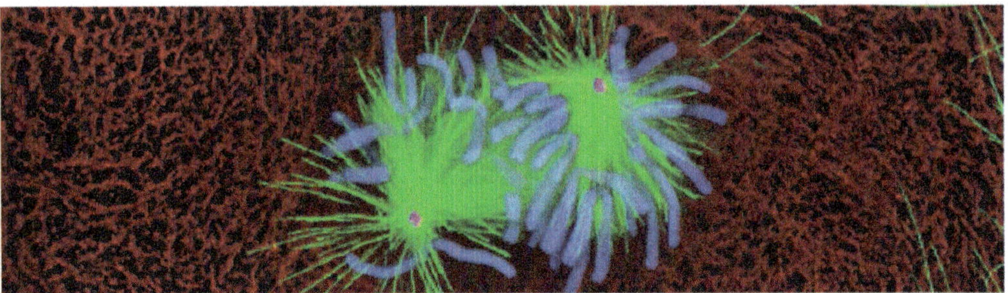

Chapter Review

Meiosis

9.1 Discovery of Meiosis

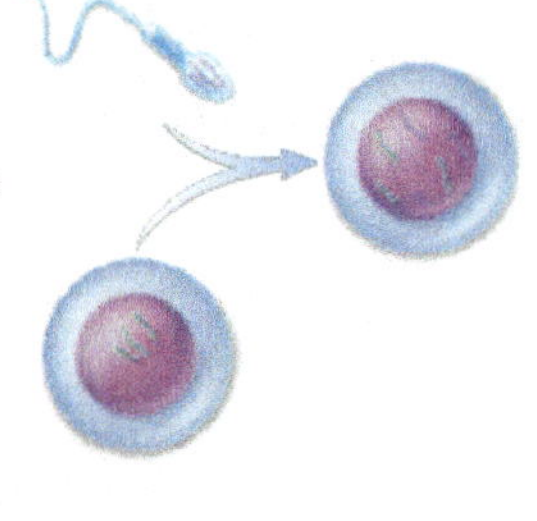

- In sexually reproducing organisms, a gamete from the male fuses with a gamete from the female in a process called fertilization, or syngamy. The number of chromosomes in gametes must be halved to maintain the correct number of chromosomes in offspring (**figure 9.1**). Organisms accomplish this through a cell division process called meiosis.
- A cell that contains two copies of each chromosome is called a diploid cell. Cells, such as gametes, that contain only one copy of each chromosome are haploid cells.
- Sexual reproduction involves meiosis, but some organisms also undergo asexual reproduction, which is reproducing by mitosis or binary fission (**figure 9.2**).

9.2 The Sexual Life Cycle

- Sexual life cycles alternate between diploid and haploid stages, with variation in the amount of time devoted to each stage. Three types of sexual life cycles exist: In many protists the majority of the life cycle is devoted to the haploid stage; in most animals the majority of the life cycle is devoted to the diploid stage; and in plants and some algae the life cycle is split more equally between a haploid stage and a diploid stage (**figure 9.3**). Germ-line cells are diploid but produce haploid gametes (**figure 9.4**).

9.3 The Stages of Meiosis

- Meiosis (**Key Biological Process, page 175**) involves two nuclear divisions, meiosis I and meiosis II, each containing a prophase, metaphase, anaphase, and telophase. Like mitosis, the DNA replicates itself during interphase, before meiosis begins. Because there are two nuclear divisions but only one round of DNA replication, the four daughter cells contain half the number of chromosomes as the parent cell.
- Prophase I is distinguished by the exchange of genetic material between homologous chromosomes, a process called crossing over. In this process, homologous chromosomes align with each other along their lengths, and sections of homologues are physically exchanged, as shown here from **figure 9.5**. This recombines the genetic information contained in the chromosomes.

- During metaphase I, microtubules in the spindle apparatus attach to homologous chromosomes, and chromosome pairs align along the metaphase plate. The alignment of the chromosomes is random, leading to the independent assortment of chromosomes into the gametes (**figure 9.6**).
- The homologous chromosomes separate during anaphase I, being pulled apart by the spindle apparatus toward their respective poles. This differs from mitosis and later in meiosis II, where sister chromatids separate in anaphase.
- In telophase I, the chromosomes cluster at the poles. This leads to meiosis II.
- Meiosis II mirrors mitosis in that it involves the separation of sister chromatids through the phases of prophase II, metaphase II, anaphase II, and telophase II. Meiosis II differs from mitosis in that there is no DNA replication before meiosis II. Homologous pairs are separated during meiosis I, such that each daughter cell, shown forming here in telophase II from **figure 9.7**, has only one-half the number of chromosomes. Also, the chromosomes in the daughter cells at the end of meiosis II are not genetically identical because of crossing over.

Comparing Meiosis and Mitosis

9.4 How Meiosis Differs from Mitosis

- Two processes that distinguish meiosis from mitosis are crossing over through synapsis and reduction division.
- When homologous chromosomes come together during prophase I, they associate with each other along their lengths, a process called synapsis (**figure 9.8*a***). Synapsis does not occur in mitosis. During synapsis, sections of homologous chromosomes are physically exchanged in crossing over. Crossing over results in daughter cells that are not genetically identical to the parent cell or to each other. In contrast, mitosis results in daughter cells that are genetically identical to the parent cell and to each other.
- In meiosis, the daughter cells contain half the number of chromosomes as the parent cell due to reduction division. As shown here from **figure 9.8*b***, reduction division occurs because meiosis contains two nuclear divisions but only one round of DNA replication during interphase.

- The primary reasons for the differences in meiosis and mitosis stem from the synapsis of homologous chromosomes in prophase I. Because of synapsis, the arms of homologous chromosomes are close enough to undergo crossing over. Synapsis also blocks the inner kinetochores from attaching to the spindle. As a result, sister chromatids do not separate during meiosis I, resulting in reduction division (**figure 9.9**).

9.5 Evolutionary Consequences of Sex

- Sexual reproduction results in the introduction of genetic variation in future generations through independent assortment, crossing over, and random fertilization.
- Independent assortment results in the distribution of chromosomes into gametes, which creates many different combinations (**figure 9.10**).
- Crossing over provides even more genetic variability in gametes, such that the genetic combinations are virtually unlimited.
- The fusion of two gametes results in new genetic combinations that were created randomly, further increasing genetic diversity.

Test Your Understanding

1. An egg and a sperm unite to form a new organism. To prevent the new organism from having twice as many chromosomes as its parents,
   a. half of the chromosomes in the new organism quickly disassemble, leaving the correct number.
   b. half of the chromosomes from the egg and half from the sperm are ejected from the new cell.
   c. the large egg contains all the chromosomes, the tiny sperm only contributes some DNA.
   d. the egg and sperm cells only have half the number of chromosomes found in the parents due to meiosis.
2. The diploid number of chromosomes in humans is 46. The haploid number is
   a. 138.
   b. 92.
   c. 46.
   d. 23.
3. In organisms that have sexual life cycles, there is a time when there are
   a. $1n$ gametes (haploid), followed by $2n$ zygotes (diploid).
   b. $2n$ gametes (haploid), followed by $1n$ zygotes (diploid).
   c. $2n$ gametes (diploid), followed by $1n$ zygotes (haploid).
   d. $1n$ gametes (diploid), followed by $2n$ zygotes (haploid).
4. Which of the following occurs in meiosis I?
   a. All chromosomes duplicate.
   b. Homologous chromosomes randomly orient themselves on the metaphase plate, called independent assortment.
   c. The duplicated sister chromatids separate.
   d. The original cell divides into four diploid cells.
5. Which of the following occurs in meiosis II?
   a. All chromosomes duplicate.
   b. Homologous chromosomes randomly separate, called independent assortment.
   c. The duplicated sister chromatids separate.
   d. Genetically identical daughter cells are produced.
6. During which stage of meiosis is crossing over initiated?
   a. prophase I
   b. anaphase I
   c. metaphase II
   d. interphase
7. Synapsis is the process whereby
   a. homologous pairs of chromosomes separate and migrate toward a pole.
   b. homologous chromosomes exchange chromosomal material.
   c. homologous chromosomes become closely associated along their lengths.
   d. the daughter cells contain half the number of chromosomes as the parent cell.
8. Crossing over is the process whereby
   a. homologous chromosomes cross over to opposite sides of the cell.
   b. homologous chromosomes exchange chromosomal material.
   c. homologous chromosomes become closely associated along their lengths.
   d. kinetochore fibers attach to both sides of a centromere.
9. Mitosis results in ____, while meiosis results in _____.
   a. cells that are genetically identical to the parent cell/haploid cells
   b. haploid cells/diploid cells
   c. four daughter cells/two daughter cells
   d. cells with half the number of chromosomes as the parent cell/cells that vary in chromosome number
10. A major consequence of sex and meiosis is that each species
    a. remains pretty much the same because the chromosomes are carefully duplicated and passed to the next generation.
    b. has a lot of genetic reassortment due to processes in meiosis II.
    c. has a lot of genetic reassortment due to processes in meiosis I.
    d. has a lot of genetic reassortment due to processes in telophase II.

Apply Your Understanding

1. **Figure 9.5** How is it that in meiosis you can end up with four "daughter cells" that are all genetically different from one another?

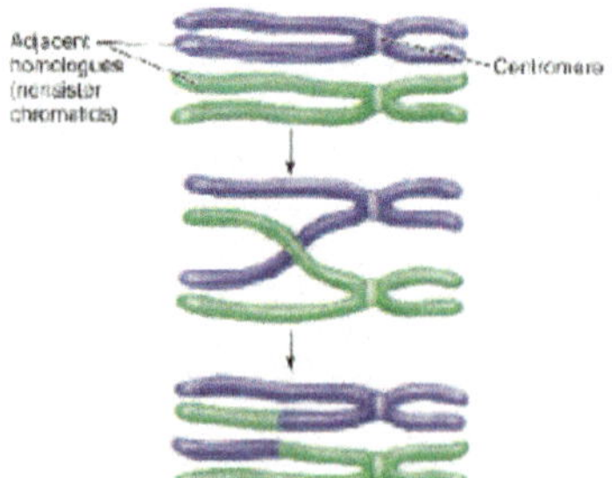

2. **Figure 9.8*a*** Referring to the homologous chromosomes shown here during prophase I, and knowing that they stay in synapsis during metaphase I, explain why it is that the sister chromatids don't separate as they do in mitosis.

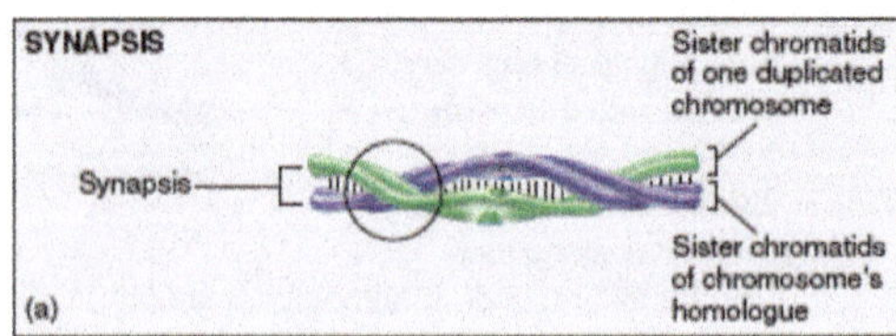

Synthesize What You Have Learned

1. It would seem that you only need one set of instructions for your body to do all the jobs it needs to carry out. So why aren't organisms simply haploid all their lives?
2. Are the gamete cells of your body haploid or diploid? Why not the alternative?
3. An organism has 56 chromosomes in its diploid stage. Indicate how many chromosomes are present in the following, and explain your reasoning:
   a. somatic cells
   b. metaphase I
   c. metaphase II
   d. gametes

10 Foundations of Genetics

Learning Objectives

Mendel

10.1 Mendel and the Garden Pea
1. Contrast the experiments of T. A. Knight and Gregor Mendel.
2. List four characteristics that made the garden pea easy for Mendel to study, and describe Mendel's experimental design.

10.2 What Mendel Observed
1. Describe what Mendel observed when crossing two contrasting traits.
2. State what percentage of F_2 individuals displayed the recessive trait and what percentage were heterozygous.

10.3 Mendel Proposes a Theory
1. State the five hypotheses of Mendel's theory.
2. Distinguish between gene and allele and between genotype and phenotype.
3. Diagram how a test cross determines the genotype of a dominant trait.

10.4 Mendel's Laws
1. State Mendel's First and Second Laws.
2. Recount the fate of Mendel's paper describing his experiments.

From Genotype to Phenotype

10.5 How Genes Influence Traits
1. Describe how genotype determines phenotype.

10.6 Some Traits Don't Show Mendelian Inheritance
1. List the five factors that can disguise Mendelian segregation.
2. Define pleiotropic effects, incomplete dominance, epistasis, and codominance.

Today's Biology: Does Environment Affect I.Q.?

Chromosomes and Heredity

10.7 Chromosomes Are the Vehicles of Mendelian Inheritance
1. Describe Morgan's surprising observation about white-eyed flies and explain how it proved the chromosomal theory of inheritance.

10.8 Human Chromosomes
1. Contrast aneuploidy with nondisjunction, monosomic with trisomic, and Klinefelter with Turner syndrome.

Human Hereditary Disorders

10.9 Studying Pedigrees
1. List the three questions asked to analyze a human pedigree.

10.10 The Role of Mutation
1. Describe the inheritance of hemophilia, sickle cell disease, Tay-Sachs disease, and Huntington's disease.

10.11 Genetic Counseling and Therapy
1. Describe three things geneticists examine in cells obtained by amniocentesis.

Inquiry & Analysis: Why Woolly Hair Runs in Families

In this pea pod, you can see the shadowy outlines of seeds that will form part of the next generation of this pea plant. While the seeds appear similar to one another, the plants they produce may differ in significant ways. This is because the gametes that produced the seeds contribute chromosomes from both parents, in effect "shuffling the deck of cards" so that a progeny plant will have some characteristics from one parent and some from the other. About 150 years ago, Gregor Mendel first described this process, before anyone knew what genes or chromosomes were. We now understand the process of heredity in considerable detail, and can begin to devise ways of treating some of the disorders that arise in people when particular genes are damaged in germ-line tissue. In this chapter you will watch as Mendel experiments with pea plants like the one above. Unlike researchers before him, Mendel carefully counted the number of each kind of pea plant his experiments produced and, looking at his results, saw a beautiful simplicity. The theory he proposed to explain it has become one of the key principles of biology.

10.1 Mendel and the Garden Pea

When you were born, many things about you resembled your mother or father. This tendency for traits to be passed from parent to offspring is called **heredity.** *Traits* are alternative forms of a character, or heritable feature. How does heredity happen? Before DNA and chromosomes were discovered, this puzzle was one of the greatest mysteries of science. The key to understanding the puzzle of heredity was found in the garden of an Austrian monastery over a century ago by a monk named Gregor Mendel (figure 10.1). Mendel used the scientific process described in chapter 1 as a powerful way of analyzing the problem. Crossing pea plants with one another, Mendel made observations that allowed him to form a simple but powerful hypothesis that accurately predicted patterns of heredity—that is, how many offspring would be like one parent and how many like the other. When Mendel's rules, introduced in chapter 1 as the theory of heredity, became widely known, investigators all over the world set out to discover the physical mechanism responsible for them. They learned that hereditary traits are instructions carefully laid out in the DNA a child receives from each parent. Mendel's solution to the puzzle of heredity was the first step on this journey of understanding and one of the greatest intellectual accomplishments in the history of science.

Figure 10.1 Gregor Mendel.
The key to understanding the puzzle of heredity was solved by Mendel, who cultivated pea plants in the garden of his monastery in Brünn, Austria.

Early Ideas About Heredity

Mendel was not the first person to try to understand heredity by crossing pea plants. Over 200 years earlier, British farmers had performed similar crosses and obtained results similar to Mendel's. They observed that in crosses between two types—tall and short plants, say—one type would disappear in one generation, only to reappear in the next. In the 1790s, for example, the British farmer T. A. Knight crossed a variety of the garden pea that had purple flowers with one that had white flowers. All the offspring of the cross had purple flowers. If two of these offspring were crossed, however, some of *their* offspring were purple and some were white. Knight noted that the purple had a "stronger tendency" to appear than white, but he did not count the numbers of each kind of offspring.

Mendel's Experiments

Gregor Mendel was born in 1822 to peasant parents and was educated in a monastery. He became a monk and was sent to the University of Vienna to study science and mathematics. Although he aspired to become a scientist and teacher, he failed his university exams for a teaching certificate and returned to the monastery, where he spent the rest of his life, eventually becoming abbot. Upon his return, Mendel joined an informal neighborhood science club, a group of farmers and others interested in science. Under the patronage of a local nobleman, each member set out to undertake scientific investigations, which were then discussed at meetings and published in the club's own journal. Mendel undertook to repeat the classic series of crosses with pea plants done by Knight and others, but this time he intended to count the numbers of each kind of offspring in the hope that the numbers would give some hint of what was going on. Quantitative approaches to science—measuring and counting—were just becoming fashionable in Europe.

Mendel's Experimental System: The Garden Pea

Mendel chose to study the garden pea because several of its characteristics made it easy to work with:

1. Many varieties were available. Mendel selected seven pairs of lines that differed in easily distinguished traits (including the white versus purple flowers that Knight had studied 60 years earlier).
2. Mendel knew from the work of Knight and others that he could expect the infrequent version of a character to disappear in one generation and reappear in the next. He knew, in other words, that he would have something to count.
3. Pea plants are small, easy to grow, produce large numbers of offspring, and mature quickly.
4. The reproductive organs of peas are enclosed within their flowers. Figure 10.2 shows a cutaway view of the flower so that you can see the anther that holds the pollen and the carpel that holds the egg. Left alone, the flowers do not open. They simply fertilize themselves with their own pollen (male gametes). To carry out a cross, Mendel had only to pry the petals apart, reach in with a scissors, and snip off the male

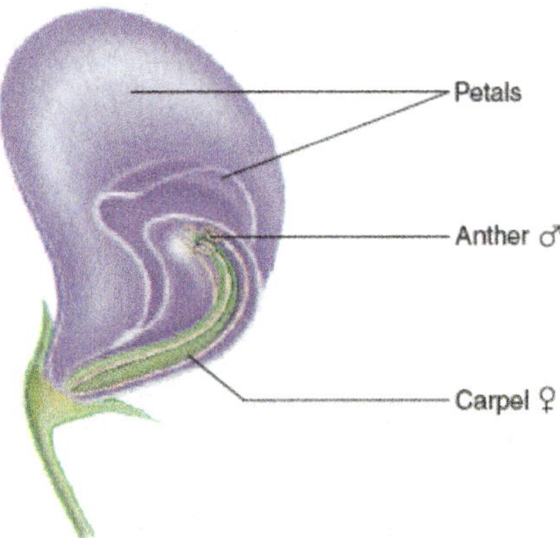

Figure 10.2 The garden pea.
Because it is easy to cultivate and because there are many distinctive varieties, the garden pea, *Pisum sativum*, was a popular choice as an experimental subject in investigations of heredity for as long as a century before Mendel's studies.

organs (anthers); he could then dust the female organs (the tip of the carpel) with pollen from another plant to make the cross.

Mendel's Experimental Design

Mendel's experimental design was the same as Knight's, only Mendel counted his plants. The crosses were carried out in three steps that are presented in the three panels in figure 10.3:

1. Mendel began by letting each variety self-fertilize for several generations. This ensured that each variety was **true-breeding,** meaning that it contained no other varieties of the trait, and so would produce only offspring of the same variety when it self-pollinated. The white flower variety, for example, produced only white flowers and no purple ones in each generation. Mendel called these lines the **P generation** (P for parental).
2. Mendel then conducted his experiment: He crossed two pea varieties exhibiting alternative traits, such as white versus purple flowers. The offspring that resulted he called the **F_1 generation** (F_1 for "first filial" generation, from the Latin word for "son" or "daughter").
3. Finally, Mendel allowed the plants produced in the crosses of step 2 to self-fertilize, and he counted the numbers of each kind of offspring that resulted in this **F_2** ("second filial") **generation.** As reported by Knight, the white flower trait reappeared in the F_2 generation, although not as frequently as the purple flower trait.

Key Learning Outcome 10.1 Mendel studied heredity by crossing true-breeding garden peas that differed in easily scored alternative traits and then allowing the offspring to self-fertilize.

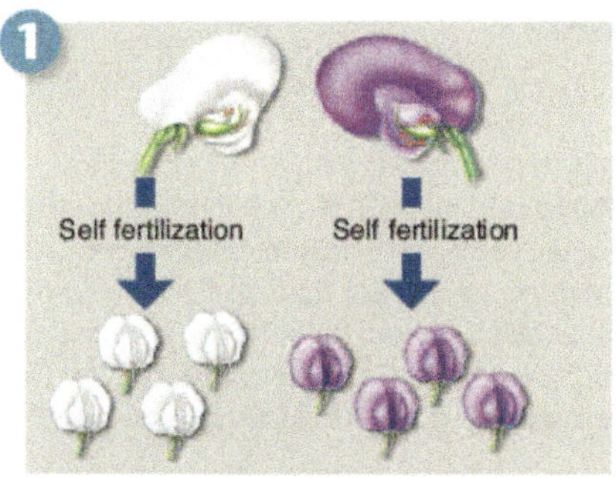

Mendel let each variety self fertilize for several generations, producing a true breeding P generation.

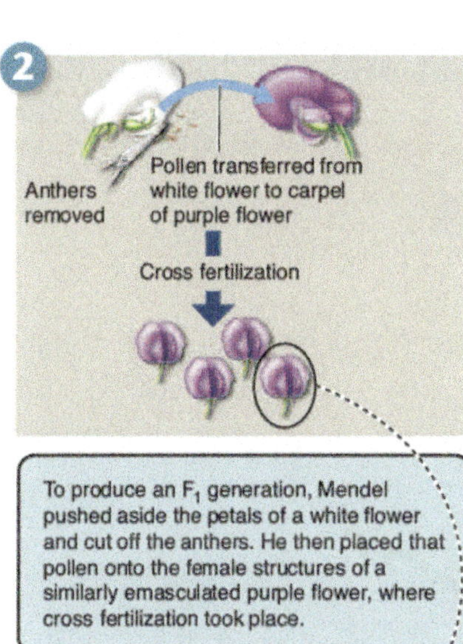

To produce an F_1 generation, Mendel pushed aside the petals of a white flower and cut off the anthers. He then placed that pollen onto the female structures of a similarly emasculated purple flower, where cross fertilization took place.

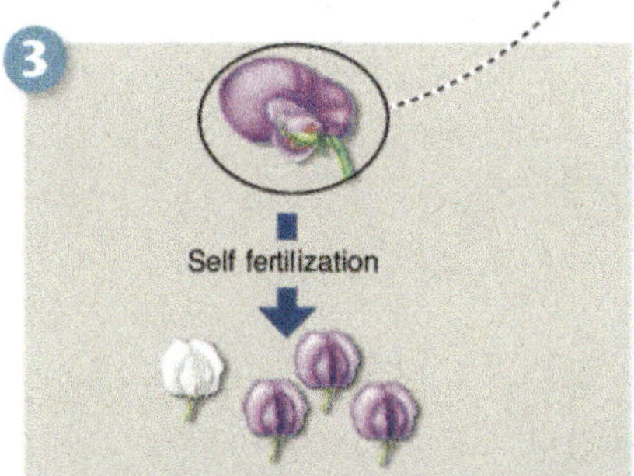

To produce an F_2 generation, Mendel let the plants in the F_1 generation self fertilize.

Figure 10.3 How Mendel conducted his experiments.

10.2 What Mendel Observed

Mendel experimented with a variety of traits in the garden pea and repeatedly made similar observations. In all, Mendel examined seven pairs of contrasting traits as shown in table 10.1. For each pair of contrasting traits that Mendel crossed he observed the same result, shown in figure 10.3, where a trait disappeared in the F_1 generation only to reappear in the F_2 generation. We will examine in detail Mendel's crosses with flower color.

The F_1 Generation

In the case of flower color, when Mendel crossed purple and white flowers, all the F_1 generation plants he observed were purple; he did not see the contrasting trait, white flowers. Mendel called the trait expressed in the F_1 plants **dominant** and the trait not expressed **recessive.** In this case, purple flower color was dominant and white flower color recessive. Mendel studied several other characters in addition to flower color, and for every pair of contrasting traits Mendel examined, one proved to be dominant and the other recessive. The dominant and recessive traits for each character he studied are indicated in table 10.1.

The F_2 Generation

After allowing individual F_1 plants to mature and self-fertilize, Mendel collected and planted the seeds from each plant to see what the offspring in the F_2 generation would look like. Mendel found (as Knight had earlier) that some F_2 plants exhibited white flowers, the recessive trait. The recessive trait had disappeared in the F_1 generation, only to reap-

TABLE 10.1 SEVEN CHARACTERS MENDEL STUDIED IN HIS EXPERIMENTS

Character			F_2 Generation	
Dominant Form	×	Recessive Form	Dominant: Recessive	Ratio
Purple flowers	×	White flowers	705:224	3.15:1 (3/4:1/4)
Yellow seeds	×	Green seeds	6,022:2,001	3.01:1 (3/4:1/4)
Round seeds	×	Wrinkled seeds	5,474:1,850	2.96:1 (3/4:1/4)
Green pods	×	Yellow pods	428:152	2.82:1 (3/4:1/4)
Inflated pods	×	Constricted pods	882:299	2.95:1 (3/4:1/4)
Axial flowers	×	Terminal flowers	651:207	3.14:1 (3/4:1/4)
Tall plants	×	Dwarf plants	787:277	2.84:1 (3/4:1/4)

Figure 10.4 Round versus wrinkled seeds.
One of the differences among varieties of pea plants that Mendel studied was the shape of the seed. In some varieties the seeds were round, whereas in others they were wrinkled.

pear in the F_2 generation. It must somehow have been present in the F_1 individuals but unexpressed!

At this stage Mendel instituted his radical change in experimental design. He *counted* the number of each type among the F_2 offspring. He believed the proportions of the F_2 types would provide some clue about the mechanism of heredity. In the cross between the purple-flowered F_1 plants, he counted a total of 929 F_2 individuals (see table 10.1). Of these, 705 (75.9%) had purple flowers and 224 (24.1%) had white flowers. Approximately one-fourth of the F_2 individuals exhibited the recessive form of the trait. Mendel carried out similar experiments with other traits, such as round versus wrinkled seeds (figure 10.4) and obtained the same result: Three-fourths of the F_2 individuals exhibited the dominant form of the character, and one-fourth displayed the recessive form. In other words, the dominant:recessive ratio among the F_2 plants was always close to 3:1.

A Disguised 1:2:1 Ratio

Mendel let the F_2 plants self-fertilize for another generation and found that the one-fourth that were recessive were true-breeding—future generations showed nothing but the recessive trait. Thus, the white F_2 individuals described previously showed only white flowers in the F_3 generation (as shown on the right in figure 10.5). Among the three-fourths of the plants that had shown the dominant trait in the F_2 generation, only one-third of the individuals were true-breeding in the F_3 generation (as shown on the left). The others showed both traits in the F_3 generation (as shown in the center)—and when Mendel counted their numbers, he found the ratio of dominant to recessive to again be 3:1! From these results Mendel concluded

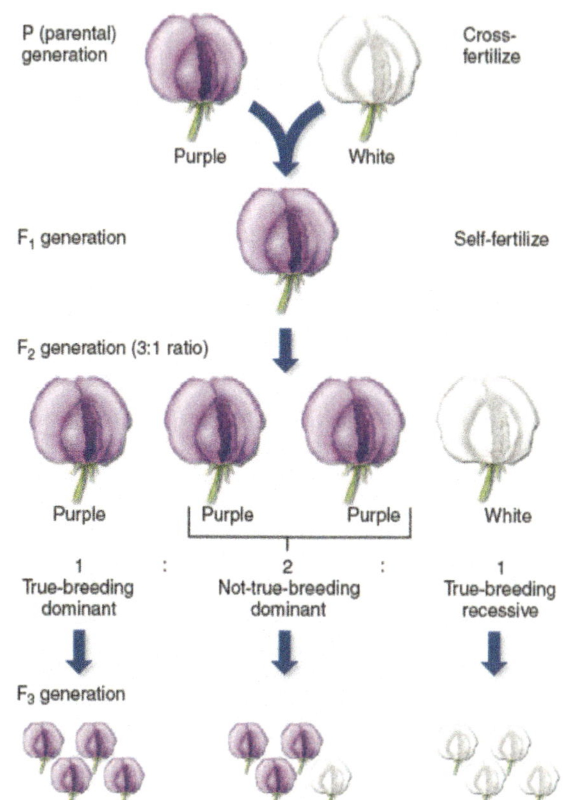

Figure 10.5 The F_2 generation is a disguised 1:2:1 ratio.
By allowing the F_2 generation to self-fertilize, Mendel found from the offspring (F_3) that the ratio of F_2 plants was one true-breeding dominant, two not-true-breeding dominant, and one true-breeding recessive.

that the 3:1 ratio he had observed in the F_2 generation was in fact a disguised 1:2:1 ratio:

1		2		1
true-breeding	**:**	**not-true-breeding**	**:**	**true-breeding**
dominant		**dominant**		**recessive**

Key Learning Outcome 10.2 **When Mendel crossed two contrasting traits and counted the offspring in the subsequent generations, he observed that all of the offspring in the first generation exhibited one (dominant) trait, and none exhibited the other (recessive) trait. In the following generation, 25% were true-breeding for the dominant trait, 50% were not-true-breeding and appeared dominant, and 25% were true-breeding for the recessive trait.**

10.3 Mendel Proposes a Theory

To explain his results, Mendel proposed a simple set of hypotheses that would faithfully predict the results he had observed. Now called Mendel's theory of heredity, it has become one of the most famous theories in the history of science. Mendel's theory is composed of five simple hypotheses:

Hypothesis 1: Parents do not transmit traits directly to their offspring. Rather, they transmit information about the traits, what Mendel called *merkmal* (the German word for "factor"). These factors act later, in the offspring, to produce the trait. In modern terminology, we call Mendel's factors **genes.**

Hypothesis 2: Each parent contains two copies of the factor governing each trait. The two copies may or may not be the same. If the two copies of the factor are the same (both encoding purple or both white flowers, for example), the individual is said to be **homozygous.** If the two copies of the factor are different (one encoding purple, the other white, for example), the individual is said to be **heterozygous.**

Hypothesis 3: Alternative forms of a factor lead to alternative traits. Alternative forms of a factor are called **alleles.** Mendel used lowercase letters to represent recessive alleles and uppercase letters to represent dominant ones. Thus, in the case of purple flowers, the dominant purple flower allele is represented as *P* and the recessive white flower allele is represented as *p*. In modern terms, we call the appearance of an individual, such as possessing white flowers, its **phenotype.** Appearance is determined by which alleles of the flower-color gene the plant receives from its parents, and we call those particular alleles the individual's **genotype.** Thus a pea plant might have the phenotype "white flower" and the genotype *pp*.

Hypothesis 4: The two alleles that an individual possesses do not affect each other, any more than two letters in a mailbox alter each other's contents. Each allele is passed on unchanged when the individual matures and produces its own gametes (egg and sperm). At the time, Mendel did not know that his factors were carried from parent to offspring on chromosomes. Figure 10.6 shows a modern view of how genes are carried on chromosomes, with homologous chromosome carrying the same genes but not necessarily the same alleles. The location of a gene on a chromosome is called its *locus* (plural, *loci*).

Hypothesis 5: The presence of an allele does not ensure that a trait will be expressed in the individual that carries it. In heterozygous individuals, only the dominant allele achieves expression; the recessive allele is present but unexpressed.

These five hypotheses, taken together, constitute Mendel's model of the hereditary process. Many traits in humans exhibit dominant or recessive inheritance similar to the traits Mendel studied in peas (table 10.2).

Analyzing Mendel's Results

To analyze Mendel's results, it is important to remember that each trait is determined by the inheritance of alleles from the parents, one allele from the mother and the other from the father. These alleles, present on chromosomes, are distributed to gametes during meiosis. Each gamete receives one copy of each chromosome, and therefore one of the alleles.

Consider again Mendel's cross of purple-flowered with white-flowered plants. Like Mendel, we will assign the symbol *P* to the dominant allele, associated with the production of purple flowers, and the symbol *p* to the recessive allele, associated with the production of white flowers. As described

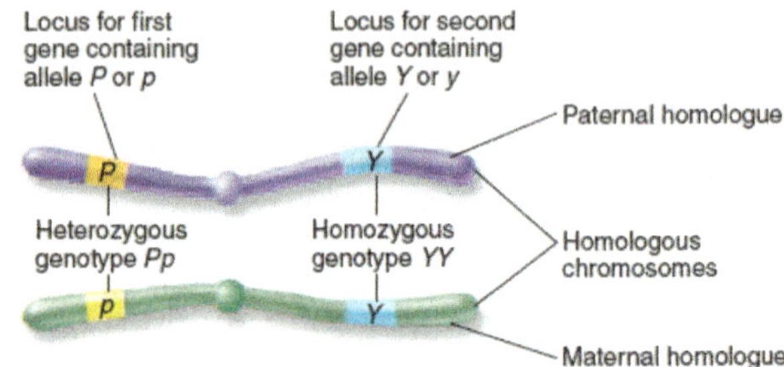

Figure 10.6 Alternative alleles of genes are located on homologous chromosomes.

TABLE 10.2 SOME DOMINANT AND RECESSIVE TRAITS IN HUMANS

Recessive Traits	Phenotypes	Dominant Traits	Phenotypes
Common baldness	M-shaped hairline receding with age	Mid-digital hair	Presence of hair on middle segment of fingers
Albinism	Lack of melanin pigmentation	Brachydactyly	Short fingers
Alkaptonuria	Inability to metabolize homogentisic acid	Phenylthiocarbamide (PTC) sensitivity	Ability to taste PTC as bitter
Red-green color blindness	Inability to distinguish red and green wavelengths of light	Camptodactyly	Inability to straighten the little finger
		Polydactyly	Extra fingers and toes

earlier, by convention, genetic traits are usually assigned a letter symbol referring to their more common forms, in this case "*P*" for purple flower color. The dominant allele is written in uppercase, as *P;* the recessive allele (white flower color) is assigned the same symbol in lowercase, *p*.

In this system, the genotype of an individual true-breeding for the recessive white-flowered trait would be designated *pp*. In such an individual, both copies of the allele specify the white-flowered phenotype. Similarly, the genotype of a true-breeding purple-flowered individual would be designated *PP*, and a heterozygote would be designated *Pp* (dominant allele first). Using these conventions, and denoting a cross between two strains with ×, we can symbolize Mendel's original cross as *pp* × *PP*.

Punnett Squares

The possible results from a cross between a true-breeding, white-flowered plant (*pp*) and a true-breeding, purple-flowered plant (*PP*) can be visualized with a **Punnett square.** In a Punnett square, the possible gametes of one individual are listed along the horizontal side of the square, while the possible gametes of the other individual are listed along the vertical side. The genotypes of potential offspring are represented by the cells within the square. Figure 10.7 walks you through the set-up of a Punnett square crossing two individual plants that are heterozygous for flower color (*Pp* × *Pp*). The genotypes of the parents are placed along the top and side and the genotypes of potential offspring appear in the cells.

The frequency that these genotypes occur in the offspring is usually expressed by a **probability.** For example, in a cross between a homozygous white-flowered plant (*pp*) and a homozygous purple-flowered plant (*PP*), *Pp* is the only possible genotype for all individuals in the F_1 generation as shown by the Punnett square on the left of figure 10.8. Because *P* is dominant to *p*, all individuals in the F_1 generation have purple flowers. When individuals from the F_1 generation are crossed, as shown by the Punnett square on the right, the probability of obtaining a homozygous dominant (*PP*) individual in the F_2 is 25% because one-fourth of the possible genotypes are *PP*. Similarly, the probability of an individual in the F_2 generation being homozygous recessive (*pp*) is 25%. Because the heterozygous genotype has two possible ways of occurring (*Pp* and *pP*), it occurs in half of the cells within the square; the probability of obtaining a heterozygous (*Pp*) individual in the F_2 is 50% (25% + 25%).

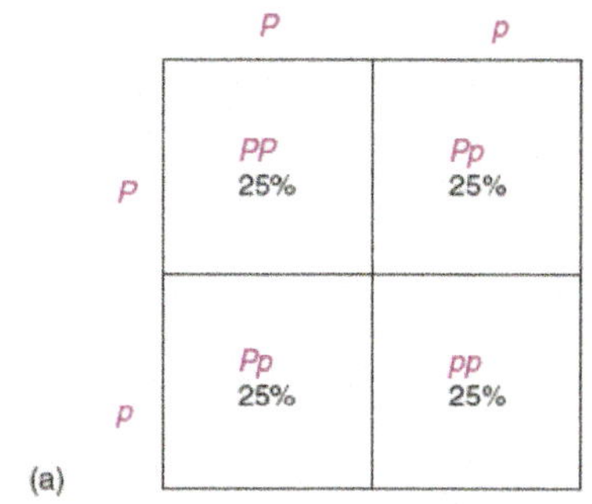

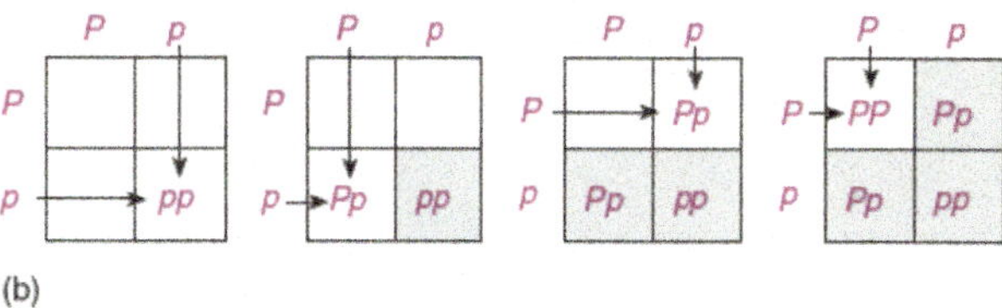

Figure 10.7 A Punnett square analysis.
(*a*) Each square represents 1/4 or 25% of the offspring from the cross. The squares in (*b*) show how the square is used to predict the genotypes of all potential offspring.

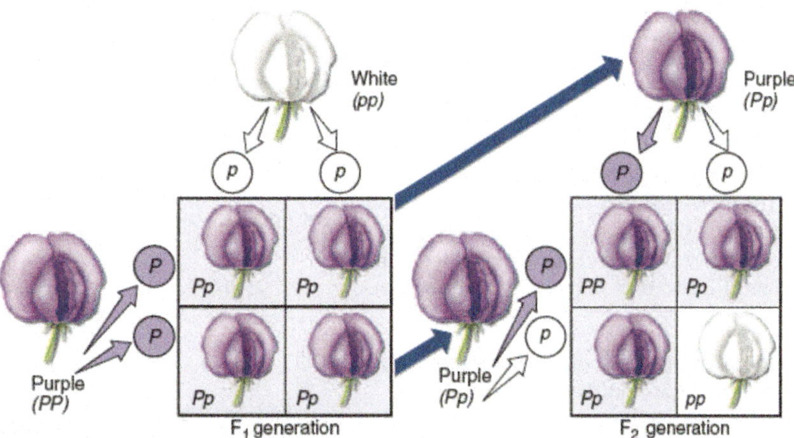

Figure 10.8 How Mendel analyzed flower color.
The only possible offspring of the first cross are *Pp* heterozygotes, purple in color. These individuals are known as the F_1 generation. When two heterozygous F_1 individuals cross, three kinds of offspring are possible: *PP* homozygotes (purple flowers); *Pp* heterozygotes (also purple flowers), which may form two ways; and *pp* homozygotes (white flowers). Among these individuals, known as the F_2 generation, the ratio of dominant phenotype to recessive phenotype is 3:1.

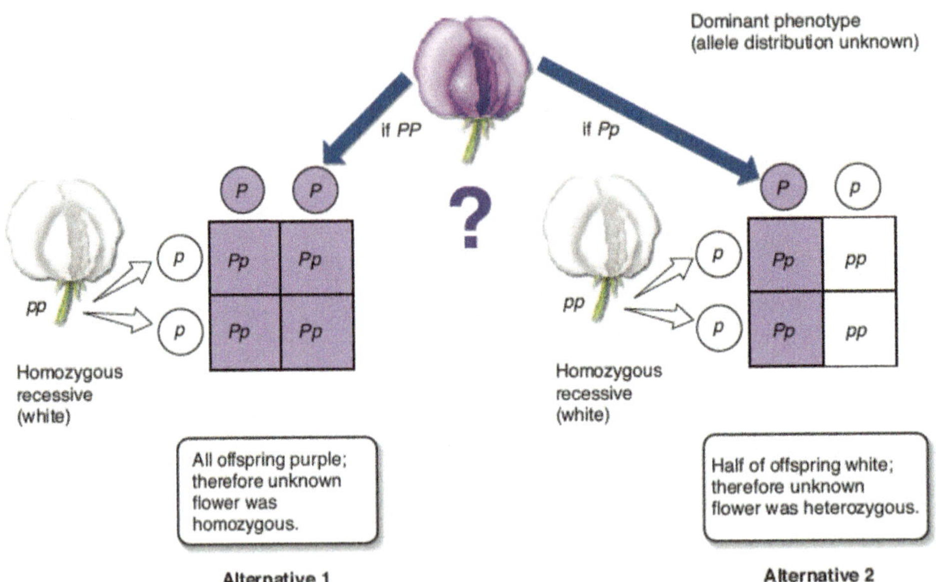

Figure 10.9 How Mendel used the testcross to detect heterozygotes.
To determine whether an individual exhibiting a dominant phenotype, such as purple flowers, was homozygous (*PP*) or heterozygous (*Pp*), Mendel devised the testcross. He crossed the individual with a known homozygous recessive (*pp*)—in this case, a plant with white flowers.

The Testcross

How did Mendel know which of the purple-flowered individuals in the F_2 generation (or the P generation) were homozygous (*PP*) and which were heterozygous (*Pp*)? It is not possible to tell simply by looking at them. For this reason, Mendel devised a simple and powerful procedure called the **testcross** to determine an individual's actual genetic composition. Consider a purple-flowered plant. It is impossible to tell whether such a plant is homozygous or heterozygous simply by looking at its phenotype. To learn its genotype, you must cross it with some other plant. What kind of cross would provide the answer? If you cross it with a homozygous dominant individual, all of the progeny will show the dominant phenotype whether the test plant is homozygous or heterozygous. It is also difficult (but not impossible) to distinguish between the two possible test plant genotypes by crossing with a heterozygous individual. However, if you cross the test plant with a homozygous recessive individual, the two possible test plant genotypes will give totally different results. To see how this works, step through a testcross of a purple-flowered plant with a white-flowered plant. Figure 10.9 shows you the two possible alternatives:

Alternative 1 (on left): Unknown plant is homozygous (*PP*). *PP* × *pp:* All offspring have purple flowers (*Pp*) as shown by the four purple squares.

Alternative 2 (on right): Unknown plant is heterozygous (*Pp*). *Pp* × *pp:* One-half of offspring have white flowers (*pp*) and one-half have purple flowers (*Pp*) as shown by the two white and two purple squares.

To perform his testcross, Mendel crossed heterozygous individuals exhibiting the dominant trait back to the parent homozygous for the recessive trait. He predicted that the dominant and recessive traits would appear in a 1:1 ratio, and that is what he observed, as you can see illustrated in alternative 2 above.

For each pair of alleles he investigated, Mendel observed phenotypic F_2 ratios of 3:1 (see table 10.1) and testcross ratios very close to 1:1, just as his model predicted.

Testcrosses can also be used to determine the genotype of an individual when two genes are involved. Mendel carried out many two-gene crosses, some of which we will soon discuss. He often used testcrosses to verify the genotypes of particular dominant-appearing F_2 individuals. Thus an F_2 individual showing both dominant traits (*A_ B_*) might have any of the following genotypes: *AABB, AaBB, AABb*, or *AaBb*. By crossing dominant-appearing F_2 individuals with homozygous recessive individuals (that is, *A_ B_* × *aabb*), Mendel was able to determine if either or both of the traits bred true among the progeny and so determine the genotype of the F_2 parent.

AABB	trait A breeds true	trait B breeds true
AaBB		trait B breeds true
AABb	trait A breeds true	
AaBb		

Key Learning Outcome 10.3 **The genes that an individual has are referred to as its genotype; the outward appearance of the individual is referred to as its phenotype. The phenotype is determined by the alleles inherited from the parents. Analyses using Punnett squares determine all possible genotypes of a particular cross. A test cross determines the genotype of a dominant trait.**

10.4 Mendel's Laws

Mendel's First Law: Segregation

Mendel's model brilliantly predicts the results of his crosses, accounting in a neat and satisfying way for the ratios he observed. Similar patterns of heredity have since been observed in countless other organisms. Traits exhibiting this pattern of heredity are called *Mendelian traits.* Because of its overwhelming importance, Mendel's theory is often referred to as Mendel's first law, or the **law of segregation.** In modern terms, Mendel's first law states that *the two alleles of a trait separate from each other during the formation of gametes, so that half of the gametes will carry one copy and half will carry the other copy.*

Mendel's Second Law: Independent Assortment

Mendel went on to ask if the inheritance of one factor, such as flower color, influences the inheritance of other factors, such as plant height. To investigate this question, he first established a series of true-breeding lines of peas that differed from one another with respect to two of the seven pairs of characteristics he had studied. He then crossed contrasting pairs of true-breeding lines. Figure 10.10 shows an experiment in which the P generation consists of homozygous individuals with round, yellow seeds (*RRYY* in the figure) that are crossed with individuals that are homozygous for wrinkled, green seeds (*rryy*). This cross produces offspring that have round, yellow seeds and are heterozygous for both of these traits (*RrYy*). Such F_1 individuals are said to be **dihybrid.** The chromosomes are then allocated to the gametes during meiosis such that there are four types of gametes for these two traits.

Mendel then allowed the dihybrid individuals to self-fertilize. If the segregation of alleles affecting seed shape and alleles affecting seed color were independent, the probability that a particular pair of seed-shape alleles would occur together with a particular pair of seed-color alleles would simply be a product of the two individual probabilities that each pair would occur separately. For example, the probability of an individual with wrinkled, green seeds appearing in the F_2 generation would be equal to the probability of an individual with wrinkled seeds (1 in 4) multiplied by the probability of an individual with green seeds (1 in 4), or 1 in 16.

In his dihybrid crosses, Mendel found that the frequency of phenotypes in the F_2 offspring closely matched the 9:3:3:1 ratio predicted by the Punnett square analysis shown in figure 10.10. He concluded that for the pairs of traits he studied, the inheritance of one trait does not influence the inheritance of the other trait, a result often referred to as Mendel's second law, or the **law of independent assortment.** We now know that this result is only valid for genes not located near one another on the same chromosome. Thus in modern terms, Mendel's second law is often stated as follows: *Genes located on different chromosomes are inherited independently of one another.*

Mendel's paper describing his results was published in the journal of his local scientific society in 1866. Unfortunately, his paper failed to arouse much interest, and his work was forgotten. Sixteen years after his death, in 1900, several investigators independently rediscovered Mendel's pioneering paper while searching the literature in preparation for publishing their own findings, which were similar to those Mendel had quietly presented more than three decades earlier.

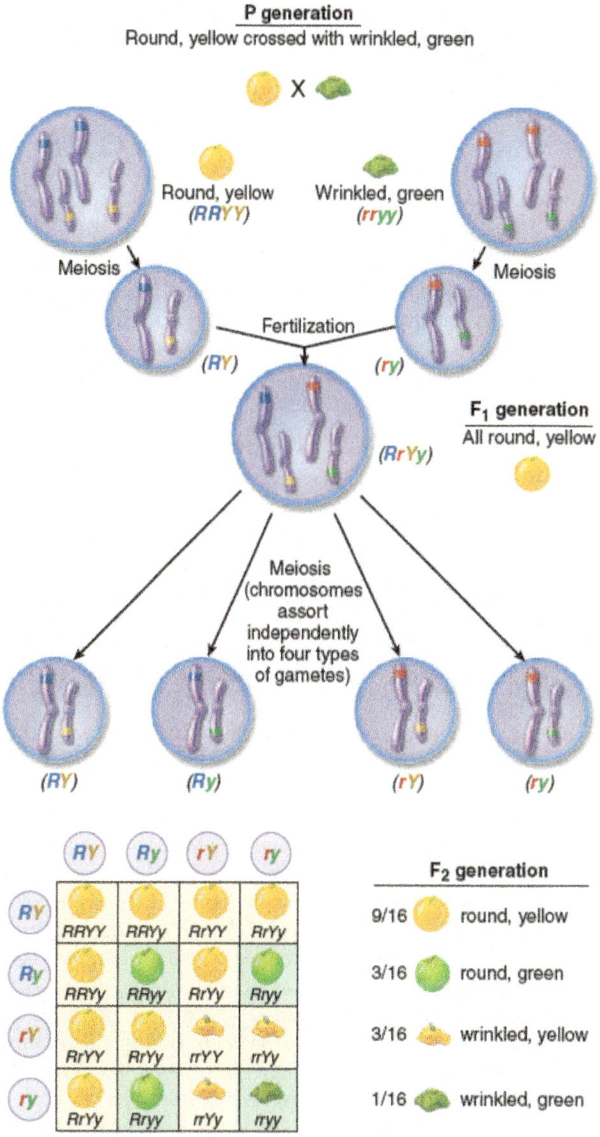

Figure 10.10 Analysis of a dihybrid cross.
This dihybrid cross shows round (*R*) versus wrinkled (*r*) seeds and yellow (*Y*) versus green (*y*) seeds. The ratio of the four possible phenotypes in the F_2 generation is predicted to be 9:3:3:1.

Key Learning Outcome 10.4 Mendel's theories of segregation and independent assortment are so well supported by experimental results that they are considered "laws."

From Genotype to Phenotype

10.5 How Genes Influence Traits

It is useful, before considering Mendelian genetics further, to gain a brief overview of how genes work. With this in mind, we will sketch, in broad strokes, a picture of how a Mendelian trait is influenced by a particular gene, how a gene can be altered by mutation, and the potential long-term evolutionary consequences of such an alteration. We will use the protein hemoglobin as our example—you can follow along on figure 10.11 starting at the bottom.

From DNA to Protein

Each body cell of an individual contains the same set of DNA molecules, called the genome of that individual. As you learned in chapter 3, DNA molecules are composed of two strands twisted about each other, each the mirror image of the other. Each strand is a long chain of nucleotide subunits that are linked together. There are four kinds of nucleotides (A, T, C, and G), and like an alphabet with four letters, the order of nucleotides determines the message encoded in the DNA of a gene.

The human genome contains 20,000 to 25,000 genes. The DNA of the human genome is parcelled out into 23 pairs of chromosomes, each chromosome containing from 1,000 to 2,000 different genes. The bands on the chromosome in figure 10.11 indicate areas that are rich in genes. You can see in the figure that the hemoglobin gene is located on chromosome 11.

At the next level in the figure, individual genes are "read" from the chromosomal DNA by enzymes that create an RNA transcript of the nucleotide sequence (except U is substituted for T). This RNA transcript of the hemoglobin (*Hb*) gene leaves the cell nucleus and acts as a work order for protein production in other parts of the cell. But, in eukaryotic cells, the RNA transcript has more information than is needed, so it is first "edited" to remove unnecessary bits before it leaves the nucleus. For example, the initial RNA gene transcript encoding the beta-subunit of the protein hemoglobin is 1,660 nucleotides long; after "editing," the resulting "messenger" RNA is 1,000 nucleotides long—you can see in the figure that the Hb mRNA is shorter than the RNA transcript of *Hb* gene.

After an RNA transcript is edited, it leaves the nucleus as messenger RNA (mRNA) and is delivered to ribosomes in the cytoplasm. Each ribosome is a tiny protein-assembly plant, and uses the sequence of the messenger RNA to determine the amino acid sequence of a particular polypeptide. In the case of beta-hemoglobin, the messenger RNA encodes a polypeptide strand of 146 amino acids.

How Proteins Determine the Phenotype

As we saw in chapter 3, polypeptide chains of amino acids, which in the figure resemble beads on a string, spontaneously fold in water into complex three-dimensional shapes. The beta-hemoglobin polypeptide folds into a compact mass that associates with three others to form an active hemoglobin protein molecule that is present in red blood cells. In the figure, each hemoglobin molecule binds oxygen (a process described fully in chapter 24) in the oxygen-rich environment of the lungs, and releases oxygen in the oxygen-poor environment of active tissues.

The oxygen-binding efficiency of the hemoglobin proteins in a person's bloodstream has a great deal to do with how well the body functions, particularly under conditions of strenuous physical activity, when delivery of oxygen to the body's muscles is the chief factor limiting the activity.

As a general rule, genes influence the phenotype by specifying the kind of proteins present in the body, which determines in large measure how that body functions.

How Mutation Alters Phenotype

A change in the identity of a single nucleotide within a gene, called a mutation, can have a profound effect if the change alters the identity of the amino acid encoded there. When a mutation of this sort occurs, the new version of the protein may fold differently, altering or destroying its function. For example, how well the hemoglobin protein performs its oxygen-binding duties depends a great deal on the precise shape that the protein assumes when it folds. A change in the identity of a single amino acid can have a drastic impact on that final shape. In particular, a change in the sixth amino acid of beta-hemoglobin from glutamic acid to valine causes the hemoglobin molecules to aggregate into stiff rods that deform blood cells into a sickle shape that can no longer carry oxygen efficiently. The resulting sickle-cell disease can be fatal.

Natural Selection for Alternative Phenotypes Leads to Evolution

Because random mutations occur in all genes occasionally, populations usually contain several versions of a gene, usually all but one of them rare. Sometimes the environment changes in such a way that one of the rare versions functions better under the new conditions. When that happens, natural selection will favor the rare allele, which will then become more common. The sickle-cell version of the beta-hemoglobin gene, rare throughout most of the world, is common in Central Africa because heterozygous individuals obtain enough functional hemoglobin from their one normal allele to get along, but are resistant to malaria, a deadly disease common there, due to their other sickle-cell allele.

Key Learning Outcome 10.5 Genes determine phenotypes by specifying the amino acid sequences, and thus the functional shapes, of the proteins that carry out cell activities. Mutations, by altering protein sequence, can change a protein's function and thus alter the phenotype in evolutionarily significant ways.

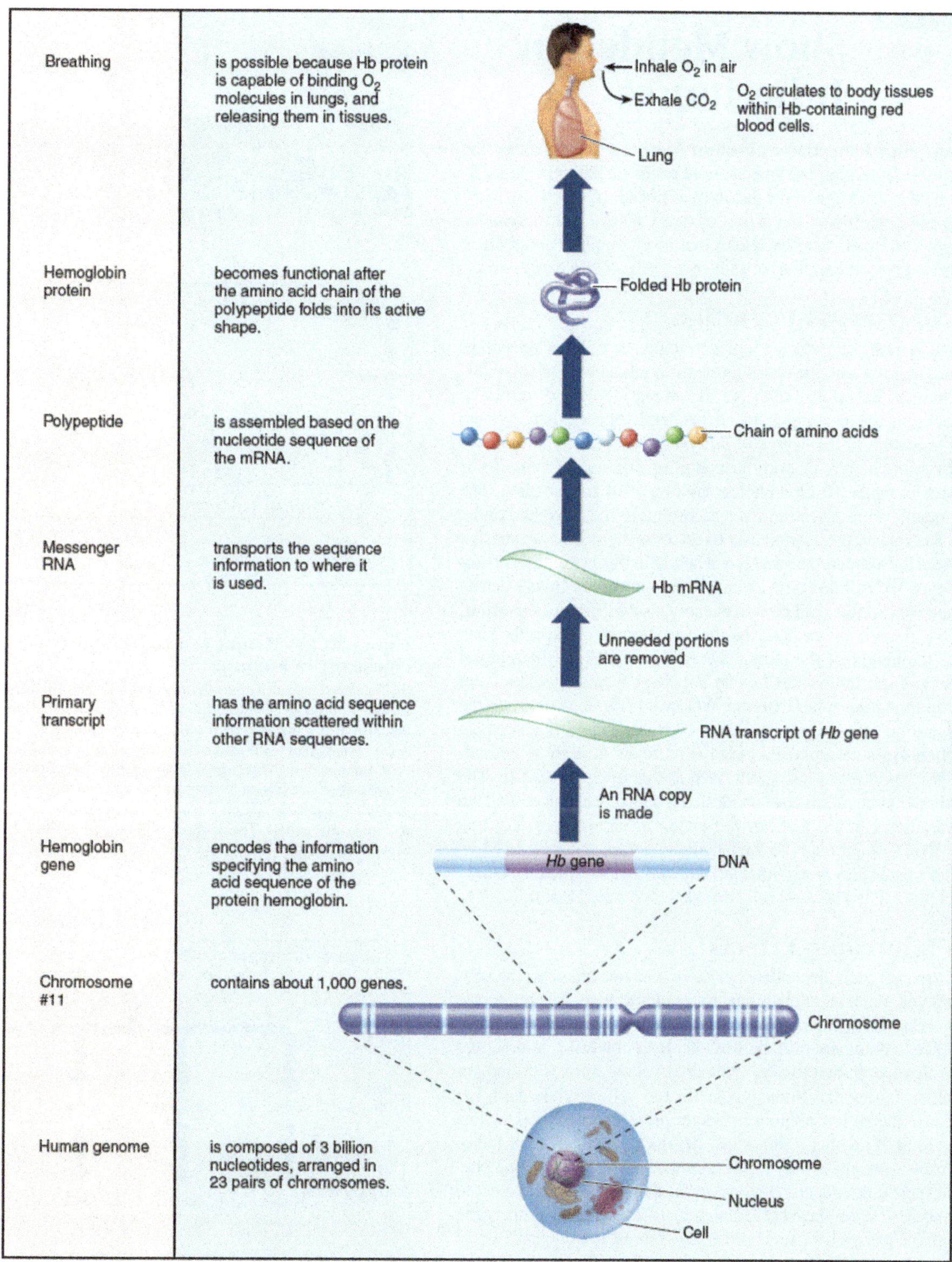

Figure 10.11 The journey from DNA to phenotype.
What an organism is like is determined in large measure by its genes. Here you see how one gene of the 20,000 to 25,000 in the human genome plays a key role in allowing oxygen to be carried throughout your body. The many steps on the journey from gene to trait are the subject of chapters 11 and 12.

10.6 Some Traits Don't Show Mendelian Inheritance

Scientists attempting to confirm Mendel's theory often had trouble obtaining the same simple ratios he had reported. Often the expression of the genotype is not straightforward. Most phenotypes reflect the action of many genes, and the phenotype can be affected by alleles that lack complete dominance, are expressed together, or influence each other's expression.

Continuous Variation

When multiple genes act jointly to influence a character such as height or weight, the character often shows a range of small differences. Because all of the genes that play a role in determining these phenotypes segregate independently of each other, we see a gradation in the degree of difference when many individuals are examined. A classic illustration of this sort of variation is seen in figure 10.12, a photograph of a 1914 college class. The students were placed in rows according to their heights, under 5 feet toward the left and over 6 feet to the right. You can see that there is considerable variation in height in this population of students. We call this type of inheritance **polygenic** (many genes) and we call this gradation in phenotypes **continuous variation.**

How can we describe the variation in a character such as the height of the individuals in figure 10.12*a*? Individuals range from quite short to very tall, with average heights more common than either extreme. What we often do is to group the variation into categories. Each height, in inches, is a separate phenotypic category. Plotting the numbers in each height category produces a histogram, such as that in figure 10.12*b*. The histogram approximates an idealized bell-shaped curve, and the variation can be characterized by the mean and spread of that curve. Compare this to the inheritance of plant height in Mendel's peas; they were either tall or dwarf, no intermediate height plants existed because only one gene controlled that trait.

(a)

Number of individuals: 0, 10, 20, 30
Height: 5'0", 5'6", 6'0"

(b)

Figure 10.12 Height is a continuously varying character in humans.

(*a*) This photograph shows the variation in height among students of the 1914 class of the Connecticut Agricultural College. Because many genes contribute to height and tend to segregate independently of each other, there are many possible combinations of those genes. (*b*) The cumulative contribution of different combinations of alleles for height forms a continuous spectrum of possible heights, in which the extremes are much rarer than the intermediate values. This is quite different from the 3:1 ratio seen in Mendel's F_2 peas.

Pleiotropic Effects

Often, an individual allele has more than one effect on the phenotype. Such an allele is said to be **pleiotropic.** When the pioneering French geneticist Lucien Cuenot studied yellow fur in mice, a dominant trait, he was unable to obtain a true-breeding yellow strain by crossing individual yellow mice with one another. Individuals homozygous for the yellow allele died, because the yellow allele was pleiotropic: One effect was yellow color, but another was a lethal developmental defect. A pleiotropic gene alteration may be dominant with respect to one phenotypic consequence (yellow fur) and recessive with respect to another (lethal developmental defect). In pleiotropy, one gene affects many characters, in marked contrast to polygeny, where many genes affect one character. Pleiotropic effects are difficult to predict, because the genes that affect a character often perform other functions we may know nothing about.

Pleiotropic effects are characteristic of many inherited disorders, such as cystic fibrosis and sickle-cell disease,

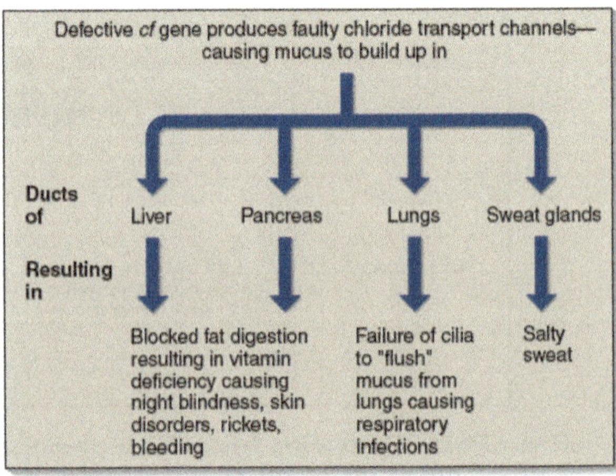

Figure 10.13 Pleiotropic effects of the cystic fibrosis gene, *cf*.

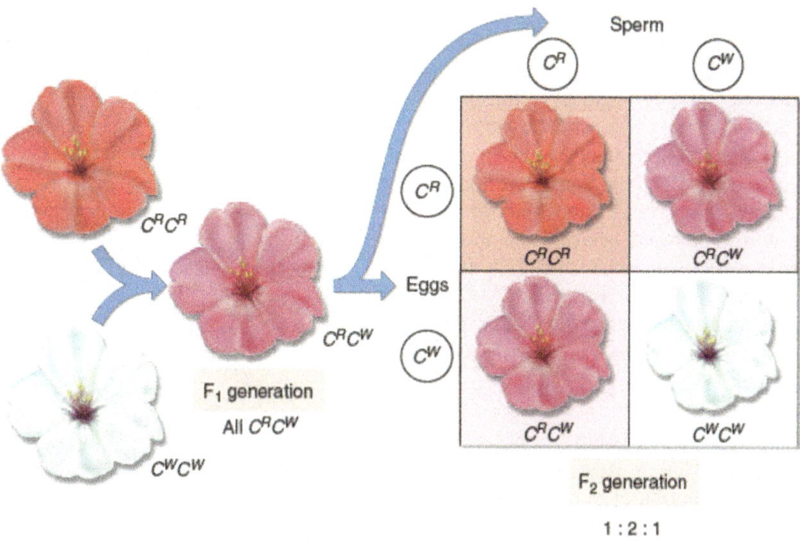

Figure 10.14 Incomplete dominance.

In a cross between a red-flowered Japanese four o'clock, genotype $C^R C^R$, and a white-flowered one ($C^W C^W$), neither allele is dominant. The heterozygous progeny have pink flowers and the genotype $C^R C^W$. If two of these heterozygotes are crossed, the phenotypes of their progeny occur in a ratio of 1:2:1 (red:pink:white).

discussed later in this chapter. In these disorders, multiple symptoms can be traced back to a single gene defect. As shown in figure 10.13, cystic fibrosis patients exhibit overly sticky mucus, salty sweat, liver and pancreas failure, and a battery of other symptoms. All are pleiotropic effects of a single defect, a mutation in a gene that encodes a chloride ion transmembrane channel. In sickle-cell disease, a defect in the oxygen-carrying hemoglobin molecule causes anemia, heart failure, increased susceptibility to pneumonia, kidney failure, enlargement of the spleen, and many other symptoms. It is usually difficult to deduce the nature of the primary defect from the range of its pleiotropic effects.

Incomplete Dominance

Not all alternative alleles are fully dominant or fully recessive in heterozygotes. Some pairs of alleles exhibit **incomplete dominance** and produce a heterozygous phenotype that is intermediate between those of the parents. For example, the cross of red- and white-flowered Japanese four o'clocks described in figure 10.14 produced red-, pink-, and white-flowered F_2 plants in a 1:2:1 ratio—heterozygotes are intermediate in color. This is different than in Mendel's pea plants that didn't exhibit incomplete dominance; the heterozygotes expressed the dominant phenotype.

Environmental Effects

The degree to which many alleles are expressed depends on the environment. Some alleles are heat-sensitive, for example. Traits influenced by such alleles are more sensitive to temperature or light than are the products of other alleles. The arctic fox in figure 10.15, for example, makes fur pigment only when the weather is warm. Can you see why

(a)

(b)

Figure 10.15 Environmental effects on an allele.

(*a*) An arctic fox in winter has a coat that is almost white, so it is difficult to see the fox against a snowy background. (*b*) In summer, the same fox's fur darkens to a reddish brown, so that it resembles the color of the surrounding tundra.

this trait would be an advantage for the fox? Imagine a fox that didn't possess this trait and was white all year round. It would be very visible to predators in the summer, standing out against its darker surroundings. Similarly, the *ch* allele in Himalayan rabbits and Siamese cats encodes a heat-sensitive version of tyrosinase, one of the enzymes mediating the production of melanin, a dark pigment. The ch version of the enzyme is inactivated at temperatures above about 33°C. At the surface of the main body and head, the temperature is above 33°C and the tyrosinase enzyme is inactive, while it is more active at body extremities such as the tips of the ears and tail, where the temperature is below 33°C. The dark melanin pigment this enzyme produces causes the ears, snout, feet, and tail to be black.

Epistasis

In some situations, two or more genes interact with each other, such that one gene contributes to or masks the expression of the other gene. This becomes apparent when analyzing dihybrid crosses involving these traits. Recall that when individuals heterozygous for two different genes mate (a dihybrid cross), offspring may display the dominant phenotype for both genes, either one of the genes, or for neither gene. Sometimes, however, an investigator cannot find four phenotype classes because two or more of the genotypes express the same phenotypes.

As was stated earlier, few phenotypes are the result of the action of one gene. Most traits reflect the action of many genes, some that act sequentially or jointly. **Epistasis** is an interaction between the products of two genes in which one of the genes modifies the phenotypic expression produced by the other. For example, some commercial varieties of corn, *Zea mays,* exhibit a purple pigment called *anthocyanin* in their seed coats, while others do not. In 1918, geneticist R. A. Emerson crossed two true-breeding corn varieties, neither exhibiting anthocyanin pigment. Surprisingly, all of the F_1 plants produced purple seeds.

When two of these pigment-producing F_1 plants were crossed to produce an F_2 generation, 56% were pigment producers and 44% were not. What was happening? Emerson correctly deduced that two genes were involved in producing pigment, and that the second cross had thus been a dihybrid cross like those performed by Mendel. Mendel had predicted 16 equally possible ways gametes could combine with each other, resulting in genotypes with a phenotypic ratio of 9:3:3:1 ($9 + 3 + 3 + 1 = 16$). How many of these were in each of the two types Emerson obtained? He multiplied the fraction that were pigment producers (0.56) by 16 to obtain 9 and multiplied the fraction that were not (0.44) by 16 to obtain 7. Thus, Emerson had a **modified ratio** of 9:7 instead of the usual 9:3:3:1 ratio. Figure 10.16 shows the results of the dihybrid cross made by Emerson. Go back and compare these results with Mendel's dihybrid cross in figure 10.10 and you can see that the F_2 genotypes in Emerson's results are consistent with what Mendel found; so why are the phenotypic ratios different?

Why Was Emerson's Ratio Modified? It turns out that in corn plants either one of the two genes that contribute to kernel color can block the expression of the other. One of the genes (*B*) produces an enzyme that permits colored pigment to be produced only if a dominant allele (*BB* or *Bb*) is present. The other gene (*A*) produces an enzyme that in its dominant form (*AA* or *Aa*) allows the pigment to be deposited on the seed coat color. Thus, an individual with two recessive alleles for gene *A* (no pigment deposition) will have white seed coats even though it is able to manufacture the pigment because it possesses dominant alleles for gene *B* (purple pigment production). Similarly, an individual with dominant alleles for gene *A* (pigment can be deposited) will also have white seed coats if it has only recessive alleles for gene *B* (pigment production) and cannot manufacture the pigment.

To produce and deposit pigment, a plant must possess at least one functional copy of each enzyme gene (*A_B_*). Of the 16 genotypes predicted by random assortment, 9 contain at least one dominant allele of both genes; they produce purple progeny and are colored darker in the Punnett square in figure 10.16. The remaining 7 genotypes lack dominant alleles at either or both loci ($3 + 3 + 1 = 7$) and so are phenotypically the same (nonpigmented—the light-color boxes in the Punnett square), giving the phenotypic ratio of 9:7 that Emerson observed.

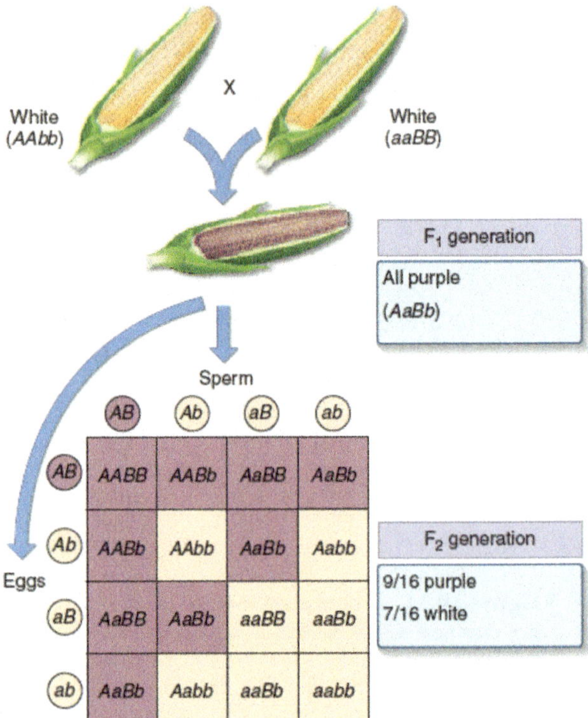

Figure 10.16 How epistasis affects kernel color. The purple pigment found in some varieties of corn is the result of two genes. Unless a dominant allele is present at each of the two loci, no pigment is expressed.

Today's *Biology*

Does Environment Affect I.Q.?

Nowhere has the influence of environment on the expression of genetic traits led to more controversy than in studies of I.Q. scores. I.Q. is a controversial measure of general intelligence based on a written test that many feel to be biased toward white middle-class America. However well or poorly I.Q. scores measure intelligence, a person's I.Q. score has been believed for some time to be determined largely by his or her genes.

How did science come to that conclusion? Scientists measure the degree to which genes influence a multigene trait by using an off-putting statistical measure called the *variance*. Variance is defined as the square of the standard deviation (a measure of the degree-of-scatter of a group of numbers around their mean value), and has the very desirable property of being additive—that is, the total variance is equal to the sum of the variances of the factors influencing it.

What factors contribute to the total variance of I.Q. scores? There are three. The first factor is variation at the gene level, some gene combinations leading to higher I.Q. scores than others. The second factor is variation at the environmental level, some environments leading to higher I.Q. scores than others. The third factor is what a statistician calls the *covariance,* the degree to which environment affects genes.

The degree to which genes influence a trait like I.Q., the *heritability* of I.Q., is given the symbol H and is defined simply as the fraction of the total variance that is genetic.

So how heritable is I.Q.? Geneticists estimate the heritability of I.Q. by measuring the environmental and genetic contributions to the total variance of I.Q. scores. The environmental contributions to variance in I.Q. can be measured by comparing the I.Q. scores of identical twins reared together with those reared apart (any differences should reflect environmental influences). The genetic contributions can be measured by comparing identical twins reared together (which are 100% genetically identical) with fraternal twins reared together (which are 50% genetically identical). Any differences should reflect genes, as twins share identical prenatal conditions in the womb and are raised in virtually identical environmental circumstances, so when traits are more commonly shared between identical twins than fraternal twins, the difference is likely genetic.

When these sorts of "twin studies" have been done in the past, researchers have uniformly reported that I.Q. is highly heritable, with values of H typically reported as being around 0.7 (a very high value). While it didn't seem significant at the time, almost all the twins available for study over the years have come from middle-class or wealthy families.

The study of I.Q. has proven controversial, because I.Q. scores are often different when social and racial groups are compared. What is one to make of the observation that I.Q. scores of poor children measure lower as a group than do scores of children of middle-class and wealthy families? This difference has led to the controversial suggestion by some that the poor are genetically inferior.

What should we make of such a harsh conclusion? To make a judgement, we need to focus for a moment on the fact that these measures of the heritability of I.Q. have all made a critical assumption, one to which population geneticists, who specialize in these sorts of things, object strongly. The assumption is that environment does not affect gene expression, so that covariance makes no contribution to the total variance in I.Q. scores—that is, that the covariance contribution to H is zero.

Studies have allowed a direct assessment of this assumption. Importantly, it proves to be flat wrong.

In November of 2003, researchers reported an analysis of twin data from a study carried out in the late 1960s. The National Collaborative Prenatal Project, funded by the National Institutes of Health, enrolled nearly 50,000 pregnant women, most of them black and quite poor, in several major U.S. cities. Researchers collected abundant data, and gave the children I.Q. tests seven years later. Although not designed to study twins, this study was so big that many twins were born, 623 births. Seven years later, 320 of these pairs were located and given I.Q. tests. This thus constitutes a huge "twin study," the first ever conducted of I.Q. among the poor.

When the data were analyzed, the results were unlike any ever reported. The heritability of I.Q. was different in different environments! Most notably, the influence of genes on I.Q. was far less in conditions of poverty, where environmental limitations seem to block the expression of genetic potential. Specifically, for families of high socioeconomic status, H = 0.72, much as reported in previous studies, but for families raised in poverty, H = 0.10, a very low value, indicating that genes were making little contribution to observed I.Q. scores. The lower a child's socioeconomic status, the less impact genes had on I.Q.

These data say, with crystal clarity, that the genetic contributions to I.Q. don't mean much in an impoverished environment.

How does poverty in early childhood affect the brain? Neuroscientists reported in 2008 that many children growing up in very poor families experience poor nutrition and unhealthy levels of stress hormones, both of which impair their neural development. This affects language development and memory for the rest of their lives.

Clearly, improvements in the growing and learning environments of poor children can be expected to have a major impact on their I.Q. scores. Additionally, these data argue that the controversial differences reported in mean I.Q. scores between racial groups may well reflect no more than poverty, and are no more inevitable.

Figure 10.17 The effect of epistatic interactions on coat color in dogs.
The coat color seen in Labrador retrievers is an example of the interaction of two genes, each with two alleles. The *E* gene determines if the pigment will be deposited in the fur, and the *B* gene determines how dark the pigment will be.

Other Examples of Epistasis In many animals, coat color is the result of epistatic interactions among genes. Coat color in Labrador retrievers, a breed of dog, is due primarily to the interaction of two genes. The *E* gene determines if dark pigment will be deposited in the fur or not. If a dog has the genotype *ee* (like the two dogs on the left in figure 10.17), no pigment will be deposited in the fur, and it will be yellow. If a dog has the genotype *EE* or *Ee* (*E_*), pigment will be deposited in the fur (like the two dogs on the right).

A second gene, the *B* gene, determines how dark the pigment will be. Dogs with the genotype *E_bb* will have brown fur and are called chocolate labs. Dogs with the genotype *E_B_* are black labs with black fur. But, even in yellow dogs, the *B* gene does have some effect. Yellow dogs with the genotype *eebb* (on the far left) will have brown pigment on their nose, lips, and eye rims, while yellow dogs with the genotype *eeB_* (the second from the left) will have black pigment in these areas. The genes for coat color in this breed have been found, and a genetic test is available to determine the coat color in a litter of puppies.

Codominance

A gene may have more than two alleles in a population, and in fact most genes possess several different alleles. Often in heterozygotes there isn't a dominant allele; instead, the effects of both alleles are expressed. In these cases, the alleles are said to be **codominant.**

Codominance is seen in the color patterning of some animals. For example, the "roan" pattern is a coloring pattern exhibited in some varieties of horses and cattle. A roan animal expresses both white and colored hairs on at least part of its body. This intermingling of the different colored hairs creates either an overall lighter color, or patches of lighter and darker colors. The roan pattern results from a heterozygous genotype, such as produced by mating of a homozygous white and homozygous colored. Could the intermediate color be the result of incomplete dominance? No. The heterozygote that receives a white allele and a colored allele does not have individual hairs that are a mix of the two colors; rather, both alleles are being expressed, with the result that the animal has some hairs that are white and some that are colored. The gray horse in figure 10.18 is exhibiting the roan pattern. It looks like it has gray hairs, but if you were able to examine its coat closely, you would see both white hairs and black hairs, giving it an overall gray color.

A human gene that exhibits more than one dominant allele is the gene that determines ABO blood type. This gene encodes an enzyme that adds sugar molecules to lipids on the surface of red blood cells. These sugars act as recognition markers for cells in the immune system and are called cell surface antigens. The gene that encodes the enzyme, designated *I*, has three common alleles: I^B, whose product adds the sugar galactose; I^A, whose product adds galactosamine; and *i*, which codes for a protein that does not add a sugar.

Different combinations of the three *I* gene alleles occur in different individuals because each person may

Figure 10.18 **Codominance in color patterning.**

This roan horse is heterozygous for coat color. The offspring of a cross between a white homozygote and a black homozygote, it expresses both phenotypes. Some of the hairs on its body are white and some are black.

be homozygous for any allele or heterozygous for any two. An individual heterozygous for the I^A and I^B alleles produces both forms of the enzyme and adds both galactose and galactosamine to the surfaces of red blood cells. Because both alleles are expressed simultaneously in heterozygotes, the I^A and I^B alleles are codominant. Both I^A and I^B are dominant over the i allele because both I^A or I^B alleles lead to sugar addition and the i allele does not. The different combinations of the three alleles produce four different phenotypes:

1. Type A individuals add only galactosamine. They are either I^AI^A homozygotes or I^Ai heterozygotes (the three darkest boxes in figure 10.19).
2. Type B individuals add only galactose. They are either I^BI^B homozygotes or I^Bi heterozygotes (the three lightest-colored boxes).
3. Type AB individuals add both sugars and are I^AI^B heterozygotes (the two intermediate-colored boxes).
4. Type O individuals add neither sugar and are ii homozygotes (the one white box in figure 10.19).

These four different cell surface phenotypes are called the **ABO blood groups.** A person's immune system can distinguish between these four phenotypes. If a type A individual receives a transfusion of type B blood, the recipient's immune system recognizes that the type B blood cells possess a "foreign" antigen (galactose) and attacks the donated blood cells, causing the cells to clump or agglutinate. This also happens if the donated blood is type AB. However, if the donated blood is type O, it contains no galactose or galactosamine antigens on the surfaces of its blood cells, and so elicits no immune response to these antigens. For this reason, the type O individual is often referred to as a "universal donor." In general, any individual's immune system will tolerate a transfusion of type O blood. Because neither galactose nor galactosamine is foreign to type AB individuals (whose red blood cells have both sugars), those individuals may receive any type of blood.

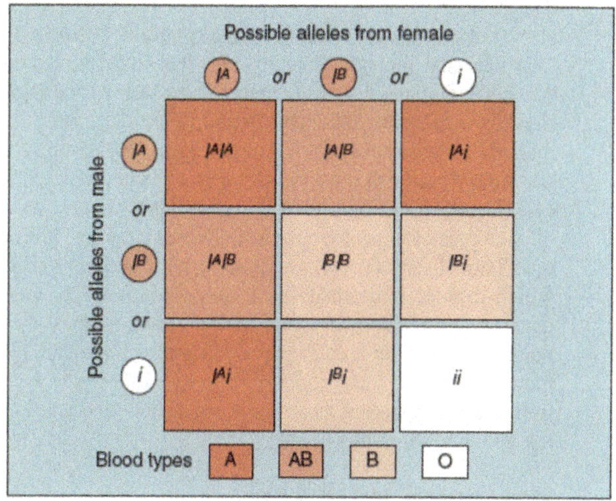

Figure 10.19 **Multiple alleles controlling the ABO blood groups.**

Three common alleles control the ABO blood groups. The different combinations of the three alleles result in four different blood type phenotypes: type A (either I^AI^A homozygotes or I^Ai heterozygotes), type B (either I^BI^B homozygotes or I^Bi heterozygotes), type AB (I^AI^B heterozygotes), and type O (ii homozygotes).

Key Learning Outcome 10.6 **A variety of factors can disguise the Mendelian segregation of alleles. Among them are continuous variation, which results when many genes contribute to a trait; pleiotropic effects, where one allele affects many phenotypes; incomplete dominance, which produces heterozygotes unlike either parent; environmental influences on the expression of phenotypes; and the interaction of more than one allele, as seen in epistasis and codominance.**

Chromosomes and Heredity

10.7 Chromosomes Are the Vehicles of Mendelian Inheritance

The Chromosomal Theory of Inheritance

In the early twentieth century it was by no means obvious that chromosomes were the vehicles of hereditary information. A central role for chromosomes in heredity was first suggested in 1900 by the German geneticist Karl Correns, in one of the papers announcing the rediscovery of Mendel's work. Soon observations that similar chromosomes paired with one another during meiosis led to the *chromosomal theory of inheritance,* first formulated by American Walter Sutton in 1902.

Several pieces of evidence supported Sutton's theory. One was that reproduction involves the initial union of only two cells, egg and sperm. If Mendel's model was correct, then these two gametes must make equal hereditary contributions. Sperm, however, contain little cytoplasm, suggesting that the hereditary material must reside within the nuclei of the gametes. Furthermore, while diploid individuals have two copies of each pair of homologous chromosomes, gametes have only one. This observation was consistent with Mendel's model, in which diploid individuals have two copies of each heritable gene and gametes have one. Finally, chromosomes segregate during meiosis, and each pair of homologues orients on the metaphase plate independently of every other pair. Segregation and independent assortment were two characteristics of the genes in Mendel's model.

Problems with the Chromosomal Theory

Investigators soon pointed out one problem with this theory, however. If Mendelian traits are determined by genes located on the chromosomes, and if the independent assortment of Mendelian traits reflects the independent assortment of chromosomes in meiosis, why does the number of traits that assort independently in a given kind of organism often greatly exceed the number of chromosome pairs the organism possesses? This seemed a fatal objection, and it led many early researchers to have serious reservations about Sutton's theory.

Morgan's White-Eyed Fly

The essential correctness of the chromosomal theory of heredity was demonstrated by a single small fly. In 1910 Thomas Hunt Morgan, studying the fruit fly *Drosophila melanogaster,* detected a mutant male fly that differed strikingly from normal fruit flies: Its eyes were white instead of red (figure 10.20).

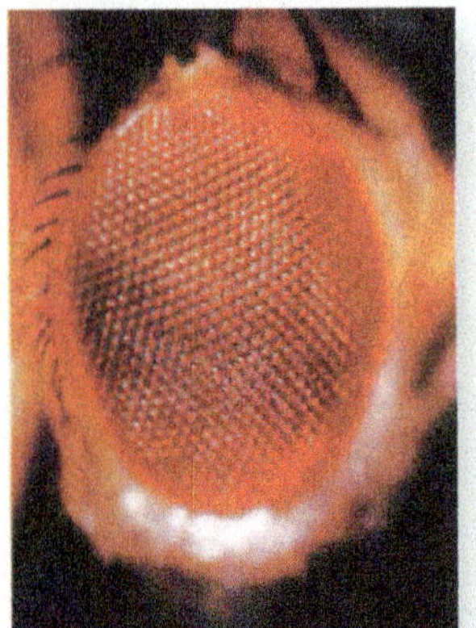
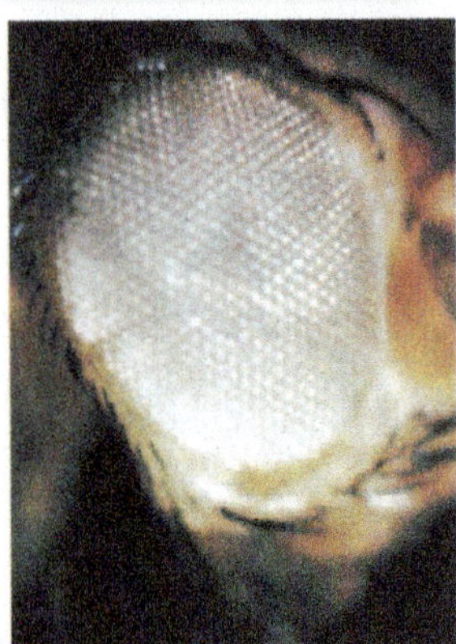

Figure 10.20 Red-eyed (wild type) and white-eyed (mutant) *Drosophila*.
The white-eye defect is hereditary, the result of a mutation in a gene located on the X chromosome. By studying this mutation, Morgan first demonstrated that genes are on chromosomes.

Morgan immediately set out to determine if this new trait would be inherited in a Mendelian fashion. He first crossed the mutant male with a normal female to see if red or white eyes were dominant. All of the F_1 progeny had red eyes, so Morgan concluded that red eye color was dominant over white. Following the experimental procedure that Mendel had established long ago, Morgan then crossed the red-eyed flies from the F_1 generation with each other. Of the 4,252 F_2 progeny Morgan examined, 782 (18%) had white eyes. Although the ratio of red eyes to white eyes in the F_2 progeny was greater than 3:1, the results of the cross nevertheless provided clear evidence that eye color segregates. However, there was something about the outcome that was strange and totally unpredicted by Mendel's theory—*all of the white-eyed F_2 flies were males!*

How could this result be explained? Perhaps it was impossible for a white-eyed female fly to exist; such individuals might not be viable for some unknown reason. To test this idea, Morgan testcrossed the female F_1 progeny with the original white-eyed male. He obtained white-eyed and red-eyed males and females in a 1:1:1:1 ratio, just as Mendelian theory predicted. Hence, a female could have white eyes. Why, then, were there no white-eyed females among the progeny of the original cross?

Sex Linkage Confirms the Chromosomal Theory

The solution to this puzzle involved sex. In *Drosophila,* the sex of an individual is determined by the number of copies of a particular chromosome, the X chromosome, that an individual possesses. A fly with two X chromosomes is a female, and a fly with only one X chromosome is a male. In males, the single X chromosome pairs in meiosis with a large, dissimilar partner called the Y chromosome. The female thus produces only X gametes, while the male produces both X and Y gametes. When fertilization involves an X sperm, the result is an XX zygote, which develops into a female; when fertilization involves a Y sperm, the result is an XY zygote, which develops into a male.

The solution to Morgan's puzzle is that the gene causing the white-eye trait in *Drosophila* resides only on the X chromosome—it is absent from the Y chromosome. (We now know that the Y chromosome in flies carries almost no functional genes.) A trait determined by a gene on the sex chromosome is said to be **sex-linked.** Knowing the white-eye trait is recessive to the red-eye trait, we can now see that Morgan's result was a natural consequence of the Mendelian assortment of chromosomes. Figure 10.21 steps you through Morgan's experiment, showing both the eye color alleles and the sex chromosomes. In this experiment, the F_1 generation all had red eyes, while the F_2 generation contained flies with white eyes—but they were all males. This at-first-surprising result happens because the segregation of the white-eye trait has a one-to-one correspondence with the segregation of the X chromosome. In other words, the white-eye gene is on the X chromosome. In humans, traits such as color-blindness (see page 208 and chapter 29) and hemophilia (a blood-clotting disease discussed later in this chapter) are sex-linked.

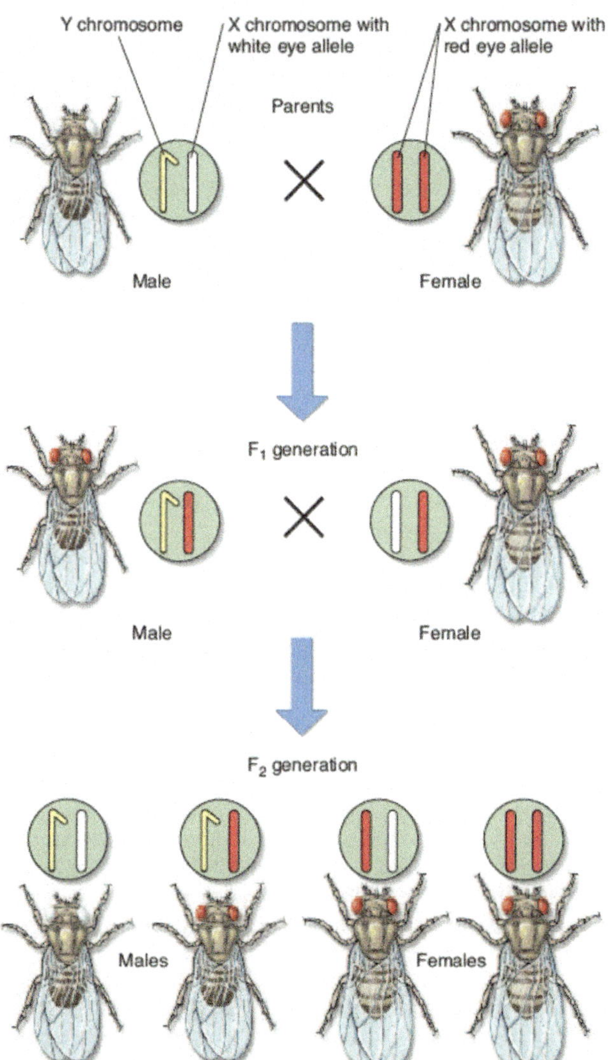

Figure 10.21 Morgan's experiment demonstrating the chromosomal basis of sex linkage.
The white-eyed mutant male fly was crossed with a normal female. The F_1 generation flies all exhibited red eyes, as expected for flies heterozygous for a recessive white-eye allele. In the F_2 generation, all of the white-eyed flies were male.

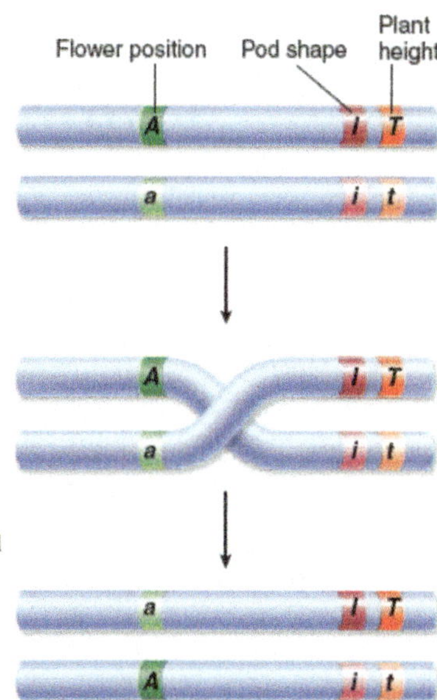

Figure 10.22 Linkage.
Genes that are located farther apart on a chromosome, like the genes for flower position (*A*) and pod shape (*I*) in Mendel's peas, will assort independently because crossing over results in recombination of these alleles. Pod shape (*I*) and plant height (*T*), however, are positioned very near each other, such that crossing over usually would not occur. These genes are said to be linked and do not undergo independent assortment.

Morgan's experiment presented the first clear evidence that the genes determining Mendelian traits reside on chromosomes, just as Sutton had proposed. Now we can see that the reason Mendelian traits assort independently is because chromosomes assort independently. When Mendel observed the segregation of alternative traits in pea plants, he was observing a reflection of the meiotic segregation of the chromosomes, which contained the characters he was observing.

If genes are located on chromosomes, you might expect that two genes on the same chromosome would segregate together. However, if the two genes are located far from each other on the chromosome, like genes *A* and *I* in figure 10.22, the likelihood of crossing over occurring between them is very high, leading to independent segregation. Conversely, the closer two genes are to each other on a chromosome, like genes *I* and *T*, the less likely it is that a cross-over event will occur between them. Genes that are located quite close to each other almost always segregate together, meaning that they are inherited together. The tendency of close-together genes to segregate together is called **linkage.**

Key Learning Outcome 10.7 **Mendelian traits assort independently because they are determined by genes located on chromosomes that assort independently in meiosis.**

10.8 Human Chromosomes

The principles of genetics first proposed by Mendel apply not only to pea plants and fruit flies, but also to you. How closely you resemble your father or mother was largely established before your birth by the chromosomes you received from them, just as meiosis in pea plants determined the segregation of Mendel's traits. But many of the alleles found in human populations demand more serious concern than the color of a pea. Some of the most devastating human disorders result from alleles specifying defective forms of proteins that have important functions in our bodies. By studying human heredity, scientists are more able to predict which disorders parents might expect to pass on to their children, and with what probability.

Although humans pass genes on to the next generation in much the same way that other organisms do, we naturally have a special curiosity about ourselves. Because we know that some illnesses are hereditary and others are not, we cannot escape concern when a member of our family becomes ill. If a family member has had a stroke, we tend to worry about our own future health because we know that the propensity to suffer strokes can be hereditary. Few parents have babies without worrying about the possibility of birth defects. Genes are also clearly involved in such conditions as diabetes, depression, and alcoholism. The way in which genes interact with the environment to produce individuals with differing personalities is the subject of continuing intensive study. Because of the importance of genes in determining the course of our lives, we are all human geneticists, interested in what the laws of genetics reveal about ourselves and our families.

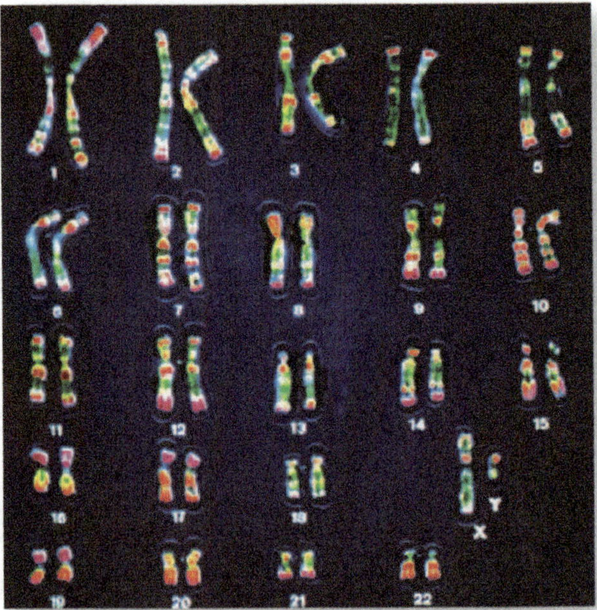

Figure 10.23 A human karyotype.
Photographs of each of the 46 chromosomes have been arranged in descending order of size. The banding patterns revealed by staining allow the investigator to identify homologues and pair them together.

Human Chromosomes

Although chromosomes were discovered more than a century ago, the exact number of chromosomes that humans possess (46) was not established until 1956, when new techniques for accurately determining the number and form of human and other mammalian chromosomes were developed.

Biologists examine human chromosomes by collecting a blood sample, adding chemicals that induce the white blood cells in the sample to divide (red blood cells have lost their nuclei and cannot divide), and then adding other chemicals that arrest cell division at metaphase. Metaphase is the stage of mitosis when the chromosomes are most condensed and thus most easily distinguished from one another. The cells are then flattened, spreading out their contents, and the individual chromosomes are separated for examination. The chromosomes are stained and photographed, and a chromosomal "portrait" called a karyotype is prepared with the photographs of the individual chromosomes. A human karyotype is presented in figure 10.23. By convention, the chromosomes in a karyotype are presented with homologues together, and chromosomes arranged in order of descending size.

Of the 23 pairs of human chromosomes you see in figure 10.23, 22 consist of members that are similar in size and morphology in both males and females. These chromosomes are called **autosomes.** In many plants and animals, including peas, fruit flies, and humans, the two members of the remaining pair—the so-called **sex chromosomes**—are unlike each other in males and similar in females. In humans, females are designated XX and males are designated XY. The Y chromosome is much smaller than the X chromosome and carries only a tenth the number of genes. Among the genes present on the Y chromosome are those that determine "maleness," and, therefore, humans who inherit the Y chromosome develop into males.

The karyotypes of individuals are often examined to detect genetic abnormalities arising from extra or lost chromosomes. For example, the human birth defect Down syndrome (discussed on the facing page) is associated with the presence of an extra copy of chromosome 21, which can be recognized easily in karyotypes, as there are 47 chromosomes rather than 46, and the extra chromosome can be identified by its banding pattern as a third copy of chromosome 21. Karyotypes of fetal cells taken before birth can reveal genetic abnormalities of this sort.

Nondisjunction

Some of the most significant human hereditary disorders arise as a result of problems with how human chromosomes sort during meiosis.

Sometimes during meiosis, sister chromatids or homologous chromosomes that paired up during metaphase remain stuck together instead of separating. The failure of chromosomes to separate correctly during either meiosis I or II is called **nondisjunction.** Nondisjunction leads to **aneuploidy,** an abnormal number of chromosomes. The nondisjunction you see in figure 10.24 occurs because the homologous pair of larger chromosomes failed to separate in anaphase I. The gametes that result from this division have unequal numbers of chromosomes. Under normal meiosis, all gametes would be expected to have two chromosomes, but as you can see, two of these gametes have three chromosomes, while two others have just one.

Almost all humans of the same sex have the same karyotype simply because other arrangements don't work well. Humans who have lost even one copy of an autosome (called **monosomics**) do not survive development. In all but a few cases, humans who have gained an extra autosome (called **trisomics**) also do not survive. However, five of the smallest chromosomes—those numbered 13, 15, 18, 21, and 22—can be present in humans as three copies and still allow the individual to survive for a time. The presence of an extra chromosome 13, 15, or 18 causes severe developmental defects, and infants with such a genetic makeup die within a few months. In contrast, individuals who have an extra copy of chromosome 21 or, more rarely, chromosome 22, usually survive to adulthood. In such individuals, the maturation of the skeletal system is delayed, so they generally are short and have poor muscle tone. Their mental development is also affected.

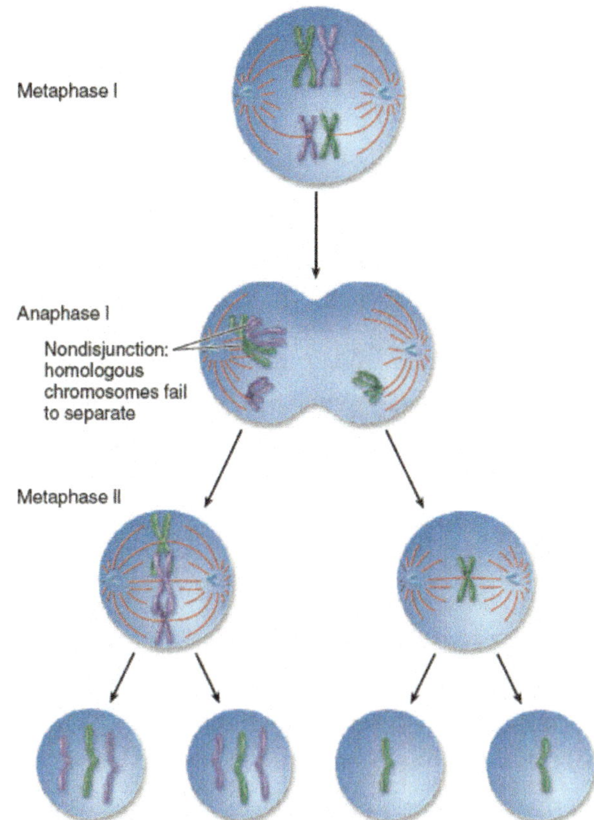

Figure 10.24 Nondisjunction in anaphase I.
In nondisjunction that occurs during meiosis I, one pair of homologous chromosomes fails to separate in anaphase I, and the gametes that result have one too many or one too few chromosomes. Nondisjunction can also occur in meiosis II, when sister chromatids fail to separate during anaphase II.

Down Syndrome The developmental defect produced by the trisomy 21 seen in figure 10.25 was first described in 1866 by J. Langdon Down; for this reason, it is called **Down syndrome.** About 1 in every 750 children exhibits Down syndrome, and the frequency is similar in all racial groups. It is much more common in children of older mothers. The graph in figure 10.26 (see next page) shows the increasing incidence in older mothers. In mothers under 30 years old, the incidence is only about 0.6 per 1,000 (or 1 in 1,500 births), while in mothers 30 to 35 years old, the incidence doubles to about 1.3 per 1,000 births (or 1 in 750 births). In mothers over 45, the risk is as high as 63 per 1,000 births (or 1 in 16 births). The reason that older mothers are more prone to Down syndrome babies is that all the eggs that a woman will ever produce are present in her ovaries by the time she is born, and as she gets older they may accumulate damage that can result in nondisjunction.

Figure 10.25 Down syndrome.
(*a*) In this karyotype of a male individual with Down syndrome, the trisomy at position 21 can be clearly seen. (*b*) A person with Down syndrome.

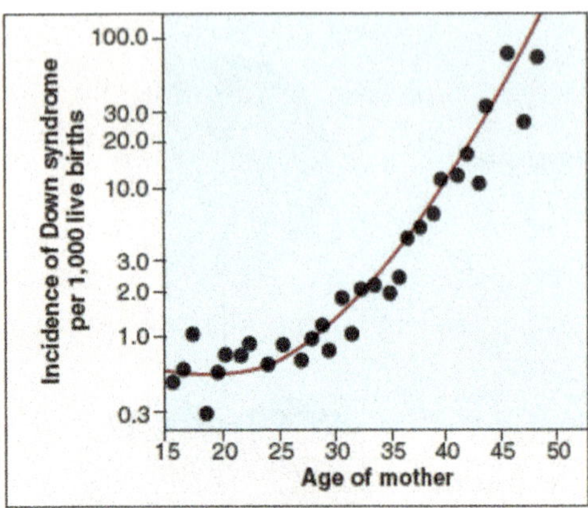

Figure 10.26 Correlation between maternal age and the incidence of Down syndrome.
As women age, the chances they will bear a child with Down syndrome increase. After a woman reaches age 35, the frequency of Down syndrome increases rapidly.

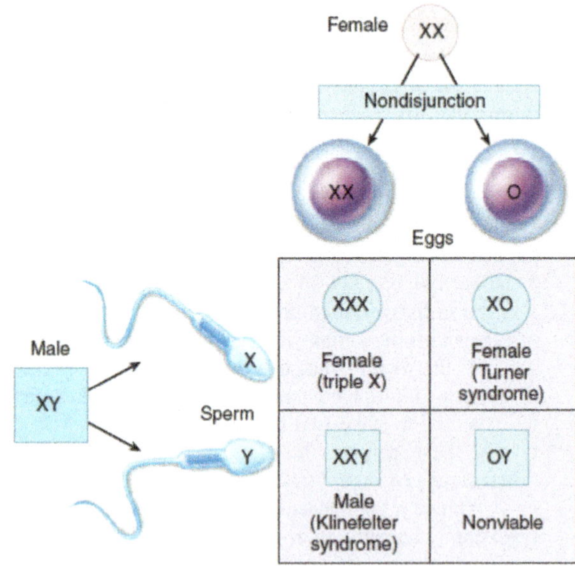

Figure 10.27 Nondisjunction of the X chromosome.
Nondisjunction of the X chromosome can produce sex chromosome aneuploidy—that is, abnormalities in the number of sex chromosomes.

Nondisjunction Involving the Sex Chromosomes

As noted, 22 of the 23 pairs of human chromosomes are perfectly matched in both males and females and are called autosomes. The remaining pair are the sex chromosomes, X and Y. In humans, as in *Drosophila* (but by no means in all diploid species), females are XX and males XY; any individual with at least one Y chromosome is male. The Y chromosome is highly condensed and bears few functional genes in most organisms. Some of the active genes the Y chromosome does possess are responsible for the features associated with "maleness." Individuals who gain or lose a sex chromosome do not generally experience the severe developmental abnormalities caused by changes in autosome numbers. Such individuals may reach maturity, but with somewhat abnormal features.

Nondisjunction of the X Chromosome When X chromosomes fail to separate during meiosis, some of the gametes that are produced possess both X chromosomes and so are XX gametes; the other gametes that result from such an event have no sex chromosome and are designated "O."

Figure 10.27 shows what happens if gametes from X chromosome nondisjunction combine with sperm. If an XX gamete combines with an X gamete, the resulting XXX zygote (in the upper left of the Punnett square) develops into a female who is taller than average but other symptoms can vary greatly. Some are normal in most respects, others may have lower reading and verbal skills, and still others are mentally retarded. If an XX gamete combines with a Y gamete (in the lower left), the XXY zygote develops into a sterile male who has many female body characteristics and, in some cases, diminished mental capacity. This condition, called *Klinefelter syndrome,* occurs in about 1 in 500 male births.

If an O gamete fuses with a Y gamete (in the lower right), the OY zygote is nonviable and fails to develop further because humans cannot survive when they lack the genes on the X chromosome. If an O gamete fuses with an X gamete (in the upper right), the XO zygote develops into a sterile female of short stature, with a webbed neck and immature sex organs that do not undergo changes during puberty. The mental abilities of XO individuals are normal in verbal learning but lower in nonverbal/math-based problem solving. This condition, called *Turner syndrome,* occurs roughly once in every 5,000 female births.

Nondisjunction of the Y Chromosome The Y chromosome can also fail to separate in meiosis, leading to the formation of YY gametes. When these gametes combine with X gametes, the XYY zygotes develop into fertile males of normal appearance. The frequency of the XYY genotype is about 1 per 1,000 newborn males.

Key Learning Outcome 10.8 The particular array of chromosomes that an individual possesses is called the karyotype. The human karyotype usually contains 23 pairs of chromosomes. Autosome loss is always lethal, and an extra autosome is with few exceptions lethal too.

10.9 Studying Pedigrees

To study human heredity, scientists look at the results of crosses that have already been made. They study family trees, or **pedigrees,** to identify which relatives exhibit a trait. Then they can often determine whether the gene producing the trait is sex-linked (that is, located on the X chromosome) or autosomal, and whether the expression of the trait is dominant or recessive. Frequently the pedigree will also help an investigator infer which individuals in a family are homozygous and which are heterozygous for the allele specifying the trait.

Analyzing a Pedigree for Albinism

Albino individuals lack all pigmentation; their hair and skin are completely white. In the United States about 1 in 38,000 Caucasians and 1 in 22,000 African-Americans are albino. In the pedigree of albinism among a family of Hopi Indians presented in figure 10.28, each symbol represents one individual in the family history, with the circles representing females and the squares, males. In this pedigree, individuals that exhibit a trait being studied—in this case, albinism—are indicated by solid symbols; heterozygote "carriers" exhibiting normal phenotypes are indicated by half-filled symbols. Marriages are represented by horizontal lines connecting a circle and a square, from which a cluster of vertical lines descend indicating the children, arranged from left to right in order of their birth.

To analyze this pedigree of albinism, a geneticist traditionally asks three questions:

1. *Is albinism sex-linked or autosomal?* If the trait is sex-linked, it is usually seen only in males; if it is autosomal, it appears in both sexes fairly equally. In the pedigree below, the proportion of affected males (4 of 12, or 33%) is reasonably similar to the proportion of affected females (8 of 19, or 42%). (When counting numbers of affected individuals in a pedigree, exclude the parents in generation I, as well as any "outsiders" who marry into the family.) From this result, it is reasonable to conclude the trait is autosomal.
2. *Is albinism dominant or recessive?* If the trait is dominant, every albino child will have an albino parent. If recessive, however, an albino child's parents can appear normal, since both parents may be heterozygous "carriers." In the pedigree below, parents of most of the albino children do not exhibit the trait, which indicates that albinism is recessive. Four children in one family *do* have albino parents. The allele is very common among the Hopi Indians, from which this pedigree was derived, and thus homozygous individuals such as these albino parents are present in the Hopis in sufficient numbers that they sometimes marry. In this family, *both* parents are albino and *all* four children are albino, which is consistent with the finding that the trait albinism is recessive, with both parents homozygous for the allele.
3. *Is the albinism trait determined by a single gene, or by several?* If the trait is determined by a single gene, then a ratio of 3:1 (normal to albino) offspring should be born to heterozygous parents (indicated by half-filled symbols), reflecting Mendelian segregation in a cross. Thus about 25% of these children should be albino. But if the trait is determined by several genes, albinism would only be present in a few percent. In this pedigree, 8 of 24 children born to heterozygotes exhibit albinism, or 33%, strongly suggesting that only one gene is segregating in these crosses.

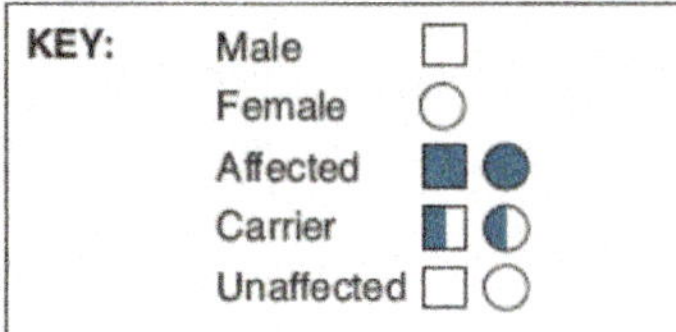

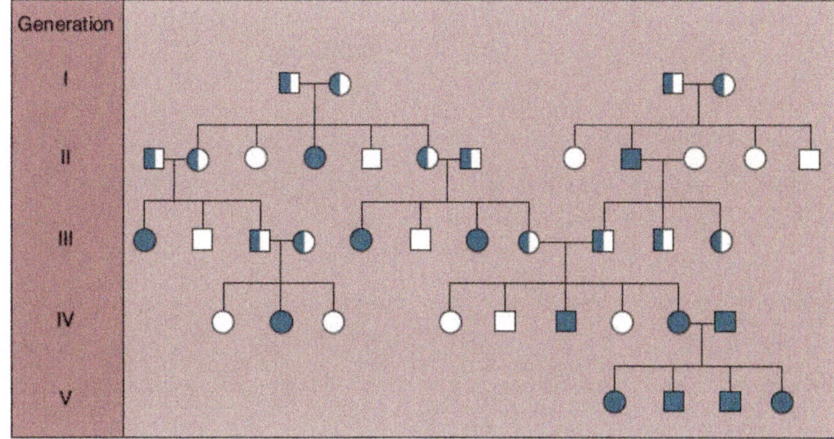

Figure 10.28 A pedigree of albinism.
In the photo, one of three girls from a Hopi Indian family (the left-most family in generation IV of the pedigree) is albino. The pedigree shows the inheritance of the gene causing albinism in this family, with the solid blue symbols indicating persons who are albino.

Analyzing a Pedigree for Color Blindness

The albinism pedigree analysis you have just examined indicates that albinism is an autosomal recessive trait controlled by a single gene. The inheritance of other human traits is studied in a similar way, although sometimes with different results. As an example, let us analyze a different trait. Red-green color blindness is an infrequent, although not rare, inherited trait in humans, affecting 5% to 9% of males. Color blindness is a group of eye disorders in which a person is not able to distinguish certain colors or shades of colors. It doesn't mean that they see only in black and white, but rather that they see colors but some different colors look the same to them. Special types of cells in the retina of the eye detect different colors of light and different shades. Recall from the discussion of the electromagnetic spectrum in chapter 6 that visible light contains different wavelengths of photons that appear as the spectrum of visible light shown here:

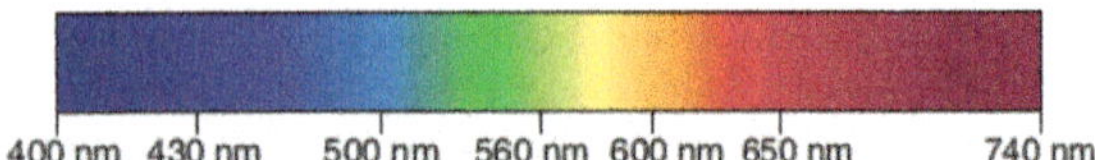

Our eyes contain three types of color receptors: one absorbs red light, one green light, and a third absorbs blue light. People with red-green color blindness have deficiencies in their ability to detect red and green light as being different, and so these colors appear the same to them. Test samples called Ishihara plates are used to determine if a person is color blind. The test plates contain different colored dots arranged to reveal a shape, often a number. People with normal vision are able to see the number while a person who is color blind for those colors is not able to see it. An Ishihara test for red-green color blindness is shown in figure 10.29.

Like albinism, a pedigree can be used to reveal the pattern of inheritance of color blindness. In the pedigree shown below, a red-green color blind man has five children with a woman who is heterozygous for the allele. Again, the solid-color symbols indicate an affected individual, in this case red-green color blind. Half-filled symbols indicate a heterozygous individual who carries the trait but does not express it.

To analyze this pedigree, you ask the same three questions as before:

1. *Is red-green color blindness sex-linked or autosomal?* Of the five affected individuals, all are male. The trait is clearly sex-linked.
2. *Is red-green color blindness dominant or recessive?* If the trait is dominant, then every color-blind child should have a color-blind parent. In this pedigree, however, that is not true in any family after that of the original male. The trait is clearly recessive.
3. *Is the red-green color blindness trait determined by a single gene?* If it is, then children born to heterozygous parents should be color-blind in about 25% of cases, reflecting a 3:1 Mendelian segregation of the trait. In this pedigree, 4 of 14, or 28%, of the children of heterozygous parents are color blind, indicating that a single gene is segregating (do not count the five children of the generation I parents because the father in this case is homozygous for the trait).

The results of this pedigree indicate that color blindness is caused by a single sex-linked, recessive gene. This doesn't mean that females can't be color blind, but in order for a female to be color blind, both X chromosomes would have to carry the color blind gene, and this only occurs in 0.5% of females.

Key Learning Outcome 10.9 The study of family trees can often reveal if an inherited trait is caused by a single gene, if that gene is located on the X chromosome, and if its alleles are recessive.

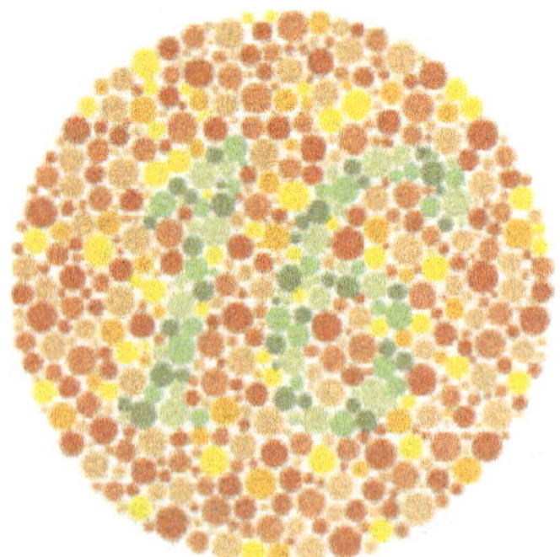

Source: This image has been reproduced from Ishihara's *Tests for Color Deficiency* published by KANEHARA TRADING INC., located in Tokyo, Japan. But tests for color deficiency cannot be conducted with this material. For accurate testing, the original plates should be used.

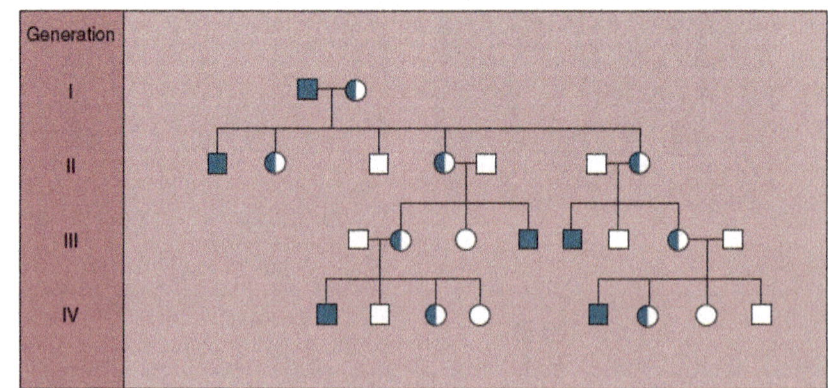

Figure 10.29 Pedigree of color blindness.
Individuals who are red-green color blind cannot see the number, as all the dots appear the same color. The pedigree traces red-green color blindness through four generations of a family.

10.10 The Role of Mutation

Hemophilia: A Sex-Linked Trait

Blood in a cut clots as a result of the polymerization of protein fibers circulating in the blood. A dozen proteins are involved in this process, and all must function properly for a blood clot to form. A mutation causing any of these proteins to lose their activity leads to a form of **hemophilia,** a hereditary condition in which the blood clots slowly or not at all.

Hemophilias are recessive disorders, expressed only when an individual does not possess any copy of the normal allele and so cannot produce one of the proteins necessary for clotting. Most of the genes that encode the blood-clotting proteins are on autosomes, but two (designated VIII and IX) are on the X chromosome. These two genes are sex-linked (see section 10.7): Any male who inherits a mutant allele will develop hemophilia because his other sex chromosome is a Y chromosome that lacks any alleles of those genes.

The most famous instance of hemophilia, often called the Royal hemophilia, is a sex-linked form that arose in the royal family of England. This hemophilia was caused by a mutation in gene IX that occurred in one of the parents of Queen Victoria of England (1819–1901). The pedigree in figure 10.30 shows that in the six generations since Queen Victoria, 10 of her male descendants have had hemophilia (the solid squares). The present British royal family has escaped the disorder because Queen Victoria's son, King Edward VII, did not inherit the defective allele, and all the subsequent rulers of England are his descendants. Three of Victoria's nine children did receive the defective allele, however, and they carried it by marriage into many of the other royal families of Europe. In this photo, Queen Victoria of England is surrounded by some of her descendants in 1894. Standing behind Victoria and wearing feathered boas are two of Victoria's granddaughters, Alice's daughter's: Princess Irene of Prussia (right), and Alexandra (left), who would soon become Czarina of Russia. Both Irene and Alexandra were also carriers of hemophilia.

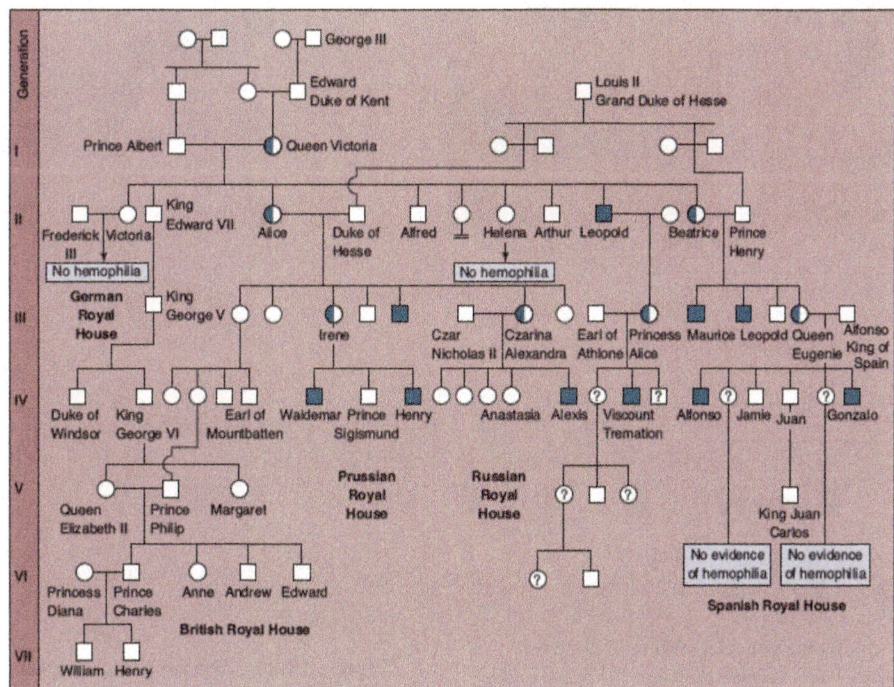

Figure 10.30 The Royal hemophilia pedigree.

Queen Victoria's daughter Alice introduced hemophilia into the Russian and Prussian royal houses, and her daughter Beatrice introduced it into the Spanish royal house. Victoria's son Leopold, himself a victim, also transmitted the disorder in a third line of descent. Half-shaded symbols represent carriers with one normal allele and one defective allele; fully shaded symbols represent affected individuals. Squares represent males; circles represent females.

Sickle-Cell Disease: A Recessive Trait

Sickle-cell disease is a recessive hereditary disorder. Its inheritance is shown in the pedigree in figure 10.31, where affected individuals are homozygous, carrying two copies of the mutated gene. Affected individuals have defective molecules of hemoglobin, the protein within red blood cells that carries oxygen. Consequently, these individuals are unable to properly transport oxygen to their tissues. The defective hemoglobin molecules stick to one another, forming stiff, rodlike structures and resulting in the formation of sickle-shaped red blood cells (figure 10.31). As a result of their stiffness and irregular shape, these cells have difficulty moving through the smallest blood vessels; they tend to accumulate in those vessels and form clots. People who have large proportions of sickle-shaped red blood cells tend to have intermittent illness and a shortened life span.

The hemoglobin in the defective red blood cells differs from that in normal red blood cells in only one of hemoglobin's 574 amino acid subunits. In the defective hemoglobin, the amino acid valine replaces a glutamic acid at a single position in the protein. Interestingly, the position of the change is far from the active site of hemoglobin where the iron-bearing heme group binds oxygen. Instead, the change occurs on the outer edge of the protein. Why then is the result so catastrophic? The sickle-cell mutation puts a very nonpolar amino acid on the surface of the hemoglobin protein, creating a "sticky patch" that sticks to other such patches—nonpolar amino acids tend to associate with one another in polar environments like water. As one hemoglobin adheres to another, chains of hemoglobin molecules form.

Individuals heterozygous for the sickle-cell allele are generally indistinguishable from normal persons. However, some of their red blood cells show the sickling characteristic when they are exposed to low levels of oxygen. The allele responsible for sickle-cell disease is particularly common among people of African descent because the sickle-cell allele is more common in Africa. About 9% of African Americans are heterozygous for this allele, and about 0.2% are homozygous and therefore have the disorder. In some groups of people in Africa, up to 45% of all individuals are heterozygous for this allele, and fully 6% are homozygous and express

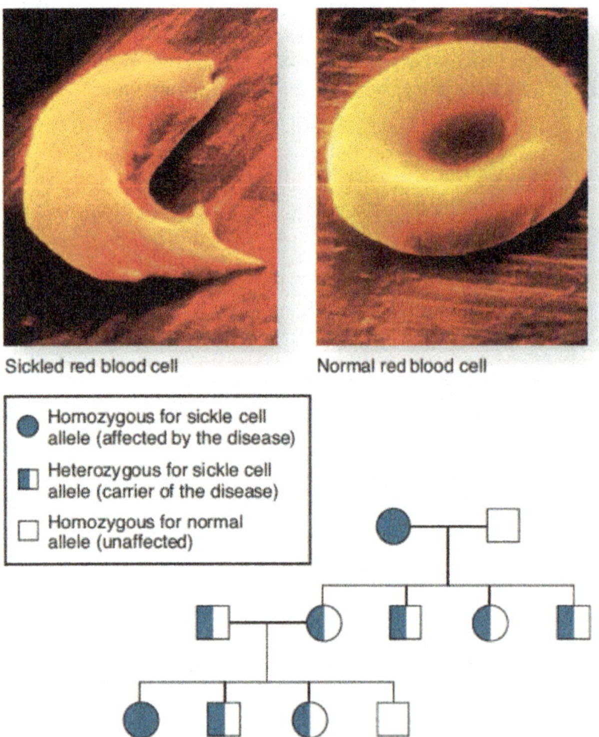

Figure 10.31 Inheritance of sickle-cell disease.
Sickle-cell disease is a recessive autosomal disorder. If one parent is homozygous for the recessive trait, all of the offspring will be carriers (heterozygotes), like the F_1 generation of Mendel's testcross. A normal red blood cell is shaped like a flattened sphere. In individuals homozygous for the sickle-cell trait, many of the red blood cells have sickle shapes.

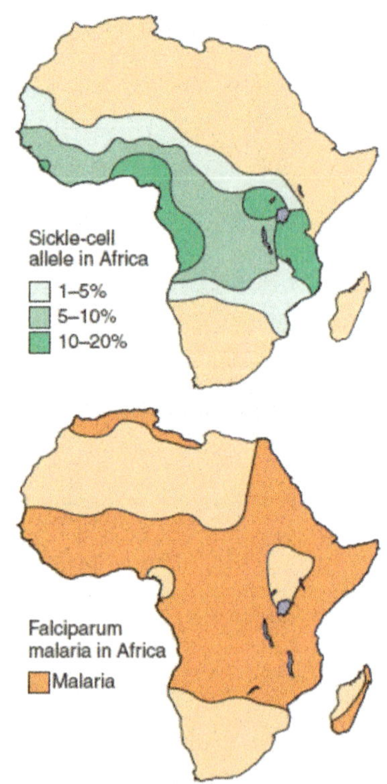

Figure 10.32 The sickle-cell allele confers resistance to malaria.
The distribution of sickle-cell disease closely matches the occurrence of malaria in central Africa. This is not a coincidence. The sickle-cell allele, when heterozygous, confers resistance to malaria, a very serious disease.

the disorder. What factors determine the high frequency of sickle-cell disease in Africa? It turns out that heterozygosity for the sickle-cell allele increases resistance to malaria, a common and serious disease in Central Africa. Comparing the two maps shown in figure 10.32, you can see that the area of the sickle-cell trait matches well with the incidence of malaria. The interactions of sickle-cell disease and malaria are discussed further in chapter 14.

Tay-Sachs Disease: A Recessive Trait

Tay-Sachs disease is an incurable hereditary disorder in which the brain deteriorates. Affected children appear normal at birth and usually do not develop symptoms until about the eighth month, when signs of mental deterioration appear. The children are blind within a year after birth, and they rarely live past five years of age.

The Tay-Sachs allele produces the disease by encoding a nonfunctional form of the enzyme hexosaminidase A. This enzyme breaks down *gangliosides,* a class of lipids occurring within the lysosomes of brain cells. As a result, the lysosomes fill with gangliosides, swell, and eventually burst, releasing oxidative enzymes that kill the cells. There is no known cure for this disorder.

Tay-Sachs disease is rare in most human populations, occurring in only 1 in 300,000 births in the United States. However, the disease has a high incidence among Jews of Eastern and Central Europe (Ashkenazi) and among American Jews, 90% of whom trace their ancestry to Eastern and Central Europe. In these populations, it is estimated that 1 in 28 individuals is a heterozygous carrier of the disease, and approximately 1 in 3,500 infants has the disease. Because the disease is caused by a recessive allele, most of the people who carry the defective allele do not themselves develop symptoms of the disease because, as shown by the middle bar in figure 10.33, their one normal gene produces enough enzyme activity (50%) to keep the body functioning normally.

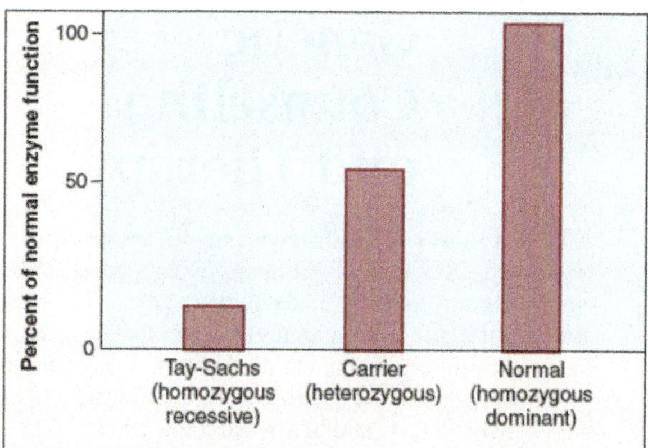

Figure 10.33 Tay-Sachs disease.
Homozygous individuals *(left bar)* typically have less than 10% of the normal level of hexosaminidase A *(right bar),* while heterozygous individuals *(middle bar)* have about 50% of the normal level—enough to prevent deterioration of the central nervous system.

Huntington's Disease: A Dominant Trait

Not all hereditary disorders are recessive. **Huntington's disease** is a hereditary condition caused by a dominant allele that causes the progressive deterioration of brain cells. Perhaps 1 in 24,000 individuals develops the disorder. Because the allele is dominant, every individual who carries the allele expresses the disorder. Nevertheless, the disorder persists in human populations because its symptoms usually do not develop until the affected individuals are more than 30 years old, and by that time most of those individuals have already had children. Consequently, as illustrated by the pedigree in figure 10.34, the allele is often transmitted before the lethal condition develops.

Key Learning Outcome 10.10 Many human hereditary disorders reflect the presence of rare (and sometimes not so rare) mutations within human populations.

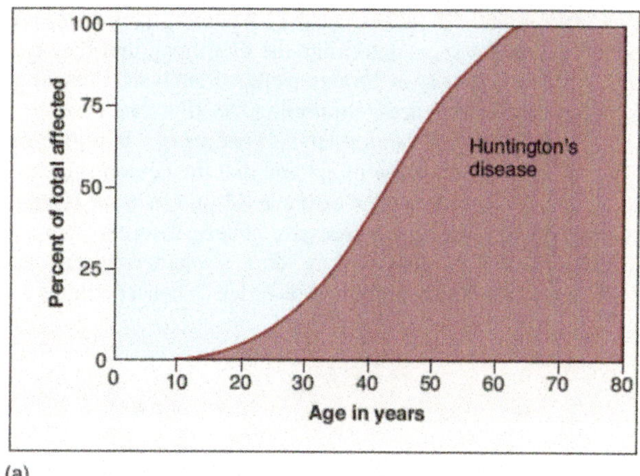

(a)

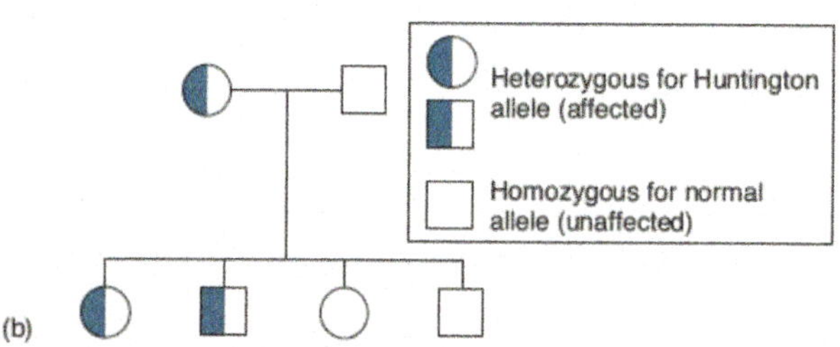

(b)

Figure 10.34 Huntington's disease is a dominant genetic disorder.
(a) Because of the late age of onset of Huntington's disease, the allele causing it persists despite being both dominant and fatal. *(b)* The pedigree illustrates how a dominant lethal allele can be passed from one generation to the next. Although the mother was affected, we can tell that she was heterozygous because if she were homozygous dominant, all of her children would have been affected. However, by the time she found out that she had the disease, she had probably already given birth to her children. In this way the trait passes on to the next generation even though it is fatal.

10.11 Genetic Counseling and Therapy

Although most genetic disorders cannot yet be cured, we are learning a great deal about them, and progress toward successful therapy is being made in many cases. However, in the absence of a cure, some parents may feel their only recourse is to try to avoid producing children with these conditions. The process of identifying parents at risk of producing children with genetic defects and of assessing the genetic state of early embryos is called **genetic counseling.** Genetic counseling can help prospective parents determine their risk of having a child with a genetic disorder and advise them on medical treatments or options if a genetic disorder is determined to exist in an unborn child.

High-Risk Pregnancies

If a genetic defect is caused by a recessive allele, how can potential parents determine the likelihood that they carry the allele? One way is through pedigree analysis, often employed as an aid in genetic counseling. As illustrated earlier in this chapter, by analyzing a person's pedigree, it is sometimes possible to estimate the likelihood that the person is a carrier for certain disorders. For example, if one of your relatives has been afflicted with a recessive genetic disorder such as cystic fibrosis, it is possible that you are a heterozygous carrier of the recessive allele for that disorder. When a pedigree analysis indicates that both parents of an expected child have a significant probability of being heterozygous carriers of a recessive allele responsible for a serious genetic disorder, the pregnancy is said to be a high-risk pregnancy. In such cases, there is a significant probability that the child will exhibit the clinical disorder.

Another class of high-risk pregnancies is that in which the mothers are more than 35 years old. As we have seen, the frequency of birth of infants with Down syndrome increases dramatically in the pregnancies of older women (see figure 10.26).

Genetic Screening

When a pregnancy is determined to be high risk, many women elect to undergo **amniocentesis,** a procedure that permits the prenatal diagnosis of many genetic disorders. Figure 10.35 shows how an amniocentesis is performed. In the fourth month of pregnancy, a sterile hypodermic needle is inserted into the expanded uterus of the mother, and a small sample of the amniotic fluid bathing the fetus is removed. Within the fluid are free-floating cells derived from the fetus; once removed, these cells can be grown in cultures in the laboratory. During amniocentesis, the position of the needle and that of the fetus are usually observed by means of **ultrasound.** The ultrasound image in figure 10.36 clearly reveals the fetus's position in the uterus. You can see its head and a hand extending up, maybe sucking its thumb. The sound waves used in ultrasound generate a live image that permits the person withdrawing the amniotic fluid to do so without damaging the fetus. In addition, ultrasound can be used to examine the fetus for signs of major abnormalities.

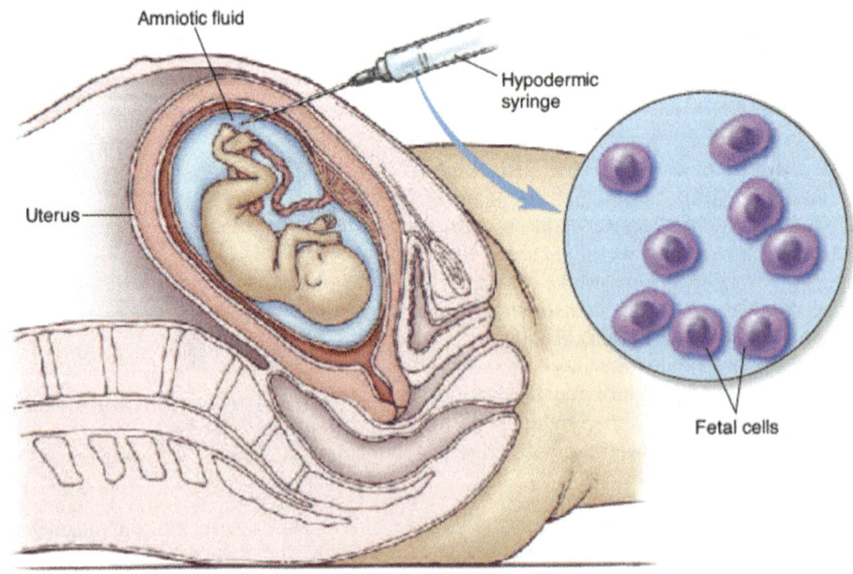

Figure 10.35 Amniocentesis.
A needle is inserted into the amniotic cavity, and a sample of amniotic fluid, containing some free cells derived from the fetus, is withdrawn into a syringe. The fetal cells are then grown in culture and their karyotype and many of their metabolic functions are examined.

In recent years, physicians have increasingly turned to another invasive procedure for genetic screening called **chorionic villus sampling.** In this procedure, the physician removes cells from the chorion, a membranous part of the placenta that nourishes the fetus. This procedure can be used earlier in pregnancy (by the eighth week) and yields results much more rapidly than does amniocentesis, but can increase the risk of miscarriage.

Genetic counselors look at three things in the cultures of cells obtained from amniocentesis or chorionic villus sampling:

1. **Chromosomal karyotype.** Analysis of the karyotype can reveal aneuploidy (extra or missing chromosomes) and gross chromosomal alterations.
2. **Enzyme activity.** In many cases, it is possible to test directly for the proper functioning of enzymes involved in genetic disorders. The lack of normal enzymatic activity signals the presence of the disorder. Thus, the lack of the enzyme responsible for breaking down phenylalanine signals PKU (phenylketonuria), the absence of the enzyme responsible for the breakdown of gangliosides indicates Tay-Sachs disease, and so forth.
3. **Genetic markers.** Genetic counselors can look for an association with known genetic markers. For sickle-cell anemia, Huntington's disease, and one form of muscular dystrophy (a genetic disorder characterized by weakened muscles), investigators have found other mutations on the same chromosomes that, by chance, occur at about the same place as the mutations that cause those disorders. By testing for the presence of these other mutations, a genetic counselor can identify individuals with a high probability of possessing the disorder-causing mutations. Finding such mutations in the first place is a little like searching for a needle in a haystack, but persistent efforts have proved successful in these three disorders. The associated mutations are detectable because they alter the length of the DNA segments that DNA-cleaving enzymes produce when they cut strands of DNA at particular places, an approach described in more detail in chapter 13.

DNA Screening

The mutations that cause hereditary defects are frequently caused by alteration of a single DNA nucleotide within a key gene. Such spot differences between the version of a gene you have and the one another person has are called "single nucleotide polymorphisms," or SNPs. With the completion of the Human Genome Project (described in detail in chapter 13), researchers have begun assembling a huge database of hundreds of thousands of SNPs. Each of us differs from the standard "type sequence" in several thousand gene-altering SNPs. Screening SNPs and comparing them to known SNP databases should soon allow genetic counselors to screen each patient for copies of genes leading to hereditary disorders such as cystic fibrosis and muscular dystrophy.

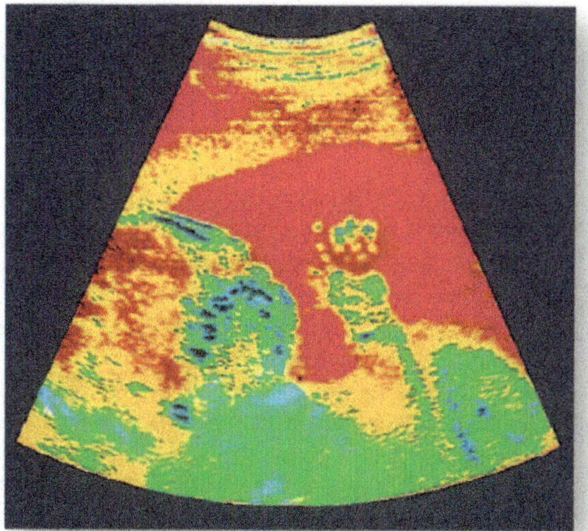

Figure 10.36 An ultrasound view of a fetus.
During the fourth month of pregnancy, when amniocentesis is normally performed, the fetus usually moves about actively. The head of the fetus (visualized in *green*) is to the *left*.

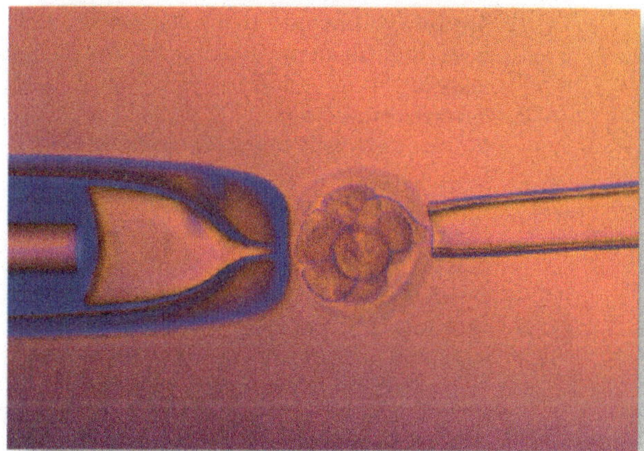

Figure 10.37 Preimplantation genetic screening.
The photograph shows a human embryo at the eight-cell stage, just before one of the eight cells is to be extracted for genetic testing by researchers.

Parents conceiving by in vitro fertilization have available a well-established screening procedure known as **preimplantation genetic screening.** In this test, the egg is fertilized outside the mother, in glassware, and allowed to divide three times, until it contains eight cells. One of the eight cells is then removed from each of several such 8-cell embryos (figure 10.37) and tested for any of 150 genetic defects. The remaining 7-cell embryos are each able to develop into normal fetuses, giving the parents the choice of identifying and implanting an embryo that is disease free.

Key Learning Outcome 10.11 **It has recently become possible to detect genetic defects early in pregnancy, allowing for appropriate planning by the prospective parents.**

INQUIRY & ANALYSIS

Why Woolly Hair Runs in Families

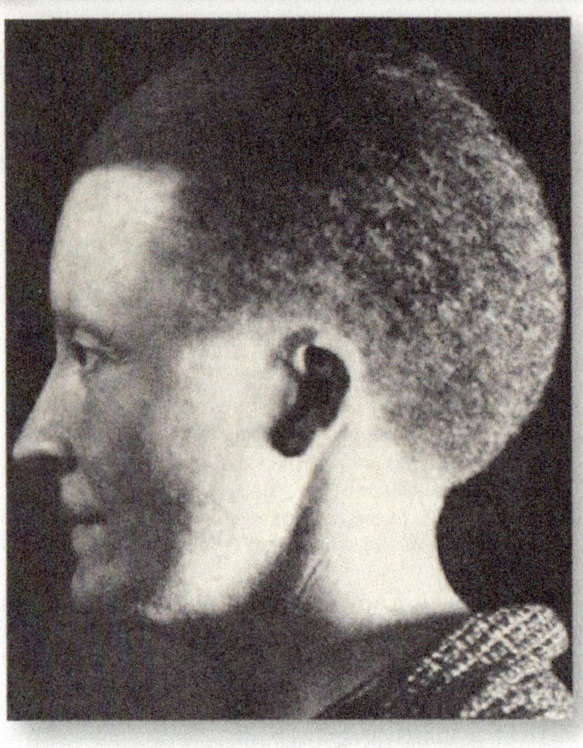

The woman in the photo on the right does not cut her hair. Her hair breaks off naturally as it grows, keeping it from getting long. Other members of her family have the same sort of hair, suggesting it is a hereditary trait. Because of its curly, fuzzy texture, this trait has been given the name "woolly hair."

While the woolly-hair trait is rare, it flares up in certain families. The extensive pedigree below (drawn curved so as to fit in the large families produced by the second and subsequent generations) records the incidence of woolly hair in five generations (the Roman numerals on the left) of a Norwegian family. As is the convention, affected individuals are indicated by solid symbols, with circles females and squares males. The pedigree below will provide you with the information you need to discover how this trait is inherited within human families.

1. **Applying Concepts** In the diagram below, how many individuals are documented? Are all of them related?
2. **Interpreting Data**
   a. Does the woolly-hair trait appear in both sexes equally?
   b. Does every woolly-hair child have a woolly-hair parent?
   c. What percentage of the offspring born to a woolly-haired parent are also woolly haired?
3. **Making Inferences**
   a. Is woolly hair sex-linked or autosomal?
   b. Is woolly hair dominant or recessive?
   c. Is the woolly-hair trait determined by a single gene, or by several?
4. **Drawing Conclusions**
   a. How many copies of the woolly-hair allele are necessary to produce a detectable change in a person's hair?
   b. Are there any woolly-hair homozygous individuals in the pedigree? Explain.

Pedigree of Woolly Hair Among a Norwegian Family

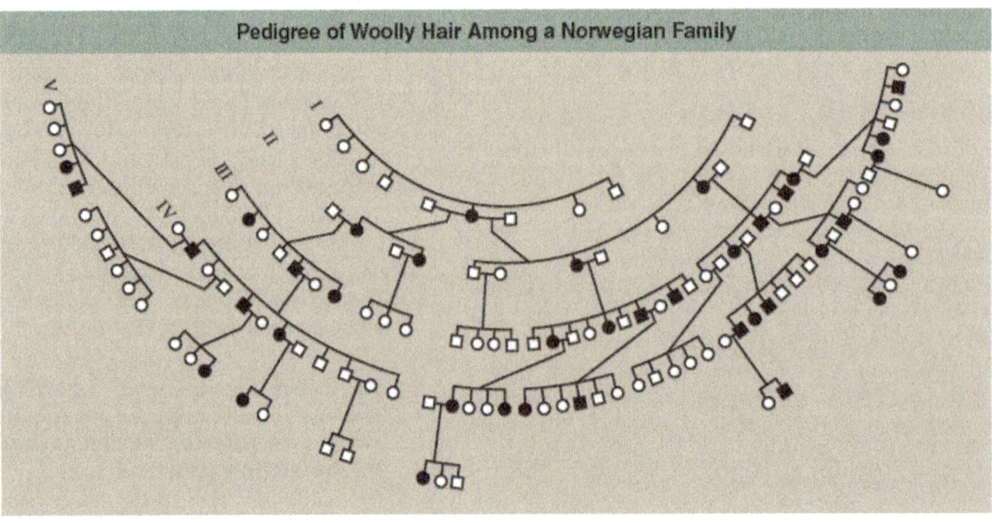

Chapter Review

Mendel

10.1 Mendel and the Garden Pea

- Mendel, shown here from **figure 10.1**, studied inheritance using the garden pea (**figure 10.2**) and the scientific method.
- Mendel used plants that were true-breeding for a particular characteristic; these plants were the P generation. He then crossed two P generation plants that expressed alternate traits (different forms of a characteristic). Their offspring were called the F_1 generation. He then allowed the F_1 plants to self-fertilize, giving rise to the F_2 generation (**figure 10.3**).

10.2 What Mendel Observed

- In Mendel's experiments, the F_1 generation plants all expressed the same alternative form, called the dominant trait. In the F_2 generation, 3/4 of the offspring expressed the dominant trait and 1/4 expressed the other trait, called the recessive trait. Mendel found this 3:1 ratio in the F_2 generation in all the seven traits he studied (**table 10.1**). Mendel then found that this 3:1 ratio was actually a 1:2:1 ratio—1 true-breeding dominant: 2 not-true-breeding dominant: 1 true-breeding recessive (**figure 10.5**).

10.3 Mendel Proposes a Theory

- Mendel's theory of heredity explains that characteristics are passed from parent to offspring as alleles (alternate forms of a characteristic), one allele inherited from each parent. If both of the alleles are the same, the individual is homozygous for the trait. If the individual has one dominant and one recessive allele, it is heterozygous for the trait. An individual's alleles are referred to as its genotype, and the expression of those alleles is its phenotype.
- A Punnett square can be used to predict the probabilities of inheriting certain genotypes and phenotypes in the offspring of a cross (**figures 10.7** and **10.8**).
- A testcross is the mating of an individual of unknown genotype with an individual that is homozygous recessive. It is done to determine if the unknown genotype is homozygous or heterozygous for the dominant trait (**figure 10.9**).

10.4 Mendel's Laws

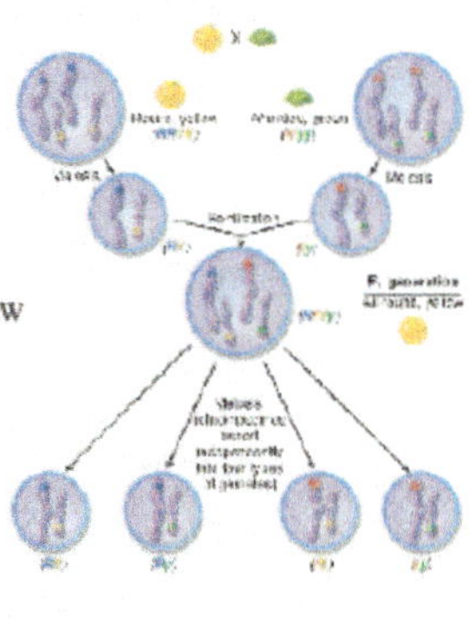

- Mendel's law of segregation states that alleles are distributed into gametes so that half of the gametes will carry one copy of a trait and the remaining gametes carry the other copy of the trait. Mendel's law of independent assortment states that the inheritance of one trait does not influence the inheritance of other traits. Genes located on different chromosomes are inherited independent of each other (**figure 10.10**).

From Genotype to Phenotype

10.5 How Genes Influence Traits

- Genes coded in DNA determine phenotype because DNA encodes the amino acid sequences of proteins, and proteins are the outward expression of genes (**figure 10.11**). Alternative forms of a gene, called alleles, result from mutations.

10.6 Some Traits Don't Show Mendelian Inheritance

- Not all traits follow the inheritance patterns outlined by Mendel. Continuous variation results when more than one gene contributes in a cumulative way to a phenotype, resulting in a continuous array of phenotypes. This pattern of inheritance is called polygenic (**figure 10.12**). Pleiotropic effects result when one gene influences more than one trait (**figure 10.13**). Incomplete dominance results when alternative alleles are not fully dominant or fully recessive such that heterozygous individuals express a phenotype that is intermediate between the dominant and recessive phenotypes (**figure 10.14**). The expression of some genes is influenced by environmental factors (**figure 10.15**), such as the changing of fur color triggered by heat-sensitive alleles. Epistasis occurs when two or more genes interact, having an additive or masking effect (**figure 10.16**) or resulting in several different phenotypes (**figure 10.17**). Codominance occurs when there isn't a dominant allele—two alleles are expressed resulting in phenotypic expression of both alleles (**figures 10.18** and **10.19**).

Chromosomes and Heredity

10.7 Chromosomes Are the Vehicles of Mendelian Inheritance

- Genes assort independently because they are located on chromosomes that assort independently during meiosis. Morgan demonstrated this using an X-linked gene in fruit flies (**figure 10.21**). However, the farther apart two genes are on a chromosome, the more likely they are to segregate independently because of crossing over (**figure 10.22**).

10.8 Human Chromosomes

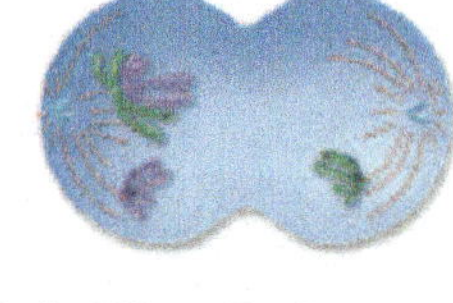

- Humans have 23 pairs of homologous chromosomes: 22 pairs of autosomes and one pair of sex chromosomes. Nondisjunction occurs when sister chromatids or homologous pairs fail to separate during meiosis (**figure 10.24**), resulting in gametes with too many or too few chromosomes. Nondisjunction of autosomes is usually fatal, Down Syndrome being an exception (**figures 10.25** and **10.26**), but the effects of nondisjunction of sex chromosomes are less severe (**figure 10.27**).

Human Hereditary Disorders

10.9 Studying Pedigrees

- By looking at pedigrees, scientists can determine various aspects about the genetics of a trait (**figures 10.28** and **10.29**).

10.10 The Role of Mutation

- Mutations can lead to genetic disorders such as hemophilia (**figure 10.30**), sickle-cell disease (**figures 10.31–10.32**), Tay-Sachs disease (**figure 10.33**), and Huntington's disease (**figure 10.34**).

10.11 Genetic Counseling and Therapy

- Some genetic disorders can be detected during pregnancy using methods such as amniocentesis (**figure 10.35**) and chorionic villus sampling.

Test Your Understanding

1. Gregor Mendel studied the garden pea plants because
   a. pea plants are small, easy to grow, grow quickly, and produce lots of flowers and seeds.
   b. he knew about studies with the garden pea that had been done for hundreds of years, and wanted to continue them, using math—counting and recording differences.
   c. he knew that there were many varieties available with distinctive characteristics.
   d. All of the above.
2. Mendel examined seven characteristics, such as flower color. He crossed plants with two different forms of a character (purple flowers and white flowers). In every case the first generation of offspring (F_1) were
   a. all purple flowers.
   b. half purple flowers and half white flowers.
   c. 3/4 purple and 1/4 white flowers.
   d. all white flowers.
3. Following question 2, when Mendel allowed the F_1 generation to self-fertilize, the offspring in the F_2 generation were
   a. all purple flowers.
   b. half purple flowers and half white flowers.
   c. 3/4 purple and 1/4 white flowers.
   d. all white flowers.
4. Mendel then studied his results, and proposed a set of hypotheses stating that parents transmit
   a. traits directly to their offspring and they are expressed.
   b. some factor, or information, about traits to their offspring and it may or may not be expressed.
   c. some factor, or information, about traits to their offspring and it will always be expressed.
   d. some factor, or information, about traits to their offspring and both traits are expressed in every generation, perhaps in a "blended" form with information from the other parent.
5. A cross between two individuals results in a ratio of 9:3:3:1 for four possible phenotypes. This is an example of a
   a. dihybrid cross.
   b. monohybrid cross.
   c. testcross.
   d. None of these is correct.
6. Human height shows a continuous variation from the very short to the very tall. Height is most likely controlled by
   a. epistatic genes.
   b. environmental factors.
   c. sex-linked genes.
   d. multiple genes.
7. In the human ABO blood grouping, the four basic blood types are type A, type B, type AB, and type O. The blood proteins A and B are
   a. simple dominant and recessive traits.
   b. incomplete dominant traits.
   c. codominant traits.
   d. sex-linked traits.
8. What finding finally determined that genes were carried on chromosomes?
   a. heat sensitivity of certain enzymes that determined coat color
   b. sex-linked eye color in fruit flies
   c. the finding of complete dominance
   d. establishing pedigrees
9. Nondisjunction
   a. occurs when homologous chromosomes or sister chromatids fail to separate during meiosis.
   b. may lead to Down syndrome.
   c. results in aneuploidy.
   d. All of the above.
10. Which of the following analyses can detect aneuploidy?
    a. enzyme activity
    b. chromosomal karyotyping
    c. pedigrees
    d. genetic markers

Apply Your Understanding: Additional Genetics Problems

1. Silky feathers in chickens is a single-gene recessive trait whose effect is to produce shiny plumage. If you had a normal-feathered bird, what would be the easiest cross to perform to determine if a bird is homozygous or heterozygous for the silky allele?
2. Among Hereford cattle there is a dominant allele called *polled;* the individuals that have this allele lack horns. Suppose you acquire a herd consisting entirely of polled cattle, and you carefully determine that no cow in the herd has horns. Some of the calves born that year, however, grow horns. You remove them from the herd and make certain that no horned adult has gotten into your pasture. Despite your efforts, more horned calves are born the next year. What is the reason for the appearance of the horned calves? If your goal is to maintain a herd consisting entirely of polled cattle, what should you do?
3. Brachydactyly is a rare human trait that causes a shortening of the length of the fingers by a third. A review of medical records reveals that the progeny of marriages between a brachydactylous person and a normal person are approximately half brachydactylous. What proportion of offspring in matings between two brachydactylous individuals would be expected to be brachydactylous?
4. Your instructor presents you with a *Drosophila* (fruit fly) with red eyes, as well as a stock of white-eyed flies and another stock of flies homozygous for the red-eye allele. You know that the presence of white eyes in *Drosophila* is caused by homozygosity for a recessive allele. How would you determine whether the single red-eyed fly was heterozygous for the white-eye allele?
5. Hemophilia is a recessive sex-linked human blood disease that leads to failure of blood to clot normally. One form of hemophilia has been traced to the royal family of England, from which it spread throughout the royal families of Europe. For the purposes of this problem, assume that it originated as a mutation either in Prince Albert or in his wife, Queen Victoria.

   a. Prince Albert did not have hemophilia. If the disease is a sex-linked recessive abnormality, how could it have originated in Prince Albert, a male, who would have been expected to exhibit sex-linked recessive traits?

   b. Alexis, the son of Czar Nicholas II of Russia and Empress Alexandra (a granddaughter of Victoria), had hemophilia, but their daughter Anastasia did not. Anastasia died, a victim of the Russian revolution, before she had any children. Can we assume that Anastasia would have been a carrier of the disease? Would your answer be different if the disease had been present in Nicholas II or in Alexandra?
6. A normally pigmented man marries an albino woman. They have three children, one of whom is an albino. What is the genotype of the father?

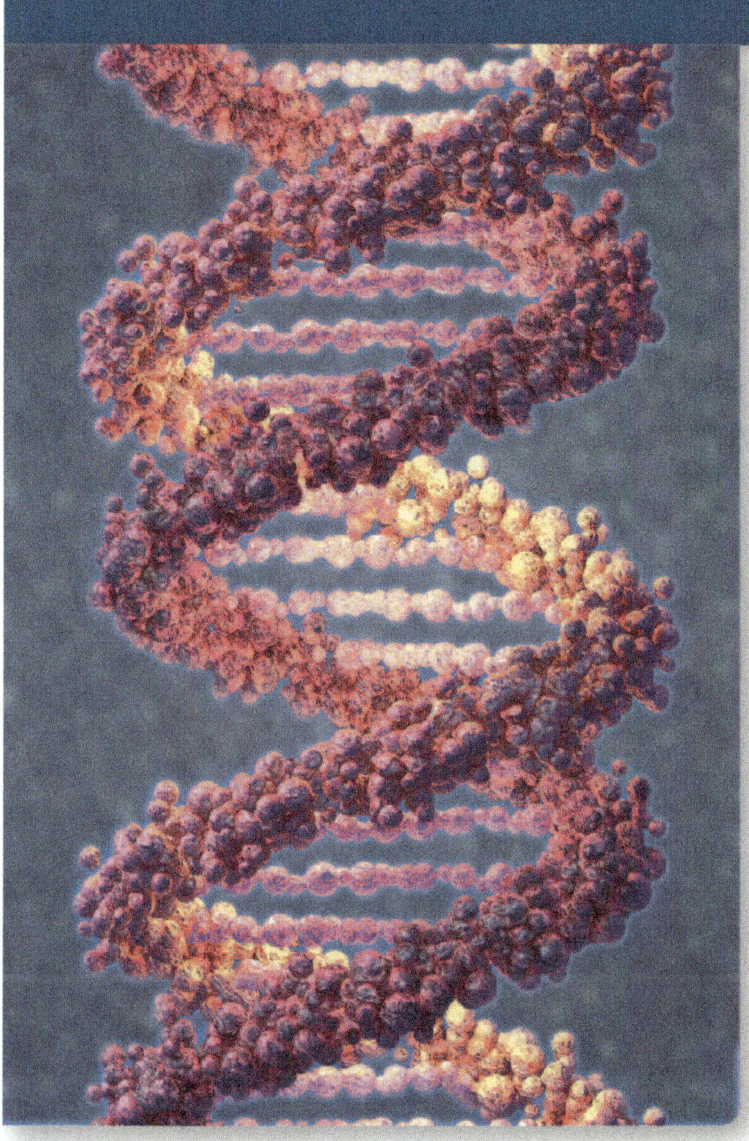

11

DNA: The Genetic Material

Learning Objectives

Genes Are Made of DNA

11.1 The Discovery of Transformation

1. Identify the two macromolecules found in chromosomes.
2. Describe Griffith's experiment demonstrating transformation.
3. Explain why Griffith's mixture of harmless dead bacteria with capsules and harmless live bacteria lacking capsules is deadly.

11.2 Experiments Identifying DNA as the Genetic Material

1. List the five ways that Avery's transforming principle resembles DNA.
2. Describe the Hershey-Chase experiment, and explain how it demonstrated that DNA is the hereditary material.

11.3 Discovering the Structure of DNA

1. Name the four DNA nucleotides, and identify which are purines.
2. Distinguish purine from pyrimidine.
3. State Chargaff's rule.
4. Describe Watson and Crick's proposed structure for the DNA molecule.
5. Explain why each nucleotide in Watson and Crick's DNA structure can form a base pair with only one of the four potential nucleotides.

DNA Replication

11.4 How the DNA Molecule Copies Itself

1. Diagram three alternative mechanisms of DNA replication.
2. Describe the Meselson-Stahl experiment, and explain how it confirms one of these mechanisms.
3. Name and describe the four stages of DNA replication.
4. Distinguish between the leading and lagging strands; define Okazaki fragments.
5. Explain the function of RNA primers and of ligases.

Altering the Genetic Message

11.5 Mutation

1. Contrast the inheritance of somatic and germ-line mutations.
2. List four molecular events that can alter the sequence of DNA.
3. Distinguish between transposition and chromosomal rearrangement.

Today's Biology: DNA Fingerprinting
Today's Biology: Tracing the DNA of Irish Kings
Biology and Staying Healthy: Protecting Your Genes

Inquiry & Analysis: Are Mutations Random or Directed by the Environment?

The realization that patterns of heredity can be explained by the segregation of chromosomes in meiosis raised a question that occupied biologists for over 50 years: What is the exact nature of the connection between hereditary traits and chromosomes? In this chapter you will examine some of the chain of experiments that have led to our current understanding of the molecular mechanisms of heredity. The experiments determining that DNA is the genetic material are among the most elegant in science. Just as in a good detective story, each conclusion has led to new questions. The intellectual path taken has not always been a straight one, the best questions not always obvious. But however erratic and lurching the course of the experimental journey, our picture of heredity has become progressively clearer, the image more sharply defined. We now understand in considerable detail how the DNA molecule copies itself, and how changes to it lead to hereditary gene mutations.

Genes Are Made of DNA

11.1 The Discovery of Transformation

The Griffith Experiment

As we learned in chapters 8, 9, and 10, chromosomes contain genes, which, in turn, contain hereditary information. However, Mendel's work left a key question unanswered: What *is* a gene? When biologists began to examine chromosomes in their search for genes, they soon learned that chromosomes are made of two kinds of macromolecules, both of which you encountered in chapter 3: **proteins** (long chains of *amino acid* subunits linked together in a string) and **DNA** (deoxyribonucleic acid—long chains of *nucleotide* subunits linked together in a string). It was possible to imagine that either of the two was the stuff that genes are made of—information might be stored in a sequence of different amino acids, or in a sequence of different nucleotides. But which one is the stuff of genes, protein or DNA? This question was answered clearly in a variety of different experiments, all of which shared the same basic design: If you separate the DNA in an individual's chromosomes from the protein, which of the two materials is able to change another individual's genes?

In 1928, British microbiologist Frederick Griffith made a series of unexpected observations while experimenting with pathogenic (disease-causing) bacteria. Figure 11.1 takes you stepwise through his discoveries. When he infected mice with a virulent strain of *Streptococcus pneumoniae* bacteria (then known as *Pneumococcus*), the mice died of blood poisoning, as you can see in panel 1. However, when he infected similar mice with a mutant strain of *S. pneumoniae* that lacked the virulent strain's polysaccharide capsule, the mice showed no ill effects, as you can see in panel 2. The capsule was apparently necessary for infection. The normal pathogenic form of this bacterium is referred to as the *S form* because it forms smooth colonies in a culture dish. The mutant form, which lacks an enzyme needed to manufacture the polysaccharide capsule, is called the *R form* because it forms rough colonies.

To determine whether the polysaccharide capsule itself had a toxic effect, Griffith injected dead bacteria of the virulent S strain into mice and as panel 3 shows, the mice remained perfectly healthy. Finally, as shown in panel 4, he injected mice with a mixture containing dead S bacteria of the virulent strain and live, capsuleless R bacteria, each of which by itself did not harm the mice. Unexpectedly, the mice developed disease symptoms and many of them died. The blood of the dead mice was found to contain high levels of live, virulent *Streptococcus* type S bacteria, which had surface proteins characteristic of the live (previously R) strain. Somehow, the information specifying the polysaccharide capsule had passed from the dead, virulent S bacteria to the live, capsuleless R bacteria in the mixture, permanently transforming the capsuleless R bacteria into the virulent S variety.

Key Learning Outcome 11.1 Hereditary information can pass from dead cells to living ones and transform them.

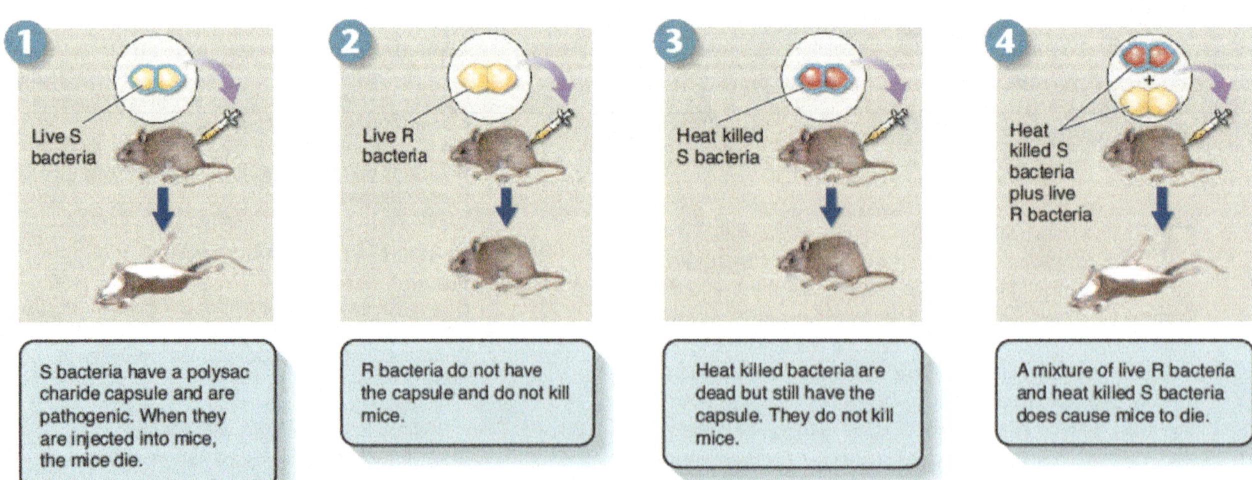

Figure 11.1 How Griffith discovered transformation.

Transformation, the movement of a gene from one organism to another, provided some of the key evidence that DNA is the genetic material. Griffith found that extracts of dead pathogenic strains of the bacterium *Streptococcus pneumoniae* can "transform" live harmless strains into live pathogenic strains.

11.2 Experiments Identifying DNA as the Genetic Material

The Avery Experiments

The agent responsible for transforming *Streptococcus* went undiscovered until 1944. In a classic series of experiments, Oswald Avery and his coworkers Colin MacLeod and Maclyn McCarty characterized what they referred to as the "transforming principle." Avery and his colleagues prepared the same mixture of dead S *Streptococcus* and live R *Streptococcus* that Griffith had used, but first they removed as much of the protein as they could from their preparation of dead S *Streptococcus*, eventually achieving 99.98% purity. Despite the removal of nearly all protein from the dead S *Streptococcus*, the transforming activity was not reduced. Moreover, the properties of the transforming principle resembled those of DNA in several ways:

Same chemistry as DNA. When the purified principle was analyzed chemically, the array of elements agreed closely with DNA.

Same behavior as DNA. In an ultracentrifuge, the transforming principle migrated like DNA; in electrophoresis and other chemical and physical procedures, it also acted like DNA.

Not affected by lipid and protein extraction. Extracting the lipid and protein from the purified transforming principle did not reduce its activity.

Not destroyed by protein- or RNA-digesting enzymes. Protein-digesting enzymes did not affect the principle's activity, nor did RNA-digesting enzymes.

Destroyed by DNA-digesting enzymes. The DNA-digesting enzyme destroyed all transforming activity.

The evidence was overwhelming. They concluded that "a nucleic acid of the deoxyribose type is the fundamental unit of the transforming principle of *Pneumococcus* Type III"—in essence, that DNA is the hereditary material.

The Hershey-Chase Experiment

Avery's result was not widely appreciated at first because most biologists still preferred to think that genes were made of proteins. In 1952, however, a simple experiment carried out by Alfred Hershey and Martha Chase was impossible to ignore. The team studied the genes of viruses that infect bacteria. These viruses attach themselves to the surface of bacterial cells and inject their genes into the interior; once inside, the genes take over the genetic machinery of the cell and order the manufacturing of hundreds of new viruses. When mature, the progeny viruses burst out to infect other cells. These bacteria-infecting viruses have a very simple structure: a core of DNA surrounded by a coat of protein.

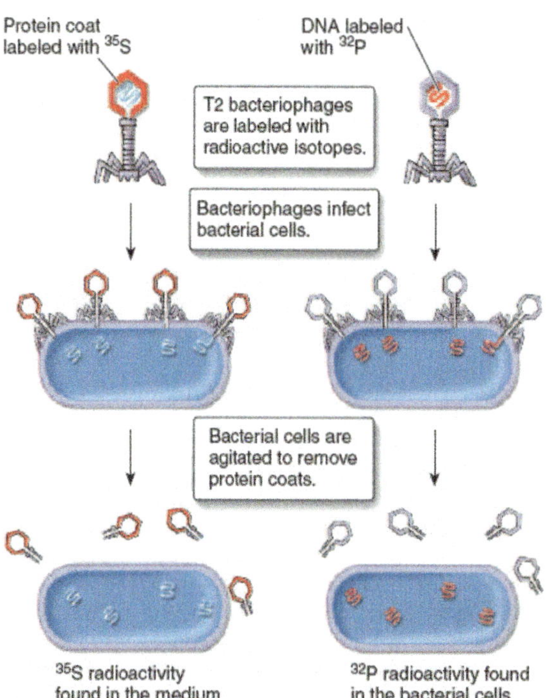

Figure 11.2 The Hershey-Chase experiment.
The experiment that convinced most biologists that DNA is the genetic material was carried out soon after World War II, when radioactive isotopes were first becoming commonly available to researchers. Hershey and Chase used different radioactive labels to "tag" and track protein and DNA. They found that when bacterial viruses inserted their genes into bacteria to guide the production of new viruses, ^{35}S radioactivity did not enter infected bacterial cells and ^{32}P radioactivity did. Clearly the virus DNA, not the virus protein, was responsible for directing the production of new viruses.

In this experiment, shown in figure 11.2, Hershey and Chase used radioactive isotopes to "label" the DNA and protein of the viruses. Radioactively tagged molecules are indicated in red in the figure. In the preparation on the right, the viruses were grown so that their DNA contained radioactive phosphorus (^{32}P); in another preparation, on the left side of the figure, the viruses were grown so that their protein coats contained radioactive sulfur (^{35}S). After the labeled viruses were allowed to infect bacteria, Hershey and Chase shook the suspensions forcefully to dislodge attacking viruses from the surface of bacteria, used a rapidly spinning centrifuge to isolate the bacteria, and then asked a very simple question: What did the viruses inject into the bacterial cells, protein or DNA? They found that the bacterial cells infected by viruses containing the ^{32}P label had labeled tracer in their interiors; cells infected by viruses containing the ^{35}S labeled tracer did not. The conclusion was clear: The genes that viruses use to specify new viruses are made of DNA and not protein.

Key Learning Outcome 11.2 Several key experiments demonstrated conclusively that DNA, not protein, is the hereditary material.

11.3 Discovering the Structure of DNA

As it became clear that DNA was the molecule that stored the hereditary information, researchers began to question how this nucleic acid could carry out the complex function of inheritance. At the time, investigators did not know what the DNA molecule looked like.

We now know that DNA is a long, chainlike molecule made up of subunits called **nucleotides.** As you can see in figure 11.3, each nucleotide has three parts: a central sugar called *deoxyribose*, a phosphate (PO_4) group, and an organic base. The sugar (the lavender pentagon structure) and the phosphate group (the yellow-circled structure) are the same in every nucleotide of DNA. However, there are four different kinds of bases: two large ones with double-ring structures, and two small ones with single rings. The large bases, called **purines,** are **A** (adenine) and **G** (guanine). The small bases, called **pyrimidines,** are **C** (cytosine) and **T** (thymine). A key observation, made by Erwin Chargaff, was that DNA molecules always had equal amounts of purines and pyrimidines. In fact, with slight variations due to imprecision of measurement, the amount of A always equals the amount of T, and the amount of G always equals the amount of C. This observation (A = T, G = C), known as **Chargaff's rule,** suggested that DNA had a regular structure.

The significance of Chargaff's rule became clear in 1953 when the British chemist Rosalind Franklin carried out an X-ray diffraction experiment. In these experiments, DNA molecules are bombarded with X-ray beams, and when individual rays encounter atoms, their paths are bent or diffracted like a thrown ball bounces off or around an object. Each atomic encounter creates a pattern on photographic film, shown in figure 11.4*a*, that looks like the ripples created by tossing a rock into a smooth lake. Franklin's results suggested that the DNA molecule had the shape of a coiled spring or a corkscrew, a form called a **helix,** with the image in the photo from the viewpoint of looking down the center of the molecule.

Franklin's work was shared with two researchers at Cambridge University, Francis Crick and James Watson, before it was published. Using Tinkertoy-like models of the bases, Watson and Crick deduced the true structure of DNA (figure 11.4*b*): The DNA molecule is a **double helix,** a winding staircase of two strands whose bases face one another (figure 11.4*c*). Chargaff's rule is a direct reflection of this structure—every bulky purine on one strand is paired with a slender pyrimidine on the other strand. Specifically, A (the blue bases) pairs with T (the orange bases), and G (the purple bases) pairs with C (the pink bases). Because hydrogen bonds, shown as dotted lines, can form between the **base pairs,** the molecule keeps a constant thickness.

Key Learning Outcome 11.3 The DNA molecule consists of two strands of nucleotides held together by hydrogen bonds between bases. The two strands wind into a double helix.

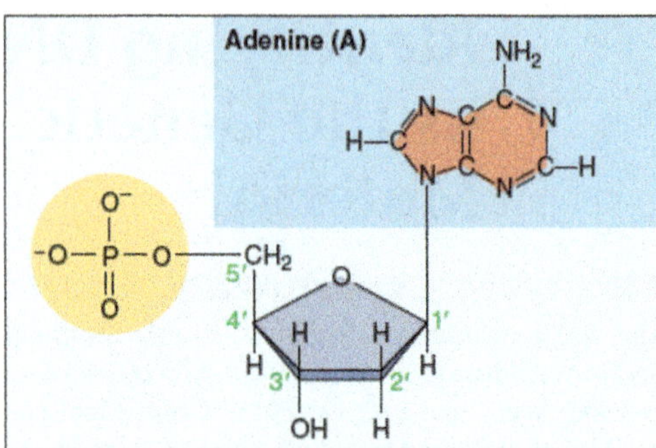

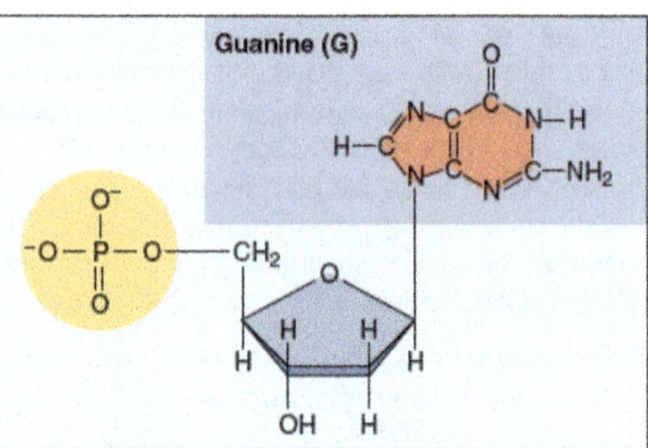

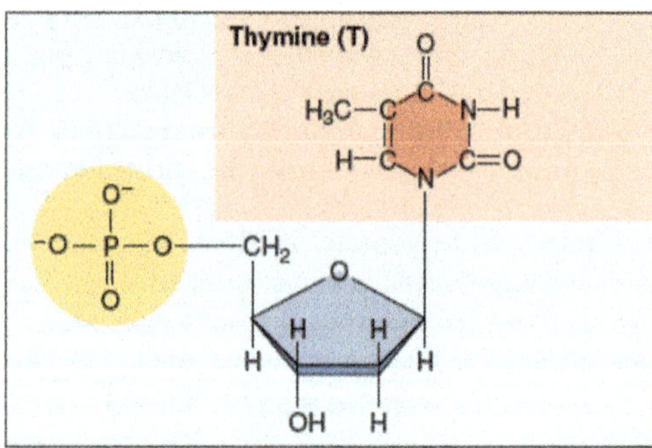

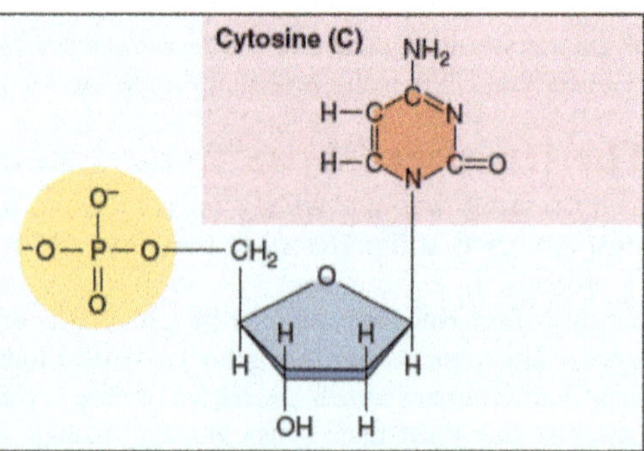

Figure 11.3 The four nucleotide subunits that make up DNA.
The nucleotide subunits of DNA are composed of three parts: a central five-carbon sugar called deoxyribose, a phosphate group, and an organic, nitrogen-containing base.

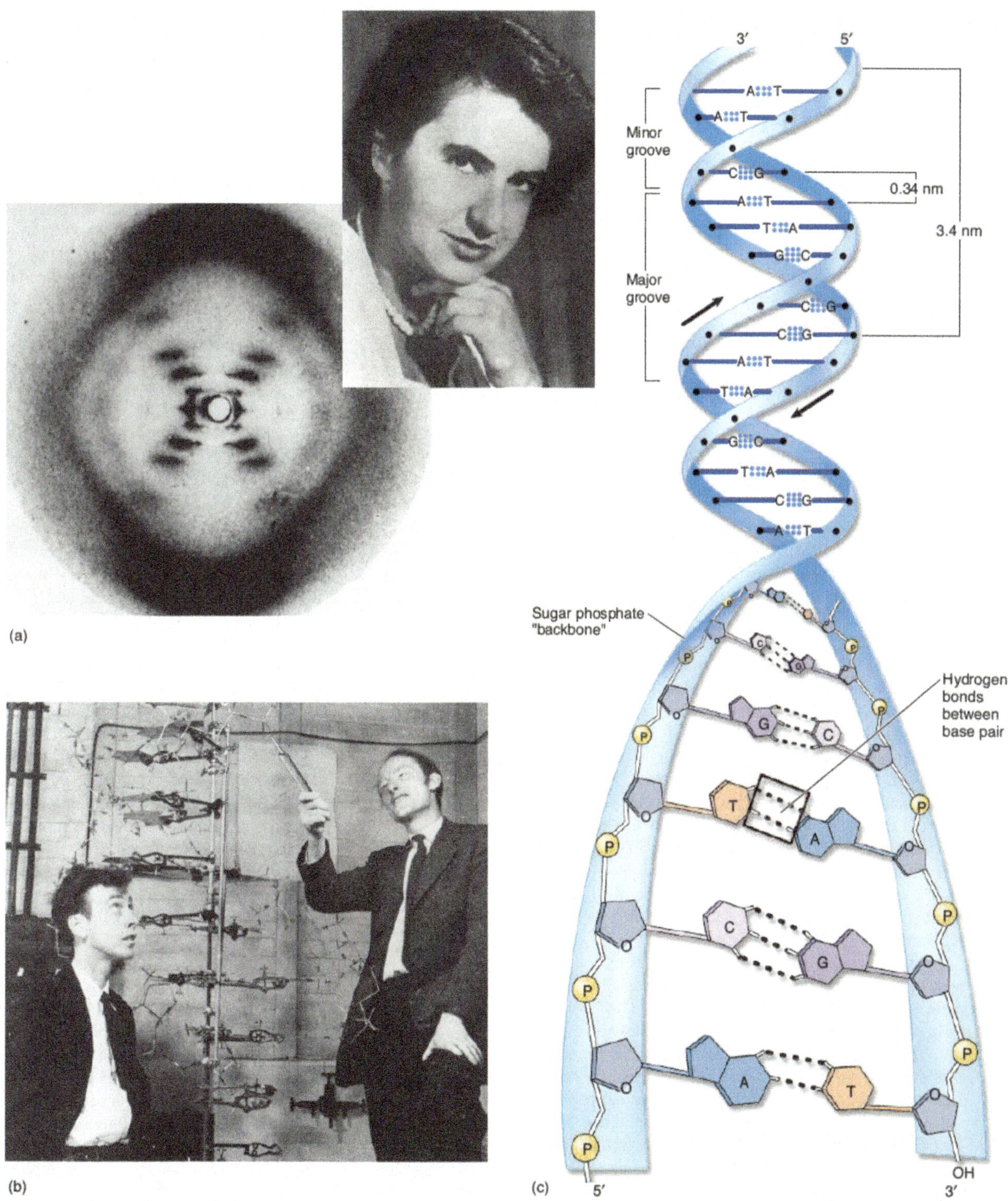

Figure 11.4 The DNA double helix.

(*a*) This X-ray diffraction photograph was made in 1953 by Rosalind Franklin (inset) in the laboratory of Maurice Wilkins. It suggested to Watson and Crick that the DNA molecule was a helix, like a winding staircase. (*b*) In 1953 Watson and Crick deduced the structure of DNA. James Watson (seated and peering up at their homemade model of the DNA molecule) was a young American postdoctoral student, and Francis Crick (pointing) was an English scientist. (*c*) The dimensions of the double helix were suggested by the X-ray diffraction studies. In a DNA duplex molecule, only two base pairs are possible: adenine (A) with thymine (T) and guanine (G) with cytosine (C). A G–C base pair has three hydrogen bonds; an A–T base pair has only two.

DNA Replication

11.4 How the DNA Molecule Copies Itself

The attraction that holds the two DNA strands together is the formation of weak hydrogen bonds between the bases that face each other from the two strands. That is why A pairs with T and not C; A can only form hydrogen bonds with T. Similarly, G can form hydrogen bonds with C but not T. In the Watson-Crick model of DNA, the two strands of the double helix are said to be *complementary* to each other. One chain of the helix can have any sequence of bases, of A, T, G, and C, but this sequence completely determines that of its partner in the helix. If the sequence of one chain is ATTGCAT, the sequence of its partner in the double helix must be TAACGTA. Each chain in the helix is a complementary mirror image of the other. This **complementarity** makes it possible for the DNA molecule to copy itself during cell division in a very direct manner. But, there are three possible alternatives as to how the DNA could serve as a template for the assembly of new DNA molecules.

First, the two strands of the double helix could separate and serve as templates for the assembly of two new strands by base pairing A with T and G with C. This is what happens in figure 11.5*a*, with the original strand colored blue and the newly formed strands red. After replicating, the original strands rejoin, preserving the original strand of DNA and forming an entirely new strand. This is called *conservative replication.*

In the second alternative, the double helix need only "unzip" and assemble a new complementary chain along each single strand. This form of DNA replication is called *semiconservative replication,* because while the sequence of the original duplex is conserved after one round of replication, the duplex itself is not. Instead, each strand of the duplex becomes part of another duplex. You can see in figure 11.5*b* that the blue strand is from the original helix and the red strand is newly formed.

In the third alternative, called *dispersive replication,* the original DNA would serve as a template for the formation of new DNA strands but the new and old DNA would be dispersed among the two daughter strands. As shown in figure 11.5*c*, each daughter strand is made up of sections of original (blue) strands and new (red) strands.

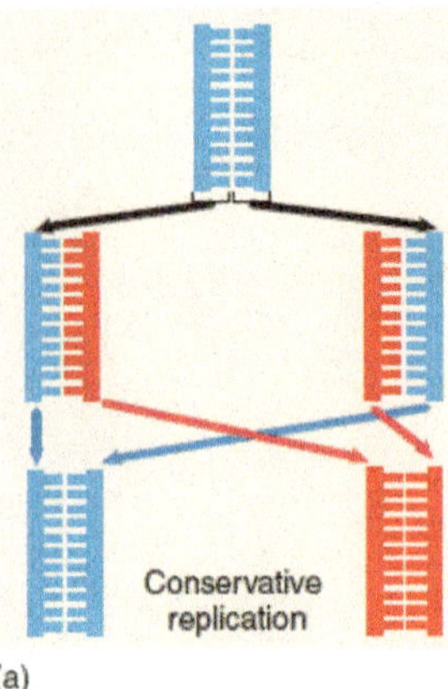

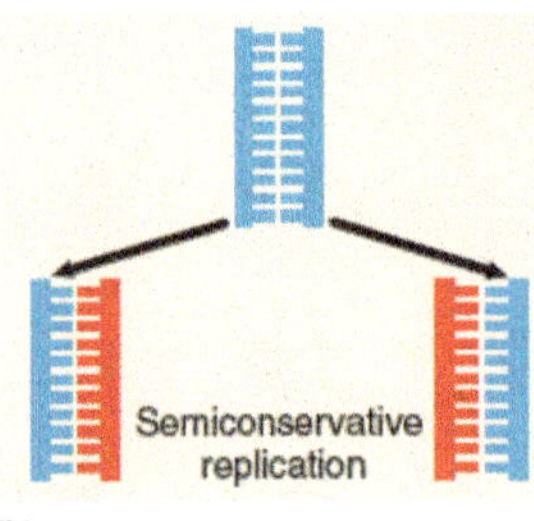

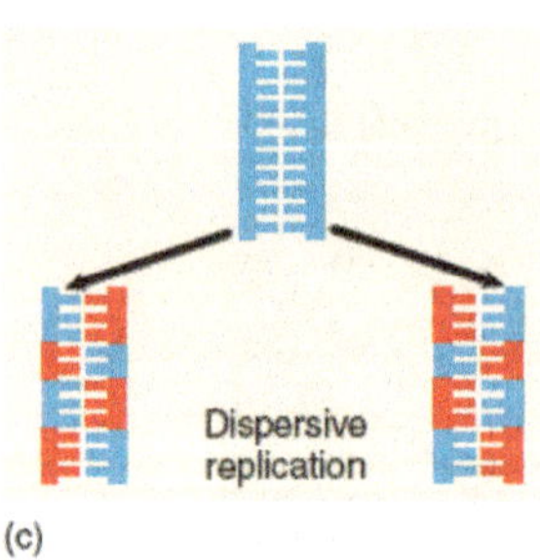

Figure 11.5 Alternative mechanisms of DNA replication.

The Meselson-Stahl Experiment

The three alternative hypotheses of DNA replication were tested in 1958 by Matthew Meselson and Franklin Stahl of the California Institute of Technology. These two scientists grew bacteria in a medium containing the heavy isotope of nitrogen, ^{15}N, which became incorporated into the bases of the bacterial DNA (the upper petri dish in figure 11.6). After several generations, samples were taken from this culture and grown in a medium containing the normal lighter isotope ^{14}N, which became incorporated into the newly replicating DNA. Bacterial samples were taken from the ^{14}N media at 20 minute intervals (❷ through ❹). DNA was extracted from all three samples and a fourth sample, ❶, that served as a control.

By dissolving the DNA they had collected in a heavy salt called cesium chloride, and then spinning the solution at very high speeds in an ultracentrifuge, Meselson and Stahl were able to separate DNA strands of different densities. The centrifugal forces caused the cesium ions to migrate toward the bottom of the centrifuge tube, creating a gradient of cesium concentration, and thus a gradation of density. Each DNA strand floats or sinks in the gradient until it reaches the position where its density exactly matches the density of the cesium there. Because ^{15}N strands are denser than ^{14}N strands, they migrate farther down the tube to a denser region of cesium.

The DNA collected immediately after the transfer was all dense, as shown in test tube ❷. However, after the bacteria completed their first round of DNA replication in the ^{14}N medium, the density of their DNA had decreased to a value intermediate between ^{14}N-DNA and ^{15}N-DNA, as shown in test tube ❸. After the second round of replication, two density classes of DNA were observed, one intermediate and one equal to that of ^{14}N-DNA, as shown in test tube ❹.

Meselson and Stahl interpreted their results as follows: After the first round of replication, each daughter DNA duplex was a hybrid possessing one of the heavy strands of the parent molecule and one light strand; when this hybrid duplex replicated, it contributed one heavy strand to form another hybrid duplex and one light strand to form a light duplex. Thus, this experiment clearly ruled out conservative and dispersive DNA replication, and confirmed the prediction of the Watson-Crick model that DNA replicates in a semiconservative manner.

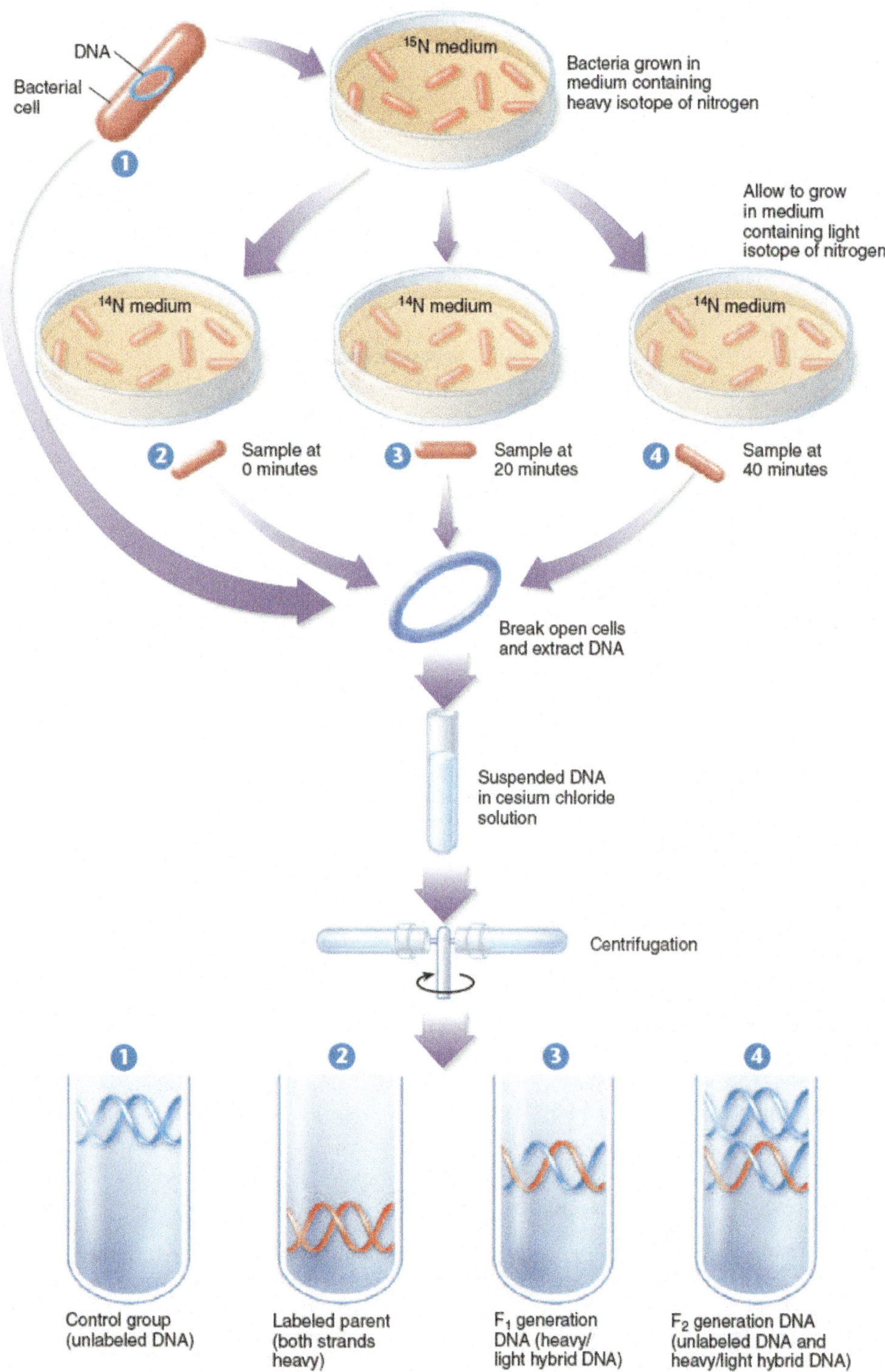

Figure 11.6 The Meselson-Stahl experiment.

Bacterial cells were grown for several generations in a medium containing a heavy isotope of nitrogen (^{15}N) and then were transferred to a new medium containing the normal lighter isotope (^{14}N). (The bacteria shown here are not drawn to scale, as tens of thousands of bacterial cells grow on even a tiny portion of a plate in culture.) At various times thereafter, samples of the bacteria were collected, and their DNA was dissolved in a solution of cesium chloride, which was spun rapidly in a centrifuge. The labeled and unlabeled DNA settled in different areas of the tube because they differed in weight. The DNA with two heavy strands settled down toward the bottom of the tube. The DNA with two light strands settled higher up in the tube. The DNA with one heavy and one light strand settled in between the other two.

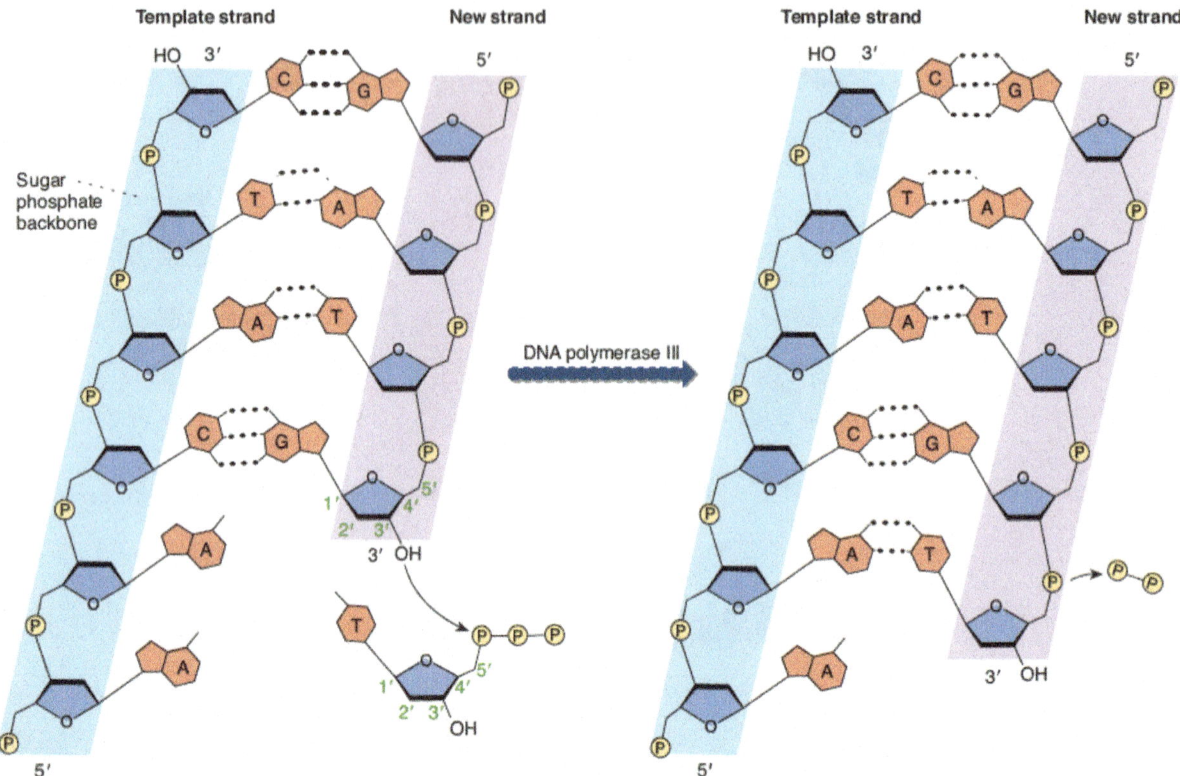

Figure 11.7 How nucleotides are added in DNA replication.
In a nucleotide, the phosphate group is attached to the 5′ carbon atom of the sugar, and an OH group is attached to the 3′ carbon atom. So, on a DNA strand, there will be a 5′ phosphate at one end of the chain and a 3′ OH at the other end. In the DNA double helix, the two strands of nucleotides pair up in opposite orientations, with one strand running 5′ to 3′ and the other running 3′ to 5′. When nucleotides are added to a growing strand of DNA by the enzyme DNA polymerase III, the first phosphate group of the incoming nucleotide attaches to the OH group on the end nucleotide of the existing strand.

How DNA Copies Itself

The copying of DNA before cell division is called **DNA replication.** This process is overseen by six proteins in prokaryotes (in eukaryotes, some of the enzymes are different). These proteins coordinate the unwinding of the DNA duplex and assembly of new complementary DNA strands by the addition of nucleotides to existing strands (figure 11.7). Here is how the process works.

1 Unwinding

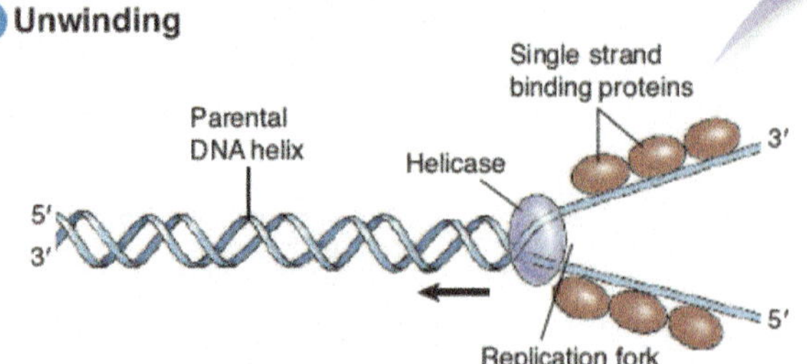

Before DNA replication can begin, an enzyme called *helicase* separates and unwinds the strands of the parental DNA. Single-strand binding proteins stabilize the single-stranded regions of DNA before they are replicated. Helicase moves up the DNA helix, unwinding as it goes.

2 Priming the Leading Strand

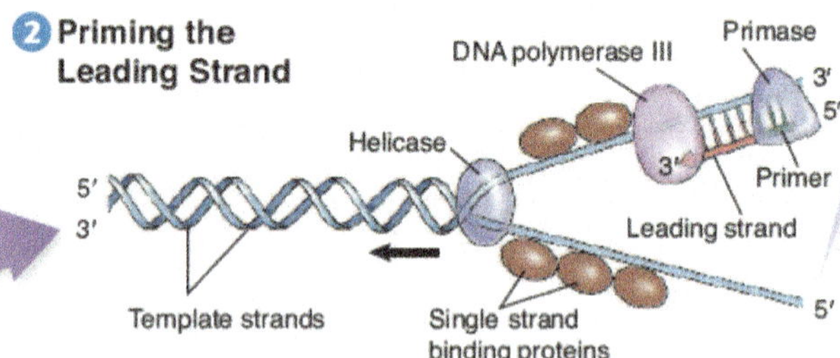

After the parental DNA duplex is unwound, an enzyme complex called *DNA polymerase III* can add nucleotides complementary to the exposed template strand of DNA. But this enzyme cannot begin a new strand; it can only add nucleotides to an existing strand. Thus, before a new strand of DNA can be built, an enzyme called *primase* must synthesize a short section of joined RNA nucleotides, called a *primer,* complementary to the single-stranded template. DNA polymerase III can then add onto the primer and assemble a complementary new strand of DNA on each old strand. One of the new strands of DNA is called the *leading strand;* it is the one that is extended in a 5′ to 3′ direction toward the replication fork.

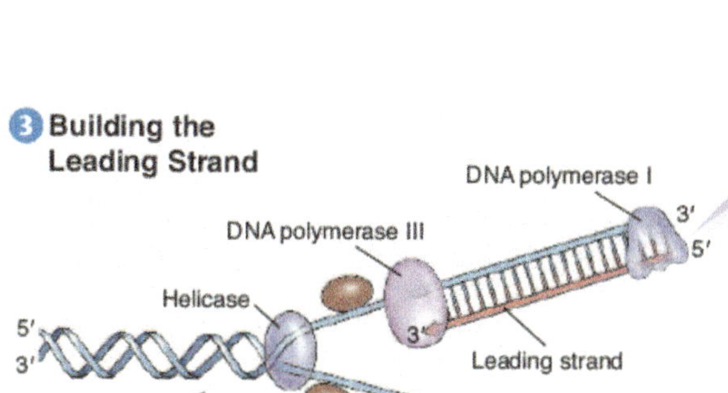

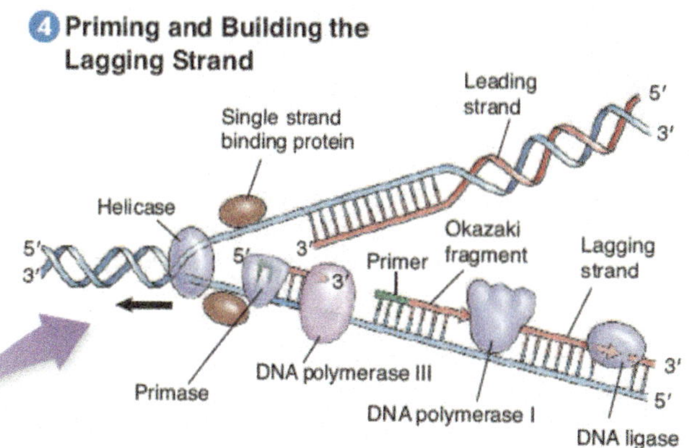

DNA polymerase III builds the leading strand by adding nucleotides to the 3′ end of the strand: the 5′ phosphate end of a nucleotide to be added attaches to the 3′ sugar end of the existing strand. DNA polymerase III moves toward the replication fork and builds the leading strand as one continuous strand. Before the leading strand can be completed, another polymerase enzyme called *DNA polymerase I* removes the RNA primer and fills in the gap with DNA nucleotides. The newly synthesized hybrid DNA can then rewind into a helix.

Because DNA polymerase III can only assemble new strands in the 5′ to 3′ direction, the other strand, called the *lagging strand,* is assembled in short 5′ to 3′ segments, moving away from the replication fork. Each lagging strand segment begins with an RNA primer, and then DNA polymerase III builds away from the replication fork until it encounters the previously synthesized section. These short stretches of newly synthesized DNA on the lagging strand are called *Okazaki fragments.* As the helix opens further, a new RNA primer is added, and DNA polymerase III must release the template it has completed and begin with the "new" template. The Okazaki fragments are then linked when DNA polymerase I removes the RNA primers and an enzyme called *DNA ligase* joins the ends of the newly synthesized segments of DNA. The entire lagging strand can only be replicated in this discontinuous fashion.

Eukaryotic chromosomes each contain a single, very long molecule of DNA (figure 11.8), one far too long to copy all the way from one end to the other with a single replication fork. Each eukaryotic chromosome is instead copied in sections of about 100,000 nucleotides, each with its own replication origin and fork.

The enormous amount of DNA that resides within the cells of your body represents a long series of DNA replications, starting with the DNA of a single cell—the fertilized egg. Living cells have evolved many mechanisms to avoid errors during DNA replication and to preserve the DNA from damage. These mechanisms of **DNA repair** proofread the strands of each daughter cell against one another for accuracy and correct any mistakes. But the proofreading is not perfect. If it were, no mistakes such as mutations would occur, no variation in gene sequence would result, and evolution would come to a halt. Mutation will be discussed in more detail in the next section and in chapter 14.

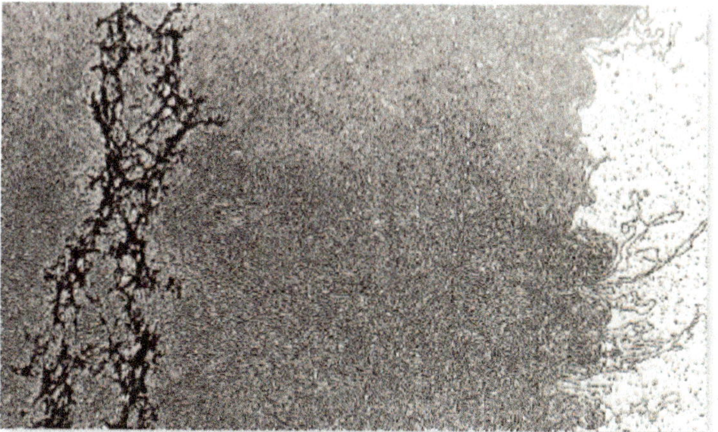

Figure 11.8 DNA of a single chromosome.

This chromosome has been relieved of most of its packaging proteins, leaving the DNA in its extended form. The residual protein scaffolding appears as the dark material on the left side of the micrograph.

Key Learning Outcome 11.4 The basis for the great accuracy of DNA replication is complementarity. DNA's two strands are complementary mirror images of each other, so either one can be used as a template to reconstruct the other.

Altering the Genetic Message

11.5 Mutation

There are two general ways in which the genetic message is altered: mutation and recombination. A change in the content of the genetic message—the base sequence of one or more genes—is referred to as a **mutation.** As you learned in the previous section, DNA copies itself by forming complementary strands along single strands of DNA when they are separated. The template strand directs the formation of the new strand. However, this replication process is not foolproof. Sometimes errors are made and these are called mutations. Some mutations alter the identity of a particular nucleotide, while others remove or add nucleotides to a gene. A change in the position of a portion of the genetic message is referred to as **recombination.** Some recombination events move a gene to a different chromosome; others alter the location of only part of a gene. The cells of eukaryotes contain an enormous amount of DNA, and the mechanisms that protect and proofread the DNA are not perfect. If they were, no variation would be generated.

Mistakes Happen

In fact, cells do make mistakes during replication, as shown in figure 11.9. And mutations can also occur because of DNA alteration by chemicals, like those in cigarette smoke, or by radiation, like the ultraviolet light from the sun or tanning beds. However, mutations are rare. In humans, sequencing the genomes of an entire family has revealed that only about 60 out of the 3 billion nucleotides of the genome are altered by mutation each generation. If changes were common, the genetic instructions encoded in DNA would soon degrade into meaningless gibberish. Limited as it might seem, the steady trickle of change that does occur is the very stuff of evolution. Every difference in the genetic messages that specify different organisms arose as the result of genetic change.

Figure 11.9 Mutation.
Fruit flies normally have one pair of wings, extending from the thorax. This fly is a *bithorax* mutant. Because of a mutation in a gene regulating a critical stage of development, it possesses two thorax segments and thus two sets of wings.

Kinds of Mutation

The message that DNA carries in its genes is the "instructions" of how to make proteins. The sequence of nucleotides in a strand of DNA translates into the sequence of amino acids that makes up a protein. This process was introduced in section 10.5 on page 194 and will be described in more detail in chapter 12. If the core message in the DNA is altered through mutation, as shown by the substitution of T (in red) for G in figure 11.10, then the protein product can also be altered, sometimes to the point where it can no longer function properly. Because mutations can occur randomly in a cell's DNA, most mutations are detrimental, just as making a random change in a computer program usually worsens performance. The consequences of a detrimental mutation may be minor or catastrophic, depending on the function of the altered gene.

Mutations in Germ-Line Tissues The effect of a mutation depends critically on the identity of the cell in which the mutation occurs. During the embryonic development of all multicellular organisms, there comes a point when cells destined to form gametes (germ-line cells) are segregated from those that will form the other cells of the body (somatic cells). Only when a mutation occurs within a germ-line cell is it passed to subsequent generations as part of the hereditary endowment of the gametes derived from that cell. Mutations in germ-line

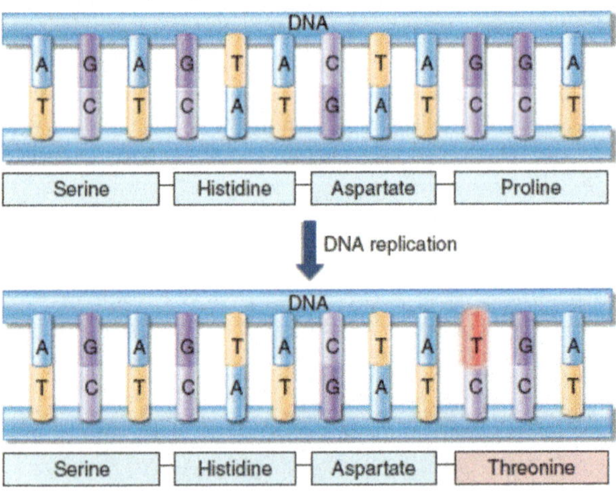

(a) Base substitution (red) in DNA: changes G to T in the DNA strand and, as a result, proline to threonine in the protein.

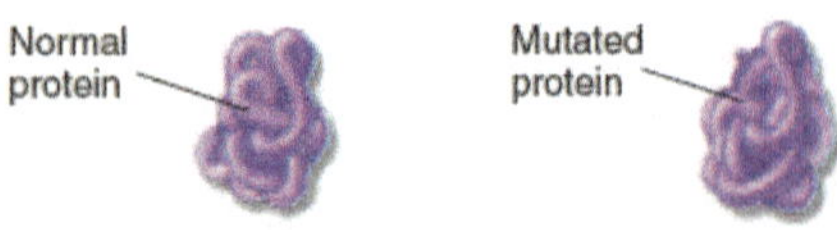

(b) The mutated protein with the amino acid substitute folds differently than the normal protein and its function will most likely be affected.

Figure 11.10 Base substitution mutation.
(*a*) Some changes in a DNA sequence can result in a change in a single amino acid. (*b*) This results in a mutated protein that may not function the same as the normal protein.

Today's *Biology*

DNA Fingerprinting

Only identical twins have exactly the same DNA sequence. All other people differ from one another at many sites. In 1985 British geneticist Alec Jeffreys took advantage of this to develop a new forensic (that is, crime scene investigation) tool, DNA fingerprinting. The procedure involves cleaving an individual's DNA into small bits, which are spread apart on a gel to yield a pattern of bands, a "DNA fingerprint" characteristic of that person. The photo on the right shows the DNA fingerprints a prosecuting attorney presented in a rape trial in 1987. They consisted of autoradiographs, parallel bars on X-ray film. Each bar represents the position of a DNA fragment produced by techniques that will be described in more detail in chapter 13. The dark lanes with many bars represent standardized controls. Two different ways of producing the DNA fragments are shown, each highlighting particular sequences. A sample had been taken from the victim within hours of her attack; from it semen was collected and the semen DNA analyzed for its patterns.

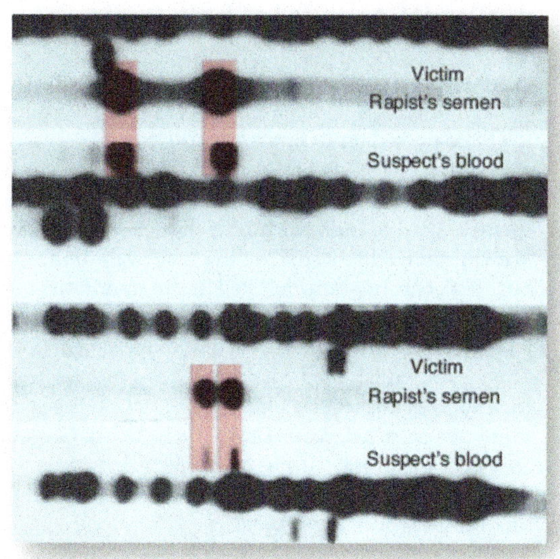

Compare the DNA fingerprint patterns of the semen to that of the suspect. You can see that the suspect's two patterns match that of the rapist, and these patterns are quite different from those of the victim. Clearly the semen collected from the rape victim and the blood sample from the suspect came from the same person. The suspect was Tommie Lee Andrews, and on November 6, 1987, the jury returned a verdict of guilty.

Since the Andrews verdict, DNA fingerprinting has been used as evidence in many court cases. DNA can be obtained at a crime scene from several different sources, such as small amounts of blood, hair, or semen. As the man who analyzed Andrews's DNA says: "It's like leaving your name, address, and social security number at the scene of the crime. It's that precise."

While some ways of detecting DNA differences highlight profiles shared by many people, others are quite rare. Using several, identity can be clearly established or ruled out. The DNA profiles of O. J. Simpson and blood samples from the murder scene of his former wife from his highly publicized and controversial murder trial in 1995 are presented on the right.

DNA fingerprinting is certainly not restricted to prosecution. It can also be used to establish innocence. More than 120 convicted people have been freed in the last 14 years using DNA evidence presented by The Innocence Project lawyers, for example.

Of course, the procedures involved in creating the DNA fingerprints and in analyzing them must be carried out properly—sloppy procedures could lead to a wrongful conviction. After widely publicized instances of questionable lab procedures, national standards have been developed.

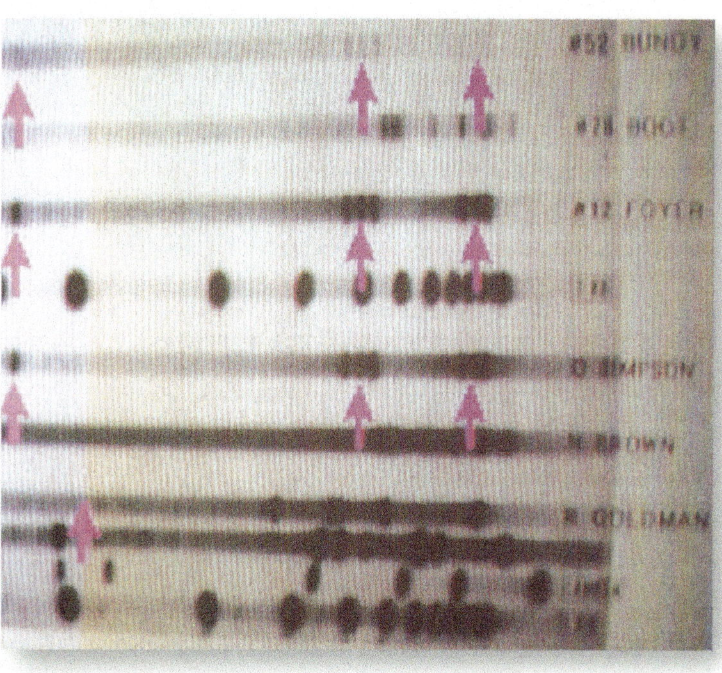

Today's *Biology*

Tracing the DNA of Irish Kings

Every time a mutation occurs in germ-line DNA—the DNA producing egg or sperm—there is the possibility that it will be passed on to future generations. However, there are a lot of "ifs": if the DNA change is not corrected by the cell's error-detecting machinery; if that particular egg or sperm is used to make a child; if that child survives and has children. Still, despite all the "ifs," we humans have over the centuries accumulated lots of mutations in our DNA. "Our DNA is a history book," geneticists say.

With the molecular tools that modern genetics provides, scientists are beginning to read that book, to trace the course of our species's history by tracking the changes that have occurred in our DNA. The National Geographic Society, for example, has been conducting a Geno-graphic Project comparing over 100,000 DNA samples from people all over the world, from Arctic Inuit Eskimos and Kenya's Masai to Australian aborigines and North American Pueblo Indians. Their hope is to create a picture of ancestral migratory routes, the historical paths people have taken as they populated the globe.

To gain the clearest possible picture of the past, gene researchers focus on DNA of the Y chromosome and the mitochondria. The Y chromosome of males does not recombine with other chromosomes. This has the effect of keeping mutations together once they occur. Similarly, the mitochondrial DNA of females is passed down from mother to child without recombination (sperm contribute no mitochondria to the fertilized egg).

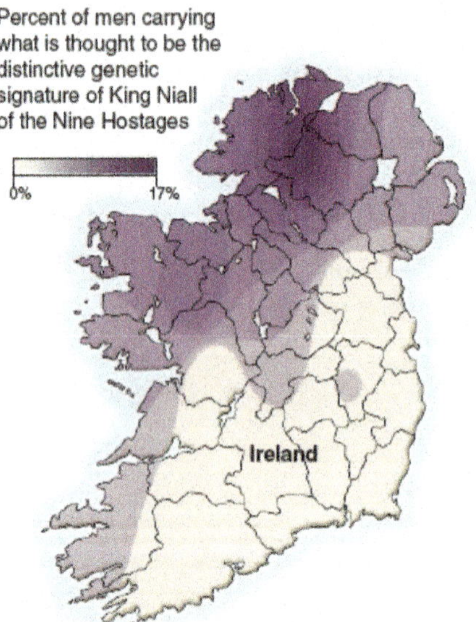

To compare large numbers of DNA samples, it is not practical to sequence the entire Y chromosome or mitochondrial DNA of each individual. Instead, investigators monitor several dozen highly variable DNA locations, short bits of the chromosome within which lots of mutations have occurred. Because DNA mutations are rare events and there has been no recombination to shuffle the changes, when the same particular combination of mutations (what gene researchers call a haplotype) occurs in two people, they almost certainly are related, having inherited their Y chromosome or mitochondrial DNA from a common ancestor. Looking at many individuals in this way, investigators can build up a picture of who is related to whom—a portrait of the past, inscribed on our genes.

To see how this works, consider a study carried out in 2006 by Daniel Bradley and colleagues at Trinity College in Dublin, Ireland. They set out to apply DNA studies to Irish history, which has always been a bit of a muddle. Writing did not become common in Ireland until 600 A.D., and little is known for sure of earlier events in Irish history. This has not, of course, prevented the Irish from preserving a rich story of those times.

Much as the British tell of a mythical King Arthur who few historians believe was a real person, so the Irish recount the tale of an Irish high king of the fifth century A.D. from whom an alarming number of Irishmen claim descent. Niall Noigiallach—Niall of the Nine Hostages—was so named because early in his reign he consolidated his power by taking hostages from the royal families of each of the five provinces that then constituted Ireland, as well as from Scotland, the Saxons, the Britons, and the Franks.

He founded a dynasty, the Ui Neill ("the descendants of Niall"), which ruled the northwest of Ireland from about 600 to 900 A.D. When the Irish took surnames around 1000 A.D., some chose names associated with the Ui Neill dynasties, names like Gallagher, Boyle, Doherty, O'Conner, Reilly, Flynn, Devlin, Donnelly, McLoughlin, Molloy, O'Rourke, and of course O'Neill (the prefix "O" is often added).

Did Niall of the Nine Hostages really exist? The DNA evidence gathered by Bradley argues yes. Some 20% of men in northwest Ireland have a distinctive genetic signature on their Y chromosomes, a haplotype carried down for over a thousand years. As you can see on the map to the left, this signature haplotype predominates in the northwest, the seat of Ui Neill power.

Indeed, wherever in the world you look (the Irish were particularly adept at migrating—over 400,000 residents of New York City claim Irish ancestry), this haplotype is much more common among Irishmen with the Ui Neill surnames than among Irishmen as a whole.

Niall is said to have had 14 sons, a large number even for those days, which might go a long way towards explaining that some 2 million men worldwide now carry his distinctive Y chromosome. Like Genghis Khan, ancestor of 16 million men in Asia, Niall of the Nine Hostages seems to have left quite a genetic footprint.

tissue are of enormous biological importance because they provide the raw material from which natural selection produces evolutionary change.

Mutations in Somatic Tissues Change can occur only if there are new, different allele combinations available to replace the old. Mutation produces new alleles, and recombination puts the alleles together in different combinations. In animals, it is the occurrence of these two processes in germ-line tissue that is important to evolution because mutations in somatic cells (somatic mutations) are not passed from one generation to the next. However, a somatic mutation may have drastic effects on the individual organism in which it occurs, because it is passed on to all of the cells that are descended from the original mutant cell. Thus, if a mutant lung cell divides, all cells derived from it will carry the mutation. Somatic mutations of lung cells are, as we shall see, the principal cause of lung cancer in humans.

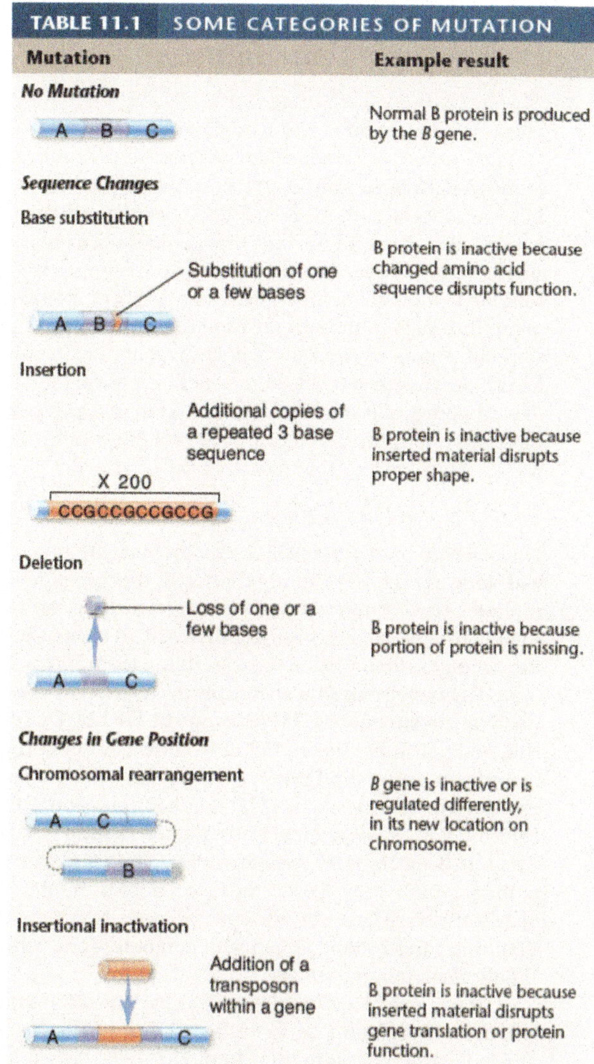

TABLE 11.1 SOME CATEGORIES OF MUTATION

Mutation		Example result
No Mutation	A B C	Normal B protein is produced by the *B* gene.
Sequence Changes		
Base substitution	Substitution of one or a few bases; A B C	B protein is inactive because changed amino acid sequence disrupts function.
Insertion	Additional copies of a repeated 3 base sequence; X 200; CCGCCGCCGCCG	B protein is inactive because inserted material disrupts proper shape.
Deletion	Loss of one or a few bases; A C	B protein is inactive because portion of protein is missing.
Changes in Gene Position		
Chromosomal rearrangement	A C; B	*B* gene is inactive or is regulated differently, in its new location on chromosome.
Insertional inactivation	Addition of a transposon within a gene; A C	B protein is inactive because inserted material disrupts gene translation or protein function.

Altering the Sequence of DNA One category of mutational changes affects the message itself, producing alterations in the sequence of DNA nucleotides (table 11.1). If alterations involve only one or a few base pairs in the coding sequence, they are called **point mutations.** Sometimes the identity of a base changes (*base substitution*), while other times one or a few bases are added (*insertion*) or lost (*deletion*). If an insertion or deletion throws the reading of the gene message out of register, a **frame-shift mutation** results. Figure 11.10 shows a base substitution mutation that results in the change of an amino acid, from proline to threonine. This could be a minor change or catastrophic. However, suppose that this had been the deletion of a nucleotide, that the cytosine base nucleotide had been skipped during replication. This would shift the register of the DNA message (imagine removing the w from this sentence, yielding "*This oulds hiftt her egistero fth eDN Amessag*") and you can see the problem. Many point mutations result from damage to the DNA caused by **mutagens,** usually radiation or chemicals. The latter class of mutations is of particular importance because modern industrial societies often release many chemical mutagens into the environment.

Changes in Gene Position Another category of mutations affects the way the genetic message is organized. In both prokaryotes and eukaryotes, individual genes may move from one place in the genome to another by **transposition** (see also page 259). When a particular gene moves to a different location, its expression or the expression of neighboring genes may be altered. In addition, large segments of chromosomes in eukaryotes may change their relative locations or undergo duplication. Such **chromosomal rearrangements** often have drastic effects on the expression of the genetic message.

The Importance of Genetic Change

All evolution begins with alterations in the genetic message that create new alleles or alter the organization of genes on chromosome. Some changes in germ-line tissue produce alterations that enable an organism to leave more offspring, and those changes tend to be preserved as the genetic endowment of future generations. Other changes reduce the ability of an organism to leave offspring. Those changes tend to be lost, as the organisms that carry them contribute fewer members to future generations. Evolution can be viewed as the selection of particular combinations of alleles from a pool of alternatives. The rate of evolution is ultimately limited by the rate at which these alternatives are generated. Genetic change through mutation and recombination provides the raw material for evolution.

Genetic changes in somatic cells do not pass on to offspring, and so they have no direct evolutionary consequence. However, changes in the genes of somatic cells can have an important immediate impact if the gene affects development or is involved with regulation of cell proliferation.

Key Learning Outcome 11.5 **Rare changes in genes, called mutations, can have significant effects on the individual when they occur in somatic tissue, but they are inherited only if they occur in germ-line tissue. Inherited changes provide the raw material for evolution.**

Biology and *Staying Healthy*

Protecting Your Genes

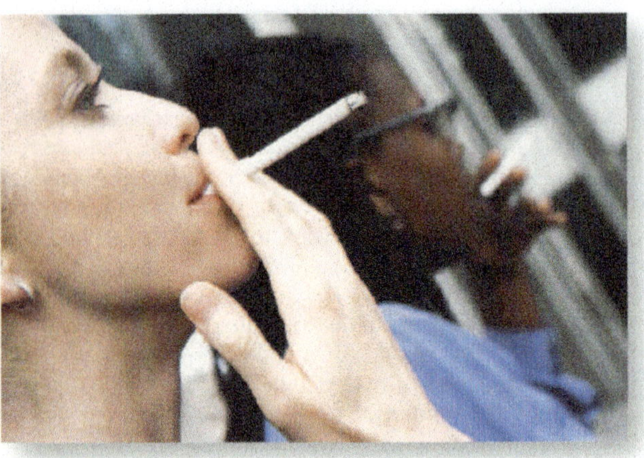

This text's discussion of changes in genes—mutations—has largely focused on heredity, how changes in the information encoded in DNA can affect offspring. It is important, however, to realize that inherited mutations occur only in germ-line tissue, in the cells that generate your eggs or sperm. Mutations in the other cells of your body, in so-called somatic tissues, are not inherited. This does not, however, mean that such mutations are not important. In fact, somatic mutations can have a disastrous impact upon your health because they can lead to cancer. Protecting the DNA of your body's cells from damaging mutation is perhaps the most important thing you can do to prolong your life. Here we will examine two potential threats.

Smoking and Lung Cancer

The association of particular chemicals in cigarette smoke with lung cancer, particularly chemicals that are potent mutagens (see chapters 8 and 24), led researchers early on to suspect that lung cancer might be caused, at least in part, by the action of chemicals on the cells lining the lung.

The hypothesis that chemicals in tobacco cause cancer was first advanced over 200 years ago in 1761 by Dr. John Hill, an English physician. Hill noted unusual tumors of the nose in heavy snuff users and suggested tobacco had produced these cancers. In 1775, a London surgeon, Sir Percivall Pott, made a similar observation, noting that men who had been chimney sweeps exhibited frequent cancer of the scrotum. He suggested that soot and tars might be responsible. These observations led to the hypothesis that lung cancer results from the action of tars and other chemicals in tobacco smoke.

It was over a century before this hypothesis was directly tested. In 1915, Japanese doctor Katsusaburo Yamagiwa applied extracts of tar to the skin of 137 rabbits every two or three days for three months. Then he waited to see what would happen. After a year, cancers appeared at the site of application in seven of the rabbits. Yamagiwa had induced cancer with the tar, the first direct demonstration of chemical carcinogenesis. In the decades that followed, this approach demonstrated that many chemicals can cause cancer.

But do these lab studies apply to people? Do tars in cigarette smoke in fact induce lung cancer in humans? In 1949, the American physician Ernst Winder and the British epidemiologist Richard Doll independently reported that lung cancer showed a strong link to the smoking of cigarettes, which introduces tars into the lungs. Winder interviewed 684 lung cancer patients and 600 normal controls, asking whether each had ever smoked. Cancer rates were 40 times higher in heavy smokers than in nonsmokers. From these studies, it seemed likely as long as 50 years ago that tars and other chemicals in cigarette smoke induce cancer in the lungs of persistent smokers. While this suggestion was resisted by the tobacco industry, the evidence that has accumulated since these pioneering studies makes a clear case, and there is no longer any real doubt. Chemicals in cigarette smoke cause cancer.

As you will learn in chapter 24 (page 525), tars and other chemicals in cigarette smoke cause lung cancer by mutating DNA, disabling genes that in normal lung cells restrain cell division. Lacking these restraints, the altered lung cells begin to divide ceaselessly, and lung cancer results. Just under 160,000 Americans died of lung cancer last year, and almost all of them were cigarette smokers.

If cigarette smoking is so dangerous, why do so many Americans smoke? Fully 23% of American men smoke, and 18% of women. Are they not aware of the danger? Of course they are. But they are not able to quit. Tobacco smoke, you see, also contains another chemical, nicotine, which is highly addictive. The nature of the addiction is discussed in detail in chapter 28 (page 592). Basically, what happens is that a smoker's brain makes physiological compensations to overcome the effects of nicotine, and once these adjustments are made the brain does not function normally without nicotine. The body's physiological response to nicotine is profound and unavoidable; there is no way to prevent addiction to nicotine with willpower.

Many people attempting to quit smoking use patches containing nicotine to help them, the idea being that providing nicotine removes the craving for cigarettes. This is true, it does—as long as you keep using the patch. Actually, using such patches simply substitutes one (admittedly less dangerous) nicotine source for another. If you are going to quit smoking, there is no way to avoid the necessity of eliminating the drug to which you are addicted, nicotine. There is no easy way out. The only way to quit is to quit.

Clearly, if you do not smoke, you should not start. Asked what three things were most important to improve Americans' health, a prominent physician replied: "Don't smoke. Don't smoke. Don't smoke."

Tanning and Skin Cancer

Almost all cells in the human body undergo cell division, replacing themselves as they wear out. Some adult cells do this quite frequently, others rarely if ever. Skin cells divide quite frequently. Exposed to a lot of wear-and-tear, they divide about every 27 days to replace dead or damaged cells. The skin sloughs off dead cells from the surface and replaces these with new cells from beneath. The average person will lose about 105 pounds of skin by the time he or she turns 70.

While skin can be damaged in many ways, the damage that seems to have the most long-term affect is caused by the sun. The skin contains cells called melanocytes that produce a pigment called melanin when exposed to UV light. Melanin produces a yellow to brown color in the skin. The type of melanin and the amount produced is genetically determined. People with darker skin types have more melanocytes and produce a melanin that is dark brown in color. Protected by UV-absorbing melanin, they almost never burn. Fair-skinned people have fewer melanocytes and produce melanin that is more yellow in color. Unprotected by melanin, these people sunburn easily and rarely tan. When cells on the body's surface are badly damaged by the sun—what we call a sunburn, the cells slough off. Recall the peeling that you experienced if you have ever had a bad sunburn.

Up until the early 20th century, a tan was a condition that people went to great lengths to avoid. A tanned body was a sign of the working class, people who had to work in the sun. The wealthy elite avoided the sun with pale skin being in fashion. All of this changed in the 1920s, when tans became a status symbol, with the wealthy able to travel to warm, sunny destinations, even in the middle of winter. That tan, bronzed glow that people would sit in the sun for hours to achieve was thought to be both healthy and attractive.

During the 1970s, doctors started to see an uptick in the number of cases of melanoma, a deadly form of skin cancer. New cases were increasing about 6% each year. Researchers proposed that UV rays from the sun were the underlying cause of this epidemic of skin cancer and warned people to avoid the sun when possible and protect themselves with sunscreen.

Malignant melanoma is the most deadly of skin cancers, although treatable if caught early. Melanoma is cancer of melanocyte cells. Melanoma lesions usually appear as shades of tan, brown, and black and often begin in or near a mole, and so changes in a mole are a symptom of melanoma. Melanoma is most prevalent in fair-skinned people, but unlike the other forms of skin cancer, it can also affect people with darker complexions.

The public has been slow to respond to warnings about avoiding sun exposure, perhaps because the cosmetic benefits of tanning are immediate while the health hazards are much delayed. The desire to achieve that tanned, bronzed body is as strong as ever.

A good tan requires regular exposure to the sun to maintain it, so indoor tanning salons have become popular. Tanning booths emit concentrated UV rays from two sides, allowing a person to tan in less time and in all weather conditions (sun, rain, snow). The indoor tanning business has grown in the United States to a $2 billion-a-year industry with an estimated 28 million Americans tanning annually.

People thought that building up a tan through the use of tanning booths would protect a person's skin from burning and would reduce the time exposed to the UV radiation, both leading to a reduced risk of skin cancer. However, recent research does not support these assumptions. A 2003 study of 106,000 Scandinavian women showed that exposure to UV rays in a tanning booth as little as once a month can increase your risk of melanoma by 55%, especially when the exposure is during early adulthood. Those women who were in their 20s and used sun lamps to tan were at the highest risk, about 150% higher than those who didn't use a tanning bed. As with other studies, fair-skinned women were at the greatest risk. In fact, tanning booths, even for those people who tan more easily, heighten the risk for skin cancer because people use the tanning booths year-round, increasing their cumulative exposure.

It is difficult to avoid the conclusion that to protect your genes you should avoid tanning booths. Like smoking cigarettes, excessive tanning is gambling with your life.

Are Mutations Random or Directed by the Environment?

Once biologists appreciated that Mendelian traits were in fact alternative versions of DNA sequences that resulted from mutations, a very important question arose and needed to be answered—are mutations random events that might happen anywhere on the DNA in a chromosome, or are they directed to some degree by the environment? Do the mutagens in cigarettes, for example, damage DNA at random locations, or do they preferentially seek out and alter specific sites such as those regulating the cell cycle?

This key question was addressed and answered in an elegant, deceptively simple experiment carried out in 1943 by two of the pioneers of molecular genetics, Salvadore Luria and Max Delbruck. They chose to examine a particular mutation that occurs in laboratory strains of the bacterium *E. coli*. These bacterial cells are susceptible to T1 viruses, tiny chemical parasites that infect, multiply within, and kill the bacteria. If 10^5 bacterial cells are exposed to 10^{10} T1 viruses, and the mixture spread on a culture dish, not one cell grows—every single *E. coli* cell is infected and killed. However, if you repeat the experiment using 10^9 bacterial cells, lots of cells survive! When tested, these surviving cells prove to be mutants, resistant to T1 infection. The question is, did the T1 virus cause the mutations, or were they present all along, too rare to be present in a sample of only 10^5 cells but common enough to be present in 10^9 cells?

To answer this question, Luria and Delbruck devised a simple experiment they called a "fluctuation test," illustrated here. Five cell generations are shown for each of four independent bacterial cultures, all tested for resistance in the fifth generation. If the T1 virus causes the mutations (top row), then each culture will have more or less the same number of resistant cells, with only a little fluctuation (that is, variation among the four). If, on the other hand, mutations are spontaneous and so equally likely to occur in any generation, then bacterial cultures in which the T1-resistance mutation occurs in earlier generations will possess far more resistant cells by the fifth generation than cultures in which the mutation occurs in later generations, resulting in wide fluctuation among the four cultures. The table presents the data they obtained for 20 individual cultures.

Number of Bacteria Resistant to T1 Virus

Culture number	Resistant colonies found	Culture number	Resistant colonies found
1	1	11	107
2	0	12	0
3	3	13	0
4	0	14	0
5	0	15	1
6	5	16	0
7	0	17	0
8	5	18	64
9	0	19	0
10	6	20	35

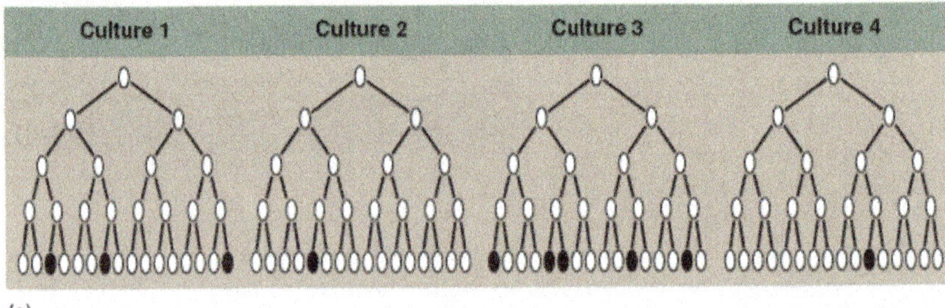

(a)

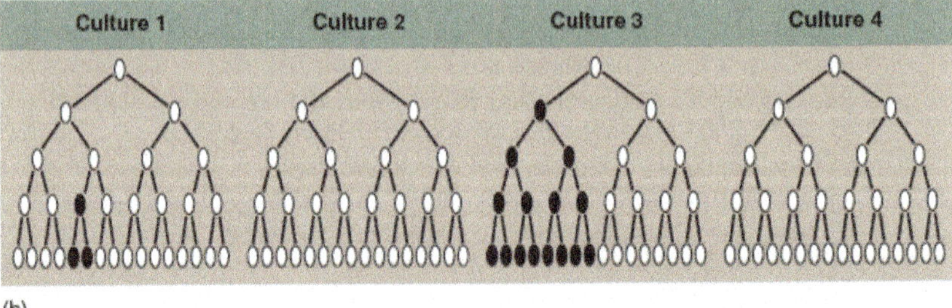

(b)

1. **Applying Concepts** Is there a dependent variable in this experiment? Explain.
2. **Interpreting Data** What is the mean number of T1-resistant colonies found in the 20 individual cultures?
3. **Making inferences**
   a. Comparing the 20 individual cultures, do the cultures exhibit similar numbers of T1-resistant bacterial cells?
   b. Which of the two alternative outcomes illustrated above, (a) or (b), is more similar to the outcome obtained by Luria and Delbruck in this experiment?
4. **Drawing Conclusions** Are these data consistent with the hypothesis that the mutation for T1 resistance among *E. coli* bacteria is caused by exposure to T1 virus? Explain.

Chapter Review

Genes Are Made of DNA

11.1 The Discovery of Transformation

- Using *Streptococcus pneumoniae* bacteria, Griffith showed that information that controls physical characteristics can be passed from one bacterium to another, even from a dead bacterium.
- By injecting mice with different strains of *S. pneumoniae*, Griffith determined that some strains were pathogenic, resulting in the mice's deaths. Bacteria of the pathogenic strains contained polysaccharide capsules (S strain), like the bacteria shown here from **figure 11.1**, while those without the coats (R strain) were nonlethal. When he mixed dead pathogenic bacteria (S), which usually would not cause death, and live nonpathogenic bacteria (R) and injected them into mice, the mice died. The dead mice contained living S strains.

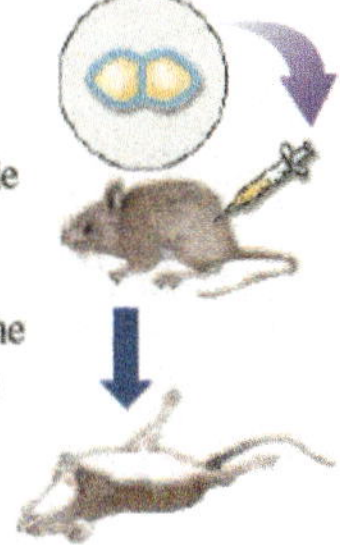

- Something passed from the dead lethal bacteria to the live nonlethal bacteria, causing them to turn deadly. Griffith did not determine whether this "transforming principle" was protein or DNA.

11.2 Experiments Identifying DNA as the Genetic Material

- Avery and colleagues showed that protein was not the source of this transformation. They replicated Griffith's experiment but removed all protein from the preparation. The virulent strain with its protein coat removed was still able to transform nonvirulent bacteria. This experimental result supported the hypothesis that DNA, not protein, was the transforming principle.
- Using bacterial viruses, Hershey and Chase showed that genes were carried on DNA and not proteins. They used two different radioactively tagged preparations, DNA in one and protein in the other (**figure 11.2**). Each preparation was used to infect bacteria. When they screened the two bacterial cultures, they discovered that the infected bacteria contained radioactively tagged DNA.

11.3 Discovering the Structure of DNA

- The structure of DNA was not known. The basic chemical components of DNA were determined to be nucleotides, like the one shown here from **figure 11.3**. Each nucleotide has a similar structure: a deoxyribose sugar attached to a phosphate group and one of four organic bases.

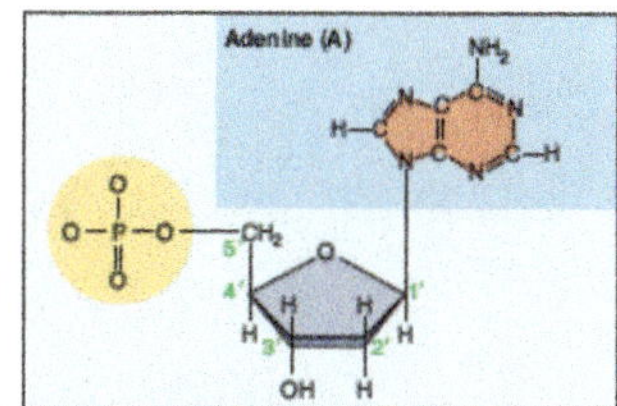

- Erwin Chargaff observed that two sets of bases are always present in equal amounts in a molecule of DNA (A nucleotides equal T nucleotides, and C nucleotides equal G nucleotides). This observation, called Chargaff's rule, gave some insight into the structure of DNA—that there was some regularity to the structure.
- Using X-ray diffraction, Rosalind Franklin was able to form a "picture" of DNA. The image suggested that the DNA molecule was coiled, a form called a helix.
- Using Chargaff's and Franklin's research, Watson and Crick determined that DNA is a double helix, two strands that are connected by base pairing between the nucleotide bases. An A nucleotide on one strand pairs with T on the other, and similarly G pairs with C (**figure 11.4**).

DNA Replication

11.4 How the DNA Molecule Copies Itself

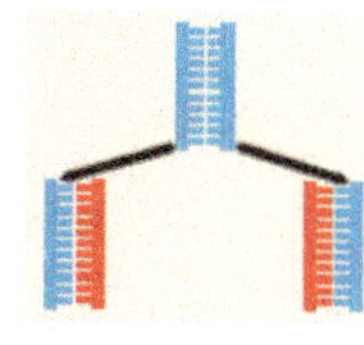

- The complementarity of DNA (A pairs with T and C with G) implies a method of replication where a single strand of DNA can serve as the template for production of another strand, but there are several ways in which this could occur (**figure 11.5**).
- Meselson and Stahl showed that DNA replicates semiconservatively, using each of the original strands as templates to form new strands. In semiconservative replication, each new strand of DNA consists of a template strand from the parent DNA and a newly synthesized strand that is complementary to the template strand (**figure 11.6**).
- In replication, the DNA molecule first unwinds by the actions of an enzyme called helicase. Each DNA strand is then copied by the actions of an enzyme called DNA polymerase. The two original strands serve as templates to the new DNA strands. DNA polymerase adds nucleotides to the new DNA strands that are complementary to the original single strands. DNA polymerase can only add on to an existing strand, and so the new strand begins after a section of nucleic acids called a primer is added. A different enzyme builds the primer. Nucleotides are added to the growing strand in a 5′ to 3′ direction (**figure 11.7**).
- The point where the DNA separates is called the replication fork. Because nucleotides can only be added onto the 3′ end of the growing DNA strand, DNA copies in a continuous manner on one of the strands, called the leading strand, and in a discontinuous manner on the other strand, called the lagging strand (**integrated art, page 224**). On the lagging strand, primers are inserted at the replication fork, and nucleotides are added in sections (**integrated art, page 225**). Before the new DNA strands rewind, the primers are removed and the DNA segments are linked together with another enzyme called DNA ligase.
- Errors can occur during the replication of DNA. The cell has many mechanisms to correct damage to the DNA or mistakes made during replication. This proofreading process compares one strand against its complementary strand and corrects errors, but this system is not foolproof.

Altering the Genetic Message

11.5 Mutation

- A mutation is a change in the nucleotide sequence of the genetic message. Mutations that change one or only a few nucleotides are called point mutations (**figure 11.10**). Some mutations occur through the movement of sections of DNA from one place to another, a process called transposition (**table 11.1**).

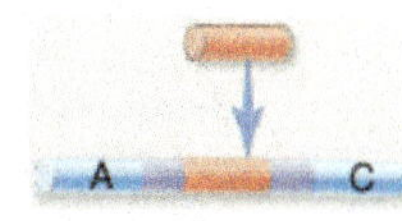

Test Your Understanding

1. In his experiments, Frederick Griffith found that
   a. hereditary information within a cell cannot be changed.
   b. hereditary information can be added to cells from other cells.
   c. mice infected with live R strains die.
   d. mice infected with heat-killed S strains die.
2. The experiment performed by Alfred Hershey and Martha Chase showed that the molecule viruses use to specify new viruses is
   a. a protein.
   b. a carbohydrate.
   c. ATP.
   d. DNA.
3. Erwin Chargaff, Rosalind Franklin, Francis Crick, and James Watson all worked on pieces of information relating to the
   a. structure of DNA.
   b. function of DNA.
   c. inheritance of DNA.
   d. mutations of DNA.
4. The four DNA nucleotides are all different in terms of
   a. their sizes.
   b. the number of hydrogen bonds they can form with their base pair.
   c. the type of nitrogen base.
   d. the type of sugar.
5. Which of the following lists the purine nucleotide bases?
   a. adenine and cytosine
   b. guanine and thymine
   c. cytosine and thymine
   d. adenine and guanine
6. If one strand of a DNA molecule has the base sequence ATTGCAT, its complementary strand will have the sequence
   a. ATTGCAT.
   b. TAACGTA.
   c. GCCATGC.
   d. CGGTACG.
7. Regarding the duplication of DNA, we now know that each double helix
   a. rejoins after replicating.
   b. splits down the middle into two single strands, and each one then acts as a template to build its complement.
   c. fragments into small chunks that duplicate and reassemble.
   d. All of these are true for different types of DNA.
8. DNA polymerase can only add nucleotides to an existing chain, so _______________ is required.
   a. a primer
   b. helicase
   c. a lagging strand
   d. a leading strand
9. Genetic messages can be altered in two ways:
   a. through semiconservative replication or conservative replication.
   b. through the chromosome or through the protein.
   c. by mutation or by recombination.
   d. by activation or by repression.
10. Mutations can occur in
    a. germ-line tissues and are passed on to future generations.
    b. somatic tissues and are passed on to future generations.
    c. germ-line tissues but not in somatic tissues.
    d. somatic tissues but not in germ-line tissues.

Apply Your Understanding

1. **Figure 11.4c** What are some of the possible problems that could occur if the C nucleotide indicated with the red arrow is accidentally replaced with an A nucleotide?

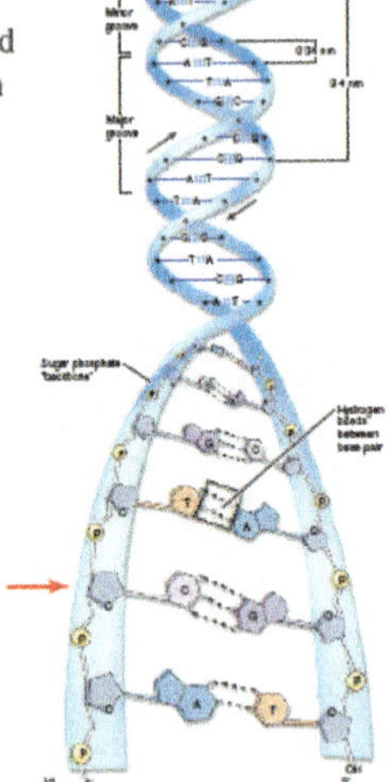

2. **Table 11.1** What types of mutations in the table result in a shift in the reading frame of the DNA? Explain.

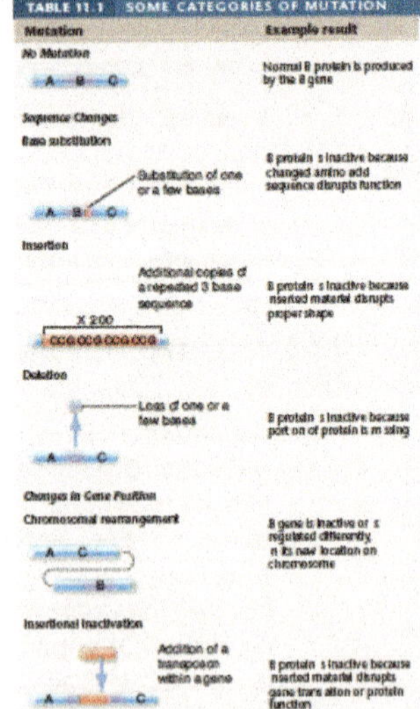

Synthesize What You Have Learned

1. The discovery that DNA is the genetic material was an experimental journey rather than a flash of insight. Highlighting individual experiments, use this journey to defend the statement attributed to Sir Isaac Newton in 1676 (though some say that Bernard of Chartres said it first, way back in about 1130!) that scientists build new ideas in science by "standing on the shoulders of giants."
2. Certain strains of bacteria are resistant to the antibiotic tetracycline, while other strains are sensitive to it. Design an experiment you would carry out to determine whether or not tetracycline resistance is an inherited trait specified by the DNA of the resistant strain.

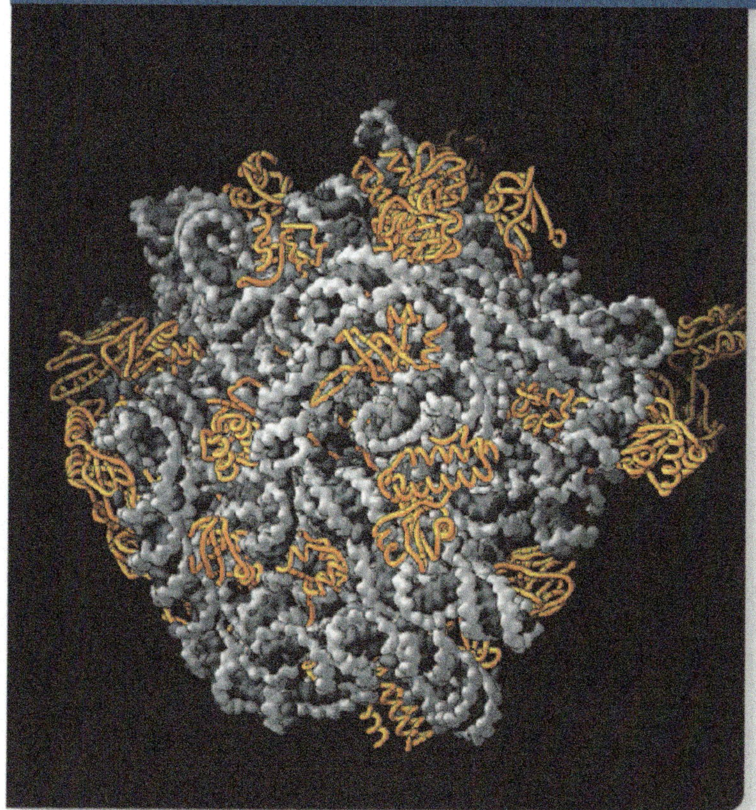

12

How Genes Work

Learning Objectives

From Gene to Protein

12.1 The Central Dogma
1. State the so-called "Central Dogma."
2. List the four kinds of RNA used in the synthesis of proteins.
3. Name the two stages of gene expression.
4. Describe the location and function of a promoter site.

12.2 Transcription
1. State the chemical difference between DNA and RNA.
2. Indicate in what chemical direction an mRNA chain is assembled during transcription.

12.3 Translation
1. Distinguish between codon and anticodon.
2. State how many codons do not code for an amino acid.
3. Describe the composition of a ribosome.
4. Contrast the roles of tRNA and activating enzymes in reading the genetic code.
5. List the order in which the A, P, and E sites of a ribosome are occupied by an amino acid.

12.4 Gene Expression
1. Compare the architecture of prokaryotic and eukaryotic genes.
2. Explain alternative splicing.
3. Compare the cellular locations of transcription and translation in prokaryotes and eukaryotes.
4. Describe the six stages of eukaryotic protein synthesis.

Regulating Gene Expression in Prokaryotes

12.5 How Prokaryotes Control Transcription
1. Define operon.
2. Diagram how the *lac* operon works.
3. Contrast the regulatory action of repressors and activators.
4. Describe how the CAP protein acts as an activator.

Regulating Gene Expression in Eukaryotes

12.6 Transcriptional Control in Eukaryotes
1. Describe the regulatory roles of histones and of DNA methylation in eukaryotes.

12.7 Controlling Transcription from a Distance
1. Distinguish between basal transcription factors and specific transcription factors.
2. Use a diagram to explain how enhancers regulate gene expression at a distance.

12.8 RNA-Level Control
1. Define RNA interference.
2. Explain how siRNA can regulate gene expression.

Biology and Staying Healthy: Silencing Genes to Treat Disease

12.9 Complex Regulation of Gene Expression
1. Name and describe six points where eukaryotic gene expression is controlled.

Inquiry & Analysis: Building Proteins in a Test Tube

Ribosomes, like the one you see here, are very complex cellular machines that assemble the polypeptide segments of proteins, using information that has been copied from genes onto RNA molecules. The ribosomes read the gene information copied onto these messenger RNA transcripts, and use it to determine the amino acid sequence of the new polypeptide that the ribosome is assembling. Each ribosome is made of over 50 different proteins (shown here in gold), as well as three chains of RNA composed of some 3,000 nucleotides (shown here in gray). It has been traditionally assumed that the proteins in a ribosome act as enzymes to catalyze the amino acid assembly process, with the RNA acting as a scaffold to position the proteins. In the year 2000 we learned the reverse to be true. Powerful X-ray diffraction studies revealed the complete detailed structure of a ribosome at atomic resolution. Unexpectedly, the many proteins of a ribosome are scattered over its surface like decorations on a Christmas tree. The role of these proteins seems to be to stabilize the many bends and twists of the RNA chains, the proteins acting like spot-welds between the RNA strands they touch. Importantly, there are no proteins on the inside of the ribosome where the chemistry of protein synthesis takes place—just twists of RNA. Thus, it is the ribosome's RNA, not its proteins, that catalyzes the joining together of amino acids! Clearly, our knowledge of how genes work is still increasing, often adjusting what seem to be fundamental concepts.

From Gene to Protein

12.1 The Central Dogma

The discovery that genes are made of DNA, discussed in chapter 11, left unanswered the question of how the information in DNA is used. How does a string of nucleotides in a spiral molecule determine if you have red hair? We now know that the information in DNA is arrayed in little blocks, like entries in a dictionary, and each block is a gene that specifies the sequence of amino acids for a polypeptide. These polypeptides form the proteins that determine what a particular cell will be like.

All organisms, from the simplest bacteria to ourselves, use the same basic mechanism of reading and expressing genes, so fundamental to life as we know it that it is often referred to as the "Central Dogma": Information passes from the genes (DNA) to an RNA copy of the gene, and the RNA copy directs the sequential assembly of a chain of amino acids. Said briefly, **DNA ⟶ RNA ⟶ protein.**

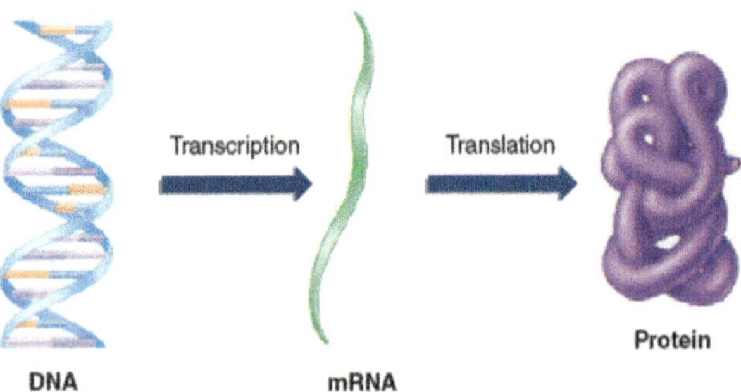

A cell uses four kinds of RNA in the synthesis of proteins: messenger RNA (mRNA), silencing RNA (siRNA), ribosomal RNA (rRNA), and transfer RNA (tRNA). These types of RNA are described in more detail later in this chapter.

The use of information in DNA to direct the production of particular proteins is called **gene expression.** Gene expression occurs in two stages: In the first stage, *transcription,* mRNA molecules are synthesized from genes within the DNA; in the second stage, *translation,* the mRNA is used to direct the production of polypeptides, the components of proteins.

Transcription: An Overview

The first step of the Central Dogma is the transfer of information from DNA to RNA, which occurs when an mRNA copy of the gene is produced. Because the DNA sequence in the gene is transcribed into an RNA sequence, this stage is called transcription. Transcription is initiated when the enzyme *RNA polymerase* binds to a special nucleotide sequence called a *promoter* located at the beginning of a gene. Starting there, the RNA polymerase moves along the strand into the gene (figure 12.1). As it encounters each DNA nucleotide, it adds the corresponding complementary RNA nucleotide to a growing mRNA strand. Thus, guanine (G), cytosine (C), thymine (T), and adenine (A) in the DNA would signal the addition of C, G, A, and uracil (U), respectively, to the mRNA.

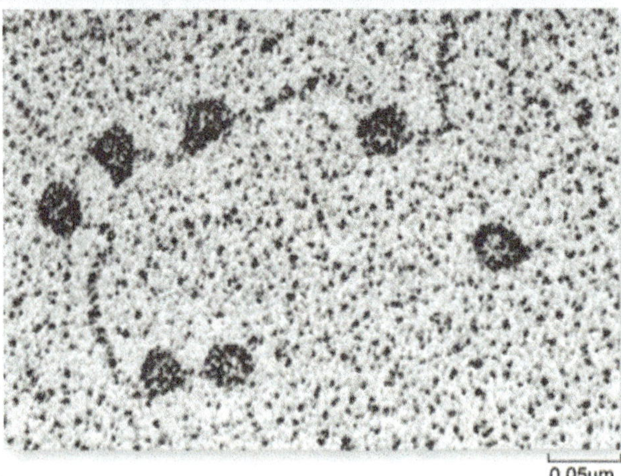

Figure 12.1 RNA polymerase.
In this electron micrograph, the dark circles are RNA polymerase molecules synthesizing RNA from a DNA template.

When the RNA polymerase arrives at a transcriptional "stop" signal at the opposite end of the gene, it disengages from the DNA and releases the newly assembled RNA chain. This chain is a complementary transcript of the gene from which it was copied.

Translation: An Overview

The second step of the Central Dogma is the transfer of information from RNA to protein, which occurs when the information contained in the mRNA transcript is used to direct the sequence of amino acids during the synthesis of polypeptides by ribosomes. This process is called translation because the nucleotide sequence of the mRNA transcript is translated into an amino acid sequence in the polypeptide. Translation begins when an rRNA molecule within the ribosome recognizes and binds to a "start" sequence on the mRNA. The ribosome then moves along the mRNA molecule, three nucleotides at a time. Each group of three nucleotides is a codeword that specifies which amino acid will be added to the growing polypeptide chain, and is recognized by a specific tRNA molecule. The ribosome continues in this fashion until it encounters a translational "stop" signal; then it disengages from the mRNA and releases the completed polypeptide.

Key Learning Outcome 12.1 The information encoded in genes is expressed in two phases: transcription, which produces an mRNA molecule whose sequence is complementary to the DNA sequence of the gene; and translation, which assembles a polypeptide.

12.2 Transcription

Just as an architect protects building plans from loss or damage by keeping them safe in a central place and issuing only blueprint copies to on-site workers, so your cells protect their DNA instructions by keeping them safe within a central DNA storage area, the nucleus. The DNA never leaves the nucleus. Instead, the process of **transcription** creates "blueprint" copies of particular genes that are sent out into the cell to direct the assembly of proteins (figure 12.2).

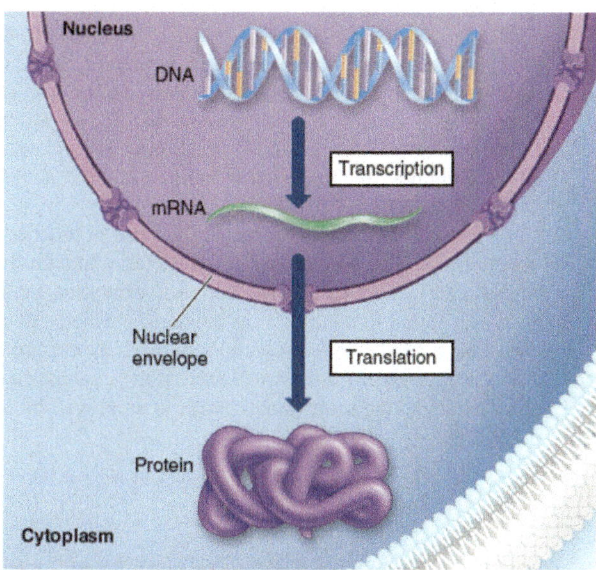

Figure 12.2 Overview of gene expression in a eukaryotic cell.

These working copies of genes are made of ribonucleic acid (RNA) rather than DNA. Recall that RNA is the same as DNA except that the sugars in RNA have an extra oxygen atom and the pyrimidine base thymine (T) is replaced by a similar pyrimidine base called uracil, U (see figure 3.9).

The Transcription Process

The RNA copy of a gene used in the cell to produce a polypeptide is called **messenger RNA (mRNA)**—it is the messenger that conveys the information from the nucleus to the cytoplasm. The copying process that makes the mRNA is called transcription—just as monks in monasteries used to make copies of manuscripts by faithfully transcribing each letter, so enzymes within the nuclei of your cells make mRNA copies of your genes by faithfully complementing each nucleotide.

In your cells, the transcriber is a large and very sophisticated protein called **RNA polymerase.** It binds to one strand of a DNA double helix at the promoter site and then moves along the DNA strand like a train engine on a track. Although DNA is double-stranded, the two strands have complementary rather than identical sequences, so RNA polymerase is only able to bind one of the two DNA strands (the one with the promoter-site sequence it recognizes). As RNA polymerase goes along the DNA strand it is copying, it pairs each nucleotide with its complementary RNA version (G with C, A with U), building an mRNA chain in the 5′ to 3′ direction as it moves along the strand (figure 12.3).

Key Learning Outcome 12.2 **Transcription is the production of an mRNA copy of a gene by the enzyme RNA polymerase.**

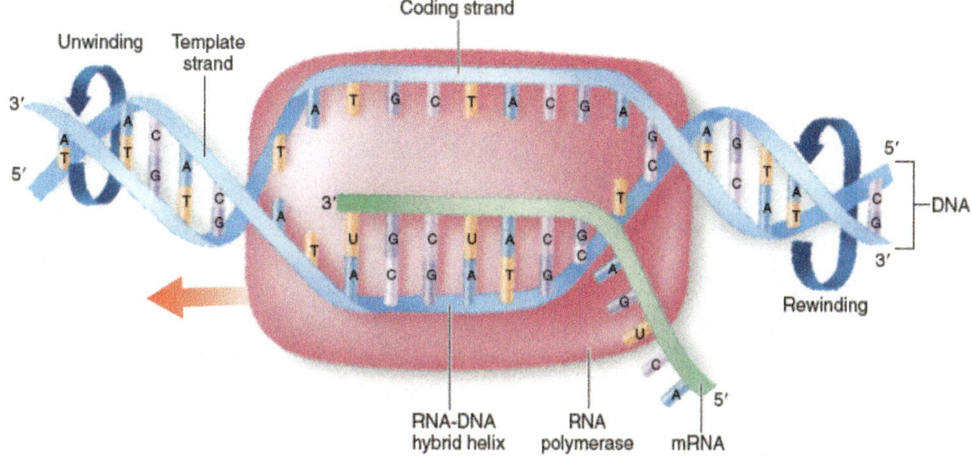

Figure 12.3 Transcription.

One of the strands of DNA functions as a template on which nucleotide building blocks are assembled into mRNA by RNA polymerase as it moves along the DNA strand.

12.3 Translation

The Genetic Code

The essence of Mendelian genetics is that information determining hereditary traits, traits passed from parent to child, is encoded information. The information is written within the chromosomes in blocks called **genes.** Genes affect Mendelian traits by directing the production of particular proteins. The essence of gene expression, of using your genes, is reading the information encoded within DNA and using that information to produce proteins.

To correctly read a gene, a cell must translate the information encoded in DNA into the language of proteins—that is, it must convert the order of the gene's nucleotides into the order of amino acids in a polypeptide, a process called **translation.** The rules that govern this translation are called the **genetic code.**

The mRNA is transcribed from a gene in a linear sequence, one nucleotide following another, beginning at the promoter. There, RNA polymerase binds to the DNA and begins its assembly of the mRNA. Transcription ends when the RNA polymerase reaches a certain nucleotide sequence that signals it to stop.

However, the mRNA is not translated in this same way. The mRNA is "read" by a ribosome in three-nucleotide units. Each three-nucleotide sequence of the mRNA is called a **codon.** Each codon, with the exception of three, codes for a particular amino acid. Biologists worked out which codon corresponds to which amino acid by trial-and-error experiments carried out in test tubes. In these experiments, investigators used artificial mRNAs to direct the synthesis of polypeptides in the tube, and then looked to see the sequence of amino acids in the newly formed polypeptides. An mRNA that was a string of UUUUUU . . . , for example, produced a polypeptide that was a string of phenylalanine (Phe) amino acids, telling investigators that the codon UUU corresponded to the amino acid Phe. The entire genetic code dictionary is presented in figure 12.4. The first letters of the codon are positioned down the left side, the second across the top, and the third down the right side. To determine the amino acid encoded by a codon, say AGC, go to "A" on the left, follow the row over to the "G" column, and go down to the C on the right. As you discover, AGC encodes the amino acid serine. Because at each position of a three-letter codon, any of the four different nucleotides (U, C, A, G) may be used, there are 64 different possible three-letter codons ($4 \times 4 \times 4 = 64$) in the genetic code.

The genetic code is universal, the same in practically all organisms. GUC codes for valine in bacteria, in fruit flies, in eagles, and in your own cells. The only exception biologists have ever found to this rule is in the way in which cell organelles that contain DNA (mitochondria and chloroplasts) and a few microscopic protists read the "stop" codons. In every other instance, the same genetic code is employed by all living things.

The Genetic Code

First Letter	Second Letter: U		C		A		G		Third Letter
U	UUU	Phenylalanine	UCU	Serine	UAU	Tyrosine	UGU	Cysteine	U
	UUC		UCC		UAC		UGC		C
	UUA	Leucine	UCA		UAA	Stop	UGA	Stop	A
	UUG		UCG		UAG	Stop	UGG	Tryptophan	G
C	CUU	Leucine	CCU	Proline	CAU	Histidine	CGU	Arginine	U
	CUC		CCC		CAC		CGC		C
	CUA		CCA		CAA	Glutamine	CGA		A
	CUG		CCG		CAG		CGG		G
A	AUU	Isoleucine	ACU	Threonine	AAU	Asparagine	AGU	Serine	U
	AUC		ACC		AAC		AGC		C
	AUA		ACA		AAA	Lysine	AGA	Arginine	A
	AUG	Methionine; Start	ACG		AAG		AGG		G
G	GUU	Valine	GCU	Alanine	GAU	Aspartate	GGU	Glycine	U
	GUC		GCC		GAC		GGC		C
	GUA		GCA		GAA	Glutamate	GGA		A
	GUG		GCG		GAG		GGG		G

Figure 12.4 The genetic code (RNA codons).

A codon consists of three nucleotides read in sequence. For example, ACU codes for threonine. The first letter, A, is in the First Letter column; the second letter, C, is in the Second Letter row; and the third letter, U, is in the Third Letter column. Most amino acids are specified by more than one codon. For example, threonine is specified by four codons, which differ only in the third nucleotide (ACU, ACC, ACA, and ACG).

Translating the RNA Message into Proteins

The final result of the transcription process is the production of an mRNA copy of a gene. Like a photocopy, the mRNA can be used without damage or wear and tear on the original. After transcription of a gene is finished, the mRNA passes out of the nucleus (in eukaryotes) and into the cytoplasm through pores in the nuclear membrane. There, translation of the genetic message occurs. In translation, organelles called **ribosomes** use the mRNA produced by transcription to direct the synthesis of a polypeptide following the genetic code.

The Protein-Making Factory Ribosomes are the polypeptide-making factories of the cell. Each is very complex, containing over 50 different proteins and several segments of **ribosomal RNA (rRNA).** Ribosomes use mRNA, the "blueprint" copies of nuclear genes, to direct the assembly of polypeptides, which are then combined into proteins.

Ribosomes are composed of two pieces, or subunits, one nested into the other like a fist in the palm of your hand. The "fist" is the smaller of the two subunits, the pink structure in figure 12.5. Its rRNA has a short nucleotide sequence exposed on the surface of the subunit. This exposed sequence is identical to a sequence called the leader region that occurs at the beginning of all genes. Because of this, an mRNA molecule binds to the exposed rRNA of the small subunit like a fly sticking to flypaper.

The Key Role of tRNA Directly adjacent to the exposed rRNA sequence are three small pockets or dents, called the A, P, and E sites, in the surface of the ribosome (shown in figure 12.5 and discussed shortly). These sites have just the right shape to bind yet a third kind of RNA molecule, **transfer RNA (tRNA).** It is tRNA molecules that bring amino acids to the ribosome used in making proteins. The tRNA molecules are chains about 80 nucleotides long. The string of nucleotides folds back on itself, forming a three-looped structure shown in figure 12.6*a*. The looped structure further folds into a compact shape shown in figure 12.6*b*, with a three-nucleotide sequence at one end (the pink loop) and an amino acid attachment site on the other end (the 3′ end).

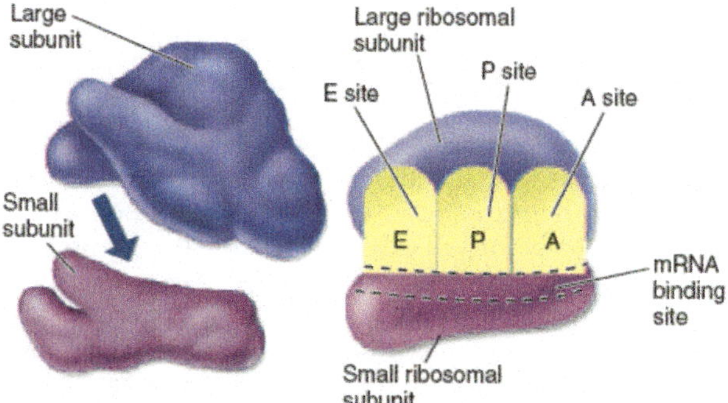

Figure 12.5 A ribosome is composed of two subunits.

The smaller subunit fits into a depression on the surface of the larger one. The A, P, and E sites on the ribosome play key roles in protein synthesis.

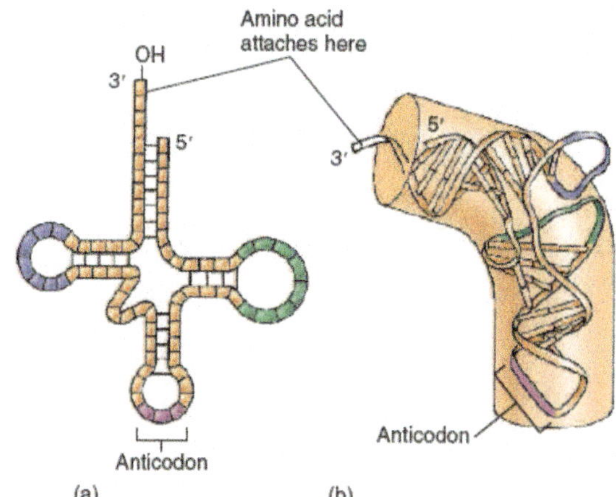

Figure 12.6 The structure of tRNA.

tRNA, like mRNA, is a long strand of nucleotides. However, unlike mRNA, hydrogen bonding occurs between its nucleotides, causing the strand to form hairpin loops, as seen in (*a*). The loops then fold up on each other to create the compact, three-dimensional shape seen in (*b*). Amino acids attach to the free, single-stranded —OH end of a tRNA molecule. A three-nucleotide sequence called the anticodon in the lower loop of tRNA interacts with a complementary codon on the mRNA.

The three-nucleotide sequence, called the **anticodon,** is very important: It is the complementary sequence to 1 of the 64 codons of the genetic code! Special enzymes, called *activating enzymes,* match amino acids in the cytoplasm with their proper tRNAs. The anticodon determines which amino acid will attach to a particular tRNA.

Because the first dent in the ribosome, called the *A site* (the attachment site where amino-acid-bearing tRNAs will bind) is directly adjacent to where the mRNA binds to the rRNA, three nucleotides of the mRNA are positioned directly facing the anticodon of the tRNA. Like the address on a letter, the anticodon ensures that an amino acid is delivered to its correct "address" on the mRNA where the ribosome is assembling the polypeptide.

Making the Polypeptide Once an mRNA molecule has bound to the small ribosomal subunit, the other larger ribosomal subunit binds as well, forming a complete ribosome. The ribosome then begins the process of translation, illustrated in the panels of the Key Biological Process illustration on the next page. Panel 1 shows how the mRNA begins to thread through the ribosome like a string passing through the hole in a doughnut. The mRNA passes through in short spurts, three nucleotides at a time, and at each burst of movement a new three-nucleotide codon on the mRNA is positioned opposite the A site in the ribosome, where a tRNA molecule first binds, as shown in panel 2.

As each new tRNA brings in an amino acid to each new codon presented at the A site, the old tRNA paired with the previous codon is passed over to the *P site* where peptide bonds form between the incoming amino acid and the growing polypeptide chain. The tRNA in the P site eventually shifts

KEY BIOLOGICAL PROCESS: Translation

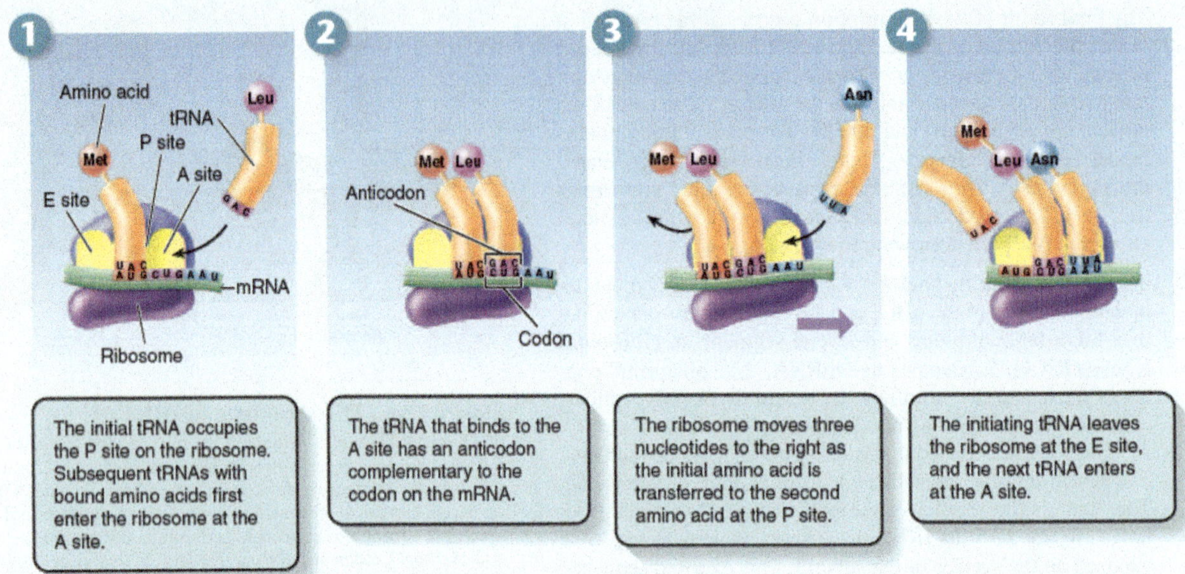

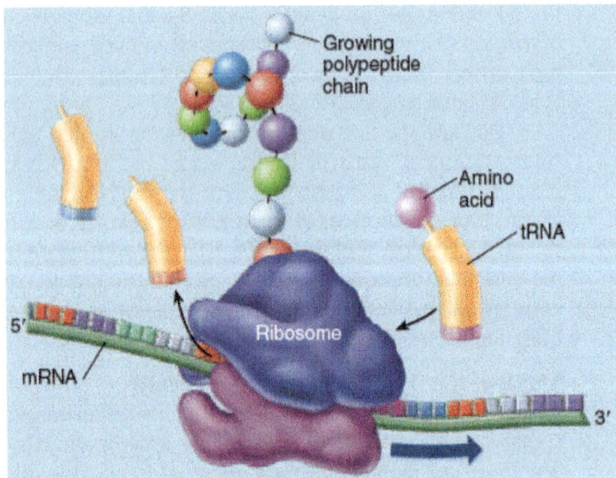

Figure 12.7 Ribosomes guide the translation process.

tRNA binds to an amino acid as determined by the anticodon sequence. Ribosomes bind the loaded tRNAs to their complementary sequences on the strand of mRNA. tRNA adds its amino acid to the growing polypeptide chain, which is released as the completed protein.

to the *E site* (the exit site), as shown in panel 3, and the amino acid it carried is attached to the end of a growing amino acid chain. The tRNA is then released (panel 4). So as the ribosome proceeds down the mRNA, one tRNA after another is selected to match the sequence of mRNA codons. In figure 12.7 you can see the ribosome traveling along the length of the mRNA, the tRNAs bringing the amino acids into the ribosome and the growing polypeptide chain extending out from the ribosome. Translation continues until a "stop" codon is encountered, which signals the end of the polypeptide. The ribosome complex falls apart, and the newly made polypeptide is released into the cell.

As explained earlier, the overall flow of genetic information, the so-called "Central Dogma," is from DNA to mRNA to protein. For example, the polypeptide that is being formed in the Key Biological Process illustration above began with the DNA nucleotide sequence TACGACTTA, which is first transcribed into the mRNA sequence AUGCUGAAU. This sequence is then translated by the tRNAs into a polypeptide composed of the amino acids methionine—leucine—asparagine.

Key Learning Outcome 12.3 The genetic code dictates how a particular nucleotide sequence specifies a particular amino acid sequence. A gene is transcribed into mRNA, which is then translated into a polypeptide. The sequence of mRNA codons dictates the corresponding sequence of amino acids in a growing polypeptide chain.

12.4 Gene Expression

The Central Dogma, discussed in section 12.1, is the same in all organisms. Figure 12.8 overviews the components needed for the key processes of DNA replication, transcription, and translation and the products that are formed in each. In general, the components are the same, the processes are the same, and the products are the same whether in prokaryotes or eukaryotes. However, there are some differences in gene expression between the two types of cells.

Architecture of the Gene

In prokaryotes, a gene is an uninterrupted stretch of DNA nucleotides whose transcript is read three nucleotides at a time to make a chain of amino acids. In eukaryotes, by contrast, genes are fragmented. In these more complex genes, the DNA nucleotide sequences encoding the amino acid sequence of a polypeptide are called **exons,** and the exons are interrupted frequently by extraneous nucleotides, "extra stuff" called **introns.** You can see them in the segment of DNA illustrated in figure 12.9; the exons are

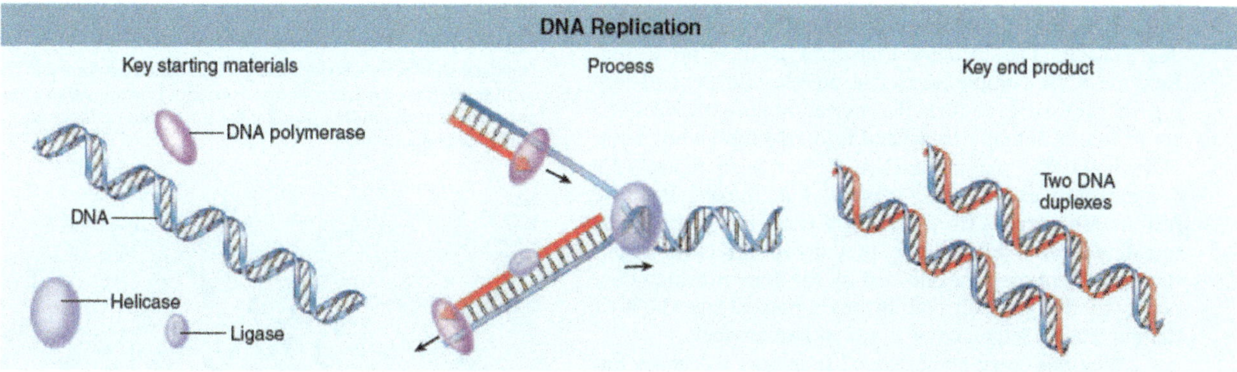

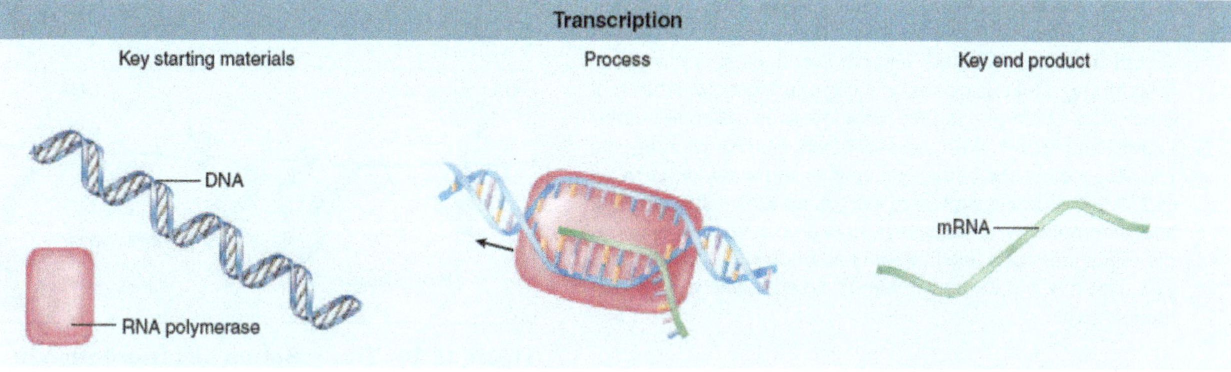

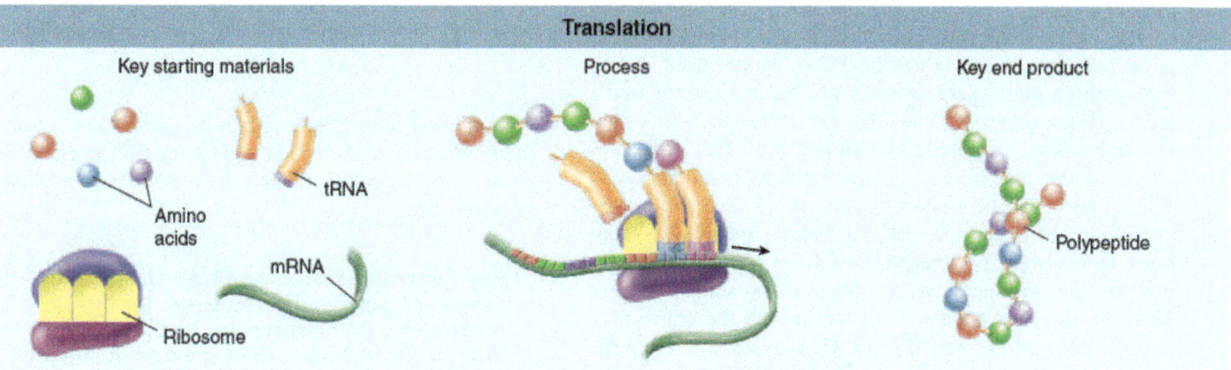

Figure 12.8 The processes of DNA replication, transcription, and translation.
These processes are generally the same in prokaryotes and eukaryotes.

the blue areas and the introns are the orange areas. Imagine looking at an interstate highway from a satellite. Scattered randomly along the thread of concrete would be cars, some moving in clusters, others individually; most of the road would be bare. That is what a eukaryotic gene is like: scattered exons embedded within much longer sequences of introns. In humans, only 1% to 1.5% of the genome is devoted to the exons that encode polypeptides, while 24% is devoted to the noncoding introns.

When a eukaryotic cell transcribes a gene, it first produces a **primary RNA transcript** of the entire gene, shown in figure 12.9 with the exons in green and the introns in orange. Enzymes add modifications called a *5′ cap* and a *3′ poly-A tail,* which protect the RNA transcript from degradation. The primary transcript is then processed. Enzyme-RNA complexes excise out the introns and join together the exons to form the shorter, mature mRNA transcript that is actually translated into an amino acid chain. Notice that the mature mRNA transcript in figure 12.9 contains only exons (green segments), no introns. Because introns are excised from the RNA transcript before it is translated into a polypeptide, they do not affect the structure of the polypeptide encoded by the gene in which they occur, despite the fact that introns represent over 90% of the nucleotide sequence of a typical human gene.

Why this crazy organization? It appears that many human genes can be spliced together in more than one way. In many instances, exons are not just random fragments, but rather functional modules. One exon encodes a straight stretch of protein, another a curve, yet another a flat place. Like mixing Tinkertoy parts, you can construct quite different assemblies by employing the same exons in different combinations and orders. With this sort of **alternative splicing,** the 25,000 genes of the human genome seem to encode as many as 120,000 different expressed messenger RNAs. It seems that added complexity in humans has been achieved not by gaining more gene parts (we have only about twice as many genes as a fruit fly), but rather by coming up with new ways to put them together.

Protein Synthesis

Protein synthesis in eukaryotes is more complex than in prokaryotes. Prokaryotic cells lack a nucleus and so there is no barrier between where mRNA is synthesized during transcription and where proteins are formed during translation. Consequently, a gene can be translated as it is being transcribed. Figure 12.10 shows how the ribosomes attach to the mRNA as it is synthesized in prokaryotes. These clusters of ribosomes on the mRNA are called *polyribosomes.* In eukaryotic cells, a nuclear membrane separates the process of transcription from translation, making protein synthesis much more complicated. Figure 12.11 on the next page walks you through the entire process. Transcription (step 1) and RNA processing (step 2) occur within the nucleus. In step 3, the mRNA travels to the cytoplasm where it binds to the ribosome. In step 4, tRNAs bind to their appropriate amino acids, which correspond to their anticodons. In steps 5 and 6, the tRNAs bring the amino acids to the ribosome and the mRNA is translated into a polypeptide.

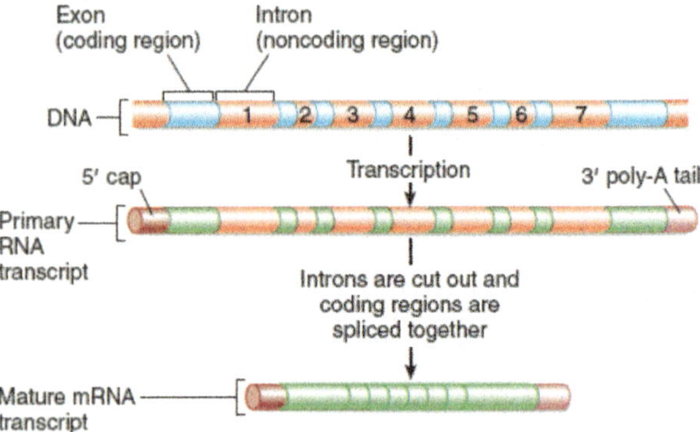

Figure 12.9 Processing eukaryotic RNA.
The gene shown here codes for a protein called ovalbumin. The ovalbumin gene and its primary transcript contain seven segments not present in the mRNA used by the ribosomes to direct the synthesis of the protein.

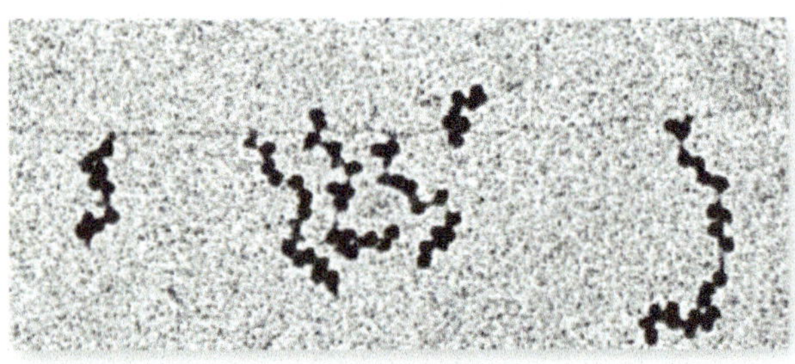

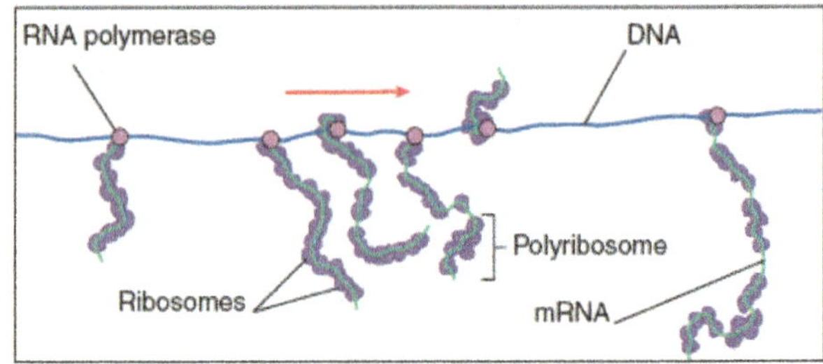

Figure 12.10 Transcription and translation in prokaryotes.
Ribosomes attach to an mRNA as it is formed, producing polyribosomes that translate the gene soon after it is transcribed.

Key Learning Outcome 12.4 The general process of gene expression is similar in prokaryotes and eukaryotes, but differences exist in the architecture of the gene and the location of transcription and translation in the cell.

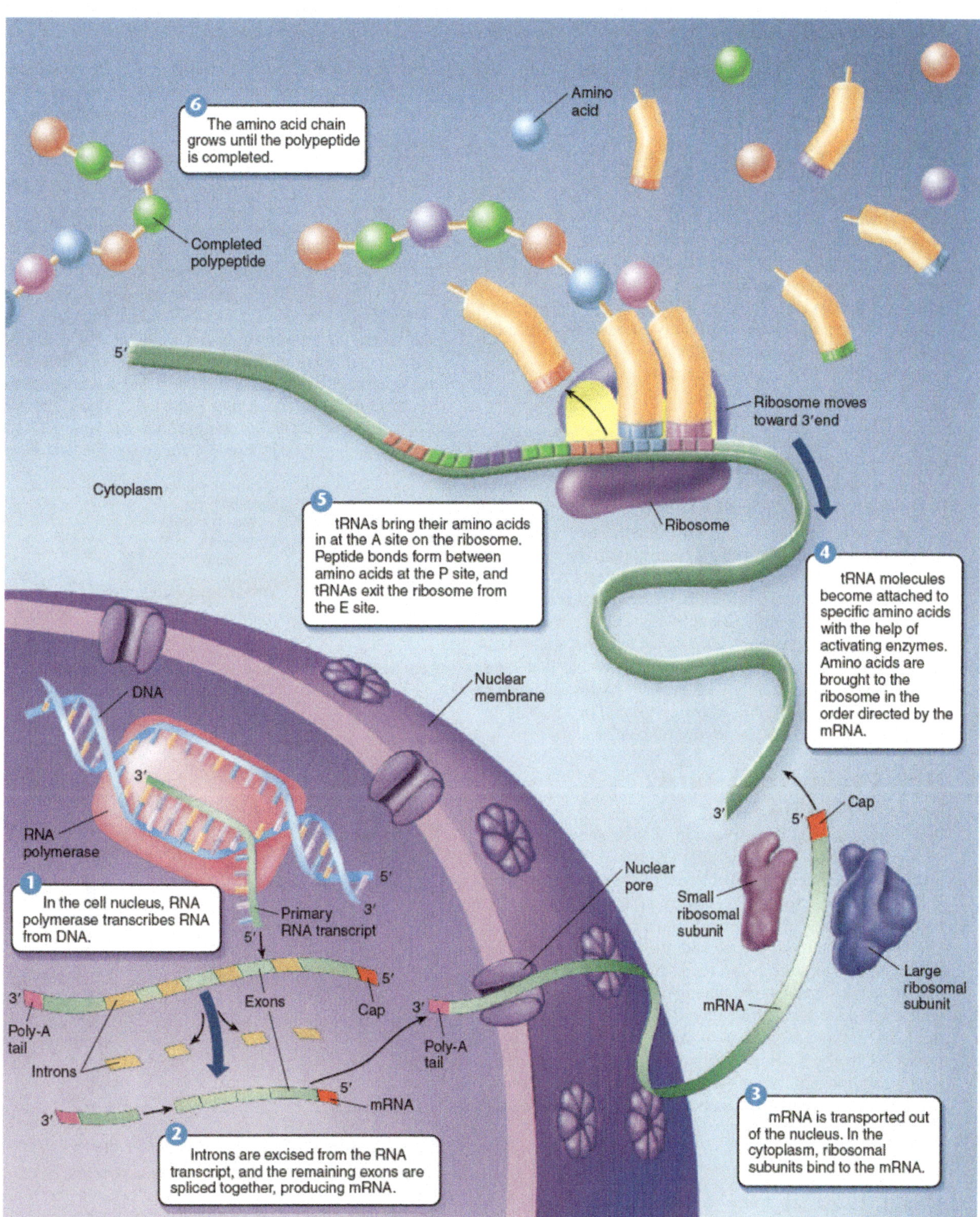

Figure 12.11 How protein synthesis works in eukaryotes.

Regulating Gene Expression in Prokaryotes

12.5 How Prokaryotes Control Transcription

Being able to translate a gene into a polypeptide is only part of gene expression. Every cell must also be able to regulate when particular genes are used. Imagine if every instrument in a symphony played at full volume all the time, all the horns blowing full blast and each drum beating as fast and loudly as it could! No symphony plays that way because music is more than noise—it is the controlled expression of sound. In the same way, growth and development are due to the controlled expression of genes, each brought into play at the proper moment to achieve precise and delicate effects.

Control of gene expression is accomplished very differently in prokaryotes than in the cells of complex multicellular organisms. Prokaryotic cells have been shaped by evolution to grow and divide as rapidly as possible, enabling them to exploit transient resources. Proteins in prokaryotes turn over rapidly. This allows them to respond quickly to changes in their external environment by changing patterns of gene expression. In prokaryotes, the primary function of gene control is to adjust the cell's activities to its immediate environment. Changes in gene expression alter which enzymes are present in the cell in response to the quantity and type of available nutrients and the amount of oxygen present. Almost all of these changes are fully reversible, allowing the cell to adjust its enzyme levels up or down as the environment changes.

How Prokaryotes Turn Genes Off and On

Prokaryotes control the expression of their genes largely by saying *when* individual genes are to be transcribed. At the beginning of each gene are special regulatory sites that act as points of control. Specific regulatory proteins within the cell bind to these sites, turning transcription of the gene off or on.

For a gene to be transcribed, the RNA polymerase has to bind to a **promoter,** a specific sequence of nucleotides on the DNA that signals the beginning of a gene. In prokaryotes, gene expression is controlled by either blocking or allowing the RNA polymerase access to the promoter. Genes can be turned off by the binding of a **repressor,** a protein that binds to the DNA blocking the promoter. Genes can be turned on by the binding of an **activator,** a protein that makes the promoter more accessible to the RNA polymerase.

Repressors

Many genes are "negatively" controlled: They are turned off except when needed. In these genes, the regulatory site is located between the place where the RNA polymerase binds to the DNA (the promoter site) and the beginning edge of the gene. When a regulatory protein called a repressor is bound to its regulatory site, called the *operator,* its presence blocks the movement of the polymerase toward the gene. Imagine if you went to sit down to eat dinner and someone was already sitting in your chair—you could not begin your meal until this person was removed from your chair. In the same way, the polymerase cannot begin transcribing the gene until the repressor protein is removed.

To turn on a gene whose transcription is blocked by a repressor, all that is required is to remove the repressor. Cells do this by binding special "signal" molecules to the repressor protein; the binding causes the repressor protein to contort into a shape that doesn't fit DNA, and it falls off, removing the barrier to transcription. A specific example demonstrating how repressor proteins work is the set of genes called the *lac* operon in the bacterium *Escherichia coli.* An **operon** is a segment of DNA containing a cluster of genes that are transcribed as a unit. The *lac* operon, shown in figure 12.12, consists of both polypeptide-encoding genes (labeled genes 1, 2,

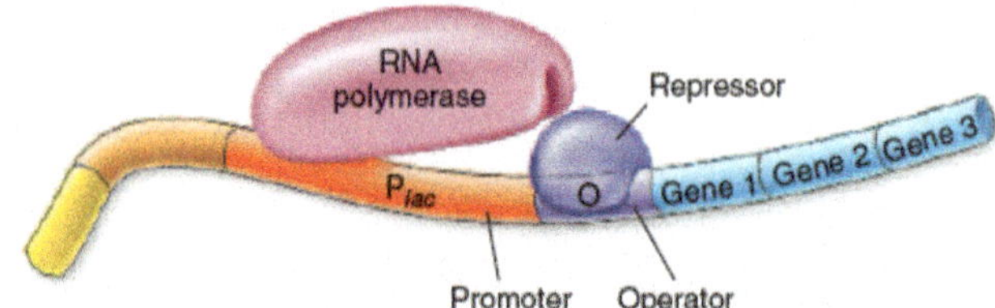

(a) *lac* operon is "repressed"

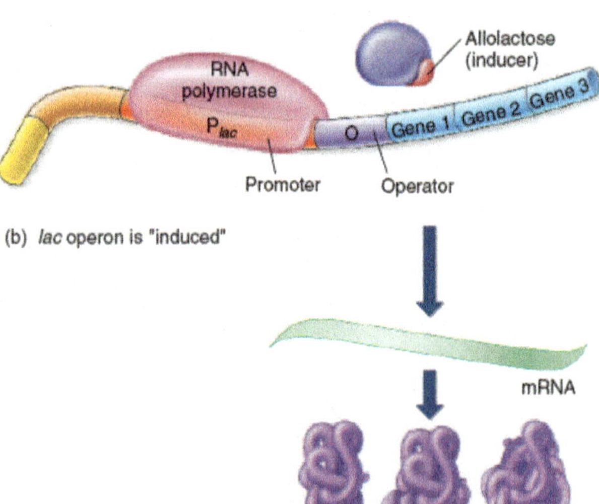

(b) *lac* operon is "induced"

Figure 12.12 How the lac operon works.
(*a*) The *lac* operon is shut down ("repressed") when the repressor protein is bound to the operator site. Because promoter and operator sites overlap, RNA polymerase and the repressor cannot bind at the same time. (*b*) The *lac* operon is transcribed ("induced") when allolactose binds to the repressor protein changing its shape so that it can no longer sit on the operator site and block polymerase binding.

and 3, which code for enzymes involved in breaking down the sugar lactose) and associated regulatory elements—the operator (the purple segment) and promoter (the orange segment). Transcription is turned off when a repressor molecule binds to the operator such that RNA polymerase cannot bind to the promoter. When *E. coli* encounters the sugar lactose, a metabolite of lactose called allolactose binds to the repressor protein and induces a twist in its shape that causes it to fall from the DNA. As you can see in figure 12.12*b*, RNA polymerase is no longer blocked, so it starts to transcribe the genes needed to break down the lactose to get energy.

Activators

Because RNA polymerase binds to a specific promoter site on one strand of the DNA double helix, it is necessary that the DNA double helix unzip in the vicinity of this site for the polymerase protein to be able to sit down properly. In many genes, this unzipping cannot take place without the assistance of a regulatory protein called an activator that binds to the DNA in this region and helps it unwind. Just as in the case of the repressor protein described previously, cells can turn genes on and off by binding "signal" molecules to the activator protein. These molecules either prevent the activator from binding to the DNA or enable it to do so. In the *lac* operon, a protein called *catabolite activator protein* (CAP) acts as an activator. CAP has to bind a signal molecule, cAMP, before it can associate with the DNA. Once the CAP/cAMP complex forms, as shown in figure 12.13, it binds to the DNA and makes the promoter more accessible to RNA polymerase.

Why bother with activators? Imagine if you had to eat every time you encountered food! Activator proteins enable a cell to cope with this sort of problem. Activators and repressors work together to control transcription. To understand how, let's consider the *lac* operon again, now shown in figure 12.14. When a bacterium encounters the sugar lactose, it may already have lots of energy in the form of glucose, as shown in panel ❶, and so does not need to break down more lactose.

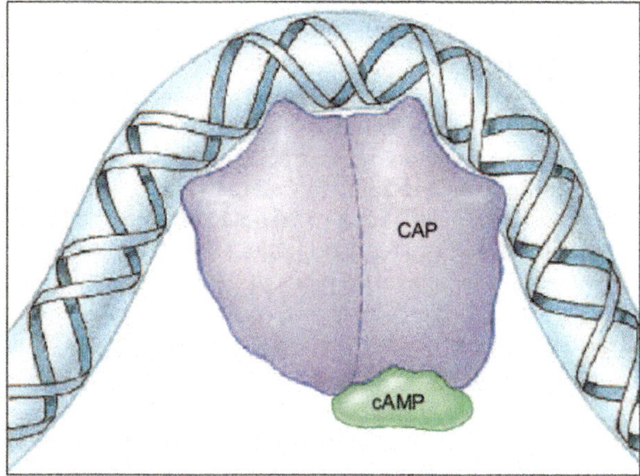

Figure 12.13 How an activator works.
Binding of the catabolite activator protein (CAP)/cAMP complex to DNA causes the DNA to bend around it. This increases the activity of RNA polymerase.

CAP can only bind and activate gene transcription when glucose levels are low. Because RNA polymerase requires the activator to function, the *lac* operon is not expressed. Also if glucose is present and lactose is absent, not only is the activator CAP unable to bind, but also a repressor blocks the promoter, as shown in panel ❷ below and in figure 12.12*a*. In the absence of both glucose and lactose, cAMP, the "low glucose" signal molecule (the green pie-shaped piece in panels ❸ and ❹) binds to CAP, and CAP is able to bind to the DNA. However, the repressor is still blocking transcription, as shown in panel ❸. Only in the absence of glucose and in the presence of lactose, the repressor is removed, the activator (CAP) is bound, and transcription proceeds, as shown in panel ❹.

Key Learning Outcome 12.5 Cells control the expression of genes by determining when they are transcribed. Some regulatory proteins block the binding of RNA polymerase, and others facilitate it.

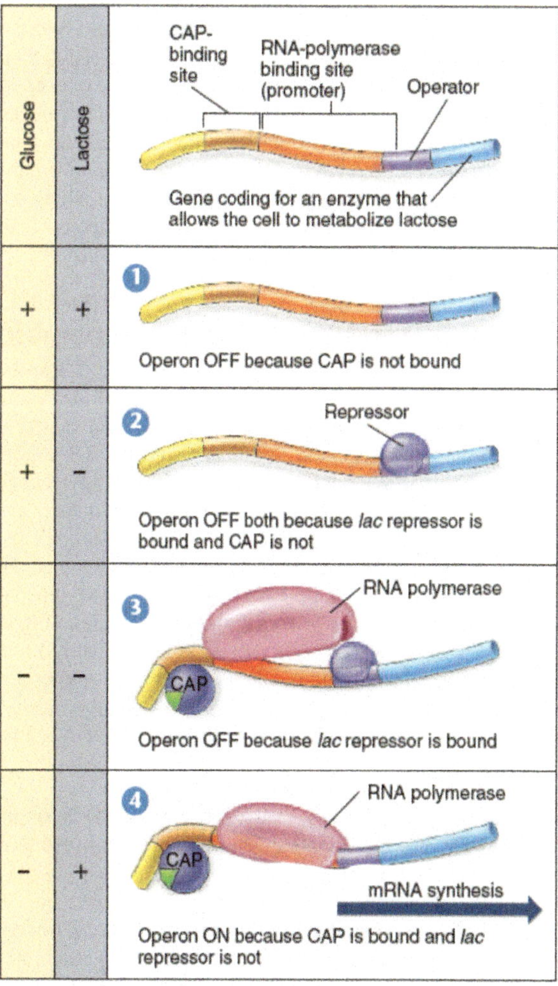

Figure 12.14 Activators and repressors at the *lac* operon.

Regulating Gene Expression in Eukaryotes

12.6 Transcriptional Control in Eukaryotes

The Goals of Gene Expression Are Different in Eukaryotes

In multicellular organisms with relatively constant internal environments, the primary function of gene control in a cell is not to respond to that cell's immediate environment, like a prokaryote does, but rather to participate in regulating the body as a whole. Some of these changes in gene expression compensate for changes in the physiological condition of the body. Others mediate the decisions that ultimately produce the body, ensuring that the right genes are expressed in the right cells at the right time during development. The growth and development of multicellular organisms entail a long series of biochemical reactions. To produce the necessary enzymes, genes are transcribed in a carefully prescribed order, each for a specified period of time, following a fixed genetic program. The one-time expression of the genes that guide such a program is fundamentally different from the reversible metabolic adjustments prokaryotic cells make to the environment, like the turning on and off of the *lac* operon. In all multicellular organisms, changes in gene expression within particular cells serve the needs of the whole organism, rather than the survival of individual cells.

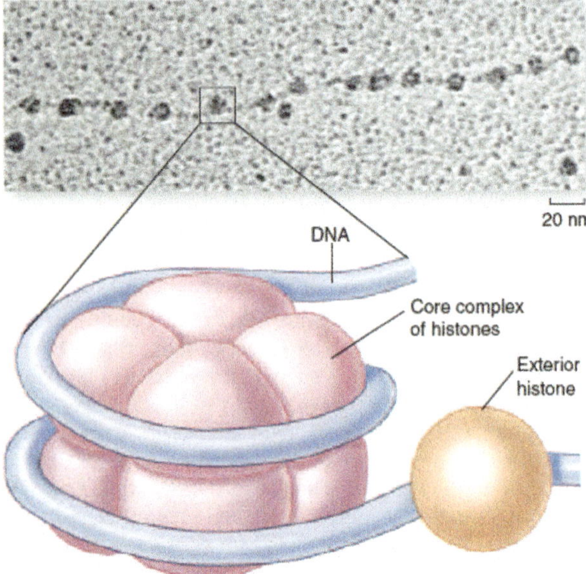

Figure 12.15 DNA coils around histones.
Within chromosomes, DNA is packaged into nucleosomes. In the electron micrograph (*top*), the DNA is partially unwound, and individual nucleosomes can be seen. In a nucleosome, the DNA double helix coils around a core complex of eight histones; one additional histone binds to the outside of the nucleosome.

The Structure of Chromosomes Can Affect Eukaryotic Gene Expression

The first hurdle faced by RNA polymerase in transcribing a eukaryotic gene is gaining access to it. The DNA of eukaryotes is packaged into chromosomes. The packaging of DNA into nucleosomes and then into higher-order chromosome structures (refer back to figure 8.4) is directly related to the control of gene expression. Chromosome structure at its lowest level is the organization of DNA and histone proteins into nucleosomes, as shown in figure 12.15. These nucleosomes may block binding of RNA polymerase and other proteins called transcription factors at the promoter. The higher-order organization of chromosomes is not completely understood. It involves modifying histones to produce a greater condensation of the chromosomal material, called *chromatin*, making promoters even less accessible for protein-DNA interactions.

DNA Methylation

Chemical methylation of the DNA was once thought to play a major role in regulating gene expression in vertebrate cells. Methylation is the process of adding a methyl group (—CH_3) to cytosine nucleotides, creating 5-methylcytosine, which is still able to participate in base pairs. Scientists observed that many inactive mammalian genes are methylated, and it was tempting to conclude that methylation caused the inactivation. However, methylation is now viewed as having a less direct role in transcriptional regulation. It appears instead to block accidental transcription of "turned-off" genes. DNA methylation thus ensures that once a gene is turned off, it stays off.

Chromatin Structure and Transcriptional Activators

As in prokaryotes, not all gene regulation in eukaryotes involves repression of transcription. In at least some instances, activation of a gene is needed before it can be transcribed. Most activating factors seem to act directly on transcription, helping the RNA polymerase to bind to the promoter. Other "coactivators" act by modifying the structure of chromatin to make DNA accessible. Recently, some coactivators have been shown to interact with histones. In these cases, it appears that transcription is increased by adding methyl groups to the histones. The methylation of histones disrupts the higher-order chromatin structure, making the DNA more accessible. This control also appears to work in the opposite way, with corepressors having been shown to remove methyl groups from the histones, making the DNA coil more tightly around the histones and restricting access to the DNA.

Key Learning Outcome 12.6 Transcriptional control in eukaryotes can be effected by the tight packaging of DNA into nucleosomes. Activators can loosen DNA's association with histones, making the DNA more accessible to RNA polymerase.

12.7 Controlling Transcription from a Distance

In eukaryotes, transcription is considerably more complex, and the amount of DNA involved in regulating eukaryotic genes is much greater.

Eukaryotic Transcription Factors

Eukaryotic transcription requires not only the RNA polymerase molecule, but also a variety of other proteins, called *transcription factors,* that interact with the polymerase.

Basal transcription factors are necessary for the assembly of a transcription apparatus and recruitment of RNA polymerase to a promoter. While these factors are required for transcription to occur, they do not increase the rate of transcription above a low rate, the so-called basal rate. These factors, colored green in figure 12.16, all come together to form the *initiation complex.* This is clearly much more complex than a bacterial RNA polymerase, which is a single enzyme protein.

The initiation complex, once assembled, will not achieve transcription at a high level without the participation of other gene-specific factors. The number and diversity of these **specific transcription factors,** colored tan in figure 12.16, is overwhelming. Multicellular organisms control which genes are expressed by regulating which specific transcription factors are available at a particular time and place.

Enhancers

While prokaryotic gene control regions, such as the operator, are positioned immediately upstream of the coding region, this is not true in eukaryotes. It turns out that far away sites called **enhancers** can have a major impact on the rate of transcription. Enhancers are nucleotide sequences where specific transcription factors, acting as activators, bind the DNA. The ability of enhancers to act over large distances is accomplished by DNA bending to form a loop. In figure 12.17, an activator binds to the yellow-colored enhancer far from the promoter. The DNA loops, bringing the activator in contact with the RNA polymerase/initiation complex so that transcription can begin.

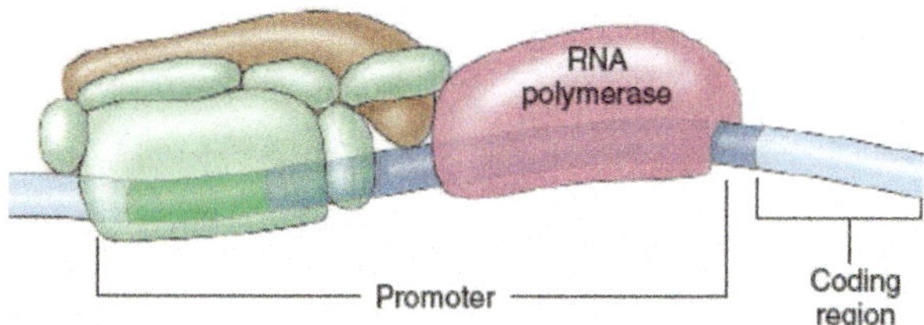

Figure 12.16 Formation of a eukaryotic initiation complex.

The basal transcription factors (*green*) bind to the promoter region of the DNA and form an initiation complex. A number of specific transcription factors (*tan*) bind to the basal transcription factor complex (the initiation complex) and together recruit the RNA polymerase molecule to the promoter.

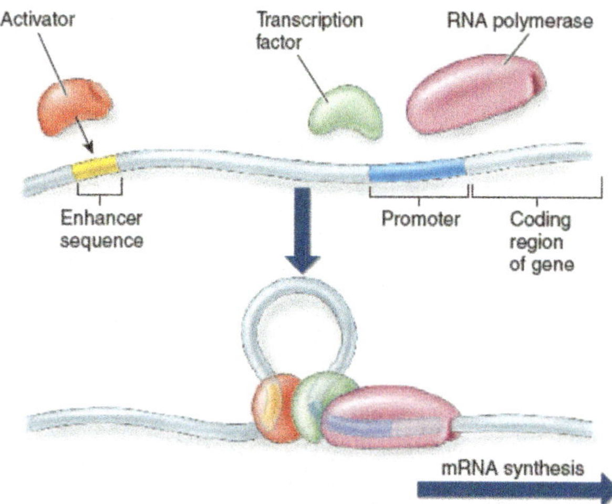

Figure 12.17 How enhancers work.

The activator binding site, or enhancer, is often located far from the gene. The binding of an activator protein brings the enhancer in contact with the gene.

Complicating the process even more, several different additional transcription factors may modulate the action of a particular specific transcription factor. These **coactivators** and **mediators** act by first binding the transcription factor, and then binding to the transcription apparatus.

Tying It All Together

How can we make sense of this extremely complicated situation? Virtually all genes that are transcribed by RNA polymerase in eukaryotes need the same group of basal factors to assemble an initiation complex, but its ultimate level of transcription depends in each instance on the other specific factors involved that make up a *transcription complex.* This kind of combined gene regulation leads to great flexibility in the control of gene expression. It provides the cell the ability to produce finely graded responses to the many environmental and developmental signals that it may receive. The eukaryotic cell achieves a higher level of control because of the interaction of a large number of protein regulatory elements (figure 12.18). This control is more sophisticated but not fundamentally different from the integration achieved by the prokaryotic *lac* operon using two regulatory proteins.

Key Learning Outcome 12.7 **Transcription factors and enhancers give eukaryotic cells great flexibility in controlling gene expression.**

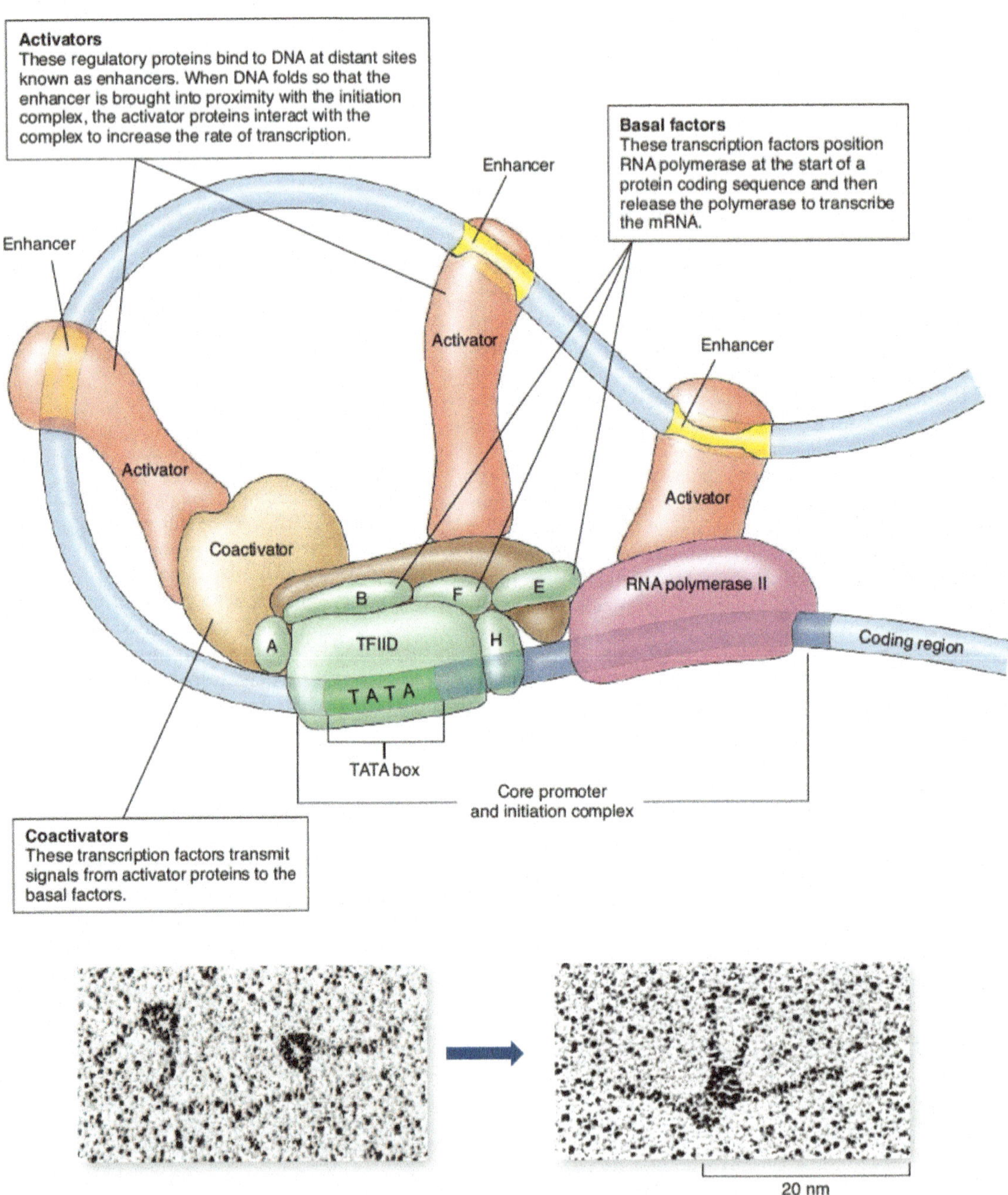

Figure 12.18 Interactions of various factors within the transcription complex.
All specific transcription factors bind to enhancer sequences that may be distant from the promoter. These proteins can then interact with the initiation complex by DNA looping to bring the factors into proximity with the initiation complex. As detailed in the text, some transcription factors, activators, can directly interact with the RNA polymerase II or the initiation complex while others require additional coactivators. The electron micrographs are of the bacterial activator NtrC. When it binds to an enhancer, you can see how this causes the DNA to loop over to a distant site where RNA polymerase is bound, activating transcription. While such enhancers are rare in prokaryotes, they are common in eukaryotes.

12.8 RNA-Level Control

Thus far we have discussed gene regulation entirely in terms of proteins that regulate the start of transcription by blocking or activating the "reading" of a particular gene by RNA polymerase. Within the last decade, however, it has become increasingly clear that RNA molecules can regulate the expression of genes, acting after transcription as a second level of control.

Discovery of RNA Interference

As will be discussed in chapter 13, the bulk of the eukaryotic genome is not translated into proteins. This finding was puzzling at first, but biologists now suspect that RNA transcripts of these regions might play an important role in gene regulation. The finding that almost all the differences between human and chimpanzee DNA occur in such regions only adds to the suspicion.

All this began to make sense in 1998, when a simple experiment was carried out, for which Americans Andrew Fire and Craig Mello later won the Nobel Prize in Physiology or Medicine in 2006. These investigators injected double-stranded RNA molecules into the nematode worm *Caenorhabditis elegans*. This resulted in the silencing of the gene whose sequence was complementary to the double-stranded RNA, and of no other gene. The investigators called this very specific effect **gene silencing,** or **RNA interference.** What is going on here? As you will learn in chapter 16, RNA viruses replicate themselves through double-stranded intermediates—at a critical stage, a virus enzyme called *reverse transcriptase* travels along the virus RNA and assembles a complementary strand. Because the life cycle of many viruses involves a double-stranded RNA stage, RNA interference may have evolved as a cellular defense mechanism against these viruses; the evolution of this adaptation would have predated the evolutionary divergence of plants and animals. Indeed, double-stranded viral RNAs can be targeted for destruction by RNA interference machinery. Without intending to do so, the nematode researchers had stumbled across this defense.

How RNA Interference Works

Investigating interference, researchers noted that in the process of silencing a gene, plants produced short RNA molecules (ranging in length from 21 to 28 nucleotides) that matched the gene being silenced. Earlier researchers were focusing on far larger messenger RNA (mRNA), transfer RNA (tRNA), and ribosomal RNA (rRNA) and had not noticed these far smaller bits, tossing them out during experiments. These small RNAs appeared to regulate the activity of specific genes.

Soon researchers found evidence of similar small RNAs in a wide range of other organisms. In the plant *Arabidopsis thaliana,* small RNAs seemed to be involved in the regulation of genes critical to early development, while in yeasts they were identified as the agents that silence genes in tightly packed regions of the genome. In the ciliated protozoan *Tetrahymena thermophila,* the loss of major blocks of DNA during development seems guided by small RNA molecules.

The first clue of how small fragments of RNA can act to regulate gene expression emerged when researchers noted that stretches of double-stranded RNA injected into *C. elegans* can dissociate. Each single strand can then form a double-stranded RNA by folding back in a hairpin loop, like the three sections of RNA shown toward the top of figure 12.19. This occurs because the two ends of the strand have a complementary nucleotide sequence. When the RNA loops, the complementary bases form base pairings that hold the strands together much as they do in the strands of a DNA duplex.

Figure 12.19 How RNA interference works. Double-stranded RNA is cut by dicer. The resulting siRNA associates with proteins forming a complex called RISC. siRNA becomes single stranded and binds to targeted mRNAs with the same or similar sequences, which blocks translation of the gene.

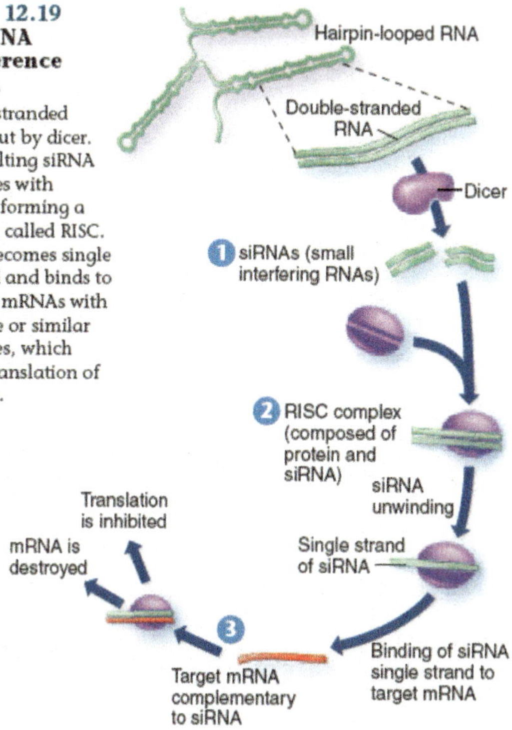

Exactly how does such a double-stranded RNA inhibit the expression of the gene from which the double-stranded RNA has been generated? In the first stage of RNA interference, an enzyme called *dicer* recognizes long, double-stranded RNA molecules and cuts them into short, small RNA segments called *siRNAs (small interfering RNAs)* 1. In the next step, the siRNAs can assemble into a ribonucleoprotein complex called *RISC (RNA Interference Silencing Complex)* 2. RISC then unwinds the siRNA duplex, which leaves one single strand of RNA that is able to bind to mRNAs complementary to it 3 and thus silence the genes that produced those mRNA molecules.

Once the siRNA has bound to mRNA, the silencing is achieved in one of two ways: Either the mRNA is inhibited by blocking its translation into protein, or the mRNA is destroyed. The choice between inhibition and destruction is thought to be governed by how closely the sequence of the siRNA matches the mRNA sequence, with destruction being the outcome for best-matched targets.

Key Learning Outcome 12.8 Small interfering RNAs, called siRNAs, are formed from double-stranded sections of RNA molecules. These siRNAs bind to mRNA molecules in the cell and block their translation.

Biology and *Staying Healthy*

Silencing Genes to Treat Disease

The recent discovery that eukaryotes control their genes by selectively "silencing" particular gene transcripts has electrified biologists, as it opens exciting possibilities for treating disease and infection. Many diseases are caused by the expression of one or more genes. AIDS, for example, requires the expression of several genes of the HIV virus. Many chronic human diseases result from excessively active genes. What if doctors could somehow shut these genes off?

The idea is simple. If you can isolate a gene involved in the disorder and determine its sequence, then in principle you could synthesize an RNA molecule with the sequence of the opposite or "anti-sense" strand. This RNA would thus have a sequence complementary to the messenger RNA produced by that gene. Introduced into cells, this synthesized RNA might be able to bind to the messenger RNA, creating a double-stranded RNA that could not be read by ribosomes. If an anti-sense therapy could be made to work and be delivered practically and inexpensively, the AIDS epidemic could be halted in its tracks. Indeed, any viral infection could be combatted in this way. Influenza is perhaps the greatest killer of all infectious diseases. A workable anti-sense therapy could provide a means of stamping out a bird flu epidemic before the virus spreads.

By far the most exciting promise of anti-sense gene silencing therapy is the possibility of practical cancer therapy. Discussed in chapter 8, cancer kills more Americans than any other disease. We now know in considerable detail how cancer comes about. It results from damage to genes that regulate the cell cycle. The great promise of RNA gene silencing therapy comes from those cancer-causing gene mutations that increase the effectiveness of one or more "divide" signals. If these mutant genes could be silenced, the cancer could be shut down.

The possibility of using complementary RNA to silence troublesome genes has gotten a huge boost in the last few years from the discovery of a unique virus defense system in eukaryotes. In order to protect themselves from RNA virus infection, cells have a complex system for detecting, attacking, and destroying viral RNA. The system takes advantage of a subtle vulnerability of the infecting virus: At some point, in order to multiply within the infected cell, the virus must express its genes—it must make complementary copies of them that can serve as messenger RNAs to direct production of virus proteins. At that point, while the viral RNA molecule is double-stranded, the virus is vulnerable to attack: At no place in the cell is double-stranded RNA usually found, so by targeting double-stranded RNAs for immediate destruction, a cell can defeat virus infections.

Silencing genes with complementary RNA, dubbed "RNA interference," offers the exciting hope that successful treatment of many diseases may be literally at our doorstep.

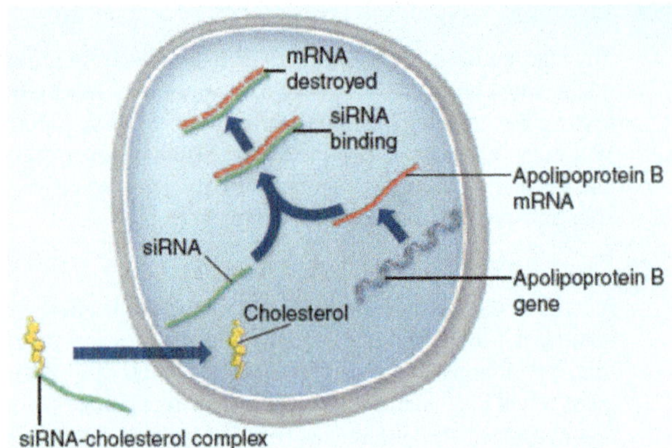

First, however, scientists must figure out how to make RNA interference therapies work. They are facing some formidable technical problems, not the least of which is to find a way to deliver the interfering RNA to, and into, the target cells. The problem is that RNA is rapidly broken down in the bloodstream, and most of the body's cells don't readily absorb it, even if it does reach them. Some researchers are attempting to package the RNA into viruses, although as you will learn in chapter 13, gene therapies that have attempted this approach can trigger an immune response and could even cause cancer. Gene therapy researchers have been seeking safer virus gene-delivery vehicles; what they learn will surely be put to good effect.

One interesting alternative approach is to modify the RNA to protect it and make it more easily taken up by cells. This work focuses on the mRNA that encodes apolipoprotein B, a molecule involved in the metabolism of cholesterol. High levels of apolipoprotein are found in people with high levels of cholesterol, associated with increased risk of coronary heart disease. Interfering RNAs that target apolipoprotein B mRNA result in destruction of the mRNA, and lower levels of cholesterol. To effectively deliver it to the body's tissues, researchers simply attached a molecule of cholesterol to each interfering RNA molecule, as shown above. Levels of apolipoprotein B were reduced 50% to 70%, and blood cholesterol levels plummeted downwards, to the same levels seen in cells from which the apolipoprotein B gene had been deleted. It is not clear if this approach will work for many other RNAs, but it looks promising.

A second major problem confronting those seeking to develop successful therapies based on RNA silencing of troublesome genes is one of specificity. It is very important that only the target gene be silenced. Before carrying out clinical trials involving large numbers of people, it is imperative that we be sure the interfering RNA will not shut down vital human genes as well as the targeted virus or cancer genes. Some studies suggest this will not be a problem, while in others a range of "off-target" genes seem to be affected. This possibility will have to be carefully evaluated for each new therapy being developed.

12.9 Complex Regulation of Gene Expression

As you have seen, the eukaryotic gene is structurally more complex than the prokaryotic gene, and the regulation of gene expression is also more complex. Eukaryotic gene expression is controlled at many stages, reviewed in figure 12.20. Chromatin structure can affect gene expression by determining if a gene will be accessible to RNA polymerase. The availability of many factors influences the rate at which a particular gene is transcribed. Once transcribed, a gene's expression can be altered by alternative splicing, or it can be silenced by RNA interference. Although gene regulation often occurs earlier in the process of gene expression, some control mechanisms act later. The availability of translational proteins can affect protein synthesis, and a protein can also be chemically modified after it is produced.

Key Learning Outcome 12.9 **A eukaryotic gene is controlled at many points during gene expression.**

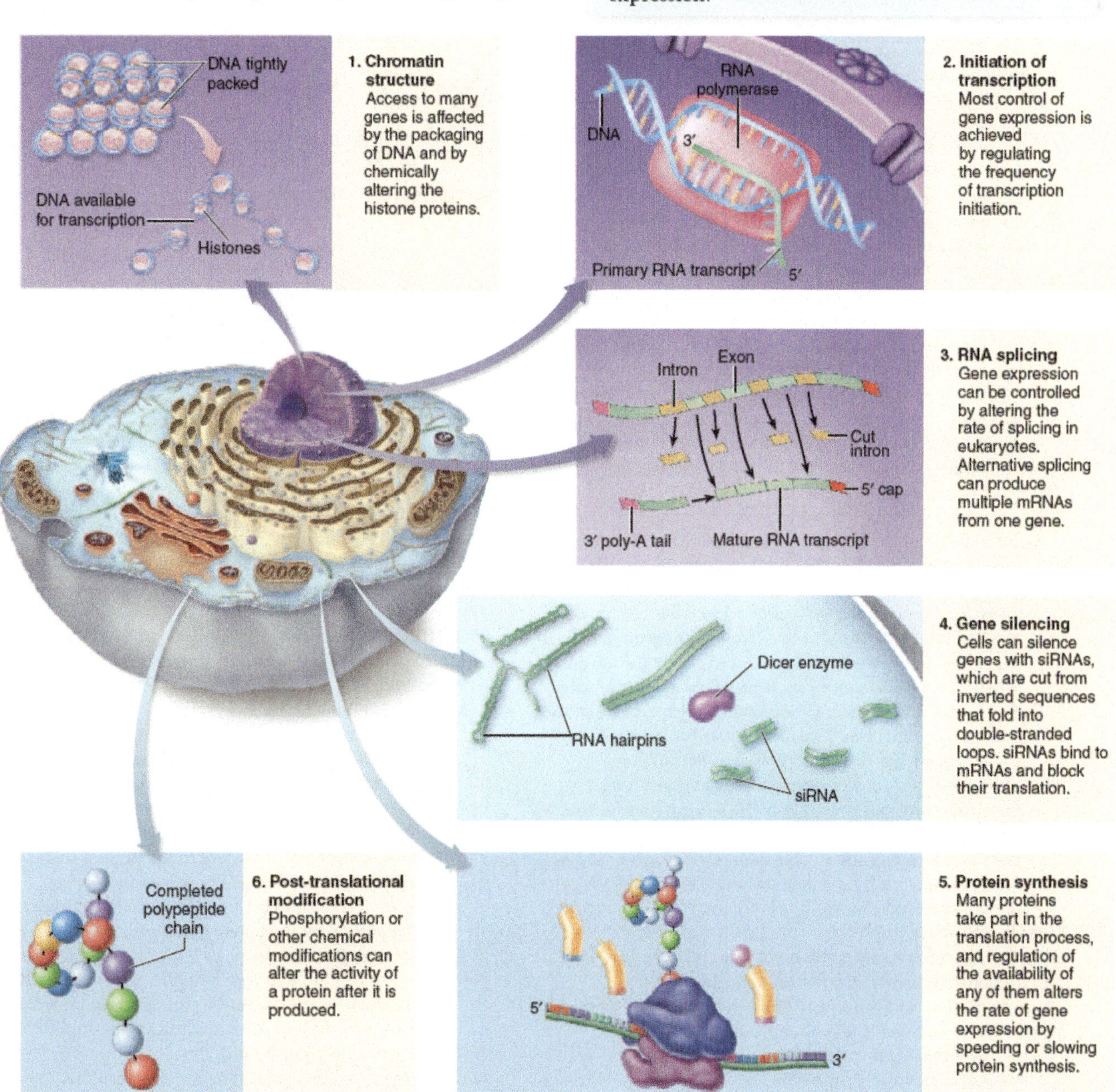

Figure 12.20 **The control of gene expression in eukaryotes.**

INQUIRY & ANALYSIS

Building Proteins in a Test Tube

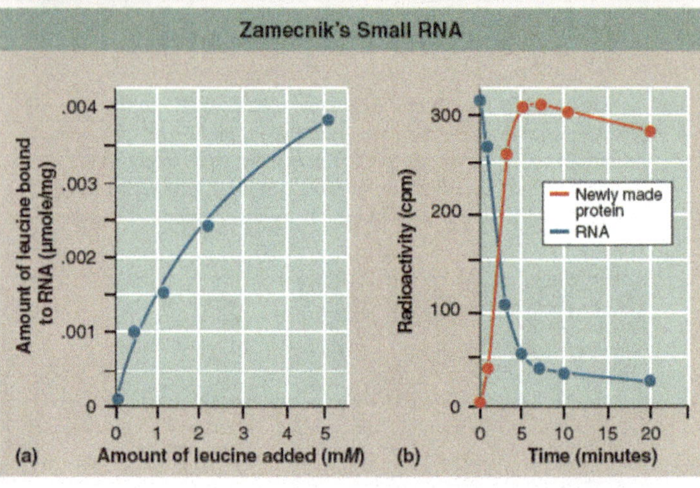

The complex mechanisms used by cells to build proteins were not discovered all at once. Our understanding came slowly, accumulating through a long series of experiments, each telling us a little bit more. To gain some sense of the incremental nature of this experimental journey, and to appreciate the excitement that each step gave, it is useful to step into the shoes of an investigator back when little was known and the way forward was not clear.

The shoes we will step into are those of Paul Zamecnik, an early pioneer in protein synthesis research. Working with colleagues at Massachusetts General Hospital in the early 1950s, Zamecnik first asked the most direct of questions: Where in the cell are proteins synthesized? To find out, they injected radioactive amino acids into rats. After a few hours, the labeled amino acids could be found as part of newly made proteins in the livers of the rats. And, if the livers were removed and checked only minutes after injection, radioactive-labeled proteins were found only associated with small particles in the cytoplasm. Composed of protein and RNA, these particles, later named ribosomes, had been discovered years earlier by electron microscope studies of cell components. This experiment identified them as the sites of protein synthesis in the cell.

After several years of trial-and-error tinkering, Zamecnik and his colleagues had worked out a "cell-free" protein-synthesis system that would lead to the synthesis of proteins in a test tube. It included ribosomes, mRNA, and ATP to provide energy. It also included a collection of required soluble "factors" isolated from homogenized rat cells that somehow worked with the ribosome to get the job done. When Zamecnik's team characterized these required factors, they found most of them to be proteins, as expected, but also present in the mix was a small RNA, very unexpected.

To see what this small RNA was doing, they performed the following experiment. In a test tube, they added various amounts of ^{14}C-leucine (that is, the radioactively labeled amino acid leucine) to the cell-free system containing the soluble factors, ribosomes, and ATP. After waiting a bit, they then isolated the small RNA from the mixture and checked it for radioactivity. You can see the results in graph (*a*).

In a follow-up experiment, they mixed the radioactive leucine-small RNA complex that this experiment had generated with cell extracts containing intact endoplasmic reticulum (that is, a cell system of ribosomes on membranes quite capable of making protein). Looking to see where the radioactive label now went, they then isolated the newly made protein as well as the small RNA [see graph (*b*)].

EXPERIMENT A, shown in graph (a)

1. **Applying Concepts** What is the dependent variable?
2. **Interpreting Data** Does the amount of leucine added to the test tube have an effect on the amount of leucine found bound to the small RNA?
3. **Making Inferences** Is the amount of leucine bound to small RNA proportional to the amount of leucine added to the mixture?
4. **Drawing Conclusions** Can you reasonably conclude from this result that the amino acid leucine is binding to the small RNA?

EXPERIMENT B, shown in graph (b)

1. **Applying Concepts** What is the dependent variable?
2. **Interpreting Data**
   a. Monitoring radioactivity for 20 minutes after the addition of the radioactive leucine-small RNA complex to the cell extract, what happens to the level of radioactivity in the small RNA (blue)?
   b. Over the same period, what happens to the level of radioactivity in the newly made protein (red)?
3. **Making Inferences** Is the same amount of radioactivity being lost from the small RNA that is being gained by the newly made protein?
4. **Drawing Conclusions**
   a. Is it reasonable to conclude that the small RNA is donating its amino acid to the growing protein? [Hint: As a result of this experiment, the small RNA was called transfer RNA.]
   b. If you were to isolate the protein from this experiment made after 20 minutes, which amino acids would be radioactively labeled? Explain.

Chapter Review

From Gene to Protein

12.1 The Central Dogma

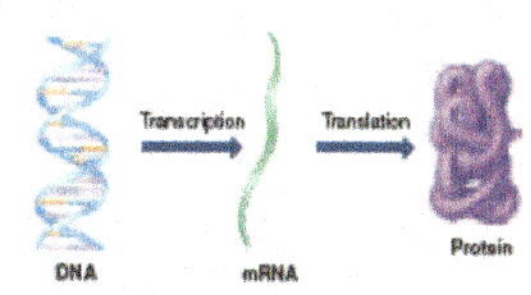

- DNA is the storage site of genetic information in the cell. The process of gene expression, DNA to RNA to protein, is called the "Central Dogma."
- Gene expression occurs in two stages using different types of RNA: transcription, where an mRNA copy is made from the DNA, and translation, where the information on the mRNA is translated into a protein using rRNA and tRNA.

12.2 Transcription

- In transcription, DNA serves as a template for mRNA synthesis by a protein called RNA polymerase. RNA polymerase binds to one strand of the DNA at a site called the promoter. RNA polymerase adds complementary nucleotides onto the growing mRNA in a way that is similar to the actions of DNA polymerase (**figure 12.3**).

12.3 Translation

- The message within the mRNA is coded in the order of the nucleotides, which are read in three-nucleotide units called codons. Each codon corresponds to a particular amino acid. The rules that govern the translation of codons on the mRNA into amino acids is called the genetic code (**figure 12.4**).
- The genetic code contains 64 codons but codes for only 20 amino acids; therefore, there is duplication. In many instances, two or more different codons encode the same amino acid.
- In translation, the mRNA carries the message to the cytoplasm. rRNA combines with proteins to form a structure called a ribosome (**figure 12.5**), the platform on which proteins are assembled. tRNAs, like one shown here from **figure 12.6**, carry amino acids to the ribosome to build the polypeptide chain.
- A ribosome contains a small rRNA subunit and a large rRNA subunit. Translation begins when the mRNA binds to the small ribosomal subunit, which triggers the binding of the large subunit to the small subunit, forming a complete ribosome.
- A ribosome moves along mRNA, while the tRNA molecules that contain anticodon sequences complementary to the codons bring amino acids to the ribosome. The amino acids add to the growing polypeptide chain (**Key Biological Process, page 240**).

12.4 Gene Expression

- Prokaryotic genes are contained within a stretch of DNA that is transcribed into mRNA and then is translated in its entirety. Eukaryotic genes are fragmented, containing coding regions called exons, and noncoding regions called introns.
- The entire eukaryotic gene is transcribed into RNA, but the introns are spliced out before translation. This editing process is shown here from **figure 12.9**.

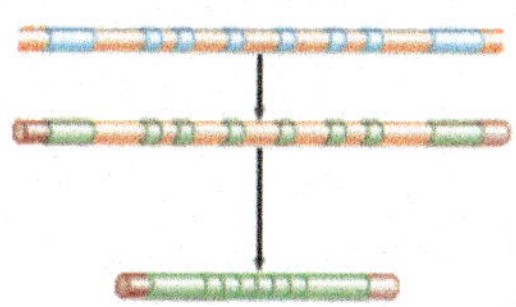

- Exons can be spliced together in different ways, a process called alternative splicing, producing different protein products from the same sections of DNA.
- In prokaryotes, transcription and translation occur simultaneously in the cytoplasm (**figure 12.10**). In eukaryotes, the RNA transcript is first produced and then processed (introns spliced out) in the nucleus. The mRNA then travels to the cytoplasm to be translated into a polypeptide (**figure 12.11**).

Regulating Gene Expression in Prokaryotes

12.5 How Prokaryotes Control Transcription

- In prokaryotic cells, genes are turned off when a repressor protein blocks the promoter, as shown here from **figure 12.12**, binding to a site called an operator.
- Some genes can be turned on only when a protein called an activator binds to the DNA and opens up the double helix so that RNA polymerase can bind to the promoter.
- The *lac* operon contains a cluster of genes that are involved in the breakdown of the sugar lactose. When the proteins produced by the *lac* operon genes are needed, an inducer molecule will bind to the repressor protein so it can't attach to the DNA, thereby freeing the promoter so that RNA polymerase can bind (**figure 12.12**).
- The *lac* operon is also controlled by an activator. The activator alters the shape of DNA, which allows the RNA polymerase to bind to the DNA (**figure 12.13**). It is only when the activator binds to the DNA and the repressor is removed that the RNA polymerase can bind to the promoter (**figure 12.14**).

Regulating Gene Expression in Eukaryotes

12.6 Transcriptional Control in Eukaryotes

- The coiling of DNA around histones (**figure 12.15**) restricts RNA polymerase's access to the DNA. Controlling gene expression may involve chemical modification of histones that make the DNA more accessible.

12.7 Controlling Transcription from a Distance

- In eukaryotic cells, transcription requires the binding of transcription factors before RNA polymerase can bind to the promoter (**figure 12.16**). Eukaryotic genes are controlled from distant locations called enhancers (**figures 12.17** and **12.18**).

12.8 RNA-Level Control

- RNA interference blocks translation. Small sections of RNA, called siRNA, bind to mRNA in the cytoplasm, which blocks the expression of the gene (**figure 12.19**).

12.9 Complex Regulation of Gene Expression

- Eukaryotic gene expression is controlled at many levels (**figure 12.20**).

Test Your Understanding

1. Which of the following is not a type of RNA?
   a. nRNA (nuclear RNA)
   b. mRNA (messenger RNA)
   c. rRNA (ribosomal RNA)
   d. tRNA (transfer RNA)
2. Each amino acid in a polypeptide is specified by
   a. an enhancer.
   b. a promoter.
   c. an rRNA molecule.
   d. a codon.
3. The three-nucleotide codon system can be arranged into ____________ combinations.
   a. 16
   b. 20
   c. 64
   d. 128
4. The process of obtaining a copy of the information in a gene as a strand of messenger RNA is called
   a. polymerase.
   b. expression.
   c. transcription.
   d. translation.
5. The site where RNA polymerase attaches to the DNA molecule to start the formation of an RNA molecule is called a(n)
   a. promoter.
   b. exon.
   c. intron.
   d. enhancer.
6. The process of taking the information on a strand of messenger RNA and building an amino acid chain, which will become all or part of a protein molecule, is called
   a. polymerase.
   b. expression.
   c. transcription.
   d. translation.
7. If an mRNA codon reads UAC, its complementary anticodon will be
   a. TUC.
   b. ATG.
   c. AUG.
   d. CAG.
8. Which of the following accurately describes gene expression in prokaryotic cells?
   a. All genes are on all the time in all cells, making the needed amino acid sequences.
   b. Some genes are always off unless a promoter turns them on.
   c. Some genes are always on unless a promoter turns them off.
   d. Some genes remain off as long as a repressor is bound.
9. Which of the following statements is correct about eukaryotic gene expression?
   a. mRNAs must have introns spliced out.
   b. mRNAs contain the transcript of only one gene.
   c. Enhancers act from a distance.
   d. All of the above.
10. Which of the following is *not* a mechanism of controlling gene expression in eukaryotic cells?
    a. blocking translation with siRNA
    b. activating an enhancer
    c. translating a gene as it is being transcribed
    d. alternative splicing of the primary RNA transcript

Apply Your Understanding

1. **Page 236** Assume that this figure is showing gene expression in a eukaryotic cell. What step is missing in the process?

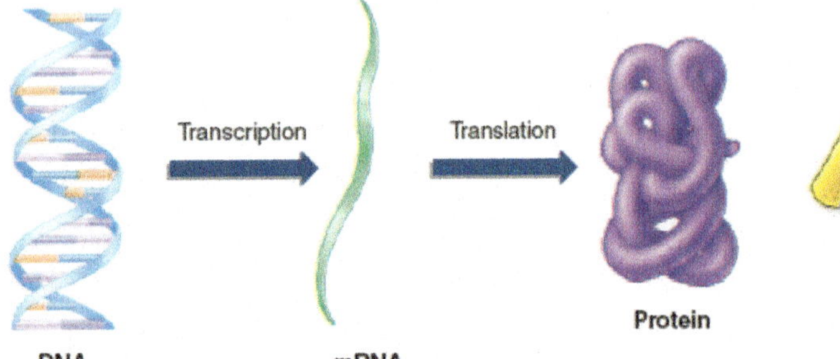

2. **Figure 12.12** Can genes 1, 2, and 3 be transcribed? Describe what would happen if an inducer molecule was added to the complex.

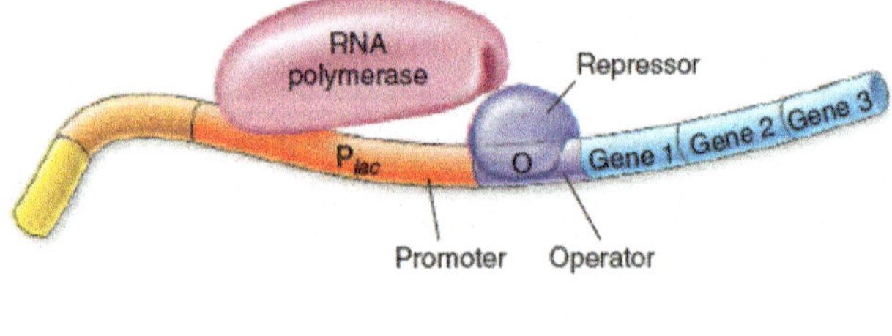

Synthesize What You Have Learned

1. On the television program *The X Files*, Agent Scully discovers an extraterrestrial life form that has a DNA genome like ours, but with a four-letter genetic code instead of the triplet genetic code that we earthly inhabitants possess. How many different amino acids would this extraterrestrial code be able to specify (assuming there were that many kinds of amino acids available on the extraterrestrial planet)? Why do you think the terrestrial code allows 64 combinations, when only 22 amino acids are common here on earth? Why do you think only 20 amino acids are common in proteins here on earth?
2. What would happen if all the genes in a cell were always active?
3. The nucleotide sequence of a hypothetical gene is:
   TACATACTTAGTTACGTCGCCCGGAAATAT
   a. What will be the sequence on the mRNA when it is transcribed?
   b. What will be the amino acid sequence of the protein when it's translated?
   c. What would happen to the amino acid chain if the highlighted nucleotide underwent a mutation and was changed to an A nucleotide?

13
The New Biology

Learning Objectives

Sequencing Entire Genomes

13.1 Genomics
1. Define genome.
2. Describe the four stages of DNA sequencing.

13.2 The Human Genome
1. State the number of base pairs in the human genome.
2. State the number of human protein-encoding genes; describe the four kinds.
3. Estimate how much of the genome is noncoding, and describe the four classes of noncoding DNA.

Genetic Engineering

13.3 A Scientific Revolution
1. Define restriction enzyme.
2. Explain how they are used to transfer genes.
3. Define cDNA, DNA fingerprinting, and PCR amplification.

Today's Biology: DNA and the Innocence Project

13.4 Genetic Engineering and Medicine
1. Describe how a piggyback vaccine is constructed and used.

13.5 Genetic Engineering and Agriculture
1. Describe how pest and herbicide resistance have been bioengineered in crop plants.
2. Assess whether GM foods are safe to eat, and whether GM crops are harmful to the environment.

Today's Biology: A DNA Timeline

The Revolution in Cell Technology

13.6 Reproductive Cloning
1. Explain the key insight of Keith Campbell that led to successful cloning.
2. Describe Wilmut's cloning of Dolly the sheep.
3. Explain the role of epigenetics in cloning success.

13.7 Stem Cell Therapy
1. Diagram the four stages of embryonic stem cell therapy.
2. Outline the ethical objections, and describe how Yamanaka's discovery suggests a way to circumvent them.

13.8 Therapeutic Use of Cloning
1. Distinguish between reproductive and therapeutic cloning.

13.9 Gene Therapy
1. Describe gene transfer therapy and its gene vector problem.

Inquiry & Analysis: Can Modified Genes Escape from GM Crops?

This sheep, Dolly, was the first animal to be cloned from a single adult cell. The lamb you see beside her is her offspring, normal in every respect. From Dolly we learn that genes are not lost during development. If a single adult cell can be induced to switch the proper combination of genes on and off, that one cell can develop into a normal adult individual. Embryonic stem cells are like this—poised to become any cell of the body as the embryo develops. It may be possible to replace damaged tissues with healthy tissue grown from a patient's own embryonic stem cells, so long as the disorder is not an inherited one. The approach has been used successfully in laboratory mice to cure a variety of disorders. However, its use in humans would require employing stem cells derived from the patient, and because this may involve destroying a human embryo the approach is controversial. Another approach, when the damaged tissue does result from a defective gene, is to repair rather than replace, using a virus to transfer a healthy gene into those tissues that lack it. In this chapter you will explore genomic screening, the application of gene technology to medicine and agriculture, reproductive cloning, stem cell tissue replacement, and gene therapy, all areas in which a revolution is reshaping biology.

Sequencing Entire Genomes

13.1 Genomics

Recent years have seen an explosion of interest in comparing the entire DNA content of different organisms, a new field of biology called **genomics.** While initial successes focused on organisms with relatively small numbers of genes, researchers have recently completed the sequencing of several large eukaryotic genomes, including our own.

The full complement of genetic information of an organism—all of its genes and other DNA—is called its **genome.** To study a genome, the DNA is first sequenced, a process that allows each nucleotide of a DNA strand to be read in order. The first genome to be sequenced was a very simple one: a small bacterial virus called ϕ-X174 (ϕ is the Greek letter phi). Frederick Sanger, inventor of the first practical way to sequence DNA, obtained the sequence of this 5,375-nucleotide genome in 1977. This was followed by the sequencing of dozens of prokaryotic genomes. The advent of automated DNA sequencing machines in recent years has made the DNA sequencing of much larger eukaryotic genomes practical, including our own (table 13.1).

Sequencing DNA

In sequencing DNA, the DNA of unknown sequence is first cut into fragments. Each DNA fragment is then copied (amplified), so there are thousands of copies of the fragment. The DNA fragments are then mixed with copies of DNA polymerase, copies of a primer (recall from chapter 11 that DNA polymerase can only add nucleotides onto an existing strand of nucleotides), a supply of the four nucleotide bases, and a supply of four different chain-terminating chemical tags. The chemical tags act as one of the four nucleotide bases in DNA synthesis, undergoing complementary base pairing. First, heat is applied to denature the double-stranded DNA fragments. The solution is then allowed to cool, allowing the primer (the lighter blue box in figure 13.1❶) to bind to a single strand of the DNA, and synthesis of the complementary strand proceeds. Whenever a chemical tag is added instead of a nucleotide base, the synthesis stops, as shown in the figure. For example, the terminating red "T" was added after three normal nucleotides and synthesis stopped. Because of the relatively low concentration of the chemical tags compared with the nucleotides, a tag that binds to G on the DNA fragment, for example, will not necessarily be added to the first G site. Thus, the mixture will contain a series of double-stranded DNA fragments of different lengths, corresponding to the different distances the polymerase traveled from the primer before a chain-terminating tag was incorporated (six are shown in ❶).

The series of fragments are then separated according to size by gel electrophoresis. The fragments become arrayed like the rungs of a ladder, each rung being one base longer than the one below it. Compare the lengths of the fragments in ❶ and their positions on the gel in ❷. The shortest fragment has only one nucleotide (G) added to the primer, so it is the lowest rung on the gel. In automated DNA sequencing, fluorescently colored chemical tags are used to label the fragments, one color for each type of nucleotide. Computers read the colors on the gel to determine the DNA sequence and display this sequence as a series of colored peaks (❸ and ❹). The development of automated sequencers in the mid-1990s has allowed the sequencing of large eukaryotic genomes. A research institute with several hundred such instruments can sequence 100 million base pairs every day, with only 15 minutes of human attention!

Key Learning Outcome 13.1 Powerful automated DNA sequencing technology is now revealing the DNA sequences of entire genomes.

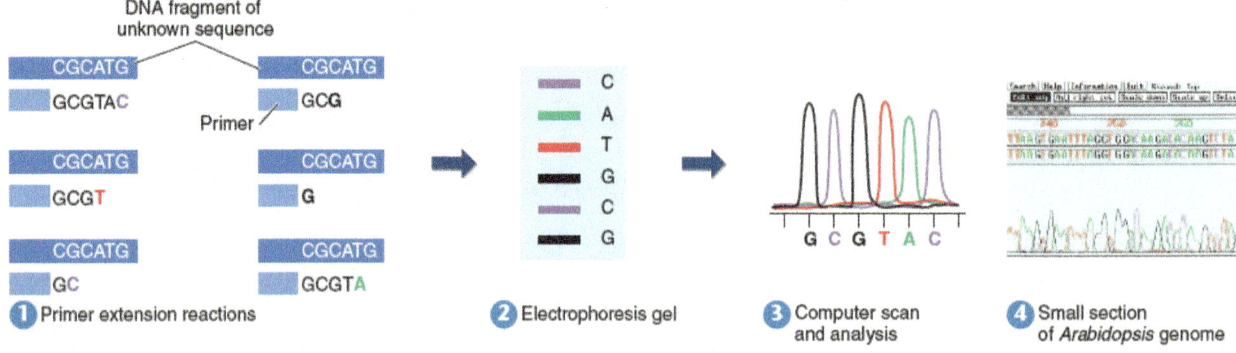

Figure 13.1 How to sequence DNA.
❶ DNA is sequenced by adding complementary bases to a single-stranded fragment. DNA synthesis stops when a chemical tag is inserted instead of a nucleotide, resulting in different sizes of DNA fragments. ❷ The DNA fragments of varying lengths are separated by gel electrophoresis, the smaller fragments migrating farther down the gel. (The boldface letters indicate the chemical tags added in step ❶ that stopped the replication process.) ❸ Computers scan the gel, from smallest to largest fragments, and display the DNA sequence as a series of colored peaks. ❹ Data from an automated DNA-sequencing run show the nucleotide sequence for a small section of the *Arabidopsis* (plant) genome.

TABLE 13.1 SOME EUKARYOTIC GENOMES

Organism		Estimated Genome Size (Mbp)	Number of Genes (×1,000)	Nature of Genome
Vertebrates				
	Homo sapiens (human)	3,200	20–25	The first large genome to be sequenced; the number of transcribable genes is far less than expected; much of the genome is occupied by repeated DNA sequences.
	Pan troglodytes (chimpanzee)	2,800	20–25	There are few base substitutions between chimp and human genomes, less than 2%, but many small sequences of DNA have been lost as the two species diverged, often with significant effects.
	Mus musculus (mouse)	2,500	25	Roughly 80% of mouse genes have a functional equivalent in the human genome; importantly, large portions of the noncoding DNA of mouse and human have been conserved; overall, rodent genomes (mouse and rat) appear to be evolving more than twice as fast as primate genomes (humans and chimpanzees).
	Gallus gallus (chicken)	1,000	20–23	One-third the size of the human genome; genetic variation among domestic chickens seems much higher than in humans.
	Fugu rubripes (pufferfish)	365	35	The *Fugu* genome is only one-ninth the size of the human genome, yet it contains 10,000 more genes.
Invertebrates				
	Caenorhabditis elegans (nematode)	97	21	The fact that every cell of *C. elegans* has been identified makes its genome a particularly powerful tool in developmental biology.
	Drosophila melanogaster (fruit fly)	137	13	*Drosophila* telomere regions lack the simple repeated segments that are characteristic of most eukaryotic telomeres. About one-third of the genome consists of gene-poor centric heterochromatin.
	Anopheles gambiae (mosquito)	278	15	The extent of similarity between *Anopheles* and *Drosophila* is approximately equal to that between human and pufferfish.
	Nematostella vectensis (sea anemone)	450	18	The genome of this cnidarian is much more like vertebrate genomes than nematode or insect genomes that appear to have become streamlined by evolution.
Plants				
	Oryza sativa (rice)	430	33–50	The rice genome contains only 13% as much DNA as the human genome, but roughly twice as many genes; like the human genome, it is rich in repetitive DNA.
	Populus trichocarpa (cottonwood tree)	500	45	This fast-growing tree is widely used by the timber and paper industries. Its genome, fifty times smaller than the pine genome, is one-third heterochromatin.
Fungi				
	Saccharomyces cerevisiae (brewer's yeast)	13	6	*S. cerevisiae* was the first eukaryotic cell to have its genome fully sequenced.
Protists				
	Plasmodium falciparum (malaria parasite)	23	5	The *Plasmodium* genome has an unusually high proportion of adenine and thymine. Scarcely 5,000 genes contain the bare essentials of the eukaryotic cell.

13.2 The Human Genome

On June 26, 2000, geneticists announced that the entire human genome had been sequenced. This effort presented no small challenge, as the human genome is huge—more than 3 billion base pairs, which is the largest genome sequenced to date. To get an idea of the magnitude of the task, consider that if all 3.2 billion base pairs were written down on the pages of this book, the book would be 500,000 pages long, and it would take you about 60 years, working eight hours a day, every day, at five bases a second, to read it all.

Reading the human genome for the first time, geneticists encountered four big surprises.

1. The Number of Genes Is Quite Low

The human genome sequence contains only 20,000 to 25,000 protein-encoding genes, only 1% of the genome. As you can see in figure 13.2, this is scarcely more genes than in a nematode worm (21,000 genes), not quite double the number in a fruit fly (13,000 genes). Researchers had confidently anticipated at least four times as many genes because over 100,000 unique messenger RNA (mRNA) molecules can be found in human cells—surely, they argued, it would take as many genes to make them.

How can human cells contain more mRNAs than genes? Recall from chapter 12 that in a typical human gene, the sequence of DNA nucleotides that specifies a protein is broken into many bits called exons, scattered among much longer segments of nontranslated DNA called introns. Imagine this paragraph was a human gene; all the occurrences of the letter "e" could be considered exons, while the rest would be noncoding introns, which make up 24% of the human genome.

When a cell uses a human gene to make a protein, it first manufactures mRNA copies of the gene, then splices the exons together, getting rid of the intron sequences in the process. Now here's the turn of events researchers had not anticipated: The exon portions of human gene transcripts are often spliced together in different ways, called *alternative splicing*. As we discussed in chapter 12, each exon is actually a module; one exon may code for one part of a protein, another for a different part of a protein. When the exon transcripts are mixed in different ways, very different protein shapes can be built.

With alternative mRNA splicing, it is easy to see how 25,000 genes can encode four times as many proteins. The added complexity of human proteins occurs because the gene parts are put together in new ways. Great music is made from simple tunes in much the same way.

2. Some Chromosomes Have Few Genes

In addition to the fragmenting of genes by the scattering of exons throughout the genome, there is another interesting "organizational" aspect of the genome. Genes are not distributed evenly over the genome. The small chromosome number 19 is packed densely with genes, transcription factors, and other functional elements. The much larger chromosome numbers 4 and 8, by contrast, have few genes, scattered like isolated hamlets in a desert. On most chromosomes, vast stretches of seemingly barren DNA fill the chromosomes between clusters rich in genes.

3. Genes Exist in Many Copy Numbers

Four different classes of protein-encoding genes are found in the human genome, differing largely in gene copy number.

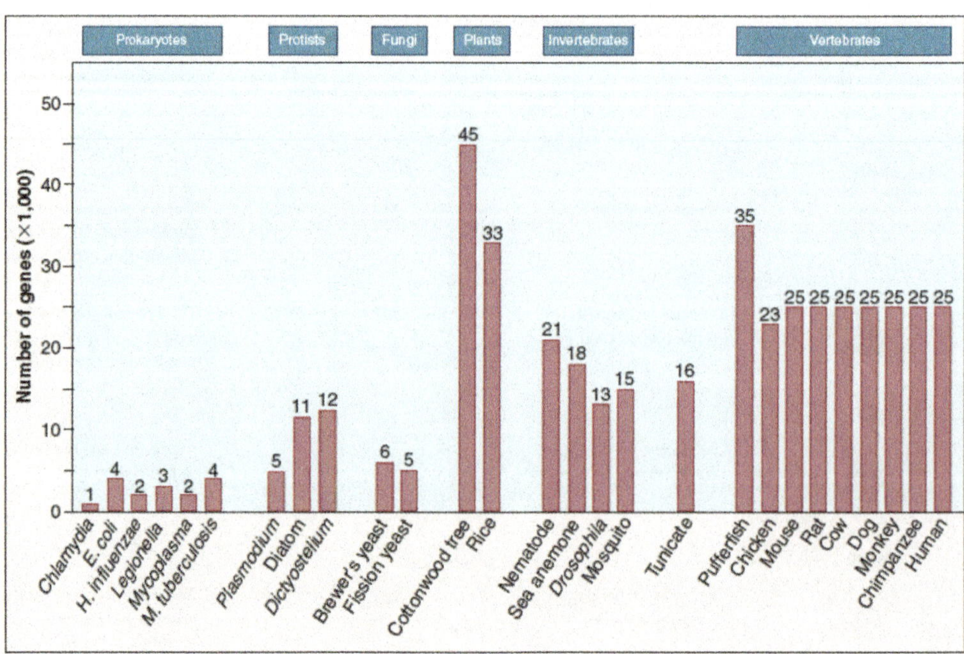

Figure 13.2 Comparing genome size.

All mammals have the same size genome, 20,000 to 25,000 protein-encoding nuclear genes. The unexpectedly larger sizes of the plant and pufferfish genomes are thought to reflect whole-genome duplications rather than increased complexity.

Single-copy genes. Many eukaryotic genes exist as single copies at a particular location on a chromosome. Mutations in these genes produce recessive Mendelian inheritance of those traits. Silent copies inactivated by mutation, called *pseudogenes,* are as common as protein-encoding genes.

Segmental duplications. Human chromosomes contain many segmental duplications, whole blocks of genes that have been copied over from one chromosome to another. Blocks of similar genes in the same order are found throughout the genome. Chromosome 19 seems to have been the biggest borrower, with blocks of genes shared with 16 other chromosomes.

Multigene families. Many genes exist as parts of multigene families, groups of related but distinctly different genes that often occur together in a cluster. Multigene families contain from three to several dozen genes. Although they differ from each other, the genes of a multigene family are clearly related in their sequences, making it likely that they arose from a single ancestral sequence.

Tandem clusters. These groups of repeated genes consist of DNA sequences that are repeated many thousands of times, one copy following another in tandem array. By transcribing all of the copies in these tandem clusters simultaneously, a cell can rapidly obtain large amounts of the product they encode. For example, the genes encoding rRNA are present in clusters of several hundred copies.

4. Most Genome DNA Is Noncoding

The fourth notable characteristic of the human genome is the startling amount of noncoding DNA it possesses. Only 1% to 1.5% of the human genome is coding DNA, devoted to genes encoding proteins. Each of your cells has about 6 feet of DNA stuffed into it, but of that, less than 1 inch is devoted to genes! Nearly 99% of the DNA in your cells seems to have little or nothing to do with the instructions that make you who you are (figure 13.3).

There are four major types of noncoding human DNA:

Noncoding DNA within genes. As discussed earlier, a human gene is made up of numerous fragments of protein-encoding information (exons) embedded within a much larger matrix of noncoding DNA (introns). Introns make up 24% of the human genome—exons only 1%!

Structural DNA. Some regions of the chromosomes remain highly condensed, tightly coiled, and untranscribed throughout the cell cycle. These portions—about 20% of the DNA—tend to be localized around the centromere, or located near the telomeres, or ends, of the chromosome.

Repeated sequences. Scattered about chromosomes are simple sequence repeats of two or three nucleotides like CA or CGG, repeated like a broken record thousands and thousands of times. These make up about 3% of the human genome. An additional 7% is devoted to other sorts of duplicated sequences. Repetitive sequences with excess C and G tend to be found in the neighborhood of genes, while A- and T-rich repeats dominate the nongene deserts. The light bands on chromosome karyotypes now have an explanation—they are regions rich in GC and genes. Dark bands signal neighborhoods rich in A and T, which are thin on genes. Chromosome 8, for example, contains many nongene areas that are indicated by dark bands, while chromosome 19 is dense with genes and so it has few dark bands.

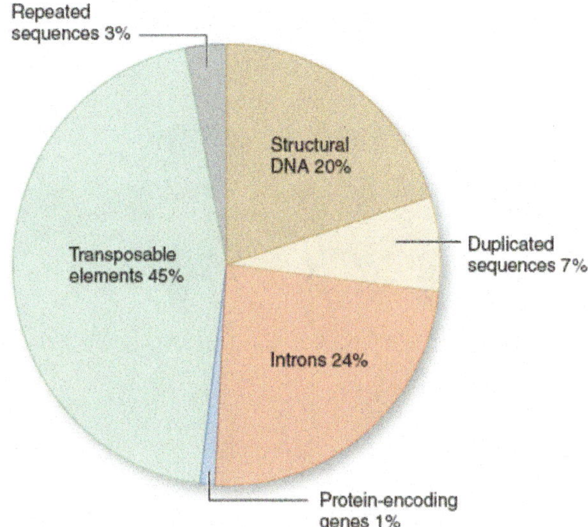

Figure 13.3 The human genome.
Very little of the human genome is devoted to protein-encoding genes, indicated by the light blue section in this pie chart.

Transposable elements. Fully 45% of the human genome consists of mobile parasitic bits of DNA called *transposable elements.* Discovered by Barbara McClintock in 1950 (she won the Nobel Prize in Physiology or Medicine for her discovery in 1983), transposable elements are bits of DNA that are able to jump from one location on a chromosome to another—tiny molecular versions of Mexican jumping beans. Because they leave a copy of themselves behind when they jump, their numbers in the genome increase as generations pass. Nested within the human genome are over half a million copies of an ancient transposable element called *Alu,* composing fully 10% of the entire human genome. Often jumping right into genes, *Alu* transpositions cause many harmful mutations.

Key Learning Outcome 13.2 The entire 3.2-billion-base pair human genome has been sequenced. Only about 1% to 1.5% of the human genome is devoted to protein-encoding genes. Much of the rest is composed of transposable elements.

13.3 A Scientific Revolution

In recent years, **genetic engineering**—the ability to manipulate genes and move them from one organism to another—has led to great advances in medicine and agriculture (figure 13.4). Most of the insulin used to treat diabetes is now obtained from bacteria that contain a human insulin gene. In late 1990, the first transfers of genes from one human to another were carried out to correct the effects of defective genes in a rare genetic immune disorder called *Severe Combined Immunodeficiency* (SCID)—often called the "Bubble Boy Disorder" based on a young boy who lived out his life in an enclosed, germ-free environment. In addition, cultivated plants and animals can be genetically engineered to resist pests, grow bigger, or grow faster.

Restriction Enzymes

The first stage in any genetic engineering experiment is to chop up the "source" DNA to get a copy of the gene you wish to transfer. This first stage is the key to successful transfer of the gene, and learning how to do it is what has led to the genetic revolution. The trick is in how the DNA molecules are cut. The cutting must be done in such a way that the resulting DNA fragments have "sticky ends" that can later be joined with another molecule of DNA.

This special form of molecular surgery is carried out by **restriction enzymes,** also called *restriction endonucleases,* which are special enzymes that bind to specific short sequences (typically four to six nucleotides long) on the DNA. These sequences are very unusual in that they are symmetrical—the two strands of the DNA duplex have the same nucleotide sequence, running in opposite directions! The sequence in figure 13.5, for example, is GAATTC. Try writing down the sequence of the opposite strand: it is CTTAAG—the same sequence, written backward. This sequence is recognized by

Curing disease. One of two young girls who were the first humans "cured" of a hereditary disorder by transferring into their bodies healthy versions of a defective gene. The transfer was successfully carried out in 1990, and twenty years later the girls remain healthy.

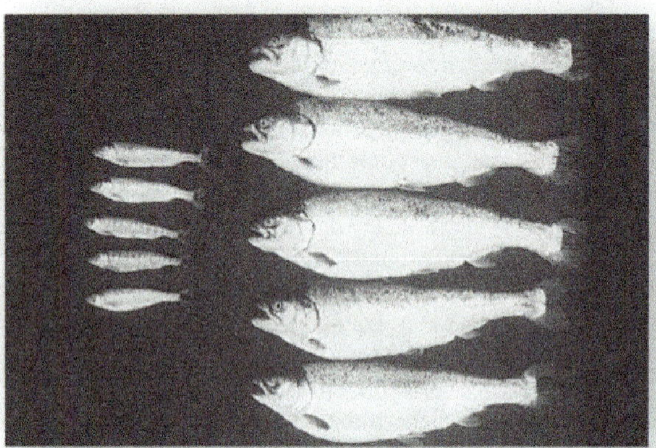

Increasing yields. The genetically engineered salmon on the *right* have shortened production cycles and are heavier than the nontransgenic salmon of the same age on the *left*.

Pest-proofing plants. The genetically engineered cotton plants on the *right* have a gene that inhibits feeding by weevils; the cotton plants on the *left* lack this gene, and produce far fewer cotton bolls.

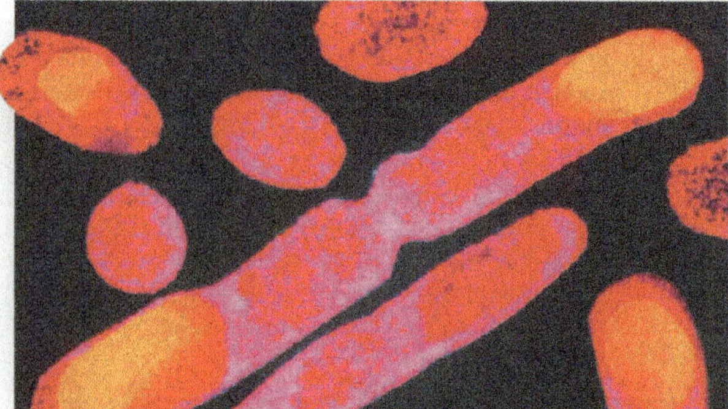

Producing insulin. The common bacteria *Escherichia coli* (*E. coli*) can be genetically engineered to contain the gene that codes for the protein insulin. The bacteria are turned into insulin-producing factories and can produce large quantities of insulin for diabetic patients. In the image above, insulin-producing sites inside genetically-altered *E. coli* cells are orange.

Figure 13.4 Examples of genetic engineering.

the restriction enzyme *Eco*RI. Other restriction enzymes recognize other sequences.

What makes the DNA fragments "sticky" is that most restriction enzymes do not make their incision in the center of the sequence; rather, the cut is made to one side. In the sequence in figure 13.5 ❶, the cut is made on both strands between the G and A nucleotides, G/AATTC. This produces a break, with short, single strands of DNA dangling from each end. Because the two single-stranded ends are complementary in sequence, they could pair up and heal the break, with the aid of a sealing enzyme—*or* they could pair with *any other DNA fragment cut by the same enzyme* because all would have the same single-stranded sticky ends. Figure 13.5 ❷ shows how DNA from another source (the orange DNA) also cut with *Eco*RI has the same sticky ends as the original source DNA. Any gene in any organism cut by the enzyme that attacks GAATTC sequences will have the same sticky ends, and can be joined to any other with the aid of a sealing enzyme called *DNA ligase* ❸.

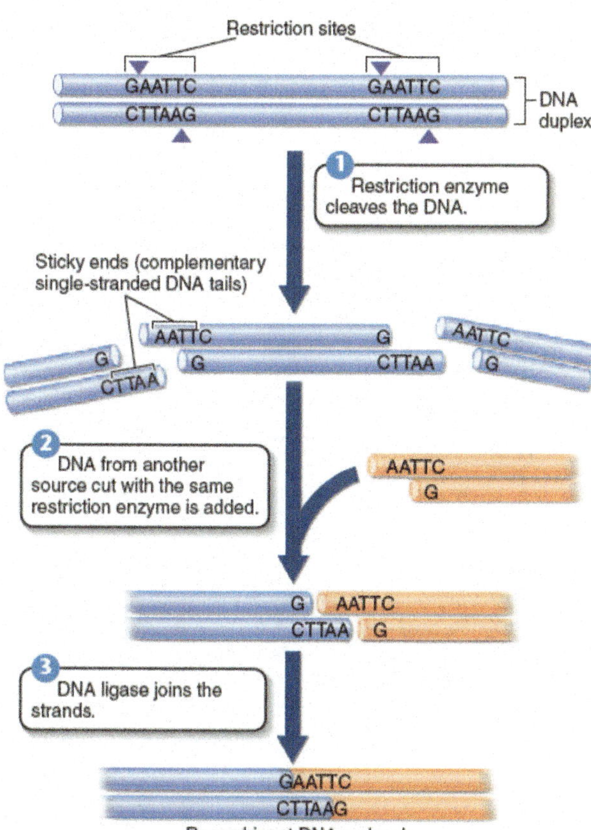

Figure 13.5 How restriction enzymes produce DNA fragments with sticky ends.

The restriction enzyme *Eco*RI always cleaves the sequence GAATTC between G and A. Because the same sequence occurs on both strands, both are cut. However, the two sequences run in opposite directions on the two strands. As a result, single-stranded tails are produced that are complementary to each other, or "sticky."

Formation of cDNA

As already mentioned, eukaryotic genes are encoded in segments called exons separated from one another by numerous nontranslated sequences called introns. The entire gene is transcribed by RNA polymerase, producing what is called the primary RNA transcript (figure 13.6). Before a eukaryotic gene can be translated into a protein, the introns must be cut out of this primary transcript. The fragments that remain are then spliced together to form the mRNA, which is eventually translated in the cytoplasm. When transferring eukaryotic genes into bacteria (discussed in section 13.4), it is necessary to transfer DNA that has had the intron information removed because bacteria lack the enzymes to carry out this processing. Bacterial genes do not contain introns. To produce eukaryotic DNA without introns, genetic engineers first isolate from the cytoplasm the processed mRNA corresponding to a particular gene. The cytoplasmic mRNA has *only* exons, properly spliced together. An enzyme called *reverse transcriptase* is then used to make a complementary DNA strand of the mRNA. A double-stranded DNA molecule is then produced. Such a version of a gene is called *complementary DNA*, or **cDNA.**

cDNA technology is being used in other ways, such as determining the patterns of gene expression in different cells. Every cell in an organism contains the same DNA, but genes are selectively turned on and off in any given cell. Researchers can see what genes are actively expressed using cDNAs.

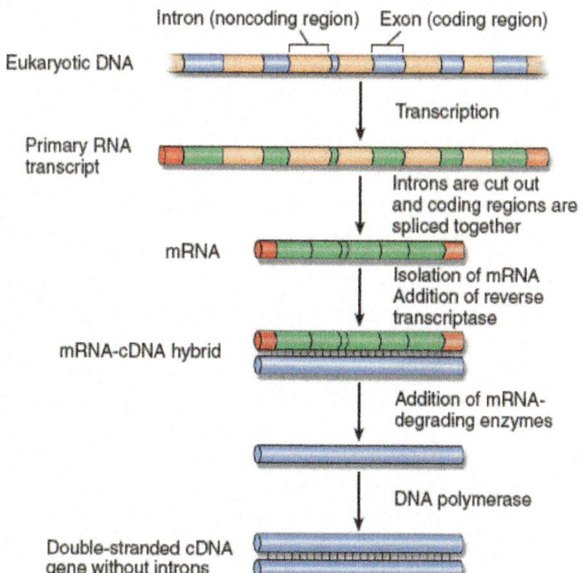

Figure 13.6 cDNA: Producing an intron-free version of a eukaryotic gene for genetic engineering.

In eukaryotic cells, a primary RNA transcript is processed into the mRNA. The mRNA is isolated and converted into cDNA.

DNA Fingerprinting and Forensic Science

DNA fingerprinting is a process used to compare samples of DNA. Just as fingerprinting revolutionized forensic evidence in the early 1900s, so DNA fingerprinting is revolutionizing it today. A hair, a minute speck of blood, a drop of semen, all can serve as sources of DNA to convict or clear a suspect.

The process of DNA fingerprinting uses probes to fish out particular sequences to be compared from the thousands of other sequences in the human genome. Because different people have different DNA sequences throughout their genome, they will have different restriction enzyme cutting sites, and so produce different-sized fragments that move to different locations on a gel. Radioactive probes bind to particular locations that occur randomly many times in the genome, "lighting up" the fragments that contain that sequence. If several different probes are used, the chance of any two individuals having the same gel pattern is less than one in a billion. DNA fragments that bind the probes are then visible on autoradiographic film as dark bands. The autoradiograph gel patterns are in essence DNA "fingerprints" that can be used in criminal investigations and other identification applications.

In the first time DNA evidence was used in a court of law, DNA probes were used to characterize DNA isolated from a rape victim's blood, from semen left by the rapist, and from the suspect's blood:

You can see that the suspect's pattern (indicated in pink) matches that of the rapist, and is not at all like that of the victim. Other probes produced similar results. On November 6, 1987, the jury returned a verdict of guilty, the first time a person in the United States was convicted of a crime based on DNA evidence. Since this verdict, DNA fingerprinting has been admitted as evidence in thousands of court cases.

PCR Amplification

Tiny DNA samples, as little as that found in a single human hair, can be magnified to millions of copies by a process called **PCR** (for **polymerase chain reaction**). The use of DNA in forensic analysis often depends critically on this PCR technology, developed in 1983 by Kary Mullis. In PCR, a double-stranded DNA fragment is heated so it becomes single stranded; each of the strands is then copied by DNA polymerase to produce two double-stranded fragments. The fragments are heated again and copied again to produce four double-stranded fragments. This cycle is repeated many times, each time doubling the number of copies, until enough copies of the DNA fragment exist for analysis (figure 13.7).

The amount of DNA found in a single hair will do. Biologists used to think that DNA was present only in the cells at the root of a hair, not in the hair shaft made of the protein keratin. But we now know follicle cells become incorporated into the growing shaft and their DNA is sealed in by the keratin, protecting it from being degraded by bacteria and fungi.

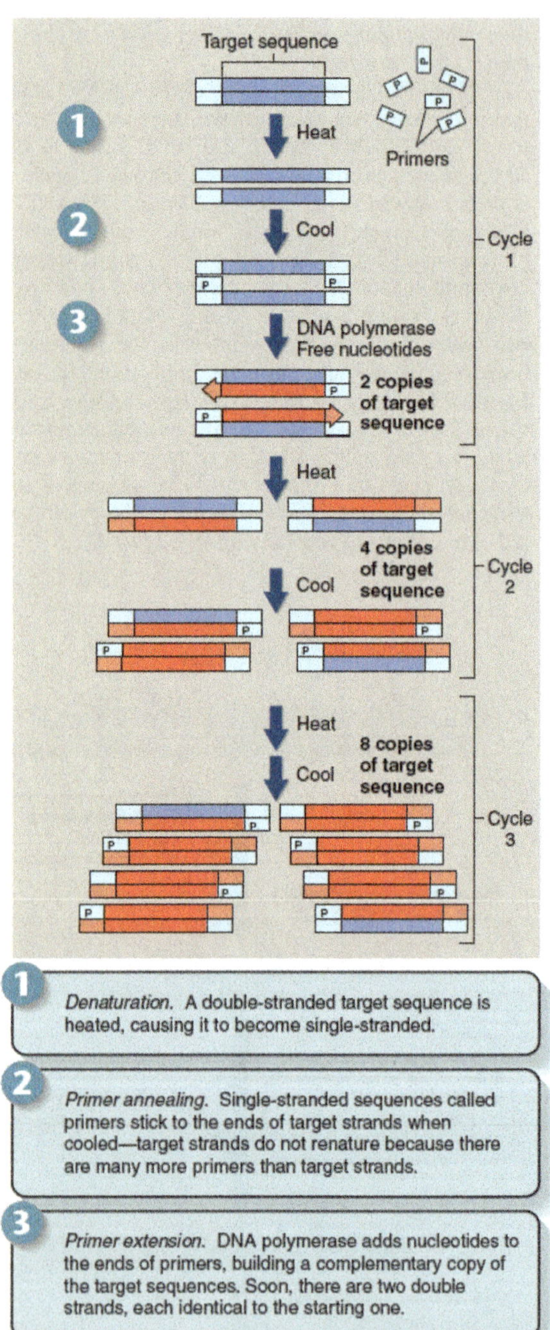

1 *Denaturation.* A double-stranded target sequence is heated, causing it to become single-stranded.

2 *Primer annealing.* Single-stranded sequences called primers stick to the ends of target strands when cooled—target strands do not renature because there are many more primers than target strands.

3 *Primer extension.* DNA polymerase adds nucleotides to the ends of primers, building a complementary copy of the target sequences. Soon, there are two double strands, each identical to the starting one.

Figure 13.7 How the polymerase chain reaction works.

Key Learning Outcome 13.3 **Restriction enzymes, the key tools that make genetic engineering possible, bind to specific short sequences of DNA and cut the DNA. This produces fragments with "sticky ends," which can be rejoined in different combinations.**

Today's *Biology*

DNA and the Innocence Project

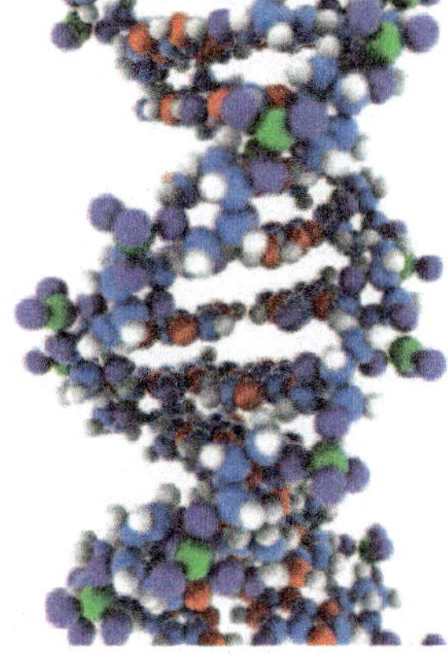

Every person's DNA is uniquely their own, a sequence of nucleotides found in no other person. Like a molecular social security number a billion digits long, the nucleotide sequence of an individual's genes can provide proof-positive identification of a rapist or murderer from DNA left at the scene of a crime—proof more reliable than fingerprints, more reliable than an eye witness, even more reliable than a confession by the suspect.

Never was this demonstrated more clearly than on May 16, 2006 in a Rochester, New York courtroom. There a judge freed convicted murderer Douglas Warney after 10 years in prison.

The murder occurred on New Year's Day in 1996. The bloody body of one William Beason, a prominent community activist, was found in his bed. Police called in all the usual suspects, in this case everyone known to be an acquaintance of the victim. Warney, an unemployed 34-year-old who had dropped out of school in the eighth grade, committed robberies, and worked as a male hustler, learned that detectives wanted to speak to him about the killing, and went to the police station for questioning. Within hours he was charged with murder.

Warney's interrogation was not recorded, but it resulted in a signed confession based on the words that the detective sergeant said Warney uttered, a confession that contained accurate details about the murder scene that had not been made public. Warney said that the victim was wearing a nightgown and had been cooking chicken in a pot, and that the murderer had used a 12-inch serrated knife. The case against him in court rested almost entirely on these vivid details. Even though Warney recanted his confession at his trial, the accuracy of his confession was damning. The details that Warney provided to the sergeant could have come only from someone who was present at the crime scene.

There were problems with his confession, brought out at trial. Three elements of the signed statement's account of that night were clearly not true. It said Warney had driven to the victim's house in his brother's car, but his brother did not own a car; it said Warney disposed of his bloody clothes after the stabbing in the garbage can behind the house, but the can, buried in snow from the day of the crime, did not contain bloody clothes; it named a relative of Warney as an accomplice, but that relative was in a secure rehabilitation center on that day.

The most difficult bit of evidence to match to the written confession was blood found at the scene that was not that of the victim. Drops of a second person's blood were found on the floor and on a towel. The difficult bit was that the blood was a different blood type than Warney's.

At trial, prosecutors pointed out that the blood could have come from the accomplice mentioned in the confession, and pounded away on the point that the details in the confession could only have come from first-hand knowledge of the crime.

After a short trial, Douglas Warney was convicted of the murder of William Beason and sentenced to 25 years.

When appeals failed, Mr. Warney in 2004 sought help from the Innocence Project, a nonprofit legal clinic that helps identify wrongly convicted individuals and secure their freedom. Set up at the Benjamin N. Cardozo School of Law in New York by lawyers Barry Scheck and Peter Neufeld, the Innocence Project specializes in using DNA technology to establish innocence.

The Innocence Project staff petitioned the court for additional DNA testing using new sensitive DNA probes, arguing that this might disclose evidence that would have resulted in a different verdict. The judge refused, ruling that the possibility that the blood found at the scene of the crime might match that of a criminal already in the state databank was "too speculative and improbable" to warrant the new tests.

Someone in the prosecutor's office must have been persuaded, however. Without notifying Warney's legal team, Project Innocence, or the court, this good Samaritan arranged for new DNA tests on the blood drops found at the crime scene. When compared to the New York State criminal DNA database, they hit a strong match. The blood was that of Eldred Johnson, in prison for slitting his landlady's throat in Utica two weeks before Beason was killed.

When confronted, the prisoner Johnson readily admitted to the stabbing of Beason. He said he was the sole killer, and had never met Douglas Warney.

This leaves the interesting question of how Warney's signed confession came to include such accurate information about the crime scene. It now appears he may have been fed critical details about the crime scene by the homicide detective leading the investigation, who has since died.

DNA testing has become a pillar of the American criminal justice system. It has provided key evidence that has established the guilt of thousands of suspects beyond any reasonable doubt—and, as you see here, also provided the evidence that our criminal justice system sometimes convicts and sentences innocent people. Over more than a decade, the Innocence Project and other similar efforts have cleared hundreds of convicted people, strong proof that wrongful convictions are not isolated or rare events. DNA testing opens a window of hope for the wrongly convicted.

13.4 Genetic Engineering and Medicine

Much of the excitement about genetic engineering has focused on its potential to improve medicine—to aid in curing and preventing illness. Major advances have been made in the production of proteins used to treat illness, and in the creation of new vaccines to combat infections.

Making "Magic Bullets"

Many illnesses occur because of gene defects that prevent our bodies from making critical proteins. Juvenile diabetes is such an illness. The body is unable to control levels of sugar in the blood because a critical protein, **insulin,** cannot be made. This failure can be overcome if the body can be supplied with the protein it lacks. The donated protein is in a very real sense a "magic bullet" to combat the body's inability to regulate itself.

Until recently, the principal problem with using regulatory proteins as drugs was in manufacturing the protein. Proteins that regulate the body's functions are typically present in the body in very low amounts, and this makes them difficult and expensive to obtain in quantity. With genetic engineering techniques, the problem of obtaining large amounts of rare proteins has been largely overcome. The cDNA of genes encoding medically important proteins are now introduced into bacteria (table 13.2). Because the host bacteria can be grown cheaply, large amounts of the desired protein can be easily isolated. In 1982, the U.S. Food and Drug Administration approved the use of human insulin produced from genetically engineered bacteria, the first commercial product of genetic engineering.

The use of genetic engineering techniques in bacteria has provided ample sources of therapeutic proteins, but the application extends beyond bacteria. Today hundreds of pharmaceutical companies around the world are busy producing other medically important proteins, expanding the use of these genetic engineering techniques. A gene added to the DNA of the mouse on the right in figure 13.8 produces human growth hormone, allowing the mouse to grow larger than its twin.

The advantage of using genetic engineering is clearly seen with **factor VIII,** a protein that promotes blood clotting. A deficiency in factor VIII leads to hemophilia, an inherited disorder (discussed in chapter 10) that is characterized by prolonged bleeding. For a long time, hemophiliacs received blood factor VIII that had been isolated from donated blood. Unfortunately, some of the donated blood had been infected with viruses such as HIV and hepatitis B, which were then unknowingly transmitted to those people who received blood transfusions. Today the use of genetically engineered factor VIII produced in the laboratory eliminates the risks associated with blood products obtained from other individuals.

TABLE 13.2 GENETICALLY ENGINEERED DRUGS

Product	Effects and Uses
Anticoagulants	Involved in dissolving blood clots; used to treat heart attack patients
Colony-stimulating factors	Stimulate white blood cell production; used to treat infections and immune system deficiencies
Erythropoietin	Stimulates red blood cell production; used to treat anemia in individuals with kidney disorders
Factor VIII	Promotes blood clotting; used to treat hemophilia
Growth factors	Stimulate differentiation and growth of various cell types; used to aid wound healing
Human growth hormone	Used to treat dwarfism
Insulin	Involved in controlling blood sugar levels; used in treating diabetes
Interferons	Disrupt the reproduction of viruses; used to treat some cancers
Interleukins	Activate and stimulate white blood cells; used to treat wounds, HIV infections, cancer, immune deficiencies

Figure 13.8 Genetically engineered human growth hormone.

These two mice are genetically identical, but the large one has one extra gene: the gene encoding human growth hormone. The gene was added to the mouse's genome by genetic engineers and is now a stable part of the mouse's genetic make-up. In humans, growth hormone is used to treat various forms of dwarfism.

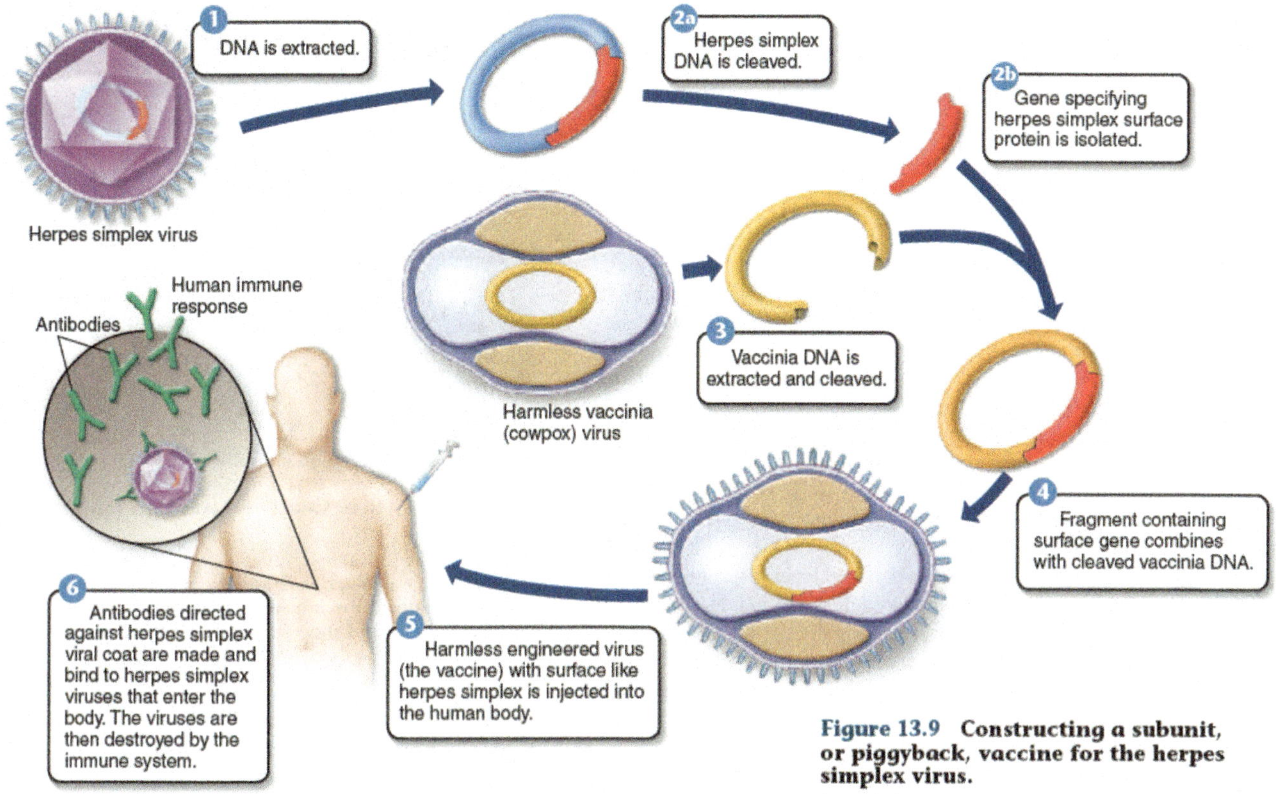

Figure 13.9 Constructing a subunit, or piggyback, vaccine for the herpes simplex virus.

Piggyback Vaccines

Another area of potential significance involves the use of genetic engineering to produce subunit vaccines against viruses such as those that cause herpes and hepatitis. Genes encoding part of the protein-polysaccharide coat of the herpes simplex virus or hepatitis B virus are spliced into a fragment of the vaccinia (cowpox) virus genome. The vaccinia virus, which is essentially harmless to humans and was used by British physician Edward Jenner more than 200 years ago in his pioneering vaccinations against smallpox, is now used as a vector to carry a viral coat gene into cultured mammalian cells. As shown in figure 13.9, the steps in constructing a subunit vaccine for herpes simplex begin with (1) extracting the herpes simplex viral DNA and (2) isolating a gene that codes for a protein on the surface of the virus. The cowpox viral DNA is extracted and cleaved (3), and the herpes gene is then combined with the cowpox DNA (4). The recombinant DNA is inserted into a cowpox virus. Many copies of the recombinant virus, which have the outside coat of a herpes virus, are produced. When this recombinant virus is injected into a human (5), the immune system produces antibodies directed against the coat of the recombinant virus (6). The person therefore develops an immunity to the virus. Vaccines produced in this way, also known as **piggyback vaccines,** are harmless because the vaccinia virus is benign, and only a small fragment of the DNA from the disease-causing virus is introduced via the recombinant virus.

In 1995, the first clinical trial began of a new kind of vaccine, called a DNA vaccine. DNA containing a viral gene is injected and taken up by cells of the body, where the gene is expressed. The infected cells trigger a cellular immune response, in which blood cells known as killer T cells attack the infected cells. The first DNA vaccines spliced an influenza virus gene encoding an internal nucleoprotein into a plasmid, which was then injected into mice. The mice developed strong cellular immune responses against influenza. The approach offers great promise.

In 2010, the first effective **cancer vaccines** were announced. A cancer vaccine is therapeutic rather than preventive, stimulating the immune system to attack a tumor in the same way invading microbes are attacked. The first cancer vaccine approved for clinical use employs proteins from prostate cancer cells to induce the immune system to attack prostate cancer tumors. Another cancer vaccine, very effective in mice but not yet approved for humans, uses a protein called alpha-lactalbumin to trigger an attack on breast cancer cells. The protein is not found on normal breast cancer cells except when women are breast-feeding. As 96% of the lifetime risk of a woman for breast cancer is after child-bearing years, post-menopausal use of this vaccine may prove to be a powerful therapy against early undetected breast cancers.

> **Key Learning Outcome 13.4** Genetic engineering has facilitated the production of medically important proteins and led to novel vaccines.

13.5 Genetic Engineering and Agriculture

Pest Resistance

An important effort of genetic engineers in agriculture has involved making crops resistant to insect pests without spraying with pesticides, a great saving to the environment. Consider cotton. Its fibers are a major source of raw material for clothing throughout the world, yet the plant itself can hardly survive in a field because many insects attack it. Over 40% of the chemical insecticides used today are employed to kill insects that eat cotton plants. The world's environment would greatly benefit if these thousands of tons of insecticide were not needed. Biologists are now in the process of producing cotton plants that are resistant to attack by insects.

One successful approach uses a kind of soil bacterium, *Bacillus thuringiensis* (Bt), that produces a protein that is toxic when eaten by crop pests, such as larvae (caterpillars) of butterflies. When the gene producing the Bt protein is inserted into the chromosomes of tomatoes, the plants begin to manufacture Bt protein. While not harmful to humans, it makes the tomatoes highly toxic to hornworms (one of the most serious pests of commercial tomato crops).

Many important plant pests also attack roots. To combat these pests, genetic engineers are introducing the *Bt* gene into different kinds of bacteria, ones that colonize the roots of crop plants. Any insects eating such roots consume the bacteria and so are lethally attacked by the Bt protein.

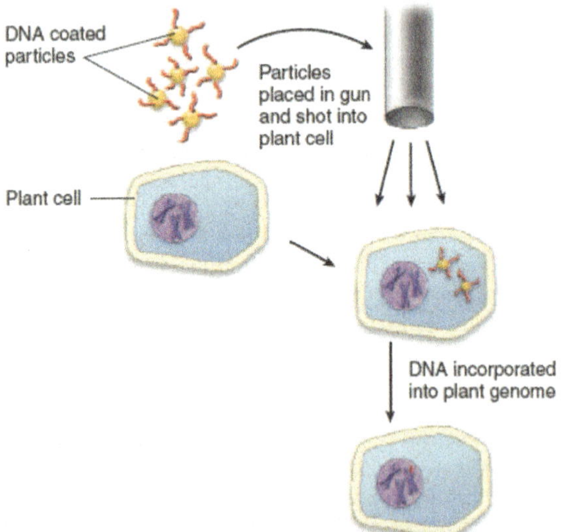

Figure 13.10 Shooting genes into cells.
A DNA particle gun, also called a gene gun, fires tungsten or gold particles coated with DNA into plant cells. The DNA-coated particles pass through the cell wall and into the cell, where the DNA is incorporated into the plant cell's DNA. The gene encoded by the DNA is expressed.

Figure 13.11 Genetically engineered herbicide resistance.
All four of these petunia plants were exposed to equal doses of an herbicide. The two on *top* were genetically engineered to be resistant to glyphosate, the active ingredient in the herbicide, whereas the two dead ones on the *bottom* were not.

Herbicide Resistance

A major advance has been the creation of crop plants that are resistant to the herbicide *glyphosate*, a powerful biodegradable herbicide that kills most actively growing plants. Glyphosate is used in orchards and agricultural fields to control weeds. Growing plants need to make a lot of protein, and glyphosate stops them from making protein by destroying an enzyme necessary for the manufacture of so-called aromatic amino acids (that is, amino acids that contain a ring structure, like phenylalanine—see figure 3.5). Humans are unaffected by glyphosate because we don't make aromatic amino acids—we obtain them from plants we eat! To make crop plants resistant to this powerful plant killer, genetic engineers screened thousands of organisms until they found a species of bacteria that could make aromatic amino acids in the presence of glyphosate. They then isolated the gene encoding the resistant enzyme and successfully introduced the gene into plants. They inserted the gene into the plants using DNA particle guns, also called gene guns. You can see in figure 13.10 how a DNA particle gun works. Small tungsten or gold pellets are coated with DNA (red in the figure) that contains the gene of interest and placed in the DNA particle gun. The DNA gun literally shoots the gene into plant cells in culture where the gene can be incorporated into the plant genome and then expressed. Plants that have been genetically engineered in this way are shown in figure 13.11. The two plants on top were genetically engineered to be resistant to glyphosate, the herbicide that killed the two plants at the bottom of the photo.

The creation of glyphosate-tolerant crops is of major benefit to the environment. Glyphosate is quickly broken down in the environment, which makes its use a great improvement over long-lasting chemical herbicides. Also, not having to plow to remove weeds reduces the loss of fertile topsoil to erosion.

More Nutritious Crops

The cultivation of genetically modified (GM) crops of corn, cotton, soybeans, and other plants (table 13.3) has become commonplace in the United States. In 2004, 85% of soybeans in the United States were planted with seeds genetically modified to be herbicide resistant. The result has been that less tillage was needed and, as a consequence, soil erosion was greatly lessened. Pest-resistant GM corn in 2004 comprised over 50% of all corn planted in the United States, and pest-resistant GM cotton comprised 81% of all cotton. In both cases, the change greatly lessens the amount of chemical pesticide used on the crops. These benefits of soil preservation and chemical pesticide reduction, while significant, have been largely bestowed upon farmers, making their cultivation of crops cheaper and more efficient.

Like the first act of a play, these developments have served mainly to set the stage for the real action, which is only now beginning to happen. The real promise of plant genetic engineering is to produce genetically modified plants with desirable traits that directly benefit the consumer.

One recent advance, nutritionally improved "golden" rice, gives us a hint of what is to come. In developing countries, large numbers of people live on simple diets that are poor sources of vitamins and minerals (what botanists called "micronutrients"). Worldwide, the two major micronutrient deficiencies are iron, which affects 1.4 billion women, 24% of the world population, and vitamin A, affecting 40 million children, 7% of the world population. The deficiencies are especially severe in developing countries where the major staple food is rice. In recent research, Swiss bioengineer Ingo Potrykus and his team at the Institute of Plant Sciences, Zurich, have gone a long way toward solving this problem. Supported by the Rockefeller Foundation and with results to be made free to developing countries, the work is a model of what plant genetic engineering can achieve.

To solve the problem of dietary iron deficiency among rice eaters, Potrykus first asked why rice is such a poor source of dietary iron. The problem, and the answer, proved to have three parts:

1. *Too little iron.* The proteins of rice endosperm have unusually low amounts of iron. To solve this problem, a ferritin gene (abbreviated as *Fe* in figure 13.12) was transferred into rice from beans. Ferritin is a protein with an extraordinarily high iron content, and so greatly increased the iron content of the rice.
2. *Inhibition of iron absorption by the intestine.* Rice contains an unusually high concentration of a chemical called phytate, which inhibits iron absorption in the intestine—it stops your body from taking up the iron in the rice. To solve this problem, a gene encoding an enzyme called phytase (abbreviated as *Pt*) that destroys phytate was transferred into rice from a fungus.
3. *Too little sulfur for efficient iron absorption.* The human body requires sulfur for the uptake of iron, and rice has very little of it. To solve this problem, a gene encoding a sulfur-rich protein (abbreviated as *S*) was transferred into rice from wild rice.

To solve the problem of vitamin A deficiency, the same approach was taken. First, the problem was identified. It turns out rice only goes partway toward making vitamin A; there are no enzymes in rice to catalyze the last four steps. To solve the problem, genes encoding these four enzymes (abbreviated A_1 A_2 A_3 A_4) were added to rice from a flower, the daffodil.

The development of transgenic rice is only the first step in the battle to combat dietary deficiencies. The added nutritional value only makes up for half a person's requirements, and many years will be required to breed the genes into lines adapted to local conditions, but it is a promising start, representative of the very real promise of genetic engineering.

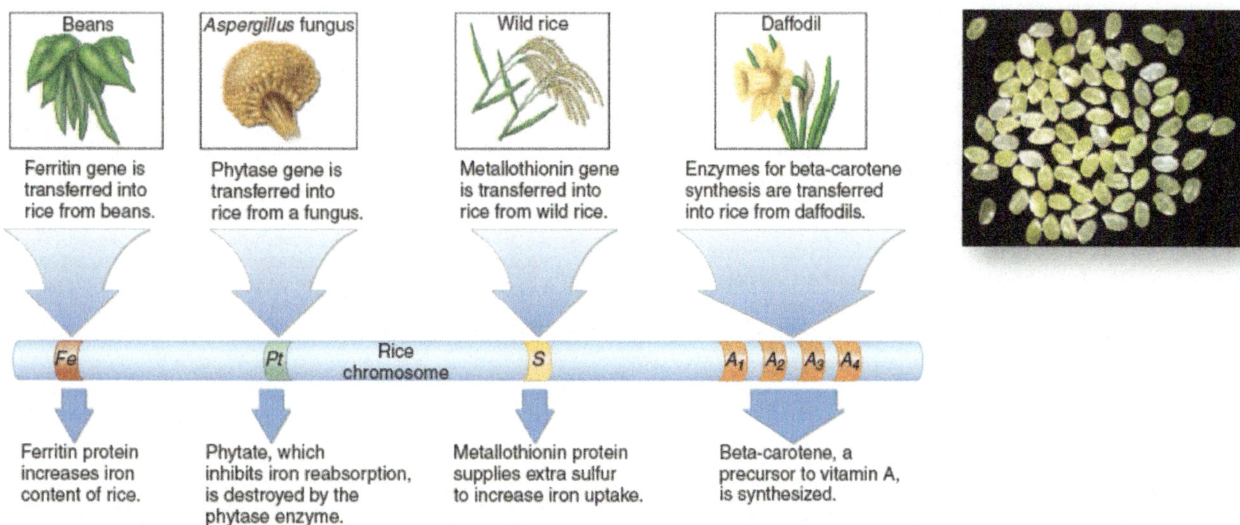

Figure 13.12 Transgenic "golden" rice.

How Do We Measure the Potential Risks of Genetically Modified Crops?

Is Eating Genetically Modified Food Dangerous? Many consumers worry that when bioengineers introduce novel genes into genetically modified (GM) crops, there may be dangerous consequences for the food we eat. The introduction of glyphosate-resistance into soybeans is an example. Could introduced proteins like the enzyme making the GM soybeans glyphosate-tolerant cause a fatal immune reaction in some people? Because the potential danger of allergic reactions is quite real, every time a protein-encoding gene is introduced into a GM crop it is necessary to carry out extensive tests of the introduced protein's allergen potential. No GM crop currently being produced in the United States contains a protein that acts as an allergen to humans. On this score, then, the risk of genetic engineering to the food supply seems to be slight.

Are GM Crops Harmful to the Environment? Those concerned about the widespread use of GM crops raise three legitimate concerns:

1. *Harm to Other Organisms.* Might pollen from Bt corn harm non-pest insects that chance to eat it? Studies suggest little possibility of harm.
2. *Resistance.* All insecticides and herbicides used in agriculture share the problem that pests eventually evolve resistance to them, in much the same way that bacterial populations evolve resistance to antibiotics. To prevent this, farmers are required to plant at least 20% non-Bt crops alongside Bt crops to provide refuges where insect populations are not under selection pressure and in this way to slow the development of resistance. As a result, despite the widespread use of Bt crops like corn, soybeans, and cotton since 1996, there are as of yet only a few cases of insects developing resistance to Bt plants in the field. Unfortunately, the same restrictions have not been required for farmers using the herbicide glyphosate, leading to a different result: By the year 2010, glyphosate-resistant weeds have been reported by upset farmers in 22 states.
3. *Gene Flow.* How about the possibility that introduced genes will pass from GM crops to their wild relatives? For the major GM crops, there is usually no potential relative around to receive the modified gene from the GM crop. There are no wild relatives of soybeans in Europe, for example. Thus there can be no gene escape from GM soybeans in Europe, any more than genes can flow from you to your pet dog or cat. However—and this is a big however—for secondary crops only now being genetically modified, studies suggest it will be difficult to prevent GM crops from interbreeding with surrounding relatives to create new hybrids.

Key Learning Outcome 13.5 **Genetic engineering affords great opportunities for progress in food production. On balance, the risks appear slight, and the potential benefits substantial.**

TABLE 13.3 GENETICALLY MODIFIED CROPS

Crop	Description
Rice 	Genes have been added to commercial rice from daffodils for vitamin A, and from beans, fungi, and wild rice to supply dietary iron; transgenic strains that are cold-tolerant are under development.
Wheat 	New strains of wheat, resistant to the herbicide glyphosate, greatly reduce the need for tilling and so reduce loss of topsoil.
Soybean 	A major animal feed crop, soybeans tolerant of the herbicide glyphosate were used in 90% of U.S. soybean acreage in 2010. Varieties are being developed that contain the *Bt* gene, to protect the crop from insect pests without chemical pesticides. The nutritional value of soybean crops is being improved by genetic engineers in several ways, including transgenic varieties with high tryptophan (soybeans are poor in this essential amino acid), reduced trans-fatty acids, and enhanced omega-3 (beneficial) fatty acids, common in fish oil but low in plants.
Corn 	Corn varieties resistant to insect pests (Bt corn) are widely planted (40% of U.S. acreage); varieties also tolerant of the herbicide glyphosate have been recently developed. Varieties that are drought resistant are being developed, as well as nutritionally improved lines with high lysine, vitamin A, and high levels of the unsaturated fat oleic acid, which reduces harmful cholesterol and so prevents clogged arteries.
Cotton 	Cotton crops are attacked by cotton bollworm, budworm, and other lepidopteran insects; more than 40% of all chemical pesticide tonnage worldwide is applied to cotton. A form of the *Bt* gene toxic to all lepidopterans but harmless to other insects has transformed cotton to a crop that requires few chemical pesticides. 81% of U.S. acreage is Bt cotton.
Peanut 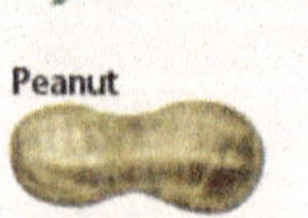	The lesser cornstalk borer causes serious damage to peanut crops. An insect-resistant variety is under development by gene engineers to control this pest.
Potato 	Verticillium wilt (a fungal disease) infects the water-conducting tissues of potatoes, reducing crop yields 40%. An antifungal gene from alfalfa reduces infections sixfold.
Canola 	Canola, a major vegetable oil and animal feed crop, is typically grown in narrow rows with little cultivation, requiring extensive application of chemical herbicides to keep down weeds. New glyphosate-tolerant varieties require far less chemical treatment. 80% of U.S. canola acreage planted is gene-modified canola.

Today's *Biology*

A DNA Timeline

2006 Japanese cell biologist Shinya Yamanaka uses only four transcription factors to reprogram adult skin cells into embryonic stem cells, opening the possibility of ethical therapeutic cloning.

In 2000, Craig Venter of Celera, President Clinton, and Francis Collins of the Human Genome Project announce the human genome.

2000 Two teams, led by Craig Venter and Francis Collins, complete draft sequences of the human genome.

1998 Andrew Fire and Craig Mello discover RNA interference, leading to a Nobel Prize only eight years later.

1996 Ian Wilmut uses the nucleus of an adult cell to successfully clone a sheep, "Dolly."

1995 Craig Venter sequences the first genome of an organism, the single-celled bacterium *Haemophilus influenzae*.

1992 Lawyers Barry Scheck and Peter Neufeld start the Innocence Project, whose efforts have cleared more than 120 wrongly convicted people through the use of DNA technology.

1985 British geneticist Alec Jeffreys invents DNA fingerprinting, the use of DNA in forensic analysis to match people to biological tissue found at crime scenes.

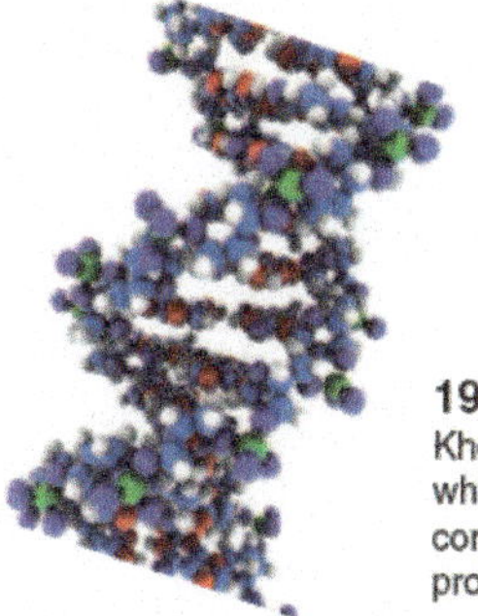

1973 Herbert Boyer and Stanley Cohen invent genetic engineering, successfully inserting an amphibian RNA gene into a different organism.

1983 Kary Mullis develops the polymerase chain reaction (PCR), allowing amplification and analysis of minute traces of DNA, such as that found in a single human hair.

1964 Marshall Nirenberg and Har Khorana break the genetic code, learning which three-letter code words of DNA correspond to each amino acid in proteins.

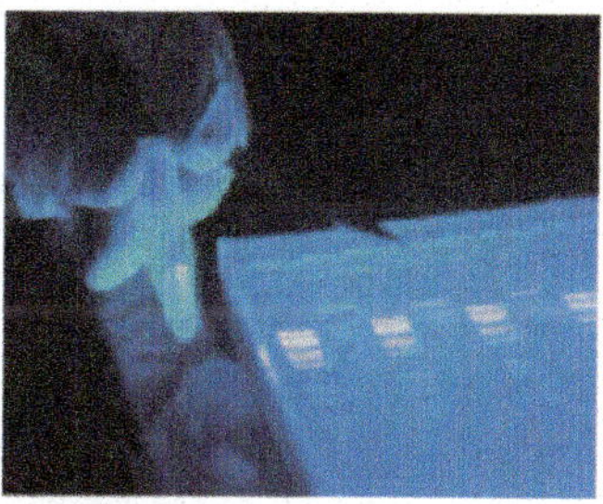

1956 Vernon Ingram shows that sickle cell disease is due to a DNA mutation leading to a single amino acid change in the protein hemoglobin.

1953 James Watson and Francis Crick propose that the DNA molecule is a double helix, each strand's nucleotide sequence complementary to the other.

1952 Alfred Hershey and Martha Chase demonstrate that viruses inject DNA into bacteria to reproduce, not protein; this experiment convinces most biologists that DNA is the genetic material.

1950 Graduate student Ray Gosling, working in the lab of British biochemist Maurice Wilkins, obtains the first clear X-ray diffraction patterns of DNA; over the next two years, Rosalind Franklin and he produce ever-clearer pictures.

1944 American biochemist Oswald Avery purifies Griffith's transforming principle, and demonstrates conclusively that it is DNA, although this conclusion was not appreciated at first.

1928 British microbiologist Frederick Griffith discovers transformation of living bacteria by material from dead ones.

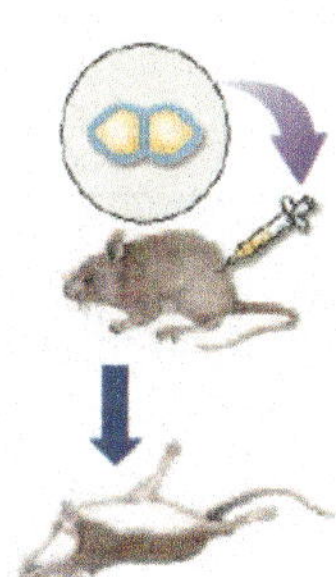

1869 German chemist Friedrich Miescher discovers DNA, called "nucleic acid" because it was isolated from sperm nuclei and is slightly acidic.

The Revolution in Cell Technology

13.6 Reproductive Cloning

One of the most active and exciting areas of biology involves recently developed approaches to manipulating animal cells. In this section, you will encounter three areas where landmark progress is being made in cell technology: reproductive cloning of farm animals, stem cell research, and gene therapy. Advances in cell technology hold the promise of revolutionizing our lives.

The idea of cloning animals was first suggested in 1938 by German embryologist Hans Spemann (called the "father of modern embryology"), who proposed what he called a "fantastical experiment": remove the nucleus from an egg cell (creating an enucleated egg) and put in its place a nucleus from another cell. When attempted many years later (figure 13.13), this experiment actually succeeded in frogs, sheep, monkeys, and many other animals. However, only donor nuclei extracted from early embryos seemed to work. After repeated failures using nuclei from adult cells, many researchers became convinced that the nuclei of animal cells become irreversibly committed to a developmental pathway after the first few cell divisions of the developing embryo.

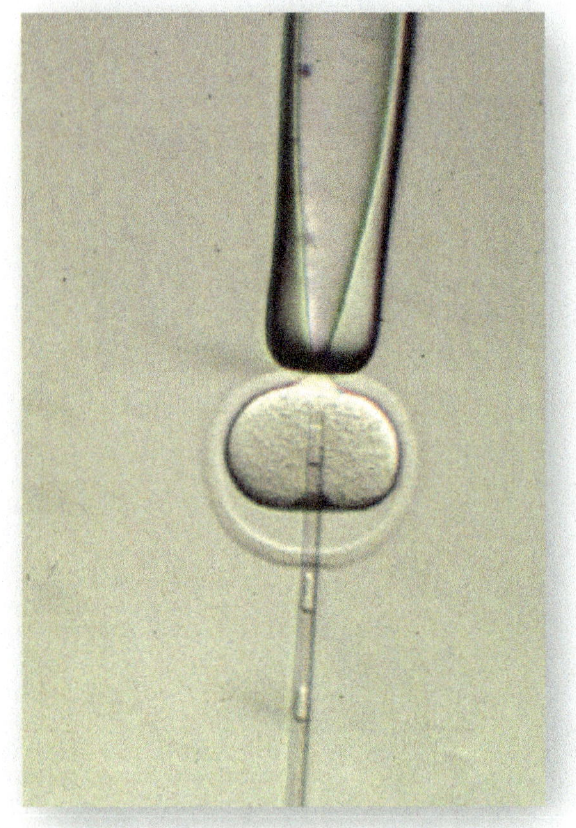

Figure 13.13 A cloning experiment.
In this photo, a nucleus is being injected from a micropipette (bottom) into an enucleated egg cell held in place by a pipette.

Wilmut's Lamb

Then, in the 1990s, a key insight was made in Scotland by geneticist Keith Campbell, a specialist in studying the cell cycle of agricultural animals. Recall from chapter 8 that the division cycle of eukaryotic cells progresses in several stages. Campbell reasoned, "Maybe the egg and the donated nucleus need to be at the same stage in the cell cycle." This proved to be a

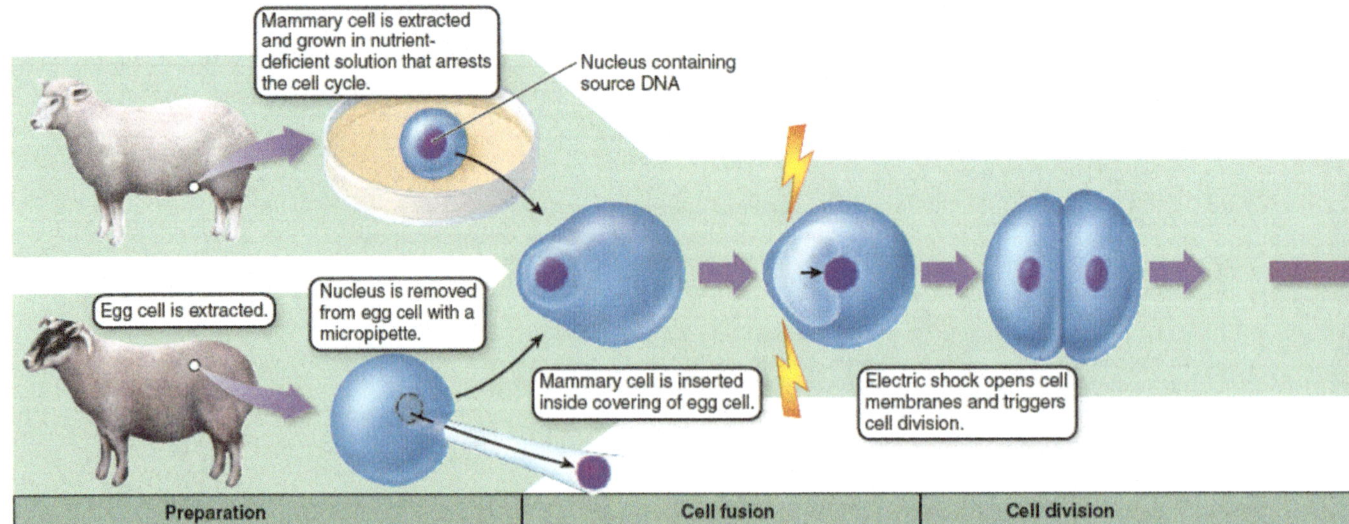

Figure 13.14 Wilmut's animal cloning experiment.

key insight. In 1994 researchers succeeded in cloning farm animals from advanced embryos by first starving the cells, so that they paused at the beginning of the cell cycle. Two starved cells are thus synchronized at the same point in the cell cycle.

Campbell's colleague Ian Wilmut then attempted the key breakthrough, the experiment that had been eluding researchers: He set out to transfer the nucleus from an adult differentiated cell into an enucleated egg, and to allow the resulting embryo to grow and develop in a surrogate mother, hopefully producing a healthy animal (figure 13.14). Approximately five months later, on July 5, 1996, the mother gave birth to a lamb. This lamb, "Dolly," was the first successful clone generated from an adult animal cell. Dolly grew into a healthy adult, and as you can see in the photo at the beginning of this chapter, she went on to have healthy offspring normal in every respect.

Progress with Reproductive Cloning

Since Dolly's birth in 1996, scientists have successfully cloned a wide variety of farm animals with desired characteristics, including cows, pigs, goats, horses, and donkeys, as well as pets like cats and dogs. Snuppy, the puppy in figure 13.15, was the first dog to be cloned. For most farm animals, cloning procedures have become increasingly efficient since Dolly was cloned. However, the development of clones into adults tends to go unexpectedly haywire. Almost none survive to live a normal life span. Even Dolly died prematurely in 2003, having lived only half a normal sheep life span.

Figure 13.15 Cloning the family pet.
This puppy named "Snuppy" is the first dog cloned. Beside him to the left, is the adult male dog that provided the skin cell from which Snuppy was cloned. The dog in the photo on the right was Snuppy's surrogate mother.

The Importance of Gene Reprogramming

What is going wrong? It turns out that as mammalian eggs and sperm mature, their DNA is conditioned by the parent female or male, a process called reprogramming. Chemical changes are made to the DNA that alter when particular genes are expressed without changing the nucleotide sequences. In the years since Dolly, scientists have learned a lot about gene reprogramming, also called **epigenetics.** Epigenetics works by blocking the cell's ability to read certain genes. A gene is locked in the off position by adding a $-CH_3$ (methyl) group to some of its cytosine nucleotides. After a gene has been altered like this, the polymerase protein that is supposed to "read" the gene can no longer recognize it. The gene has been shut off.

We are only beginning to learn how to reprogram human DNA, so any attempt to clone a human is simply throwing stones in the dark, hoping to hit a target we cannot see. For this and many other reasons, human reproductive cloning is regarded as highly unethical.

Key Learning Outcome 13.6 Although recent experiments have demonstrated the possibility of cloning animals from adult tissue, the cloning of farm animals often fails for lack of proper epigenetic reprogramming.

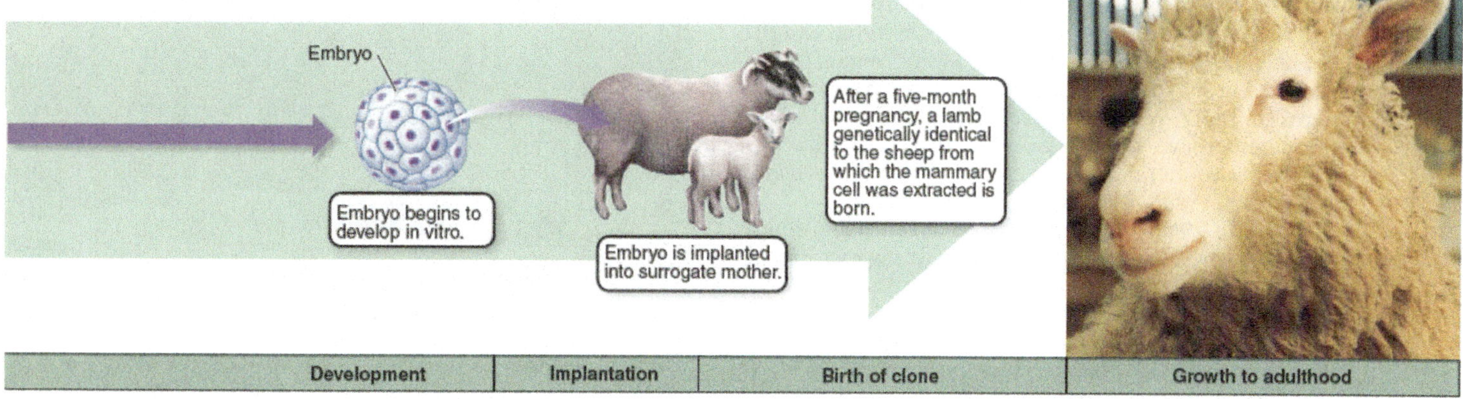

13.7 Stem Cell Therapy

You can see a mass of human embryonic stem cells in figure 13.16. Many are **totipotent**—able to form any body tissue, and even an entire adult animal. What is an embryonic stem cell, and why is it totipotent? To answer this question, we need to consider for a moment where an embryo comes from. At the dawn of a human life, a sperm fertilizes an egg to create a single cell destined to become a child. As development commences, that cell begins to divide, producing after four divisions a small mass of 16 **embryonic stem cells.** Each of these embryonic stem cells has all of the genes needed to produce a normal individual.

As development proceeds, some of these embryonic stem cells become committed to forming specific types of tissues, such as nerve tissues, and, after this step is taken, cannot ever produce any other kind of cell. In the case of nerve tissue, they are then called *nerve stem cells.* Others become specialized to produce blood cells, others to produce muscle tissue, and still others to form the other tissues of the body. Each major tissue is formed from its own kind of tissue-specific **adult stem cell.** Because an adult stem cell forms only that one kind of tissue, it is not totipotent.

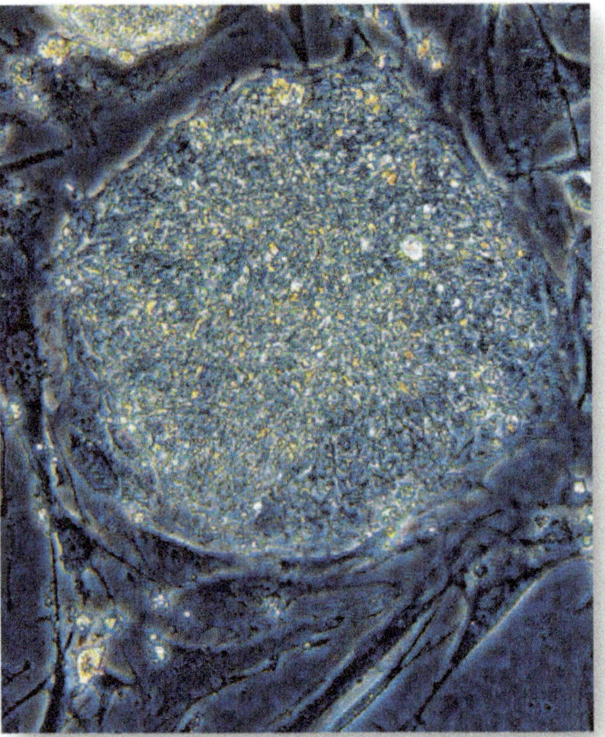

Figure 13.16 Human embryonic stem cells (×20). This mass is a colony of undifferentiated human embryonic stem cells growing in tissue culture and surrounded by fibroblasts (elongated cells) that serve as a "feeder layer."

Using Stem Cells to Repair Damaged Tissues

Embryonic stem cells offer the exciting possibility of restoring damaged tissues. To understand how, follow along in figure 13.18. A few days after fertilization, an embryonic stage called the *blastocyst* forms ❶. Embryonic stem cells are harvested from its inner cell mass or from cells of the embryo at a later stage ❷. These embryonic stem cells can be grown in tissue culture as seen in figure 13.16, and in principle be induced to form any type of tissue in the body ❸. The resulting healthy tissue can then be injected into the patient where it will grow and replace damaged tissue ❹. Alternatively, where possible, adult stem cells can be isolated and when injected back into the body, can form certain types of tissue cells.

Both adult and embryonic stem cell transfer experiments have been carried out successfully in mice. Adult blood stem cells have been used to cure leukemia. Heart muscle cells grown from mouse embryonic stem cells have successfully replaced the damaged heart tissue of a living mouse. In other experiments, damaged spinal neurons have been partially repaired. DOPA-producing neurons of mouse brains, whose loss is responsible for Parkinson's disease, have been successfully replaced with embryonic stem cells, as have islet cells of the pancreas, whose loss leads to juvenile diabetes.

Because the course of development is broadly similar in all mammals, these experiments in mice suggest exciting possibilities for stem cell therapy in humans. The hope is that individuals with Parkinson's disease, like Michael J. Fox (figure 13.17), might be partially or fully cured with stem cell therapy. As you might imagine, work proceeds intensively in this field of research.

There are ethical objections to using embryonic stems cells but new experimental results hint at ways around this

Figure 13.17 Promoting a cure for Parkinson's. Michael J. Fox, with whom you may be familiar as a star of the *Back to the Future* film series and the TV show *Family Ties*, is a victim of Parkinson's disease, and a prominent spokesman for those who suffer from it. Here you see him testifying before the U.S. Senate (along with fellow advocate Mary Tyler Moore) on the need for vigorous efforts to support research seeking a cure.

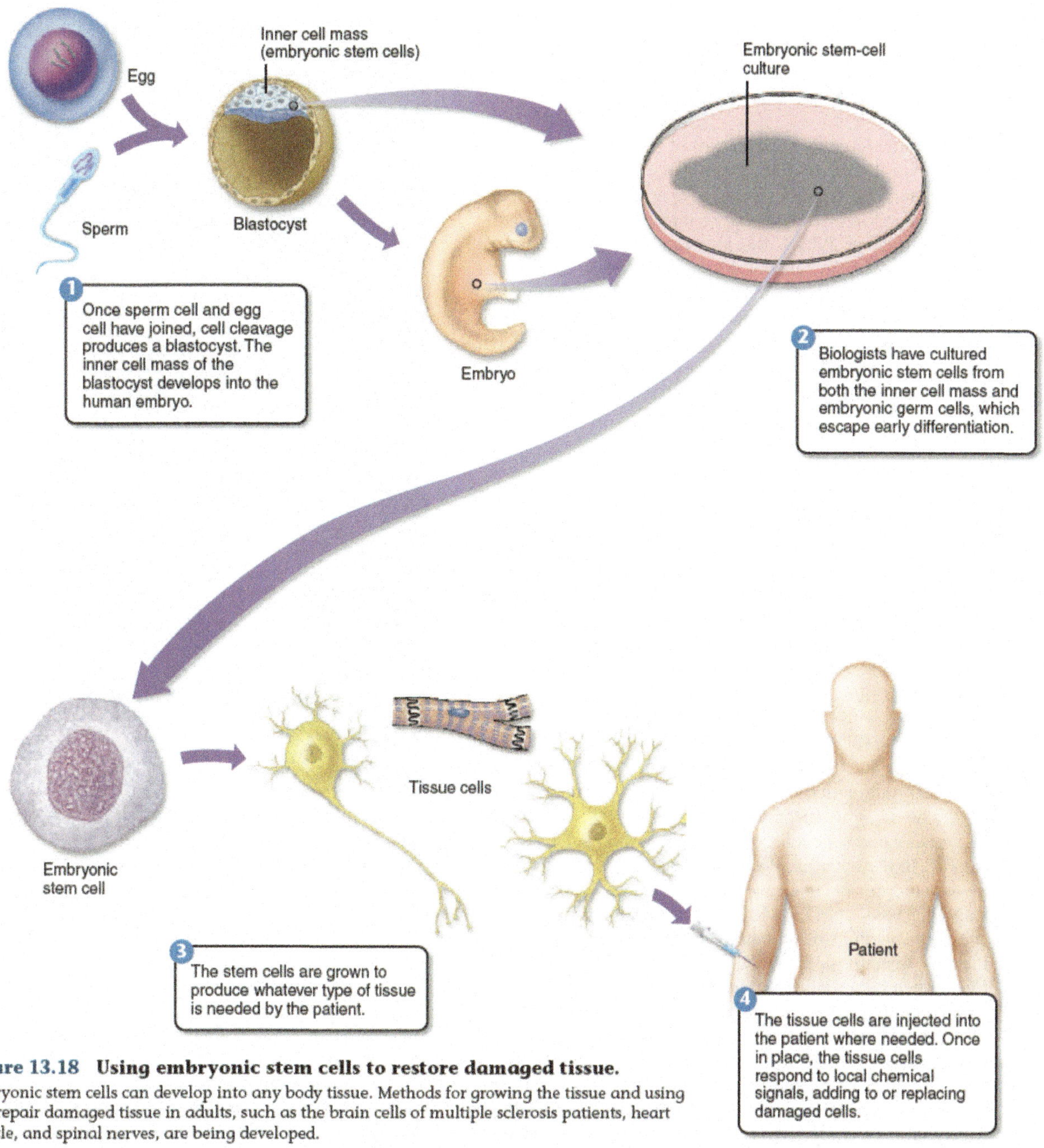

Inner cell mass (embryonic stem cells)

Egg

Sperm

Blastocyst

Embryonic stem-cell culture

Embryo

1 Once sperm cell and egg cell have joined, cell cleavage produces a blastocyst. The inner cell mass of the blastocyst develops into the human embryo.

2 Biologists have cultured embryonic stem cells from both the inner cell mass and embryonic germ cells, which escape early differentiation.

Embryonic stem cell

Tissue cells

Patient

3 The stem cells are grown to produce whatever type of tissue is needed by the patient.

4 The tissue cells are injected into the patient where needed. Once in place, the tissue cells respond to local chemical signals, adding to or replacing damaged cells.

Figure 13.18 Using embryonic stem cells to restore damaged tissue.
Embryonic stem cells can develop into any body tissue. Methods for growing the tissue and using it to repair damaged tissue in adults, such as the brain cells of multiple sclerosis patients, heart muscle, and spinal nerves, are being developed.

ethical maze. In 2007, researchers in two independent laboratories reported that they had engineered embryonic stemlike cells from normal adult human skin cells. The cells they created were pluripotent—they could differentiate into many different cell types. Whether pluripotency extends to totipotency is still being investigated. How were these cells transformed? The essential clue came six years earlier, when fusing adult cells with embryonic stem cells transformed the adult cells into pluripotent cells, as if factors had been transmitted to the adult cells that conferred pluripotency. Then, in a crucial advance in 2006, Japanese cell biologist Shinya Yamanaka introduced into adult human skin cells not the entire contents of an embryonic stem cell, but just the genes for four transcription factors. Once inside, these four factors induced a series of events that converted the adult cell to pluripotency. In effect, he had found a way to reprogram the adult cells to be embryonic stem cells. From proof-of-principle in a laboratory culture dish to actual medical application is still a leap, but the possibility is exciting.

Key Learning Outcome 13.7 Human adult and embryonic stem cells offer the possibility of replacing damaged or lost human tissues.

13.8 Therapeutic Use of Cloning

While exciting, the therapeutic uses of stem cells to cure leukemia, type I diabetes, Parkinson's disease, damaged heart muscle, and injured nerve tissue were all achieved in experiments carried out using strains of mice without functioning immune systems. Why is this important? Because had these mice possessed fully functional immune systems, they almost certainly would have rejected the implanted stem cells as foreign. Humans with normal immune systems might well refuse to accept transplanted stem cells simply because they are from another individual. For such stem cell therapy to work in humans, this problem needs to be addressed and solved.

Cloning to Achieve Immune Acceptance

Early in 2001, a research team at the Rockefeller University reported a way around this potentially serious problem. Their solution? They first isolated skin cells from a mouse, then using the same procedure that created Dolly, they created a 120-cell embryo from them. The embryo was then destroyed, its embryonic stem cells harvested and cultured (figure 13.19) for transfer to replace injured tissue. This procedure is called **therapeutic cloning.** Therapeutic cloning and the procedure that was used to create Dolly, called **reproductive cloning,** are contrasted in figure 13.20. You can see that steps 1 through 5 are essentially the same for both procedures, but the two methods proceed differently after that. In reproductive cloning, the blastocyst from step 5 is implanted in a surrogate mother in step 6a, developing into a baby that is genetically identical to the nucleus donor, step 7a. In therapeutic cloning, by contrast, stem cells from the blastocyst of step 5 are removed and grown in culture, step 6. These stem cells are developed into particular tissue types, such as pancreatic islet cells in step 7, and can then be injected or transplanted into a patient that needs them, such as a diabetic patient, where the new islet cells can begin producing insulin.

Therapeutic cloning, or, more technically, *somatic cell nuclear transfer,* successfully addresses the key problem that must be solved before embryonic stem cells can be used to repair damaged human tissues, which is immune acceptance. Because stem cells are cloned from the body's own tissues in therapeutic cloning, they pass the immune system's "self" identity check, and the body readily accepts them.

Gene Reprogramming to Achieve Immune Acceptance

In therapeutic cloning, the cloned embryo is destroyed to obtain embryonic stem cells. What is the moral standing of a six-day human embryo? Considering it a living individual, many people regard therapeutic cloning to be ethically unacceptable. Recent research discussed on the previous page suggests an alternative approach that avoids this problem: reprogramming adult cells into embryonic stem cells by introducing just a few genes into the adult cells. The genes are so-called transcription factors, turning on key genes that act to reverse the "shut off" epigenetic changes that have occurred during development of the adult cells. Human applications, if even possible, are probably far into the future, but the possibility of reprogramming adult cells is exciting.

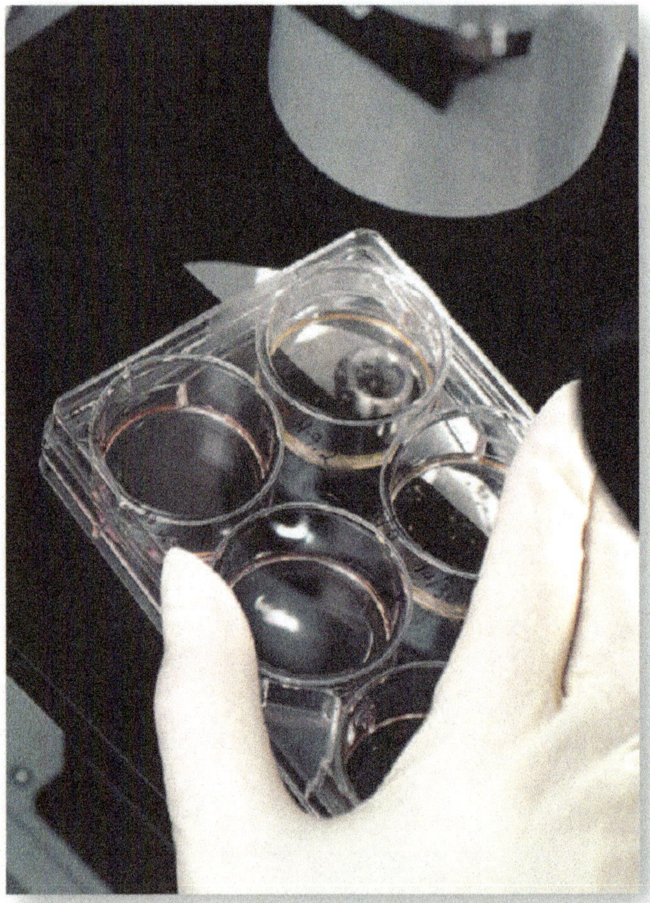

Figure 13.19 Embryonic stem cells growing in cell culture.
Embryonic stem cells derived from early human embryos will grow indefinitely in tissue culture. When transplanted, they can sometimes be induced to form new cells of the adult tissue into which they have been placed. This suggests exciting therapeutic uses.

Key Learning Outcome 13.8 Therapeutic cloning involves initiating blastocyst development from a patient's tissue using nuclear transplant procedures, then using the blastocyst's embryonic stem cells to replace the patient's damaged or lost tissue. Gene reprogramming of adult tissue cells may allow a less controversial approach.

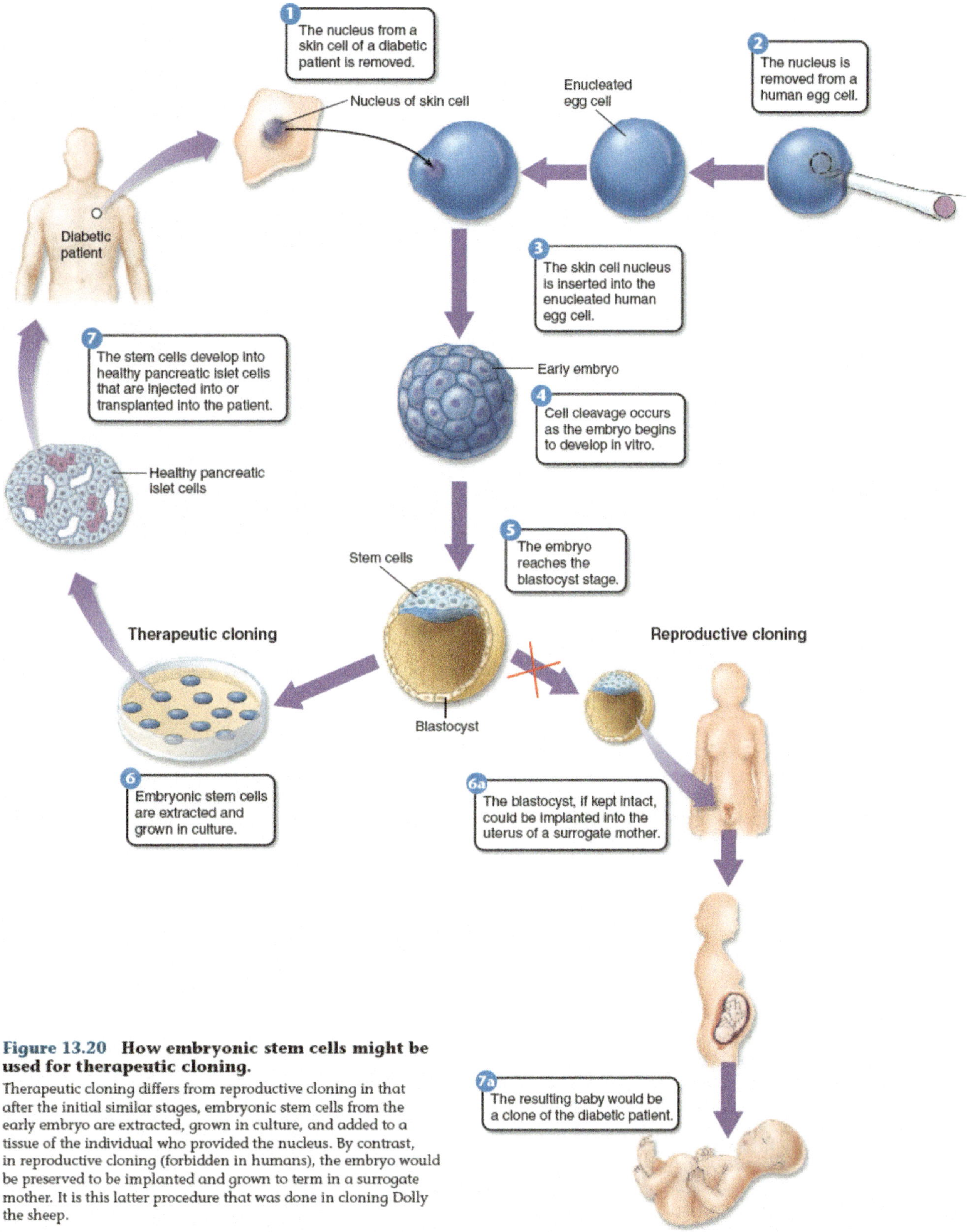

Figure 13.20 How embryonic stem cells might be used for therapeutic cloning.

Therapeutic cloning differs from reproductive cloning in that after the initial similar stages, embryonic stem cells from the early embryo are extracted, grown in culture, and added to a tissue of the individual who provided the nucleus. By contrast, in reproductive cloning (forbidden in humans), the embryo would be preserved to be implanted and grown to term in a surrogate mother. It is this latter procedure that was done in cloning Dolly the sheep.

13.9 Gene Therapy

The third major advance in cell technology involves introducing "healthy" genes into cells that lack them. For decades scientists have sought to cure often-fatal genetic disorders like cystic fibrosis, muscular dystrophy, and multiple sclerosis by replacing the defective gene with a functional one.

Early Success

A successful **gene transfer therapy** procedure was first demonstrated in 1990 (see section 13.3). Two girls were cured of a rare blood disorder due to a defective gene for the enzyme adenosine deaminase. Scientists isolated working copies of this gene and introduced them into bone marrow cells taken from the girls. The gene-modified bone marrow cells were allowed to proliferate, then were injected back into the girls. The girls recovered and stayed healthy. For the first time, a genetic disorder was cured by gene therapy.

The Rush to Cure Cystic Fibrosis

Researchers quickly set out to apply the new approach to one of the big killers, cystic fibrosis. The defective gene, labelled *cf*, had been isolated in 1989. Five years later, in 1994, researchers successfully transferred a healthy *cf* gene into a mouse that had a defective one—they in effect had cured cystic fibrosis in a mouse. They achieved this remarkable result by adding the *cf* gene to a virus that infected the lungs of the mouse, carrying the gene with it "piggyback" into the lung cells. The virus chosen as the "vector" was adenovirus (the red viruses in figure 13.21), a virus that causes colds and is very infective of lung cells. To avoid any complications, the lab mice used in the experiment had their immune systems disabled.

Very encouraged by these well-publicized preliminary trials with mice, several labs set out in 1995 to attempt to cure cystic fibrosis by transferring healthy copies of the *cf* gene into human patients. Confident of success, researchers added the human *cf* gene to adenovirus then administered the gene-bearing virus into the lungs of cystic fibrosis patients. For eight weeks the gene therapy did seem successful, but then disaster struck. The gene-modified cells in the patients' lungs came under attack by the patients' own immune systems. The "healthy" *cf* genes were lost and with them any chance of a cure.

Problems with the Vector

Other attempts at gene therapy met with similar results, eight weeks of hope followed by failure. In retrospect, although it was not obvious then, the problem with these early attempts seems predictable. Adenovirus causes colds. Do you know anyone who has never had a cold? When you get a cold, your body produces antibodies to fight off the infection, and so all of us have antibodies directed against adenovirus. We were introducing therapeutic genes in a vector our bodies are primed to destroy.

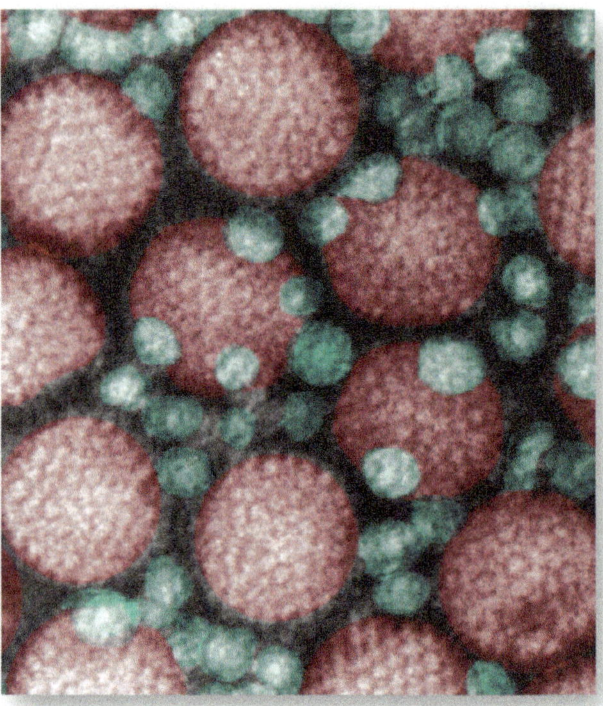

Figure 13.21 Adenovirus and AAV vectors (×200,000).

Adenovirus, the *red* virus particles above, has been used to carry healthy genes in clinical trials of gene therapy. Its use as a vector is problematic, however. AAV, the much smaller *bluish-green* virus particles seen in association with adenovirus here, lacks the many problems of adenovirus and is a much more promising gene transfer vector.

A second serious problem is that when the adenovirus infects a cell, it inserts its DNA into the human chromosome. Unfortunately, it does so at a random location. This means that the insertion events could cause mutations—if the viral DNA inserts into the middle of a gene, it could inactivate that gene. Because the spot where the adenovirus inserts is random, some of the mutations that result can be expected to cause cancer, certainly an unacceptable consequence.

A More Promising Vector

Researchers are now investigating a much more promising vector, a tiny virus called *adeno-associated virus* (AAV—the smaller bluish-green viruses in figure 13.21) that has only two genes. To create a vector for gene transfer, researchers remove both of the AAV genes. The shell that remains is still quite infective and can carry human genes into patients. AAV does not elicit a strong immune response—cells infected with AAV are not eliminated by a patient's immune system. Importantly, AAV enters human DNA far less frequently than adenovirus, and so is less likely to produce cancer-causing mutations.

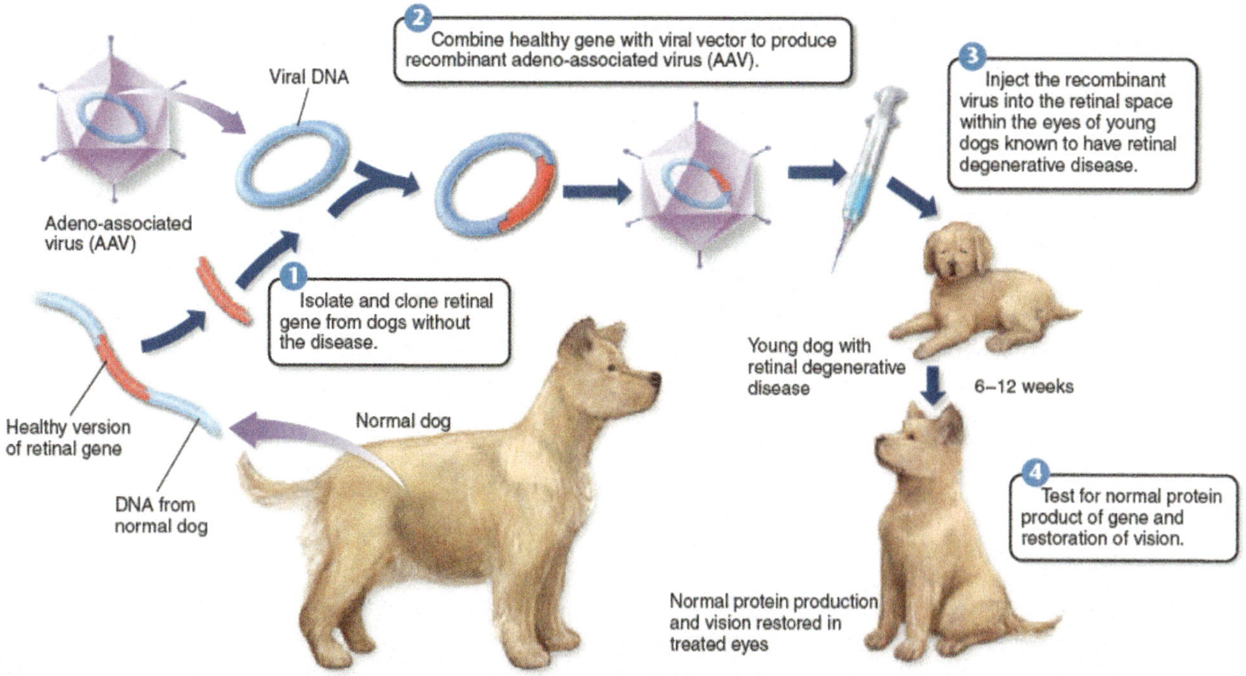

Figure 13.22 Using gene therapy to cure a retinal degenerative disease in dogs.
Researchers were able to use genes from healthy dogs to restore vision in dogs blinded by an inherited retinal degenerative disease. This disease also occurs in human infants and is caused by a defective gene that leads to early vision loss, degeneration of the retinas, and blindness. In the gene therapy experiments, genes from dogs without the disease were inserted into 3-month-old dogs that were known to carry the defective gene and that had been blind since birth. Six weeks after the treatment, the dogs' eyes were producing the normal form of the gene's protein product, and by three months, tests showed that the dogs' vision was restored.

Success with New Vectors

In 1999, AAV successfully cured anemia in rhesus monkeys. In monkeys, humans, and other mammals, red blood cell production is stimulated by a protein called *erythropoietin* (EPO). People with a type of anemia caused by low red blood cell counts, like dialysis patients, get regular injections of EPO. Using AAV to carry a souped-up EPO gene into the monkeys, scientists were able to greatly elevate their red blood cell counts, curing the monkeys of anemia—and they stayed cured.

A similar experiment using AAV cured dogs of a hereditary disorder leading to retinal degeneration and blindness. These dogs had a defective gene that produced a mutant form of a protein associated with the retina of the eye and were blind. Recombinant viral DNA was made using a healthy version of the gene, shown in steps ❶ and ❷ in figure 13.22. Injection of AAV bearing the needed gene into the fluid-filled compartment behind the retina, step ❸, restored sight in the dogs, step ❹. This procedure was recently tried on human patients with some success.

In 2003, gene therapy clinical trials attempting to cure severe combined immune deficiency (SCID) were halted when 5 of the 20 patients in the trial developed leukemia. Apparently the vector had contained a small segment of DNA homologous to a leukemia-causing human gene. When the vector inserted there, the leukemia-causing genes were activated.

Researchers stripped out the leukemia-causing segment of the vector, and launched yet another series of new gene therapy clinical trials. In 2009, a team used an improved vector to successfully treat 12 patients suffering from Leber's congenital blindness. All patients had some improvement in eyesight. In another study reported in 2009, two patients were treated for a rare, fatal brain disease called adrenoleukodystrophy (ALD), the disease featured in the film "Lorenzo's Oil." The treatment stopped the progression of the disease in its tracks. Three years later, the patients remain stable and can attend school.

In 2010, researchers began a new SCID trial with the improved vector, encouraged by the fact that all the patients in the 2003 trial who did not develop leukemia were completely cured of SCID. Trials are also underway for a wide variety of other disorders.

Key Learning Outcome 13.9 In principle, it should be possible to cure hereditary disorders like cystic fibrosis by transferring a healthy gene into the cells of affected tissues. Early attempts using adenovirus vectors were not often successful. New virus vectors avoid the problems of earlier vectors and offer promise of gene transfer therapy cures.

Can Modified Genes Escape from GM Crops?

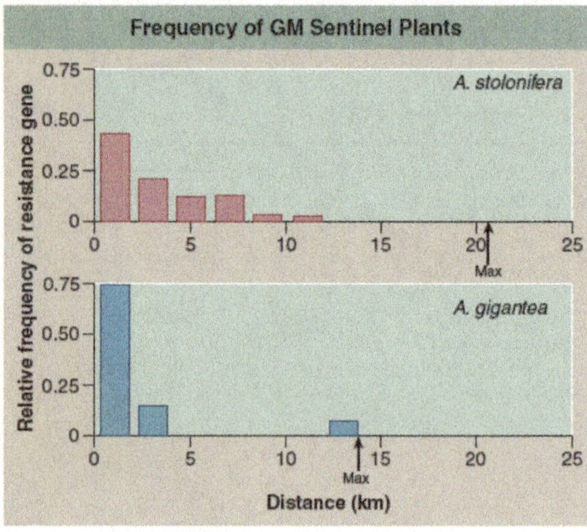

On page 268, the question of whether gene flow of GM crops posed a problem to the environment was discussed. A field experiment conducted in 2004 by the Environmental Protection Agency assessed the possibility that introduced genes could pass from genetically modified golf course grass to other plants. Investigators introduced a gene conferring herbicide resistance (the EPSP synthetase gene for resistance to glyphosate) into golf course bentgrass, *Agrostis stolonifera,* and then looked to see if the gene passed from the GM grass to other plants of the same species, and also if it passed to other related species.

The map below displays the setup of this elaborate field study. A total of 178 *A. stolonifera* plants were placed outside the golf course, many of them downwind. An additional 69 bentgrass plants were found to be already growing downwind, most of them the related species *A. gigantea*. Seeds were collected from each of these plants, and the DNA of resulting seedlings tested for the presence of the gene introduced into the GM golf course grass. In the graph, the upper red histogram (a **histogram** is a "bar graph" that sorts data into a series of discontinuous categories, the value of each bar representing the number of individuals in a category, or, as in this case, the average value of entries in that category) presents the relative frequency with which the gene was found in *A. stolonifera* plants located at various distances from the golf course. The lower blue histogram does the same for *A. gigantea* plants.

1. **Applying Concepts**
   a. Reading a Histogram. Does the gene conferring resistance to herbicide pass to other plants of this species, *A. stolonifera*? To individuals of the related species *A. gigantea*?
   b. What is the maximal distance over which the herbicide resistance gene is transferred to other plants of this species? Of the related species? What are these distances, expressed in miles?
2. **Interpreting Data**
   a. What general statement can be made about the effect of distance on the likelihood that the herbicide resistance gene will pass to another plant?
   b. Are there any significant differences in the gene flow to individuals of *A. stolonifera* and to individuals of the related species *A. gigantea?*
3. **Making Inferences** What mechanism do you propose to account for this gene flow?
4. **Drawing Conclusions** Is it fair to conclude that genetically modified traits can pass from crops to other plants? What qualifications would you place on your conclusion?

Chapter Review

Sequencing Entire Genomes

13.1 Genomics

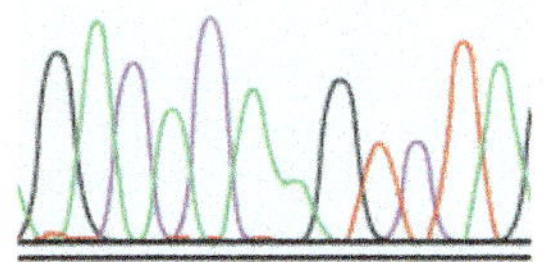

- The genetic information of an organism, its genes and other DNA, is called its genome. The sequencing and study of genomes is an area of biology called genomics (**table 13.1**).
- The sequencing of entire genomes, a once long and tedious process, has been made faster and easier with automated systems (**figure 13.1**).

13.2 The Human Genome

- The human genome contains about 20,000 to 25,000 genes (**figure 13.2**), far less than what was expected based on the number of unique mRNA molecules present in our cells.
- Genes are organized in different ways in the genome, with nearly 99% of the human genome containing noncoding segments of DNA (**figure 13.3**).

Genetic Engineering

13.3 A Scientific Revolution

- Genetic engineering is the process of moving genes from one organism to another. It is having a major impact on medicine and agriculture (**figure 13.4**).
- Restriction enzymes are a special kind of enzyme that binds to short sequences of DNA and cuts them at specific locations. When two different molecules of DNA are cut with the same restriction enzyme, sticky ends form, allowing the segments of different DNA to be joined (**figure 13.5**).
- Before transferring a eukaryotic gene into a bacterial cell, the intron regions must be removed. This is accomplished with the use of cDNA, a complementary copy of the gene using the processed mRNA to make double-stranded DNA that doesn't contain the introns (**figure 13.6**).
- DNA fingerprinting is a process using probes to compare two samples of DNA. The probes bind to the DNA samples, creating restriction patterns that can be compared.
- The polymerase chain reaction (PCR) is a procedure used to amplify small amounts of DNA (**figure 13.7**).

13.4 Genetic Engineering and Medicine

- Genetic engineering is used in the production of medically important proteins used to treat illnesses (**table 13.2**).
- Vaccines are developed using genetic engineering. A gene that encodes a viral protein of a pathogenic virus is inserted into the DNA of a harmless virus that serves as a vector, as shown here from **figure 13.9**. The vector carrying the recombinant DNA is injected into a human. The vector infects the body, replicates, and the recombinant DNA is translated producing the viral proteins. The body elicits an immune response against the proteins, which protects the person from an infection by the pathogenic virus in the future.

13.5 Genetic Engineering and Agriculture

- Genetic engineering has been used in crop plants to make them more cost effective to grow (**figure 13.11**) or more nutritious (**figure 13.12**). However, GM plants are a source of controversy because of potential dangers that may result from the genetic manipulation of crop plants.

The Revolution in Cell Technology

13.6 Reproductive Cloning

- Ian Wilmut succeeded in cloning a sheep by synchronizing the donated nucleus and the egg cell to the same stage of the cell cycle (**figure 13.14**).
- Other animals have been successfully cloned, but problems and complications arise, often causing premature death. The problems with cloning appear to be caused by the lack of modifications that need to be made to the DNA, which turns certain genes on or off, a process called epigenetic reprogramming.

13.7 Stem Cell Therapy

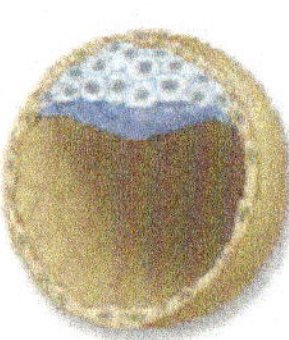

- Embryonic stem cells are totipotent cells, which are cells that are able to divide and develop into any type of cell in the body or develop into an entire individual. These cells are present in the early embryo. Because of the totipotent nature of embryonic stem cells, they could be used to replace tissues lost or damaged due to accident or disease (**figure 13.18**).

13.8 Therapeutic Use of Cloning

- The use of embryonic stem cells to replace damaged tissue has one major drawback: tissue rejection. The embryonic stem cells are treated as foreign cells by the patient's body and are rejected. Therapeutic cloning could alleviate this problem.
- Therapeutic cloning is the process whereby a cell from an individual who has lost tissue function is cloned, producing an embryo that is genetically identical to the person. Embryonic stem cells are then harvested from the cloned embryo and injected into the same individual. The embryonic stem cells regrow the lost or damaged tissue without eliciting an immune response (**figure 13.20**). However, this procedure, like others using embryonic stem cells, is controversial. Adult cells epigenetically reprogrammed to behave like embryonic stem cells offers great promise of a more acceptable treatment.

13.9 Gene Therapy

- Using gene therapy, a patient with a genetic disorder is cured by replacing a defective gene with a "healthy" gene. In theory, this should work—but early attempts to cure cystic fibrosis failed because of immunological reactions to the adenovirus vector used to carry the healthy genes into the patient.
- The recent focus of gene therapy has been to identify a vector that avoids the problems encountered with the adenovirus vector. Promising results in experiments using a parvovirus called adeno-associated virus (AAV) has scientists hopeful that new vectors will eliminate the problems seen with adenovirus (**figure 13.22**).

Test Your Understanding

1. The total amount of DNA in an organism, including all of its genes and other DNA, is its
   a. heredity.
   b. genetics.
   c. genome.
   d. genomics.
2. A possible reason why humans have such a small number of genes as opposed to what was anticipated by scientists is that
   a. humans don't need more than 25,000 genes to function.
   b. the exons used to make a specific mRNA can be rearranged to form different proteins.
   c. the sample size used to sequence the human genome was not big enough, so the number of genes estimated could be low.
   d. the number of genes will increase as scientists find out what all of the noncoding DNA actually does.
3. A protein that can cut DNA at specific DNA base sequences is called a
   a. DNase.
   b. DNA ligase.
   c. restriction enzyme.
   d. DNA polymerase.
4. Complementary DNA or cDNA is produced by
   a. inserting a gene into a bacterial cell.
   b. exposing the mRNA of the desired eukaryotic gene to reverse transcriptase.
   c. exposing the source DNA to restriction enzymes.
   d. exposing the source DNA to a probe.
5. Which of the following statements is correct?
   a. DNA fingerprinting is not admissible in court.
   b. DNA fingerprinting can prove with 100% certainty that two samples of DNA are from the same person.
   c. DNA fingerprinting becomes more and more reliable as more probes are used.
   d. No two people will ever have the same restriction pattern.
6. Using drugs produced by genetically engineered bacteria allows
   a. the drug to be produced in far larger amounts than in the past.
   b. humans to permanently correct the effect of a missing gene from their own systems.
   c. humans to cure cystic fibrosis.
   d. All of the above.
7. Some of the advantages to using genetically modified organisms in agriculture include
   a. increased yield.
   b. maintaining current nutritive value.
   c. mass-producing proteins.
   d. curing genetic diseases.
8. Which of the following is *not* a concern about the use of genetically modified crops?
   a. possible danger to humans after consumption
   b. insecticide resistance developing in pest species
   c. gene flow into natural relatives of GM crops
   d. harm to the crop itself from mutations
9. One of the main biological problems with replacing damaged tissue through the use of embryonic stem cells is
   a. immunological rejection of the tissue by the patient.
   b. that stem cells may not target appropriate tissue.
   c. the time needed to grow sufficient amounts of tissue.
   d. that genetic mutation of chosen stem cells may cause future problems.
10. In gene therapy, healthy genes are placed into animal cells that have defective genes by using
    a. a DNA particle gun.
    b. micropipettes (needles).
    c. viruses.
    d. Cells are not modified genetically. Instead, healthy tissue is grown and transplanted into the patient.

Apply Your Understanding

1. **Figure 13.1** Can you sequence the unknown section of DNA with the DNA fragments obtained from the following DNA analysis?

2. **Figure 13.14** Your friend Thomas wants to know why scientists don't just take an egg cell, with its own nucleus intact, and shock it to begin cell division. How do you answer him?

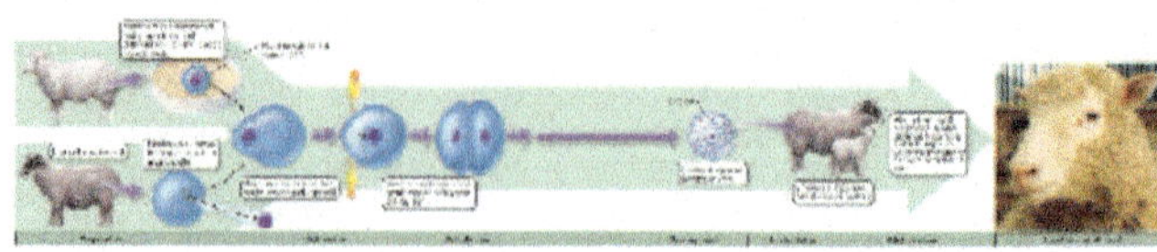

Synthesize What You Have Learned

1. The goal behind therapeutic cloning is to replace tissue that is damaged due to an accident or nongenetic disease. Do you think therapeutic cloning as shown in figure 13.20 would work to replace tissue damage caused by genetic disorders?
2. If a person has a genetic disease such as cystic fibrosis, the hope is that we will be able to use gene therapy to cure him or her. When that happens, will the patient no longer be able to pass the *cf* gene on to his or her children?
3. Much of the technology for producing GM foods is owned by multinational corporations, which seek to maintain intellectual ownership of their creations. As one example, Monsanto Corporation requires farmers to sign contracts for glyphosate-tolerant soybeans that prevent the farmers from saving seed for replanting the next year. The company has aggressively brought suit against violators. On the one hand, companies need to be able to profit from their products, and the development costs of GM foods are enormous. Without potential profit, future GM crops will not be developed. On the other hand, in many highly populated regions of the world, people who face famine when their crops fail simply cannot afford to pay the price of seeds every year. How would you want to see this challenging issue handled?

These four finches live on the Galápagos Islands, a cluster of volcanic islands far out to sea off the coast of South America. All descendants of a single ancestral migrant, blown to the islands from the mainland long ago, the Galápagos finches gave Darwin valuable clues about how natural selection shapes the evolution of species. The two upper finches are ground finches, their different beaks adapting them to eat different-sized seeds. The finch on the left consumes smaller, slender seeds. The stouter beak of the finch on the right enables it to crack open larger, drier seeds. On the lower left is a woodpecker finch, a kind of tree finch that carries around a cactus spine, which it uses to probe for insects in deep crevices. On the lower right is a warbler finch that like its namesake, eats crawling insects. Each of these species utilizes food resources differently. Their different ways of interacting with the community within which they live generate the selective pressures that shape the evolution of groups like Darwin's finches.

14

Evolution and Natural Selection

Learning Objectives

Evolution

14.1 Darwin's Voyage on HMS Beagle

The great diversity of life on earth—ranging from bacteria to elephants and roses—is the result of a long process of **evolution,** the change that occurs in organisms's characteristics through time. In 1859, the English naturalist Charles Darwin (1809–82; figure 14.1) first suggested an explanation for why evolution occurs, a process he called *natural selection.* Biologists soon became convinced Darwin was right and now consider evolution one of the central concepts of the science of biology. In this chapter, we examine Darwin and evolution in detail, as the concepts we encounter will provide a solid foundation for your exploration of the living world.

The theory of evolution proposes that a population can change over time, sometimes forming a new species. A **species** is a population or group of populations that possess similar characteristics and can interbreed and produce fertile offspring. This famous theory provides a good example of how a scientist develops a hypothesis—in this case, a hypothesis of how evolution occurs—and how, after much testing, the hypothesis is eventually accepted as a theory.

Charles Robert Darwin was an English naturalist who, after 30 years of study and observation, wrote one of the most famous and influential books of all time. This book, *On the Origin of Species by Means of Natural Selection, or The Preservation of Favoured Races in the Struggle for Life,* created a sensation when it was published, and the ideas Darwin expressed in it have played a central role in the development of human thought ever since.

In Darwin's time, most people believed that the various kinds of organisms and their individual structures resulted from direct actions of the Creator. Species were thought to be specially created and unchangeable over the course of time. In contrast to these views, a number of earlier philosophers had presented the view that living things must have changed during the history of life on earth. Darwin proposed a concept he called natural selection as a coherent, logical explanation for this process. Darwin's book, as its title indicates, presented a conclusion that differed sharply from conventional wisdom. Although his theory did not challenge the existence of a Divine Creator, Darwin argued that this Creator did not simply create things and then leave them forever unchanged. Instead, Darwin's God expressed Himself through the operation of natural laws that produced change over time—evolution.

The story of Darwin and his theory begins in 1831, when he was 22 years old. The small British naval vessel HMS *Beagle* that you see in figure 14.2 was about to set sail on a five-year navigational mapping expedition around the coasts of South America. The red arrows in figure 14.3 indicate the route taken by HMS *Beagle*. The young (26-year-old) captain of HMS *Beagle,* unable by British naval tradition to have social contact with his crew, and anticipating a voyage that would last many years, wanted a gentleman companion, someone to talk to. Indeed, the *Beagle*'s previous skipper had broken down and shot himself to death after three solitary years away from home.

Figure 14.1 The theory of evolution by natural selection was proposed by Charles Darwin.
This rediscovered photograph appears to be the last ever taken of the great biologist. It was taken in 1881, the year before Darwin died.

On the recommendation of one of his professors at Cambridge University, Darwin, the son of a wealthy doctor and very much a gentleman, was selected to serve as the captain's companion, primarily to share his table at mealtime during every shipboard dinner of the long voyage. Darwin paid his own expenses, and even brought along a manservant.

Darwin took on the role of ship's naturalist (the official naturalist, a man named Robert McKormick, left the ship before the first year was out). During this long voyage, Darwin had the chance to study a wide variety of plants and animals on continents and islands and in distant seas. He was able to explore the biological richness of the tropical forests, examine the extraordinary fossils of huge extinct mammals in Patagonia at the southern tip of South America, and observe the remarkable series of related but distinct forms of life on the **Galápagos Islands.** Such an opportunity clearly played an important role in the development of his thoughts about the nature of life on earth.

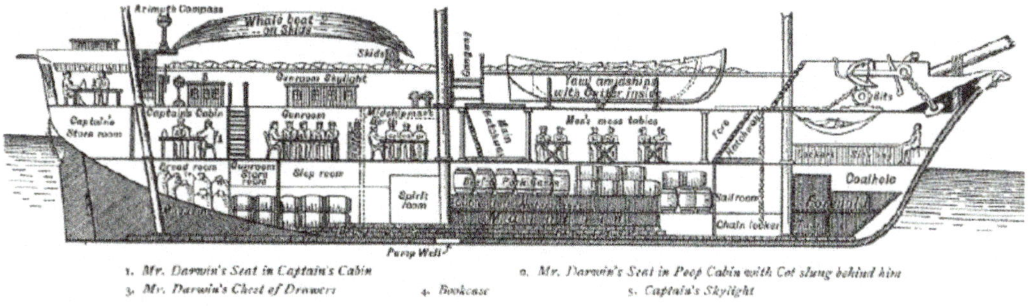

Figure 14.2 Cross section of HMS *Beagle*.
HMS *Beagle,* a 10-gun brig of 242 tons, only 90 feet in length, had a crew of 74 people! After he first saw the ship, Darwin wrote to his college professor Henslow: "The absolute want of room is an evil that nothing can surmount."

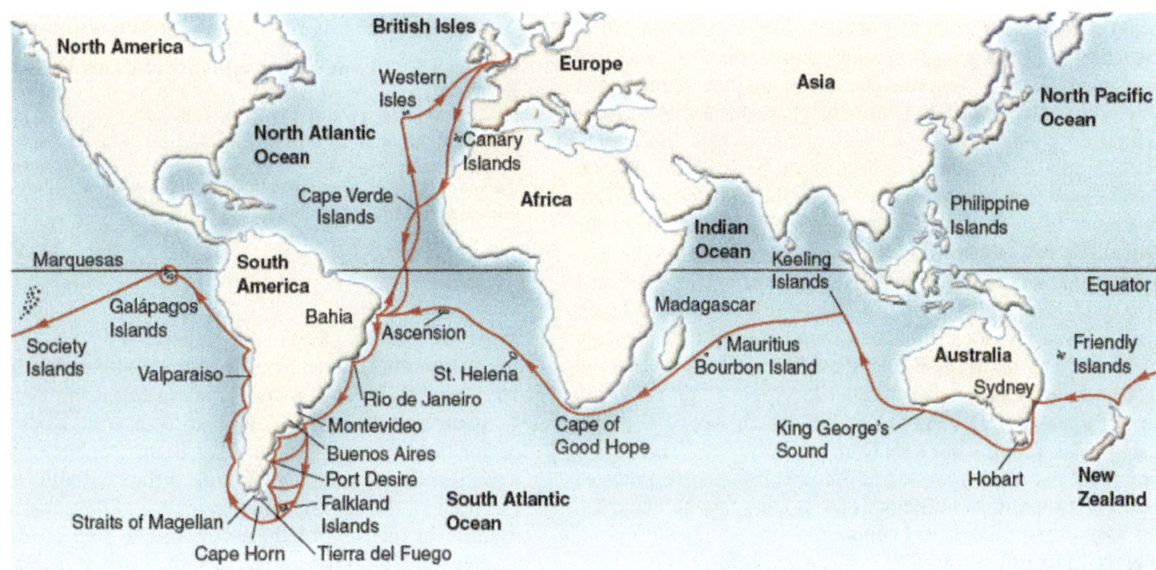

Figure 14.3 The five-year voyage of HMS *Beagle*.
Although the ship sailed around the world, most of its time was spent exploring the coasts and coastal islands of South America, such as the Galápagos Islands. Darwin's studies of the animals of these islands played a key role in the eventual development of his theory of evolution by means of natural selection.

When Darwin returned from the voyage at the age of 27, he began a long period of study and contemplation. During the next 10 years, he published important books on several different subjects, including the formation of oceanic islands from coral reefs and the geology of South America. He also devoted eight years of study to barnacles, a group of small marine animals with shells that inhabit rocks and pilings, eventually writing a four-volume work on their classification and natural history. In 1842, Darwin and his family moved out of London to a country home at Down, in the county of Kent. In these pleasant surroundings, Darwin lived, studied, and wrote for the next 40 years.

Key Learning Outcome 14.1 **Darwin was the first to propose natural selection as the mechanism of evolution that produced the diversity of life on earth.**

14.2 Darwin's Evidence

One of the obstacles that had blocked the acceptance of any theory of evolution in Darwin's day was the incorrect notion, widely believed at that time, that the earth was only a few thousand years old. The discovery of thick layers of rocks, evidences of extensive and prolonged erosion, and the increasing numbers of diverse and unfamiliar fossils discovered during Darwin's time made this assertion seem less and less likely. The great geologist Charles Lyell (1797–1875), whose *Principles of Geology* (1830) Darwin read eagerly as he sailed on HMS *Beagle*, outlined for the first time the story of an ancient world in which plant and animal species were constantly becoming extinct while others were emerging. It was this world that Darwin sought to explain.

What Darwin Saw

When HMS *Beagle* set sail, Darwin was fully convinced that species were immutable, meaning that they were not subject to being changed. Indeed, it was not until two or three years after his return that he began to seriously consider the possibility that they could change. Nevertheless, during his five years on the ship, Darwin observed a number of phenomena that were of central importance to him in reaching his ultimate conclusion. For example, in the rich fossil beds of southern South America, he observed fossils of the extinct armadillo shown on the right in figure 14.4. They were surprisingly similar in form to the armadillos that still lived in the same area, shown on the left. Why would similar living and fossil organisms be in the same area unless the earlier form had given rise to the other? Later, Darwin's observations would be strengthened by the discovery of other examples of fossils that show intermediate characteristics, pointing to successive change.

Repeatedly, Darwin saw that the characteristics of similar species varied somewhat from place to place. These geographical patterns suggested to him that organismal lineages change gradually as individuals move into new habitats. On the Galápagos Islands, 900 kilometers (540 miles) off the coast of Ecuador, Darwin encountered a variety of different finches on the islands. The 14 species, although related, differed slightly in appearance. Darwin felt it most reasonable to assume all these birds had descended from a common ancestor blown by winds from the South American mainland several million years ago. Eating different foods, on different islands, the species had changed in different ways, most notably in the size of their beaks. The larger beak of the ground finch in the upper left of figure 14.5 is better suited to crack open the large seeds it eats. As the generations descended from the common ancestor, these ground finches changed and adapted, what Darwin referred to as "descent with modification"—evolution.

In a more general sense, Darwin was struck by the fact that the plants and animals on these relatively young volcanic islands resembled those on the nearby coast of South America. If each one of these plants and animals had been created independently and simply placed on the Galápagos Islands, why didn't they resemble the plants and animals of islands with similar climates, such as those off the coast of Africa, for example? Why did they resemble those of the adjacent South American coast instead?

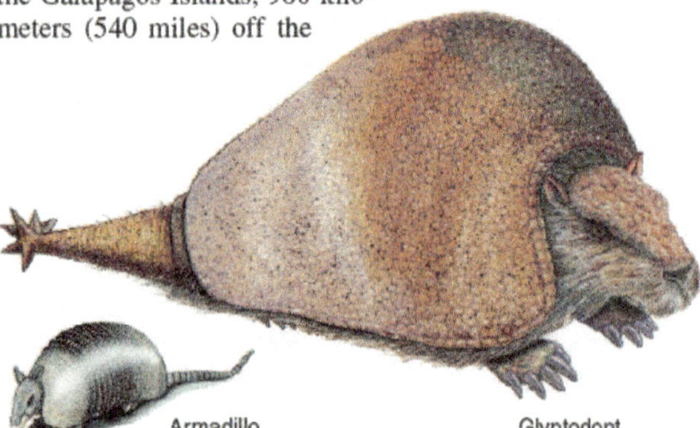

Figure 14.4 Fossil evidence of evolution.
The now-extinct glyptodont was a large 2,000-kilogram South American armadillo (about the size of a small car), much larger than the modern armadillo, which weighs an average of about 4.5 kilograms and is about the size of a house cat. The similarity of fossils such as the glyptodonts to living organisms found in the same regions suggested to Darwin that evolution had taken place.

Figure 14.5 Four Galápagos finches and what they eat.
Darwin observed 14 different species of finches on the Galápagos Islands, differing mainly in their beaks and feeding habits. These four finches eat very different food items, and Darwin surmised that the very different shapes of their beaks represented evolutionary adaptations improving their ability to do so.

Key Learning Outcome 14.2 The fossils and patterns of life that Darwin observed on the voyage of HMS *Beagle* eventually convinced him that evolution had taken place.

14.3 The Theory of Natural Selection

It is one thing to observe the results of evolution but quite another to understand how it happens. Darwin's great achievement lies in his formulation of the hypothesis that evolution occurs because of natural selection.

Darwin and Malthus

Of key importance to the development of Darwin's insight was his study of Thomas Malthus's *Essay on the Principle of Population* (1798). In his book, Malthus pointed out that populations of plants and animals (including human beings) tend to increase geometrically, while the ability of humans to increase their food supply increases only arithmetically. A geometric progression is one in which the elements increase by a constant factor; the blue line in figure 14.6 shows the progression 2, 6, 18, 54, . . . and each number is three times the preceding one. An arithmetic progression, in contrast, is one in which the elements increase by a constant difference; the red line shows the progression 2, 4, 6, 8, . . . and each number is two greater than the preceding one.

Because populations increase geometrically, virtually any kind of animal or plant, if it could reproduce unchecked, would cover the entire surface of the world within a surprisingly short time. Instead, population sizes of species remain fairly constant year after year, because death limits population numbers. Malthus's conclusion provided the key ingredient that was necessary for Darwin to develop the hypothesis that evolution occurs by natural selection.

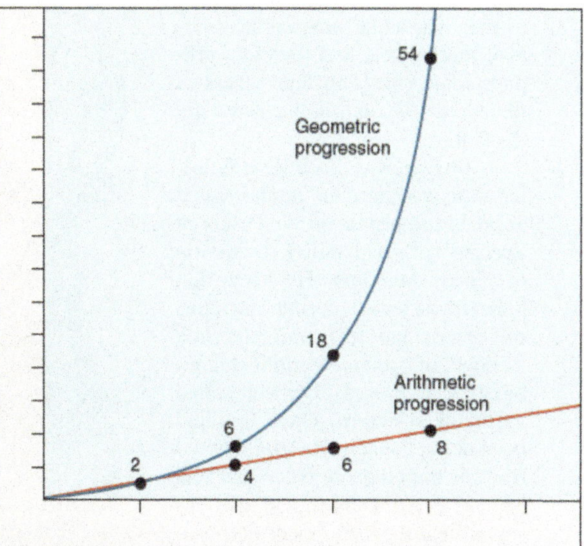

Figure 14.6 Geometric and arithmetic progressions. An arithmetic progression increases by a constant difference (for example, units of 1 or 2 or 3), while a geometric progression increases by a constant factor (for example, by 2 or by 3 or by 4). Malthus contended that the human growth curve was geometric, but the human food production curve was only arithmetic. Can you see the problems this difference would cause?

Natural Selection

Sparked by Malthus's ideas, Darwin saw that although every organism has the potential to produce more offspring than can survive, only a limited number actually do survive and produce further offspring. Many examples appear in nature. Sea turtles, for instance, will return to the beaches where they hatched to lay their eggs. Each female will lay about 100 eggs. The beach could be covered with thousands of hatchlings, like in figure 14.7, trying to make it to water's edge. Less than 10% will actually reach adulthood and return to this beach to reproduce. Darwin combined his observation with what he had seen on the voyage of HMS *Beagle,* as well as with his own experiences in breeding domestic animals, and made an important association: Those individuals that possess physical, behavioral, or other attributes that help them live in their environment are more likely to survive than those that do not have these characteristics. By surviving, they gain the opportunity to pass on their favorable characteristics to their offspring. As the frequency of these characteristics increases in the population, the nature of the population as a whole will gradually change. Darwin called this process **natural selection.** The driving force he identified has often been referred to as *survival of the fittest.* However, this is not to say the biggest or the strongest always survive. These characteristics may be favorable in one environment but less favorable in another.

Figure 14.7 Sea turtle hatchlings. These newly-hatched sea turtles make their way to the ocean from their nests on the beach. Thousands of eggs may be laid on a beach during a spawning, but less than 10% will survive to adulthood. Natural predators, human egg poachers, and environmental challenges prevent the majority of offspring from surviving. As Darwin observed, sea turtles produce more offspring than will actually survive to reproduce.

The organisms that are "best suited" to their particular environment survive more often, and therefore produce more offspring than others in the population, and in this sense are the "fittest."

Darwin was thoroughly familiar with variation in domesticated animals and began *On the Origin of Species* with a detailed discussion of pigeon breeding. He knew that breeders selected certain varieties of pigeons and other animals, such as dogs, to produce certain characteristics, a process Darwin called **artificial selection.** Once this had been done, the animals would breed true for the characteristics that had been selected. Darwin had also observed that the differences purposely developed between domesticated races or breeds were often greater than those that separated wild species. Domestic pigeon breeds, for example, show much greater variety than all of the hundreds of wild species of pigeons found throughout the world. Such relationships suggested to Darwin that evolutionary change could occur in nature too. Surely if pigeon breeders could foster such variation by "artificial selection," nature through environmental pressures could do the same, playing the breeder's role in selecting the next generation—a process Darwin called *natural selection*.

Figure 14.8 Darwin greets his monkey ancestor.

In his time, Darwin was often portrayed unsympathetically, as in this drawing from an 1874 publication.

Darwin's theory provides a simple and direct explanation of biological diversity, or why animals are different in different places—because habitats differ in their requirements and opportunities, the organisms with characteristics favored locally by natural selection will tend to vary in different places. As we will discuss later in this chapter in section 14.9, there are five evolutionary forces that can affect biological diversity, although natural selection is the only evolutionary force that produces *adaptive* changes.

Darwin Drafts His Argument

Darwin drafted the overall argument for evolution by natural selection in a preliminary manuscript in 1842. After showing the manuscript to a few of his closest scientific friends, however, Darwin put it in a drawer and for 16 years turned to other research. No one knows for sure why Darwin did not publish his initial manuscript—it is very thorough and outlines his ideas in detail. Some historians have suggested that Darwin was wary of igniting public, and even private, criticism of his evolutionary ideas—there could have been little doubt in his mind that his theory of evolution by natural selection would spark controversy. Others have proposed that Darwin was simply refining his theory, although there is little evidence he altered his initial manuscript in all that time.

Wallace Has the Same Idea

The stimulus that finally brought Darwin's theory into print was an essay he received in 1858. A young English naturalist named Alfred Russel Wallace (1823–1913) sent the essay to Darwin from Malaysia; it concisely set forth the theory of evolution by means of natural selection, a theory Wallace had developed independently of Darwin. Like Darwin, Wallace had been greatly influenced by Malthus's 1798 book. Colleagues of Wallace, knowing of Darwin's work, encouraged him to communicate with Darwin. After receiving Wallace's essay, Darwin arranged for a joint presentation of their ideas at a seminar in London. Darwin then completed his own book, expanding the 1842 manuscript that he had written so long ago, and submitted it for publication.

Publication of Darwin's Theory

Darwin's book appeared in November 1859 and caused an immediate sensation. Although people had long accepted that humans closely resembled apes in many characteristics, the possibility that there might be a direct evolutionary relationship was unacceptable to many. Darwin did not actually discuss this idea in his book, but it followed directly from the principles he outlined. In a subsequent book, *The Descent of Man,* Darwin presented the argument directly, building a powerful case that humans and living apes have common ancestors. Many people were deeply disturbed with the suggestion that human beings were descended from the same ancestor as apes, and Darwin's book on evolution caused him to become a victim of the satirists of his day—the cartoon in figure 14.8 is a vivid example. Darwin's arguments for the theory of evolution by natural selection were so compelling, however, that his views were almost completely accepted within the intellectual community of Great Britain after the 1860s.

Key Learning Outcome 14.3 The fact that populations do not really expand geometrically implies that nature acts to limit population numbers. The traits of organisms that survive to produce more offspring will be more common in future generations—a process Darwin called natural selection.

14.4 The Beaks of Darwin's Finches

Darwin's Galápagos finches played a key role in his argument for evolution by natural selection. He collected 31 specimens of finches from three islands when he visited the Galápagos Islands in 1835. Darwin, not an expert on birds, had trouble identifying the specimens. He believed by examining their beaks that his collection contained wrens, "gross-beaks," and blackbirds.

The Importance of the Beak

Upon Darwin's return to England, ornithologist John Gould examined the finches. Gould recognized that Darwin's collection was in fact a closely related group of distinct species, all similar to one another except for their beaks. In all, 14 species are now recognized, 13 from the Galápagos and one from far-distant Cocos Island. The ground finches with the larger beaks in figure 14.9 feed on seeds that they crush in their beaks, whereas those with narrower beaks eat insects, including the warbler finch (named for its resemblance to a mainland bird). Other species include fruit and bud eaters, and species that feed on cactus fruits and the insects they attract; some populations of the sharp-beaked ground finch even include "vampires" that creep up on seabirds and use their sharp beaks to drink their blood. Perhaps most remarkable are the tool users, like the woodpecker finch you see in the upper left of the figure, that picks up a twig, cactus spine, or leaf stalk, trims it into shape with its beak, and then pokes it into dead branches to pry out grubs.

Figure 14.10 A gene shapes the beaks of Darwin's finches.
A cell signalling molecule called "bone morphogenic protein 4" (BMP4) has been shown by DNA researchers to tailor the shape of the beak in Darwin's finches.

The differences in the beaks of Darwin's finches are due to differences in the genes of the birds. When biologists compare the DNA of large ground finches (with stout beaks for cracking large seeds) to the DNA of small ground finches (with more slender beaks), the only growth factor gene that is different in the DNA of the two species is *BMP4* (figure 14.10 and figure 1.15). The difference is in how the gene is used. The large ground finches, with larger beaks, make more BMP4 protein than do the small ground finches.

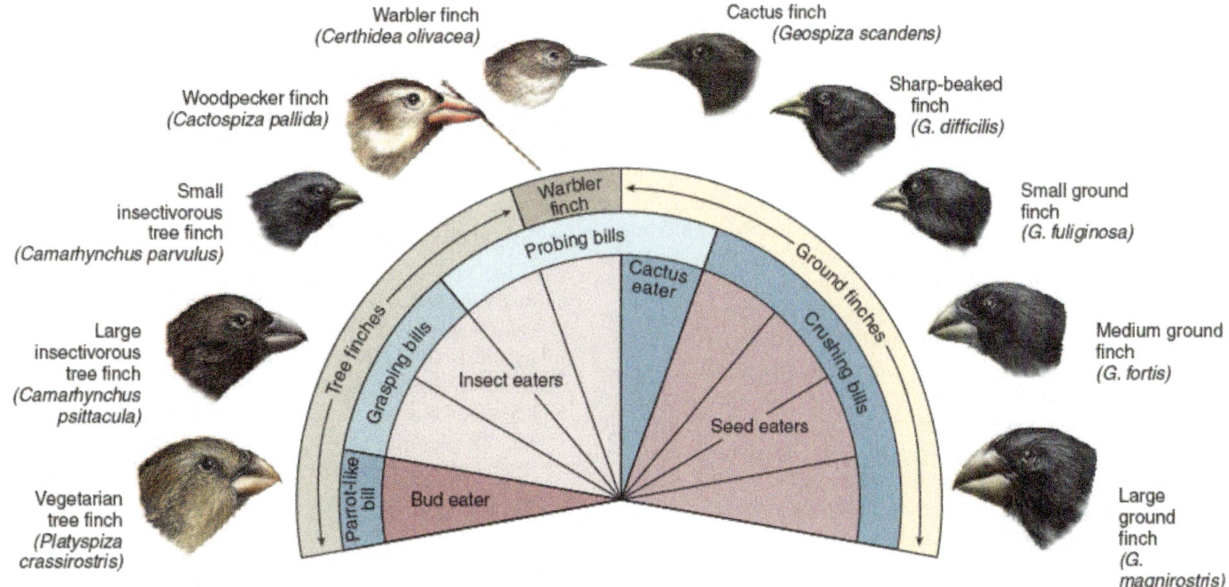

Figure 14.9 A diversity of finches on a single island.
Ten species of Darwin's finches from Isla Santa Cruz, one of the Galápagos Islands. The 10 species show differences in beaks and feeding habits. These differences presumably arose when the finches arrived and encountered habitats lacking small birds. Scientists concluded that all of these birds derived from a single common ancestor.

The correspondence between the beaks of the 14 finch species and their food source immediately suggested to Darwin that evolution had shaped them:

"Seeing this gradation and diversity of structure in one small, intimately related group of birds, one might really fancy that from an original paucity of birds in this archipelago, one species has been taken and modified for different ends."

Checking to See if Darwin Was Right

If Darwin's suggestion that the beak of an ancestral finch had been "modified for different ends" is correct, then it ought to be possible to see the different species of finches acting out their evolutionary roles, each using its beak to acquire its particular food specialty. The four species that crush seeds within their beaks, for example, should feed on different seeds, with those with stouter beaks specializing on harder-to-crush seeds.

Many biologists visited the Galápagos after Darwin, but it was 100 years before any tried this key test of his hypothesis. When the great naturalist David Lack finally set out to do this in 1938, observing the birds closely for a full five months, his observations seemed to contradict Darwin's proposal! Lack often observed many different species of finch feeding together on the same seeds. His data indicated that the stout-beaked species and the slender-beaked species were feeding on the very same array of seeds.

We now know that it was Lack's misfortune to study the birds during a wet year, when food was plentiful. The size of the finch's beak is of little importance in such flush times; slender and stout beaks work equally well to gather the abundant tender small seeds. Later work revealed a very different picture during dry years, when few seeds are available.

A Closer Look

Starting in 1973, Peter and Rosemary Grant of Princeton University and generations of their students have studied the medium ground finch, *Geospiza fortis*, on a tiny island in the center of the Galápagos called Daphne Major. These finches feed preferentially on small, tender seeds, abundantly available in wet years. The birds resort to larger, drier seeds that are harder to crush when small seeds are hard to find. Such lean times come during periods of dry weather, when plants produce few seeds, large or small.

By carefully measuring the beak shape of many birds every year, the Grants were able to assemble for the first time a detailed portrait of evolution in action. The Grants found that beak depth changed from one year to the next in a predictable fashion. During droughts, plants produced few seeds, and all available small seeds quickly were eaten, leaving large seeds as the major remaining source of food. As a result, birds with large beaks survived better, because they were better able to break open these large seeds. Consequently, the average beak depth of birds in the population increased the next year because this next generation included offspring of the large-beaked birds that survived. The offspring of the surviving "dry year" birds had larger beaks, an evolutionary response which led to the peaks you see in the graph in figure 14.11. The reason there are peaks and not plateaus is that the average beak size decreased again when wet seasons returned because the larger beak size was no longer more favorable when seeds were plentiful and so smaller-beaked birds survived to reproduce.

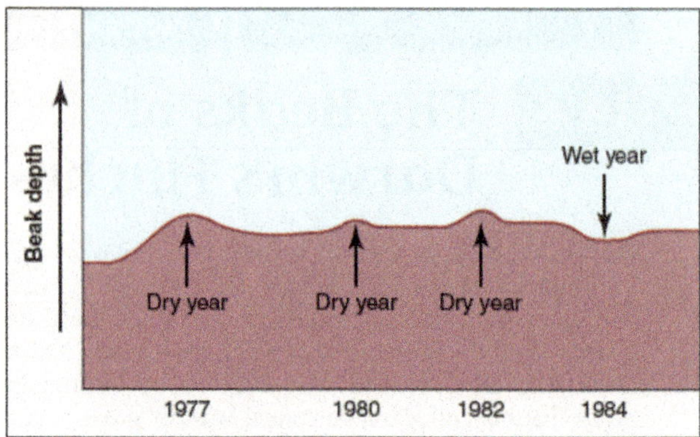

Figure 14.11 Evidence that natural selection alters beak size in *Geospiza fortis*.

In dry years, when only large, tough seeds were available, the mean beak size increased. In wet years, when many small seeds were available, smaller beaks became more common.

Could these changes in beak dimension reflect the action of natural selection? An alternative possibility might be that the changes in beak depth do not reflect changes in gene frequencies but rather are simply a response to diet, with poorly fed birds having stouter beaks. To rule out this possibility, the Grants measured the relation of parent beak size to offspring beak size, examining many broods over several years. The depth of the beak was passed down faithfully from one generation to the next, suggesting the differences in beak size indeed reflected gene differences.

Support for Darwin

If the year-to-year changes in beak depth can be predicted by the pattern of dry years, then Darwin was right after all—natural selection influences beak size based on available food supply. In the study discussed here, birds with stout beaks have an advantage during dry periods, for they can break the large, dry seeds that are the only food available. When small seeds become plentiful once again with the return of wet weather, a smaller beak proves a more efficient tool for harvesting smaller seeds.

Key Learning Outcome 14.4 In Darwin's finches, natural selection adjusts the shape of the beak in response to the nature of the food supply, adjustments that are occurring even today.

14.5 How Natural Selection Produces Diversity

Darwin believed that each Galápagos finch species had adapted to the particular foods and other conditions on the particular island it inhabited. Because the islands presented different opportunities, a cluster of species resulted. Presumably, the ancestor of Darwin's finches reached these newly formed islands before other land birds, so that when it arrived, all of the niches where birds occur on the mainland were unoccupied. A *niche* is what a biologist calls the way a species makes a living—the biological (that is, other organisms) and physical (climate, food, shelter, etc.) conditions with which an organism interacts as it attempts to survive and reproduce. As the new arrivals to the Galápagos moved into vacant niches and adopted new lifestyles, they were subjected to diverse sets of selective pressures. Under these circumstances, the ancestral finches rapidly split into a series of populations, some of which evolved into separate species.

The phenomenon by which a cluster of species change, as they occupy a series of different habitats within a region, is called *adaptive radiation*. Figure 14.12 shows how the 14 species of Darwin's finches on the Galápagos Islands and Cocos Island are thought to have evolved. The ancestral population, indicated by the base of the brackets, migrated to the islands about 2 million years ago and underwent adaptive radiation giving rise to the 14 different species.

The descendants of the original finches that reached the Galápagos Islands now occupy many different kinds of habitats on the islands. The 14 species that inhabit the Galápagos Islands and Cocos Island occupy four types of niches:

1. **Ground finches.** There are six species of *Geospiza* ground finches. Most of the ground finches feed on seeds. The size of their beaks is related to the size of the seeds they eat. Some of the ground finches feed primarily on cactus flowers and fruits and have longer, larger, more pointed beaks.
2. **Tree finches.** There are five species of insect-eating tree finches. Four species have beaks that are suitable for feeding on insects. The woodpecker finch has a chisel-like beak. This unique bird carries around a twig or a cactus spine, which it uses to probe for insects in deep crevices.
3. **Vegetarian finch.** The very heavy beak of this bud-eating bird is used to wrench buds from branches.
4. **Warbler finches.** These unusual birds play the same ecological role in the Galápagos woods that warblers play on the mainland, searching continually over the leaves and branches for insects. They have a slender, warblerlike beak.

Key Learning Outcome 14.5 Darwin's finches, all derived from one similar mainland species, have radiated widely on the Galápagos Islands, filling unoccupied niches in a variety of ways.

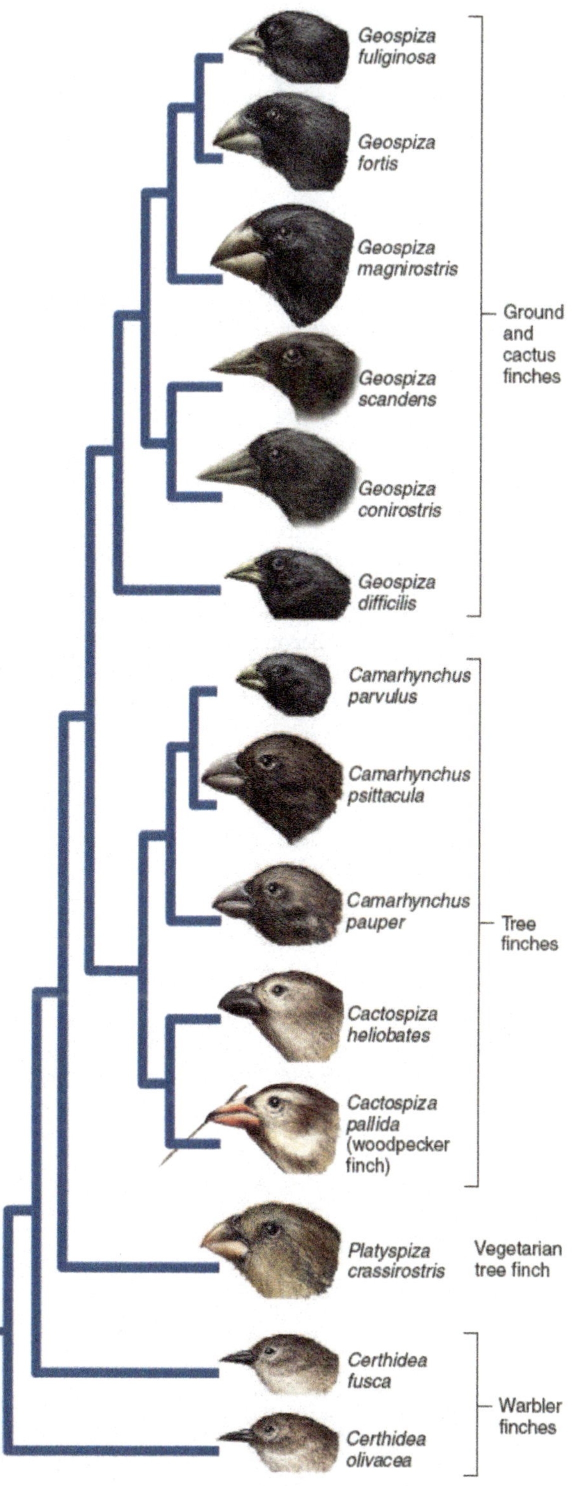

Figure 14.12 An evolutionary tree of Darwin's finches.

This family tree was constructed by comparing DNA of the 14 species. Their position at the base of the finch tree suggests that warbler finches were among the first adaptive types to evolve in the Galápagos.

14.6 The Evidence for Evolution

The evidence that Darwin presented in *The Origin of Species* to support his theory of evolution was strong. We will now examine other lines of evidence supporting Darwin's theory, including information revealed by examining fossils, anatomical features, and molecules such as DNA and proteins.

The Fossil Record

The most direct evidence of macroevolution is found in the fossil record. **Fossils** are the preserved remains, tracks, or traces of once-living organisms. Fossils are created when organisms become buried in sediment. The calcium in bone or other hard tissue mineralizes, and the surrounding sediment eventually hardens to form rock. Most fossils are, in effect, skeletons. In the rare cases when fossils form in very fine sediment, feathers may also be preserved. When remains are frozen or become suspended in amber (fossilized plant resin), however, the entire body may be preserved. The fossils contained in layers of sedimentary rock reveal a history of life on earth.

By dating the rock in which a fossil occurs, like the one in figure 14.13, we can get an accurate idea of how old the fossil is. Rocks are dated by measuring the amount of certain radioisotopes in the rock. A radioisotope will break down, or decay, into other isotopes or elements. This occurs at a constant rate and so the amount of a radioisotope present in the rock is an indication of the rock's age.

Using Fossils to Test the Theory of Evolution

If the theory of evolution is correct, then the fossils we see preserved in rock should represent a history of evolutionary change. The theory makes the clear prediction that a parade of successive changes should be seen, as first one change occurs and then another. If the theory of evolution is not correct, on the other hand, then such orderly change is not expected.

Figure 14.13 Dinosaur fossil of *Parasaurolophus*.

To test this prediction, we follow a logical procedure:

1. *Assemble a collection of fossils of a particular group of organisms.* You might, for example, gather together a collection of fossil titanotheres, a hoofed mammal that lived between about 50 million and 35 million years ago.
2. *Date each of the fossils.* In dating the fossils, it is important to make no reference to what the fossil is like. Imagine it as being concealed in a black box of rock, with only the box being dated.
3. *Order the fossils by their age.* Without looking in the "black boxes," place them in a series, beginning with the oldest and proceeding to the youngest.
4. *Now examine the fossils.* Do the differences between the fossils appear jumbled, or is there evidence of successive change as evolution predicts? You can judge for yourself in figure 14.14. During the 15 million years spanned by this collection of titanothere fossils, the small, bony protuberance located above the nose 50 million years ago evolved in a series of continuous changes into relatively large blunt horns.

It is important not to miss the key point of the result you see illustrated in figure 14.14: Evolution is an observation, not a conclusion. Because the dating of the samples is independent of what the samples are like, *successive change through time is a data statement*. While the statement that evolution

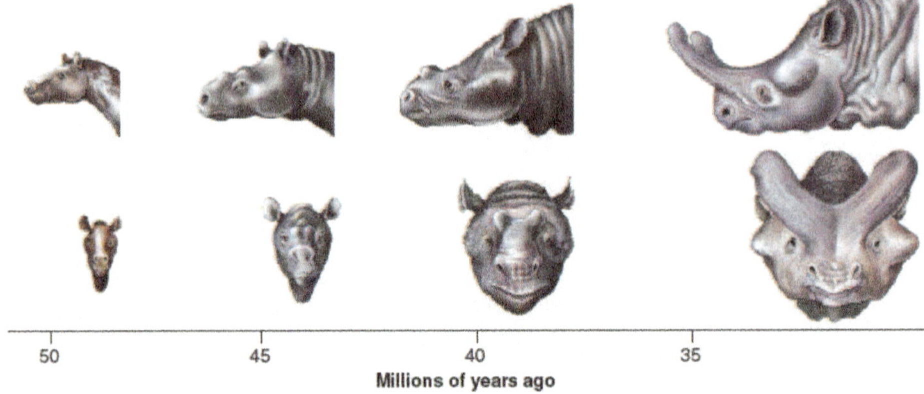

Figure 14.14 Testing the theory of evolution with fossil titanotheres.

Here you see illustrated changes in a group of hoofed mammals known as titanotheres, which lived between about 50 million and 35 million years ago. During this time, the small, bony protuberance located above the nose 50 million years ago evolved into relatively large, blunt horns.

Today's *Biology*

Darwin and Moby Dick

Moby Dick, the white whale hunted by Captain Ahab in Melville's novel, was a sperm whale. One of the ocean's great predators, a large sperm whale is a voracious meat-eater that may span over 60 feet and weigh 50 tons. A sperm whale is not a fish, though. Unlike the great white shark in *Jaws*, a whale has hairs (not many), and a female whale has milk-producing mammary glands with which it feeds its young. A sperm whale is a mammal, just as you are! This raises an interesting question. If Darwin is right about the fossil record reflecting life's evolutionary past, then fossils tell us mammals evolved from reptiles on land at about the time of the dinosaurs. How did they end up back in the water?

The evolutionary history of whales has long fascinated biologists, but only in recent years have fossils been discovered that reveal the answer to this intriguing question. A series of discoveries now allows biologists to trace the evolutionary history of the most colossal animals ever to live on earth back to their beginnings at the dawn of the Age of Mammals. Whales, it turns out, are the descendants of four-legged land mammals that reinvaded the sea some 50 million years ago, much as seals and walruses are doing today. It's pretty startling to realize that Moby Dick's evolutionary ancestor lived on the steppes of Asia and looked like a modest-sized pig a few feet long and weighing perhaps 50 pounds, with four toes on each foot.

From what land mammal did whales arise? Researchers had long speculated that it might be a hoofed meat-eater with three toes known as a *mesonychid*, related to rhinoceroses. Subtle clues suggested this—the arrangement of ridges on the molar teeth, the positioning of the ear bones in the skull. But findings announced in 2001 reveal these subtle clues to have been misleading. Ankle bones from two newly described 50 million-year-old whale species discovered by Philip Gingerich of the University of Michigan are those of an artiodactyl, a four-toed mammal related to hippos, cattle, and pigs. Even more recently, Japanese researchers studying DNA have discovered unique genetic markers shared today only by whales and hippos.

Biologists now conclude that whales, like hippos, are descended from a group of early four-hoofed mammals called *anthracotheres*, modest-sized grazing animals with a piggish appearance abundant in Europe and Asia 50 million years ago.

In Pakistan in 1994, biologists discovered its descendant, the oldest known whale. The fossil was 49 million years old, had four legs, each with four-toed feet and a little hoof at the tip of each toe. Dubbed *Ambulocetus* (walking whale), it was sharp-toothed and about the size of a large sea lion. Analysis of the minerals in its teeth reveal it drank fresh water, so like a seal it was not yet completely a marine animal. Its nostrils were on the end of the snout, like a dog's.

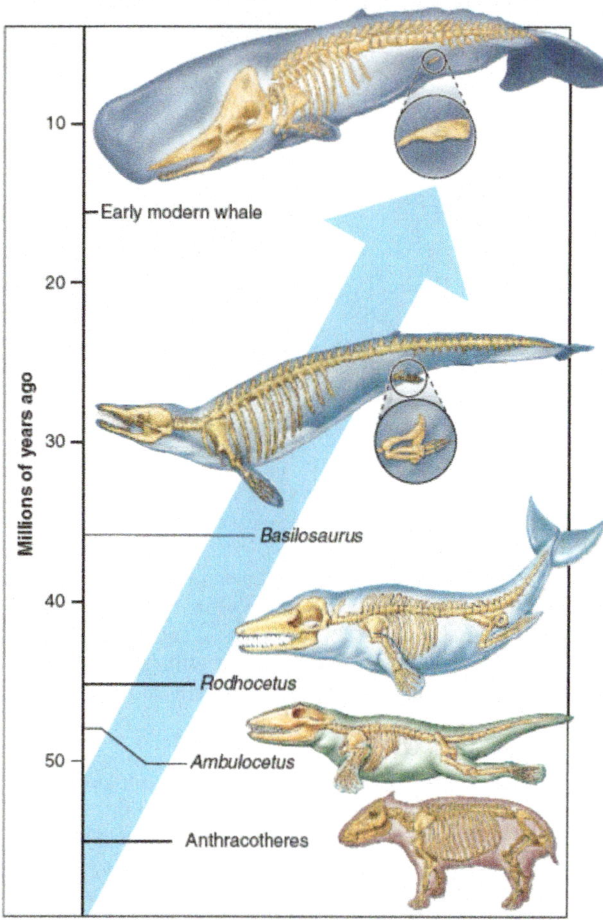

Appearing in the fossil record a few million years later is *Rodhocetus,* also seal-like but with smaller hind limbs and the teeth of an ocean water drinker. Its nostrils are shifted higher on the skull, halfway towards the top of the head.

Almost 10 million years later, about 37 million years ago, we see the first representatives of *Basilosaurus,* a giant 60-foot-long serpentlike whale with shrunken hind legs still complete down to jointed knees and toes.

The earliest modern whales appear in the fossil record 15 million years ago. The nostrils are now in the top of the head, a "blowhole" that allows it to break the surface, inhale, and resubmerge without having to stop or tilt the head up. The hind legs are gone, with vestigial tiny bones remaining that are unattached to the pelvis. Still, today's whales retain all the genes used to code for legs—occasionally a whale is born having sprouted a leg or two.

So it seems to have taken 35 million years to evolve a whale from the piglike ancestor of a hippopotamus—intermediate steps preserved in the fossil record for us to see. Darwin, who always believed that gaps in the vertebrate fossil record would eventually be filled in, would have been delighted.

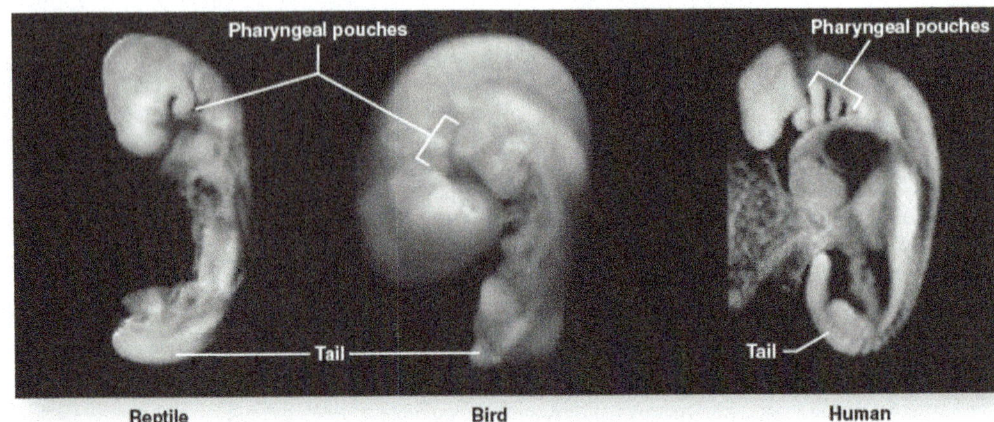

Figure 14.15 Embryos show our early evolutionary history. These embryos, representing various vertebrate animals, show the primitive features that all vertebrates share early in their development, such as pharyngeal pouches and a tail.

is the result of natural selection is a theory advanced by Darwin, the statement that macroevolution has occurred is a factual observation.

Many other examples illustrate this clear confirmation of the key prediction of Darwin's theory. The evolution of today's large, single-hoof horse with complex molar teeth from a much smaller four-toed ancestor with much simpler molar teeth is a familiar and clearly documented instance.

The Anatomical Record

Much of the evolutionary history of vertebrates can be seen in the way in which their embryos develop. Figure 14.15 shows three different embryos early in development, and as you can see, all vertebrate embryos have pharyngeal pouches (that develop into gill slits in fish); also every vertebrate embryo has a long bony tail, even if the tail is not present in the fully developed animal. These relict developmental forms strongly suggest that all vertebrates share a basic set of developmental instructions.

As vertebrates have evolved, the same bones are sometimes still there but put to different uses, their presence betraying their evolutionary past. For example, the forelimbs of vertebrates are all **homologous structures;** that is, although the structure and function of the bones have diverged, they are derived from the same body part present in a common ancestor. You can see in figure 14.16 how the bones of the forelimb have been modified for different functions. The yellow- and purple-colored bones, which correspond in humans to the bones of the forearm and wrist and fingers, respectively, are modified to make up the wings in the bat, the full leg of the horse, and the paddle in the fin of the porpoise.

Not all similar features are homologous. Sometimes features found in different lineages come to resemble each other as a result of parallel evolutionary adaptations to similar environments. This form of evolutionary change is referred to as *convergent evolution,* and these similar-looking features are called **analogous structures.** For example, the wings of birds, pterosaurs, and bats are analogous structures, modified through natural selection to serve the same function and

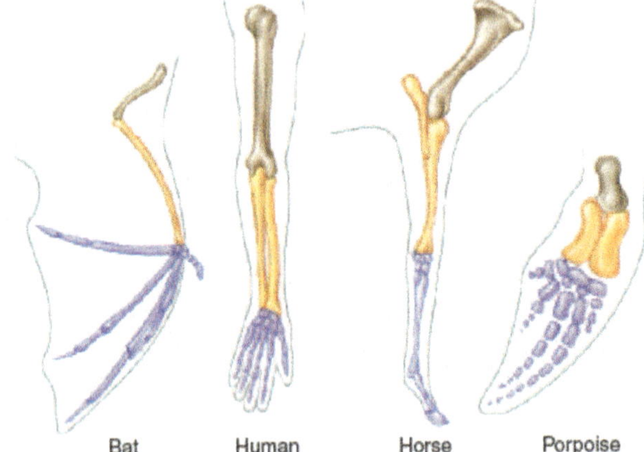

Figure 14.16 Homology among vertebrate limbs. Homologies among the forelimbs of four mammals show the ways in which the proportions of the bones have changed in relation to the particular way of life of each organism. Although considerable differences can be seen in form and function, the same basic bones are present in each forelimb.

therefore look the same (figure 14.17). Similarly, the marsupial mammals of Australia evolved in isolation from placental mammals, but similar selective pressures have generated very similar kinds of animals.

Sometimes structures are put to no use at all! In living whales, which evolved from hoofed mammals, the bones of the pelvis that formerly anchored the two hind limbs are all that remain of the rear legs, unattached to any other bones and serving no apparent purpose (the reduced pelvic bone can be seen in the figure in the "Today's Biology" reading on the previous page). Another example of what are called *vestigial organs* is the human appendix. The great apes, our closest relatives, have an appendix much larger than ours attached to the gut tube, which holds bacteria used in digesting the cellulose cell walls of the plants eaten by these primates. The human appendix is a vestigial version of this structure that now serves no function in digestion (although it may have acquired an alternate function in the lymphatic system).

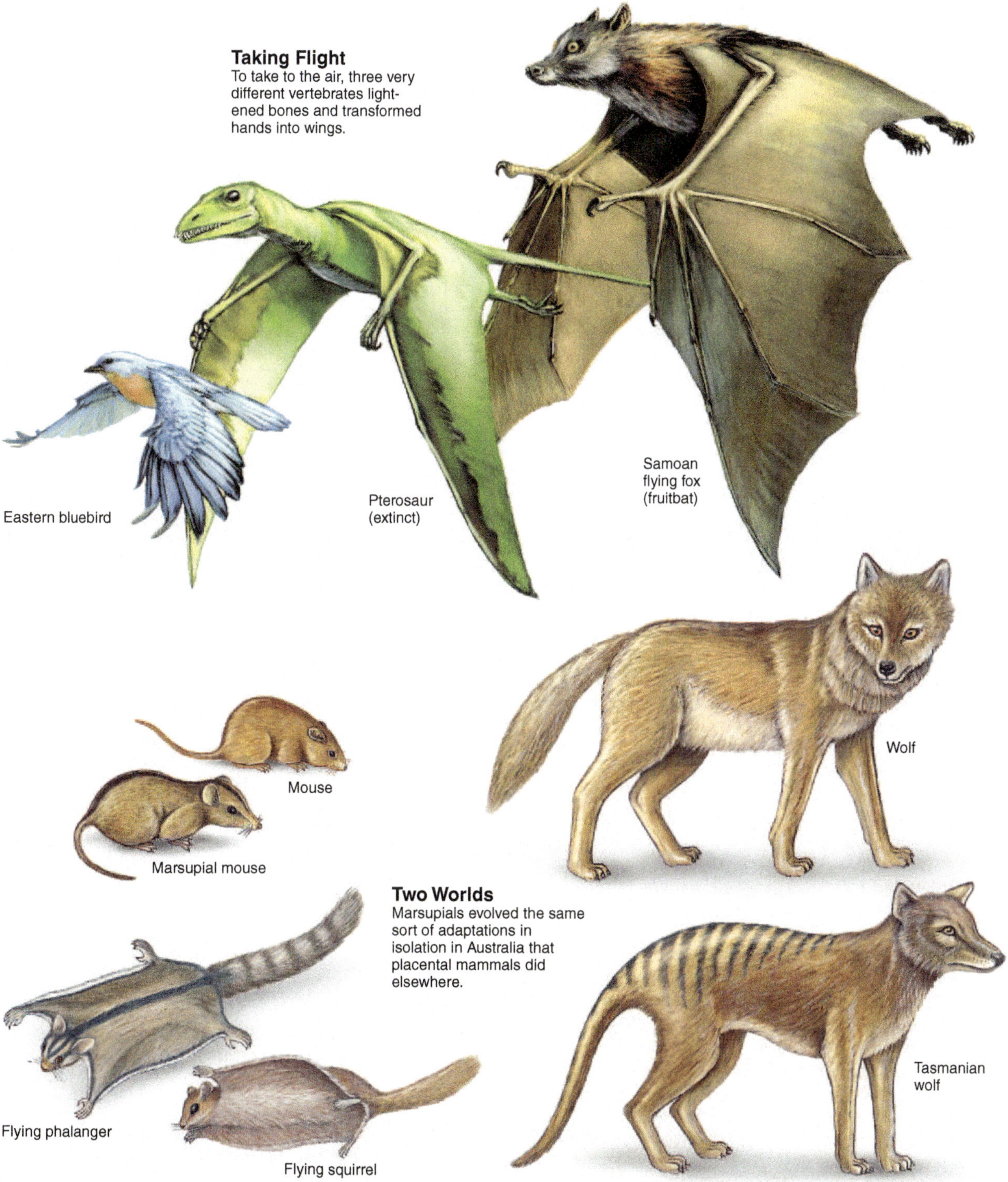

Figure 14.17 Convergent evolution: many paths to one goal.

Over the course of evolution, form often follows function. Members of very different animal groups often adapt in similar fashions when challenged by similar opportunities. These are but a few of many examples of such convergent evolution. The flying vertebrates represent mammals (bat), reptiles (pterosaur), and birds (bluebird). The three pairs of terrestrial vertebrates each contrast a North American placental mammal with an Australian marsupial one.

Source: "Taking Flight" image from The New York Times, *December 15, 1988.* The New York Times. *Reprinted with permission.*

The Molecular Record

Traces of our evolutionary past are also evident at the molecular level. We possess the same set of color vision genes as our ancestors, only more complex, and we employ pattern formation genes during early development that all animals share. Indeed, if you think about it, the fact that organisms have evolved from a series of simpler ancestors implies that a record of evolutionary change is present in the cells of each of us, in our DNA. According to evolutionary theory, new alleles arise from older ones by mutation and come to predominance through favorable selection. A series of evolutionary changes thus implies a continual accumulation of genetic changes in the DNA. From this you can see that evolutionary theory makes a clear prediction: Organisms that are more distantly related should have accumulated a greater number of evolutionary differences than two species that are more closely related.

This prediction is now subject to direct test. Recent DNA research allows us to directly compare the genomes of different organisms. The result is clear: For a broad array of vertebrates, the more distantly related two organisms are, the greater their genomic difference. This research is described later in this chapter on pages 298 and 299.

This same pattern of divergence can be clearly seen at the protein level. Comparing the hemoglobin amino acid sequence of different species with the human sequence in figure 14.18, you can see that species more closely related to humans have fewer differences in the amino acid structure of their hemoglobin. Macaques, primates closely related to humans, have fewer differences from humans (only 8 different amino acids) than do more distantly related mammals like dogs (which have 32 different amino acids). Nonmammalian terrestrial vertebrates differ even more, and marine vertebrates are the most different of all. Again, the prediction of evolutionary theory is strongly confirmed.

Molecular Clocks This same pattern is seen when the DNA sequence of an individual gene is compared over a much broader array of organisms. One well-studied case is the mammalian *cytochrome c* gene (cytochrome *c* is a protein that plays a key role in oxidative metabolism). Figure 14.19 compares the time when two species diverged on the *x* axis to the number of differences in their *cytochrome c* gene on the *y* axis. To practice using this data set, go back about 75 million years ago to find a common ancestor for humans and rodents—in that time there have been about 60 base substitutions in cytochrome *c*. This graph reveals a very important finding: Evolutionary changes appear to accumulate in cytochrome *c* at a constant rate, as indicated by the straightness of the blue line connecting the points. This constancy is sometimes referred to as a molecular clock. Most proteins for which data are available appear to accumulate changes over time in this fashion, although different proteins can evolve at very different rates.

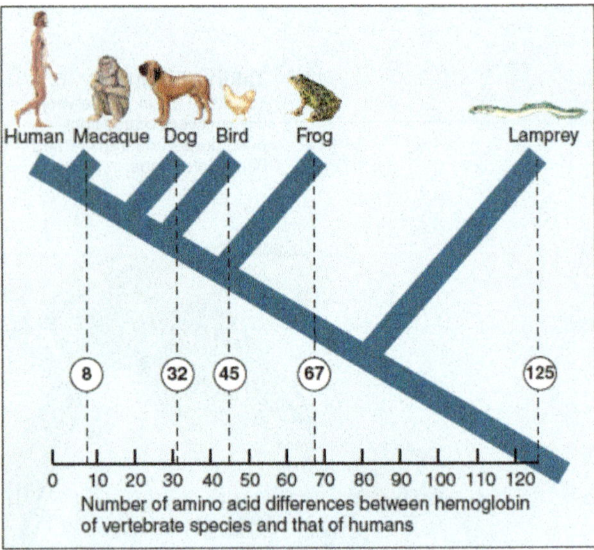

Figure 14.18 Molecules reflect evolutionary divergence.
The greater the evolutionary distance from humans (as revealed by the *blue* evolutionary tree based on the fossil record), the greater the number of amino acid differences in the vertebrate hemoglobin polypeptide.

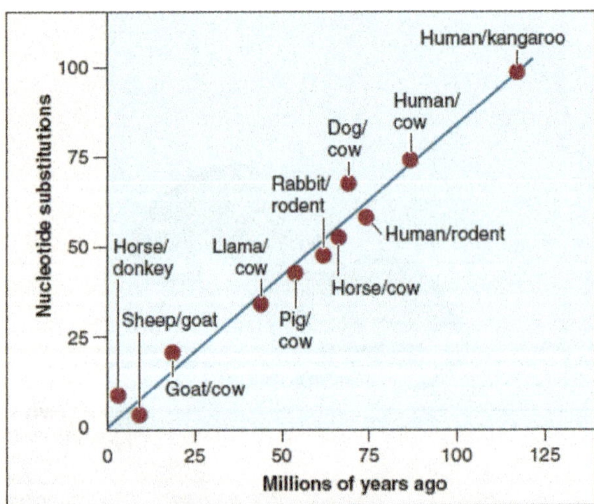

Figure 14.19 The molecular clock of cytochrome *c*.
When the time since each pair of organisms presumably diverged is plotted against the number of nucleotide differences in cytochrome *c*, the result is a straight line, suggesting that the *cytochrome c* gene is evolving at a constant rate.

Key Learning Outcome 14.6 **The fossil record provides a clear record of successive evolutionary change. Comparative anatomy also offers evidence that evolution has occurred. Finally, the genetic record exhibits successive evolution, the DNA of organisms accumulating increasing numbers of changes over time.**

14.7 Evolution's Critics

Of all the major ideas of biology, evolution is perhaps the best known to the general public because many people mistakenly believe that evolution represents a challenge to their religious beliefs. A person can have a spiritual faith in God and still be an excellent scientist—and evolutionist. Because Darwin's theory of evolution is the subject of often-bitter public controversy, we will examine the objections of evolution's critics in detail to see why there is such a disconnect between science and public opinion.

History of the Controversy

An Old Conflict Immediately after publication of *The Origin of Species,* English clergymen attacked Darwin's book as heretical; Gladstone, England's prime minister and a famous statesman, condemned it. The book was defended by Thomas Huxley and other scientists, who gradually won over the scientific establishment, and by the turn of the century evolution was generally accepted by the world's scientific community.

The Fundamentalist Movement By the 1920s the teaching of evolution had become frequent enough in American public schools to alarm conservative critics of evolution who saw Darwinism as a threat to their Christian beliefs. Between 1921 and 1929, fundamentalists introduced bills outlawing the teaching of evolution in 37 state legislatures. Four passed: Tennessee, Mississippi, Arkansas, and Texas.

Civil rights groups used the case of high school teacher John Scopes to challenge the Tennessee law within months of its being passed in 1925. The trial attracted national attention—you might have seen it portrayed in the film *Inherit the Wind.* Scopes, who had indeed violated the new law, lost.

After the 1920s, there were few other attempts to pass state laws preventing the teaching of evolution. Only one bill was introduced between 1930 and 1963. Why? Because Darwin's fundamentalist critics had succeeded quietly in winning their way. Textbooks published throughout the 1930s ignored evolution, new editions of texts removing the words *evolution* and *Darwin* from their indices. In the 1920–29 period, for example, the average number of words about the evolution of humans in 93 secondary school texts was 1,339; in the 1930–39 period it had dropped to 439. As late as 1950–59 it was 614. To quote biologist Ernst Mayr, "The word EVOLUTION simply disappeared from American schoolbooks."

These antievolution laws remained on the books for many years. Then in 1965 teacher Susan Epperson was convicted of teaching evolution under the 1928 Arkansas law. In 1968 the United States Supreme Court found the Arkansas antievolution law to be unconstitutional; the 1920s laws were soon repealed.

Russian advances in space in the early 1960s created a public outcry for better American science education. New biology textbooks reintroduced evolution, and gave it renewed emphasis. The average number of words per text devoted to the evolution of man, for example, rose from 614 in the period 1950–59 to 8,977 in the period 1960–69. By the 1970s, evolution again formed the core of most biology schoolbooks.

The Scientific Creationism Movement Again alarmed by the prevalence of evolution in public school biology classes, Darwin's critics took a new approach. It began with a proposal by the Institute for Creationism Research in 1964, which said, "Creationism is just as much a science as is evolution, and evolution is just as much a religion as is creation." This proposal has become known as *creationism science*. It was soon followed by the introduction of legislation in state legislatures mandating that "all theories of origins be accorded equal time." Creationism was represented as being as much a scientific theory as evolution, to which students had a right to be exposed.

In 1981 the state legislatures of Arkansas and Louisiana passed "equal-time" bills into law. The Louisiana equal-time law requiring "balanced treatment of creation-science and evolution-science in public schools" was struck down by the Supreme Court in 1987, which judged that creation science is not, in fact, science but rather a religious view that has no place in public science classrooms.

Local Action In the following decades, Darwin's critics began to substitute the school board for the legislature. Unlike most European countries, which set their school curricula through a central education ministry, U.S. education is highly decentralized, with elected education boards setting science standards at the state and local levels. These standards determine the content of statewide assessment tests, and have a major impact on what is taught in classrooms.

Critics of Darwin have run successfully for seats on local and state education boards across the United States, and from these positions have begun to alter standards to lessen the impact of evolution in the classroom. Great publicity followed the removal of evolution from the Kansas state standards in 1999 and again in 2005, but in many other states the same effect has been achieved more quietly. Only 22 states today mandate the teaching of natural selection, for example. Four states fail to mention evolution at all.

Intelligent Design In recent years, critics of Darwin have begun new attempts to combat the teaching of evolution in the classroom, arguing before state and local school boards that life is too complex for natural selection and so must reflect intelligent design. They go on to argue that this "theory of intelligent design" (ID) should be presented in the science classroom as an alternative to the theory of evolution.

Scientists object strongly to dubbing ID a scientific theory. The essence of science is seeking explanations in what can be observed, tested, replicated by others, and possibly falsified. Explanations that cannot be tested and potentially rejected simply aren't science. If someone invokes a nonnatural cause—a supernatural force—in their research, and you decide to test it, can you think of any way to do so such that it could be falsified? Supernatural causation is not science.

The source of sharp public controversy, intelligent design has been overwhelmingly rejected by the scientific community, which does not regard intelligent design to be science at all, but rather thinly disguised creationism, a religious view that has no place in the science classroom.

Arguments Advanced by Darwin's Critics

Critics of evolution have raised a variety of objections to Darwin's theory of evolution by natural selection:

1. **Evolution is not solidly demonstrated.** *"Evolution is just a theory,"* critics point out, as if theory meant lack of knowledge, some kind of guess. Scientists, however, use the word theory in a very different sense than the general public does (see section 1.8). Theories are the solid ground of science, supported with much experimental evidence and that of which we are most certain. Few of us doubt the theory of gravity because it is "just a theory."
2. **The intelligent design argument.** *"The organs of living creatures are too complex for a random process to have produced."* This classic "argument from design" was first proposed nearly 200 years ago by William Paley in his book *Natural Theology*—the existence of a clock is evidence of the existence of a clockmaker, Paley argues. Similarly, Darwin's critics argue that organs like the mammalian ear are too complex to be due to blind evolution. There must have been a designer. Biologists do not agree. The intermediates in the evolution of the mammalian ear are well documented in the fossil record. These intermediate forms were each favored by natural selection because they each had value—being able to amplify sound a little is better than not being able to amplify it at all. Complex structures like the mammalian ear evolved as a progression of slight improvements. Nor is the solution always optimal. The vertebrate eye, for example, is poorly designed.

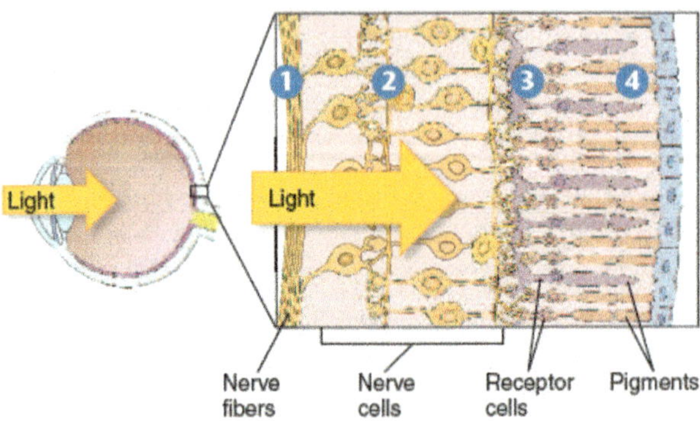

The visual pigments in a vertebrate eye that are stimulated by light are embedded in the retinal tissue, facing backward to the direction of the light. The light has to pass through nerve fibers ❶, nerve cells ❷, and receptor cells ❸, before reaching the pigments ❹. No intelligent designer would design an eye backwards!
3. **There are no fossil intermediates.** *"No one ever saw a fin on the way to becoming a leg,"* critics claim, pointing to the many gaps in the fossil record in Darwin's day. Since then, however, most fossil intermediates in vertebrate evolution have indeed been found. A clear line of fossils now traces the transition between fish and amphibians, between reptiles and mammals, and between apes and humans. The animal that produced the fossil shown below is an extinct lobe-finned fish (genus *Tiktaalik*) that lived approximately 375 million years ago. Coined by its discoverer as a "fishopod," it clearly has some characteristics that are fishlike, similar to fish that lived about 380 million years ago, and others that are more like early tetrapods, which lived about 365 million years ago. *Tiktaalik* appears to be a transitional animal, between fish and amphibians.

4. **Evolution violates the second law of thermodynamics.** *"A jumble of soda cans doesn't by itself jump neatly into a stack—things become more disorganized due to random events, not more organized."* Biologists point out that this argument ignores what the second law really says: Disorder increases in a closed system, which the earth most certainly is not. Energy enters the biosphere from the sun, fueling life and all the processes that organize it.
5. **Natural selection does not imply evolution.** *"No scientist has come up with an experiment where fish evolve into frogs and leap away from predators."* Is microevolution (evolution within a species) the mechanism that has produced macroevolution (evolution among species)? Most biologists that have studied the problem think so. The differences between breeds produced by artificial selection—such as chihuahuas, dachshunds, and greyhounds—are more distinctive than differences between wild canine species. Laboratory selection experiments with insects easily create forms that cannot interbreed and thus would in nature be considered different species. Thus, production of radically different forms has indeed been observed, repeatedly.
6. **Life could not have evolved in water.** *"Because the peptide bond does not form spontaneously in water, amino acids could never have spontaneously linked together to form proteins; nor is there any chemical reason why biological proteins contain only the L-isomer and not the D-isomer."* Both of these contentions are valid, but do not require rejecting evolution. Rather, they suggest that the early evolution of life took place on a surface rather than in solution. Amino acids link up spontaneously on the surface of clays, for example, which can have a shape that selects the L-isomer.

The Irreducible Complexity Fallacy

The century-and-a-half-old "intelligent design" argument of William Paley has been recently articulated in a new molecular guise by Lehigh University biochemistry professor Michael Behe. In his 1996 book *Darwin's Black Box: The Biochemical Challenge to Evolution,* Behe argues that the intricate molecular machinery of our cells is so elaborate, our body processes so interconnected, that they cannot be explained by evolution from simpler stages in the way that Darwinists explain the evolution of the mammalian ear. The molecular machinery of the cell is "irreducibly complex." Behe defines an irreducibly complex system as "a single system composed of several well-matched, interacting parts that contribute to the basic function, wherein the removal of any one of the parts causes the system to effectively cease functioning." Each part plays a vital role. Remove just one, Behe emphasizes, and cell molecular machinery cannot function.

As an example of such an irreducibly complex system, Behe describes the series of more than a dozen blood clotting proteins that act in our body to cause blood to clot around a wound. Take out any step in the complex cascade of reactions that leads to coagulation of blood, says Behe, and your body's blood would leak out from a cut like water from a ruptured pipe. Remove a single enzyme from the complementary system that confines the clotting process to the immediate vicinity of the wound, and all your lifeblood would harden. Either condition would be fatal. The need for *all* the parts of such complex systems to work leads directly to Behe's criticism of Darwin's theory of evolution by natural selection. Behe writes that "irreducibly complex systems cannot evolve in a Darwinian fashion." If dozens of different proteins all must work correctly to clot blood, how could natural selection act to fashion any one of the individual proteins? No one protein does anything on its own, just as a portion of a watch doesn't tell time. Behe argues that, like Paley's watch, the blood clotting system must have been designed all at once, as a single functioning machine.

What's wrong with Behe's argument, as evolutionary scientists have been quick to point out, is that each part of a complex molecular machine does not evolve by itself, despite Behe's claim that it must. The several parts evolve together, in concert, precisely because evolution acts on the system, not its parts. That's the fundamental fallacy in Behe's argument. Natural selection can act on a complex system because at every stage of its evolution, the system functions. Parts that improve function are added, and, because of later changes, eventually become essential, in the same way that the second rung of a ladder becomes essential once you have added a third.

The mammalian blood clotting system, for example, has evolved in stages from much simpler systems. By comparing the amino acid sequences of the many proteins, biochemist Russell Doolittle has estimated how long it has been since each protein evolved. You can see what he has learned in figure 14.20. The core of the vertebrate clotting system, called the "common pathway" (highlighted in blue), evolved at the dawn of the vertebrates approximately 600 million years ago, and is found today in lampreys, the most primitive fish. As vertebrates evolved, proteins were added to the clotting system, improving its efficiency. The so-called *extrinsic pathway* (highlighted in pink), triggered by substances released from damaged tissues, was added 500 million years ago. Each step in the pathway amplifies what goes before, so adding the extrinsic pathway greatly increases the amplification and thus the sensitivity of the system. Fifty million years later, a third component was added, the so-called *intrinsic pathway* (highlighted in tan). It is triggered by contact with the jagged surfaces produced by injury. Again, amplification and sensitivity were increased to ultimately end up with blood clots formed by the cross linking of fibrin (highlighted in green). At each stage as the clotting system evolved to become more complex, its overall performance came to depend on the added elements. Mammalian clotting, which utilizes all three pathways, no longer functions if any one of them is disabled. Blood clotting has become "irreducibly complex"—as the result of Darwinian evolution. Behe's claim that complex cellular and molecular processes can't be explained by Darwinism is wrong. Indeed, examination of the human genome reveals that the cluster of blood clotting genes arose through duplication of genes, with increasing amounts of change. The evolution of the blood clotting system is an observation, not a surmise. Its irreducible complexity is a fallacy.

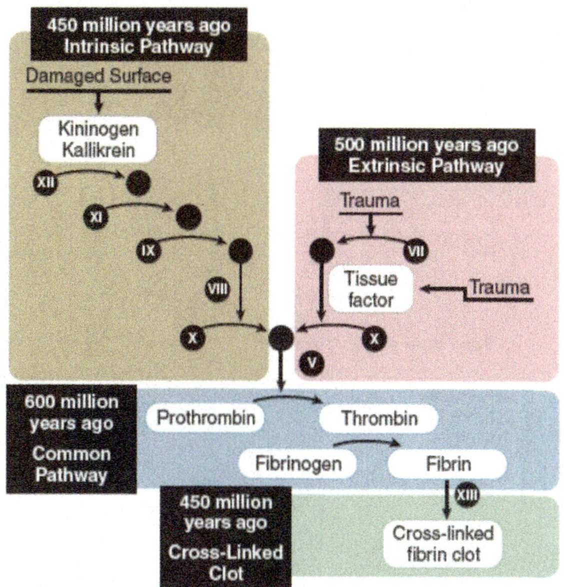

Figure 14.20 How blood clotting evolved.
The blood clotting system evolved in steps, with new proteins adding on to the preceding step.

Key Learning Outcome 14.7 Darwin's theory of evolution, while accepted overwhelmingly by scientists, has its objectors. Their criticisms are without scientific merit.

A Closer Look

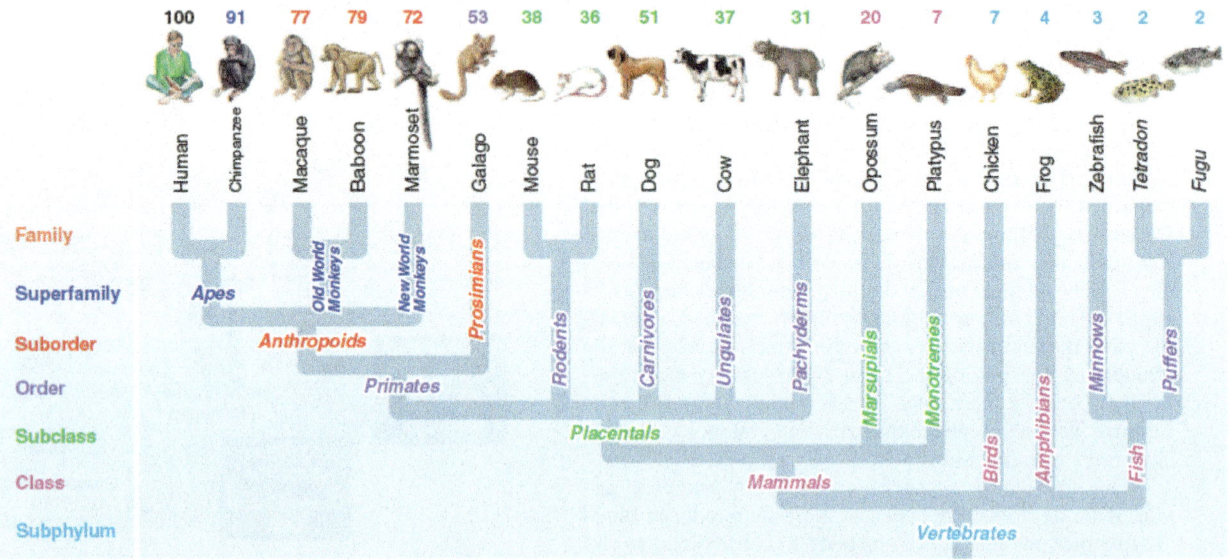

Putting Intelligent Design to the Test

In the spring of 2006 the South Carolina Board of Education rejected a state panel's proposal to change high school standards by calling on students to critically analyze evolution. The Board stated it felt the proposal was a ploy to promote the avoidance of teaching evolution. Similar proposals to add a requirement that students critically analyze evolution had been rejected earlier in the year by the Utah and Ohio Boards of Education, and are currently under consideration in several other states.

What are we to make of this? Surely no scientist can object to critically analyzing any theory. That is what science is all about, seeking explanations for what can be observed, tested, replicated, and possibly falsified. Indeed, biologists claim that Darwin's theory of evolution has been subjected to as much critical analysis as any theory in the history of science.

So why the objection to this change in high school standards? Because many scientists and teachers, apparently including the South Carolina Board of Education, feel the change is simply intended to promote the teaching of a nonscientific alternative to evolution in classrooms.

This distinction between an assertion that can be tested and one that cannot goes to the very nature of science. Actually, nothing makes this difference more clear cut than the critical analysis so sought after by South Carolina's critics of evolution. So let's do it. Let's put Darwin to the test.

As explained earlier in the chapter, if Darwin's assertion is correct, that organisms evolved from ancestral species, then we should be able to track evolutionary changes in our DNA. The variation that we see between species reflects adaptations to environmental challenges, adaptations that result from changes in DNA. Therefore, a series of evolutionary changes should be reflected in an accumulation of genetic changes in the DNA. This hypothesis, that evolutionary changes reflect accumulated changes in DNA, leads to the following prediction: Two species that are more distantly related (for example, humans and mice) should have accumulated a greater number of evolutionary differences than two species that are more closely related (say, humans and chimpanzees).

So have they? Let's compare vertebrate species to see. The "family tree" above shows how biologists believe 18 different vertebrate species are related. Apes and monkeys, because they are in the same order (primates), are considered more closely related to each other than either are to members of another order, such as mice and rats (rodents).

The wealth of genomes (a genome is all the DNA that an organism possesses) that have been sequenced since completion of the human genome project allows us to directly compare the DNA of these 18 vertebrates. To reduce the size of the task, investigators at the National Human Genome Research Institute working at the University of California, Santa Cruz, focused on 44 so-called ENCODE regions scattered around the vertebrate genomes. These regions, corresponding to 30 Mb (megabase, or million bases) or roughly 1% of the total human genome, were selected to be representative of the genome as a whole, containing protein-encoding genes as well as noncoding DNA.

For each vertebrate species, the investigators determined the similarity of its DNA to that of humans—that is, the percent of the nucleotides in that organism's 44 ENCODE regions that match those of the human genome.

You can see the result in each instance presented as a number above the picture of each organism on the vertebrate family tree. As Darwin's theory predicts, the closer the relatives, the less the genomic difference we see. The chimpanzee genome is more like the human genome (91%) than the monkey genomes are (72 to 79%). Furthermore, these five genomes, all in the primate order, are more like each other than any are to those of another order, such as rodents (mouse and rat).

In general, as you proceed through the taxonomic categories of the vertebrate family tree from very distant relatives on the right (some in the same class as humans) to very close ones on the left (in the same family), you can see clearly that genomic similarity increases as taxonomic distance decreases—just as Darwin's theory predicts. The prediction of evolutionary theory is solidly confirmed.

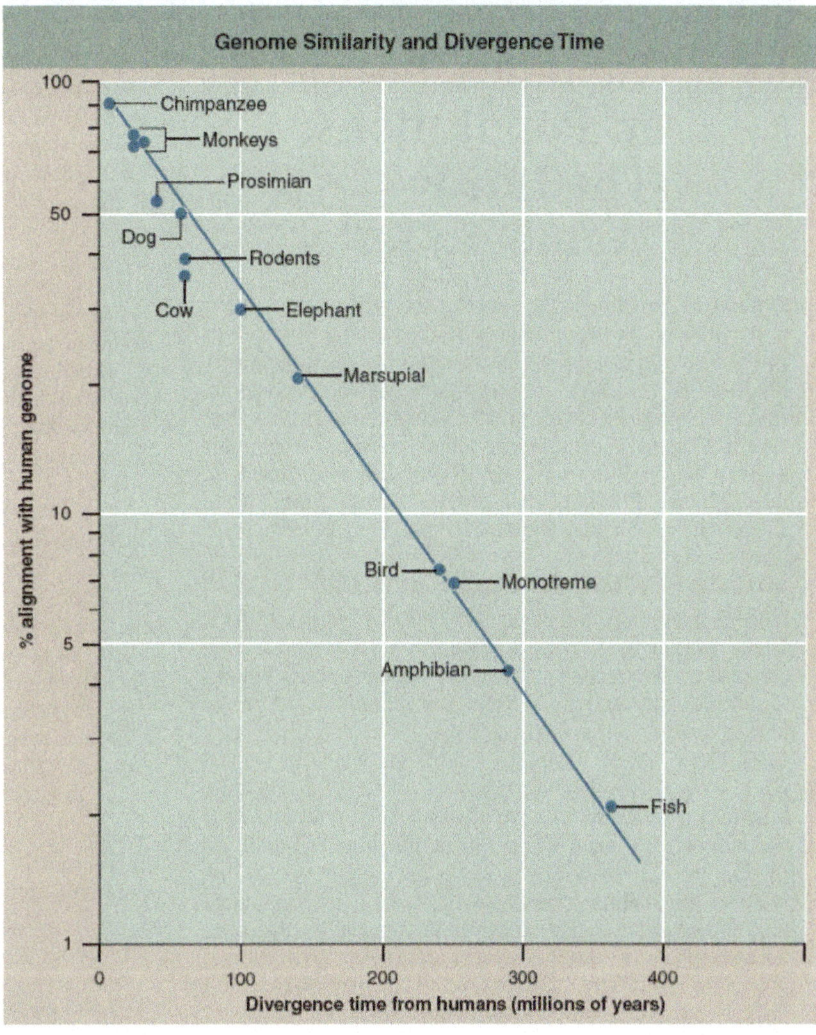

The analysis does not have to stop here. The evolutionary history of the vertebrates is quite well known from fossils, and because many of these fossils have been independently dated using tools such as radioisotope dating, it is possible to recast the analysis in terms of concrete intervals of time, and assess directly whether or not vertebrate genomes accumulate more differences over longer periods of time as Darwin's theory predicts.

For each of the 18 vertebrates being analyzed, the graph above plots genomic similarity—how alike the DNA sequence of the vertebrate's ENCODE regions are to those of the human genome—against divergence time (that is, how many millions of years have elapsed since that vertebrate and humans shared a common ancestor in the fossil record). Thus the last common ancestor shared by chickens and humans was an early reptile called a *dicynodont* that lived some 250 million years ago; since then the genomes of the two species have changed so much that only 7% of their ENCODE sequences are still the same.

The result seen in the graph is striking and very clear: Over their more than 300 million year history, vertebrates have accumulated more and more genetic change in their DNA. "Descent with modification" was Darwin's definition of evolution, and that is exactly what we see in the graph. The evolution of the vertebrate genome is not a theory, but an observation.

The wealth of data made available by the human genome project has allowed a powerful test of Darwin's prediction. The conclusion to which the test leads us is that evolution is an observed fact, clearly revealed in the DNA of vertebrates.

This is the sort of critical analysis that science requires, and that the theory of evolution has again passed. Anyone suggesting that a nonscientific alternative to evolution, such as Intelligent Design, offers an alternative scientific explanation to evolution is welcome to subject it to the same sort of critical analysis you have seen employed here. Can you think of a way to do so? It is precisely because the assertion of intelligent design cannot be critically analyzed—it does not make any testable prediction—that it is not science and has no place in science classrooms.

How Populations Evolve

14.8 Genetic Change in Populations: The Hardy-Weinberg Rule

Population genetics is the study of the properties of genes in populations. Genetic variation within natural populations could not be explained by Darwin and his contemporaries. The way in which meiosis produces genetic segregation among the progeny of a hybrid had not yet been discovered. And, although Mendel performed his experiments during this same time period, his work was largely unknown. Selection, scientists then thought, should always favor an optimal form, and so tend to eliminate variation.

Hardy-Weinberg Equilibrium

Indeed, variation within populations puzzled many scientists; **alleles** (alternative forms of a gene) that were dominant were believed to drive recessive alleles out of populations, with selection favoring an optimal form. The solution to the puzzle of why genetic variation persists was developed in 1908 by G. H. Hardy and W. Weinberg. Hardy and Weinberg studied **allele frequencies** (the proportion of alleles of a particular type in a population) in a population's *gene pool,* which is the sum of all of the genes in a population, including all alleles in all individuals. Hardy and Weinberg pointed out that in a large population in which there is random mating, and in the absence of forces that change allele frequencies, the original genotype proportions remain constant from generation to generation. Dominant alleles do not, in fact, replace recessive ones. Because their proportions do not change, the genotypes are said to be in **Hardy-Weinberg equilibrium.**

The Hardy-Weinberg rule is viewed as a baseline to which the frequencies of alleles in a population can be compared. If the allele frequencies are not changing (they are in Hardy-Weinberg equilibrium), the population is not evolving. If, however, allele frequencies are sampled at one point in time and they differ greatly from what would be expected under Hardy-Weinberg equilibrium, then the population is undergoing evolutionary change.

Hardy and Weinberg came to their conclusion by analyzing the frequencies of alleles in successive generations. The **frequency** of something is defined as the proportion of individuals with a certain characteristic, compared to the entire population. Thus, in the population of 1,000 cats shown in figure 14.21, there are 840 black cats and 160 white cats. To determine the frequency of black cats, divide 840 by 1,000 (840/1,000), which is 0.84. The frequency of white cats is 160/1,000 = 0.16.

Knowing the frequency of the phenotype, one can calculate the frequency of the genotypes and alleles in the population. By convention, the frequency of the more common of two alleles (in this case *B* for the black allele) is designated by the letter p and that of the less common allele (*b* for the white allele) by the letter q. Because there are only two alleles, the sum of p and q must always equal 1 ($p + q = 1$).

In algebraic terms, the Hardy-Weinberg equilibrium is written as an equation. For a gene with two alternative alleles *B* (frequency p) and *b* (frequency q), the equation looks like this:

$$p^2 + 2pq + q^2 = 1$$

p^2	+	$2pq$	+	q^2	=	1
Individuals homozygous for allele *B*		**Individuals heterozygous for alleles *B* and *b***		**Individuals homozygous for allele *b***		

You will notice that not only does the sum of the alleles add up to 1 but so does the sum of the frequencies of genotypes.

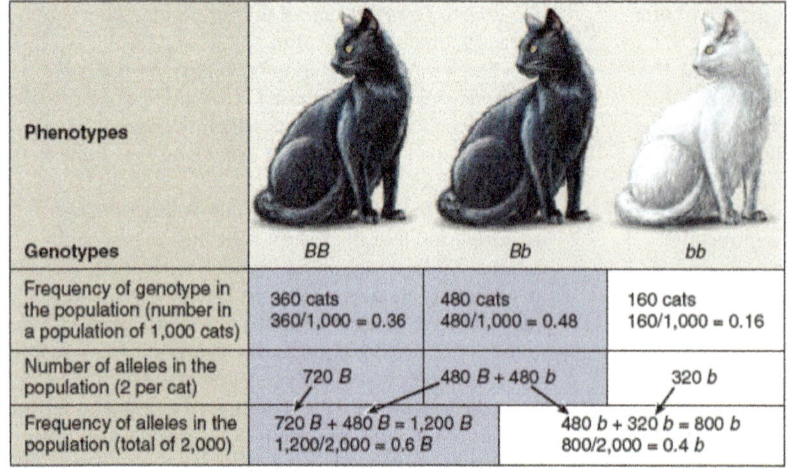

Phenotypes			
Genotypes	*BB*	*Bb*	*bb*
Frequency of genotype in the population (number in a population of 1,000 cats)	360 cats 360/1,000 = 0.36	480 cats 480/1,000 = 0.48	160 cats 160/1,000 = 0.16
Number of alleles in the population (2 per cat)	720 *B*	480 *B* + 480 *b*	320 *b*
Frequency of alleles in the population (total of 2,000)	720 *B* + 480 *B* = 1,200 *B* 1,200/2,000 = 0.6 *B*		480 *b* + 320 *b* = 800 *b* 800/2,000 = 0.4 *b*

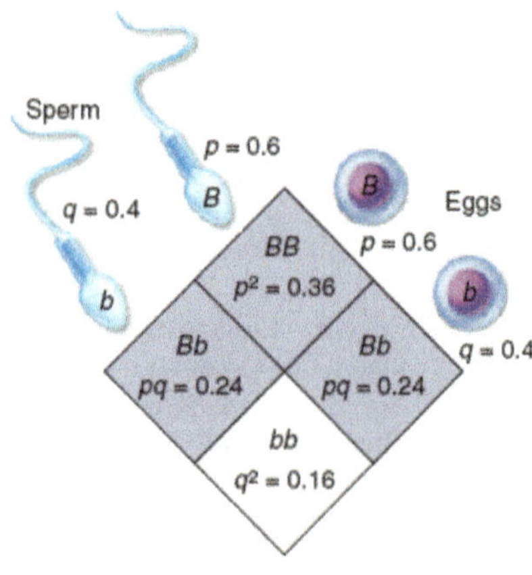

Figure 14.21 Calculating allele frequencies at Hardy-Weinberg equilibrium. The example here is a population of 1,000 cats, in which 160 are white and 840 are black. White cats are *bb*, and black cats are *BB* or *Bb*.

Knowing the frequencies of the alleles in a population doesn't reveal whether the population is evolving or not. We need to look at future generations to determine this. Using the allele frequencies calculated for our population of cats, we can predict what the genotypic and phenotypic frequencies will be in future generations. The Punnett square shown on the right in figure 14.21 is constructed with allele frequencies of 0.6 for the *B* allele and 0.4 for the *b* allele, taken from the bottom row of the chart. It might help you to consider these frequencies as percentages, with a 0.6 representing 60% of the population and 0.4 representing 40% of the population. According to the Hardy-Weinberg rule, 60% of the sperm in the population will carry the *B* allele (indicated as $p = 0.6$ in the Punnett square) and 40% of the sperm will carry the *b* allele ($q = 0.4$). When these are crossed with eggs carrying the same allele frequencies (60% or $p = 0.6$ *B* allele and 40% or $q = 0.4$ *b* allele), the predicted genotypic frequencies can be simply calculated. The genotypic ratio for *BB*, in the upper quadrant, equals the frequency of *B* (0.6) multiplied by the frequency of *B* (0.6) or ($0.6 \times 0.6 = 0.36$). So, if the population is not evolving, the genotypic ratio for *BB* would stay the same, and 0.36 or 36% of the cats in future generations would be homozygous dominant (*BB*) for coat color. Likewise, 0.48 or 48% of the cats would be heterozygous *Bb* ($0.24 + 0.24 = 0.48$), and 0.16 or 16% of the cats would be homozygous recessive *bb*.

Hardy-Weinberg Assumptions

The Hardy-Weinberg rule is based on certain assumptions. The equation on page 300 is true only if the following five assumptions are met:

1. The size of the population is very large or effectively infinite.
2. Individuals mate with one another at random.
3. There is no mutation.
4. There is no input of new copies of any allele from any extraneous source (such as migration from a nearby population) or losses of copies of alleles through emigration (individuals leaving the population).
5. All alleles are replaced equally from generation to generation (natural selection is not occurring).

Hardy-Weinberg: A Null Hypothesis

Many populations, and most human populations, are large and randomly mating with respect to most traits (a few traits affecting appearance undergo strong sexual selection in humans). Thus, many populations are similar to the ideal population envisioned by Hardy and Weinberg. For some genes, however, the observed proportion of heterozygotes does not match the value calculated from the allele frequencies. When this occurs, it indicates that something is acting on the population to alter one or more of the genotypic frequencies, whether it is selection, nonrandom mating, migration, or some other factor. Viewed in this light, Hardy-Weinberg can be viewed as a *null hypothesis*. A null hypothesis is a prediction that is made stating there will be no differences in the parameters being measured. If over several generations, the genotypic frequencies in the population do not match those predicted by the Hardy-Weinberg equation, the null hypothesis would be rejected and the assumption made that some force is acting on the population to change the frequencies of alleles. The factors that can affect the frequencies of alleles in a population are discussed in detail later in this chapter.

Case-Study: Cystic Fibrosis in Humans

How valid are the predictions made by the Hardy-Weinberg equation? For many genes, they prove to be very accurate. As an example, consider the recessive allele responsible for the serious human disease cystic fibrosis. This allele (q) is present in Caucasians in North America at a frequency of 0.022. What proportion of Caucasian North Americans, therefore, is expected to express this trait? The frequency of double-recessive individuals (q^2) is expected to be:

$$q^2 = 0.022 \times 0.022 = 0.00048$$

which equals 0.48 in every 1,000 individuals or about 1 in every 2,000 individuals, very close to real estimates.

What proportion is expected to be heterozygous carriers? If the frequency of the recessive allele q is 0.022, then the frequency of the dominant allele p must be $p = 1 - q$ or:

$$p = 1 - 0.022 = 0.978$$

The frequency of heterozygous individuals ($2pq$) is thus expected to be:

$$2 \times 0.978 \times 0.022 = 0.043$$

It is estimated that 12 million individuals in the United States are carriers of the cystic fibrosis allele. In a population of 292 million people, that is a frequency of 0.041, very close to projections using the Hardy-Weinberg equation. However, if the frequency of the cystic fibrosis allele in the United States were to change, this would suggest that the population is no longer following the assumptions of the Hardy-Weinberg rule. For example, if prospective parents who were carriers of the allele chose not to have children, the frequency of the allele would decrease in future generations. Mating would no longer be random, because those carrying the allele would not mate. Consider another scenario. If gene therapies were developed that were able to cure the symptoms of cystic fibrosis, patients would survive longer and would have more of an opportunity to reproduce. This would increase the frequency of the allele in future generations. An increase could also result from an influx of the allele into the population by migration, if the allele were more frequent among individuals migrating into the country.

Key Learning Outcome 14.8 **In a large, randomly-mating population that fulfills the other Hardy-Weinberg assumptions, allele frequencies can be expected to be in Hardy-Weinberg equilibrium. If they are not, then the population is undergoing evolutionary change.**

14.9 Agents of Evolution

Many factors can alter allele frequencies. But only five alter the proportions of homozygotes and heterozygotes enough to produce significant deviations from the proportions predicted by the Hardy-Weinberg rule:

1. Mutation
2. Nonrandom mating
3. Genetic drift
4. Migration
5. Selection

Mutation

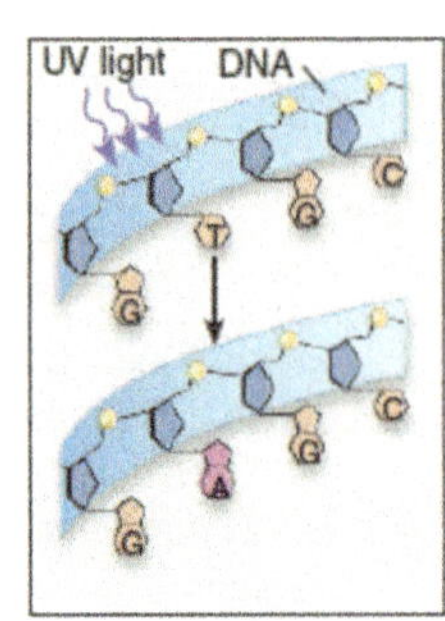

A **mutation** is a change in a nucleotide sequence in DNA. For example, a T nucleotide could undergo a mutation and be replaced with an A nucleotide. Mutation from one allele to another obviously can change the proportions of particular alleles in a population. But mutation rates are generally too low to significantly alter Hardy-Weinberg proportions of common alleles. Many genes mutate 1 to 10 times per 100,000 cell divisions. Some of these mutations are harmful, while others are neutral or, even rarer, beneficial. Also, the mutations must affect the DNA of the germ cells (egg and sperm), or the mutation will not be passed on to offspring. The mutation rate is so slow that few populations are around long enough to accumulate significant numbers of mutations. However, no matter how rare, mutation is the ultimate source of genetic variation in a population.

Nonrandom Mating

Individuals with certain genotypes sometimes mate with one another either more or less commonly than would be expected on a random basis, a phenomenon known as **nonrandom mating.** One type of nonrandom mating is **sexual selection,** choosing a mate often based on certain physical characteristics. Another type of nonrandom mating is inbreeding, or mating with relatives, such as in the self-fertilization of a flower. Inbreeding increases the proportions of individuals that are homozygous because no individuals mate with any genotype but their own. As a result, inbred populations contain more homozygous individuals than predicted by the Hardy-Weinberg rule. For this reason, populations of self-fertilizing plants consist primarily of homozygous individuals, whereas outcrossing plants, which interbreed with individuals different from themselves, have a higher proportion of heterozygous individuals. Nonrandom mating alters genotype frequencies but not allele frequencies. The allele frequencies remain the same—the alleles are just distributed differently among the offspring.

Genetic Drift

In small populations, the frequencies of particular alleles may be changed drastically by chance alone. In an extreme case, individual alleles of a given gene may all be represented in few individuals, and some of the alleles may be accidentally lost if those individuals fail to reproduce or die. This loss of individuals and their alleles is due to random events rather than the fitness of the individuals carrying those alleles. This is not to say that alleles are always lost with genetic drift, but allele frequencies appear to change randomly, as if the frequencies were drifting; thus, random changes in allele frequencies is known as **genetic drift.** A series of small populations that are isolated from one another may come to differ strongly as a result of genetic drift.

When one or a few individuals migrate and become the founders of a new, isolated population at some distance from their place of origin, the alleles that they carry are of special significance in the new population. Even if these alleles are rare in the source population, they will become a significant fraction of the new population's genetic endowment. This is called the **founder effect.** As a result of the founder effect, rare alleles and combinations often become more common in new, isolated populations. The founder effect is particularly important in the evolution of organisms that occur on oceanic islands, such as the Galápagos Islands that Darwin visited. Most of the kinds of organisms that occur in such areas were probably derived from one or a few initial founders. In a similar way, isolated human populations are often dominated by the genetic features that were characteristic of their founders, particularly if only a few individuals were involved initially (figure 14.22).

Even if organisms do not move from place to place, occasionally their populations may be drastically reduced in size. This may result from flooding, drought,

Figure 14.22
The founder effect.
This Amish woman is holding her child, who has Ellis-van Creveld syndrome. The characteristic symptoms are short limbs, dwarfed stature, and extra fingers. This disorder was introduced in the Amish community by one of its founders in the 18th century and persists to this day because of reproductive isolation.

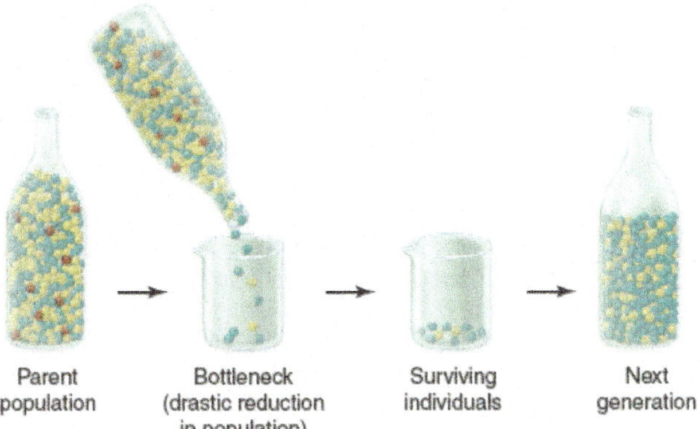

Figure 14.23 Genetic drift: a bottleneck effect.
The parent population contains roughly equal numbers of green and yellow individuals and a small number of red individuals. By chance, the few remaining individuals that contribute to the next generation are mostly green. The bottleneck occurs because so few individuals form the next generation, as might happen after an epidemic or a catastrophic storm.

earthquakes, and other natural forces or from progressive changes in the environment. The surviving individuals constitute a random genetic sample of the original population. Such a restriction in genetic variability has been termed the **bottleneck effect** (figure 14.23). The very low levels of genetic variability seen in African cheetahs today is thought to reflect a near-extinction event in the past.

Migration

Migration is defined in genetic terms as the movement of individuals between populations. It can be a powerful force, upsetting the genetic stability of natural populations. Migration includes movement of individuals into a population, called *immigration*, or the movement of individuals out of a population, called *emigration*. If the characteristics of the newly arrived individuals differ from those already there, and if the newly arrived individuals adapt to survive in the new area and mate successfully, then the genetic composition of the receiving population may be altered.

Sometimes migration is not obvious. Subtle movements include the drifting of gametes of plants, or of the immature stages of marine organisms, from one place to another. For example, a bee can carry pollen from a flower in one population to a flower in another population. By doing this, the bee may be introducing new alleles into a population. However it occurs, migration can alter the genetic characteristics of populations and cause a population to be out of Hardy-Weinberg equilibrium. Thus, migration can cause evolutionary change. The magnitude of effects of migration is based on two factors: (1) the proportion of migrants in the population, and (2) the difference in allele frequencies between the migrants and the original population. The actual evolutionary impact of migration is difficult to assess, and depends heavily on the selective forces prevailing at the different places where the populations occur.

Selection

As Darwin pointed out, some individuals leave behind more progeny than others, and the likelihood they will do so is affected by their inherited characteristics. The result of this process is called **selection** and was familiar even in Darwin's day to breeders of horses and farm animals. In so-called **artificial selection,** the breeder selects for the desired characteristics. For example, mating larger animals with each other produces offspring that are larger. In **natural selection,** Darwin suggested the environment plays this role, with conditions in nature determining which kinds of individuals in a population are the most fit (meaning individuals that are best suited to their environment; see section 14.3) and so affecting the proportions of genes among individuals of future populations. The environment imposes the conditions that determine the results of selection and, thus, the direction of evolution (figure 14.24).

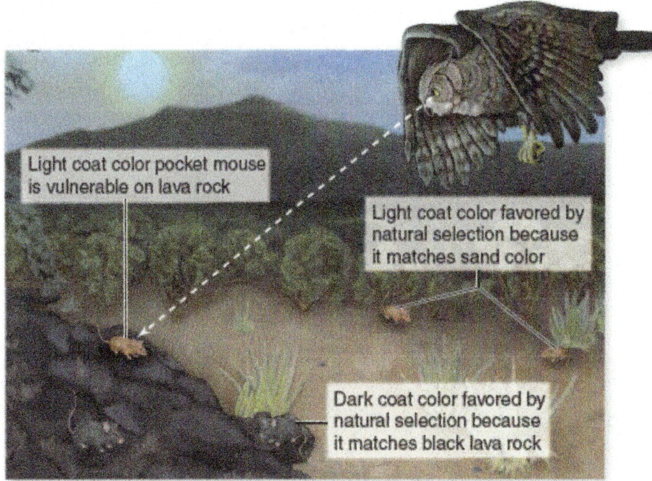

Figure 14.24 Selection for coat color in mice.
In the American southwest, ancient lava flows have produced black rock formations that contrast starkly with the surrounding light-colored desert sand. Populations of many species of animals occurring on these rocks are dark in color, whereas sand-dwelling populations are much lighter. For example, in pocket mice, selection favors coat color that matches their surroundings. The close match between coat color and background color camouflages the mice and provides protection from avian predators. These mice are very visible when placed in the opposite habitats.

Forms of Selection

Selection operates in natural populations of a species as skill does in a football game. In any individual game, it can be difficult to predict the winner because chance can play an important role in the outcome. But over a long season, the teams with the most skillful players usually win the most games. In nature, those individuals best suited to their environments tend to win the evolutionary game by leaving the most offspring, although chance can play a major role in the life of any one individual. While you cannot predict the fate of any one individual, or any one coin toss, it is possible to predict which kind of individual will tend to become more common in populations of a species, as it is possible to predict the proportion of heads after many coin tosses.

In nature, many traits, perhaps most, are affected by more than one gene. The interactions between genes are typically complex, as you saw in chapter 10. For example, alleles of many different genes play a role in determining human height (see figure 10.12). In such cases, selection operates on all the genes, influencing most strongly those that make the greatest contribution to the phenotype. How selection changes the population depends on which genotypes are favored. Three types of natural selection have been identified: stabilizing selection, disruptive selection, and directional selection.

Stabilizing Selection

When selection acts to eliminate both extremes from an array of phenotypes—for example, eliminating larger and smaller body sizes—the result is an increase in the frequency of the already common intermediate phenotype (such as a midsized body). This is called **stabilizing selection:**

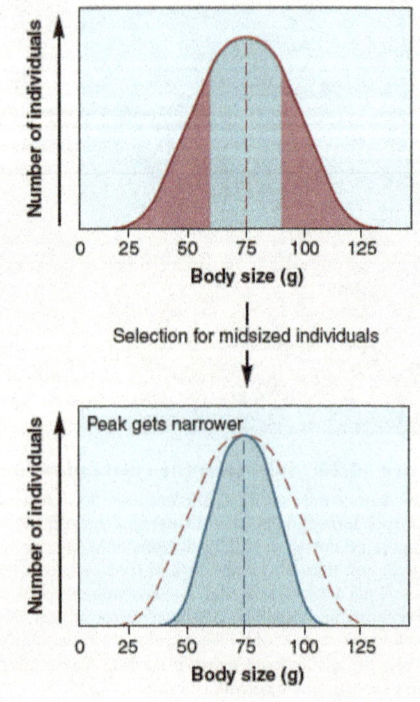

In a classic study carried out after an "uncommonly severe storm of snow, rain, and sleet" on February 1, 1898, 136 starving English sparrows were collected and brought to the laboratory of H. C. Bumpus at Brown University in Providence, Rhode Island. Of these, 64 died and 72 survived. Bumpus took standard measurements on all the birds. He found that among males, the surviving birds tended to be bigger, as one might expect from the action of directional selection (discussed later). However, among females, the birds that survived were those that were more average in size. Among the female birds that perished were many more individuals that had extreme measurements, either very large or very small.

In Bumpus' quaint phrasing, "The process of selective elimination is most severe with extremely varying individuals no matter in what directions the variation may occur. It is quite as dangerous to be conspicuously above a certain standard of organic excellence as it is to be conspicuously below the standard. It is the *type* that nature favors."

In the Bumpus study, selection had acted most strongly against these "extreme-sized" female birds. Stabilizing selection does not change which phenotype is the most common of the population—the average-sized birds were already the most common phenotype—but rather makes it even more common by eliminating extremes. In effect, selection is operating to prevent change away from the middle range of values.

Many examples similar to Bumpus's female sparrows are known. For example, in humans, infants with intermediate weight at birth have the highest survival rate:

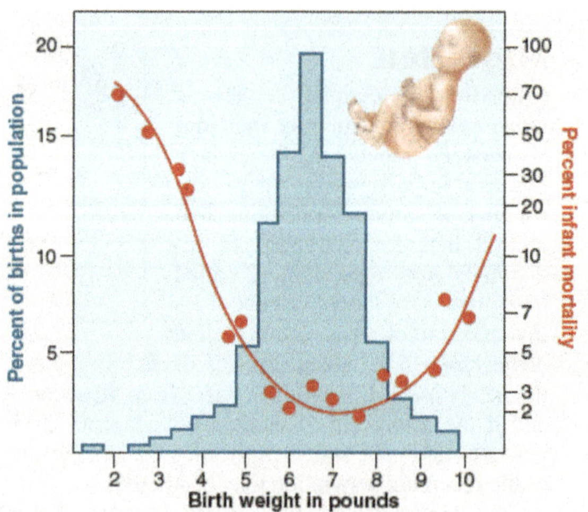

More specifically, the death rate among human babies is lowest at an intermediate birth weight between 7 and 8 pounds indicated by the red line in the graph above, which consists of data compiled from U.S. birth records over many years. The intermediate weights are also the most common in the population, indicated by the blue area. Larger and smaller babies both occur less frequently and have a greater tendency to die at or near birth. In a similar way, chickens eggs of intermediate weight have the highest hatching success.

Disruptive Selection

In some situations, selection acts to eliminate the intermediate type, resulting in the two more extreme phenotypes becoming more common in the population. This type of selection is called **disruptive selection:**

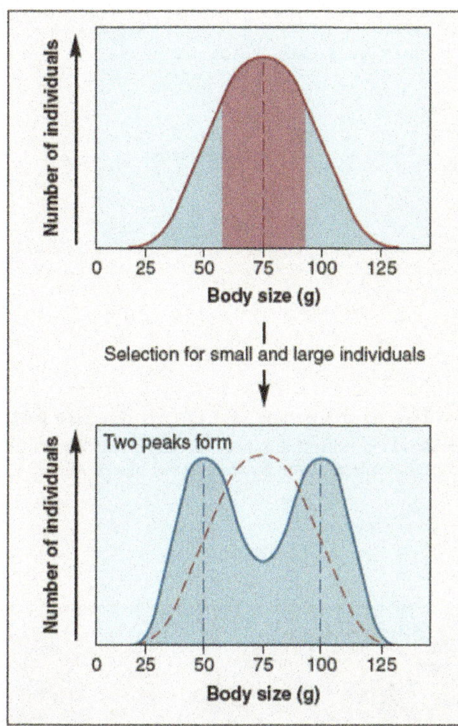

A clear example is the different beak sizes of the African black-bellied seedcracker finch *Pyrenestes ostrinus*. Populations of these birds contain individuals with large and small beaks, but very few individuals with intermediate-sized beaks. As their name implies, these birds feed on seeds, and the available seeds fall into two size categories: large and small. Only large-beaked birds, like the one on the left in the figure below, can open the tough shells of large seeds, whereas birds with the smallest beaks, like the one on the right, are more adept at handling small seeds. Birds with intermediate-sized beaks are at a disadvantage with both seed types: unable to open large seeds and too clumsy to efficiently process small seeds. Consequently, selection acts to eliminate the intermediate phenotypes, in effect partitioning the population into two phenotypically distinct groups.

Directional Selection

In other situations, selection acts to eliminate one extreme from an array of phenotypes, resulting in the other extreme phenotype becoming more common in the population. This form of selection is called **directional selection:**

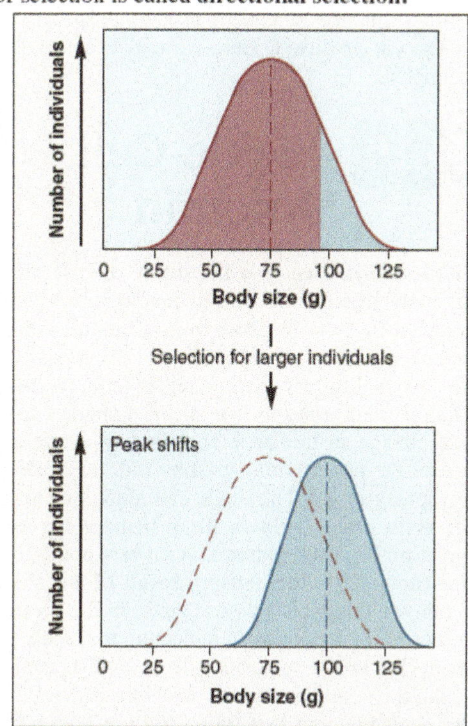

For example, in the experiment below, flies (*Drosophila*) that moved toward light were eliminated from the population, and only flies that moved away from light were used as parents for the next generation. After 20 generations of selected matings, flies that flew toward light were far less frequent in the population.

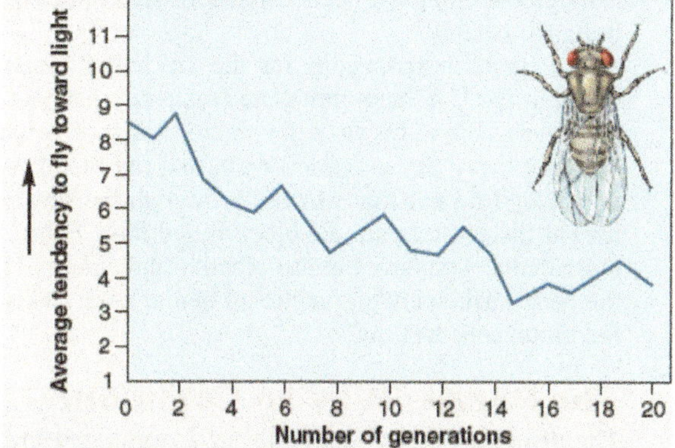

Key Learning Outcome 14.9 Five evolutionary forces have the potential to significantly alter allele and genotype frequencies in populations: mutation, nonrandom mating, genetic drift, migration, and selection. Selection can favor intermediate values, or one or both extremes.

Adaptation Within Populations

In the time since Darwin suggested the pivotal role of natural selection in evolution, many examples have been found in which natural selection is clearly acting to change the genetic makeup of species, just as Darwin predicted. Here we will examine three examples.

14.10 Sickle-Cell Anemia

Sickle-cell disease is a hereditary disease affecting hemoglobin molecules in the blood. It was first detected in 1904 in Chicago in a blood examination of an individual complaining of tiredness. You can see the original doctor's report in figure 14.25. The disorder arises as a result of a single nucleotide change in the gene encoding β-hemoglobin, one of the key proteins used by red blood cells to transport oxygen. The sickle-cell mutation changes the sixth amino acid in the β-hemoglobin chain (position B6) from glutamic acid (very polar) to valine (nonpolar). The unhappy result of this change is that the nonpolar *valine* at position B6, protruding from a corner of the hemoglobin molecule, fits nicely into a nonpolar pocket on the opposite side of another hemoglobin molecule; the nonpolar regions associate with each other. As the two-molecule unit that forms still has both a B6 valine and an opposite nonpolar pocket, other hemoglobins clump on, and long chains form as in figure 14.26*a*. The result is the deformed "sickle-shaped" red blood cell you see in figure 14.26*b*. In normal everyday hemoglobin, by contrast, the polar amino acid *glutamic acid* occurs at position B6. This polar amino acid is not attracted to the nonpolar pocket, so no hemoglobin clumping occurs, and cells are normal shaped as in figure 14.26*c*.

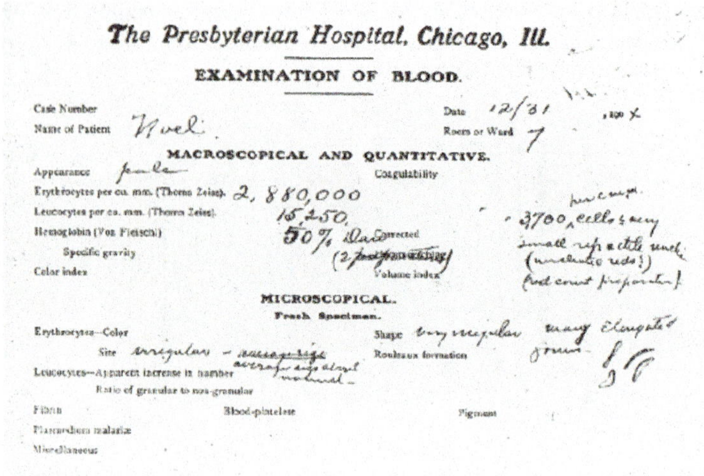

The Presbyterian Hospital, Chicago, Ill.

EXAMINATION OF BLOOD.

Case Number
Name of Patient Noel
Date 12/31, 190 4
Room or Ward 7

MACROSCOPICAL AND QUANTITATIVE.

Appearance pale
Coagulability
Erythrocytes per cu. mm. (Thoma Zeiss). 2,880,000
Leucocytes per cu. mm. (Thoma Zeiss). 15,250
Hemoglobin (Von Fleischl) 50%
Specific gravity
Color index
Volume index

MICROSCOPICAL.

Fresh Specimen.

Erythrocytes—Color
Size irregular
Shape very irregular
Roulesux formation
Leucocytes—Apparent increase in number
Ratio of granular to non-granular
Fibrin
Blood-platelets
Pigment
Plasmodium malariæ
Miscellaneous

Figure 14.25 The first known sickle-cell disease patient. Dr. Ernest Irons's blood examination report on his patient Walter Clement Noel, December 31, 1904, described his oddly shaped red blood cells.

Persons homozygous for the sickle-cell genetic mutation in the β-hemoglobin gene frequently have a reduced life span. This is because the sickled form of hemoglobin does not carry oxygen atoms well, and red blood cells that are sickled do not flow smoothly through the tiny capillaries but instead jam up and block blood flow. Heterozygous individuals, who have both a defective and a normal form of the gene, make enough functional hemoglobin to keep their red blood cells healthy.

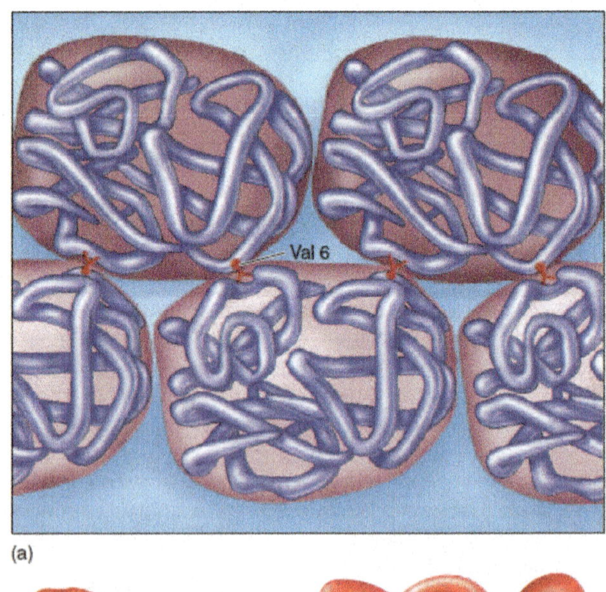

(a)

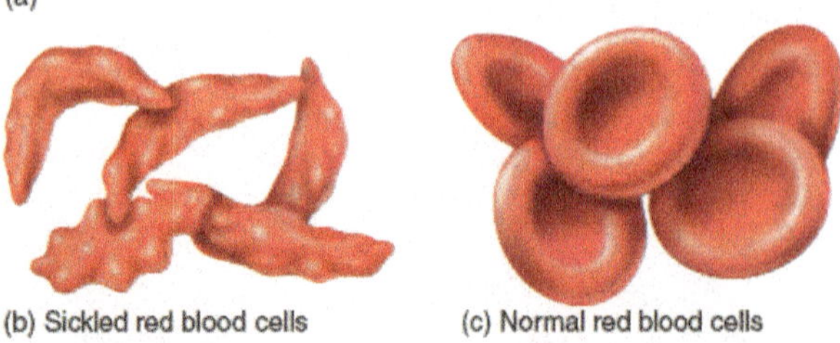

(b) Sickled red blood cells

(c) Normal red blood cells

Figure 14.26 Why the sickle-cell mutation causes hemoglobin to clump.

The Puzzle: Why So Common?

The disorder is now known to have originated in Central Africa, where the frequency of the sickle-cell allele is about 0.12. One in 100 people is homozygous for the defective allele and develops the fatal disorder. Sickle-cell disease affects roughly two African Americans out of every thousand but is almost unknown among other racial groups.

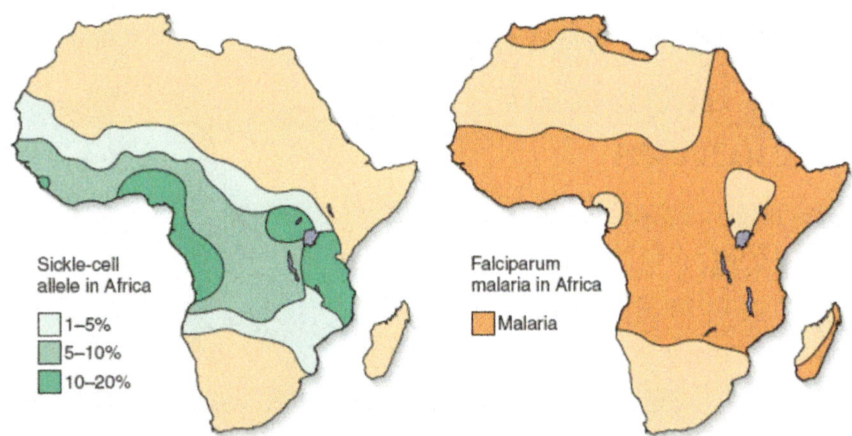

Figure 14.27 How stabilizing selection maintains sickle-cell disease.
The diagrams show the frequency of the sickle-cell allele (*left*) and the distribution of falciparum malaria (*right*). Falciparum malaria is one of the most devastating forms of the often fatal disease. As you can see, its distribution in Africa is closely correlated with that of the allele of the sickle-cell characteristic.

If Darwin is right, and natural selection drives evolution, then why has natural selection not acted against the defective allele in Africa and eliminated it from the human population there? Why is this potentially fatal allele instead very common there?

The Answer: Stabilizing Selection

The defective allele has not been eliminated from Central Africa because people who are heterozygous for the sickle-cell allele are much less susceptible to malaria, one of the leading causes of death in Central Africa. Examine the maps in figure 14.27, and you will see the relationship between sickle-cell disease and malaria clearly. The map on the left shows the frequency of the sickle-cell allele, the darker green areas indicating a 10% to 20% frequency of the allele. The map on the right indicates the distribution of malaria in dark orange. Clearly, the areas that are colored in darker green on the left map overlap many of the dark orange areas in the map on the right. Even though the population pays a high price—the many individuals in each generation who are homozygous for the sickle-cell allele die—the deaths are far fewer than would occur due to malaria if the heterozygous individuals were not malaria resistant. One in 5 individuals (20%) are heterozygous and survive malaria, while only 1 in 100 (1%) are homozygous and die of sickle-cell disease. Similar inheritance patterns of the sickle-cell allele are found in other countries frequently exposed to malaria, such as areas around the Mediterranean, India, and Indonesia. Natural selection has favored the sickle-cell allele in Central Africa and other areas hit by malaria because the payoff in survival of heterozygotes more than makes up for the price in death of homozygotes. This phenomenon is an example of **heterozygote advantage.**

Stabilizing selection (also called *balancing selection*) is thus acting on the sickle-cell allele: (1) Selection tends to eliminate the sickle-cell allele because of its lethal effects on homozygous individuals, and (2) selection tends to favor the sickle-cell allele because it protects heterozygotes from malaria. Like a manager balancing a store's inventory, natural selection increases the frequency of an allele in a species as long as there is something to be gained by it, until the cost balances the benefit.

Stabilizing selection occurs because malarial resistance counterbalances lethal sickle-cell disease. Malaria is a tropical disease that has essentially been eradicated in the United States since the early 1950s, and stabilizing selection has not favored the sickle-cell allele here. Africans brought to America several centuries ago have not gained any evolutionary advantage in all that time from being heterozygous for the sickle-cell allele. There is no benefit to being resistant to malaria if there is no danger of getting malaria anyway. As a result, the selection against the sickle-cell allele in America is not counterbalanced by any advantage, and the allele has become far less common among African Americans than among native Africans in Central Africa.

Stabilizing selection is thought to have influenced many other human genes in a similar fashion. The recessive *cf* allele causing cystic fibrosis is unusually common in northwestern Europeans. People that are heterozygous for the *cf* allele are protected from the dehydration caused by cholera, and the *cf* allele may provide protection against typhoid fever too. Apparently, the bacterium causing typhoid fever uses the healthy version of the CFTR protein (see page 80) to enter the cells it infects, but it cannot use the cystic fibrosis version of the protein. As with sickle-cell disease, heterozygotes are protected.

Key Learning Outcome 14.10 The prevalence of sickle-cell disease in African populations is thought to reflect the action of natural selection. Natural selection favors individuals carrying one copy of the sickle-cell allele, because they are resistant to malaria, common in Africa.

14.11 Peppered Moths and Industrial Melanism

The peppered moth, *Biston betularia,* is a European moth that rests on tree trunks during the day. Until the mid-19th century, almost every captured individual of this species had light-colored wings. From that time on, individuals with dark-colored wings increased in frequency in the moth populations near industrialized centers, until they made up almost 100% of these populations. Dark individuals had a dominant allele that was present but very rare in populations before 1850. Biologists soon noticed that in industrialized regions where the dark moths were common, the tree trunks were darkened almost black by the soot of pollution. Dark moths were much less conspicuous resting on them than light moths were. In addition, air pollution that was spreading in the industrialized regions had killed many of the light-colored lichens on tree trunks, making the trunks darker.

Selection for Melanism

Can Darwin's theory explain the increase in the frequency of the dark allele? Why did dark moths gain a survival advantage around 1850? An amateur moth collector named J. W. Tutt proposed in 1896 what became the most commonly accepted hypothesis explaining the decline of the light-colored moths. He suggested that light forms were more visible to predators on sooty trees that have lost their lichens. Consequently, birds ate the peppered moths resting on the trunks of trees during the day. The dark forms, in contrast, were at an advantage because they were camouflaged (figure 14.28). Although Tutt initially had no evidence, British ecologist Bernard Kettlewell tested the hypothesis in the 1950s by rearing populations of peppered moths with equal numbers of dark and light individuals. Kettlewell then released these populations into two sets of woods: one, near heavily polluted Birmingham, the other, in unpolluted Dorset. Kettlewell set up traps in the woods to see how many of both kinds of moths survived. To evaluate his results, he had marked the released moths with a dot of paint on the underside of their wings, where birds could not see it.

In the polluted area near Birmingham, Kettlewell trapped 19% of the light moths, but 40% of the dark ones. This indicated that dark moths had a far better chance of surviving in these polluted woods where the tree trunks were dark. In the relatively unpolluted Dorset woods, Kettlewell recovered 12.5% of the light moths but only 6% of the dark ones. This indicated that where the tree trunks were still light-colored, light moths had a much better chance of survival. Kettlewell later solidified his argument by placing dead moths on trees and filming birds looking for food. Sometimes the birds actually passed right over a moth that was the same color as its background.

Industrial Melanism

Industrial melanism is a term used to describe the evolutionary process in which darker individuals come to predominate over lighter individuals since the industrial revolution as a result of natural selection. The process is widely believed to have taken place because the dark organisms are better concealed from their predators in habitats that have been darkened by soot and other forms of industrial pollution, as suggested by Kettlewell.

Dozens of other species of moths have changed in the same way as the peppered moth in industrialized areas throughout Eurasia and North America, with dark forms becoming more common from the mid-19th century onward as industrialization spread.

Figure 14.28 Tutt's hypothesis explaining industrial melanism.

Color variants of the peppered moth (*Biston betularia*). Tutt proposed that the dark moth is more visible to predators on unpolluted trees (*top*), while the light moth is more visible to predators on bark blackened by industrial pollution (*bottom*).

Selection Against Melanism

As of the second half of the 20th century, with the widespread implementation of pollution controls, these trends are reversing, not only for the peppered moth in many areas in England but also for many other species of moths throughout the northern continents. These examples provide some of the best-documented instances of changes in allelic frequencies of natural populations as a result of natural selection due to specific factors in the environment.

In England, the air pollution promoting industrial melanism began to reverse following enactment of Clean Air legislation in 1956. Beginning in 1959, the *Biston* population at Caldy Common outside Liverpool has been sampled each year. The frequency of the melanic (dark) form dropped from a high of 94% in 1960 to a low of 19% in 1995 (figure 14.29). Similar reversals have been documented at numerous other locations throughout England. The drop correlates well with a drop in air pollution, particularly with tree-darkening sulfur dioxide and suspended particulates.

Interestingly, the same reversal of industrial melanism appears to have occurred in America during the same time that it was happening in England. Industrial melanism in the American subspecies of the peppered moth was not as widespread as in England, but it has been well documented at a rural field station near Detroit. Of 576 peppered moths collected there from 1959 to 1961, 515 were melanic, a frequency of 89%. The American Clean Air Act, passed in 1963, led to significant reductions in air pollution. Resampled in 1994, the Detroit field station peppered moth population had only 15% melanic moths! The moths in Liverpool and Detroit, both part of the same natural experiment, exhibit strong evidence of natural selection.

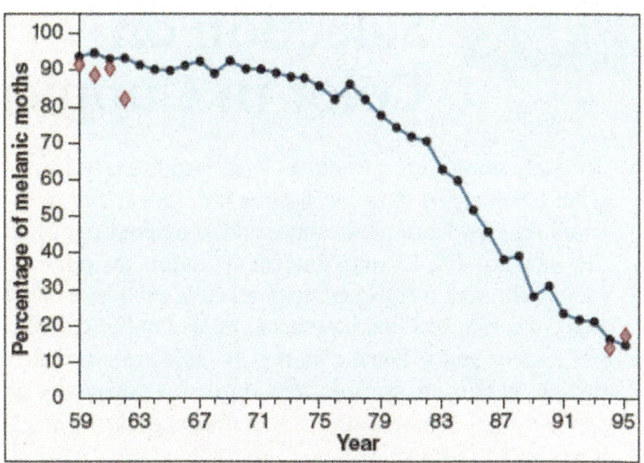

Figure 14.29 Selection against melanism.
The circles indicate the frequency of melanic *Biston betularia* moths at Caldy Common in England, sampled continuously from 1959 to 1995. Red diamonds indicate frequencies of melanic *B. betularia* in Michigan from 1959 to 1962 and from 1994 to 1995.

Reconsidering the Target of Natural Selection

Tutt's hypothesis, widely accepted in the light of Kettlewell's studies, is currently being reevaluated. The problem is that the recent selection against melanism does not appear to correlate with changes in tree lichens. At Caldy Common, the light form of the peppered moth began its increase in frequency long before lichens began to reappear on the trees. At the Detroit field station, the lichens never changed significantly as the dark moths first became dominant and then declined over the last 30 years. In fact, investigators have not been able to find peppered moths on Detroit trees at all, whether covered with lichens or not. Wherever the moths rest during the day, it does not appear to be on tree bark. Some evidence suggests they rest on leaves on the treetops, but no one is sure.

The action of selection may depend on other differences between light and dark forms of the peppered moth as well as their wing coloration. Researchers report, for example, a clear difference in their ability to survive as caterpillars under a variety of conditions. Perhaps natural selection is also targeting the caterpillars rather than the adults. While we can't yet say exactly what the targets of selection are, researchers are actively investigating this, one of the best-documented instances of natural selection in action.

Natural Selection for Melanism in Mice

Melanism is not restricted to insects. Cats and many other mammals have melanic forms that are subject to natural selection in much the same way as moths. The coat color of desert pocket mice that live on differently colored rock habitats provides a clear-cut example of natural selection acting on melanism. In Arizona and New Mexico, these small, wild pocket mice live in isolated black volcanic lava beds and the pale soils between them. Melanin synthesis during hair development of pocket mice is regulated by the receptor gene *MC1R*. Mutations that disable *MC1R* lead to melanism. Such mutations are dominant alleles, so whenever they are present in a population, dark pocket mice are seen. When wild populations of pocket mice were surveyed by biologists from the University of Arizona, there was a striking correlation between coat color and the color of the rock on which the population of pocket mice lived.

As you can see in the two upper photographs, the close match between coat color and background color gives the mice cryptic protection from avian predators, particularly owls. These mice are very visible when placed in the opposite habitats (lower photos).

Key Learning Outcome 14.11 **Natural selection has favored the dark form of the peppered moth in areas subject to severe air pollution, perhaps because on darkened trees they are less easily seen by moth-eating birds. Selection has in turn favored the light form as pollution has abated.**

14.12 Selection on Color in Guppies

To study evolution, biologists have traditionally investigated what has happened in the past, sometimes many millions of years ago. To learn about dinosaurs, a paleontologist looks at dinosaur fossils. To study human evolution, an anthropologist looks at human fossils and, increasingly, examines the "family tree" of mutations that have accumulated in human DNA over millions of years. For the biologists taking this traditional approach, evolutionary biology is similar to astronomy and history, relying on observation rather than experiment to examine ideas about past events.

Nonetheless, evolutionary biology is not entirely an observational science. Darwin was right about many things, but one area in which he was mistaken concerns the pace at which evolution occurs. Darwin thought that evolution occurred at a very slow, almost imperceptible, pace. However, in recent years many case studies have demonstrated that, in some circumstances, evolutionary change can occur rapidly. Consequently, it is possible to establish experimental studies to test evolutionary hypotheses. Although laboratory studies on fruit flies and other organisms have been common for more than 50 years, it has only been in recent years that scientists have started conducting experimental studies of evolution in nature. One excellent example of how observations of the natural world can be combined with rigorous experiments in the lab and in the field concerns research on the guppy, *Poecilia reticulata.*

Guppies Live in Different Environments

The guppy is a popular aquarium fish because of its bright coloration and prolific reproduction. In nature, guppies are found in small streams in northeastern South America and the nearby island of Trinidad. In Trinidad, guppies are found in many mountain streams. One interesting feature of several streams is that they have waterfalls. Amazingly, guppies and some other fish are capable of colonizing portions of the stream above the waterfall. The killifish, *Rivulus hartii,* is a particularly good colonizer; apparently on rainy nights, it will wriggle out of the stream and move through the damp leaf litter. Guppies are not so proficient, but they are good at swimming upstream. During flood seasons, rivers sometimes overflow their banks, creating secondary channels that move through the forest. During these occasions, guppies may be able to move upstream and invade pools above waterfalls. By contrast, not all species are capable of such dispersal and thus are only found in these streams below the first waterfall. One species whose distribution is restricted by waterfalls is the pike cichlid, *Crenicichla alta,* a voracious predator that feeds on other fish, including guppies.

Because of these barriers to dispersal, guppies can be found in two very different environments. The guppies you see living in pools just below the waterfalls in figure 14.30 are faced with predation by the pike cichlid. This substantial risk keeps rates of survival relatively low. By contrast, in similar pools just above the waterfall, the only predator present is the killifish, which only rarely preys on guppies. Guppy populations above and below waterfalls exhibit many differences. In the high-predation pools, male guppies exhibit the drab coloration you see in the guppies below the waterfall in figure 14.30. Moreover, they tend to reproduce at a younger age and attain relatively smaller adult sizes. By contrast, male fish above the waterfall in the figure display gaudy colors that they use to court females. Adults mature later and grow to larger sizes.

These differences suggest the function of natural selection. In the low-predation environment, males display gaudy colors and spots that help in mating. Moreover, larger males are most successful at holding territories and mating with females, and larger females lay more eggs. Thus, in the absence of predators, larger and more colorful fish may have produced more offspring, leading to the evolution of those traits. In pools below the waterfall, however, natural selection would favor different traits. Colorful males are likely to attract the attention of the pike cichlid, and high predation

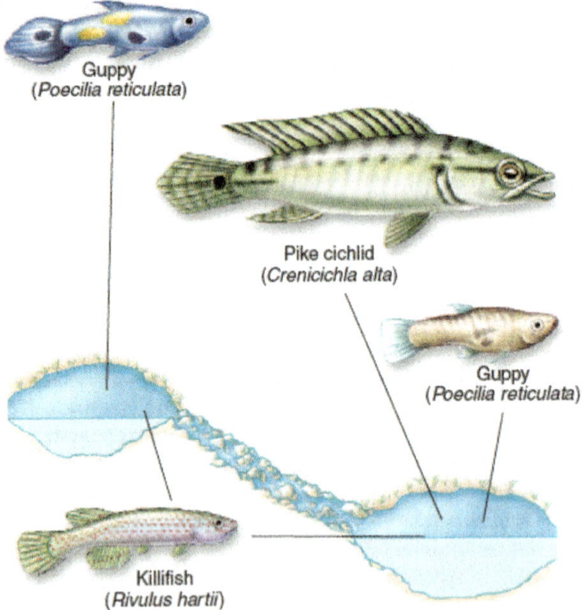

Figure 14.30 The evolution of protective coloration in guppies.
In pools below waterfalls where predation is high, male guppies (*Poecilia reticulata*) are drab colored. In the absence of the highly predatory pike cichlid (*Crenicichla alta*), male guppies in pools above waterfalls are much more colorful and attractive to females. The killifish (*Rivulus hartii*) is also a predator but only rarely eats guppies. The evolution of these differences in guppies can be experimentally tested.

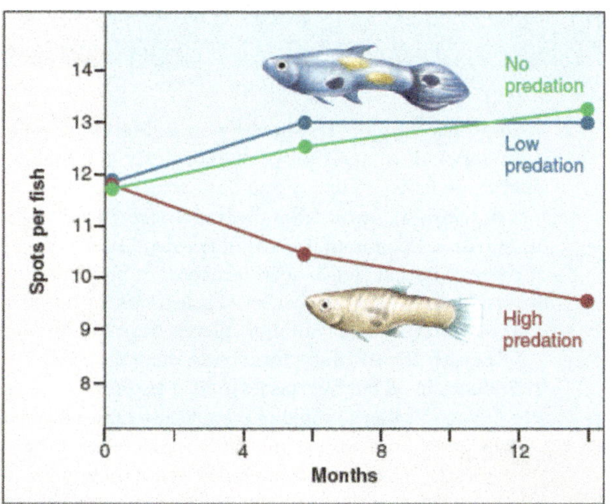

Figure 14.31 Evolutionary change in spot number.
Guppies raised in low-predation or no predation environments in laboratory greenhouses had a greater number of spots, whereas selection in more dangerous environments, like the pools with the highly predatory pike cichlid, led to less conspicuous fish. The same results are seen in field experiments conducted in pools above and below waterfalls (photo).

rates mean that most fish live short lives; thus, individuals that are more drab and shunt energy into early reproduction, rather than into growth to a larger size, are likely to be favored by natural selection.

The Experiments

Although the differences between guppies living above and below the waterfalls suggest that they represent evolutionary responses to differences in the strength of predation, alternative explanations are possible. Perhaps, for example, only very large fish are capable of swimming upstream past the waterfall to colonize pools. If this were the case, then a founder effect would occur in which the new population was established solely by individuals with genes for large size.

Laboratory Experiment The only way to rule out such alternative possibilities is to conduct a controlled experiment. John Endler, now of the University of California, Santa Barbara, conducted the first experiments in large pools in laboratory greenhouses. At the start of the experiment, a group of 2,000 guppies was divided equally among 10 large pools. Six months later, pike cichlids were added to four of the pools and killifish to another four, with the remaining two pools left to serve as "no predator" controls. Fourteen months later (which corresponds to 10 guppy generations), the scientists compared the populations. You can see their results in figure 14.31. The guppies in the killifish pool (the blue line) and control pools (the green line) were notably large, brightly colored fish with about 13 colorful spots per individual. In contrast, the guppies in the pike cichlid pools (the red line) were smaller and drab in coloration, with a reduced number of spots (about 9 per fish). These results clearly suggest that predation can lead to rapid evolutionary change, but do these laboratory experiments reflect what occurs in nature?

Field Experiment To find out, Endler and colleagues—including David Reznick, now at the University of California, Riverside—located two streams that had guppies in pools below a waterfall, but not above it (see photograph in figure 14.31). As in other Trinidadian streams, the pike cichlid was present in the lower pools, but only the killifish was found above the waterfalls. The scientists then transplanted guppies to the upper pools and returned at several-year intervals to monitor the populations. Despite originating from populations in which predation levels were high, the transplanted populations rapidly evolved the traits characteristic of low-predation guppies: They matured late, attained greater size, and had brighter colors. Control populations in the lower pools, by contrast, continued to be drab and matured early and at smaller sizes. Laboratory studies confirmed that the differences between the populations were the result of genetic differences. These results demonstrate that substantial evolutionary change can occur in less than 12 years. More generally, these studies indicate how scientists can formulate hypotheses about how evolution occurs and then test these hypotheses in natural conditions. The results give strong support to the theory of evolution by natural selection.

Key Learning Outcome 14.12 Experiments can be conducted in nature to test hypotheses about how evolution occurs. Such studies reveal that natural selection can lead to rapid evolutionary change.

Author's *Corner*

Are Bird-Killing Cats Nature's Way of Making Better Birds?

Death is not pretty, early in the morning on the doorstep. A small dead bird was left at our front door one morning, lying by the newspaper as if it might at any moment fly away. I knew it would not. Like other birds before it, it was a gift to our household by Feisty, a cat who lives with us. Feisty is a killer of birds, and every so often he leaves one for us, like rent.

We have four cats, and the other three, true housecats, would not know what to do with a bird. Feisty is different, a long-haired gray Persian with the soul of a hunter. While the other three cats sleep safely in the house with us, Feisty spends most nights outside, prowling.

Feisty's nocturnal donations are not well received by my family. More than once it has been suggested, as we donate the bird to the trashman, that perhaps Feisty would be happier living in the country.

As a biologist I try to take a more scientific view. I tell my girls that getting rid of Feisty is unwarranted because hunting cats like Feisty actually help birds, in a Darwinian sort of way. Like an evolutionary quality control check, I explain, predators ensure that only those individuals of a population that are better-suited to their environment contribute to the next generation, by the simple expedient of removing the lesser-suited. By taking the birds who are least able to escape predation—the sick and the old—Feisty culls the local bird population, leaving it on average a little better off.

That's what I tell my girls. It all makes sense, from a biological point of view, and it is a story they have heard before, in movies like *Never Cry Wolf*, and *The Lion King*. So Feisty is given a reprieve, and survives to hunt another night.

What I haven't told my girls is how little evidence actually backs up this pretty defense of Feisty's behavior. My explanation may be couched in scientific language, but without proof this "predator-as-purifier" tale is no more than a hypothesis. It might be true, and then again it might not. By such thin string has Feisty's future with our family hung.

Recently the string became a strong cable. Two French biologists put the hypothesis I had been using to defend Feisty to the test. To my great relief, it was supported.

Drs. Anders Møller and Johannes Erritzoe of the Université Pierre et Marie Curie in Paris devised a simple way to test the hypothesis. They compared the health of birds killed by domestic cats like Feisty with that of birds killed in accidents such as flying into glass windows or moving cars. Glass windows do not select for the weak or infirm—a healthy bird flies into a glass window and breaks its neck just as easily as a sickly bird. If cats are actually selecting the less-healthy birds, then their prey should include a larger proportion of sickly individuals than those felled by flying into glass windows.

How can we know what birds are sickly? Drs. Møller and Erritzoe examined the size of the dead bird's spleens. The size of its spleen is a good indicator of how healthy a bird is. Birds experiencing a lot of infections, or harboring a lot of parasites, have smaller spleens than healthy birds.

They examined 18 species of birds, more than 500 individuals. In all but two species (robins and goldcrests) they found that the spleens of birds killed by cats were significantly smaller than those killed accidentally. We're not splitting hairs here, talking about some minor statistical difference. Spleens were on average a third smaller in cat-killed birds. In five bird species (blackcaps, house sparrows, lesser whitethroats, skylarks, and spotted flycatchers), the spleens of birds pounced on by cats were less than half the size of those killed by flying at speed into glass windows or moving cars.

As a control to be sure that additional factors were not operating, the Paris biologists checked for other differences between birds killed by cats and birds killed accidentally. Weight, sex, and wing length, all of which you could imagine might be important, were not significant. Cat-killed birds had, on average, the same weight, proportion of females, and wing length as accident-killed birds.

One other factor did make a difference: age. About 50% of the birds killed accidentally were young, while fully 70% of the birds killed by cats were. Apparently it's not quite so easy to catch an experienced old codger as it is a callow youth.

So Feisty was just doing Darwin's duty, I pleaded, informing my girls that the birds he catches would soon have died anyway. But a dead bird on a doorstep argues louder than any science, and they remained unconvinced.

They are my daughters, and thus not ones to give in without a fight. Scouring the Internet, they assembled this counter-argument: Predatory house cats not unlike Feisty, as well as feral cats (domesticated cats that have been abandoned to the wild), are causing major problems for native bird populations of England, New Zealand, and Australia, as well as here in the United States. Although house cats like Feisty have the predatory instincts of their ancestors, they seem to lack the restraint that their wild relatives have. Most wild cats hunt only when hungry, but pet and feral cats seem to "love the kill," not killing for food but for sport.

So Darwin and I lost this argument. It seems I must restrict Feisty's hunting expeditions after all. While a little pruning may benefit a bird population, wholesale slaughter only devastates it. I will always see a lion whenever I look at Feisty on the prowl, but it will be a lion restricted to indoor hunting.

14.13 The Biological Species Concept

A key aspect of Darwin's theory of evolution is his proposal that adaptation (microevolution) leads ultimately to large-scale changes leading to species formation and higher taxonomic groups (macroevolution). The way natural selection leads to the formation of new species has been thoroughly documented by biologists, who have observed the stages of the species-forming process, or **speciation,** in many different plants, animals, and microorganisms. Speciation usually involves successive change: First, local populations become increasingly specialized; then, if they become different enough, natural selection may act to keep them that way.

Before we can discuss how one species gives rise to another, we need to understand exactly what a species is. The evolutionary biologist Ernst Mayr coined the **biological species concept,** which defines species as "groups of actually or potentially interbreeding natural populations which are reproductively isolated from other such groups."

In other words, the biological species concept says that a species is composed of populations whose members mate with each other and produce fertile offspring—or would do so if they came into contact. Conversely, populations whose members do not mate with each other or who cannot produce fertile offspring are said to be **reproductively isolated** and, thus, members of different species.

What causes reproductive isolation? If organisms cannot interbreed or cannot produce fertile offspring, they clearly belong to different species. However, some populations that are considered to be separate species can interbreed and produce fertile offspring, but they ordinarily do not do so under natural conditions. They are still considered to be reproductively isolated in that genes from one species generally will not be able to enter the gene pool of the other species. Table 14.1 summarizes the steps at which barriers to successful reproduction may occur. Such barriers are termed **reproductive isolating mechanisms** because they prevent genetic exchange between species. We will first discuss *prezygotic isolating mechanisms,* those that prevent the formation of zygotes. Then we will examine *postzygotic isolating mechanisms,* those that prevent the proper functioning of zygotes after they have formed.

Even though the definition of what constitutes a species is of fundamental importance to evolutionary biology, this issue has still not been completely settled and is currently the subject of considerable research and debate. For example, the biological species concept has had a number of problems. Plants of different species cross fertilize and produce fertile hybrids at much higher frequencies than first thought. Hybridization is common enough to cast doubt about whether reproductive isolation is the only force maintaining the integrity of plant species.

TABLE 14.1 ISOLATING MECHANISMS

Mechanism	Description
Prezygotic Isolating Mechanisms	
Geographic isolation	Species occur in different areas, which are often separated by a physical barrier such as a river or mountain range.
Ecological isolation	Species occur in the same area, but they occupy different habitats. Survival of hybrids is low because they are not adapted to either environment of their parents.
Temporal isolation	Species reproduce in different seasons or at different times of the day.
Behavioral isolation	Species differ in their mating rituals.
Mechanical isolation	Structural differences between species prevent mating.
Prevention of gamete fusion	Gametes of one species function poorly with the gametes of another species or within the reproductive tract of another species.
Postzygotic Isolating Mechanisms	
Hybrid inviability or infertility	Hybrid embryos do not develop properly, hybrid adults do not survive in nature, or hybrid adults are sterile or have reduced fertility.

Key Learning Outcome 14.13 **A species is generally defined as a group of similar organisms that does not exchange genes extensively with other groups in nature.**

14.14 Isolating Mechanisms

Prezygotic Isolating Mechanisms

Geographical Isolation This mechanism is perhaps the easiest to understand. Species that exist in different areas are not able to interbreed. The two populations of flowers in the first panel of table 14.1 are separated by a mountain range and so would not be capable of interbreeding.

Ecological Isolation Even if two species occur in the same area, they may utilize different portions of the environment and thus not hybridize because they do not encounter each other, like the lizards in the second panel of table 14.1. One lives on the ground and the other in the trees. Another example in nature is the ranges of lions and tigers in India. Their ranges overlapped until about 150 years ago. Even when they did overlap, however, there were no records of natural hybrids. Lions stayed mainly in the open grassland and hunted in groups called prides; tigers tended to be solitary creatures of the forest. Because of their ecological and behavioral differences, lions and tigers rarely came into direct contact with each other, even though their ranges overlapped thousands of square kilometers. Figure 14.32 shows that hybrids are possible; the tigon shown in figure 14.32c is a hybrid of a lion and tiger. These matings do not occur in the wild but can happen in artificial environments such as zoos.

Temporal Isolation *Lactuca graminifolia* and *L. canadensis,* two species of wild lettuce, grow together along roadsides throughout the southeastern United States. Hybrids between these two species are easily made experimentally and are completely fertile. But such hybrids are rare in nature because *L. graminifolia* flowers in early spring and *L. canadensis* flowers in summer. This is called temporal isolation and is shown in the third panel in table 14.1. When the blooming periods of these two species overlap, as they do occasionally, the two species do form hybrids, which may become locally abundant.

Behavioral Isolation In chapter 37, we will consider the often elaborate courtship and mating rituals of some groups of animals, which tend to keep these species distinct in nature even if they inhabit the same places. This behavioral isolation is discussed in the fourth panel of table 14.1. For example, mallard and pintail ducks are perhaps the two most common freshwater ducks in North America. In captivity, they produce completely fertile offspring, but in nature they nest side-by-side and rarely hybridize.

Mechanical Isolation Structural differences that prevent mating between related species of animals and plants is called mechanical isolation and is shown in panel five of table 14.1.

(a)

(b)

(c)

Figure 14.32 Lions and tigers are ecologically isolated.

The ranges of lions and tigers used to overlap in India. However, lions and tigers do not hybridize in the wild because they utilize different portions of the habitat. (a) Tigers are solitary animals that live in the forest, whereas (b) lions live in open grassland. (c) Hybrids, such as this tigon, have been successfully produced in captivity, but hybridization does not occur in the wild.

Figure 14.33 Postzygotic isolation in leopard frogs. Numbers indicate the following species in the geographic ranges shown: (1) *Rana pipiens;* (2) *Rana blairi;* (3) *Rana sphenocephala;* (4) *Rana berlandieri.* These four species resemble one another closely in their external features. Their status as separate species was first suspected when hybrids between them were found to produce defective embryos in the laboratory. Subsequent research revealed that the mating calls of the four species differ substantially, indicating that the species have both pre- and postzygotic isolating mechanisms.

Flowers of related species of plants often differ significantly in their proportions and structures. Some of these differences limit the transfer of pollen from one plant species to another. For example, bees may pick up the pollen of one species on a certain place on their bodies; if this area does not come into contact with the receptive structures of the flowers of another plant species, the pollen is not transferred.

Prevention of Gamete Fusion In animals that shed their gametes directly into water, eggs and sperm derived from different species may not attract one another. Many land animals may not hybridize successfully because the sperm of one species may function so poorly within the reproductive tract of another that fertilization never takes place. In plants, the growth of pollen tubes may be impeded in hybrids between different species. In both plants and animals, the operation of such isolating mechanisms prevents the union of gametes even following successful mating. The sixth panel in table 14.1 discusses this isolating mechanism.

Postzygotic Isolating Mechanisms

All of the factors we have discussed up to this point tend to prevent hybridization. If hybrid matings do occur, and zygotes are produced, many factors may still prevent those zygotes from developing into normally functioning, fertile individuals. Development in any species is a complex process. In hybrids, the genetic complements of two species may be so different that they cannot function together normally in embryonic development. For example, hybridization between sheep and goats usually produces embryos that die in the earliest developmental stages.

Figure 14.33 shows four species of leopard frogs (genus *Rana*) and their ranges throughout North America. It was assumed for a long time that they constituted a single species. However, careful examination revealed that although the frogs appear similar, successful mating between them is rare because of problems that occur as the fertilized eggs develop. Many of the hybrid combinations cannot be produced even in the laboratory. Examples of this kind, in which similar species have been recognized only as a result of hybridization experiments, are common in plants.

Even if hybrids survive the embryo stage, however, they may not develop normally. If the hybrids are weaker than their parents, they will almost certainly be eliminated in nature. Even if they are vigorous and strong, as in the case of the mule, a hybrid between a female horse and a male donkey, they may still be sterile and thus incapable of contributing to succeeding generations. Sterility may result in hybrids because the development of sex organs may be abnormal, because the chromosomes derived from the respective parents may not pair properly, or from a variety of other causes.

Key Learning Outcome 14.14 Prezygotic isolating mechanisms lead to reproductive isolation by preventing the formation of hybrid zygotes. Postzygotic mechanisms lead to the failure of hybrid zygotes to develop normally, or they prevent hybrids from becoming established in nature.

INQUIRY & ANALYSIS

Does Natural Selection Act on Enzyme Polymorphism?

The essence of Darwin's theory of evolution is that, in nature, selection favors some gene alternatives over others. Many studies of natural selection have focused on genes encoding enzymes because populations in nature tend to possess many alternative alleles of their enzymes (a phenomenon called *enzyme polymorphism*). Often investigators have looked to see if weather influences which alleles are more common in natural populations. A particularly nice example of such a study was carried out on a fish, the mummichog (*Fundulus heteroclitus*), which ranges along the East Coast of North America. Researchers studied allele frequencies of the gene encoding the enzyme lactate dehydrogenase, which catalyzes the conversion of pyruvate to lactate. As you learned in chapter 7, this reaction is a key step in energy metabolism, particularly when oxygen is in short supply. There are two common alleles of lactate dehydrogenase in these fish populations, with allele *a* being a better catalyst at lower temperatures than allele *b*.

In an experiment, investigators sampled the frequency of allele *a* in 41 fish populations located over 14 degrees of latitude, from Jacksonville, Florida (31° North), to Bar Harbor, Maine (44° North). Annual mean water temperatures change 1° C per degree change in latitude. The survey is designed to test a prediction of the hypothesis that natural selection acts on this enzyme polymorphism. If it does, then you would expect that allele *a*, producing a better "low-temperature" enzyme, would be more common in the colder waters of the more northern latitudes. The graph on the right presents the results of this survey. The points on the graph are derived from pie chart data such as shown for 20 populations in the map (a **pie chart diagram** assigns a slice of the pie to each variable; the size of the slice is proportional to the contribution made by that variable to the total). The blue line on the graph is the line that best fits the data (a **"best-fit" line**, also called a **regression line**, is determined statistically by a process called *regression analysis*).

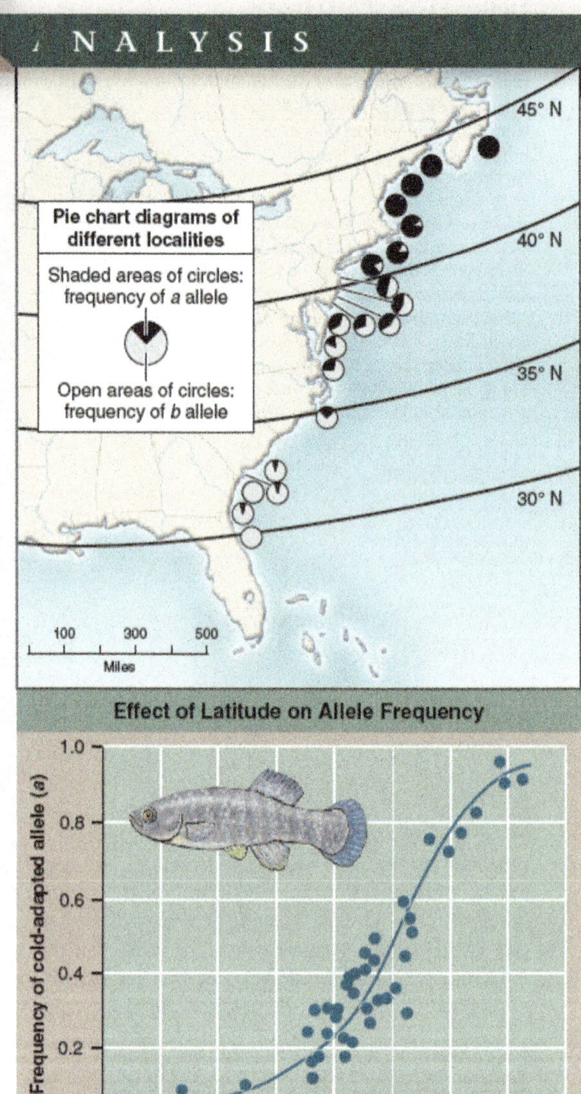

1. **Applying Concepts**
   a. Variable. In the graph, what is the dependent variable?
   b. Reading pie charts. In the fish population located at 35° N latitude, what is the frequency of the *a* allele? Locate this point on the graph.
   c. Analyzing a continuous variable. Compare the frequency of allele *a* among fish captured in waters at 44° N latitude with the frequency among fish captured at 31° N latitude. Is there a pattern? Describe it.
2. **Interpreting Data** At what latitude do fish populations exhibit the greatest variability in allele *a* frequency?
3. **Making Inferences**
   a. Are fish populations in cold waters at 44° N latitude more or less likely to contain heterozygous individuals than fish populations in warm waters at 31° N latitude? Why this difference, or lack of it?
   b. Where along this latitudinal gradient in the frequency of allele *a* would you expect to find the highest frequency of heterozygous individuals? Why?
4. **Drawing Conclusions** Are the differences in population frequencies of allele *a* consistent with the hypothesis that natural selection is acting on the alleles encoding this enzyme? Explain.
5. **Further Analysis** If you were to release fish captured at 32° N into populations located at 44° N, so that the local population now had equal frequencies of the two alleles, what would you expect to happen in future generations? How might you test this prediction?

Chapter Review

Evolution

14.1 Darwin's Voyage on HMS *Beagle*

- The theory of evolution through natural selection, a theory proposed by Darwin, is overwhelmingly accepted by scientists and is considered to be the backbone of the science of biology.

14.2 Darwin's Evidence

- Darwin observed fossils in South America of extinct species that resembled living species (**figure 14.4**). On the Galápagos Islands, Darwin observed finches that differed slightly in appearance between islands but resembled finches found on the South American mainland (**figure 14.5**).

14.3 The Theory of Natural Selection

- Key to Darwin's hypothesis was the observation by Malthus that the food supply limits population growth. A population grows only as large as that which can live off of the available food that limits geometric growth of a population (**figure 14.6**).
- Using Malthus's observations and his own, Darwin proposed that individuals that are better suited to their environments survive to produce offspring, gaining the opportunity to pass their characteristics on to future generations, what Darwin called natural selection.

Darwin's Finches: Evolution in Action

14.4 The Beaks of Darwin's Finches

- By observing the different sizes and shapes of beaks in the closely related finches of the Galápagos Islands (**figure 14.9**), and correlating the beaks with the types of food consumed, Darwin concluded that the birds's beaks were modified from an ancestral species based on the food available, each suited to its food supply. Scientists have identified a gene, *BMP4,* that is expressed differently in birds with different shaped beaks.

14.5 How Natural Selection Produces Diversity

- The 14 species of finches found on the islands off the coast of South America descended from a mainland species that adapted to different niches, a process called adaptive radiation (**figure 14.12**).

The Theory of Evolution

14.6 The Evidence for Evolution

- The evidence for evolution includes the fossil record. The titanothere and its ancestors are known only from the fossil record (**figure 14.14**). The fossil record reveals organisms that are intermediate in form.
- The evidence for evolution also includes the anatomical record, which reveals similarities in structures between species (**figure 14.15**). Homologous structures are similar in structure but differ in their functions (**figure 14.16**). Analogous structures are similar in function but differ in their underlying structure.
- The molecular record traces changes in the genomes and proteins of species over time (**figures 14.18** and **14.19**).

14.7 Evolution's Critics

- Darwin's theory of evolution through natural selection has always had its critics. Their criticisms of evolution, however, are without scientific merit (**integrated art, page 296** and **figure 14.20**).

How Populations Evolve

14.8 Genetic Change in Populations: The Hardy-Weinberg Rule

- If a population follows the five assumptions of Hardy-Weinberg, the frequencies of alleles within the population will not change (**figure 14.21**). However, if a population is small, has selective mating, experiences mutations or migration, or is under the influence of natural selection, the allele frequencies will be different from those predicted by the Hardy-Weinberg Rule.

14.9 Agents of Evolution

- Five factors act on populations to change their allele and genotype frequencies (**integrated art, pages 302–303**). Mutations are changes in DNA. Nonrandom mating occurs when individuals seek out mates based on certain traits. Genetic drift is the random loss of alleles in a population due to chance occurrences, not due to fitness (**figures 14.22** and **14.23**). Migrations are the movements of individuals or alleles into or out of a population. Selection occurs when individuals with certain traits leave more offspring because their traits allow them to better respond to the challenges of their environment (**figure 14.24**).
- Selection can act on the genes in that population in several different ways (**integrated art, pages 304–305**). Stabilizing selection tends to reduce extreme phenotypes. Disruptive selection tends to reduce intermediate phenotypes. Directional selection tends to reduce one extreme phenotype from the population.

Adaptation Within Populations

14.10 Sickle-Cell Anemia

- Sickle-cell disease is an example of heterozygote advantage, where individuals who are heterozygous for a trait tend to survive better in areas with malaria (**figures 14.26** and **14.27**).

14.11 Peppered Moths and Industrial Melanism

- Natural selection favors dark-colored (melanic) organisms in areas of heavy pollution or in other cases of background matching (**figures 14.28** and **14.29**).

14.12 Selection on Color in Guppies

- Experiments have shown evolutionary change in guppy populations due to natural selection (**figures 14.30** and **14.31**).

How Species Form

14.13 The Biological Species Concept

- The biological species concept states that a species is a group of organisms that mate with each other and produce fertile offspring, or would do so if in contact with each other. If they cannot mate, or mate but cannot produce fertile offspring, they are said to be reproductively isolated (**table 14.1**).

14.14 Isolating Mechanisms

- There are two types of isolating mechanisms: prezygotic and postzygotic. Prezygotic isolating mechanisms prevent the formation of a hybrid zygote. Postzygotic isolating mechanisms prevent normal development of a hybrid zygote or result in sterile offspring (**figure 14.33**).

Test Your Understanding

1. Darwin was greatly influenced by Thomas Malthus, who pointed out that
   a. food supplies increase geometrically.
   b. populations increase arithmetically.
   c. populations are capable of geometric increase, yet remain at constant levels.
   d. the food supply usually increases faster than the population that depends on it.
2. Darwin proposed that individuals with traits that help them live in their immediate environment are more likely to survive and reproduce than individuals without those traits. He called this
   a. natural selection.
   b. arithmetic progression.
   c. the theory of evolution.
   d. geometric progression.
3. A great deal of research has been done on Darwin's finches over the last 70 years. The research
   a. seems to often contradict Darwin's original ideas.
   b. seems to agree with Darwin's original ideas.
   c. does not show any clear patterns that support or refute Darwin's original ideas.
   d. suggests a different explanation for the evolution of finches.
4. One of the major sources of evidence for evolution is in the comparative anatomy of organisms. Features that look different but have similar structural origin are called
   a. homologous structures.
   b. analogous structures.
   c. vestigial structures.
   d. equivalent structures.
5. A large group of organisms lives in a large, stable ecosystem. There is no competition for resources. Individuals show no mate preferences. All organisms appear to be identical except for a few individuals in the most recent generation of offspring that exhibit a different fur coat color and pattern. The ecosystem and population are geographically isolated from other populations of the same organism. Which Hardy-Weinberg assumption seems to have been violated?
   a. large population size
   b. random mating within the population
   c. no mutation within the population
   d. no input of new alleles from outside or loss of alleles
6. A population of 1,000 individuals has 200 individuals who show a homozygous recessive phenotype and 800 individuals who express the dominant phenotype. What is the frequency of homozygous recessive individuals in this population?
   a. 0.20
   b. 0.30
   c. 0.45
   d. 0.55
7. A chance event occurs that causes a population to lose some individuals (they died)—hence, a loss of alleles in the population results from
   a. mutation.
   b. migration.
   c. selection.
   d. genetic drift.
8. Selection that causes one extreme phenotype to be more frequent in a population is an example of
   a. disruptive selection.
   b. stabilizing selection.
   c. directional selection.
   d. equivalent selection.
9. A key element of Ernst Mayr's biological species concept is
   a. homologous isolation.
   b. divergent isolation.
   c. convergent isolation.
   d. reproductive isolation.
10. Which of the following is *not* a prezygotic isolating mechanism?
    a. behavioral isolation
    b. ecological isolation
    c. hybrid infertility
    d. None of the above.

Apply Your Understanding

1. **Pages 304-305** Because of prolonged drought, the trees on an island are producing nuts that are much smaller with thicker and harder shells. What would you predict would happen to birds that depend on the nuts for food? What type of selection will result?

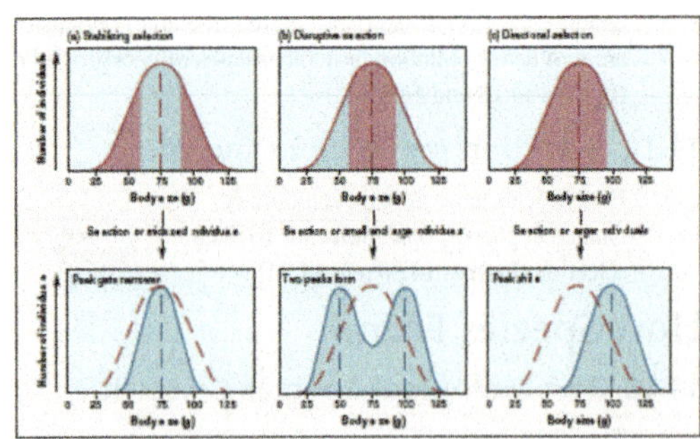

2. **Table 14.1** A very heavy rainstorm floods a mountain river, changing its course and digging a deep canyon through the soft soils of the meadow in the valley downstream. How might mice populations on the two sides of the valley be affected?

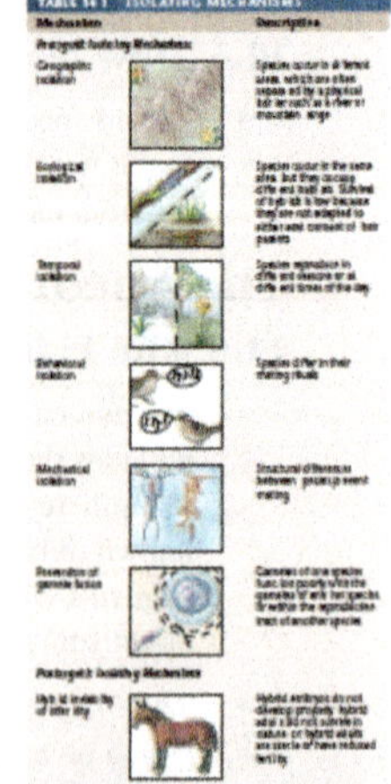

Synthesize What You Have Learned

1. Can natural selection occur among genetically identical clones? Explain your reasoning.
2. The evolutionary pathways of some groups of related organisms seem to have moved from larger to smaller organisms over time, such as the glyptodont to the armadillo, the mammoth to the elephant. Yet other groups of related organisms have exhibited a trend of increased sizes of species over time, such as the tiny eohippus to the horse. Explain how this can occur.
3. In a courtroom in 2005, biologist Ken Miller criticized the claims of intelligent design. After noting that 99.9% of the organisms that have ever lived on earth are now extinct, he said that "an intelligent designer who designed things, 99.9% of which didn't last, certainly wouldn't be very intelligent." Evaluate Miller's criticism.

Glossary

Terms & Concepts

A

absorption (L. *absorbere,* to swallow down) The movement of water and substances dissolved in water into a cell, tissue, or organism.
acid Any substance that dissociates to form H^+ ions when dissolved in water. Having a pH value less than 7.
acoelomate (Gr. *a,* not + *koiloma,* cavity) A bilaterally symmetrical animal not possessing a body cavity, such as a flatworm.
actin (Gr. *actis,* ray) One of the two major proteins that make up myofilaments (the other is myosin). It provides the cell with mechanical support and plays major roles in determining cell shape and cell movement.
action potential A single nerve impulse. A transient all-or-none reversal of the electrical potential across a neuron membrane. Because it can activate nearby voltage-sensitive channels, an action potential propagates along a nerve cell.
activation energy The energy a molecule must acquire to undergo a specific chemical reaction.
activator A regulatory protein that binds to the DNA and makes it more accessible for transcription.
active transport The transport of a solute across a membrane by protein carrier molecules to a region of higher concentration by the expenditure of chemical energy. One of the most important functions of any cell.
adaptation (L. *adaptare,* to fit) Any peculiarity of structure, physiology, or behavior that promotes the likelihood of an organism's survival and reproduction in a particular environment.
adenosine triphosphate (ATP) A molecule composed of ribose, adenine, and a triphosphate group. ATP is the chief energy currency of all cells. Cells focus all of their energy resources on the manufacture of ATP from ADP and phosphate, which requires the cell to supply 7 kilocalories of energy obtained from photosynthesis or from electrons stripped from foodstuffs to form 1 mole of ATP. Cells then use this ATP to drive endergonic reactions.
adhesion (L. *adhaerere,* to stick to) The molecular attraction exerted between the surfaces of unlike bodies in contact, as water molecules to the walls of the narrow tubes that occur in plants.
aerobic (Gr. *aer,* air + *bios,* life) Oxygen-requiring.
allele (Gr. *allelon,* of one another) One of two or more alternative forms of a gene.
allele frequency The relative proportion of a particular allele among individuals of a population. Not equivalent to gene frequency, although the two terms are sometimes confused.
allosteric interaction (Gr. *allos,* other + *stereos,* shape) The change in shape that occurs when an activator or repressor binds to an enzyme. These changes result when specific, small molecules bind to the enzyme, molecules that are not substrates of that enzyme.
alternation of generations A reproductive life cycle in which the multicellular diploid phase produces spores that give rise to the multicellular haploid phase and the multicellular haploid phase produces gametes that fuse to give rise to the zygote. The zygote is the first cell of the multicellular diploid phase.
alveolus, *pl.* alveoli (L. *alveus,* a small cavity) One of the many small, thin-walled air sacs within the lungs in which the bronchioles terminate.
amniotic egg An egg that is isolated and protected from the environment by a more or less impervious shell. The shell protects the embryo from drying out, nourishes it, and enables it to develop outside of water.
anaerobic (Gr. *an,* without + *aer,* air + *bios,* life) Any process that can occur without oxygen. Includes glycolysis and fermentation. Anaerobic organisms can live without free oxygen.
anaphase In mitosis and meiosis II, the stage initiated by the separation of sister chromatids, during which the daughter chromosomes move to opposite poles of the cell; in meiosis I, marked by separation of replicated homologous chromosomes.
angiosperms The flowering plants, one of five phyla of seed plants. In angiosperms, the ovules at the time of pollination are completely enclosed by tissues.
anterior (L. *ante,* before) Located before or toward the front. In animals, the head end of an organism.
anther (Gr. *anthos,* flower) The part of the stamen of a flower that bears the pollen.
antibody (Gr. *anti,* against) A protein substance produced by a B cell lymphocyte in response to a foreign substance (antigen) and released into the bloodstream. Binding to the antigen, antibodies mark them for destruction by other elements of the immune system.
anticodon The three-nucleotide sequence of a tRNA molecule that is complementary to, and base pairs with, an amino acid-specifying codon in mRNA.
antigen (Gr. *anti,* against + *genos,* origin) A foreign substance, usually a protein, that stimulates lymphocytes to proliferate and secrete specific antibodies that bind to the foreign substance, labeling it as foreign and destined for destruction.
apical meristem (L. *apex,* top + Gr. *meristos,* divided) In vascular plants, the growing point at the tip of the root or stem.
aposematic coloration An ecological strategy of some organisms that "advertise" their poisonous nature by the use of bright colors.
appendicular skeleton (L. *appendicula,* a small appendage) The skeleton of the limbs of the human body containing 126 bones.
archaea A group of prokaryotes that are among the most primitive still in existence, characterized by the absence of peptidoglycan in their cell walls, a feature that distinguishes them from bacteria.
asexual Reproducing without forming gametes. Asexual reproduction does not involve sex. Its outstanding characteristic is that an individual offspring is genetically identical to its parent.
association neuron A nerve cell found only in the CNS that acts as a functional link between sensory neurons and motor neurons. Also called interneuron.
atom (Gr. *atomos,* indivisible) A core (nucleus) of protons and neutrons surrounded by an orbiting cloud of electrons. The chemical behavior of an atom is largely determined by the distribution of its electrons, particularly the number of electrons in its outermost level.
atomic number The number of protons in the nucleus of an atom. In an atom that does not bear an electric charge (that is, one that is not an ion), the atomic number is also equal to the number of electrons.
autonomic nervous system (Gr. *autos,* self + *nomos,* law) The motor pathways that carry commands from the central nervous system to regulate the glands and nonskeletal muscles of the body. Also called the involuntary nervous system.
autosome (Gr. *autos,* self + *soma,* body) Any of the 22 pairs of human chromosomes that are similar in size and morphology in both males and females.
autotroph (Gr. *autos,* self + *trophos,* feeder) An organism that can harvest light energy from the sun or from the oxidation of inorganic compounds to make organic molecules.
axial skeleton The skeleton of the head and trunk of the human body containing 80 bones.

axon (Gr., axle) A process extending out from a neuron that conducts impulses away from the cell body.

B

bacterium, *pl.* bacteria (Gr. *bakterion,* dim. of *baktron,* a staff) The simplest cellular organism. Its cell is smaller and prokaryotic in structure, and it lacks internal organization.
basal body In eukaryotic cells that contain flagella or cilia, a form of centriole that anchors each flagellum.
base Any substance that combines with H^+ ions thereby reducing the H^+ ion concentration of a solution. Having a pH value above 7.
Batesian mimicry After Henry W. Bates, English naturalist. A situation in which a palatable or nontoxic organism resembles another kind of organism that is distasteful or toxic. Both species exhibit warning coloration.
B cell A lymphocyte that recognizes invading pathogens much as T cells do, but instead of attacking the pathogens directly, it marks them with antibodies for destruction by the nonspecific body defenses.
bilateral symmetry (L. *bi,* two + *lateris,* side; Gr. *symmetria,* symmetry) A body form in which the right and left halves of an organism are approximate mirror images of each other.
binary fission (L. *binarius,* consisting of two things or parts + *fissus,* split) Asexual reproduction of a cell by division into two equal, or nearly equal, parts. Bacteria divide by binary fission.
binomial system (L. *bi,* twice, two + Gr. *nomos,* usage, law) A system of nomenclature that uses two words. The first names the genus, and the second designates the species.
biomass (Gr. *bios,* life + *maza,* lump or mass) The total weight of all of the organisms living in an ecosystem.
biome (Gr. *bios,* life + *-oma,* mass, group) A major terrestrial assemblage of plants, animals, and microorganisms that occur over wide geographical areas and have distinct characteristics. The largest ecological unit.
buffer A substance that takes up or releases hydrogen ions (H^+) to maintain the pH within a certain range.

C

calorie (L. *calor,* heat) The amount of energy in the form of heat required to raise the temperature of 1 gram of water 1 degree Celsius.
calyx (Gr. *kalyx,* a husk, cup) The sepals collectively. The outermost flower whorl.
cancer Unrestrained invasive cell growth. A tumor or cell mass resulting from uncontrollable cell division.
capillary (L. *capillaris,* hairlike) A blood vessel with a very small diameter. Blood exchanges gases and metabolites across capillary walls. Capillaries join the end of an arteriole to the beginning of a venule.
carbohydrate (L. *carbo,* charcoal + *hydro,* water) An organic compound consisting of a chain or ring of carbon atoms to which hydrogen and oxygen atoms are attached in a ratio of approximately 1:2:1. A compound of carbon, hydrogen, and oxygen having the generalized formula $(CH_2O)_n$ where n is the number of carbon atoms.
carcinogen (Gr. *karkinos,* cancer + -gen) Any cancer-causing agent.
cardiovascular system (Gr. *kardia,* heart + L. *vasculum,* vessel) The blood circulatory system and the heart that pumps it. Collectively, the blood, heart, and blood vessels.
carpel (Gr. *karpos,* fruit) A leaflike organ in angiosperms that encloses one or more ovules.
carrying capacity The maximum population size that a habitat can support.
catabolism (Gr. *katabole,* throwing down) A process in which complex molecules are broken down into simpler ones.
catalysis (Gr. *katalysis,* dissolution + *lyein,* to loosen) The enzyme-mediated process in which the subunits of polymers are positioned so that their bonds undergo chemical reactions.
catalyst (Gr. *kata,* down + *lysis,* a loosening) A general term for a substance that speeds up a specific chemical reaction by lowering the energy required to activate or start the reaction. An enzyme is a biological catalyst.
cell (L. *cella,* a chamber or small room) The smallest unit of life. The basic organizational unit of all organisms. Composed of a nuclear region containing the hereditary apparatus within a larger volume called the cytoplasm bounded by a lipid membrane.
cell cycle The repeating sequence of growth and division through which cells pass each generation.
cellular respiration The process in which the energy stored in a glucose molecule is released by oxidation. Hydrogen atoms are lost by glucose and gained by oxygen.
central nervous system The brain and spinal cord, the site of information processing and control within the nervous system.
centromere (Gr. *kentron,* center + *meros,* a part) A constricted region of the chromosome joining two sister chromatids, to which the kinetochore is attached.
chemical bond The force holding two atoms together. The force can result from the attraction of opposite charges (ionic bond) or from the sharing of one or more pairs of electrons (a covalent bond).
chemiosmosis The cellular process responsible for almost all of the adenosine triphosphate (ATP) harvested from food and for all the ATP produced by photosynthesis.
chemoautotroph An autotrophic bacterium that uses chemical energy released by specific inorganic reactions to power its life processes, including the synthesis of organic molecules.
chiasma, *pl.* chiasmata (Gr. a cross) In meiosis, the points of crossing over where portions of chromosomes have been exchanged during synapsis. A chiasma appears as an X-shaped structure under a light microscope.
chloroplast (Gr. *chloros,* green + *plastos,* molded) A cell-like organelle present in algae and plants that contains chlorophyll (and usually other pigments) and is the site of photosynthesis.
chromatid (Gr. *chroma,* color + L. *-id,* daughters of) One of two daughter strands of a duplicated chromosome that is joined by a single centromere.
chromatin (Gr. *chroma,* color) The complex of DNA and proteins of which eukaryotic chromosomes are composed.
chromosome (Gr. *chroma,* color + *soma,* body) The vehicle by which hereditary information is physically transmitted from one generation to the next. In a eukaryotic cell, long threads of DNA that are associated with protein and that contain hereditary information.
cilium, *pl.* cilia (L. eyelash) Refers to flagella, which are numerous and organized in dense rows. Cilia propel cells through water. In human tissue, they move water or mucus over the tissue surface.
cladistics A taxonomic technique used for creating hierarchies of organisms based on derived characters that represent true phylogenetic relationship and descent.
class A taxonomic category ranking below a phylum (division) and above an order.
clone (Gr. *klon,* twig) A line of cells, all of which have arisen from the same single cell by mitotic division. One of a population of individuals derived by asexual reproduction from a single ancestor. One of a population of genetically identical individuals.
codominance In genetics, a situation in which the effects of both alleles at a particular locus are apparent in the phenotype of the heterozygote.
codon (L. code) The basic unit of the genetic code. A sequence of three adjacent nucleotides in DNA or mRNA that codes for one amino acid or for polypeptide termination.
coelom (Gr. *koilos,* a hollow) A body cavity formed between layers of mesoderm and in which the digestive tract and other internal organs are suspended.
coenzyme A cofactor of an enzyme that is a nonprotein organic molecule.
coevolution (L. *co-,* together + *e-,* out + *volvere,* to fill) A term that describes the long-term evolutionary adjustment of one group of organisms to another.
commensalism (L. *cum,* together with + *mensa,* table) A symbiotic relationship in which one species benefits while the other neither benefits nor is harmed.
community (L. *communitas,* community, fellowship) The populations of different species that live together and interact in a particular place.
competition Interaction between individuals for the same scarce resources. Intraspecific competition is competition between individuals of a single species. Interspecific competition is competition between individuals of different species.

competitive exclusion The hypothesis that if two species are competing with one another for the same limited resource in the same place, one will be able to use that resource more efficiently than the other and eventually will drive that second species to extinction locally.
complement system The chemical defense of a vertebrate body that consists of a battery of proteins that insert in bacterial and fungal cells, causing holes that destroy the cells.
concentration gradient The concentration difference of a substance as a function of distance. In a cell, a greater concentration of its molecules in one region than in another.
condensation The coiling of the chromosomes into more and more tightly compacted bodies begun during the G_2 phase of the cell cycle.
conjugation (L. *conjugare,* to yoke together) An unusual mode of reproduction in unicellular organisms in which genetic material is exchanged between individuals through tubes connecting them during conjugation.
consumer In ecology, a heterotroph that derives its energy from living or freshly killed organisms or parts thereof. Primary consumers are herbivores; secondary consumers are carnivores or parasites.
cortex (L. bark) In vascular plants, the primary ground tissue of a stem or root, bounded externally by the epidermis and internally by the central cylinder of vascular tissue. In animals, the outer, as opposed to the inner, part of an organ, as in the adrenal, kidney, and cerebral cortexes.
cotyledon (Gr. *kotyledon,* a cup-shaped hollow) Seed leaf. Monocot embryos have one cotyledon, and dicots have two.
countercurrent flow In organisms, the passage of heat or of molecules (such as oxygen, water, or sodium ions) from one circulation path to another moving in the opposite direction. Because the flow of the two paths is in opposite directions, a concentration difference always exists between the two channels, facilitating transfer.
covalent bond (L. *co-,* together + *valare,* to be strong) A chemical bond formed by the sharing of one or more pairs of electrons.
crossing over An essential element of meiosis occurring during prophase when nonsister chromatids exchange portions of DNA strands.
cuticle (L. *cutis,* skin) A very thin film covering the outer skin of many plants.
cytokinesis (Gr. *kytos,* hollow vessel + *kinesis,* movement) The C phase of cell division in which the cell itself divides, creating two daughter cells.
cytoplasm (Gr. *kytos,* hollow vessel + *plasma,* anything molded) A semifluid matrix that occupies the volume between the nuclear region and the cell membrane. It contains the sugars, amino acids, proteins, and organelles (in eukaryotes) with which the cell carries out its everyday activities of growth and reproduction.
cytoskeleton (Gr. *kytos,* hollow vessel + *skeleton,* a dried body) In the cytoplasm of all eukaryotic cells, a network of protein fibers that supports the shape of the cell and anchors organelles, such as the nucleus, to fixed locations.

D

deciduous (L. *decidere,* to fall off) In vascular plants, shedding all the leaves at a certain season.
dehydration reaction Water-losing. The process in which a hydroxyl (OH) group is removed from one subunit of a polymer and a hydrogen (H) group is removed from the other subunit, linking the subunits together and forming a water molecule as a by-product.
demography (Gr. *demos,* people + *graphein,* to draw) The statistical study of population. The measurement of people or, by extension, of the characteristics of people.
density The number of individuals in a population in a given area.
deoxyribonucleic acid (DNA) The basic storage vehicle or central plan of heredity information. It is stored as a sequence of nucleotides in a linear nucleotide polymer. Two of the polymers wind around each other like the outside and inside rails of a circular staircase.
depolarization The movement of ions across a cell membrane that wipes out locally an electrical potential difference.
deuterostome (Gr. *deuteros,* second + *stoma,* mouth) An animal in whose embryonic development the anus forms from or near the blastopore, and the mouth forms later on another part of the blastula. Also characterized by radial cleavage.
dicot Short for dicotyledon; a class of flowering plants generally characterized by having two cotyledons, netlike veins, and flower parts in fours or fives.
diffusion (L. *diffundere,* to pour out) The net movement of molecules to regions of lower concentration as a result of random, spontaneous molecular motions. The process tends to distribute molecules uniformly.
dihybrid (Gr. *dis,* twice + L. *hibrida,* mixed offspring) An individual heterozygous for two genes.
dioecious (Gr. *di,* two + *eikos,* house) Having male and female flowers on separate plants of the same species.
diploid (Gr. *diploos,* double + *eidos,* form) A cell, tissue, or individual with a double set of chromosomes.
directional selection A form of selection in which selection acts to eliminate one extreme from an array of phenotypes. Thus, the genes promoting this extreme become less frequent in the population.
disaccharide (Gr. *dis,* twice + *sakcharon,* sugar) A sugar formed by linking two monosaccharide molecules together. Sucrose (table sugar) is a disaccharide formed by linking a molecule of glucose to a molecule of fructose.
disruptive selection A form of selection in which selection acts to eliminate rather than favor the intermediate type.
diurnal (L. *diurnalis,* day) Active during the day.
division Traditionally, a major taxonomic group of the plant kingdom comparable to a phylum of the animal kingdom. Today divisions are called phyla.
dominant allele An allele that dictates the appearance of heterozygotes. One allele is said to be dominant over another if an individual heterozygous for that allele has the same appearance as an individual homozygous for it.
dorsal (L. *dorsum,* the back) Toward the back, or upper surface. Opposite of ventral.
double fertilization A process unique to the angiosperms, in which one sperm nucleus fertilizes the egg and the second one fuses with the polar nuclei. These two events result in the formation of the zygote and the primary endosperm nucleus, respectively.

E

ecdysis (Gr. *ekdysis,* stripping off) The shedding of the outer covering or skin of certain animals. Especially the shedding of the exoskeleton by arthropods.
ecology (Gr. *oikos,* house + *logos,* word) The study of the relationships of organisms with one another and with their environment.
ecosystem (Gr. *oikos,* house + *systema,* that which is put together) A community, together with the nonliving factors with which it interacts.
ectoderm (Gr. *ecto,* outside + *derma,* skin) One of three embryonic germ layers that forms in the gastrula; giving rise to the outer epithelium and to nerve tissue.
ectothermic Referring to animals whose body temperature is regulated by their behavior or their surroundings.
electron A subatomic particle with a negative electrical charge. The negative charge of one electron exactly balances the positive charge of one proton. Electrons orbit the atom's positively charged nucleus and determine its chemical properties.
electron transport chain A collective term describing the series of membrane-associated electron carriers embedded in the inner mitochondrial membrane. It puts the electrons harvested from the oxidation of glucose to work driving proton-pumping channels.
electron transport system A collective term describing the series of membrane-associated electron carriers embedded in the thylakoid membrane of the chloroplast. It puts the electrons harvested from water molecules and energized by photons of light to work driving proton-pumping channels.
element A substance that cannot be separated into different substances by ordinary chemical methods.
emergent properties Novel properties in the hierarchy of life that were not present at the simpler levels of organization.

endergonic (Gr. *endon,* within + *ergon,* work) Reactions in which the products contain more energy than the reactants and require an input of usable energy from an outside source before they can proceed. These reactions are not spontaneous.
endocrine gland (Gr. *endon,* within + *krinein,* to separate) A ductless gland producing hormonal secretions that pass directly into the bloodstream or lymph.
endocrine system The dozen or so major endocrine glands of a vertebrate.
endocytosis (Gr. *endon,* within + *kytos,* cell) The process by which the edges of plasma membranes fuse together and form an enclosed chamber called a vesicle. It involves the incorporation of a portion of an exterior medium into the cytoplasm of the cell by capturing it within the vesicle.
endoderm (Gr. *endon,* outside + *derma,* skin) One of three embryonic germ layers that forms in the gastrula; giving rise to the epithelium that lines internal organs and most of the digestive and respiratory tracts.
endoplasmic reticulum (ER) (L. *endoplasmic,* within the cytoplasm + *reticulum,* little net) An extensive network of membrane compartments within a eukaryotic cell; attached ribosomes synthesize proteins to be exported.
endoskeleton (Gr. *endon,* within + *skeletos,* hard) In vertebrates, an internal scaffold of bone or cartilage to which muscles are attached.
endosperm (Gr. *endon,* within + *sperma,* seed) A nutritive tissue characteristic of the seeds of angiosperms that develops from the union of a male nucleus and the polar nuclei of the embryo sac. The endosperm is either digested by the growing embryo or retained in the mature seed to nourish the germinating seedling.
endosymbiotic (Gr. *endon,* within + *bios,* life) theory A theory that proposes how eukaryotic cells arose from large prokaryotic cells that engulfed smaller ones of a different species. The smaller cells were not consumed but continued to live and function within the larger host cell. Organelles that are believed to have entered larger cells in this way are mitochondria and chloroplasts.
endothermic The ability of animals to maintain an elevated body temperature using their metabolism.
energy The capacity to bring about change, to do work.
enhancer A site of regulatory protein binding on the DNA molecule distant from the promoter and start site for a gene's transcription.
entropy (Gr. *en,* in + *tropos,* change in manner) A measure of the disorder of a system. A measure of energy that has become so randomized and uniform in a system that the energy is no longer available to do work.
enzyme (Gr. *enzymos,* leavened; from *en,* in + *zyme,* leaven) A protein capable of speeding up specific chemical reactions by lowering the energy required to activate or start the reaction but that remains unaltered in the process.
epidermis (Gr. *epi,* on or over + *derma,* skin) The outermost layer of cells. In vertebrates, the nonvascular external layer of skin of ectodermal origin; in invertebrates, a single layer of ectodermal epithelium; in plants, the flattened, skinlike outer layer of cells.
epistasis (Gr. *epistasis,* a standing still) An interaction between the products of two genes in which one modifies the phenotypic expression produced by the other.
epithelium (Gr. *epi,* on + *thele,* nipple) A thin layer of cells forming a tissue that covers the internal and external surfaces of the body. Simple epithelium consists of the membranes that line the lungs and major body cavities and that are a single cell layer thick. Stratified epithelium (the skin or epidermis) is composed of more complex epithelial cells that are several cell layers thick.
erythrocyte (Gr. *erythros,* red + *kytos,* hollow vessel) A red blood cell, the carrier of hemoglobin. Erythrocytes act as the transporters of oxygen in the vertebrate body. During the process of their maturation in mammals, they lose their nuclei and mitochondria, and their endoplasmic reticulum is reabsorbed.
estrus (L. *oestrus,* frenzy) The period of maximum female sexual receptivity. Associated with ovulation of the egg. Being "in heat."
estuary (L. *aestus,* tide) A partly enclosed body of water, such as those that often form at river mouths and in coastal bays, where the salinity is intermediate between that of saltwater and freshwater.
ethology (Gr. *ethos,* habit or custom + *logos,* discourse) The study of patterns of animal behavior in nature.
euchromatin (Gr. *eu,* true + *chroma,* color) Chromatin that is extended except during cell division, from which RNA is transcribed.
eukaryote (Gr. *eu,* true + *karyon,* kernel) A cell that possesses membrane-bounded organelles, most notably a cell nucleus, and chromosomes whose DNA is associated with proteins; an organism composed of such cells. The appearance of eukaryotes marks a major event in the evolution of life, as all organisms on earth other than bacteria and archaea are eukaryotes.
eumetazoan (Gr. *eu,* true + *meta,* with + *zoion,* animal) A "true animal." An animal with a definite shape and symmetry and nearly always distinct tissues.
eutrophic (Gr. *eutrophos,* thriving) Refers to a lake in which an abundant supply of minerals and organic matter exists.
evaporation The escape of water molecules from the liquid to the gas phase at the surface of a body of water.
evolution (L. *evolvere,* to unfold) Genetic change in a population of organisms over time (generations). Darwin proposed that natural selection was the mechanism of evolution.
exergonic (L. *ex,* out + Gr. *ergon,* work) Any reaction that produces products that contain less free energy than that possessed by the original reactants and that tends to proceed spontaneously.
exocytosis (Gr. *ex,* out of + *kytos,* cell) The extrusion of material from a cell by discharging it from vesicles at the cell surface. The reverse of endocytosis.
exoskeleton (Gr. *exo,* outside + *skeletos,* hard) An external hard shell that encases a body. In arthropods, comprised mainly of chitin.
experiment The test of a hypothesis. An experiment that tests one or more alternative hypotheses and those that are demonstrated to be inconsistent with experimental observation are rejected.

F

facilitated diffusion The transport of molecules across a membrane by a carrier protein in the direction of lowest concentration.
family A taxonomic group ranking below an order and above a genus.
feedback inhibition A regulatory mechanism in which a biochemical pathway is regulated by the amount of the product that the pathway produces.
fermentation (L. *fermentum,* ferment) A catabolic process in which the final electron acceptor is an organic molecule.
fertilization (L. *ferre,* to bear) The union of male and female gametes to form a zygote.
fitness The genetic contribution of an individual to succeeding generations, relative to the contributions of other individuals in the population.
flagellum, *pl.* flagella (L. *flagellum,* whip) A fine, long, threadlike organelle protruding from the surface of a cell. In bacteria, a single protein fiber capable of rotary motion that propels the cell through the water. In eukaryotes, an array of microtubules with a characteristic internal 9 + 2 microtubule structure that is capable of vibratory but not rotary motion. Used in locomotion and feeding. Common in protists and motile gametes. A cilium is a short flagellum.
food web The food relationships within a community. A diagram of who eats whom.
founder effect The effect by which rare alleles and combinations of alleles may be enhanced in new populations.
frequency In statistics, defined as the proportion of individuals in a certain category, relative to the total number of individuals being considered.
fruit In angiosperms, a mature, ripened ovary (or group of ovaries) containing the seeds.

G

gamete (Gr. wife) A haploid reproductive cell. Upon fertilization, its nucleus fuses with that of another gamete of the opposite sex. The resulting diploid cell (zygote) may develop into a new diploid individual, or in some protists and fungi, may undergo meiosis to form haploid somatic cells.
gametophyte (Gr. *gamete,* wife + *phyton,* plant) In plants, the haploid (n), gamete-producing generation, which alternates with the diploid ($2n$) sporophyte.

ganglion, ***pl.*** **ganglia (Gr. a swelling)** A group of nerve cells forming a nerve center in the peripheral nervous system.
gastrulation The inward movement of certain cell groups from the surface of the blastula.
gene (Gr. ***genos,*** **birth, race)** The basic unit of heredity. A sequence of DNA nucleotides on a chromosome that encodes a polypeptide or RNA molecule and so determines the nature of an individual's inherited traits.
gene expression The process in which an RNA copy of each active gene is made, and the RNA copy directs the sequential assembly of a chain of amino acids at a ribosome.
gene frequency The frequency with which individuals in a population possess a particular gene. Often confused with allele frequency.
genetic code The "language" of the genes. The mRNA codons specific for the 20 common amino acids constitute the genetic code.
genetic drift Random fluctuations in allele frequencies in a small population over time.
genetic map A diagram showing the relative positions of genes.
genetics (Gr. ***genos,*** **birth, race)** The study of the way in which an individual's traits are transmitted from one generation to the next.
genome (Gr. ***genos,*** **offspring + L.** ***oma,*** **abstract group)** The genetic information of an organism.
genomics The study of genomes as opposed to individual genes.
genotype (Gr. ***genos,*** **offspring +** ***typos,*** **form)** The total set of genes present in the cells of an organism. Also used to refer to the set of alleles at a single gene locus.
genus, ***pl.*** **genera (L. race)** A taxonomic group that ranks below a family and above a species.
germination (L. ***germinare,*** **to sprout)** The resumption of growth and development by a spore or seed.
gland (L. ***glandis,*** **acorn)** Any of several organs in the body, such as exocrine or endocrine, that secrete substances for use in the body. Glands are composed of epithelial tissue.
glomerulus (L. a little ball) A network of capillaries in a vertebrate kidney, whose walls act as a filtration device.
glycolysis (Gr. ***glykys,*** **sweet +** ***lyein,*** **to loosen)** The anaerobic breakdown of glucose; this enzyme-catalyzed process yields two molecules of pyruvate with a net of two molecules of ATP.
golgi complex Flattened stacks of membrane compartments that collect, package, and distribute molecules made in the endoplasmic reticulum.
gravitropism (L. ***gravis,*** **heavy +** ***tropes,*** **turning)** The response of a plant to gravity, which generally causes shoots to grow up and roots to grow down.
greenhouse effect The process in which carbon dioxide and certain other gases, such as methane, that occur in the earth's atmosphere transmit radiant energy from the sun but trap the longer wavelengths of infrared light, or heat, and prevent them from radiating into space.
guard cells Pairs of specialized epidermal cells that surround a stoma. When the guard cells are turgid, the stoma is open; when they are flaccid, it is closed.
gymnosperm (Gr. ***gymnos,*** **naked +** ***sperma,*** **seed)** A seed plant with seeds not enclosed in an ovary. The conifers are the most familiar group.

H

habitat (L. ***habitare,*** **to inhabit)** The place where individuals of a species live.
half-life The length of time it takes for half of a radioactive substance to decay.
haploid (Gr. ***haploos,*** **single +** ***eidos,*** **form)** The gametes of a cell or an individual with only one set of chromosomes.
Hardy-Weinberg equilibrium After G. H. Hardy, English mathematician, and G. Weinberg, German physician. A mathematical description of the fact that the relative frequencies of two or more alleles in a population do not change because of Mendelian segregation. Allele and genotype frequencies remain constant in a random-mating population in the absence of inbreeding, selection, or other evolutionary forces. Usually stated as: If the frequency of allele A is p and the frequency of allele a is q, then the genotype frequencies after one generation of random mating will always be $(p + q)^2 = p^2 + 2pq + q^2$.
Haversian canal After Clopton Havers, English anatomist. Narrow channels that run parallel to the length of a bone and contain blood vessels and nerve cells.
helper T cell A class of white blood cells that initiates both the cell-mediated immune response and the humoral immune response; helper T cells are the targets of the AIDS virus (HIV).
hemoglobin (Gr. ***haima,*** **blood + L.** ***globus,*** **a ball)** A globular protein in vertebrate red blood cells and in the plasma of many invertebrates that carries oxygen and carbon dioxide.
herbivore (L. ***herba,*** **grass +** ***vorare,*** **to devour)** Any organism that eats only plants.
heredity (L. ***heredis,*** **heir)** The transmission of characteristics from parent to offspring.
heterochromatin (Gr. ***heteros,*** **different +** ***chroma,*** **color)** That portion of a eukaryotic chromosome that remains permanently condensed and therefore is not transcribed into RNA. Most centromere regions are heterochromatic.
heterokaryon (Gr. ***heteros,*** **other +** ***karyon,*** **kernel)** A fungal hypha that has two or more genetically distinct types of nuclei.
heterotroph (Gr. ***heteros,*** **other +** ***trophos,*** **feeder)** An organism that does not have the ability to produce its own food. *See also* autotroph.
heterozygote (Gr. ***heteros,*** **other +** ***zygotos,*** **a pair)** A diploid individual carrying two different alleles of a gene on its two homologous chromosomes.
hierarchical (Gr. ***hieros,*** **sacred +** ***archos,*** **leader)** Refers to a system of classification in which successively smaller units of classification are included within one another.
histone (Gr. ***histos,*** **tissue)** A complex of small, very basic polypeptides rich in the amino acids arginine and lysine. A basic part of chromosomes, histones form the core around which DNA is wrapped.
homeostasis (Gr. ***homeos,*** **similar +** ***stasis,*** **standing)** The maintaining of a relatively stable internal physiological environment in an organism or steady-state equilibrium in a population or ecosystem.
homeotherm (Gr. ***homeo,*** **similar +** ***therme,*** **heat)** An organism, such as a bird or mammal, capable of maintaining a stable body temperature.
hominid (L. ***homo,*** **man)** Human beings and their direct ancestors. A member of the family Hominidae. *Homo sapiens* is the only living member.
homologous chromosome (Gr. ***homologia,*** **agreement)** One of the two nearly identical versions of each chromosome. Chromosomes that associate in pairs in the first stage of meiosis. In diploid cells, one chromosome of a pair that carries equivalent genes.
homology (Gr. ***homologia,*** **agreement)** A condition in which the similarity between two structures or functions is indicative of a common evolutionary origin.
homozygote (Gr. ***homos,*** **same or similar +** ***zygotos,*** **a pair)** A diploid individual whose two copies of a gene are the same. An individual carrying identical alleles on both homologous chromosomes is said to be homozygous for that gene.
hormone (Gr. ***hormaein,*** **to excite)** A chemical messenger, often a steroid or peptide, produced in a small quantity in one part of an organism and then transported to another part of the organism, where it brings about a physiological response.
hybrid (L. ***hybrida,*** **the offspring of a tame sow and a wild boar)** A plant or animal that results from the crossing of dissimilar parents.
hybridization The mating of unlike parents of different taxa.
hydrogen bond A molecular force formed by the attraction of the partial positive charge of one hydrogen atom of a water molecule with the partial negative charge of the oxygen atom of another.
hydrolysis reaction (Gr. ***hydro,*** **water +** ***lyse,*** **break)** The process of tearing down a polymer by adding a molecule of water. A hydrogen is attached to one subunit and a hydroxyl to the other, which breaks the covalent bond. Essentially the reverse of a dehydration reaction.
hydrophilic (Gr. ***hydro,*** **water +** ***philic,*** **loving)** Describes polar molecules, which form hydrogen bonds with water and therefore are soluble in water.

hydrophobic (Gr. *hydro,* water + *phobos,* hating) Describes nonpolar molecules, which do not form hydrogen bonds with water and therefore are not soluble in water.
hydroskeleton (Gr. *hydro,* water + *skeletos,* hard) The skeleton of most soft-bodied invertebrates that have neither an internal nor an external skeleton. They use the relative incompressibility of the water within their bodies as a kind of skeleton.
hypertonic (Gr. *hyper,* above + *tonos,* tension) A cell that contains a higher concentration of solutes than its surrounding solution.
hypha, *pl.* hyphae (Gr. *hyphe,* web) A filament of a fungus. A mass of hyphae comprises a mycelium.
hypothalamus (Gr. *hypo,* under + *thalamos,* inner room) The region of the brain under the thalamus that controls temperature, hunger, and thirst and that produces hormones that influence the pituitary gland.
hypothesis (Gr. *hypo,* under + *tithenai,* to put) A proposal that might be true. No hypothesis is ever proven correct. All hypotheses are provisional—proposals that are retained for the time being as useful but that may be rejected in the future if found to be inconsistent with new information. A hypothesis that stands the test of time—often tested and never rejected—is called a theory.
hypotonic (Gr. *hypo,* under + *tonos,* tension) A solution surrounding a cell that has a lower concentration of solutes than does the cell.

I

inbreeding The breeding of genetically related plants or animals. In plants, inbreeding results from self-pollination. In animals, inbreeding results from matings between relatives. Inbreeding tends to increase homozygosity.
incomplete dominance The ability of two alleles to produce a heterozygous phenotype that is different from either homozygous phenotype.
independent assortment Mendel's second law: The principle that segregation of alternative alleles at one locus into gametes is independent of the segregation of alleles at other loci. Only true for gene loci located on different chromosomes or those so far apart on one chromosome that crossing over is very frequent between the loci.
industrial melanism (Gr. *melas,* black) The evolutionary process in which a population of initially light-colored organisms becomes a population of dark organisms as a result of natural selection.
inflammatory response (L. *inflammare,* to flame) A generalized nonspecific response to infection that acts to clear an infected area of infecting microbes and dead tissue cells so that tissue repair can begin.
integument (L. *integumentum,* covering) The natural outer covering layers of an animal. Develops from the ectoderm.
interneuron A nerve cell found only in the CNS that acts as a functional link between sensory neurons and motor neurons. Also called association neuron.
internode The region of a plant stem between nodes where stems and leaves attach.
interoception (L. *interus,* inner + Eng. *[re]ceptive*) The sensing of information that relates to the body itself, its internal condition, and its position.
interphase That portion of the cell cycle preceding mitosis. It includes the G_1 phase, when cells grow, the S phase, when a replica of the genome is synthesized, and a G_2 phase, when preparations are made for genomic separation.
intron (L. *intra,* within) A segment of DNA transcribed into mRNA but removed before translation. These untranslated regions make up the bulk of most eukaryotic genes.
ion An atom in which the number of electrons does not equal the number of protons. An ion carries an electrical charge.
ionic bond A chemical bond formed between ions as a result of the attraction of opposite electrical charges.
ionizing radiation High-energy radiation, such as X rays and gamma rays.
isolating mechanisms Mechanisms that prevent genetic exchange between individuals of different populations or species.
isotonic (Gr. *isos,* equal + *tonos,* tension) A cell with the same concentration of solutes as its environment.
isotope (Gr. *isos,* equal + *topos,* place) An atom that has the same number of protons but different numbers of neutrons.

J

joint The part of a vertebrate where one bone meets and moves on another.

K

karyotype (Gr. *karyon,* kernel + *typos,* stamp or print) The particular array of chromosomes that an individual possesses.
kinetic energy The energy of motion.
kinetochore (Gr. *kinetikos,* putting in motion + *choros,* chorus) A disk of protein bound to the centromere to which microtubules attach during cell division, linking chromatids to the spindle.
kingdom The chief taxonomic category. This book recognizes six kingdoms: Archaea, Bacteria, Protista, Fungi, Animalia, and Plantae.

L

lamella, *pl.* lamellae (L. a little plate) A thin, platelike structure. In chloroplasts, a layer of chlorophyll-containing membranes. In bivalve mollusks, one of the two plates forming a gill. In vertebrates, one of the thin layers of bone laid concentrically around the Haversian canals.
ligament (L. *ligare,* to bind) A band or sheet of connective tissue that links bone to bone.
linkage The patterns of assortment of genes that are located on the same chromosome. Important because if the genes are located relatively far apart, crossing over is more likely to occur between them than if they are close together.
lipid (Gr. *lipos,* fat) A loosely defined group of molecules that are insoluble in water but soluble in oil. Oils such as olive, corn, and coconut are lipids, as well as waxes, such as beeswax and earwax.
lipid bilayer The basic foundation of all biological membranes. In such a layer, the nonpolar tails of phospholipid molecules point inward, forming a nonpolar zone in the interior of the bilayers. Lipid bilayers are selectively permeable and do not permit the diffusion of water-soluble molecules into the cell.
littoral (L. *litus,* shore) Referring to the shoreline zone of a lake or pond or the ocean that is exposed to the air whenever water recedes.
locus, *pl.* loci (L. place) The position on a chromosome where a gene is located.
loop of Henle After F. G. J. Henle, German anatomist. A hairpin loop formed by a urine-conveying tubule when it enters the inner layer of the kidney and then turns around to pass up again into the outer layer of the kidney.
lymph (L. *lympha,* clear water) In animals, a colorless fluid derived from blood by filtration through capillary walls in the tissues.
lymphatic system An open circulatory system composed of a network of vessels that function to collect the water within blood plasma forced out during passage through the capillaries and to return it to the bloodstream. The lymphatic system also returns proteins to the circulation, transports fats absorbed from the intestine, and carries bacteria and dead blood cells to the lymph nodes and spleen for destruction.
lymphocyte (Gr. *lympha,* water + Gr. *kytos,* hollow vessel) A white blood cell. A cell of the immune system that either synthesizes antibodies (B cells) or attacks virus-infected cells (T cells).
lyse (Gr. *lysis,* loosening) To disintegrate a cell by rupturing its plasma membrane.

M

macromolecule (Gr. *makros,* large + L. *moliculus,* a little mass) An extremely large molecule. Refers specifically to carbohydrates, lipids, proteins, and nucleic acids.
macrophage (Gr. *makros,* large + *-phage,* eat) A phagocytic cell of the immune system able to engulf and digest invading bacteria, fungi, and other microorganisms, as well as cellular debris.
marrow The soft tissue that fills the cavities of most bones and is the source of red blood cells.
mass flow The overall process by which materials move in the phloem of plants.
mass number The mass number of an atom consists of the combined mass of all of its protons and neutrons.

meiosis (Gr. *meioun,* to make smaller) A special form of nuclear division that precedes gamete formation in sexually reproducing eukaryotes. It results in four haploid daughter cells.
Mendelian ratio After Gregor Mendel, Austrian monk. Refers to the characteristic 3:1 segregation ratio that Mendel observed, in which pairs of alternative traits are expressed in the F_2 generation in the ratio of three-fourths dominant to one-fourth recessive.
menstruation (L. *mens,* month) Periodic sloughing off of the blood-enriched lining of the uterus when pregnancy does not occur.
meristem (Gr. *merizein,* to divide) In plants, a zone of unspecialized cells whose only function is to divide.
mesoderm (Gr. *mesos,* middle + *derma,* skin) One of the three embryonic germ layers that form in the gastrula. Gives rise to muscle, bone, and other connective tissue; the peritoneum; the circulatory system; and most of the excretory and reproductive systems.
mesophyll (Gr. *mesos,* middle + *phyllon,* leaf) The photosynthetic parenchyma of a leaf, located within the epidermis. The vascular strands (veins) run through the mesophyll.
metabolism (Gr. *metabole,* change) The process by which all living things assimilate energy and use it to grow.
metamorphosis (Gr. *meta,* after + *morphe,* form + *osis,* state of) Process in which form changes markedly during postembryonic development—for example, tadpole to frog or larval insect to adult.
metaphase (Gr. *meta,* middle + *phasis,* form) The stage of mitosis characterized by the alignment of the chromosomes on a plane in the center of the cell.
metastasis, *pl.* metastases (Gr. to place in another way) The spread of cancerous cells to other parts of the body, forming new tumors at distant sites.
microevolution (Gr. *mikros,* small + L. *evolvere,* to unfold) Refers to the evolutionary process itself. Evolution within a species. Also called adaptation.
microtubule (Gr. *mikros,* small + L. *tubulus,* little pipe) In eukaryotic cells, a long, hollow cylinder about 25 nanometers in diameter and composed of the protein tubulin. Microtubules influence cell shape, move the chromosomes in cell division, and provide the functional internal structure of cilia and flagella.
mimicry (Gr. *mimos,* mime) The resemblance in form, color, or behavior of certain organisms (mimics) to other more powerful or more protected ones (models), which results in the mimics being protected in some way.
mitochondrion, *pl.* mitochondria (Gr. *mitos,* thread + *chondrion,* small grain) A tubular or sausage-shaped organelle 1 to 3 micrometers long. Bounded by two membranes, mitochondria closely resemble the aerobic bacteria from which they were originally derived. As chemical furnaces of the cell, they carry out its oxidative metabolism.
mitosis (Gr. *mitos,* thread) The M phase of cell division in which the microtubular apparatus is assembled, binds to the chromosomes, and moves them apart. This phase is the essential step in the separation of the two daughter cell genomes.
mole (L. *moles,* mass) The atomic weight of a substance, expressed in grams. One mole is defined as the mass of 6.0222×10^{23} atoms.
molecule (L. *moliculus,* a small mass) The smallest unit of a compound that displays the properties of that compound.
monocot Short for monocotyledon; flowering plant in which the embryos have only one cotyledon, the flower parts are often in threes, and the leaves typically are parallel-veined.
monomers (Gr. *mono,* single + *meris,* part) Simple molecules that can join together to form polymers.
monosaccharide (Gr. *monos,* one + *sakcharon,* sugar) A simple sugar.
morphogenesis (Gr. *morphe,* form + *genesis,* origin) The formation of shape. The growth and differentiation of cells and tissues during development.
motor endplate The point where a neuron attaches to a muscle. A neuromuscular synapse.
multicellularity A condition in which the activities of the individual cells are coordinated and the cells themselves are in contact. A property of eukaryotes alone and one of their major characteristics.
muscle (L. *musculus,* mouse) The tissue in the body of humans and animals that can be contracted and relaxed to make the body move.
muscle cell A long, cylindrical, multinucleated cell that contains numerous myofibrils and is capable of contraction when stimulated.
muscle spindle A sensory organ that is attached to a muscle and sensitive to stretching.
mutagen (L. *mutare,* to change) A chemical capable of damaging DNA.
mutation (L. *mutare,* to change) A change in a cell's genetic message.
mutualism (L. *mutuus,* lent, borrowed) A symbiotic relationship in which both participating species benefit.
mycelium, *pl.* mycelia (Gr. *mykes,* fungus) In fungi, a mass of hyphae.
mycology (Gr. *mykes,* fungus) The study of fungi. A person who studies fungi is called a mycologist.
mycorrhiza, *pl.* mycorrhizae (Gr. *mykes,* fungus + *rhiza,* root) A symbiotic association between fungi and plant roots.
myofibril (Gr. *myos,* muscle + L. *fibrilla,* little fiber) An elongated structure in a muscle fiber, composed of myosin and actin.
myosin (Gr. *myos,* muscle + *in,* belonging to) One of two protein components of myofilaments. (The other is actin.)

N

natural selection The differential reproduction of genotypes caused by factors in the environment. Leads to evolutionary change.
nematocyst (Gr. *nema,* thread + *kystos,* bladder) A coiled, threadlike stinging structure of cnidarians that is discharged to capture prey and for defense.
nephron (Gr. *nephros,* kidney) The functional unit of the vertebrate kidney. A human kidney has more than 1 million nephrons that filter waste matter from the blood. Each nephron consists of a Bowman's capsule, glomerulus, and tubule.
nerve A bundle of axons with accompanying supportive cells, held together by connective tissue.
nerve impulse A rapid, transient, self-propagating reversal in electrical potential that travels along the membrane of a neuron.
neuromodulator A chemical transmitter that mediates effects that are slow and longer lasting and that typically involve second messengers within the cell.
neuromuscular junction The structure formed when the tips of axons contact (innervate) a muscle fiber.
neuron (Gr. nerve) A nerve cell specialized for signal transmission.
neurotransmitter (Gr. *neuron,* nerve + L. *trans,* across + *mitere,* to send) A chemical released at an axon tip that travels across the synapse and binds a specific receptor protein in the membrane on the far side.
neurulation (Gr. *neuron,* nerve) The elaboration of a notochord and a dorsal nerve cord that marks the evolution of the chordates.
neutron (L. *neuter,* neither) A subatomic particle located within the nucleus of an atom. Similar to a proton in mass, but as its name implies, a neutron is neutral and possesses no charge.
neutrophil An abundant type of white blood cell capable of engulfing microorganisms and other foreign particles.
niche (L. *nidus,* nest) The role an organism plays in the environment; realized niche is the niche that an organism occupies under natural circumstances; fundamental niche is the niche an organism would occupy if competitors were not present.
nitrogen fixation The incorporation of atmospheric nitrogen into nitrogen compounds, a process that can be carried out only by certain microorganisms.
nocturnal (L. *nocturnus,* night) Active primarily at night.
node (L. *nodus,* knot) The place on the stem where a leaf is formed.
node of Ranvier After L. A. Ranvier, French histologist. A gap formed at the point where two Schwann cells meet and where the axon is in direct contact with the surrounding intercellular fluid.
nondisjunction The failure of homologous chromosomes to separate in meiosis I. The cause of Down syndrome.
nonrandom mating A phenomenon in which individuals with certain genotypes sometimes mate with one another more commonly than would be expected on a random basis.

notochord (Gr. *noto,* back + L. *chorda,* cord) In chordates, a dorsal rod of cartilage that forms between the nerve cord and the developing gut in the early embryo.
nucleic acid A nucleotide polymer. A long chain of nucleotides. Chief types are deoxyribonucleic acid (DNA), which is double-stranded, and ribonucleic acid (RNA), which is typically single-stranded.
nucleosome (L. *nucleus,* kernel + *soma,* body) The basic packaging unit of eukaryotic chromosomes, in which the DNA molecule is wound around a ball of histone proteins. Chromatin is composed of long strings of nucleosomes, like beads on a string.
nucleotide A single unit of nucleic acid, composed of a phosphate, a five-carbon sugar (either ribose or deoxyribose), and a purine or a pyrimidine.
nucleolus A region inside the nucleus where rRNA and ribosomes are produced.
nucleus (L. *a kernel,* dim. Fr. *nux,* nut) A spherical organelle (structure) characteristic of eukaryotic cells. The repository of the genetic information that directs all activities of a living cell. In atoms, the central core, containing positively charged protons and (in all but hydrogen) electrically neutral neutrons.

O

oocyte (Gr. *oion,* egg + *kytos,* vessel) A cell in the outer layer of the ovary that gives rise to an ovum. A primary oocyte is any of the 2 million oocytes a female is born with, all of which have begun the first meiotic division.
operon (L. *operis,* work) A cluster of functionally related genes transcribed onto a single mRNA molecule. A common mode of gene regulation in prokaryotes; it is rare in eukaryotes other than fungi.
order A taxonomic category ranking below a class and above a family.
organ (L. *organon,* tool) A complex body structure composed of several different kinds of tissue grouped together in a structural and functional unit.
organelle (Gr. *organella,* little tool) A specialized compartment of a cell. Mitochondria are organelles.
organism Any individual living creature, either unicellular or multicellular.
organ system A group of organs that function together to carry out the principal activities of the body.
osmoconformer An animal that maintains the osmotic concentration of its body fluids at about the same level as that of the medium in which it is living.
osmoregulation The maintenance of a constant internal solute concentration by an organism, regardless of the environment in which it lives.
osmosis (Gr. *osmos,* act of pushing, thrust) The diffusion of water across a membrane that permits the free passage of water but not that of one or more solutes. Water moves from an area of low solute concentration to an area with higher solute concentration.
osmotic pressure The increase of hydrostatic water pressure within a cell as a result of water molecules that continue to diffuse inward toward the area of lower water concentration (the water concentration is lower inside than outside the cell because of the dissolved solutes in the cell).
osteoblast (Gr. *osteon,* bone + *blastos,* bud) A bone-forming cell.
osteocyte (Gr. *osteon,* bone + *kytos,* hollow vessel) A mature osteoblast.
outcross A term used to describe species that interbreed with individuals other than those like themselves.
oviparous (L. *ovum,* egg + *parere,* to bring forth) Refers to reproduction in which the eggs are developed after leaving the body of the mother, as in reptiles.
ovulation The successful development and release of an egg by the ovary.
ovule (L. *ovulum,* a little egg) A structure in a seed plant that becomes a seed when mature.
ovum, *pl.* ova (L. egg) A mature egg cell. A female gamete.
oxidation (Fr. *oxider,* to oxidize) The loss of an electron during a chemical reaction from one atom to another. Occurs simultaneously with reduction. Is the second stage of the 10 reactions of glycolysis.
oxidative metabolism A collective term for metabolic reactions requiring oxygen.
oxidative respiration Respiration in which the final electron acceptor is molecular oxygen.

P

parasitism (Gr. *para,* beside + *sitos,* food) A symbiotic relationship in which one organism benefits and the other is harmed.
parthenogenesis (Gr. *parthenos,* virgin + Eng. *genesis,* beginning) The development of an adult from an unfertilized egg. A common form of reproduction in insects.
partial pressures (P) The components of each individual gas—such as nitrogen, oxygen, and carbon dioxide—that together constitute the total air pressure.
pathogen (Gr. *pathos,* suffering + Eng. *genesis,* beginning) A disease-causing organism.
pedigree (L. *pes,* foot + *grus,* crane) A family tree. The patterns of inheritance observed in family histories. Used to determine the mode of inheritance of a particular trait.
peptide (Gr. *peptein,* to soften, digest) Two or more amino acids linked by peptide bonds.
peptide bond A covalent bond linking two amino acids. Formed when the positive (amino, or NH_2) group at one end and a negative (carboxyl, or COOH) group at the other end undergo a chemical reaction and lose a molecule of water.
peristalsis (Gr. *peri,* around + *stellein,* to wrap) The rhythmic sequences of waves of muscular contraction in the walls of a tube.
pH Refers to the concentration of H^+ ions in a solution. The numerical value of the pH is the negative of the exponent of the molar concentration. Low pH values indicate high concentrations of H^+ ions (acids), and high pH values indicate low concentrations (bases).
phagocyte (Gr. *phagein,* to eat + *kytos,* hollow vessel) A cell that kills invading cells by engulfing them. Includes neutrophils and macrophages.
phagocytosis (Gr. *phagein,* to eat + *kytos,* hollow vessel) A form of endocytosis in which cells engulf organisms or fragments of organisms.
phenotype (Gr. *phainein,* to show + *typos,* stamp or print) The realized expression of the genotype. The observable expression of a trait (affecting an individual's structure, physiology, or behavior) that results from the biological activity of proteins or RNA molecules transcribed from the DNA.
pheromone (Gr. *pherein,* to carry + [hor] mone) A chemical signal emitted by certain animals as a means of communication.
phloem (Gr. *phloos,* bark) In vascular plants, a food-conducting tissue basically composed of sieve elements, various kinds of parenchyma cells, fibers, and sclereids.
phosphodiester bond The bond that results from the formation of a nucleic acid chain in which individual sugars are linked together in a line by the phosphate groups. The phosphate group of one sugar binds to the hydroxyl group of another, forming an—O—P—O bond.
photon (Gr. *photos,* light) The unit of light energy.
photoperiodism (Gr. *photos,* light + *periodos,* a period) A mechanism that organisms use to measure seasonal changes in relative day and night length.
photorespiration A process in which carbon dioxide is released without the production of ATP or NADPH. Because it produces neither ATP nor NADPH, photorespiration acts to undo the work of photosynthesis.
photosynthesis (Gr. *photos,* light + *-syn,* together + *tithenai,* to place) The process by which plants, algae, and some bacteria use the energy of sunlight to create from carbon dioxide (CO_2) and water (H_2O) the more complicated molecules that make up living organisms.
phototropism (Gr. *photos,* light + *trope,* turning to light) A plant's growth response to a unidirectional light source.
phylogeny (Gr. *phylon,* race, tribe) The evolutionary relationships among any group of organisms.
phylum, *pl.* phyla (Gr. *phylon,* race, tribe) A major taxonomic category, ranking above a class.
physiology (Gr. *physis,* nature + *logos,* a discourse) The study of the function of cells, tissues, and organs.
pigment (L. *pigmentum,* paint) A molecule that absorbs light.
pili (pilus) Short flagella that occur on the cell surface of some prokaryotes.

pinocytosis (Gr. *pinein,* to drink + *kytos,* cell) A form of endocytosis in which the material brought into the cell is a liquid containing dissolved molecules.
pistil (L. *pistillum,* pestle) Central organ of flowers, typically consisting of ovary, style, and stigma; a pistil may consist of one or more fused carpels and is more technically and better known as the gynoecium.
plankton (Gr. *planktos,* wandering) The small organisms that float or drift in water, especially at or near the surface.
plasma (Gr. form) The fluid of vertebrate blood. Contains dissolved salts, metabolic wastes, hormones, and a variety of proteins, including antibodies and albumin. Blood minus the blood cells.
plasma membrane A lipid bilayer with embedded proteins that control the cell's permeability to water and dissolved substances.
plasmid (Gr. *plasma,* a form or something molded) A small fragment of DNA that replicates independently of the bacterial chromosome.
platelet (Gr. dim of *plattus,* flat) In mammals, a fragment of a white blood cell that circulates in the blood and functions in the formation of blood clots at sites of injury.
pleiotropy (Gr. *pleros,* more + *trope,* a turning) A gene that produces more than one phenotypic effect.
polarization The charge difference of a neuron so that the interior of the cell is negative with respect to the exterior.
polar molecule A molecule with positively and negatively charged ends. One portion of a polar molecule attracts electrons more strongly than another portion, with the result that the molecule has electron-rich (–) and electron-poor (+) regions, giving it magnetlike positive and negative poles. Water is one of the most polar molecules known.
pollen (L. fine dust) A fine, yellowish powder consisting of grains or microspores, each of which contains a mature or immature male gametophyte. In flowering plants, pollen is released from the anthers of flowers and fertilizes the pistils.
pollen tube A tube that grows from a pollen grain. Male reproductive cells move through the pollen tube into the ovule.
pollination The transfer of pollen from the anthers to the stigmas of flowers for fertilization, as by insects or the wind.
polygyny (Gr. *poly,* many + *gyne,* woman, wife) A mating choice in which a male mates with more than one female.
polymer (Gr. *polus,* many + *meris,* part) A large molecule formed of long chains of similar molecules called subunits.
polymerase chain reaction (PCR) A process by which DNA polymerase is used to copy a sequence of DNA repeatedly, making millions of copies of the same DNA.
polymorphism (Gr. *polys,* many + *morphe,* form) The presence in a population of more than one allele of a gene at a frequency greater than that of newly arising mutations.
polynomial system (Gr. *polys,* many + [bi] nomial) Before Linnaeus, naming a genus by use of a cumbersome string of Latin words and phrases.
polyp A cylindrical, pipe-shaped cnidarian usually attached to a rock with the mouth facing away from the rock on which it is growing. Coral is made up of polyps.
polypeptide (Gr. *polys,* many + *peptein,* to digest) A general term for a long chain of amino acids linked end to end by peptide bonds. A protein is a long, complex polypeptide.
polysaccharide (Gr. *polys,* many + *sakcharon,* sugar) A sugar polymer. A carbohydrate composed of many monosaccharide sugar subunits linked together in a long chain.
population (L. *populus,* the people) Any group of individuals of a single species, occupying a given area at the same time.
posterior (L. *post,* after) Situated behind or farther back.
potential difference A difference in electrical charge on two sides of a membrane caused by an unequal distribution of ions.
potential energy Energy with the potential to do work. Stored energy.
predation (L. *praeda,* prey) The eating of other organisms. The one doing the eating is called a predator, and the one being consumed is called the prey.
primary growth In vascular plants, growth originating in the apical meristems of shoots and roots, as contrasted with secondary growth; results in an increase in length.
primary plant body The part of a plant that arises from the apical meristems.
primary producers Photosynthetic organisms, including plants, algae, and photosynthetic bacteria.
primary structure of a protein The sequence of amino acids that makes up a particular polypeptide chain.
primordium, *pl.* primordia (L. *primus,* first + *ordiri,* begin) The first cells in the earliest stages of the development of an organ or structure.
productivity The total amount of energy of an ecosystem fixed by photosynthesis per unit of time. Net productivity is productivity minus that which is expended by the metabolic activity of the organisms in the community.
prokaryote (Gr. *pro,* before + *karyon,* kernel) A simple organism that is small, single-celled, and has little evidence of internal structure.
promoter An RNA polymerase binding site. The nucleotide sequence at the end of a gene to which RNA polymerase attaches to initiate transcription of mRNA.
prophase (Gr. *pro,* before + *phasis,* form) The first stage of mitosis during which the chromosomes become more condensed, the nuclear envelope is reabsorbed, and a network of microtubules (called the spindle) forms between opposite poles of the cell.
protein (Gr. *proteios,* primary) A long chain of amino acids linked end to end by peptide bonds. Because the 20 amino acids that occur in proteins have side groups with very different chemical properties, the function and shape of a protein is critically affected by its particular sequence of amino acids.
protist (Gr. *protos,* first) A member of the kingdom Protista, which includes unicellular eukaryotic organisms and some multicellular lines derived from them.
proton A subatomic particle in the nucleus of an atom that carries a positive charge. The number of protons determines the chemical character of the atom because it dictates the number of electrons orbiting the nucleus and available for chemical activity.
protostome (Gr. *protos,* first + *stoma,* mouth) An animal in whose embryonic development the mouth forms at or near the blastopore. Also characterized by spiral cleavage.
protozoa (Gr. *protos,* first + *zoion,* animal) The traditional name given to heterotrophic protists.
pseudocoel (Gr. *pseudos,* false + *koiloma,* cavity) A body cavity similar to the coelom except that it forms between the mesoderm and endoderm.
punctuated equilibrium A hypothesis of the mechanism of evolutionary change that proposes that long periods of little or no change are punctuated by periods of rapid evolution.

Q

quaternary structure of a protein A term to describe the way multiple protein subunits are assembled into a whole.

R

radial symmetry (L. *radius,* a spoke of a wheel + Gr. *summetros,* symmetry) The regular arrangement of parts around a central axis so that any plane passing through the central axis divides the organism into halves that are approximate mirror images.
radioactivity The emission of nuclear particles and rays by unstable atoms as they decay into more stable forms. Measured in curies, with 1 curie equal to 37 billion disintegrations a second.
radula (L. scraper) A rasping, tonguelike organ characteristic of most mollusks.
recessive allele An allele whose phenotype effects are masked in heterozygotes by the presence of a dominant allele.
recombination The formation of new gene combinations. In bacteria, it is accomplished by the transfer of genes into cells, often in association with viruses. In eukaryotes, it is accomplished by reassortment of chromosomes during meiosis and by crossing over.
reducing power The use of light energy to extract hydrogen atoms from water.
reduction (L. *reductio,* a bringing back; originally, "bringing back" a metal from its oxide) The gain of an electron during a chemical reaction from one atom to another. Occurs simultaneously with oxidation.

reflex (L. *reflectere,* to bend back) An automatic consequence of a nerve stimulation. The motion that results from a nerve impulse passing through the system of neurons, eventually reaching the body muscles and causing them to contract.
refractory period The recovery period after membrane depolarization during which the membrane is unable to respond to additional stimulation.
renal (L. *renes,* kidneys) Pertaining to the kidney.
repression (L. *reprimere,* to press back, keep back) The process of blocking transcription by the placement of the regulatory protein between the polymerase and the gene, thus blocking movement of the polymerase to the gene.
repressor (L. *reprimere,* to press back, keep back) A protein that regulates transcription of mRNA from DNA by binding to the operator and so preventing RNA polymerase from attaching to the promoter.
resolving power The ability of a microscope to distinguish two points as separate.
respiration (L. *respirare,* to breathe) The utilization of oxygen. In terrestrial vertebrates, the inhalation of oxygen and the exhalation of carbon dioxide.
resting membrane potential The charge difference that exists across a neuron's membrane at rest (about 70 millivolts).
restriction endonuclease A special kind of enzyme that can recognize and cleave DNA molecules into fragments. One of the basic tools of genetic engineering.
restriction fragment-length polymorphism (RFLP) An associated genetic mutation marker detected because the mutation alters the length of DNA segments.
retrovirus (L. *retro,* turning back) A virus whose genetic material is RNA rather than DNA. When a retrovirus infects a cell, it makes a DNA copy of itself, which it can then insert into the cellular DNA as if it were a cellular gene.
ribonucleic acid (RNA) A nucleic acid that contains the sugar ribose and the pyrimidine uracil and that is used in protein production; includes mRNA, tRNA, rRNA, and siRNA.
ribose A five-carbon sugar.
ribosome A cell structure composed of protein and RNA that translates RNA copies of genes into protein.
RNA interference A type of gene silencing in which mRNA is prevented from being translated; small interfering RNAs (siRNAs) have been found to bind to mRNA and target its degradation or block its translation
RNA polymerase The enzyme that transcribes RNA from DNA.

S

saltatory conduction A very fast form of nerve impulse conduction in which the impulses leap from node to node over insulated portions.
sarcoma (Gr. *sarx,* flesh) A cancerous tumor that involves connective or hard tissue, such as muscle.
sarcomere (Gr. *sarx,* flesh + *meris,* part of) The fundamental unit of contraction in skeletal muscle. The repeating bands of actin and myosin that appear between two Z lines.
sarcoplasmic reticulum (Gr. *sarx,* flesh + *plassein,* to form, mold; L. *reticulum,* network) The endoplasmic reticulum of a muscle cell. A sleeve of membrane that wraps around each myofilament.
scientific creationism A view that the biblical account of the origin of the earth is literally true, that the earth is much younger than most scientists believe, and that all species of organisms were individually created just as they are today.
secondary growth In vascular plants, growth that results from the division of a cylinder of cells around the plant's periphery. Secondary growth causes a plant to grow in diameter.
secondary structure of a protein The folding and bending of a polypeptide chain, which is held in place by hydrogen bonds.
second messenger An intermediary compound that couples extracellular signals to intracellular processes and also amplifies a hormonal signal.
seed A structure that develops from the mature ovule of a seed plant. Contains an embryo and a food source surrounded by a protective coat.
selection The process by which some organisms leave more offspring than competing ones and their genetic traits tend to appear in greater proportions among members of succeeding generations than the traits of those individuals that leave fewer offspring.
self-fertilization The transfer of pollen from an anther to a stigma in the same flower or to another flower of the same plant.
sepal (L. *sepalum,* a covering) A member of the outermost whorl of a flowering plant. Collectively, the sepals constitute the calyx.
septum, *pl.* septa (L. *saeptum,* a fence) A partition or cross-wall, such as those that divide fungal hyphae into cells.
sex chromosomes In humans, the X and Y chromosomes, which are different in the two sexes and are involved in sex determination.
sex-linked characteristic A genetic characteristic that is determined by genes located on the sex chromosomes.
sexual reproduction Reproduction that involves the regular alternation between syngamy and meiosis. Its outstanding characteristic is that an individual offspring inherits genes from two parent individuals.
shoot In vascular plants, the aboveground parts, such as the stem and leaves.
sieve cell In the phloem (food-conducting tissue) of vascular plants, a long, slender sieve element with relatively unspecialized sieve areas and with tapering end walls that lack sieve plates. Found in all vascular plants except angiosperms, which have sieve-tube members.
soluble Refers to polar molecules that dissolve in water and are surrounded by a hydration shell.
solute The molecules dissolved in a solution. *See also* solution, solvent.
solution A mixture of molecules, such as sugars, amino acids, and ions, dissolved in water.
solvent The most common of the molecules in a solution. Usually a liquid, commonly water.
somatic cells (Gr. *soma,* body) All the diploid body cells of an animal that are not involved in gamete formation.
somite A segmented block of tissue on either side of a developing notochord.
species, *pl.* species (L. kind, sort) A level of taxonomic hierarchy; a species ranks next below a genus.
sperm (Gr. *sperma,* sperm, seed) A sperm cell. The male gamete.
spindle The mitotic assembly that carries out the separation of chromosomes during cell division. Composed of microtubules and assembled during prophase at the centrioles of the dividing cell.
spore (Gr. *spora,* seed) A haploid reproductive cell, usually unicellular, that is capable of developing into an adult without fusion with another cell. Spores result from meiosis, as do gametes, but gametes fuse immediately to produce a new diploid cell.
sporophyte (Gr. *spora,* seed + *phyton,* plant) The spore-producing, diploid ($2n$) phase in the life cycle of a plant having alternation of generations.
stabilizing selection A form of selection in which selection acts to eliminate both extremes from a range of phenotypes.
stamen (L. thread) The part of the flower that contains the pollen. Consists of a slender filament that supports the anther. A flower that produces only pollen is called staminate and is functionally male.
steroid (Gr. *stereos,* solid + L. *ol,* from oleum, oil) A kind of lipid. Many of the molecules that function as messengers and pass across cell membranes are steroids, such as the male and female sex hormones and cholesterol.
steroid hormone A hormone derived from cholesterol. Those that promote the development of the secondary sexual characteristics are steroids.
stigma (Gr. mark) A specialized area of the carpel of a flowering plant that receives the pollen.
stoma, *pl.* stomata (Gr. mouth) A specialized opening in the leaves of some plants that allows carbon dioxide to pass into the plant body and allows water vapor and oxygen to pass out of them.
stratum corneum The outer layer of the epidermis of the skin of the vertebrate body.
substrate (L. *substratus,* strewn under) A molecule on which an enzyme acts.
substrate-level phosphorylation The generation of ATP by coupling its synthesis to a strongly exergonic (energy-yielding) reaction.
succession In ecology, the slow, orderly progression of changes in community composition that takes place through time. Primary succession occurs in nature on bare substrates, over long periods of time. Secondary succession occurs when a climax community has been disturbed.

sugar Any monosaccharide or disaccharide.
surface tension A tautness of the surface of a liquid, caused by the cohesion of the liquid molecules. Water has an extremely high surface tension.
surface-to-volume ratio Describes cell size increases. Cell volume grows much more rapidly than surface area.
symbiosis (Gr. ***syn,*** **together with +** ***bios,*** **life)** The condition in which two or more dissimilar organisms live together in close association; includes parasitism, commensalism, and mutualism.
synapse (Gr. ***synapsis,*** **a union)** A junction between a neuron and another neuron or muscle cell. The two cells do not touch. Instead, neurotransmitters cross the narrow space between them.
synapsis (Gr. ***synapsis,*** **contact, union)** The close pairing of homologous chromosomes that occurs early in prophase I of meiosis. With the genes of the chromosomes thus aligned, a DNA strand of one homologue can pair with the complementary DNA strand of the other.
syngamy (Gr. ***syn,*** **together with +** ***gamos,*** **marriage)** Fertilization. The union of male and female gametes.

T

taxonomy (Gr. ***taxis,*** **arrangement +** ***nomos,*** **law)** The science of the classification of organisms.
T cell A type of lymphocyte involved in cell-mediated immune responses and interactions with B cells. Also called a T lymphocyte.
tendon (Gr. ***tenon,*** **stretch)** A strap of connective tissue that attaches muscle to bone.
tertiary structure of a protein The three-dimensional shape of a protein. Primarily the result of hydrophobic interactions of amino acid side groups and, to a lesser extent, of hydrogen bonds between them. Forms spontaneously.
test cross A cross between a heterozygote and a recessive homozygote. A procedure Mendel used to further test his hypotheses.
theory (Gr. ***theorein,*** **to look at)** A well-tested hypothesis supported by a great deal of evidence.
thigmotropism (Gr. ***thigma,*** **touch +** ***trope,*** **a turning)** The growth response of a plant to touch.
thorax (Gr. a breastplate) The part of the body between the head and the abdomen.
thylakoid (Gr. ***thylakos,*** **sac +** ***-oides,*** **like)** A flattened, saclike membrane in the chloroplast of a eukaryote. Thylakoids are stacked on top of one another in arrangements called grana and are the sites of photosystem reactions.
tissue (L. ***texere,*** **to weave)** A group of similar cells organized into a structural and functional unit.
trachea, ***pl.*** **tracheae (L. windpipe)** In vertebrates, the windpipe.
tracheid (Gr. ***tracheia,*** **rough)** An elongated cell with thick, perforated walls that carries water and dissolved minerals through a plant and provides support. Tracheids form an essential element of the xylem of vascular plants.
transcription (L. ***trans,*** **across +** ***scribere,*** **to write)** The first stage of gene expression in which the RNA polymerase enzyme synthesizes an mRNA molecule whose sequence is complementary to the DNA.
translation (L. ***trans,*** **across +** ***latus,*** **that which is carried)** The second stage of gene expression in which a ribosome assembles a polypeptide, using the mRNA to specify the amino acids.
translocation (L. ***trans,*** **across +** ***locare,*** **to put or place)** In plants, the process in which most of the carbohydrates manufactured in the leaves and other green parts of the plant are moved through the phloem to other parts of the plant.
transpiration (L. ***trans,*** **across +** ***spirare,*** **to breathe)** The loss of water vapor by plant parts, primarily through the stomata.
transposon (L. ***transponere,*** **to change the position of)** A DNA sequence carrying one or more genes and flanked by insertion sequences that confer the ability to move from one DNA molecule to another. An element capable of transposition (the changing of chromosomal location).
trophic level (Gr. ***trophos,*** **feeder)** A step in the movement of energy through an ecosystem.
tropism (Gr. ***trop,*** **turning)** A plant's response to external stimuli. A positive tropism is one in which the movement or reaction is in the direction of the source of the stimulus. A negative tropism is one in which the movement or growth is in the opposite direction.
turgor pressure (L. ***turgor,*** **a swelling)** The pressure within a cell that results from the movement of water into the cell. A cell with high turgor pressure is said to be turgid.

U

unicellular Composed of a single cell.
urea (Gr. ***ouron,*** **urine)** An organic molecule formed in the vertebrate liver. The principal form of disposal of nitrogenous wastes by mammals.
urine (Gr. ***ouron,*** **urine)** The liquid waste filtered from the blood by the kidneys.

V

vaccination The injection of a harmless microbe into a person or animal to confer resistance to a dangerous microbe.
vacuole (L. ***vacuus,*** **empty)** A cavity in the cytoplasm of a cell that is bound by a single membrane and contains water and waste products of cell metabolism. Typically found in plant cells.
van der Waals forces Weak chemical attractions between atoms that can occur when atoms are very close to each other.
variable Any factor that influences a process. In evaluating alternative hypotheses about one variable, all other variables are held constant so that the investigator is not misled or confused by other influences.
vascular bundle In vascular plants, a strand of tissue containing primary xylem and primary phloem. These bundles of elongated cells conduct water with dissolved minerals and carbohydrates throughout the plant body.
vascular cambium In vascular plants, the meristematic layer of cells that gives rise to secondary phloem and secondary xylem. The activity of the vascular cambium increases stem or root diameter.
ventral (L. ***venter,*** **belly)** Refers to the bottom portion of an animal. Opposite of dorsal.
vertebrate An animal having a backbone made of bony segments called vertebrae.
vesicle (L. ***vesicula,*** **a little (ladder)** Membrane-enclosed sacs within eukaryotic cells.
vessel element In vascular plants, a typically elongated cell, dead at maturity, that conducts water and solutes in the xylem.
villus, ***pl.*** **villi (L. a tuft of hair)** In vertebrates, fine, microscopic, fingerlike projections on epithelial cells lining the small intestine that serve to increase the absorptive surface area of the intestine.
vitamin (L. ***vita,*** **life +** ***amine,*** **of chemical origin)** An organic substance that the organism cannot synthesize, but is required in minute quantities by an organism for growth and activity.
viviparous (L. ***vivus,*** **alive +** ***parere,*** **to bring forth)** Refers to reproduction in which eggs develop within the mother's body and young are born free-living.
voltage-gated channel A transmembrane pathway for an ion that is opened or closed by a change in the voltage, or charge difference, across the cell membrane.

W

water vascular system The system of water-filled canals connecting the tube feet of echinoderms.
whorl A circle of leaves or of flower parts present at a single level along an axis.
wood Accumulated secondary xylem. Heartwood is the central, nonliving wood in the trunk of a tree. Hardwood is the wood of dicots, regardless of how hard or soft it actually is. Softwood is the wood of conifers.

X

xylem (Gr. ***xylon,*** **wood)** In vascular plants, a specialized tissue, composed primarily of elongate, thick-walled conducting cells, that transports water and solutes through the plant body.

Y

yolk (O.E. ***geolu,*** **yellow)** The stored substance in egg cells that provides the embryo's primary food supply.

Z

zygote (Gr. ***zygotos,*** **paired together)** The diploid ($2n$) cell resulting from the fusion of male and female gametes (fertilization).

Index

A

B

C

Q

R

S

T

U

V

W

X

Y

Z

CPSIA information can be obtained
at www.ICGtesting.com
Printed in the USA
LVOW02s1117120116
470265LV00014B/344/P

9 780692 345955